Encyclopedia of Tec
Innovation Manag

Encyclopedia of Technology and Innovation Management

Edited by
V. K. Narayanan and Gina Colarelli O'Connor

A John Wiley and Sons, Ltd, Publication

This edition first published in 2010

Registered office
John Wiley & Sons Ltd, The Atrium, Southern Gate, Chichester, West Sussex, PO19 8SQ, United Kingdom

For details of our global editorial offices, for customer services and for information about how to apply for permission to reuse the copyright material in this book please see our website at www.wiley.com

ISBN 978-1-405-16049-0 (H/B)

A catalogue record for this book is available from the British Library.

Project Management by OPS Ltd, Great Yarmouth, Norfolk
Typeset in 9.5/11pt Ehrhardt
Printed in Great Britain by CPI Antony Rowe, Chippenham, Wiltshire

Contents

Board of Advisors

Preface

Technology and Innovation Management (TIM) is emerging as a field of its own. It has historically been scattered across multiple disciplinary fields, and indeed, in many universities, courses on the subject can be found in schools of management, engineering, and even science. Likewise, the richness of the research agenda in the field has drawn scholars from many disciplines, including economics, psychology, sociology, and all functional areas of business. Defining the field is challenging due not merely to the complexity of the subject matter, but also because of its highly dynamic nature. But the concept of an encyclopedia of Technology and Innovation Management is surely a cue that scholars perceive a new field coalescing.

In 2004, Rosemary Nixon approached one of the coeditors (V. K. Narayanan) with the idea of this encyclopedia as a volume in the Blackwell encyclopedia series, and Cary Cooper who was a major advisor for that series was very encouraging to move ahead. As a result of a merger, this volume is now published by John Wiley & Sons Ltd. Knowing that this is a daunting task, Narayanan sought the help of Michael Hitt who identified Gina O'Connor who was willing and able to undertake the collaboration in editing this volume. Gina not only understood technology, but also brought marketing and project management experience to the endeavor. Together, they shared a passion for technology from different perspectives: strategy, marketing, and project management.

Our first task was to constitute a Board of Advisors; their names are listed on p. vii. We solicited two kinds of assistance from the board. First, we wanted them to critique the outline of topics thus preventing major gaps in our coverage of the topics in this volume. Second, we asked them for referrals, seeking prospective authors for various topics. The board's help eased our task considerably, and we want to record our deep gratitude for their assistance.

We generated the topics in three major steps. First, one of our then doctoral students (Yi Yang, currently a professor in the University of Massachusetts-Lowell) identified key words and authors from (1) major textbooks on Management of Technology listed on the Technology & Innovation Management Division website of the Academy of Management, and (2) articles written in major technology journals from 2001 to 2004 (*Research Policy* and *IEEE Transactions in Engineering Management*), and also in major management journals (*Strategic Management Journal* and *Academy of Management Journal*) . The keywords were tabulated by frequency and a frequently mentioned keyword was included in our original list of concepts. The author list gave us an initial roster of scholars to contact for entries about various concepts. Second, the editors sorted the chosen key words into clusters identifying major and minor concepts and identifying key words within them. Third, the draft of the major and minor concepts was circulated among the board members for their feedback, which led to revisions of the original list.

In order to highlight the dynamism of the field we included two sets of topics that are likely to be obsolete within three to five years: Emerging Technologies and Innovation at the National Level. The final structure of the encyclopedia reflects both the relatively enduring concepts and the elements in flux.

The topics formed the basis for contacting prospective authors, either directly or indirectly with the assistance of the board. Cumulatively, the authors represent the major contributors to the field of management of technology. Both editors spent considerable time reading the entries, making suggestions and proposing editorial changes, and in general streamlining and standardizing the entries to develop a coherent encyclopedia.

Inconsistencies with respect to terminology and definitions remain, but they are in our opinion a reflection of the diversity and vibrancy of the field. Rather than portraying TIM as a monolith, our approach has been to display the diversity of thought. In other words, when faced with the choice of insisting on consistency for its own sake, we have opted instead to allow the authors to describe their point of view. The editorial task was daunting and time-consuming, but necessary. We can say with confidence that we learned a lot in the process, not merely about the art of editing, but about the field itself, through reading the entries of these authors.

An encyclopedia has many uses, but one that is often overlooked is something that struck us in the midst of the editing process—the use of these entries in the classroom. As we read the entries, we knew they represented up-to-date and easy-to-read material that could be assigned to students as and when relevant. Indeed this thought formed the basis of a professional development workshop at the 2009 Academy of Management Conference in Chicago. Both of us plan to use these entries in our classroom.

We owe a significant note of gratitude to many individuals. First, to our authors, without whom this work would not have seen the light of the day. Second, to the board members who helped us in the critical early stages of the project and then along the way. Third, we owe a special note of gratitude to Yi Yang for research assistance in identifying the topics. Finally, to Rosemary, a pleasant, patient, soft-spoken editor who managed us well.

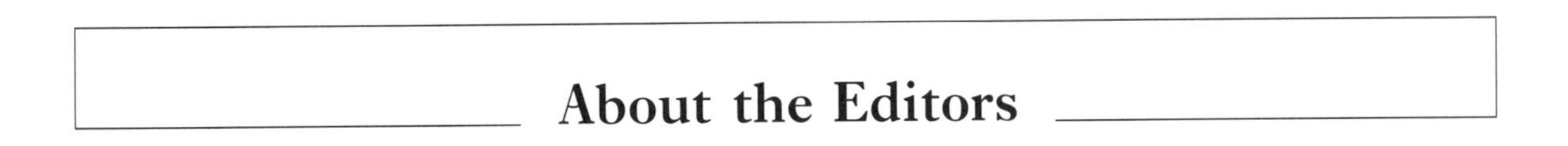

About the Editors

V. K. Narayanan

is Stubbs Professor of Strategy & Entrepreneurship and Associate Dean for Research at *LeBow College of Business, Drexel University, Philadelphia, PA, U.S.A.* His research focuses on: strategy formulation and implementation; organization design; technology strategy, innovation, and corporate entrepreneurship; knowledge management and competitor intelligence.

Gina Colarelli O'Connor

is Associate Professor of Marketing in the *Lally School of Management & Technology at Rensselaer Polytechnic Institute, Troy, NY, U.S.A.* Her research focuses on: radical innovation in large mature firms; corporate entrepreneurship; dynamic capabilities; new-market creation; and technology commercialization.

List of Contributors

Allan N. Afuah, *University of Michigan*
Matthew Allen, *Manchester Business School, U.K.*
Murugan Anandarajan, *Drexel University*
Bay Arinze, *Drexel University*
Gloria Barczak, *Northeastern University*
Heidi M. J. Bertels, *Stevens Institute of Technology*
Yi-Yu Chen, *New Jersey City University*
Susan K. Cohen, *University of Pittsburgh*
Julian Cooper, *University of Birmingham, U.K.*
Robert G. Cooper, *McMaster University*
Jeffrey G. Covin, *Indiana University*
James A. Cunningham, *National University of Ireland, Galway*
Erwin Danneels, *Worcester Polytechnic Institute*
Jenny Darroch, *Claremont Graduate University*
Donna Marie De Carolis, *Drexel University*
C. Anthony Di Benedetto, *Temple University*
Thomas Durand, *École Centrale Paris, France*
John E. Ettlie, *Rochester Institute of Technology*
George F. Farris, *Rutgers University*
Steven W. Floyd, *University of St. Gallen, Switzerland*
Kayano Fukuda, *Japan Science and Technology Agency*
J. Luis Galán, *University of Seville, Spain*
Rosanna Garcia, *Northeastern University*
Hans Georg Gemünden, *Technical University Berlin, Germany*
William Golden, *National University of Ireland, Galway*
Chittibabu Govindarajulu, *Delaware State University*
Jeffrey H. Greenhaus, *Drexel University*
Abbie Griffin, *University of Utah*
Trudy Heller, *Executive Education for the Environment*
Frederic C. Hamilton, *Olin College of Engineering*
Mei-Chih Hu, *Feng Chia University, Taiwan*
Mariann Jelinek, *College of William and Mary*
Shalini Khazanchi, *Rochester Institute of Technology*
Winston Koh, *Singapore Management University*
Peter A. Koen, *Stevens Institute of Technology*
Rishikesha T. Krishnan, *Indian Institute of Management Bangalore*
Donald F. Kuratko, *Indiana University*
Bárbara Larrañeta, *Pablo Olavide University, Spain*
Christoph H. Loch, *INSEAD, Fontainebleau, France*
Ta-Jung Lu, *National Chung-Hsing University, Taiwan*
John A. Mathews, *Macquarie University Sydney, Australia*
Judy Matthews, *Queensland University of Technology, Australia*
Robert McGowan, *University of Denver*
Iliana Payano Mejia, *Rochester Institute of Technology*
Shreefal Mehta, *Rensselaer Polytechnic Institute*
Morgan P. Miles, *Georgia Southern University*
William L. Miller, *4G Innovation LLC*
Ram Mudambi, *Temple University*
V. K. Narayanan, *Drexel University*
Edward Nelling, *DrexelUniversity*
Gina C. O'Connor, *Rensselaer Polytechnic Institute*
Paul Olk, *University of Denver*
Lois Peters, *Rensselaer Polytechnic Institute*
Phillip Phan, *Johns Hopkins University*
Kwaku O. Prakah Asante, *Ross School of Business*
Mark P. Rice, *Babson College*
Patricia Robak, *Drexel University*
Joshua L. Rosenbloom, *University of Kansas and National Bureau of Economic Research*
Steve Russell, *Siemens Corporate Research, Inc.*
Søren Salomo, *Danish Technical University*
Don Scott-Kemmis, *Australian National University*
Kenneth L. Simons, *Rensselaer Polytechnic Institute*
Svenja C. Sommer, *HEC, Paris, France*
Ashish Sood, *Emory University*
J. C. Spender, *ESADE and Lund University, Sweden*
Clyde D. Stoltenberg, *Wichita State University*
Tim Swift, *St. Joseph's University*
Gerard J. Tellis, *University of Southern California*
Mark Tribbitt, *Drexel University*
Patrick van der Duin, *Delft University of Technology, The Netherlands*
Daniele Virgillito, *University of Catania, Italy*
Jessica M. Walker, *Rochester Institute of Technology*
Judith Walls, *University of Michigan*
Jong-Wen Wann, *National Chung-Hsing University, Taiwan*
Chihiro Watanabe, *Tokyo Institute of Technology, Japan*
Christy H. Weer, *Radford University*
Carola Wolf, *University of St. Gallen, Switzerland*
Shaker A. Zahra, *University of Minnesota*
Maliha Zaman, *Drexel University*
Paschalina (Lilia) Ziamou, *The City University of New York*

Part One

Technology-specific Concepts

1

Technology: Discourse and Possibility

J. C. Spender

ESADE and Lund University

Technology is a puzzle despite its evident impact on our lives. It penetrates and structures space and time via the Internet, travel, global warming, the world's rapid financial interactions, in off-shore supply chains, and increasingly within us, as drugs and prosthetics. But we have a hard time identifying precisely what this "it" is, to grasp technology. Is "it" more than artifacts; iPods, offshore drilling rigs, or hybrid cars? Or is it a generic method paralleling the scientific method? Is it an option, an imperative, a distinct mode of human existence or merely peripheral (Heidegger, 1977)? Ellul, for instance, treats technology as an autonomous domain of human activity, sprung free of our control by our Original Sin; now, monster-like, it pursues its own imperatives and shapes us into "mass man" (Ellul, 1967). We now fear the automobile and its impact (Ladd, 2008). Against dark views we have others more comforting; technology as tools to increase our productivity towards our chosen purposes; under our control, ready-to-hand, and morally neutral (Lancaster, 1966; Mansfield, 1996; Mowery and Rosenberg, 1989). More complex are Victorian notions of technology as the means to realize our dominance over the Primitive, to free us from our natural condition; i.e., technology as the construction of our artificial world; genetic engineering rather than the fruits of the field and forest, video-gaming rather than schoolyards (Passmore, 1974; White, 1972). This variety of framings should make us suspicious of simple models; yet in this most technological of ages, it seems essential to clarify technology's nature. Ferre, summarizing, wonders whether technology should be conceived as (a) material (hardware), (b) the embodiment of scientific knowledge, (c) the extension of our natural abilities, (d) the artificial aspects of our world, or (e) man's extension of Nature (Ferre, 1988).

Rather than establish technology's essence, one approach is to explore what it is not. It is not science, nor is it Nature unalloyed, nor abstract thought, and so on. Looking for relevant dichotomies, philosophers of technology distinguish their topic from the philosophy of science (Feenberg, 1996; Ihde, 1979, 1993; Mitcham, 1994), differentiating the abstractions of theory from the practical contextualities of technology's "being-in-the-world" (Heidegger, 1977) noting the different domains of thought and action. But before we go far down this road we must decide whether we are looking for Universalist notions or ones contextualized by our own theories. Organizational and management theorists have long recognized that our definitions of technology may not work for artists, philosophers, engineers, or ethicists whose intellectual enterprise differs from ours. Perhaps technology is just another puzzle behind the unresolved contradictions of our languages that show the cutting edges of our theorizing—economic versus organizational, maximizing versus power exercizing, equilibrium versus non-equilibrium (Cyert and March, 1963; Gibbons, 2005). Perhaps it is just the doing, the practice of bridging these conceptual differences, fitting into, say, transaction cost analysis by bridging contrasting ways of modeling management's world (Williamson, 1975)? On the one hand, technology as capital, a costly tool that leverages productivity, shifts the production function, and is evaluated in terms of ROI in both make or buy modes. This is an economist's "use-based" definition, duly bound by the market demand for the technology-applying organization's output. Or we might define technology as an administrator's tool, limiting others' guile and the "impactedness" of organizational life by collecting and distributing information for decision makers. This would be an organizational theorist's definition, a mechanism for greater control. Or we might strike a "critical" attitude, seeing technology as an investment against people rather than in them that leads to workers becoming "de-skilled" (Braverman, 1974), that its anti-humanist *ethos* accelerates the disenchantment of organizational life. Or, a marketing view, how technology likewise limits the range of products and services, shaping the customer's choices.

The method here is to look at technology via the interactions that are the focus of our theorizing. Technology is then either (a) external, available in a specialized market of productivity or control-enhancing tools, or (b) endogenous, emerging in the workplace, leading to team and organizational-level learning and to novel goods and services (Nonaka and Takeuchi, 1995; Romer, 1990). Noting their patterns of power and motivation, we see how organizations are transformed by technological change; for example, as production is automated and workers establish control, countering senior executives' administrative power. Likewise, external competition is reshaped.

Technology's social context

No doubt many managers think of technology in terms of cost, profitability, control, or product market strategy. But that is not the way technology is framed by contemporaries such as Weick (Weick, 1990) or Orlikowski (Orlikowski, 1992, 2000). They reach beyond concerns with ROI, control, and competitive advantage to ask questions like "why is this technology the way it is rather than otherwise?" or "why is this technology viewed as it is by those whose activity it shapes?" These questions seem deeper, seeing technology as embedded in the world and in people's interactions, especially between those who choose a technology and those whose work is directly shaped by the technology chosen.

SCOT (Social Construction of Technology) (Bijker, Hughes, and Pinch, 1987) and ANT (Actor Network Theory) (Law and Hassard, 1999) theorists surface the social and institutional influences over seemingly pure engineering (MacKenzie and Wajcman, 1999). The SCOT angle is that technologies reflect the economic, social, and political contexts in which they come into being, combined with the science they appropriate. ANT bears more directly on sociology, arguing that society emerges between individuals whose interactions are technologically mediated; so society cannot be understood independently of the technologies that shape its practices. The general conclusion is that technology cannot be analyzed without considering the non-technological aspects of its context; it is "in-the-world" and contextualized, not abstract. Brought into the social realm, technology is inevitably "institutionalized"—"infused with meaning beyond the strict technical requirements at hand" (Selznick, 1957, p. 17)—which happens because the relationship between the technology and its users is disjoint, under-determined, with a degree of "interpretive flexibility", its practical meaning remaining "up for grabs". Technological choices lead to collisions of interests, challenging managers to "align" their technological appropriations with the organization's goals, processes, routines, and culture. Even then, the technology's inherent flexibility remains a threat to management's control.

Weick, pursuing his "sense-making" agenda, points to the perceptual and psychological phenomena that might help managers and theorists grasp how a technology acquires its meaning and practical impact. Technology is not self-evident; on the contrary it is "equivocal". Lacking any essential meaning, it means nothing except to those who interact with it. Weick's special interest is in how a technology may surprise users into errors, some catastrophic (Weick and Sutcliffe, 2001). What it means may be more to do with these interactions, and the emotions they generate in users, than with what the equipment or system's designers or selectors had in mind. Orlikowski likewise borrows from Giddens' structuration theory to theorize the interplay as technology impacts practice and practice impacts technology, and focuses on why managers' projects succeed or fail. But to argue that situations are under-determined and might therefore not turn out as planned is not the same as trying to explain, *ex post*, why they turned out as they did—and therein, of course, lies managers' struggle with technology. Unlike the philosophers, concerned with technology's impact on society generally, managers intend to harness it to their organization's goals; but what of the employees? Admitting different people might have different views about a technology they are interacting with raises questions about "who wants to know?" rather than "what is this technology?" Orlikowski's notion of "technology-as-practice"—how a technology eventually presents as a stable and institutionalized way of being—moves towards identifying the relevant agents, whether physical, individual, or group. But which of these can explain or predict success or failure? Is this technological relativism with all explanations equally valid or "up for grabs"? Likewise, though the notion of practice is intriguing, things get horrid as we extend the notion of the practicing agent from individuals to the organization's objects, procedures, equipment, institutions, and so forth, the full panoply of ANT (Latour, 1996; Law and Hassard, 1999).

Giddens' "structuration" theory attributes "interpretive flexibility" to society itself as various social arrangements are enacted and the consequences perceived lead to new arrangements (Bryant and Jary, 1991). Organization theorists like Weick and Orlikowski adopt these dialectical notions, suggesting the interactions of a technology's human and non-human aspects, its history and perceived future, give it "transience", leaving it without any deep fundamental nature, never more than what it appears to be at a particular time, a temporary synthesis. Then our obsession with causality tempts us to grant technology or society higher status as an independent variable. Social constructionists grant society higher status, without explaining how society came to be the way it is, even while granting technology the power to shape it as, say, rice-growing in a constricted geography is said to have shaped Japanese culture. The contrary is to grant technology higher status, politics, and self-interest than colliding with technology's scientific truths. Thus the QWERTY keyboard, we are told, is a triumph of convenience and institutionalization over technical efficiency.

Where does such analysis lead? Nowhere, perhaps, though the story seems interesting. Its principal point is methodological, that as we try to ground definitions of technology—whether abstract, like a compression algorithm, or physical, like aspirin, or tool-like, a ski-lift—within a complex and under-determined web of social interaction, we lose control of the discussion. Having presumed society's influence on the penetration, application, or evolution of "a technology"—whether that is institutional, cultural, scientific, organizational, or otherwise—the theorizing is un-moored until we see a theory of that determining context. More precisely, it is all very well to categorize a technology project's outcomes as displaying "inertia", "application", or "change" but can we ever understand the interpretive, technological, and institutional conditions of the context enough to know what will happen and why (Orlikowski, 2000, p. 422)? Only when we know society completely and can anticipate its changes. Likewise invoking the term "practice" but failing to identify the specific agent of that practice or its limits, we miss how contested are the intentions of the individuals or collectives whose practices intersect in any particular organization, and thus how different the answers might be to the question of "what does this agent mean by 'the technology'?" Attempts to define or analyze technology in terms of the interaction between people and objects or systems leads us into epistemological incoherence; a bait and switch. Exploiting our sense of understanding society, the explanatory base is subtly shifted from technology, the problematic, to a socially framed "technology-in-practice" or psychological cycle of function and arousal. But can the result count as an explanation? What is excluded by these notions? We get no grounding; not only is the explanation "up for grabs", it moves precious close to "anything goes" (Feyerabend, 1993).

Absent a robust theory of society, one alternative to a socially grounded theory of technology is a technologically penetrated theory of social interaction at the macro, organizational, or work-team level. This is more or less where the dispute between Habermas and Marcuse leads (Feenberg, 1996, 1999, p. 151). The former argues that behind all theories of society lies the universality of human rationality—the latter argues that such rationality is socially and historically contingent. Both allow technology as an articulation of rationality, inter-subjective, outside the person that then shapes human interaction. Here we note different rationalities, as Weber contrasted "instrumental rationality" (*zweckrationalität*) with "value-based rationality" (*wertrationalität*). It follows that questions like "why is this technology as it is" go well beyond mere contests of individual intention, interest, and power to embrace society's history and technology's path dependence emphasized by the SCOT or ANT theorists (Arthur, 1989).

Technology's objectivity and language

Clearly technology is more than its artifacts; beige boxes, wind generators, software, etc.—for these have no inherent meaning. Our responses to these artifacts determine their meaning and their impact. Yet we speak about technology as "objective" and independent of us, perhaps to hide how our responses vary. But as alternative rationalities come into view we surface the dialectical struggle between the inter-subjective rationalities we think are embedded in the technology and those of the social life they shape, the contest that so excites Ellul and the other anti-technologists. Technological rationality is advanced against the social practice–based alternatives, the contextualized rationalities advanced to explain why society is the way it is. But can we cut through this muddle?

Instead of presuming a technology can ever shed or be cleansed of the contextualities and interests that brought it into being, technology implies a distinct domain of human activity, one currently privileged (Heidegger, 1977). So the real puzzle about technology is the why and the how of this privileging for it shapes our sense-making. Readers of *Practical Mechanics* or *Radio Electronics* aside, it seems technology has no language of its own. Our failure to understand it springs from and is reflected in the lack of axioms that would underpin its own idiosyncratic language and make it comprehensible. Absent these, technology has borrowed. Engineering language obscures because it is about the properties of materials and the design and production of artifacts, not about using them or understanding their impact on our modes of life. But science gives us language that Edison could not, though grounded in causality and focused on cause and effect rather than social practice. While our forefathers were familiar with the sacred books and might quote extensively from them when discussing social concerns, today's generations are more likely to use the language of physical chemistry to discuss, say, global warming or the Green Revolution, this age's concerns. Our society is remarkable in that technology has moved on from *techne*, the Aristotelian form of knowing demonstrable in practice, to appropriate the language of science as its rhetorical mode, to conceal, perhaps, that it is as socio-historically contingent as any other domain of activity, economic, religious perhaps, or political. This seizure lies behind the idea of a Technological Age, for the language then dominates the public discourse.

Social studies of science have shown science's practices bear little relation to the classroom mythology of rigor and objectivity (Latour, 1987). Technology is no more solid or rational for its claim to be scientific. Its deeper contingent nature, that it could have been otherwise, gets hidden behind a rhetoric of scientific objectivity, just as the rhetoric of economics has evolved to convince others of the acceptability of its assumptions (McCloskey, 1998). An overly science-driven view of technology makes it impossible for us to understand "pre-scientific technology", a *techne* unframed in scientific language, of which there are many examples—military (stirrup and rifle), marine (lateen sails and compass), medical (acupuncture and the dentistry of Ancient Egypt), and managerial (the organization itself) (De Landa, 1991; White, 1964). So the question remains, "is the transient dynamic stability we treat as a technology anything more than a stabilization of the power discourses that shape social practice?" Here technology is an instrument of social power, to be controlled just as colonialists controlled the language of their subjects. It is crucial to deny it any privileged status; it is just another mode of social discourse, albeit more widespread and influential than, say, 300 years ago when religion dominated. Yet we succumb and treat technological language as objective, secular and authoritative, "evidence-based" and organized using the methods of science rather than the hit-and-miss practices Thomas Edison adopted to develop the incandescent lamp and DC technology for urban electrification.

So the more tractable questions are about how the language of technology, its rhetoric, has acquired its status and influence, pushing the moral and political issues inherent in all human activity out of sight, only to re-emerge, as they must, as the problematic for new subfields such as CSR (corporate social responsibility), or business or medical ethics. As we sense the power relations behind the language we frame the political struggle between high-tech, low-tech, and "appropriate technology" approaches, and their concern with technological colonization (Hazeltine and Bull, 1999). Partly this is an overhang from 19th-century colonialism and the struggle between those who saw natural science as "pure science" and the social sciences, if sciences at all, as poor cousins. Part is the impact that technology has on our lives, seeming to present the irresistible facts of a situation and squeeze out other discourses. Part is the professionalization and complexity of technology today, the huge educational investments necessary to comprehend what we see and puzzle out its social and moral implications. Ironically, technology has advanced in power precisely as its discourse has become less comprehensible; we marvel at its effects, having lost sight of how it supports or denies our choices—be they of diet, travel, leisure, or communication. Feeling powerless, we concede it higher status.

How can we bring technology and its impacts to heel rather than be trampled beneath them, as Ellul and Marcuse warned? How can we respect its achievements and benefits but tame its power over us? Positivism has not served us well here, for it prioritizes talk of the "real" that positivists presume exists independently of us and into which the natural sciences inquire for its universal truths. Those who treat technology as the real embedded in the social, endlessly interacting with other forms of life in processes of "structuration", try to leverage off the distinction between positivism and interpretivism, between objectivity and subjectivity. From the realist point of view, perhaps, the technology project failed because its design was faulty or inappropriate to the task; a redesign is indicated. From an interpretive point of view, perhaps, the project failed because its users made the "wrong" interpretation, suggesting that control and rationality can be restored by better communications, training, or incentives. But these two explanations never converge until we arrive at the Archimedean point of total knowledge of our universe and its causal machinery. Our real condition is elsewhere, so these approaches suggest the wrong questions. They leave us with understanding the "it" of technology as the impact of the fruits of others' explorations of the real on us. Which leaves us out of the analysis, and this is the deficiency that contemporary theorists of technology attempt to correct but cannot without a positivistic theory of the social that can converge with the chosen language of science.

A constructivist approach

Once other epistemologies are brought to bear the questions asked change. The implicit model of the human agent (whether her/his axiomatic attribute be rationality, power-seeking, emotion, self-maximizing, religiosity, etc.) is the key. We cannot critique and escape the rationalist rhetoric that supports technology's present status without also critiquing the axiomatic Model of Man which prioritizes rationality as the basis both for action and explanation in a world presumed to be rationally constructed and, consequently, fully comprehensible. We reveal something utterly dierent about technology from a constructivist position. It presents organizations as socio-economic arrangements under constant reconstruction and technology similarly, dematerializing both. Technology is impact rather than artifact. Instead of organizations having a distinct existence or ontology, they become ongoing patterns of interaction between people and other human or inorganic agents as they produce and consume. Likewise any technology-in-use appears as the social practices of producing and consuming—not at all the materialist notions that spring first to mind, the beige boxes. As we seize technology within a dynamic discourse of influence that actualizes social power, giving it neither false realism nor privileged access to Nature, we render it every bit as recursive as society itself. We are no longer able to distinguish "technology" from "organization" in any fundamental way, for organization is a hugely important technology too. Technology no longer impacts "the organization"; it merely identifies one class of the many influences over organizing processes.

But switching to a constructivist epistemology seems to do little more than take us back to the relativism of competing rationalities until we see the human agent as also being constructed. Just as organizing processes are shaped by technology, so are agents (Vygotsky, 1978). People become what they do as the recursive processes link agents—with technology-as-language as the medium. Society means some agents have the power over others and that is how technology enters the social. The pseudo-objective language of science masks this. Consider the CAFE standards, the legalities government uses to pressure the automobile industry to advance their "mileage and emissions technology". The resulting computer-controlled combustion and catalytic exhaust management technology makes no social sense abstracted into the science lab where engines can be built that offer staggeringly high mileage and low emissions—the impression we might get of the CAFE initiative. On the contrary, "fleet mileage" and emissions targets apply to the driving conditions that exist in practice, those deeply implicated in U.S. society, in what people need and expect of their transportation (Kay, 1997). To recognize how much we have been shaped by the automobile industry's decisions is to be shocked at how much power it has over our lives and who we have become. Thus each technology's artifacts are "boundary objects" to these social processes (Star, 1991). They act as the symbols, sacred objects, and ritualized processes of a science-based belief system we have privileged, suggesting some truth beyond priestly power.

Once we see the language of technology is not about "reality" or science, but is an exercise in social power, we are led to think how its rhetoric is constructed and warranted. Aristotelian rhetoric was based on the alternative modes of human persuasion—*logos*, *ethos*, and *pathos*; the first is an appeal to rationality, the second to the social relations between speaker and audience, and the third an appeal to the emotion that is the spur to action. In our hyper-rationalist age *ethos* and *pathos* are hidden, suggesting the language of technology arises at the junction of the three fundamental rational modes of human knowing. For Habermas these are indicated as the objective physical world, the social world of people, and the subjective world of feelings (Feenberg, 1999, p. 158). There are other variants; Barnard assumed the physical, social, and psychological (Barnard, 1968) while Luhmann posited the social, psychological, and the present (Luhmann, 1995). Yet a rationality-based model of the individual is implicit in all. Weick's analysis is rich in that it lies within this three-way framework, implying the meaning of a technology emerges recursively through the interaction of the social and the physical, mediated by the agent's emotion—nothing much to do with the quasi-causal models that some find in Giddens' structuration.

Adopting a constructive epistemology displaces Rational Man from this discourse. Instead we call on Agentic Man, one who constructs both world and self. Explanations of power and process are then grounded in the interacting agents' intents. For instance, ANT networks stabilize as the various agents' intents and practices coalesce into transient quietude rather than as their quasi-scientific rationalities play out to an equilibrium solution. Technology can then be captured as interplaying agents, constrained by history and material and social circumstance; perhaps physical, like carbon fiber, or social norms, important to the SCOT history of bicycles. A constructive explanation's grounding always lies in the particular agents, how they see themselves and the world they imagine—flying Wright brothers, Roosevelt's Panama Canal, the "computer for the rest of us". Technology deployed as an instrument of power to hide the intentions of the agents providing and choosing it, only appears based on rationality when others' interests have been silenced. Deconstructing the rhetoric around a technology that shapes and facilitates our practice helps us recapture our agency, bringing it into our life-world (Critchley, 1999). To speak of being driven by technology is to legitimate silencing others' agency. While one might protest and say, hey, antibiotics are real, they cure, that is just science. That we use them, that is power.

So what is the "it" of technology? This chapter argues that at its most basic "it" is the appearance of a culturally legitimized discourse around how some shape the lives of others through artifacts and ritualized processes, a seemingly de-politicized modernist form of power. The appeal is not to a transcendent Being, but to Nature and the extended possibilities revealed by, say, bronze weaponry, Salk's vaccine, or Microsoft's Vista. Technology as the rhetoric of its impacts on and meanings for us, rather than as the scientific objectivity in artifacts and systems that stand apart from us, brings it into the networks of social, economic, psychological, and political power that dynamically shape our condition. Of course, all language, being inter-subjective and standing outside us, has a mask of objectivity, leaving its practical implications problematic. But ultimately technology's meaning comes from us and not, as some would assert, from any correspondence to the positivist's real.

Bibliography

Arthur, W. B. (1989). "Competing technologies, increasing returns, and lock-in by historical events." *Economic Journal*, **99**, 116–131.

Barnard, C. I. (1968). *The Functions of the Executive* (30th Anniversary Edition). Cambridge, MA: Harvard University Press.

Bijker, W. E., Hughes, T. P., and Pinch, T. J. (Eds.). (1987). *The Social Construction of Technological Systems: New Directions in the Sociology and History of Technology*. Cambridge, MA: MIT Press.

Braverman, H. (1974). *Labor and Monopoly Capital: The Degradation of Work in the Twentieth Century*. New York: Monthly Review Press.

Bryant, C. G., and Jary, D. (1991). *Giddens' Theory of Structuration: A Critical Appreciation*. London: Routledge.

Critchley, S. (1999). *The Ethics of Deconstruction: Derrida and Levinas*. Edinburgh, U.K.: Edinburgh University Library.

Cyert, R. M., and March, J. G. (1963). *A Behavioral Theory of the Firm*. Englewood Cliffs, NJ: Prentice-Hall.

De Landa, M. (1991). *War in the Age of Intelligent Machines*. New York: Swerve Editions.

Ellul, J. (1967). *Technological Society*. New York: Random House.

Feenberg, A. (1996). "Marcuse or Habermas: Two critiques of technology." *Inquiry*, **39**, 45–70.

Feenberg, A. (1999). *Questioning Technology*. London: Routledge.

Ferre, F. (1988). *Philosophy of Technology*. Englewood Cliffs, NJ: Prentice-Hall.

Feyerabend, P. (1993). *Against Method* (Third Edition). London: Verso.

Gibbons, R. (2005). "Four formal(izable) theories of the firm?" *Journal of Economic Behavior & Organization*, **58**, 200–245.

Hazeltine, B., and Bull, C. (1999). *Appropriate Technology: Tools, Choices and Implications*. San Diego, CA: Academic Press.

Heidegger, M. (1977). *The Question concerning Technology*. New York: Harper & Row.

Ihde, D. (1979). *Technics and Praxis: A Philosophy of Technology*. Boston, MA: Reidel.

Ihde, D. (1993). *Philosophy of Technology: An Introduction*. New York: Paragon House.

Kay, J. H. (1997). *Asphalt Nation: How the Automobile Took over America and How We Can Take It Back*. Berkeley, CA: University of California Press.

Ladd, B. (2008). *Autophobia: Love and Hate in the Automotive Age*. Chicago, IL: University of Chicago Press.

Lancaster, K. (1966). "Change and innovation in the technology of consumption." *American Economic Review*, **56**(2), 14.

Latour, B. (1987). *Science in Action: How to Follow Scientists and Engineers through Society*. Cambridge, MA: Harvard University Press.

Latour, B. (1996). *Aramis or the Love of Technology*. Cambridge, MA: Harvard University Press.

Law, J., and Hassard, J. (Eds.). (1999). *Actor Network Theory and After*. Oxford, U.K.: Blackwell.

Luhmann, N. (1995). *Social Systems* (J. Bednarz and D. Baecker, Trans.). Stanford, CA: Stanford University Press.

MacKenzie, D., and Wajcman, J. (Eds.). (1999). *The Social Shaping of Technology* (Second Edition). Buckingham, U.K.: Open University Press.

Mansfield, E. (1996). "Contributions of new technology to the economy." In: B. L. R. Smith and C. E. Barfield (Eds.), *Technology, R & D, and the Economy* (pp. 114–139). Washington, DC: Brookings Institution.

McCloskey, D. N. (1998). *The Rhetoric of Economics* (Second Edition). Madison, WI: University of Wisconsin Press.

Mitcham, C. (1994). *Thinking through Technology: The Path between Engineering and Philosophy*. Chicago, IL: University of Chicago Press.

Mowery, D. C., and Rosenberg, N. (1989). *Technology and the Pursuit of Economic Growth*. Cambridge, U.K.: Cambridge University Press.

Nonaka, I., and Takeuchi, H. (1995). *The Knowledge-creating Company: How Japanese Companies Create the Dynamics of Innovation*. New York: Oxford University Press.

Orlikowski, W. J. (1992). "The duality of technology: Rethinking the concept of technology in organizations." *Organization Science*, **3**, 398–427.

Orlikowski, W. J. (2000). "Using technology and constituting structures: A practice lens for studying technology in organizations. *Organization Science*, **11**, 404–428.

Passmore, J. A. (1974). *Man's Responsibility for Nature: Ecological Problems and Western Traditions*. London: Duckworth.

Romer, P. M. (1990). "Endogenous technological change." *Journal of Political Economy*, **98**(5, Supplement), S71–S102.

Selznick, P. (1957). *Leadership in Administration: A Sociological Interpretation*. New York: Harper & Row.

Star, S. L. (1991). "Power, technology and the phenomenon of conventions: On being allergic to onions." In: J. Law (Ed.), *A Sociology of Monsters: Essays on Power, Technology and Domination* (pp. 26–56). London: Routledge.

Vygotsky, L. S. (1978). *Mind in Society: The Development of Higher Psychological Processes*. Cambridge, MA: Harvard University Press.

Weick, K. E. (1990). "Technology as equivoque: Sensemaking in new technologies." In: P. S. Goodman *et al.* (Eds.), *Technology and Organizations* (pp. 1–44). San Francisco, CA: Jossey-Bass.

Weick, K. E., and Sutcliffe, K. M. (2001). *Managing the Unexpected: Assuring High Performance in an Age of Complexity*. San Francisco, CA: Jossey-Bass.

White, L. (1964). *Medieval Technology and Social Change*. Oxford: Oxford University Press.

White, L. (1972). "The historical roots of our ecological crisis." In: C. Mitcham and R. Mackey (Eds.), *Philosophy and Technology: Readings in the Philosophical Problems of Technology* (pp. 259–265). New York: Free Press.

Williamson, O. E. (1975). *Markets and Hierarchies: Analysis and Antitrust Implications*. New York: Free Press.

2

Technology Evolution

Joshua L. Rosenbloom

University of Kansas and National Bureau of Economic Research

Technology evolution refers to changes in production processes or institutional arrangements that make it possible with a fixed set of resources to produce either (1) a greater quantity of a given product or service or (2) to produce new or qualitatively superior products or services. Technology evolution is the primary cause of rising living standards in modern economies, and the divergence of technological capabilities across countries is the chief reason for international differences in living standards.

Table 2.1 provides a concrete illustration of the impact of technology evolution on living standards in the U.S. during the 20th century. The first column shows the labor time required by an average worker to earn enough to purchase the specified product in 1895, while the second column shows the amount of time required in 2000. To facilitate comparison the third column shows the ratio of these two figures—the larger the ratio the greater the increase in labor productivity (and purchasing power) that has taken place.

The data in Table 2.1 suggest several important points about the effects of technology evolution. First, the quantity of goods that the average worker can consume has increased dramatically. This is reflected in column 3 which shows the proportionate increase in the quantity of each item that can be purchased with an hour of labor time. The median increase in productivity for this somewhat arbitrary selection of goods is between five- and eight-fold. Second, there is considerable variation in the increase in productivity across the different items. While the number of bicycles that can be purchased with an hour of work increased by a factor of 36, it increased only 2.2 times for a Steinway piano. There is even one item in the list, the sterling silver teaspoon, which has become more expensive. These differences reflect the differential impacts of the application of modern technologies—such as mass production—and changes in the cost of raw materials on individual items.

Table 2.1. Time required to earn selected items, 1895–2000

	Labor time required to earn (in hours)		
Commodity	*1895*	*2000*	*Productivity multiple*
One-speed bicycle	260	7.2	36.1
Horatio Alger (6 vols.)	21	0.6	35.0
100-piece dinner set	44	3.6	12.2
Cushioned office chair	24	2.0	12.0
Hair brush	16	2.0	8.0
Cane rocking chair	8	1.6	5.0
Solid gold locket	28	6.0	4.7
Encyclopedia Britannica	140	33.8	4.1
Steinway piano	2,400	1,107.6	2.2
Sterling silver teaspoon	26	34.0	0.8

Source: DeLong (2000, p. 5).

A third point is raised by what is not in Table 2.1. Because we can only compare productivity for items that are similar in 1895 and 2000, we cannot directly observe the impact of the introduction of new or qualitatively different goods. A moment's contemplation suggests that this list is quite long, including modern medical care, computers, MP3 players, television, and stainless steel flatware. While some of these additions serve entirely new purposes, the stainless steel flatware emphasizes the point that some new products can provide lower cost alternatives to existing items. In 1895 silver was the only rust-free flatware available; today stainless steel provides much the same service as silver at a fraction of the cost.

A more comprehensive measure of productivity improvement is provided by calculating the increase in Gross Domestic Product (GDP) per hour of labor input. Over the same period covered by Table 2.1, GDP per worker in the U.S. increased—adjusting for inflation—from $13,700 to $65,500, while the number of hours worked per year by the typical worker fell by more than one-third. Thus the average productivity of an American worker has multiplied by a factor of nearly 7 over the last century due to technological advances. Since this calculation does not account for the many ways in which our lives are enriched by new products and services not available in 1895, the seven-fold improvement is, if anything, a conservative assessment of the increase in well-being that the technology evolution has produced (DeLong, 2000).

The beginning of sustained economic growth

Technology evolution is as old as human history. For most of this time, however, the pace of change remained quite gradual. Until the late 18th century, rising productivity supported a growing population but produced little long-run advance in living standards. In the short run, episodes of below-average population growth correlated with rising living standards, and periods of more rapid population growth coincided with falling living standards. But in the long run there was little discernible trend in living standards.

Beginning in Britain sometime between 1760 and 1800, however, the pace of technological change began to accelerate. The introduction of the steam engine, new metallurgical techniques, and advances in mechanization combined with the introduction of factory methods of production fueled a rapid increase in productivity in the manufacture of cotton textiles and other products. As a result incomes began to rise at the same time that population growth accelerated.

The technologies of the British Industrial Revolution spread relatively quickly to other countries in Western Europe, the United States, and Canada. Meanwhile a stream of new innovations—railroads, electricity, synthetic dyes, better machine tools—contributed to an acceleration in the pace of economic change. By the early 20th century sustained growth had become the norm in these economies rather than stasis. The remarkable nature of this transformation is emphasized by the divergence between the West and the rest of the world. Although Japan, South Korea, Taiwan, Singapore, and Hong Kong have by now joined the ranks of modern, developed economies, the gulf in economic performance today is far wider than it was 250 years ago.

The nature of technological creativity

The technological creativity on which the modern era's sustained economic growth is based derives from the interaction of two distinct but complementary processes that Joseph Schumpeter called invention and innovation. Schumpeterian invention is the discovery of new knowledge about natural phenomena. It is, primarily, the consequence of a "struggle between mind and matter" to gain insight about how the world works (Mokyr, 1990, p. 10). Innovation in Schumpeter's terminology is the application of the existing stock of knowledge in new combinations and new ways to meet some human need. Schumpeter placed relatively little emphasis on invention, arguing that innovation was the primary source of economic advance. "Innovation is quite possible," he wrote, "without anything we should identify as invention and invention does not necessarily induce innovation, but produces of itself no economically relevant effect at all" (Schumpeter, 1939, p. 84).

The distinction Schumpeter drew between invention and innovation is conceptually important, since the factors that influence invention are likely to be somewhat different from those that affect innovation. But we should not relegate invention to secondary status. It is true that in the short run most economic progress derives from the application of existing knowledge in new ways, but without additions to the stock of basic knowledge, opportunities for innovation would eventually run into diminishing returns, and the pace of change would slow.

How technology evolves

Technology appears to evolve in two distinct ways, through gradual, incremental modifications in existing products and processes, and through discontinuous leaps in technology caused by the introduction of entirely revolutionary new innovations. While it is tempting to emphasize the introduction of revolutionary technologies like the railroad, electricity, the automobile, or the computer as the primary drivers of technology evolution, closer study suggests that the impact of these major new technologies would be far less dramatic without the accretion of small, almost invisible improvements to the original technologies. Indeed few if any of the inovations we would characterize today as revolutionary appeared so momentous at the time they were first introduced (Rosenberg, 1996).

One important reason for this is that radically new technologies are often quite primitive when they are first introduced, and this fact substantially limits their usefulness. In the case of the steam engine, for example, the earliest variants converted only a small fraction of the heat energy they consumed into mechanical effort. As a result they were only economical when located close to a source of fuel, limiting their usefulness to raising water from deep coal mines. At the beginning of the 18th century no one would have predicted the uses to which the steam engine would eventually be put, and it is worth recalling that it took several generations of largely anonymous improvement in techniques of manufacture and modification in design to reach a point where steam became competitive as a power source for factories. And it took more than a century from initial introduction until steam engines could be used to power railroads or ocean-going vessels.

The computer offers another example. The earliest electronic computers were developed in the 1940s. But these devices were large, prone to frequent failure, and extremely costly to operate. Given these limitations, Thomas Watson, Sr., the president of IBM forecast that worldwide demand for their services could be met by only a handful of the devices. Neither he nor anyone else at the time could have foreseen the ubiquity of computers today.

A second factor limiting the revolutionary impact of new technologies and making it difficult to forecast their ultimate impacts arises because of the interdependence of different technologies. What we see as an integrated technology, when fully developed, is often better understood as a system of mutually interacting innovations. Fully exploiting innovation *A* may require the development of technology *B*. If technology *B* does not exist, or is available only in a primitive state when technology *A* is first developed then the potential of technology *A* may not be immediately obvious.

An illustration of the way in which the value of a particular innovation is affected by other innovations is provided by the laser. Today lasers are used in a wide variety of applications, including long-distance transmission of voice and data, and to record and play back music, video, and other types of data. None of these applications could have been apparent at the time lasers were first developed. The use of lasers in communication required the development of fiber-optic cables capable of transmitting their signals, while the application of lasers in data storage and playback depended on a host of innovations in microprocessors, computers, and recording media.

On the one hand, these examples indicate that technology evolution is dependent on both incremental innovations and the emergence of revolutionary new technologies. Neither could exist without the other. On the other hand, they also suggest that technology evolution is inherently uncertain.

Long swings and the evolution of technology

Closely related to the question of how technology evolves is the issue of the timing of innovation. A number of economists have documented cycles in the pace of economic growth of approximately 50 to 60 years in duration consisting of alternating periods of faster and slower rates of economic growth. These fluctuations are often referred to as Kondratiev waves after the Russian economist who was one of the first to call attention to the evidence of long swings in a number of key economic indicators. Joseph Schumpeter (1939) argued that these variations in the pace of economic growth reflected cycles in the pace of innovation. According to Schumpeter periods of more rapid growth were initiated by the emergence of major new technologies or the temporal clustering of innovations, while slower growth cycles reflected periods in which the opportunities of these innovations had been largely exhausted.

The existence of Kondratiev waves, and their relationship to the timing of innovation remain controversial topics. Given the small number of cycles that can be observed in the available data, critics have pointed out that it may be premature to conclude that variations in the pace of growth reflect any sort of recurrent phenomenon. Instead they may simply reflect the impact of exogenous shocks to the economy. Similarly they have questioned the strength of the evidence linking variations in the pace of growth to innovations. Nonetheless, there are a number of scholars who continue to argue for the importance of long waves of innovation (see, e.g., Freeman and Soete, 1997, pp. 19–21; Perez, 2002).

The sources of technology evolution

Where do new technologies come from? Many new innovations arise through a process of gradual, incremental modification. Much of this creativity comes from processes that have been characterized as learning-by-doing and learning-by-using. Kenneth Arrow (1962) introduced the notion of learning-by-doing to describe technology evolution that takes place at the manufacturing stage and consists of increasing skill in production. Through careful observation of production processes and incremental improvement arising out of these observations substantial advances in productivity are possible. Nathan Rosenberg (1982, ch. 6) identified a parallel process of learning-by-using, in which technology evolution arises out of experience gained by users of complex products, such as modern aircraft or computers. On the one hand, knowledge that users of complex products gain from experience helps them to identify more clearly which features of a product are most valuable and thus lead to improvements in design that are embodied in future production. On the other hand, experience with the use and maintenance of complex products may lead to changes in how they are used or maintained that result in cost savings or increased revenue without the need for modification of the product itself.

Despite the importance of incremental improvements, however, no amount of incremental innovation can give rise to dramatic new technologies. There is, for example, no way to arrive at the automobile through incremental improvement to the horse and buggy. It might seem that such discontinuous breakthroughs would be minimally influenced by economic considerations, but there are still important feedbacks between economic activity and invention. Scientific insight begins with the perception of a problem—an area in which understanding is incomplete, and where opportunities for advance may exist. Resolution of this problem hinges on the state of existing knowledge and systematic study, which lead ultimately to a new understanding. Economic incentives influence this process through their role in helping to identify problems in need of solution. Much of the basic science of thermodynamics, for example, emerged out of efforts to better understand the factors limiting the efficiency of steam engines. Similarly much of the science of solid-state physics emerged out of the work of scientists employed at Bell Laboratories who were motivated by the problem of increasing the reliability of the telephone system (Rosenberg, 1982, pp. 141–159).

In modern, capitalist economies innovation has become an essential function for most businesses. Rather than competing with one another to produce homogeneous products at the lowest price, important sectors of modern capitalist economies are characterized by small numbers of producers who are motivated to seek a degree of monopoly power through innovation. This is the dynamic of creative destruction that Joseph Schumpeter (1942, pp. 83–86) identified as the "essential fact about capitalism." Each innovation creates a degree of market power that rewards the innovator with supernormal profits. But the innovator knows that this advantage is temporary, lasting only until the next innovation arrives. As a result commercial survival has come to rely on the regularization of the search for new innovations, and innovation has emerged as the chief form of competition for many businesses (Baumol, 2002).

The diffusion curve

So far this chapter has focused on the forces generating new technologies. But much technological change can occur without new technologies. At any point in time, most individuals and businesses are operating behind the technological frontier. Consequently the diffusion of innovations into widespread use is an extremely important topic.

Typically, the diffusion of new technologies follows an S-shaped, or logistic, diffusion curve. Figure 2.1 reproduces a number of illustrative examples plotting the number of users of selected technologies over time. As these diffusion curves suggest, the rate of adoption can vary considerably from one innovation to another. Table 2.2 makes this point even more clearly, showing the time in years from the invention of a variety of products (which may precede commercial introduction) to the time when one-quarter of the population had adopted them.

Research on diffusion has offered a variety of frameworks to understand the origins of the diffusion curve as well as the factors governing its shape. Early efforts to understand diffusion grew out of the work of sociologists and anthropologists, and focused on the role of communication and information flow between individuals and communities. The premise of much of this work is that, once developed, a new technology is superior to the existing alternatives with which it competes. Therefore any delay in adoption occurs because potential users lack information, and are uncertain about whether the new technology will actually benefit them. The rate of diffusion then depends primarily on (a) the mechanism by which information is spread from user to user and (b) variation in the willingness of individuals to adopt novel and potentially risky new technologies.

In contrast to this focus on information, economists have tended to attribute the diffusion curve mainly to heterogeneity in the population of potential users. In this framework, individuals adopt the new technology as soon as it becomes optimal for them to do so. Because potential adopters differ in one or more characteristics that affect the value of the innovation to them, however, not all of them will find the new technology immediately superior to the existing technology. In this case the S-shaped diffusion curve arises because of shifts in the cost–benefit calculation due to incremental improvement in the technology and/or changes in the characteristics of potential adopters.

Whichever framework one adopts the central empirical question in the study of technology diffusion concerns the factors that determine the speed with which a new technology is adopted. According to Hall (2005) the determinants of the rate of diffusion can be organized into three categories. The first set of factors center around the benefits that the new technology conveys relative to the

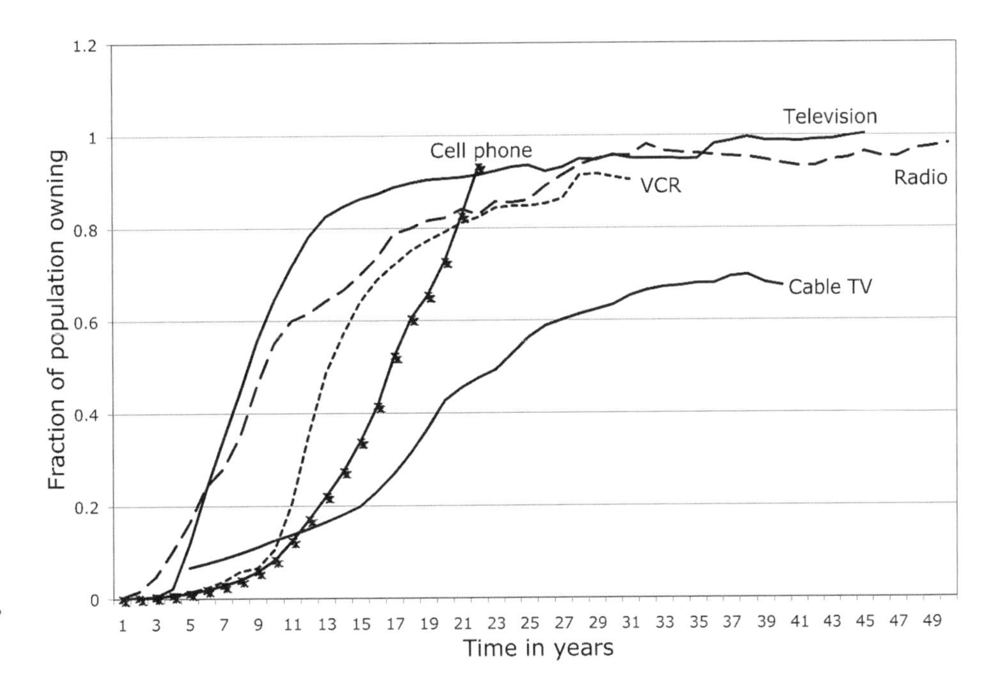

Figure 2.1.
Source: Carter *et al.* (2006, series Ae 1–28, Dg 34–45, Dg 103–105, Dg 117–130); U.S. Census Bureau (2007).

alternatives. These reflect both the intrinsic benefits of the technology and network interactions between users. Where competing standards exist, or the likely dominant standard is not clear, adopters may delay their choice to avoid choosing the "wrong" technology, thus slowing the rate of diffusion. The second set of factors concern the costs of adoption, including direct costs of acquisition plus investments in other complementary equipment or training. The higher the cost of adoption, the more slowly potential users are likely to adopt the new technology. The third cluster of influences affecting the rate of diffusion involve market size and structure, and the regulatory environment in which decisions are made. Highly concentrated markets may encourage speedy adoption by reducing network effects, or they may slow adoption if dominant firms prefer to preserve the value of existing assets. The impact of regulation is similarly ambiguous. Regulators may require firms to adopt a new technology, thus speeding diffusion; or, as in the case of the U.S. cellphone industry, regulators may impose constraints that slow the adoption of new technologies.

Table 2.2. Years from invention to diffusion to one-quarter of the population.

Product	*Year invented*	*Years to diffuse to one-quarter of the population*
Electricity	1873	46
Telephone	1876	35
Automobile	1896	55
Airplane	1903	64
Radio	1906	22
Television	1926	26
VCR	1952	34
Microwave oven	1953	30
Personal computer	1975	16
Cellular phone	1983	13
Internet	1991	7

Source: Federal Reserve Bank of Dallas (1996, p. 14).

Path dependence

It is convenient to separate discussion of the emergence of new technologies from their diffusion, but in reality these phenomena are often interrelated. Recently a good deal of attention has been devoted to the way in which small "accidents" early in the development of a technology can influence the way in which the technology ultimately develops. For this type of "path-dependent" development to occur three conditions must be met. There must be: (1) some kind of technical interrelatedness between individual users of the technology, (2) increasing returns to scale in the choice of technology, and (3) costs to switching between alternative technologies (David, 1986). The prototypical example of path dependence documented by Paul David (1986) concerns the development of the QWERTY keyboard.

According to David the arrangement of keys on typewriters and subsequently computer keyboards is sub-optimal from a purely technical perspective. Thus its persistence constitutes a puzzle for economists who are inclined to believe that technologies are chosen to maximize profits. The answer, according to David, is that the initial selection of the QWERTY arrangement was dictated by constraints imposed by early typewriter designs. Although the technological limitations that led to the QWERTY arrangement were quickly overcome, the keyboard arrangement they had dictated became locked in because of the interrelatedness between decisions by typewriter manufacturers about keyboard layouts and decisions by typists about which keyboard layout to learn, and was reinforced by the economies of standardization on a single layout. Once this standard became established the stock of typists and typewriters made the cost of switching prohibitive and perpetuated the inefficient arrangement of keys into the present.

Path dependence is most likely to arise in "network" industries, where the benefits of adopting a particular technology depend on the choices made by other individuals. For example, the value of instant messaging (IM) technologies depends on the number of other potential users who can be reached using a particular IM system. When competing IM technologies are incompatible with one another, potential adopters will have to consider not only the features of the particular systems, but how many of the people with whom they wish to communicate have already adopted the same system. As a result, the choices of early adopters will influence the evaluation of later adopters and may result in the emergence of a single dominant technology.

The history of technology evolution offers a number of examples that appear to fit the requirements for path-dependent diffusion to occur. In addition to the QWERTY keyboard, other examples include the competition between Windows and Macintosh operating systems for personal computers, the choice between VHS and Beta recording formats for videocassette recorders (VCRs), and the current competition between Blu-Ray and HD DVD formats for recording high-definition video disks.

In all of the above-mentioned cases, the choice of technology has been arrived at by an essentially unregulated process of market competition between alternative technologies, each being promoted by its own developers. In some cases a single *de facto* standard emerges. In others the market may be able to support a small number of alternative technologies. Alternatively, where the need for coordination is paramount, a standard-setting body may be established to enforce coordination on different parties.

Path dependence may also manifest itself in the realm of product design, where complex products involve the bundling of a variety of interrelated technologies. In these circumstances during the early phases of development there may be many different design variants. But over time one particular configuration often emerges as a dominant design (Utterback, 1996, ch. 2). This particular configuration comes in effect to define how the particular product is supposed to look and operate. Although the configuration may arise initially out of technologically imposed constraints, it may eventually become a constraint on future innovation.

Path dependence is an interesting feature of technology evolution because it implies that the choice of technology may depend on the distant past, and that the result may be inferior to other alternatives. As such path dependence poses a significant challenge to the conventional economic view that competition tends to move the economy toward the most efficient allocation of resources.

While accepting the theoretical possibility of path-dependent evolution, some critics have argued that it is unlikely to arise in reality. Among the most vocal critics of David's arguments are Liebowitz and Margolis (1990). They have offered empirical evidence challenging David's assertion that the QWERTY keyboard layout is inefficient, and argued on theoretical grounds that path dependence is unlikely because the existence of a superior technology creates opportunities to profit that will encourage firms to find ways to internalize the network externalities. For example, those who will benefit from a superior technology can subsidize early adopters or provide low-cost content to build up a larger user base. This appears to have been Apple's strategy in establishing the dominance of its iPod music player. By providing content at subsidized rates it has been able to establish a music format that is incompatible with other systems and lock in its dominance in the market for portable music players.

The strength of the Liebowitz and Margolis argument depends critically on the continued development of competing technologies, however. If the early dominance of a particular technological choice results in the abandonment of efforts to develop alternatives, it is possible that potentially more promising technologies may be abandoned without ever having the opportunity to be developed to a point where their superiority could be observed. In other cases the costs of switching may simply be too high to allow a significant change. One can, for example, view the co-evolution of the automobile and land use patterns in the United States as an example of this latter type of path dependence. The sunk costs of the current arrangements would make it very difficult to shift to a greater reliance on mass transit and higher density settlement patterns.

Closely related to the issue of path dependence is the establishment of standards. One way to deal with the problems of coordination that network externalities create is through the establishment of standards. In some areas, cooperative standard-setting bodies have emerged to insure a degree of order in technological choices. This is true, for example, in the development of Internet protocols. In other cases, however, *de facto* standards emerge through competition in the marketplace.

Examples of market-determined standards include the VHS videocassette format and the Windows operating system for personal computers.

The geography of technological progress

Historically, technology evolution has been characterized by a high degree of spatial clustering. New inventions and innovations occur in geographically localized clusters, and diffuse into wider use only gradually. This is apparent in the international divergence of technology since the Industrial Revolution. But it also operates within countries, where certain regions (e.g., the Silicon Valley area in California) tend to specialize in producing innovations.

In part, clustering appears to reflect aspects of the way in which knowledge is produced and communicated. Much technological knowledge is not subject to codification and formal communication. Instead it remains tacit, and only incompletely reflected in written and graphic explanations (Nelson and Wright, 1992). In the 19th century, for example, the early development of the American textile industry can be traced to the migration of skilled British mechanics with first-hand knowledge of the production and operation of the machinery used in spinning cotton yarn. Even today there are manifestations of this localization of knowledge transfer. Studies of the rate at which later patents cite earlier ones show that patents produced in the same state are more likely to be cited than similar patents produced elsewhere.

Learning-by-doing and learning-by-using constitute another mechanism through which technological progress tends to be localized. In the early phases of development, as an innovation is evolving rapidly, there needs to be a significant degree of two-way interaction between technology producers and technology users. As a result the two groups tend to co-locate within a relatively confined geographic region. For example, in the early 19th century U.S. textile production clustered in New England where key innovations in textile machinery were introduced rather than locating near the sources of raw materials in the southern region of the country. Similarly, although early automobile-manufacturing activities were spread across much of the northeastern and Midwestern U.S., the industry very quickly became concentrated around Detroit where several of the key innovators in the industry happened to be located.

Scholars have noted that this clustering represents the first phase of a characteristic technology life-cycle. As a technology matures, it becomes increasingly standardized and the need for frequent interaction between technology producers and technology users diminishes. As a result the value of locating production close to the sources of innovation diminishes and manufacturing is likely to move to areas where labor costs are lower. Thus in the late 19th century the U.S. textile industry migrated from relatively high-wage New England to the South. Now the industry has again relocated to even less expensive manufacturing locations overseas.

Another explanation for spatial clustering arises from economies of agglomeration. One important source of cost savings when economic activity is concentrated in a particular location is the benefit that workers possessing highly specialized skills and employers seeking such workers get from the size of the labor market. As the number of workers possessing specialized skills increases it becomes easier for innovating firms to find workers with the particular skills they require. By the same token, the concentration of employers at a single location increases the probability that workers will find employment that matches their skills. A second source of agglomeration economies arises from the ability of larger markets to support the development of specialized providers of inputs and services needed by innovating firms. These include legal and financial services as well as rapid prototyping and fabrication. Finally, because of the interdependent nature of production, highly skilled workers may find that working with other comparably skilled workers raises their productivity.

The existence of agglomeration economies is generally self-reinforcing. As in the case of path-dependent technology evolution, positive feedbacks mean that initially small locational advantages tend to be magnified over time. Once a particular location develops an advantage *vis-à-vis* other locations, it will tend to grow even larger, and its advantage will increase. Only when these attractions begin to be balanced by other rising costs—such as land prices—will this effect be moderated.

Technology transfer

The spatial clustering of innovation and the resulting geographic variation in technological evolution creates the potential for significant gains through the transfer of technology from more to less advanced countries or regions. Information about new products or processes of production is a non-rival good. That is, the use of this information by one economic actor does not diminish the ability of other economic actors to use it. Once it has been produced the knowledge does not need to be produced again. The only cost is the expense of communicating this knowledge. Yet historical experience reveals that the pace of technology transfer is highly uneven.

By the middle of the 19th century there already existed a highly developed international flow of information about production techniques between developed economies. Although the United States was a net importer of innovations from more advanced European countries throughout the 19th century, by the early 1850s it had begun to forge ahead in a number of areas, including the production of firearms, clocks, and other items requiring precision manufacturing of standardized parts. These achievements quickly captured European attention, and led to a number of delegations of skilled mechanics touring U.S. factories and working to import American ideas into Europe. Since this time the speed and density of these information flows has increased substantially.

Manufacturing techniques developed in Europe and the United States have flowed much more slowly to countries in other parts of the world, however. The reasons for these disparities in the transfer of technology are profoundly important since the transfer of more advanced technologies holds the promise of substantially improving living standards among the world's poor. The fact that progress has been limited and the poverty of some nations remains intractable suggests the complexity of the problem. Although scholars of this topic cannot offer easily implemented policy advice they have made progress in identifying the factors that influence the international movement of technology (Nelson and Wright, 1992).

To begin, we have already noted that communicating technological knowledge at a distance is far from costless. But this cannot be the entire explanation. A second important characteristic influencing the pace of technology transfer is the ability and willingness of receiving countries to borrow technology. Countries that have been successful in importing advanced technologies are characterized by investments in human capital that have prepared their workforce to adopt new technologies; and equally important they are open to borrowing ideas from abroad. In the case of Japan, the opening to and adoption of Western technologies was a very conscious decision; and one that contrasts with the Chinese effort in the 19th century to close themselves off from contact with the West.

A third factor affecting the pace of technology transfer is the co-evolution of technologies. The usefulness and value of many technologies is highly dependent on their interaction with other technologies. Relatedly, demand for particular types of goods is closely tied to income levels and other characteristics. Consequently it is not possible to transfer technologies piecemeal. An illustration of this phenomenon is provided by the simultaneous development of techniques of mass-production and the emergence of the modern, vertically integrated, multi-division corporate form of business organization in the United States around the turn of the 20th century. That these innovations emerged first in the United States reflects the fact that they were well adapted to the large, politically and economically integrated, and relatively wealthy U.S. economy. Europeans were well aware of these innovations and were not prevented from adopting them at the time. That they did not was due largely to the fact that they sold to smaller markets characterized by more unequal income distribution. After the Second World War, however, when the creation of the European Community and efforts to reduce trade barriers combined with transportation improvements to substantially expand European firms' access to markets, American manufacturing and management techniques spread quickly across Western Europe.

Conclusion

Technology evolution occurs through the expansion of the stock of useful knowledge and its application to the fulfillment of human needs. This is a process that is as old as human history. But the pace of change has accelerated dramatically over the past three centuries. Scholarship seeking to understand this phenomenon can be divided roughly into two themes. The first concerns the production of inventions and innovations. The central questions here concern the factors that determine the pace and characteristics of innovation, including the uneven distribution of innovation in space and time. The second set of concerns relates to the diffusion of innovations once they have been developed. Within this general theme studies have focused on the speed of diffusion, the role of path dependence and factors that either encourage or inhibit the international transfer of technology.

Many scholars of the subject would argue that technology evolution is the defining characteristic of modern capitalist economies. Understanding of precisely how capitalism and technology evolution are connected remains incomplete, but given the importance of technology evolution to rising living standards and the solution of social problems continued study of these issues appears essential.

References

Arrow, K. (1962). "The economic implications of learning by doing." *Review of Economic Studies*, **29**(2), 155–173.

Baumol, W. J. (2002). *The Free-Market Innovation Machine: Analyzing the Growth Miracle of Capitalism*. Princeton, NJ: Princeton University Press.

David, P. A. (1986). "Understanding the economics of QWERTY: The necessity of history." In: W. N. Parker (Ed.), *Economic History and the Modern Economist*. Oxford, U.K.: Basil Blackwell.

Carter, S. B. *et al.* (Eds.). (2006). *Historical Statisics of the United States* (Millenial Edition). Cambridge, U.K.: Cambridge University Press.

DeLong, J. B. (2000). *Cornucopia: The Pace of Economic Growth in the Twentieth Century* (NBER Working Paper No. 7602). Cambridge, MA: National Bureau of Economic Research.

Federal Reserve Bank of Dallas (1996). *Annual Report: The Economy at Light Speed*. Dallas, TX: Federal Reserve Bank of Dallas.

Freeman, C., and Soete, L. (1997). *The Economics of Industrial Innovation*. Cambridge, MA: MIT Press.

Hall, B. H. (2005). "Innovation and diffusion." In: J. Fagerberg, D. C. Mowery, and R. R. Nelson (Eds.), *Oxford Handbook of Innovation*. Oxford, U.K.: Oxford University Press.

Liebowitz, S. J., and Margolis, S. E. (1990). "The fable of the keys." *Journal of Law and Economics*, **33**(1), 1–25.

Mokyr, J. (1990). *The Lever of Riches: Technological Creativity and Economic Progress*. New York: Oxford University Press.

Nelson, R. N., and Wright, G. (1992). "The rise and fall of American technological leadership: The postwar era in historical perspective." *Journal of Economic Literature*, **33**(1), 1931–1964.

Perez, C. (2002). *Technological Revolutions and Finance Capital: The Dynamics of Bubbles and Golden Ages*. London: E. Elgar.

Rosenberg, N. (1982). *Inside the Black Box: Technology and Economics*. Cambridge, U.K.: Cambridge University Press.

Rosenberg, N. (1996). "Uncertainty and technological change." In: R. Landau, T. Taylor, and G. Wright (Eds.), *The Mosaic of Economic Growth*. Stanford, CA: Stanford University Press, pp. 334–356.

Schumpeter, J. A. (1939). *Business Cycles* (2 vols.). New York: McGraw-Hill.

Schumpeter, J. A. (1942). *Capitalism, Socialism and Democracy*. New York: Harper & Brothers.

U.S. Census Bureau (2007). *Statistical Abstract of the United States*. On-line resource *http://www.census.gov/compendia/statab/*

Utterback, J. M. (1996). *Mastering the Dynamics of Innovation*. Boston, MA: Harvard Business School Press.

3

Technology Transition

Ashish Sood and Gerard J. Tellis*[†]

*Emory University and [†]University of Southern California

Introduction

Many marketers think market segmentation is the most important engine of growth. (Yankelovich and Meer, 2006). On the contrary, it is technological change that is perhaps the most powerful engine of growth. Numerous examples can be cited from industry to support this claim. Technological change enabled the growth of Microsoft from a fledgling company to the colossus of the computer industry. The emergence of new classes of products (e.g., Internet-enabled products, Walkman, washing machines, etc.) suggests that technology creates new growth markets and fuels the growth of new brands. Finally, the meteoric rise of Amazon and Dell demonstrates how technological change propels small outsiders into market leaders.

Technology transition

However, firms cannot gain from technological change if they do not understand the phenomenon well. New product development and major investments in research depend upon a correct understanding of technological evolution in general and transition between leading technologies for various markets in particular. A central practical problem that faces managers is when to shift investments from the old to the new technology.

We discuss three important dimensions of technology transition: shape of technological evolution, relative performance of competing technologies during the transition, and dimensions of technological competition. In particular, we discuss the following questions:

- How do new technologies evolve?
- How do rival technologies compete?
- What are the transitions between technological changes?
- What are the performance dimensions of competition?

Currently, the main sources of answers to all these questions are limited findings in technology management literature (Anderson and Tushman, 1990, 1991; Foster, 1986; Sahal, 1981; Utterback, 1994). These sources promote a theory commonly known as "the theory of S-curve." This chapter examines this commonly accepted model of technological evolution and presents alternative perspectives.

Shape of technological evolution

Technology literature has coalesced around two aspects of the evolution of technologies: a strong consensus has developed about the phenomenon itself, while a consensus is emerging about the major explanation or theory for this phenomenon. Regarding the phenomenon, prior research suggests that technologies evolve through an initial period of slow growth, followed by one of fast growth culminating in a plateau (Foster, 1986; Sahal, 1981; Utterback, 1994). When plotted against time, the performance resembles an S-curve (see Figure 3.1).

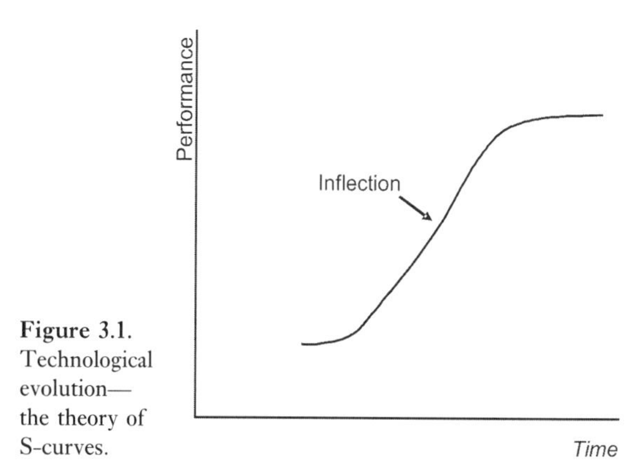

Figure 3.1. Technological evolution—the theory of S-curves.

Regarding the explanation of the S-curve, the field does not enjoy a single, strong, and unified theory of technological evolution. However, an emerging, and probably the most compelling explanation revolves around the dynamics of firms and researchers as the technology evolves through the three major stages of introduction, growth, and maturity.

In the introduction stage, a new technological platform initially makes slow progress in performance. Two reasons may explain this. First, the technology is not well known and may not attract the attention of researchers. Second, certain basic but important bottlenecks need to be overcome before any new technological platform can be translated into practical and meaningful improvements in product performance.

With continued research, the technological platform crosses a threshold after which it makes rapid progress and enters the growth stage. Three reasons may account for this change. First, the emergence of a dominant standard (Utterback, 1974), product characteristics, and consumer preferences coalesce on the new standard. Second, a large number of researchers attracted by the publicity of the new technology initiate research and produce improvements. Third, increase in sales of products that translate into greater support for research.

After a period of rapid improvement in performance, prior research suggests that the new technology reaches a maturity stage, a period when progress occurs very slowly or reaches a ceiling (Brown, 1992; Chandy and Tellis, 1998; Foster, 1986; Utterback, 1994) for various reasons. First, researchers' focus changes from product to process innovation. Second, fears of obsolescence or cannibalization leads firms to invest less in the new technology. Third, progress in the technology reaches limits of scale or system complexity.

Belief in this premise is so strong that it has almost become a law in the strategy literature. While the extant literature suggests that technological evolution follows an S-curve, it does not indicate the slope of this S-curve, the duration of the stages, or the timing or steepness of the turning points. However, there is scattered empirical support for this premise and limited theoretical support for various aspects of the S-shape curve (James and Sood, 2005; Sood and Tellis, 2005).

A recent study by Sood and Tellis (2005) suggests that the theory in this area has been partly confounded by the use of circular definitions. Many terms used to describe innovations such as revolutionary, disruptive, discontinuous, or breakthrough (Freeman, 1974; Garcia and Calantone, 2002; Schumpeter, 1939; Tushman and Anderson, 1986) are intrinsically problematic because they define an innovation in terms of its effects rather than its attributes. If the definitions are then used to predict market outcomes (e.g., new entrants displacing incumbents from disruptive technologies), researchers run the risk of asserting premises that are true by definition.

To avoid such circularity, Sood and Tellis (2005) define three types of technological change: platform, component, and design based on the intrinsic characteristics of the technology. A platform innovation is defined as the emergence of an entirely new technology based on scientific principles distinctly different from those of existing technologies. For example, the compact disk used a new platform—laser optics—to write and read data, whereas the prior technology used magnetism. A component innovation is defined as one using new parts or materials in the same technological platform (e.g., magnetic tape, floppy disk, and zip disk differ by use of components, although all are based on the platform of magnetic recording). A design innovation is defined as a reconfiguration of the linkages and layout of components within the same technological platform; for example, the changes in floppy disks from 14 to 8 inches, to 5.25 inches, to 3.5 inches, and to 2.5 inches, although all based on the platform of magnetic recording (Christensen 1992a, b, 1993). Within any platform innovation, performance improves due to innovations either in components or design or both.

Based on a dataset of 23 technologies drawn from five industries—data transfer, computer memory, desktop printers, display monitors, pharmaceuticals, and electrical lighting, Sood and Tellis (2005) report sparse support for the hypothesis that the path of technological evolution resembles an S-curve with either a visual examination of the plots or a more formal test using nonlinear regression techniques. In a majority of technologies, they find long periods of static performance interspersed with abrupt improvements in performance. These plots suggest a series of irregular step functions better approximated with multiple S-curves than a single S-curve. Across these step functions within a technology, estimates of growth rate and especially performance at maturity differ substantially.

The critical importance of these results is the following: An analyst expecting an S-shape curve would conclude that the periods of static performance meet the hypothesis and that the technology has matured at the upper asymptote, when indeed it has not. Substantial improvements in performance after the first plateau suggest the gravity of the error in abandoning the old technology prematurely.

Technological transition and performance of competing technologies

Do the paths of two technologies ever cross? If so how many times? Foster (1986) and Christensen (1997) postulate the following chain of events in the evolution of competing technologies. Sometime in the life of an old

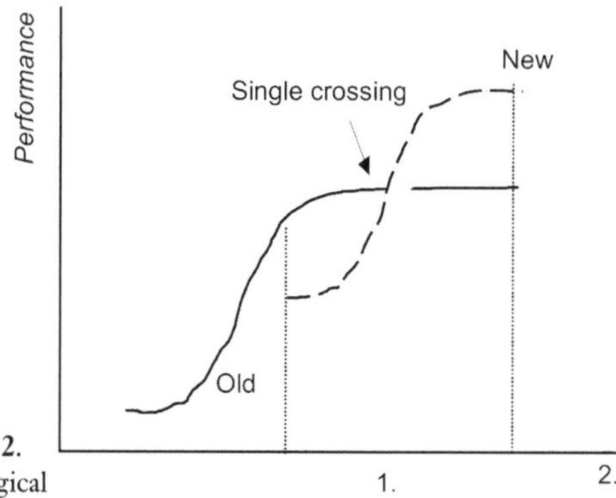

Figure 3.2. Technological evolution.

technology, a new technology emerges and makes slow progress on the primary dimension. Later it enters its growth phase and improves rapidly. In contrast, the old technology improves at a much slower rate. As a result, the new technology crosses the old technology in performance only once. This crossing of the old technology is also a signal of the end of its efficient progress (see Figure 3.2).

Numerous authors have derived strong managerial implications based on this premise (Christensen, 1997; Foster, 1986). They warn that even though managers might be able to squeeze out improvement in performance from a mature technology, the improvement is typically costly, short-lived, and small. Thus the primary recommendation of existing literature is that managers quit a maturing technology and embrace a new one to stay competitive.

However, Sood and Tellis (2005) report that a majority of new technologies perform better than the old technology, right from the time they were introduced (see Figure 3.2). Also many new technologies never improve over the old technology, while others enjoy brief spells of dominance over the old technology before the old technology regains dominance. This unexpected pattern of evolution results in three distinct types of crossings between any pair of successive technologies—no crossing at all, multiple crossings, and single crossing (see Figure 3.3).

In summary, the final status of each technology cannot be determined solely from the direction of the attack or timing of introduction. As such, it might be fatal for an incumbent to scan for competition only among technologies performing worse than its current technology. Moreover, managers expecting a single crossing are likely to be quite surprised and may make unwise decisions.

Dimensions of technological competition

Certain secondary dimensions become important as technology evolves. Progress occurs systematically along the first dimension, then moves to the second, then to the third, and so on. These dimensions form the bases of inter-technological competition. They also form the bases by which consumers choose among rival technologies or products.

The literature also suggests that the basis for such competition is quite standard and occurs in the same form across markets. For example, Christensen (1999) points out four generic dimensions of inter-technological competition: functionality, reliability, convenience, and cost. Product functionality is the primary attribute on which consumers choose products in that category. Similarly, Moore (1991) suggests that products start competing on consistent performance, or higher reliability, after subsequent innovations increase functionality beyond a certain point. Christensen (1997) suggests that after product functionality and reliability requirements are satisfied, firms become more willing to customize product designs to meet customers' specific requirements, such as convenience. Abernathy and Clark (1985) propose that the product becomes a commodity and progress occurs through price reductions once the technology has progressed up the S-curve sufficiently on functionality, reliability, and convenience. The occurrence of such generic dimensions can be important in guiding firms about the path of evolution and the direction of the next competitive attack.

Sood and Tellis (2005) suggest another perspective on the transitions between dimensions of technological competition. They suggest that each platform technology offers a completely new secondary dimension of competition while still competing on the primary dimension (see Table 3.1). For example, consider four successive technologies in monitors: CRT, LCD, plasma, and OLED. CRT monitors were initially introduced on the basis of resolution. Each subsequent technology was inferior in resolution at the time of introduction, but introduced a new important secondary dimension: resolution, compactness, screen size, and efficacy.

Technologies that excel in a particular dimension cater to particular segments that value that dimension. When the mass market focuses on one old or new dimension, niches, interested in the other dimensions, might still survive. For example, thermal printers are a popular choice in printing high-resolution pictures.

Conclusions

We can summarize the following findings of prior research on technology transition. First, using the S-curve to predict whether the performance of a technology is risky and may be misleading for two reasons: one, most of the technologies do not even demonstrate an S-shape performance curve. Two, several technologies show

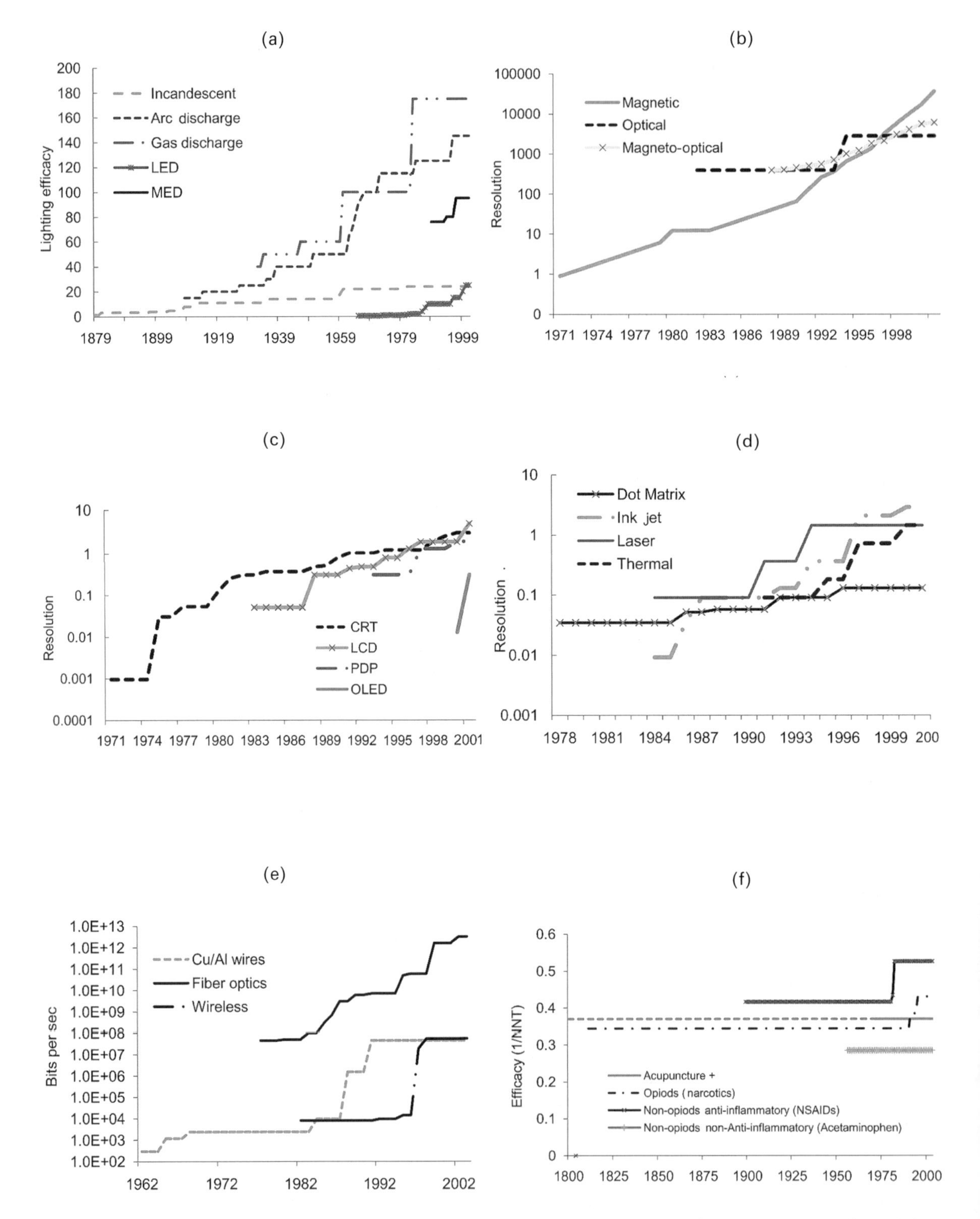

Figure 3.3. Technological evolution in six categories: (a) external lighting, (b) desktop memory*, (c) display monitors*, (d) desktop printers*, (e) data transfer*, (f) analgesics. *Note:* *performance on *y* axis is in log scale; $^{+}$accurate performance records of efficacy of acupuncture not available prior to 1971.

Table 3.1. Emergence of secondary dimensions of competition.

Market	*Secondary dimensions*
External lighting	Brightness, color rendition, light efficacy, compactness, life
Desktop memory	Capacity, reliability, life
Display monitors	Resolution, compactness, screen size, efficacy
Desktop printers	Resolution, graphics, speed, continuous color rendition
Data transfer	Transfer speed, bandwidth, connectivity/mobility
Analgesics	Analgesia, reaction speed, targeted-action, risk–benefit balance

multiple S-curves suggesting that a technology can show fresh growth after a period of slow or no improvement.

Second, the continuous emergence of new technologies and the steady growth of most technologies suggest that relying on the *status quo* is deadly for any firm. Moreover, technological progress is occurring at an ever-increasing pace. As such, paranoia rather than complacency is healthy.

Third, the attack from below remains a viable threat. Many new technologies start by offering low performance but later threaten old technologies by improving at a much faster rate. On the other hand, new technologies can perform better than old technologies even at the time of introduction. This fact heightens the threat of competition.

Fourth, another threat to incumbents is the emergence of secondary dimensions of competition. Old technologies may be completely vulnerable to these dimensions. Faced with such threats, incumbents need research to identify technological solutions to improve the value of the old technology aswell as to identify market segments that value the contributions of the old technology.

Fifth, first-mover advantages may not be lasting since entrants introduced even more innovations than incumbent firms. However, even if incumbents fail to introduce a particular new technology, all is not lost. They need not throw in the towel and divert all resources to the new technology. Many old technologies demonstrate high levels of improvement even after being dormant and static for many decades, and in some cases regain dominance. In contrast, a misplaced belief in the theory of S-curves may become a self-fulfilling prophecy and lead to the premature demise of an old technology.

References

Abernathy, W. J., and Utterback, J. M. (1978). "Patterns of industrial innovation." *Technology Review*, **80**(7), 40–47.

Abernathy, W. J., and Clark, K. B. (1985). "Innovation: Mapping the winds of creative destruction." *Research Policy*, **14**(1), 3–22.

Anderson, P., and Tushman, M. L. (1990). "Technological discontinuities and dominant designs: A cyclical model of technological change." *Administrative Science Quarterly*, **35**(4), 604–633.

Anderson, P., and Tushman, M. L. (1991). "Managing through cycles of technological change." *Research Technology Management*, **34**(3), 26–31.

Brown, Rick (1992). "Managing the 'S' curves of innovation." *Journal of Consumer Marketing*, **9**(1), 61–73.

Chandy, R. K., and Tellis, G. J. (1998). "Organizing for radical product innovation: The overlooked role of willingness to cannibalize." *Journal of Marketing Research*, **35**(4), 474–487.

Christensen, C. M. (1992a). "Exploring the limits of the technology S-curve, Part I: Component technologies." *Production and Operations Management*, 334–357.

Christensen, C. M. (1992b). "Exploring the limits of the technology S-curve, Part II: Architectural technologies." *Production and Operations Management*, 358–366.

Christensen, C. M. (1993). "The rigid disk-drive industry: A history of commercial and technological turbulence." *Business History Review*, **67**(4), 531–588.

Christensen, C. M. (1997). *The Innovator's Dilemma: When New Technologies Cause Great Firms to Fail*. Boston, MA: Harvard Business School Press.

Christensen, C. M. (1999). *Innovation and the General Manager*. Boston, MA: Irwin/McGraw-Hill.

Foster, R. (1986). *Innovation: The Attacker's Advantage*. New York: Summit Books.

Freeman, C. (1974). *The Economics of Industrial Innovation*. London: Pinter.

Garcia, R., and Calantone, R. (2002). "A critical look at technological innovation typology and innovativeness terminology: A literature review." *Journal of Product Innovation Management*, **19**(2), 10–32.

James, G. and Sood, A. (2005). "Performing hypothesis tests on the shape of functional data." *Computational Statistics and Data Analysis*, **50**(1).

Moore, G. A. (1991). *Crossing the Chasm: Marketing and Selling High-tech Goods to Mainstream Customers*. New York: HarperBusiness.

Sahal, D. (1981). "Alternative conceptions of technology." *Research Policy*, **10**(1), 2–24.

Schumpeter, J. A. (1939). *Business Cycles: A Theoretical, Historical, and Statistical Analysis of the Capitalist Process*. New York: McGraw-Hill.

Sood, A. and Tellis, G. J. (2005). *The S-Curve of Technological Evolution: Strategic Law or Self-Fulfilling Prophecy* (Working Paper No. 04-116). Boston, MA: Marketing Science Institute.

Tushman, M. L., and Anderson, P. (1986). "Technological discontinuities and organizational environments." *Administrative Science Quarterly*, **31**(3), 439–465.

Utterback, J. M. (1974). "Mastering the dynamics of innovation." *Science*, New Series **183**(4125), 620–626

Utterback, J. M. (1994). *Mastering the Dynamics of Innovation.* Boston, MA: Harvard Business School Press.

Yankelovich, D., and Meer, D. (2006). "Rediscovering market segmentation." *Harvard Business Review*, **84**(2), 122–131.

4
Technology Intelligence

Thomas Durand

Ecole Centrale Paris

4.1 Scope of technology intelligence[1]

Technology intelligence (TI) is a broad term that includes the gathering and compiling of technical information, developing technology foresight, monitoring the advancement of science and its anticipated consequences for subsequent technology development. It includes

1. Scanning of technological options potentially important for the future, technology assessment, technology mapping, and roadmaps.
2. Competitive intelligence gathering to follow or even anticipate technological choices made by competitors and to assess the volume and focus of their R&D investments in new technologies while monitoring their overall portfolio of technologies.
3. Exploitation of databases of patents and other IPR (intellectual property rights).
4. Survey of technology markets where technologies are transferred between organizations, monitoring of technology-based strategic moves (partnerships and alliances, mergers and acquisitions), etc. Although TI activities may apply at several levels (e.g., the firm and business unit level, the industry level, or even at the national policy level for the national system of innovation—Durand, 1996), the case of TI at the firm level is primarily considered here. Technology intelligence is thus the technical side of business intelligence. TI feeds into the strategy development process, identifying what is/could become possible (as far as technological feasibility is concerned) and what is/could become attractive (as far as technology performance and costs are concerned).

By extension, technology intelligence indirectly covers the theme of organizational knowledge, capabilities, and competence. As technology is one of the forms of competence in and around organizations, technology intelligence is about understanding technology dynamics, technology positions, and technical competence in a competitive context to build and implement informed strategies.

In this chapter we will summarize the objectives, functions, and tools of TI in that order.

4.2 Objectives of technology intelligence–gathering

TI serves two objectives: strategy and innovation. From a strategic viewpoint, TI aims at ensuring that the strategy development process properly takes technology into account. This means assessing the current technology portfolios of competitors compared with the firm's portfolio, anticipating and monitoring the dynamics of technology in the environment, and ensuring that the firm prepares for the technologies and technical competence that will be necessary for its future activities (i.e., those activities that are targeted through the strategy). In this sense the firm should no longer be seen as only or primarily a portfolio of product segments, but also as a portfolio of technologies and technical competencies (Figure 4.1).

[1] When the word was first introduced, "technology" formally meant the study of techniques as much as sociology is the study of the social. However, over time the word has been used to describe the technical itself, but in a specific sense: technology is a technique that may be explained by science, at least in part. In other words, a technique is essentially empirical and thus local while technology is a technique for which the fundamental mechanisms are partially understood through science. This means that technologists can anticipate whether a technology may or may not be extrapolated to other contexts and/or recombined to other technologies, thus offering a rich potential for new applications, while techniques are bound to remain limited as extrapolations and recombinations may be obtained only after real experimentations—a costly and lengthy process. Ansoff (1987) claims that the explosion of new technologies during the second half of the 20th century in fact stems from this potential of recombination which technologies demonstrate, via their scientific foundation.

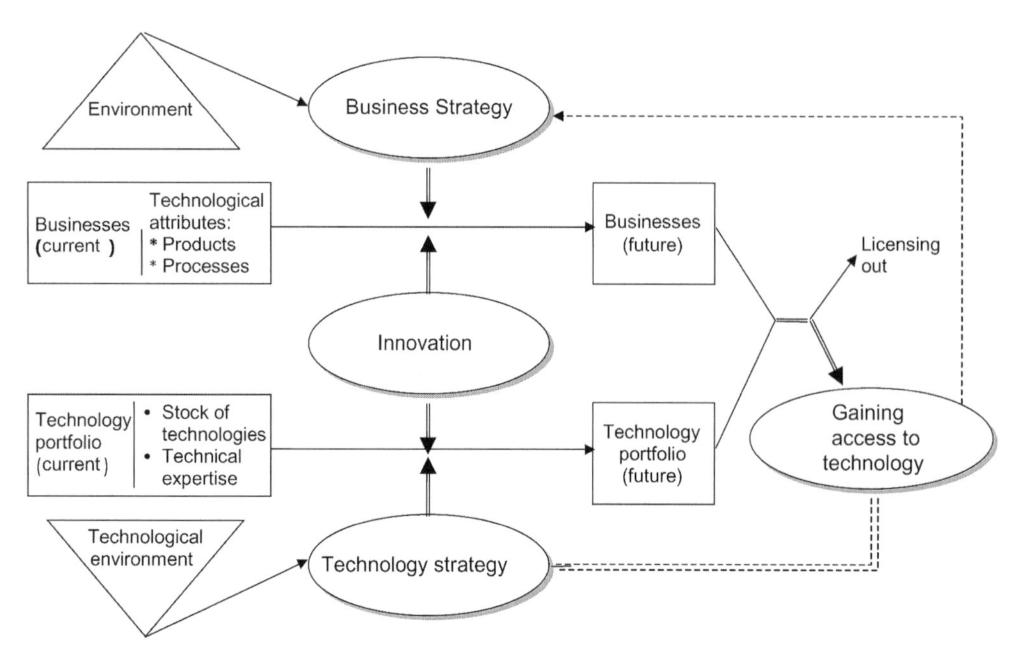

Figure 4.1.
Source: Durand in EITIM (2004).

TI also serves the innovation process. Business intelligence and technology intelligence feed strategy while contributing to the generation and/or collection of ideas. These ideas may come from within the organization but also from the outside; hence the relevance of intelligence activities that scan the outside. The ideas collected may trigger potential innovations to serve market needs and internal processes (the upper part of Figure 4.2). The ideas may also lead to technical developments (the lower part of the diagram) which could subsequently contribute to innovation projects.

In other words, TI feeds into both the strategy process and the innovation process.

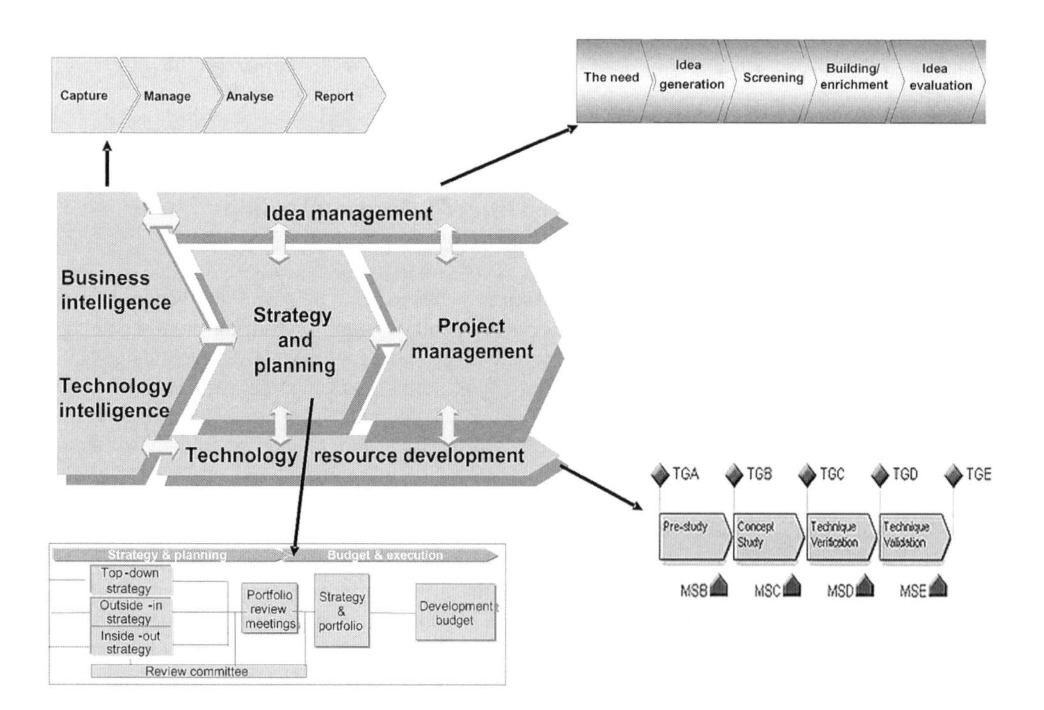

Figure 4.2.
Source: Adapted from Tetrapak.

4.3 Functions of TI

TI involves a set of activities, each of which is summarized below.

4.3.1 TI surveys technologies

Technologies deal with offerings for clients (products and services) as well as internal processes (manufacturing and logistics). Technologies also relate to soft technologies (Durand and Dubreuil, 2001), organizational innovations, or even business models (e.g., information technologies for call centers, or Internet-based distribution). In this sense, TI is about technology and its attributes. Technology may be assessed according to a variety of attributes: cost, performance (itself a multidimensional concept), functionalities, and sub-functionalities fulfilled. However, a discussion of attributes cannot be completely separated from needs. In fact, when assessing a future technological option, it is necessary to describe the types of usages and contexts of use for that technology. In other words, when identifying and assessing a new technology, TI *de facto* addresses the needs which the technology might serve. When technology intelligence exercises are conducted to identify technologies which may be key for the future, it appears that a majority of the items identified are in fact key needs and not just key technologies (e.g., "noise reduction technology" signals a need, not a technology; similarly "miniaturization" is a need while "nanotechnologies" are potential solutions).

Technologies and functional needs are thus two sides of the same coin (Durand, 2003). This is illustrated in Table 4.1 where this technology/functional need duality is further expanded. The table shows a six-column grid designed to enumerate the items identified as potential key technologies. The grid clarifies the definition, the scope, and content of each item. The table should be read starting from the italicized part (*Micro-encapsulation* in this illustration) which operates as the flag that symbolizes the key item selected, while the elements in the other columns essentially describe and illustrate the key item. In that sense, a key technology presented in the grid is both the flag (i.e., the italicized item) and the entire corresponding line of the matrix. The left part of the grid deals with the demand side (with one or several functional needs fulfilled—column 3; examples of applications—column 2; and industries downstream, i.e. typical sectors of the economy where the applications may be encountered, and thus where the corresponding technology may be used—column 1). The right part of the grid characterizes the technology (the technical solution to the functional need—column 4; the most critical aspects of the technology that may be a bottleneck, preventing full deployment of the technology, thus indicating where new developments or breakthroughs are needed—column 5; typical scientific fields where research is needed to improve the technology—column 6).

4.3.2 TI aims at surveying both current and future technologies

Beyond assessing current technologies, TI covers new technological developments and potential options for the future. The literature has divided the life cycle of technologies into phases: mature or base technologies (those readily available with a variety of external suppliers), key technologies (those which bring today a potential source of competitive advantage), pacing technologies (those currently under development which are believed to be potential future key technologies), and emerging technologies (those for which scientific progress suggests that they are promising options for the future). These categories were popularized by consultants such as ADL (Floyd, 1997; Van der Eerden and Saelens, 1991). TI covers the entire span: surveying suppliers of mature technologies; monitoring the performance, cost, and use of key technologies; identifying and assessing pacing technologies wherever they are being developed; and conducting foresight exercises to identify new technological options that may stem from upstream scientific

Table 4.1. Key technology foresight: a grid to qualify the items selected.

Industries (*column 1*)	*Example of use* (*column 2*)	*Function fulfilled* (*column 3*)	*Technology* (*column 4*)	*Critical techno points* (*column 5*)	*Scientific domains* (*column 6*)
Pharmaceuticals, Cosmetics	Drug administration	Controlled release of substance	*Micro-encapsulation*	Molecule cage	Molecular chemistry
Waste management	Confining pollutants	Controlled confinement			

Source: Durand (2003).

research activities. Thus, technology intelligence analysts monitor the dynamics of technology development, dissemination, and obsolescence. It is about understanding the stage of technology maturity, potential lock-in situations and paradigmatic shifts. One may actually conceptualize the grid of Table 4.1 from the evolutionist perspective (Dosi, 1982; Durand, 1992; Nelson and Winter, 1977): in early immature situations, when no dominant technology has yet appeared, technological uncertainty is high. Beyond the current dominant technology, several competing options may be considered for the future to fulfill a generic need, but most are still to be developed and it is thus difficult to assess which one may eventually win, if any. In such instances, technology intelligence is logically bound to identify the generic functional needs to be fulfilled (Table 4.1, column 3), listing the many competing technological options considered for the future (column 4). At this stage it may be too early to use column 5 to specify the technological bottlenecks of each option. The case of capturing and storing CO_2 illustrates this situation. This environmental need is clearly recognized but many competing technologies are being considered, while none has taken the lead yet. It is thus legitimate to flag this key need in column 3, while using column 4 to list the many technological options at hand. When a promising technology emerges as potentially dominant for some market segments (e.g., dominant design and dominant process—Abernathy and Utterback, 1975), it is possible to flag the emerging dominant technology, thus going beyond identification of the generic need. The flag (i.e., the italicized element of the line in Table 4.1) for the item selected may thus shift from column 3 to column 4. The critical technological difficulties behind the technology item may then be listed under column 5. Conversely, it should be noted that additional applications of the technology may have appeared along the way, thus providing a longer list (for column 3) of functional needs fulfilled by the key technology identified. This may be illustrated by the case of fuel cells. Fuel cells are being developed to generate energy for cars, buses, and trucks, but applications are also being considered for housing as well as mobile telephones.

4.3.3 TI involves tracking competitors

TI is finally about identifying which competitors may be better positioned to benefit from technological change, be it incremental change or disruptive (Christensen, 1997). Technological change may favor the incumbents if technological innovation is trapped within the rails of the technological trajectory in the same paradigm (Dosi, 1982). Conversely, if radical innovation strikes through a paradigmatic shift, thus shaking the entire industry out, technological change may benefit those new entrants having a portfolio of technologies and a competence base that better fits the requirements of the arising new dominant technology. It is a matter of bridging the competence gap (Durand and Guerra Viera, 1997); that is, bridging the gap between the existing portfolio of competencies and the set of competencies required by the new technology. This is where the concept of competence and "transilience" (Abernathy and Clark, 1985; Durand, 1992)—that is, the competencies that will resist through the transition—may help assess the potential impact of anticipated technological change. In addition, TI is about assessing the speed at which the new technology may disseminate, when it may occur, and the sequence of cannibalization of market segments over time.

In sum, TI is about understanding current and future technologies and competence, thus including scientific activities upstream, while covering functional needs downstream as well. TI is also about assessing potential lock-in situations on a specific technology trajectory in a given paradigm, and conversely anticipating potential radical innovations and disruptions, and assessing the timing, speed, and extent of the dissemination of the change, as well as the most likely beneficiaries.

4.4 Sources for TI

Information can be gathered from a variety of sources. As in any other intelligence activity, no single source will consistently bring the most significant pieces of information. Instead, it is important to access and combine different sources to cross-check data. I will discuss four major sources.

1. Publications represent a first accessible set of information. These include academic scientific journals, doctoral theses, technical and engineering reviews, professional technical magazines, as well as reports from government agencies or industry associations. In other words, publications span from peer-reviewed articles to the so-called "gray literature". The quality and reliability of the contents obviously varies accordingly. Yet, TI needs to scan through all of the available literature, with the provision that each source is weighted according to the relative value attached to the quality of its content.
2. Patent databases are a second interesting source for technical information. Typically, firms or individuals filing patents announce to the world the content of their invention. From that, they expect to be granted a monopoly on the technology for 20 years. This means that patents contain detailed descriptions of technologies which may or may not have been put to work. Exploiting the information available in patent databases may thus be extremely relevant for TI activities. However, not all technologies are protected via patents. In some industries, secrecy is preferred to

patents, especially when it comes to manufacturing processes, as patents are believed to be circumnavigated after only a few years (Mansfield, 1985). In addition, some players choose to file many patents around an invention, as a trick to lure competitors away from the real technology which is hidden within a foggy set of many other options. In addition, when exploiting patent databases, one should be careful to go beyond simply following the ongoing flow of new patents being granted. Instead, it is important to (re)-set the stage of the history of patents filed over the previous years on a given topic, analyzing the filing policies of the participating firms, reconstructing the linkages between patents (and technologies) via referencing among patents, and thus identifying the technological trajectories explored by the various players active in the business. This background work may then help interpret the flow of incoming patents, the corresponding technologies, and the underlying technical competences that support them. In turn, this may then help understand the technology strategy of competitors. This is typical of TI activities.

3. A third major source of information stems from so-called technology experts. This is a tricky business as the validation of individual expertise is seldom certified. In addition, some of the recognized experts may have a vested interest in the technological options which they advocate. Yet, discussions with the key individuals who are believed to bear significant pieces of the knowledge available in an industry usually prove extremely useful. This requires interviewing protocols where the data stemming from the literature and the databases have been exploited and where the experts are confronted with these data and asked to explain and justify their views. This also means challenging them with arguments heard from other experts. As in any controversy, structured discussions often prove to bring light on the matter at hand. This means that interviews with experts—and even more so with the best experts—should come after all other sources have been investigated and exploited.
4. A fourth source of information for TI is made up of an array of qualitative, informal information gathering. This includes interviewing suppliers (e.g., suppliers of key equipment), participating at fairs and conferences, collecting information from internal R&D staff who had contact with their peers from competing firms or from public research at home or abroad, reviewing the technical partnerships and alliances being announced, etc. In this context, use of the Internet is obvious. Yet, it should be stressed that the quality of information circulating on the web is highly uncertain and variable. Nevertheless, the Internet offers new tools to reach out for diffuse technical sources that may prove relevant. As an example, electronic marketplaces (e.g., InnoCentive) offer intermediation services between potential respondents worldwide and firms looking for a technological solution to a specific need that they present (while remaining anonymous). A predefined amount of money is promised by the firm posting its problem, should an appropriate technical solution that fits their need be submitted via the website. This is a typical new way for firms to innovate and to conduct TI activities (e.g., before launching costly and lengthy technical developments of their own). Conversely, the shopping lists shown on these electronic marketplaces represent a new source for TI for those who are interested to hear about unsolved problems in their industry.

In sum, TI scans through publications and patent databases, while calling upon technical experts and additional informal sources (suppliers, fairs, conferences, etc.). It should be stressed, however, that the firm is surrounded by a set of partners (clients, suppliers, distributors, universities, consultants, and other similar partners) who have historically built professional and social links into the organization. These are the usual suspects who constitute the ecosystem of the firm. The ecosystem nurtures the firm (as information flows into the company from its partners) but generates biases, if not myopia (as information is in fact filtered and distorted by these historical partners who indirectly bring their own cognitive and empirical limits into the firm—unintentionally and sometimes intentionally). As the ecosystem results from connections established with external partners over the years to solve problems of the past, it is important for TI activities to reach out for other sources of information, beyond the ecosystem itself. This may mean revisiting the ecosystem of the firm and questioning whether some key players should not be approached to enrich the set of partners from which the firm nurtures itself. In that sense, TI also deals with identifying targets for alliances and partnerships to renew the ecosystem, thus providing new sources of learning for the firm.

4.5 Methods of TI

TI activities rely upon two main types of methods, surveys and foresight.

1. Surveys are classical ways of collecting information from the categories of sources listed above. A specific topic is to be investigated (e.g., the future of fuel cells or the technology for mobile telephony in 2020). A task force is created. The literature is reviewed extensively. Ongoing activities in research labs around the world are monitored via research reports, publication scanning, exploratory interviews, company reports, and similar sources. Participants in the study are sent to key

conferences and fairs to get a sense of what the community is anticipating. Along the way, a set of key individuals, referred to as technical experts in the field, are identified. In addition, a set of potential candidate technologies are listed and documented (expected performance, potential costs, limits, pros and cons). This will serve as the background material to conduct in-depth interviews with those who were identified as technical experts. What will emerge from those interviews are the views widely shared among experts but also controversies, major questions, and dilemmas. These will constitute the core of the TI contribution (see below for deliverables and formats).

2. Foresight methods are more specific and formalized. Foresight is about future studies (i.e., studies about the futures—plural). A foresight exercise aims at creating a set of scenarios which gives a feel for the variety of potential futures. If strategy is about how to reach a desirable future, foresight is about attempting to grasp the diversity of potential future contexts and outcomes. It should be stressed that foresight is not equivalent to forecasting (Godet, 2001; Martin and Irvine, 1989; Salo, 2001; Wack, 1985a, b). Unlike forecasting, foresight is not about predicting the most probable future. Rather, foresight is about describing the variety of potential futures, in order to allow stakeholders to prepare for this variety and to contribute to shape outcomes in the direction they wish. In this sense, foresight is an input for strategy. This applies to technological foresight, as part of TI.

4.6 Tools of TI

The foresight tool box contains several methods (Godet, 2001), some of which are worth mentioning here.

1. The Delphi technique has been widely used since its development at the Rand Corporation in the 1950s. It consists of a sequence of rounds of expert judgments about a specific question such as "When could we see technology *X* become commercially viable?" Answers from each expert are then communicated to all other participating experts. On that basis, each of them is requested to revisit the matter and respond once again. The Delphi technique may be seen as a survey technique and it has been primarily used as a forecasting tool but it may serve foresight purposes as well: instead of looking for a consensus, the process may exploit the unconventional views expressed by some of the experts, inviting those experts who significantly depart from average answers to explain why they disagree with their peers. This may then help shape alternate scenarios, apart from the opinion of the mainstream.
2. *Scenario building*. This may be done directly by a group through the description of potential visions they might have about the future. It may also be done by calling upon other techniques such as the two described below. The idea behind the use of scenarios is to invite strategists to test the relevance of their strategic options against each scenario, thus preparing decision makers for the variety of potential futures and outcomes.

Structural analysis is a technique which analyzes the cross-impacts of variables within a system as a way to identify those variables which drive the system, those that operate as relays, and finally dependent variables. Once a foresight issue is raised (e.g., "what could be the future(s) of nuclear energy?"), a first step is to define the contour of the system to be studied. Variables that describe the current and future states of the system are then identified (e.g., in our example "world consumption of energy", "oil price trends", "public acceptance of nuclear technologies", or "availability of uranium", etc.). The technique then needs to systematically assess the influence of each variable on one another via a matrix where the variables appear both in lines and columns. The more influence a variable has on other variables—this appears as the sum of a line—the more driving power it has. The more a variable is influenced by other variables—this appears as the sum of a column—the more dependent it is in the system. As this analysis takes place over time, participants involved in the process share their knowledge and understanding of the issue at hand, and clarify their topics of agreement and disagreement. As a result, they are progressively put in a position to build and shape potential scenarios for the future of the system.

Another foresight technique, known as *Mactor*, operates in similar ways but concentrates on stakeholders and the families of players in the system. The analysis first describes the power and strategic means of influence which each family of players can exert on the others. Second, it identifies the main challenges facing the participants in the system. Third, it aims at assessing when two families of players taken together have shared (or conflicting) interests on each of the challenges identified. As a result, the analysis leads to a map of convergence/divergence among players in the system. As for structural analysis, this whole process again finally leads to the building of scenarios for the future of the system.

In short, foresight techniques aim at describing potential states of the future via sharply differentiated scenarios intended to help strategists open up their thinking to unconventional potential outcomes. This is quite relevant for TI as technologies often unfold in unexpected ways.

4.7 Deliverables, techniques, and format

The format of the deliverables of TI is important as communication and sharing are essential to feed into both the strategy process and the innovation process.

1. The grid presented earlier as Table 4.1 illustrates a first type of TI deliverables, by which potentially promising technologies are described through both their technical dimensions and their potential applications. This was discussed above.
2. A more detailed format for an output of TI is a dual-technology tree (Durand, 1992). This in fact visualizes part of the hierarchies of design and processes discussed by Clark (1985). The idea is to start from a generic functional need to identify all past, current, and foreseeable technologies capable of fulfilling that need. Figure 4.3 presents an example of a dual-technology tree for the treatment of diabetes, with corresponding insulin

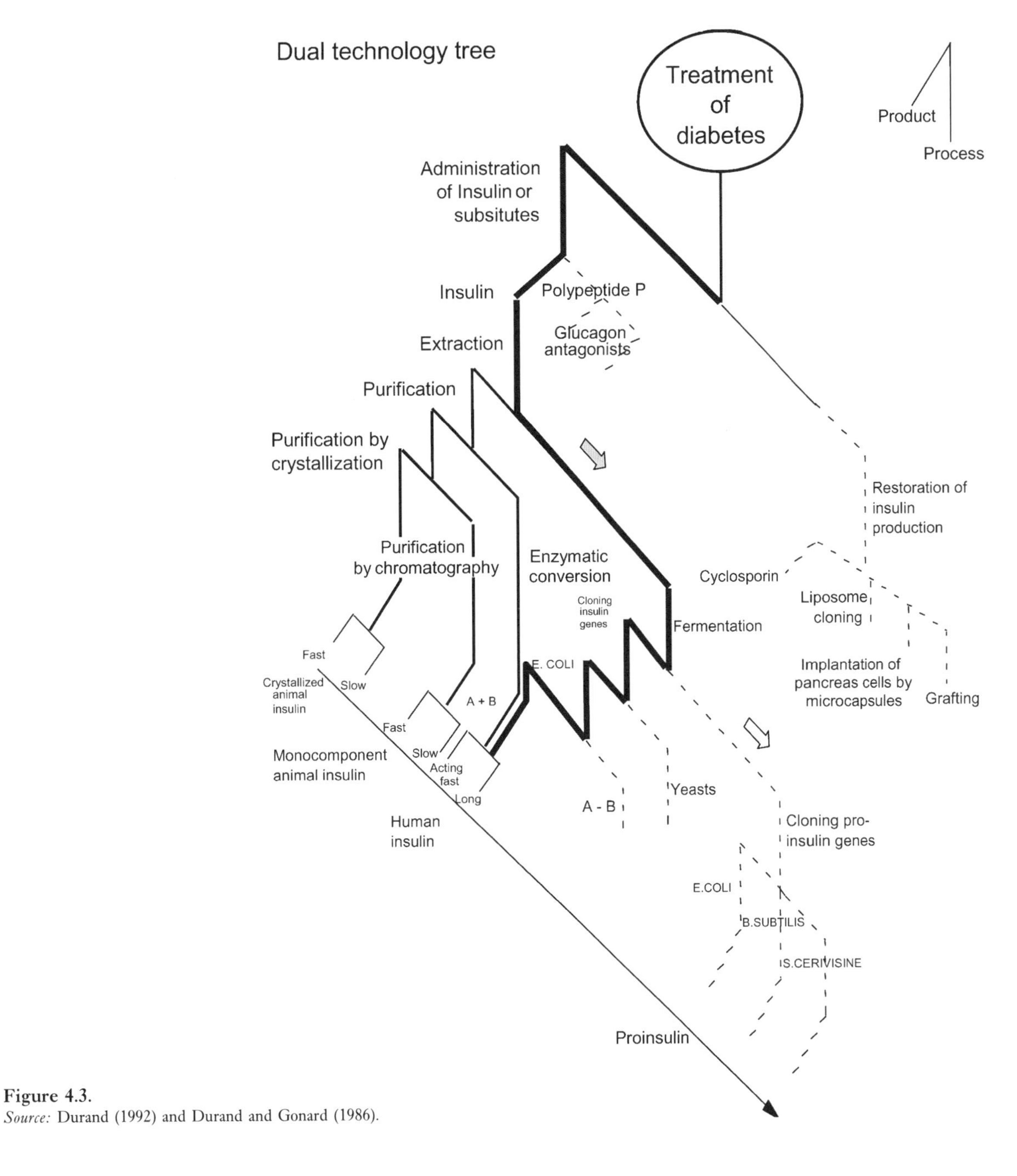

Figure 4.3.
Source: Durand (1992) and Durand and Gonard (1986).

technologies. A technological option is presented as a branch in the tree. In addition, fundamental choices (radically different ways of fulfilling the need) are presented high up in the tree, while simple variations are shown in the lower parts of the branches. This hierarchy relates to the fit between the competence base of the firms and the specific requirements of each technology. The tree shows product design options (horizontally) as well as processes (vertically), illustrating how intricately interwoven process and product design choices are; hence the use of the word "dual" to qualify the technology tree. Finally, the tree shows past technologies as normal lines on the left, today's current dominant technology in the bold lines in the middle or so, and potential technologies for the future as dotted lines on the right of the tree. This representation from left to right thus gives a sense of the past and current trajectories to help envision future technological options in context.

This form of representation actually relates to real options theory, as branches are in fact potential technological options for the future. One can further monitor competitors' activities on the map to keep track of their technological moves (R&D investments on some of the options, patent filing, acquisition of technologies, partnerships with university labs, etc.). The dual-technology tree may also be seen as a map of a territory partially known, but with unknown zones to be explored via scientific and technological search. The convention and format of the map may be adapted according to contexts; yet, a major point is that TI should deliver some forms of maps to inform strategists and decision makers about potential future technological options. This leads to a more common type of map: the roadmap.

3. A roadmap is a relevant tool to ensure consistency over time for the main components of strategic plans. Typically, a technology roadmap integrates the plan for product developments and launches together with market trends on the one hand and technology and competence development (or acquisition) on the other hand. Roadmaps are thus a multilayer tool to integrate those various plans into a framework ensuring time consistency. Roadmaps are not just the list of product launches in the next few years (see Figure 4.4). The availability of technologies, and more deeply that of technical competence and other related resources, should not be taken for granted just because the marketing department presents a plan for launches of new offerings over the next years. In that sense, drawing a multilayer roadmap is a form of consistency check. But it works the other way around as well: if new products and services may indeed be pulled by market demands, new opportunities may also be made available through technological development pushing for new applications. In that sense, TI may contribute to enrich roadmaps as far as the technology and technical competence layers are concerned.
4. Another deliverable of TI activities deals with the targeting of technology options which the firm would wish to control (i.e., to define which options to pursue actively, which to monitor at minimal cost, and which to abandon). For example, when Kodak was faced with digital photography in the 1980s, it would have made sense to start hiring electronic engineers in the company very early on, instead of denying the technological threat

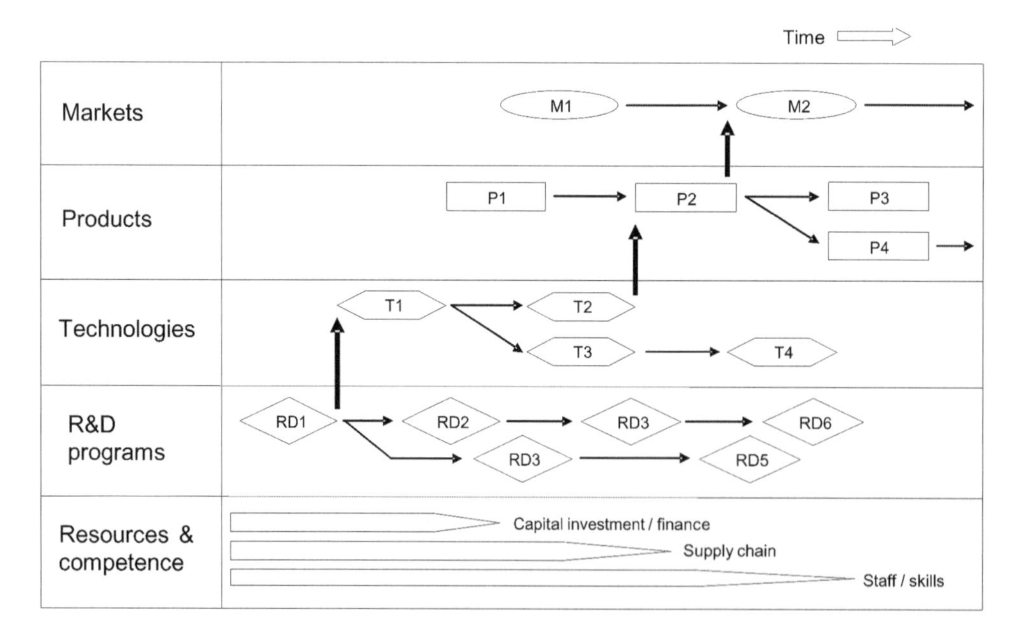

Figure 4.4. *Source:* Phaal, Farrukh, and Probert in EITIM (2004).

arguing that chemical images would remain of better quality for years ahead. Some competence in electronics would have helped the subsequent transition when it became clear that numerical photography was winning in the market. The idea is thus to select technical options that actually cover the span of competence which might be useful in the future, should any of the foreseeable technologies win the race, even the least probable. It is not easy to forecast the future of technology, but it is possible to identify the technical competence behind technologies, and it may be wise to build those pieces of competence which might turn out to help a business move fast, should an unexpected radical change occur around a particular new technological paradigm. In this sense, TI may help place bets and buy insurance contracts via competence building, just in case. This justifies the reference to real options.

Once technological options are targeted, another TI task still remains: TI is expected to contribute to inform about ways to access the technical competence needed to control technologies. This may be done internally via R&D or externally, at least in part, via R&D collaboration or acquisitions (buying technologies or acquiring firms mastering the targeted technical competence). It may require resisting the NIH (not invented here) syndrome whereby internal players, particularly R&D, do not like importing technologies from outside the firm, but TI may contribute to identifying external sources of competence, assessing the value of a piece of technology or a license, deciding whether a technology may be too strategic to share the development with a partner or a supplier, etc. These are typical deliverables expected from TI activities.

4.8 The players in the TI process

In some companies, technology is taken care of by the Vice President (VP) for R&D. They have budget and R&D staff. They program the R&D projects according to strategic priorities and marketing demands, while adding a bit of scientific and technological push for part of the budget. In the process, the R&D VPs (and their expert staff) are considered to be the most knowledgeable in the company as far as technology is concerned, especially when it comes to new technological options for the future. They are thus requested to give their opinion on some of the strategically important decisions where technology plays a role (while not necessarily being called upon to assess the technical value of a target in case of acquisitions—often a source of major frustration for many R&D VPs). As a result, in these companies, technology intelligence is primarily conducted in R&D departments. Yet, many other members of the firm conduct bits and pieces of TI: the legal department may have a patent analyst surveying patent databases; business development may gather technical information on some of the target firms they look at for potential acquisitions; engineering and manufacturing may learn about new industrial processes at technical seminars; the purchasing and sourcing department may identify new technologies via suppliers; etc. This is why TI requires a central unit to coordinate the disseminated bits and pieces of information gathered through everyday actions in various parts of the organization.

Things may be clearer when a CTO (chief technology officer) position exists. The CTO task is to conduct TI activities and strategize about technologies to support the decision-making process when it comes to technology and innovation throughout the company (selecting the best technological options and norms for a set of market applications, specifying priorities and targets for R&D programs, buying and selling technology, recommending acquisitions of firms with an attractive portfolio of technologies, launching new ventures, and making spin-off decisions, etc.). In that sense, CTOs are the logical organizers of TI activities in firms where such a position exists.

Yet another organizational issue relates to the players involved in the TI process. R&D VPs take care of R&D and cover technology matters unless a CTO is in charge (in some firms the CTO is simply a new title for the position of R&D VP). But the promotion and management of innovation is another extremely important related theme that needs to be covered. More precisely, it turns out that R&D VPs or CTOs are not well positioned to be in charge of innovation. Instead, innovation is a managerial task for all layers of management. However, TI is a typical input in the innovation process as discussed above. TI helps triggering creativity and assessing the technical feasibility and value of ideas collected. In this sense, TI needs to relate explicitly with the management lines, where innovation is to be taken care of. This is probably the most delicate and tricky part of TI.

4.9 Conclusion and managerial implications

Technology intelligence development requires a systematic activity of surveillance and monitoring. It is an essential task of management to have the organization scan the environment, searching for scientific breakthroughs and new technological options, assessing the associated risks and potential benefits, monitoring strategic moves by competitors to follow which technology they select, exchanging with suppliers about their own perspectives, listening to clients to test new ideas, collect their suggestions, and validate functional needs, given what may now become feasible due to the potential offered by new technological options, etc.

TI calls upon methods (surveys and foresights), tapping a variety of sources to deliver inputs for the strategy process (when it comes to decide which technologies to use in the business: new offerings/internal processes) and for the innovation process (both as a source of novel ideas and for assessing the potential value of innovation proposals). TI activities require the contribution of several parts of the organization, under a central monitoring system, typically from R&D or preferably a CTO. The deliverables stemming from TI activities can be formatted in a number of ways (list and description of promising technologies, dual-technology trees, roadmaps, make/cooperate/buy assessments to access the technologies and technical competence) but a key point is to make sure that TI outputs are being properly communicated and directed to inform the decision-making processes whenever technology is involved.

Finally, external technology intelligence should be complemented systematically by close monitoring of the internal technical competence base. The challenge is indeed to continuously adapt the competence base of the organization to the requirements which external technological evolutions impose on the firm. To a large extent, this is thus a process of reaction and adaptation. Yet, the firm also needs to adopt a proactive stance to try and innovate faster than its competitors; for example, recombining some of its technologies, resources, and capabilities in new, refreshing ways, possibly calling upon external resources to fully deploy its innovations.

References

Abernathy, W. J., and Clark, K. B. (1985). "Innovation: Mapping the winds of creative destruction." *Research Policy*, **14**, 3–22.

Abernathy, W. J., and Utterback, J. (1975). "A dynamic model of process and product innovation." *Omega*, 3(6), 639–656.

Ansoff, I. (1987). "Strategic management of technology." *Journal of Business Strategy*, 7(3), 28–39.

Barney, J. B. (1986a). "Strategic factor markets: Expectations, luck and business dtrategy." *Management Science*, **32**, 1231–1241,

Barney, J. B. (1986b). "Organizational culture: Can it be a source of sustained competitive advantage?" *Academy of Management Review*, **11**.

Christensen C. (1997). *The Innovator's Dilemma*. Cambridge, MA: Harvard Business School Press.

Clark, K. B. (1985). "The interaction of design hierarchies and market concepts in technological evolution." *Research Policy*, **4**, 235–251

Dosi, G. (1982). "Technological paradigm and technological trajectories." *Research Policy*, **11**, 149–162.

Durand, Th. (1992). "Dual technological trees: Assessing the intensity and strategic significance of technological change." *Research Policy*.

Durand, Th. (1996). "National management of technology and innovation: Integrating the firm's perspective into government policies." In: H. Thomas and D. O'Neal (Eds.), *Strategic Integration*. New York: John Wiley & Sons.

Durand, Th. (1998). "The alchemy of competence." In: C. K. Prahalad, G. Hamel, D. O'Neil, and H. Thomas (Eds.), *Strategic Flexibility: Managing in a Turbulent Environment*. New York: John Wiley & Sons.

Durand, Th. (2003). "12 lessons drawn from key technologies 2005, the French Technology Foresight Exercise." *Journal of Forecasting*, March.

Durand, Th., and Dubreuil, M. (2001). "Humanizing the future: Science and soft technologies." *Foresight, Journal of Future Studies, Strategic Thinking and Policies*, **3**(4), August, 285–295.

Durand, Th., and Gonard, Th. (1986). "Stratégies technologiques: le cas de l'insuline." *Revue Française de Gestion*.

Durand, Th., and Guerra-Viera, S. (1997). "Competence-based strategies when facing innovation. But what is competence?" In: H. Thomas, D. O'Neal, and R. Alvarado (Eds.), *Strategic Discovery: Competing in New Arenas*. New York: John Wiley & Sons.

EITIM (2004). "Bringing technology and innovation into the boardroom." In: Th. Durand, Granstrand, Herstatt, Nagel, Probert, Tomlin, and Tschirky (Eds.), *Strategy, Innovation and Competences for Business Value* (European Institute for Technology and Innovation Management). Basingstoke, U.K.: Palgrave Macmillan.

Floyd, C. (1997). *Managing Technology for Corporate Success*. Aldershot, U.K.: Gower Publishing.

Godet, M. (2001). *Creating Futures: Scenario Planning as a Strategic Management Tool*. Paris: Economica.

Grant, R. M. (1996). "Prospering in dynamically-competitive environments: Organizational capability as knowledge integration." *Organization Science*, **7**(4), July/August.

Hamel, G., and Heene, A. (Eds.). (1994). *Competence-Based Competition*. Chichester, U.K.: John Wiley & Sons.

Mansfield, E. (1985). "How rapidly does new industrial technology leak out?" *Journal of Industrial Economics*, **34**(2), December, 217–223

Martin, B.R., and Irvine, J. (1989). *Research Foresight*. London: Pinter.

Nelson, R., and Winter, S. (1977). "In search of a useful theory of innovation." *Research Policy*, **6**, 36–76.

Penrose, E. (1959). *The Theory of the Growth of the Firm*. Oxford, U.K.: Blackwell

Prahalad, C. K., and Hamel, G. (1990). "The core competence of the corporation." *Harvard Business Review*, 79–91.

Prahalad, C. K., and Hamel, G. (1994). *Competing for the Future*. Cambridge, MA: Harvard Business School Press.

Rowe, G., and Wright, G. (1999). "The Delphi technique as a forecasting tool: Issues and analysis." *International Journal of Forecasting*, **15**(4), October.

Rumelt, R. P. (1995). "Inertia and transformation." In: C. A. Montgomery (Ed.), *Resource-based and Evolutionary Theories of the Firm*. Boston, MA: Kluwer Academic.

Salo, A. (2001). "Incentives in technology foresight." *International Journal of Technology Management*, **21**(7/8).

Van der Eerden, C., and Saelens, F. H. (1991). "The use of science and technology indicators in strategic planning." *Long Range Planning*, **24**(3), 18–25.

Wack, P. (1985a). "Scenarios: Uncharted waters ahead." *Harvard Business Review*, September/October.

Wack, P. (1985b). "Scenarios: Shooting the rapids." *Harvard Business Review*, November/December.

Wernerfelt, B. (1984). "A resource-based view of the firm." *Strategic Management Journal*, **5**, 171–180.

5 Technology Forecasting

Steve Russell

Siemens Corporate Research, Inc.

Introduction

Technology changes in response to inventions, desires, and market conditions. In order to make the best-considered investments of talent and capital, the forecasting of these technical changes is of ever-increasing global importance. Technology consists of machinery and procedures combined with directed human knowledge, so the prediction of the changing characteristics of these components involves an intersection of concerns and specialties. The flow of innovation is often predicted using applied analytic methodologies such as equations and graphs. Where less well-instrumented data can be obtained, committees of experts may be convened to collectively determine likely trends. The operative value in the resulting projections has been to retain competitive stature and to seize the lead in emerging markets.

The forecasting of changes in technology relies on a sound understanding of the nature of technical systems and the forces that promote variations in design. Technology has many forms, ranging from material items like vehicles to systems like computer software. Biotechnology is an example of a significantly different new area of innovation which is impacting society. Some technological change comes from accidental discoveries, some from directed searches for solutions, and some comes from recombining existing components in novel ways and for new applications. After a new technological item is available, its penetration into general use often faces several obstacles. Existing products, competitive approaches, and varying degrees of skepticism can slow acceptance. Networks of promotion, delivery, and support need adjustments for easier access for the new introduction. Given this, recurring drivers and barriers have been found to have similar features and timeframes. For emergent offerings, then, companies look for the most important factors to help them plan for their developments and deployments, and to look ahead for correlated or competing products that can affect market adoption.

It is worth noting the distinctions between forecasting and related topics such as prediction, rule-based extrapolation, and planning. Technological forecasting has many similarities with familiar activities like forecasting the weather. On the other hand, predicting a winner in a political contest relies on the informal opinions of experienced pollsters and commentators—but without the repeatable patterns and deeper foundations seen in most forecasts. Some forecasts of business conditions are phrased so as to resemble hard sciences like physics. Their statements of necessary changes are, however, not actually deterministic and are usually more guidelines in thought rather than detailed forecasts. Planning is the activity that uses forecasts, or some of the less precise methods of guessing at future conditions, with the intention of taking specific subsequent actions.

This chapter deals with approaches which are used by business stakeholders to understand likely technological developments. Popular types of forecasting in technological arenas will be reviewed, including proven and accepted predictive methods that companies have institutionalized. For instance, trends expressed in data and graphs are fundamental in sharing views of recent product successes and promising near-term variations. Registration of patent-protected innovations is a central source for monitoring patterns of emerging possibilities and their early implementations. In addition to numeric methods, there are approaches for extracting and evaluating opinions from experts who have pivotal insight or influence into coming developments. Recently, social networks and personalized Internet information systems have emerged, accelerating certain types of technical progress.

A brief history of technology prediction below is followed by an overview of the frameworks for forecasting. Separate attention is given to higher level changes in systems versus smaller alterations in sub-components.

Data sources are considered, and the selection of trusted information. Methods of data analysis are then examined as well as the application of human expertise. Finally, there is a brief assessment of some changes in technology forecasting that may occur in the coming years.

Historical background

Forecasting changes in technology was of interest almost as soon as technology growth became important. For industries and governments that depend on forecasts for planning, a disciplined approach to technology prediction only emerged toward the middle of the last century. By the 1960s and 1970s many enterprises were employing technology forecasts. Data became more standardized and available. Technical and secretarial staffs were assigned to dig through published material to uncover suggestive patterns. Books and papers were produced to focus on targeted technology developments.

Patents were filed in increasing numbers and it often required physical presence at a patent office to dig for patterns and insights. More specialists mining the patent data resulted in common views on the degree of novelty and the importance of a submitted idea or device. Projecting follow-on developments from the recent trends was supported by standards in filing and capturing the key data in summarized tables for shared reviews. Once computer systems became more widespread and digital data transfer matured, the number of analysts rapidly grew. The breadth of areas of interest and the depth of analysis provided sounder foundations for many types of forecasting.

The wisdom of collected experts was harnessed to derive projections based on a wealth of real-world experience. Committees of specialists were used to tackle forecasts in areas where data alone were not sufficient. Methodologies of group decision making and idea extraction were evolved to enhance the collaborative sessions. With better data and applied expertise, forecasting became more of a discipline.

Frameworks for forecasting

For a sound forecast, good input data are instrumental in supporting analytic processing. Expert decisions also benefit from richer and diverse information sources. So, the acquisition and interpretation of trustworthy and well-structured data is a key element in technology forecasting. Group dynamics can be tuned as well, to enhance the power of collective knowledge. Reviews of prior technological advances round out the picture, revealing repeated patterns in the winnowing and fruition processes that lead from ideas to common usage.

There are differences in the characteristics of certain types of innovations that affect the types of data to collect and the nature of the subsequent analyses. A technical change may be incremental or may be fundamentally new. Also, the change may be a portion of an existing assembly, or an entirely novel system may be developed. For instance, consider the following table:

	Existing	*Novel*
Component	Better filament	Fluorescent
	Improved algorithm	Quantum code
System	Electric car	Nanotech
	Semantic search	Virtual worlds

Innovations may be part of an existing product like an electric light bulb. The component changed may be a variation on current technologies such as a filament or may be a new way to accomplish the same function such as a fluorescent element. Other innovations may involve a larger scale system such as a recombination and refitting of current parts into a new overall assembly like an electric car drive train. Or, an entirely new field of systematic changes may have been introduced such as nanotechnology. These distinctions between existing versus novel, and incremental versus systematic, have also become important in computer systems—involving algorithms as well as entire technology approaches like quantum computing.

There are specific ways in which data are derived on technological change, and in which assessments of the innovations are made. These methods include identifying trusted information banks, and performing standard analyses that are framed in commonly changed features and organization schemes of current devices. Acquiring and storing the information is also a structured process, which may involve organized scans through data sources and techniques for guessing at the inner workings of competing or novel devices such as reverse-engineering. These methods of gathering input material for forecasts are considered next.

Form specification and intelligence gathering

A composite technological item such as an electronic circuit or complex computer application can be broken down into constituent parts with distinct functions. Innovations in the sub-components typically proceed at distinct rates. Further, a given component has its own particular set of connections and dependencies on other components and on outside influences like desires for reducing the dependence on a scarce resource. These specifications of form help to direct the focus of intelligence gathering for forecasting. Technology mapping in this way finds core sub-categories, definitions, and boundaries which are specified and agreed upon by the various experts and decision influencers, based in part on their particular roles and competencies.

After making an agreed-upon map of the most important items to track for ongoing forecasts, a matrix of internal and perhaps also external dependencies for these items is filled in. This morphological breakdown can also indicate more extensive component system linkages and compatibility factors. Structured inquiries focus on pivotal sub-part performance capability metrics to populate the technology impact table. Then, the dependencies and rates of likely change are assigned appropriate numerical rankings and probability values. Of most interest are the places where changes in a single component can be leveraged to improve the overall positive characteristics of the product—such as durability, cost, energy needs, safety, capacity, or simplified maintenance. Such mapping methods have been systematically applied to focus the forecasting of developments in electronic gate arrays and sensors.

In certain industries intelligence gathering has settled into accepted patterns with a series of recommended steps. For instance, the Herring Model is a complete and repeatable gathering of useful data for extrapolations. There are five steps in this method:

(1) Assess needs. The first activity is to determine the problem areas where decisions are to be made, and the sources of knowledge required.
(2) Plan. Then, a method is derived for the timely acquisition of information and how best to utilize it.
(3) Collect. The desired information is then assembled from the identified sources, which can involve iterative passes through databases, interviews of colleagues and other specialists, and digital searches of Internet information.
(4) Analyze. The retrieved information is structured into linked databases which are organized to support reliable decision making.
(5) Presentation. The results are conveyed to interested parties in the most effective yet unbiased ways in support of investment decisions.

Scanning

Computerized scanning of document repositories, conference proceedings, journals, government reports, and internal company document content databases is a widely used approach to highlight important changes in direction and general patterns of innovation. Tagging and metadata are becoming more common in specifying the meaning and connections among documents, providing even more avenues for mining the pertinent sources of technological change. Non-textual content is also being scanned for key features, such as computer-aided drawing and manufacturing (CAD/CAM) materials, medical image scans, photographs, and even video and sound files. Processing textual material in papers has become more complex. For example, published reports and academic papers can contain terse abstracts and conclusions, and diagrams and photos which are difficult for computers to interpret in a consistent way. Natural language processing and image feature analysis developments are steadily improving, enabling forecasters to acquire extra trends from these document features.

In many technical sectors, a holistic approach to forecasting involves a flowchart-like series of steps, where information scanning, for example, follows technology mapping (detailed above). The mapping (Figure 5.1) starts by specifying the key features and modifiable components to monitor, then the computerized scanning is better directed toward those features that are most influential. Internet searches, paper compilations, and

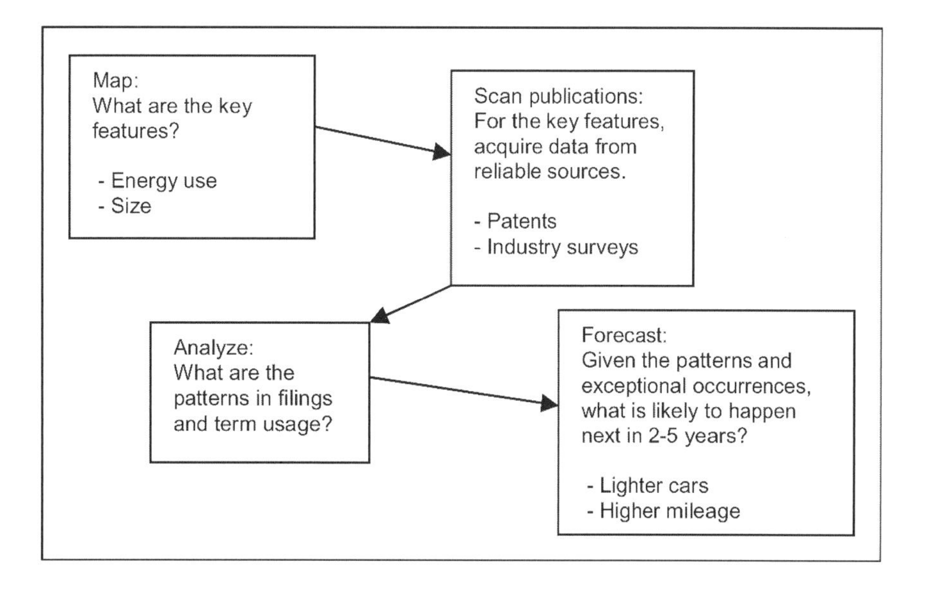

Figure 5.1

interviews of key actors are thereby framed according to the previously agreed-upon factors that distinguish the domain in question.

Data sources

Information on technological progress exists in many forms and locations. From patents to papers to corporate archives, there are variations in the reliability and consistency of such inputs. So, a key step in beginning to understand forecasting is to understand the data that it depends upon. An important note here is the accelerated connectivity of worldwide data systems, and the continuous access to vast and growing repositories of important information.

Patents

Among the better structured and monitored data sources used in technology forecasting are the official filings of inventions. Scanning through recent innovations can provide ideas for similar extensions of form and functionality. In addition, it is necessary to be aware of earlier art or competing innovations to avoid costly lawsuits and wasted investments. Patent records are readily available online—for example, in the U.S. at *http://www.uspto.gov/* Such government websites have tools to help users browse the databases for similar types of developments.

Patents have formal structures and distinct sections for describing the innovation. For instance, with some identifying material removed, below is an excerpt of a Canadian patent for a separable electrical connector:

> "... the electrical connector comprises a receptacle body, indicated generally at 10, fabricated as an integral molding ..., preferably a glass-filled polyester material ... The body is defined by a ... a top wall 18, and a bottom wall 20. A plurality of spaced apertures or passageways ... as most clearly shown in FIG. 3 ... the passageways are oriented in the body to form two parallel rows ... staggered with respect to each other ..."

The filing gives citations of previous related patents that enable or differ from this one:

> "... the prior art discloses a socket contact member which is capable of performing a wiping ... broadly referred to in U.S. Patent Number ... to ... As shown therein, two surfaces on the contact perform a wiping action ..."

The terms such as "receptacle body" and "polyester material" can be used to group together similar patents and the temporal flow of patent activity for similar purposes. The numbers of patents with similar or identical terms can help to gauge the depth of inventor interests in the area and the commonly involved components which are used in innovative ways. Tracing the citations for many such patents can reveal which ones are most influential, at least in terms of being noted as citings in subsequent invention filings. The core patents and their sources then help to establish primacy and authority patterns that may extend to future inventions and their order or dependencies. In addition to the statistics compiled on patents by official sites, there are companies that provide targeted data extracts derived from those files. There are evolving sets of tools that are used for the extraction, storage, analytics, and presentation of results for forecasting purposes. Normally an analyst is concerned with a subset of patents and submissions in their particular company's marketplace.

Patents are linked to one another in large part by their citations. Citing prior art is required for approval in most filings, referencing the preceding inventions and developments that the current filing depends upon. Tracking the linkages and progenies of developments is a kind of path analysis that can be used to understand how technological changes emerge and mature. The valuation of a patent is also linked to its preceding work and the references in subsequent innovations to it.

Other data sources

Corporate records are key assets for a firm, with a wealth of history of successful product introductions. Although the formats and styles of expression change over the years, previous company innovations can be systematically assembled for targeted studies that can support the next wave of marketable inventions. External reports are valuable from competing and related firms, and from government documents. In some cases, professional firms are relied upon as authorities in framing the discussion of upcoming technology improvements.

Parsing the sentences in a technical paper or magazine article can give better interpretations of the author's meaning and highlight the major themes regarding his or her assertions. In technology fields, these changing themes are often the central issues involved in current and upcoming innovations and their acceptance. Articles on cellphones, for instance, use different terminologies depending on whether the audience is a set of technical developers or end consumers. Taxonomies of technical terms are more structured and change more slowly than the free-form expressions in consumer evaluations. Repeatedly used terms can be displayed as tag clouds like the examples below, where the size or shading of the term indicates its likely importance. Each displayed word or phrase can be clicked to get a definition or a set of other sites that use the term. Trees or taxonomies constructed from term co-occurrences can help to organize the concepts in the field using so-called folksonomies of actual word use and idea connections.

Technology terms	*Consumer terms*
bandwidth handset	Weight size
durability luminance	IM keyboard cost
LCD pixels battery	plan color
chip chassis	**GPS** camera backlit pixels
FCC database recovery	battery storage
	web PDA provider game
	music

Forecasting methodologies

In simple cases, there is a linear development of a technology over time. This may be a steady percentage increase each month, say, in the size reduction for a type of component or its sales volume, or perhaps a doubling every quarter in the power of a certain sort of assembly. In such cases, extending the straight-line relationship into the future is often a quick and dirty way to look a bit ahead at the likely future state of affairs. Such a relationship applies under the assumption that the driving forces and constraints will remain roughly the same. Since conditions often diverge due to one or more factors these projections should only be used for, say, a year or two ahead.

New technologies more generally catch on slowly, then speed up to wider acceptance, and plateau out at some level of adoption—a so-called S-curve. There are other patterns more recently of interest such as the hype cycle for visualizing the adoption of new technologies, where a peak of interest is followed by a decline as the novelty wears off. The adoption pattern in, say, a new type of television set may be symmetric, with early growth rates which are duplicated in a reverse manner in the late-stage leveling of adoption. The Fischer–Pry logistic curve is used in many such forecast situations. Or, the growth portion may be more or less rapid than the final leveling off. The alternate Gompertz model can be useful in fitting this sort of trend, especially as more adoption data come in (Figure 5.2).

Other nonlinear relationships may be seen in the data on technological change such as polynomial curves or even sinusoidal curves for seasonal influences. Several input factors may be considered as well, in multi-dimensional graphs and tables and in matrix tabulations of what influences what and to what degree. Very detailed mathematical models and simulations can then be constructed. Their predictions can be improved with appropriate applications of the combined skills of mathematicians and business specialists.

For multiple-input factors where the data are organized into cross-tabulation tables, a set of known sub-components may be reviewed for possible rearrangements. The dependencies of sub-elements upon one another are thereby clarified. Minor innovations in localized parts can then be applied in order to make products with novel functionalities—such as in integrated computer circuit chips. The morphology of the current computer chip can thus help to indicate directions for further profitable changes. Also this approach can reveal possible road-blocks—say, due to combined heating problems. For numeric dependencies between factors such as failure rates or increased costs, the usefulness of component-to-component relationship mapping depends on collecting all possible data on the relationship cells—tested under a range of conditions where possible. Where successively more detailed breakouts are desired, the evaluation can be done in successive refinements as indicated in Figure 5.3.

Shown below is an example of a morphological analysis matrix for evolving component parameters. In this case the overall unit consists of some type of power source and a set of gears and some kind of status indicator. The successive refinements of each sub-component area are laid out in the table, where refined analyses would fill in the degree of dependency, the probability and timeframes for each

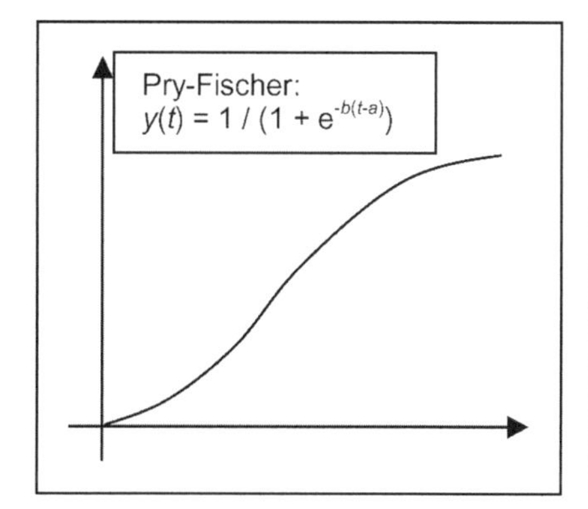

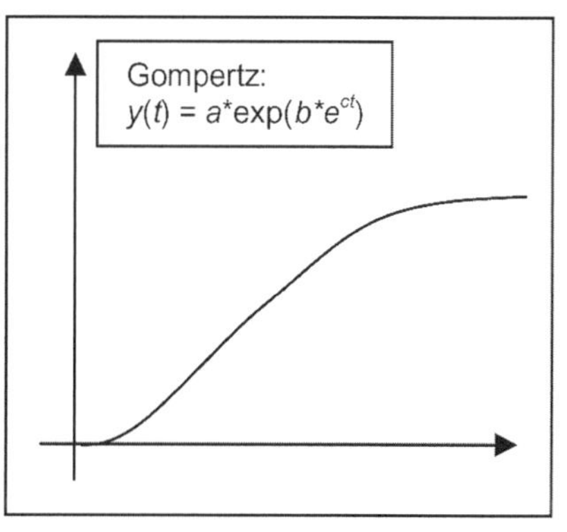

Figure 5.2

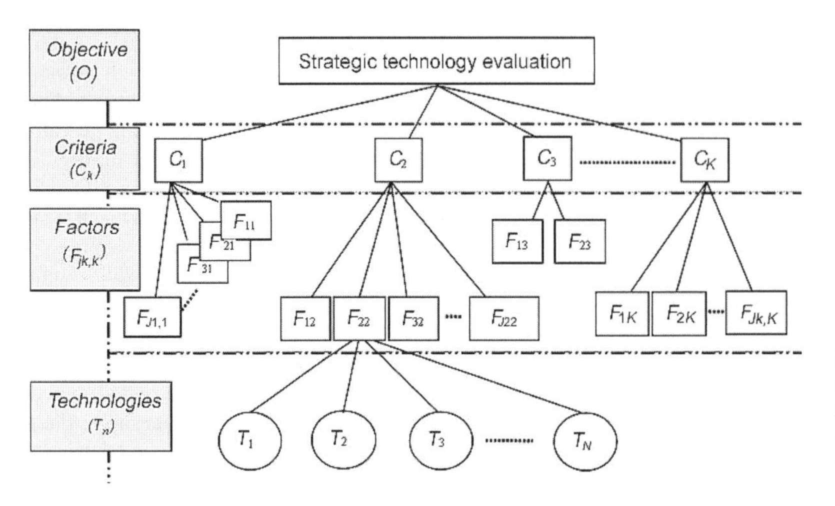

Figure 5.3. Hierarchical model for evaluating emerging technologies. *http://www.isahp.org/2005Proceedings/Papers/GerdsiN_KocaogluTechnologyRoadmapping.pdf*

change, and other measures useful in predicting development difficulty and market penetration. As mentioned above, more intensive technology mapping is useful in areas like digital circuit design, with successively more detailed sub-sub-component reviews and assessments.

Alternatives	*First*	*Second*	*Third*
Key parameters			
Power source	Mechanical	Battery	Solar
Gear type	Simple	Pinion	—
Gauge	Light	Dial	Display

Technological development often follows recurring patterns in a manner that can reveal structural principles and rules. The TRIZ model (from a Russian acronym for Theory of Inventor's Problem Solving) derives such patterns in detailed steps from patent filings and other sources on the expected evolution of product types and assemblies of components. TRIZ uses algorithms based on models of inventing new systems and iteratively refining existing products. The formulation of design problems, methods of systems analysis, and failure analysis are typically involved. A collage of tools and providers exist for various portions of this wide-ranging approach.

There are situations where data are not sufficient, so human experts are used. The Delphi method is a standard methodology for engaging independent specialists who fill out questionnaires in several rounds. There is an anonymous assessment of the inputs after each round by a facilitator, and participants are encouraged to update their answers based on the previous round's results. The responses usually converge to agreed-upon answers and recommendations. Consensus is a goal, guided by explanations for reasoning and differing outlooks. The Delphi method has been found to result in forecasts which are more accurate than an individual expert or unstructured group would provide (Figure 5.4).

Variations of the Delphi method have been widely used, such as team-based methodologies that allow each participant to express their own opinions publicly, with variations in the convergence methods. Also, structured sessions similar to brainstorming may be employed as in the Nominal Group Conferencing method. A broader sampling of ideas and opinions can also be obtained using structured or unstructured text-based surveys administered in paper or digital forms.

Problems with these group predictions can be seen even among the most technical and conscientious participants. In a recent space shuttle disaster, the analysis of the flawed decision processes found instances of uncritical acceptance of the opinions of certain more dominant team members, and the relative silencing of dissenting opinions in a few critical cases. Investigations of the sociological processes of group dynamics in such cases have uncovered a need for more care in employing a team-of-experts approach.

Social and economic factors

Economic forces affect technological advances by proving the funding lifelines for explorations and for the implementation and improvement stages. These financial forces often result from the expectations and desires of sectors of consumers. If a development is wanted badly enough and improvements are very important concerns such as in green environmental areas, there is pressure on the body of inventors who may deliver these advances. Sometimes the need and direction are clear such as in advances by a key competitor or adversary. In other cases, it becomes apparent that influential consumer groups will respond well if specific technical changes are made such as in efficiency or product size.

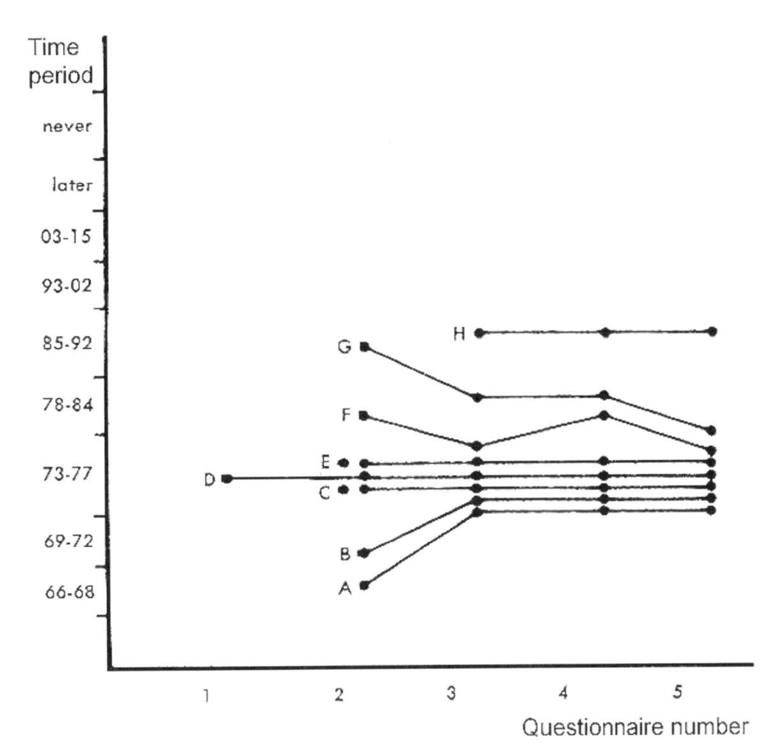

Figure 5.4. The behavior of one experimental Delphi panel on a single question: the estimated date of an anticipated event. The three panelists in the middle retained their original opinion. The two "early" and two "late" panelists revised their initial opinions to converge toward the middle. One member, holding an extreme position, neither influenced the remainder of the panel nor was influenced by it. This result is typical of panel behavior in a Delphi sequence.
http://www.airpower.maxwell.af.mil/airchronicles/aureview/1969/mar-apr/martino.html

Simulating the future

Computer modeling is continually maturing, and better standardized simulations have become routine in specific areas such as military planning and spaceflight preparation. For computationally complex cases, forecasting can benefit from such sets of simulated loads, responses, and even breakdowns. Technological advances may also involve complex interdependencies of circumstances and supply chains. Simulations with visual aids and more natural interfaces are enabling a new degree of foresight and clarity where technical advances may arise.

There are large-scale changes in technology overall and in the surrounding society that can significantly impact the process of technology forecasting. These mega-trends can change the very ground rules for prediction and may involve changes in the basic nature of the technologies that are to be forecast. For example, there has been a profound recent shift in developed countries toward a service economy. Between 70% and 80% of economic value is tied up in services as opposed to traditional "material" fields like agriculture, mining, manufacturing, and transportation. Fields like experience-rich coffee vendors, software games, and recreation often involve less technical invention than social and fashionable innovations. Technological advances aligned with services will thereby be more desired in arenas such as computer algorithms for mining changing customer tastes.

Internet-based enterprises and retail stores have become centers of investment and employment, draining dollars and thoughtful workers from older fields of familiar technologies like automobiles and electronics. Conceptualizations of technological change based on, say, manufacturing and military metaphors have been less directly successful in clarifying future developments in these lucrative areas. The technical changes in consumer products like GPS units, mobile phones, and massively multi-player games involve systems and components where patents are filed and huge investments of funds are made. Advances in software, computation and communication components, and end-user systems align and evolve in novel ways here, so the approaches for previous technical arenas need to be carefully adapted. The market factors of adoption and demand curves are unique here also, with social and personal factors which are less familiar to many technology specialists. Geographic, political, and economic factors matter more, as well as difficult-to-assess influences like a globally leveling playing field of competition for talent and resources. Environmental impacts are an increasingly serious concern as well. So, it is advisable for forecasters in the emerging areas of highest profit to consider cross-disciplinary training in non-engineering arenas including business operations and financing. Then, the matrices and functional graphs above could better include non-mechanical and non-electronic inputs in a disciplined way.

Not to be overlooked are the serendipitous advances of science where a chance accident can reveal a new principle. For example, the antibacterial drug penicillin was discovered by taking advantage of an accidental mold contamination of a culture. The alertness and skillsets that come into play here include a more open mind, some extra time for reflection, and a broad set of training. Labs and even business settings can prepare in small ways for unexpected technical improvements by reducing the emphasis on extremely targeted efficiency and performance reviews that solely focus on carefully assigned tasks and roles. Technological forecasts can incorporate such business investments that promote more innovations across the enterprise or market in question.

Trends in forecasting and future directions

Technology in some fields such as internal combustion engines has slowed in its growth and diversity, whereas in other areas, like biotech, rapid developments are occurring. Genomic, proteomic, and metabolic networks are being engineered to create new lifeforms and profitable therapies. Nano-scale analyses of these biosystems are leading to novel units for building micro-components such as highly structured buckyball carbon molecules. In such new areas, technology growth can vary greatly in its nature.

Additionally, the continued exponential increase in computing power has enabled sophisticated analyses of ever-larger datasets. New algorithms for explaining time series of data such as support vector machines and Markov Chain processes are used more frequently. Probabilistic approaches are routine, including statistical Monte Carlo data inputs and strategic game computations. Arguing against the diversifying field of expertise and the deep complexity of technical components is the need for repeatability. Durable and maintainable components need to be clear to next-generation workers who will modify and upgrade these systems. So, the drive for divergence in fundamental features and combinations will be leavened by drives for simplicity and uniform standards.

Finally, there is very rapid growth in text and social network mining. The results here can help to lead researchers to the most trustworthy sources of information and to track the accelerated rates of technological change when key mavens are influential in a given arena. Also, platforms are emerging that offer extended simulation capabilities such as virtual reality and virtual-world platforms. Such immersive telepresence environments are already providing unprecedented ways to rapidly share information across the planet and to encourage richer ways of experiencing and applying data for activities like forecasting.

Closing comments

Technology forecasting is a relatively mature discipline, at least in certain established technical areas. Similar analytic steps and data inputs are seen in many of the assessments and in the presentation of findings. Where known components are recombined into new functional combinations, the path and pace of technical advance can be relatively steady and can be forecast with some near-term accuracy. Fields such as transportation, electronics, and computers have sub-areas that are well forecast by carefully reviewing recent patterns of innovation and patents. Where expertise has proven to be the best way in the past of looking at the directions of change, stable groups of specialists can ensure continued forecast accuracy. For instance, in mining, petrochemicals, and certain agriculture sectors, human experts have been relied upon for outlooks and for planning the best company or government actions. Many research findings and consultancy groups exist for ready application in these cases. However, in more recent fields such as molecular medicine, the trends and trusted experts have not been reliably established. Here, and in emerging arenas such as digital or mixed-reality worlds the technical projections must necessarily be more speculative. Although there will still be forecasts in these areas, the user of any future condition assertions must use the results with care.

In closing, the forecasting of technological change has proven of great value, particularly to larger enterprises and government units. The anticipated changes may be thought to occur necessarily from current conditions, or these changes may be considered to result from the sustained desires and decisions of influential industry and consumer sectors. In either case, investments in time, personnel, and funds are decisions affected by the forecasts. Skills needed to improve the forecasts and the actions they induce include statistical maturity, data familiarity, and a deep sense of the forces and frictions in the field in question. Some of these skills require academic training, while in other cases years of field-immersive experience are required. For more agile predictions in emerging fields, broader training and social sensibilities are needed, especially as technology moves beyond its physical and electronic roots. The overall prescription is to at least virtually apprentice by reviewing recent pertinent forecasts, and to expand our awareness to poise for any novel developments of importance to our market area.

Bibliography

Anon. (2008). *Technological Forecasting: Judgment-based Technological Forecasting Techniques—Morphological Analysis*. New York: John Wiley & Sons; *http://www.wiley.com/college/dec/meredith298298/resources/addtopics/addtopic_s_02l.html*

Bellis, M. (2008). "Patent application tips: Example of a Canadian patent for a separable electrical connector"; *http://inventors.about.com/od/patentsbasics/a/electricalconn_3.htm*

Börner, K., Chaomei Chen, and Boyack, K. W. (2003). "Visualizing knowledge domains"; *http://www.scimaps.org/nypl/readings/Borner2003visknow.pdf*

Bright, J. R. (1994). *Practical Technology Forecasting*. Austin, TX: Technology Futures.

FreePatentsOnline. (2008). "Automated method of circuit analysis"; *http://www.freepatentsonline.com/6288393.html*

Gerdsri, N. (2002). "Analytical approach to building a technology development envelope (TDE) for roadmapping of emerging technologies"; *http://www.d.umn.edu/~honchen/an_analytical_approach_to_building_a_technology_development_envelope_for_roadmapping_of_emerging_technologies.pdf*

Hall, B. H., Jaffe, A., and Trajtenberg, M. (2002). "Market value and patent citations"; *http://emlab.berkeley.edu/users/bhhall/papers/HallJaffeTrajtenberg_RJEjan04.pdf*

Mann, D. L. (2001). "Better technology forecasting using systematic innovation methods"; *http://www.systematic-innovation.com/Articles/Missing/Mar06-Better%20Technology%20Forecasting%20Using%20Systematic%20Innovation%20Methods.pdf*

Martino, J. P. (1969). "Forecasting the progress of technology"; *http://www.airpower.maxwell.af.mil/airchronicles/aureview/1969/mar-apr/martino.html*

Millet, S. M., and Honton, E. J. (1991). *A Manager's Guide to Technology Forecasting and Strategy Analysis Methods*. Columbus, OH: Battelle Press.

Porter, A. L., Roper, A. T., Mason, T. W., Rossini, F. A., and Banks, J. (1991). *Forecasting and Management of Technology*. Malden, MA: Wiley-Interscience.

Pottengeryz, W. M., and Ting Hao Yangx. (2001). "Detecting emerging concepts in textual data mining"; *http://dimacs.rutgers.edu/~billp/pubs/SIAMETD.pdf*

Ritchey, T. (2008). "General morphological analysis: A general method for non-quantified modeling"; *http://www.swemorph.com/ma.html*

Sherden, W. A. (1999). *The Fortune Sellers: The Big Business of Buying and Selling Predictions*. New York: John Wiley & Sons.

Slocum, M. S., and Lundberg, C. O. (2001). "Technology forecasting: From emotional to empirical"; *http://www.inventioneeringco.com/Documents/Technology%20Forcasting%20Paper1.pdf*

Suryadi, K., Ridwan, A. S., Dou, H., and Purnama, A. (1999). "Technology forecasting in competitive intelligence: The use of patents analysis"; *http://isdm.univ-tln.fr/PDF/isdm3/isdm3a17_suryadi.pdf*

Technology Futures, Inc. (2009). "Five views of the future"; *http://www.tfi.com/rescon/five_views.html*

U.S. Patent and Trademark Office; *http://www.uspto.gov/*

Vanston, L. K., and Hodges, R. L. (2004). "Technology forecasting for telecommunications"; *http://www.tfi.com/pubs/w/pdf/telektronikk_peer.pdf*

6
Disruptive Technology

Erwin Danneels

Worcester Polytechnic Institute

A disruptive technology is a specific type of technological innovation. In essence, a disruptive technology initially appeals only to a marginal market segment, but because of its improvement over time it can eventually satisfy the mainstream market. As a consequence, entrant firms that supported the disruptive technology displace incumbent firms that supported the prior technology.

The term "disruptive technology" was coined by Harvard Business School professor Clayton Christensen, and was widely popularized through his best-selling book *The Innovator's Dilemma* (1997). Despite the popular appeal of the concept, the essential characteristics of a disruptive technology remain ill-defined, and discussion about disruptive technology has often been muddled.

The concept of disruptive technology and its associated mechanisms are very complex. They consist of several conceptual building blocks, including (a) trajectories of performance, (b) market segments, (c) incumbency, and (d) allocation of resources.

A technology performance trajectory portrays the changes (improvements) of one or multiple technologies over time. Graphically, it contains the passage of time on the *x*-axis and the relevant performance metrics on the *y*-axis. A demand performance trajectory portrays the changing levels of performance required by various market segments over time. Graphically, it contains the passage of time on the *x*-axis and the performance levels required by distinct markets on the *y*-axis.

A disruptive technology initially has lower performance relative to the established technology on dimensions relevant to the mainstream market, but has higher performance on dimensions valued by a marginal market. A disruptive technology initially serves a marginal market and not the mainstream market. However, over time the performance of the disruptive technology increases to a level that can satisfy the requirements of the mainstream market. Eventually the disruptive technology displaces the established technology. Often incumbent firms serving the mainstream market dismiss the relevance of the disruptive technology, and are displaced by entrant firms that developed it. Hence, disruption refers to both the displacement of the established technology by the disruptive technology, and to the displacement of incumbent firms by entrants. The mechanism of disruption provides an explanation for why established firms often fail in the face of technological changes.

Understanding the mechanism of disruption requires joint consideration of the trajectories of performance offered by technological alternatives (disruptive technology versus established technology) and the trajectories of performance demanded by various market segments (mainstream versus marginal markets). Disruptive technologies introduce a new trajectory of performance relative to an existing technology, whereas sustaining technologies improve the performance along established dimensions of performance. For instance, decreases in the graininess of traditional film improved picture quality, whereas digital imaging enabled electronic sharing of images. A disruptive technology, therefore, changes the performance metrics along which firms compete (Danneels, 2004). These trajectories of performance supplied and demanded by alternate technologies and markets can be plotted on a diagram. Disruption occurs when the trajectory of performance provided by the disruptive technology intersects with the trajectory of performance demanded in the mainstream market.

As mentioned, a disruptive technology initially appeals only to a marginal market. For example, excavators based on hydraulic technology could initially only be used to dig utility trenches in residential construction (Christensen, 1997). This marginal market values dimensions of performance on which the disruptive technology does excel. This is often a small, low-end, remote, or emerging market, and hence unattractive to incumbents who require large, high-margin markets to feed growth imperatives. Initially, disruptive technologies do not satisfy the minimum requirement along the performance metric most valued by the mainstream market and thus are considered inappropriate by incumbents for satisfying the needs of their customers. However, the performance of the disruptive technology increases over time, and eventually the performance level it offers meets the minimum level demanded by the mainstream market (Adner, 2002).

One of Christensen's most influential recommendations is that incumbents should set up a separate organization for venturing into disruptive technology. His recommendation follows logically from his explanations for the failure of incumbent firms, which are twofold. First, the resource allocation process tends to pull resources away from disruptive technology efforts to serve current customers, and therefore a spin-off with its own protected, dedicated resources is required (Christensen and Bower, 1996). Second, the disruptive technology may not fit with the mainstream organization's resources, processes, and values. For instance, he argues that it is necessary to match the size of the organization to the size of the opportunity, such that managers can get excited over the initially small market for disruptive technologies.

Later work by Christensen and colleagues made additions to the initial formulation. For instance, Christensen (2000, p. 192) stated that "disruptive technologies are typically simpler, cheaper, and more reliable and convenient than established technologies." The most recent version of the framework makes a distinction between low-end disruptions and new-market disruptions (Christensen and Raynor, 2003). A new-market disruption is "an innovation that enables a larger population of people who previously lacked the money or skill now to begin buying and using a product" (Christensen and Raynor, 2003, p. 102).

The notion of disruptive technology has been the subject of intense scholarly debate (e.g., JPIM, 2006). Much of the discussion focuses on the following issues (Danneels, 2004):

(1) the role of marketing competence in managing technological change;
(2) the success of emerging firms versus incumbent firms;
(3) whether a dedicated autonomous organization is needed to develop disruptive technology or whether incumbents can be ambidextrous; and
(4) the potential to identify disruptive technology *ex ante*.

Most empirical work has been in the form of very well-documented and thorough case studies of particular industries, but the extent to which findings from these case studies generalize across industries has not yet been addressed. Some of Christensen's claims regarding the failure of incumbents in the face of technological transition in his focal industry, the hard disk–drive industry, have been disputed (e.g., Chesbrough, 1999, 2003; King and Tucci, 2002; McKendrick, Doner, and Haggard, 2000).

The notion of disruptive technology has captured the imaginations of scholars and practitioners, but despite its popularity it remains poorly understood.

References

Adner, R. (2002). "When are technologies disruptive? A demand-based view of the emergence of competition." *Strategic Management Journal*, **23**(8), 667–688.

Chesbrough, H. (1999). "Arrested development: The experience of European hard-disk-drive firms in comparison with U.S. and Japanese firms." *Journal of Evolutionary Economics*, **9**(3), 287–329.

Chesbrough, H. (2003). "Environmental influences upon firm entry into new submarkets: Evidence from the worldwide hard-disk-drive industry conditionally." *Research Policy*, **32**(4), 403–421.

Christensen, C. M. (1997). *The Innovator's Dilemma: When New Technologies Cause Great Firms to Fail.* Boston, MA: Harvard Business School Press.

Christensen, C. M. (2000). *The Innovator's Dilemma: When New Technologies Cause Great Firms to Fail* (Second Edition). Boston, MA: Harvard Business School Press.

Christensen, C. M., and Bower, J. L. (1996). "Customer power, strategic investment, and the failure of leading firms." *Strategic Management Journal*, **17**(3), 197–218.

Christensen, C. M., and Raynor, M. E. (2003). *The Innovator's Solution: Creating and Sustaining Successful Growth.* Boston, MA: Harvard Business School Press.

Danneels, E. (2004). "Disruptive technology reconsidered: A critique and research agenda." *Journal of Product Innovation Management*, **21**(4), 246–258.

JPIM. (2006). "Dialogue on the effects of disruptive technology on firms and industries." *Journal of Product Innovation Management*, **23**(1): 2–55.

King, A. A., and Tucci, C. L. (2002). "Incumbent entry into new market niches: The role of experience and managerial choice in the creation of dynamic capabilities." *Management Science*, **48**(2), 171–186.

McKendrick, D. G., Doner, R. F., and Haggard, S. (2000). *From Silicon Valley to Singapore: Location and Competitive Advantage in the Hard-disk-drive Industry*. Stanford, CA: Stanford University Press.

7

Intellectual Property

Clyde D. Stoltenberg
Wichita State University

7.1 Introduction

The term "intellectual property" itself embodies the tension that underlies the concept as it has evolved, largely in the West. "Intellectual" suggests the mind's thought and creativity. "Property" embodies a fundamental legal concept of something that can be owned, controlled, and/or traded or exchanged through contract. Protection of intellectual property is designed to further innovation by granting economic incentives to authors, creators, and inventors. Thus, it is fundamentally tied to an economy that is market-driven.

However, maintaining the appropriate balance between creator protection and public access to knowledge in an "information society", with its rapidly evolving technology, is a tricky proposition. As one commentator has put it, "when ownership of intellectual property is challenged, it is not questioned on the grounds that intellectual property should be a public good, but rather to what degree it can be property" (Halbert, 1999). This is the tension that informs managers dealing with intellectual property. This essay will begin addressing the topic, first, by describing the categories of intellectual property. Second, the history and evolution of the categories of intellectual property will be summarized. Informed by this discussion, the chapter will conclude by addressing some of the more significant current intellectual property issues relevant to contemporary management practices and theory.

7.2 Categories

Patent, copyright, and trademark represent the three most common categories of intellectual property; they are typically protected by statute at the national level. Another layer of regulation covering these three categories emerged when the World Trade Organization (WTO) was created. Along with the WTO, the Uruguay round of negotiations under the General Agreement on Tariffs (GATT) produced an agreement on Trade-Related Aspects of Intellectual Property Rights (TRIPS), which included substantive provisions defining and articulating the rights associated with various forms of intellectual property. Because WTO member states' substantive intellectual property laws are to conform to TRIPS, some harmonization of national statutes has occurred as a result. In addition, the concept of "trade secrets" has developed as a common law principle. Finally, issues relating to computer-related technology, which straddle patent and copyright, will be discussed.

7.2.1 Patent

In the U.S. a patent can be obtained for a period of 20 years (from the date the application is filed) for inventions that are "new, useful and non-obvious" to those knowledgeable in the field of the invention. The comparable TRIPS formulation makes patentable something that is "new, involves an inventive step and is capable of industrial application" (TRIPS, Art. 27(1)). Patentable subject matter includes machines, manufactured products, compositions of matter, plants, designs, and processes (including business methods). Ideas, abstract principles, and other similar mental processes, however, can *not* be patented.

A patent owner has the right to exclude others from making, using, offering to sell, selling, or importing the inventions as defined by the claims of the patent. In exchange for this protection, the inventor must disclose in the patent application all aspects of the invention for which the benefits of patent protection are sought. The impact of this publication requirement is to disclose new and useful information that may facilitate further advancement in the art by others. Thus, an improvement of an existing invention which meets the requirements of patentability can itself be patented.

7.2.2 Copyright

In the U.S. a copyright protects "a work of original authorship fixed in a tangible medium of expression." There are eight categories of copyrightable works: literary, dramatic, musical, choreographic, pictorial (including graphic and sculptural works), motion pictures, sound recordings, and architectural works. These categories are interpreted broadly. For example, computer programs may be registered as "literary works". Protection commences at the time a work is created and lasts for the author's life plus 70 years after the author's death. In the case of a work made for hire (owned by a business), the duration is 95 years from date of first publication or 120 years from date of creation, whichever expires first.

TRIPS similarly limits copyright protection to expressions, thus excluding ideas, concepts, and procedures. TRIPS does, however, extend protection to "compilations of data or other material, whether in machine-readable or other form which by reason of the selection or arrangement of their contents constitute intellectual creations" (TRIPS, Art. 10(2)). This includes such items as databases and customer lists. TRIPS also required WTO member states to accede to the Berne Convention, which provides protection in 155 countries for works first published in the U.S.

Under U.S. law a copyright owner has the exclusive right to (1) reproduce the work, (2) distribute the work, (3) display or perform the work publicly, and (4) make derivative works or translations (Copyright Act of 1976, Sec. 106). The owner's rights are not unlimited, however. The Act provides for three exemptions: (1) under the "first sale doctrine" the purchaser of a copyrighted work has the right to resell that copy. (2) The "fair use doctrine" allows use of small portions of a copyrighted work to create another work for purposes of criticism, news reporting, teaching, and scholarly and research activities. (3) Material in the public domain is not protected.

Contrary to the case of patent, registration is not necessary to secure copyright protection. However, certain statutory remedies in infringement cases are conditioned on registration. Moreover, registation provides *prima facic* evidence of the copyrightability of the subject matter of the work, date of creation, time and place of first publication, and identity of the owner. Similarly, although the Berne Convention does not require a notice of copyright on all publicly distributed copies to avoid the work being considered in the public domain, it is still desirable to include such a notice.

7.2.3 Trademark

A trademark is any word, name, symbol, device, or any combination thereof used by a business to distinguish its goods or services from those offered by others. Within limits, trademarks can even take the form of colors as part of a design, images of real or fictitious individuals, sounds, smells, and three-dimensional configurations. To be eligible for protection (for a duration of 10 years in the U.S., successively renewable for like periods), a mark must be distinctive, whether inherently or as a result of recognition by the relevant public after a period of use. The more distinctive, unique, and well-known a mark is, the stronger it is and the more extensive its scope of legal protection. The spectrum of distinctiveness ranges from fanciful to arbitrary to suggestive to descriptive to generic, with fanciful trademarks the most protected and generic terms not protectable at all. Trademarks cannot consist of immoral, descriptive, or scandalous matter; national flags, coats of arms, or other insignia; a name, portrait, or signature identifying a particular living individual; or mere description.

For the most part, TRIPS trademark provisions are comparable with the U.S. approach. TRIPS establishes the period of protection at 7 years "renewable indefinitely" (TRIPS, Art. 18). TRIPS provides that trademark rights can be lost through abandonment or non-use. In the U.S. a finding of abandonment requires evidence of both non-use and intent to abandon; mere lapse of time alone will typically not constitute abandonment. TRIPS expressly provides for cancellation of a trademark registration if the work is unused for a period of 3 years, unless the reason for non-use was due to events beyond the registrant's control (TRIPS, Art. 19). TRIPS recognizes the importance of geographic or country-of-origin indications by providing for prevention of their false use "where a given quality, reputation, or other characteristic of the good is essentially attributable to its geographic origin" (TRIPS, Art. 22). Finally, TRIPS expressly protects "independently created industrial designs" for a period of 10 years (TRIPS, Sec. 4).

U.S. trademark law provides remedies for infringement and dilution. Infringement is based on "likelihood of confusion", which the Lanham Act defines as "any false designation of origin, false or misleading description . . . or representation of fact, which . . . is likely to cause confusion, or to cause mistake, or to deceive as to the origin, sponsorship, or approval of his of her goods, services, or commercial activities by another person" (Lanham Act, Sec. 43(a)). In addition, sufficiently famous and distinctive marks may be protected against dilution, defined as "the lessening of the capacity of a famous mark to identify and distinguish goods or services, regardless of the presence or absence of competition between the owner of the famous mark and other parties, or likelihood of confusion, mistake or deception" (Lanham Act, Sec. 43).

Just as in the case of copyright, a trademark does not have to be registered to be valid. The first to use or qualify a mark within a geographic area has priority. However, registration is desirable for confirming dates and locations of use as a matter of record. Moreover, applications are now accepted based on "intent to use", consistent with the laws of many other countries allowing application for registration without prior use. Also as in the case of copyright, a trademark notice is not required to achieve validity. However, if a trademark owner fails to give notice by use of the encircled R, profits or damages cannot be recovered in an infringement action under the statutes unless the infringer had actual knowledge

7.2.4 Trade secrets

In the U.S. protection for trade secrets evolved as a part of common law. Essentially, a trade secret can be anything that is useful or advantageous to a business but not generally known or easily ascertainable by others in the trade. To qualify, the trade secret must have business value and the owner must have made reasonable effort to maintain its secrecy (Unfair Trade Secrets Act, Sec. 1(4)). The latter can be demonstrated by the existence of written confidentiality contracts with employees, precautionary security measures, and nondisclosure provisions in licenses. The subject matter of trade secrets consists of either technical information or business information that derives economic value from the fact that it is secret and gives the proprietor some competitive advantage.

TRIPS provisions defining the parameters of trade secrets are consistent with the principles just articulated (TRIPS, Sec. 7). Of interest to licensors of intellectual property rights, however, is a provision granting countries the right to prohibit certain contractual restrictions that have "adverse effects on trade and may impede the transfer of technology." Specifically targeted are restrictions found in exclusive grantback clauses and clauses that prohibit challenges to the validity of the licensor's rights (TRIPS, Art. 40).

Remedies by trade secret proprietors are predicated on the commission of some wrongful or unethical act. Such behavior would include industrial spying or espionage, bribery of employees, or fraud. It is not wrongful to discover a competitor's trade secret through independent invention, reverse-engineering, observation of the item embodying the trade secret in public use or on public display or from public literature.

The usual remedies available to those whose trade secrets have been wrongfully appropriated include injunctions against disclosure or use of the trade secret, damages, lost profits, disgorgement of profits gained from using the trade secret, and return or destruction of items containing or using the trade secret. Some jurisdictions provide for criminal penalties, and this has been extended to the federal level by the Economic Espionage Act of 1996, which extends beyond government secrets to include business trade secrets. It is important to remember that trade secret law is wholly separate and apart from patent and copyright law.

7.2.5 Computers and technology

Computers and related technology developments have heightened the significance of intellectual property. As the value created by developing country economies has shifted from manufacturing to "knowledge creation", the value of information has increased at the same time that rapidly emerging technologies have made it harder to protect. While many proprietors of information are putting an increased premium on its control, individual creativity is finding new ways to gain access to it.

Intellectual property protection in this environment is breaking down some of the traditional boundaries between the categories described above. Computer-related technology draws on both patent and copyright regulation for protection. While manufacturing methods and processes are clearly protectable under patent law, software poses unique questions. Legal precedent bars patent protection for mathematical formulas, laws of nature, and mental processes, but software that implements algorithms and principles of nature can be protected. The traditional categories of copyright are not a perfect fit for software, which includes both source code and object code. Extending a source of protection designed for literary and artistic works to what is essentially scientific technology used to control a machine is somewhat akin to fitting a square peg into a round hole. However, software is copyrightable as "a work of original ownership fixed in a tangible medium of expression."

As difficult as it is to draw lines for purposes of applying intellectual property regulation to computer-related technology, developments in medical and biotechnology create even more profound dilemmas. The U.S. Supreme Court held in 1980 that a genetically engineered microbe capable of breaking down components of crude oil was patentable. Following a determination by the U.S. Board of Patent Appeals that plant or animal material may be patentable, the Commissioner determined that non-naturally occurring, non-human multicellular living organisms, including animals, are patentable unless "the broadest reasonable interpretation of the claimed invention as a whole encompasses a human being." Patents can be procured for making a modification to a naturally occurring substance or lifeform, or isolating or purifying such a substance. The resulting range of patentable subject matter has grown to include human, animal, and plant genes, novel nucleotide and oligonucleotide molecules, transgenic plants and animals, and genetic therapies. The implications for laboratories and research institutions are enormous.

7.3 History of intellectual property development

Having summarized the basic forms of intellectual property recognized by the current state of regulation and introduced some aspects of the impact of recent technological developments, it may be instructive to step back and examine the origins of these concepts. An understanding of the evolution of the institutional framework may help to inform our understanding of some of the current issues and controversies which will be explored at the end of this chapter. Our current state of intellectual property regulation is largely a phenomenon of the developed Western world. However, it is important to remember that the current state of things is the result of centuries of evolution; the impact of newly emerging technologies in their day had a similar effect on existing traditions as we are seeing in our own time. Major dislocations of earlier understandings are not as new or revolutionary as we may think. The fact that our intellectual property traditions are largely a product of the developed Western world also has significant implications for the level of economic development occurring in parts of the world that encompass very different traditions, cultures, and ways of thinking about the underlying concepts of intellectual property. This is particularly true with respect to Asia, which itself embodies differing cultures in its various regions, particularly when comparing East Asia with South Asia.

7.3.1 The West

Although it is commonly understood that inventive and literary property were essentially defenseless during classical times, traces of intellectual property concepts can be traced back to that period. Short-term monopolies (e.g., 1 year for unusual and outstanding dishes created by confectioners and cooks in Sybaris) were granted in Greece as an inducement for invention. Although the Romans were not favorably disposed toward monopolies, they did enact various incentives, such as exemption from civil duties for artisans in certain trades regarded as important at the time. It has been suggested that the low esteem attached to manual labor, along with the impact of the institution of slavery (discounting any need to lighten or multiply a laborer's efforts by mechanical means) accounted for minimal intellectual property development during the classical period, in spite of the flowering of scientific inquiry in Greece. The disarray characterizing the early Middle Ages following the decline of the Roman Empire, similarly would not suggest any significant institutional impetus for protecting intellectual property during that period.

Patent

This began to change, almost simultaneously in the Italian states and England, during the 13th and 14th centuries. Reflecting the rise of medieval guilds governing member activities in various trades and professions, Venice issued a decree in 1297 requiring any physician who developed a medicine based on his own secret to keep it within the guild, and prohibiting guild members from prying into it. This followed by some decades Henry III's issuance in England of letters patent confirming a grant by the Commune of Bordeaux, then under English rule, for a 15-year exclusive privilege to make fabrics of various colors using Flemish, French, and English cloth manufacturing methods. The later Middle Ages witnessed the extension of a variety of exclusive grants in France, the German principalities, and Eastern Europe. The first actual patents of invention on record in the world were granted by Florence in 1421 for a new kind of boat and by Venice in 1443 for building in each of its districts flour mills that operated without water. Venice adopted the world's first general patent statute in 1474, offering protection to inventors for their inventions in all fields. Because the statute required novelty in Venice, it admitted importation franchises in addition to true patents of invention and gave rise to the world's first known patent system. Perhaps the most famous recipient of a Venetian patent was Galileo, who received a patent in 1594 for a device that raised water and irrigated land. This development influenced patent progress elsewhere on the European continent.

An important aspect of policy in Venice, other continental states, and England was their approach to building new domestic industries by encouraging the immigration of foreign artisans. From the early 14th century on, English rulers issued letters of protection to induce foreign artisans to come to England and bring their skills and knowledge with them. Many of the foreign artisans so induced to bring their skills to England were not actually inventors, however. It fell to Queen Elizabeth I to issue England's first actual patent in 1559 for an invention. Even after the Statute of Monopolies was passed in 1624 during the reign of James I, however, inventors were not assured of the right to a patent based on objective criteria; they remained humble petitioners of the sovereign's grace, in contrast to Venetian patent grants conferred on inventors as of right. It was not until the Patent Law Amendment Act was adopted by Parliament in 1852 that a positive statutory basis existed in England for the type of letters patent specifically granted for invention.

Copyright

It was not until the introduction of the printing press in Europe in the 15th century that the rewards of publishing assumed a proportion prompting consideration of granting protection to authors and publishers. Venice, again, took the lead in promoting the resulting new trade. Between 1469 and 1517, it issued, without special statutory basis or regulation, a series of privileges relating to books and printing that included importation franchises, monopolies, and exclusion of foreign competition. The monopolies typically granted an exclusive license to print or sell an entire class of books for a specified term. These early formulations essentially ignored rights of authorship. One of the early privileges, however, was granted to the historiographer of the Venetian Republic to exercise exclusive control over the publication of his work. Although this grant did not expressly confer on the author ownership of the book as property, his rights of exclusion could be interpreted as the practical equivalent. Censorship by the authorities crept into the law beginning in 1515, and regulation of the publishing industry was tightened with a Council decree of 1548–1549 ordering the establishment of a guild into which all printers and booksellers were to be organized.

This emphasis on control also characterized England, where early regulation was less geared toward protecting author/publisher interests than regulating the political and religious ideas that could be communicated to so many as a result of the new technology. In England the sovereign established royal restrictions on the right to print, including (1) issuance of royal grants and patents conferring an exclusive right to print certain books, (2) a requirement that each copy carry the author's and printer's name, and (3) required submission of a printed copy of each work to the sovereign's private collection. Queen Mary I established the Stationers' Company of London in 1557; it was a trade association of printers and writers which possessed a royal patent to print books for resale in England. Without its approval, no one could print books for resale. Commentators have noted the similarity of this framework to the system established in Venice in 1548–1549. Sovereign control of the licensing of authors, printers, publishers, and booksellers continued through the 17th century.

It was not until Queen Anne's reign that a statute was adopted (in 1709–1710) protecting authors' rights for the first time in history. This Statute of Anne encouraged "learned Men to compose and write useful books" by giving authors a right against infringement for a term of 14 years, renewable for a like term, from date of first publication. For previously published books, the term was set at 21 years. The statute provided as penalty for infringement forfeiture of all books and sheets concerned and a fine of one penny per sheet, the total sum to be divided between Crown and plaintiff. Collection of the statutory penalty was conditioned on registration of each book with the Company of Stationers. Eventually, a question arose as to whether authors' property rights in their work were controlled exclusively by the statute or whether they also had an independent common law literary property in their works in perpetuity. In his important 1769 decision in *Millar v. Taylor*, Lord Mansfield wrote that the property right was perpetual and founded upon the common law rather than upon the statute, which did not affect this common law property right but simply conferred a greater measure of protection for a specificed period after publication. In 1774 the House of Lords overruled this decision in the case of *Donaldson v. Becket*. This case drew a fine line by holding that although there was a perpetual common law copyright, the Statute of Anne terminated the common law right of action to enforce it. Accordingly, the author's perpetual common law copyright was limited by the Statute as to remedies and protection.

Trademark

The origin of trademark predates both patent and copyright, as reflected in the marks on ancient Egyptian clay pots identifying the maker. This identification made it possible to hold the maker responsible for defects in the merchandise. By Roman times, literature contains references to makers' marks on goods as varied as cheese, wine, lamps, medicine, ointment, metallic ornaments, and glass vessels. By the Middle Ages, merchants, under the control of guilds, overcame prohibitions against advertising by using merchants' marks as a quality standard that the guilds guaranteed to customers. Typically, guild rules required every article produced by one of its members to bear both the guild symbol and the individual artisan's mark; the former informed buyers that the goods were not contraband, and the latter held the artisan responsible for poor craftsmanship. It was not until the 19th century that the concept of a trademark was regarded as a property right of some value, worthy of legal protection and rights of transferability.

The colonial period in America was influenced by the regulations that had evolved in Europe. The colonies provided incentives to establish or stimulate industries by awarding exclusive grants and other benefits and by enacting corresponding legislation. Some variations in detail evolved from colony to colony in their competition with one another. During the Constitutional Convention in 1787, the drafters recognized the abuses in granting monopolies on the British model. Deliberations on formulating a system to protect authors and inventors so they would have the exclusive rights to their works culminated in the language of Article I, Section 8, Clause 8 of the U.S. Constitution, giving Congress the power "to promote the Progress of Science and useful Arts, by securing for limited Times to Authors and Inventors the exclusive Right to their respective Writings and Discoveries." This provision represents the first time in the history of the world that a constitutional document recognized individuals' property rights in the product of their intellect and acknowledged that it was in the interest of progress to protect those rights for a certain period of time. It was not until 1870 that Congress enacted federal legislation to regulate trademarks by virtue of its authority to regulate interstate commerce.

7.3.2 The rest of the world

The evolution of intellectual property regulation, culminating in the harmonizing impact of the WTO's TRIPS regime, is largely the handiwork of what we now know as the developed countries of Western Europe and North America. From the above discussion, it is apparent that many different policies were reflected in concepts of patent, copyright, and trademark at different periods in time in different places. The intersection between regulation of competition and regulation of intellectual property is reflected in this history of institutional evolution. In England, monopolies were conferred by the sovereign as much to curry favor and preserve power as to encourage creativity and innovation. Using intellectual property to encourage invention is actually a relatively recent approach in the longer stream of history. We need to remember this as we examine both the impact of new technologies on traditional concepts of intellectual property and how other countries with different cultural backgrounds and at different stages of economic development regard these concepts.

Intellectual property as understood in the developed counties of the West comes out of a context of belief systems and thinking about property and contract that does not necessarily characterize the rest of the world. In traditional China, where the highest form of praise was to copy the works of a master so that the most highly trained connoisseur could not tell the student's painting from the master's, one might expect a different way of thinking about intellectual property. Countries whose economies may be either less developed or developing are in a relatively even harder position today to catch up than today's developed countries were when they were industrializing because the technological gap is that much greater. Looking at post-WWII development in East and Southeast Asia, from Japan to Vietnam, one can see similar patterns moving from very lax enforcement of foreign intellectual property rights to more and more rigorous enforcement as domestic entities invest in their own development of intellectual property. Indeed, one can see many comparable practices in the U.S. during the 19th century. Commentators have pointed out that "what is happening in China today happens in most newly capitalist countries, as new technologies, expanding markets and wily entrepreneurs overwhelm systems of control designed for rural areas" (Mihm, 2007). Coupled with the fact that all of today's formerly centrally planned economies moving toward greater market orientation were until recently guided by a Marxist ideology that regarded intellectual innovation as a product of society, one can see how large a role education must play in changing underlying beliefs affecting national policy toward foreign intellectual property claims.

7.4 Current and emerging issues

Globalization and the harmonization of basic underlying intellectual property substantive law resulting from TRIPS have had an enormous impact. For the first time in the history of intellectual property regulation, as can be seen from the historical discussion above, there is less difference from one country to another than ever before in the content of intellectual property rules. At the same time, the competing interests of various constituent groups affected by intellectual property remain no less diverse. Coupled with the ever-increasing rapidity of technological change that is both (1) blurring the distinction between traditional modes of delivering creative work to its consumers as convergence proceeds unabated and (2) facilitating creation of subject matter not imagined when our intellectual property rules were developed, the environment of change and uncertainty in this area becomes self-evident. Even the harmonization of underlying substantive concepts of what can be patented, copyrighted, or trademarked and length of protection has not limited variations in the mechanics of registration and procedures for enforcement. The Paris Convention (International Convention for the Protection of Industrial Property, 1883) provides a framework for giving foreign trademark and patent applications from other signatory countries the same treatment and priority as domestic applicants, but applications are made to national or regional authorities rather than to a central depository. The more recent Patent Cooperation Treaty of 1978 simplifies filing of, searching for, and publication of international patent applications and provides for filing a single, uniform international application for protection in member states, but individual national applications must follow within 30 months. Only the EU provides a consolidated multinational patent application. The Berne Convention of 1886 firmly established the rights of authors as a principle of international law but left it primarily to national laws to determine the degree of protection required.

Nor has TRIPS resolved tensions between intellectual property and environmental/biodiversity issues, regulation of competition and the provision of pharmaceutical products to the less developed world. The WTO's Doha Declaration of 2001 linked the issues of biotechnology, biodiversity, and traditional knowledge, and mandated the TRIPS Council to both be guided by TRIPS objectives and take development into account. This has been easier said than done. The tension with competition regulation inherent in the monopoly rights conferred by intellectual property law has been demonstrated by the EU's September 2007 ruling that Microsoft's refusal to provide interoperability information to competitors and its decision to tie its Windows Media Player software to the purchase of Windows operating systems amounted to an abuse of a dominant position in violation of Article 82 of the Treaty of Rome. Microsoft had argued that forcing it to hand over information about its technology to its rivals would breach its intellectual property rights. As developing countries are issuing compulsory licenses for life-saving drugs beyond the capacity of their citizens to afford, developed country representatives have warned that neither TRIPS nor the Doha Declaration justify a systematic policy of applying compulsory licenses wherever medicines exceed certain prices. Though an agreement was reached at the 2003 Cancun WTO meeting, it only provided for countries with manufacturing capacity to produce generic versions of patent drugs for domestic use and to import specific generic quantities of otherwise patent drugs.

The Bayh–Dole Act in the U.S. provides a model for transfering exclusive control over many government-funded inventions to universities and businesses operating with federal contracts for the purpose of further development and commercialization. This legislation arose from the fact that the U.S. government had commercially licensed only about 5% of its accumulated 30,000 patents. The contracting universities and businesses are then permitted to exclusively license the inventions to other parties while the government retains "march-in" rights to license the invention to a third party where it determines the invention is not being made available to the public on a reasonable basis. The Bayh–Dole Act is an example of success in one country for expanding the use of government-controlled intellectual property. In other areas, where rapidly developing technologies are leapfrogging the provisions of existing regulation, such as franchisors' use of intranet and Internet websites, data privacy, data security, and encryption, to name only several, the competing interests of contending constituent groups make resolution difficult at the national level, much less globally. It may be helpful to approach these conflicting interests with an appreciation of how contending positions have been dealt with historically in the development of intellectual property regulation.

Bibliography

Bugbee, B. (1967). *Genesis of American Patent and Copyright Law*. Washington, D.C.: Public Affairs Press.

DiMatteo, L., and Dhooge, L. (2006). *International Business Law: A Transactional Approach*. Mason, OH: Thomson/West.

Finkelstein, W., and Sims, J. (Eds.) (2005). *The Intellectual Property Handbook: A Practical Guide for Franchise, Business and IP Counsel*. Chicago, IL: ABA Publishing.

Foster, F., and Shook, R. (1993). *Patents, Copyrights, and Trademarks*. New York: John Wiley & Sons.

Halbert, D. (1999). *Intellectual Property in the Information Age: The Politics of Expanding Ownership Rights*. Westport, CN: Quorum Books.

Mihm, S. (2007). "China's business practices mirror 19th-century U.S." *Boston Globe*, August 26.

Moore, A. (2004). *Intellectual Property and Information Control: Philosophic Foundations and Contemporary Issues*. New Brunswick, NJ: Transaction Publishers.

Peeters, C., and van Pottelsberghe de la Potterie, B. (Eds.). (2006). *Economic and Management Perspectives on Intellectual Property Rights*. New York: Palgrave Macmillan.

Schaffer, R., Earle, B., and Agusti, F. (2005). *International Business Law and Its Environment*. Mason, OH: Thomson/Southwestern.

8

Options and the Analysis of Technology Projects

Edward Nelling

Drexel University

Basic concepts

The traditional method of analyzing corporate projects, referred to as capital budgeting, involves discounted cash flow computations to determine the project's net present value (NPV). This computation is done using the expected cash flows associated with a project, and works well for projects with characteristics that do not change over time. In practice, the management of projects is dynamic in nature, and managers have flexibility to make decisions as information is received over the life of a project. The actual cash flow that is realized in each year is usually greater than or less than the expected value, reflecting changing market conditions and managerial decisions. Many projects, especially those with a significant technological component, are characterized by a high degree of flexibility. This flexibility cannot easily be reflected within discounted cash flow analysis and an NPV computation. A more appropriate method for the evaluation of technology projects is *real-options* analysis.

Real options analysis involves the application of option-pricing theory to the valuation of physical assets such as corporate projects. Option-pricing theory has been well developed and widely applied to financial assets such as stocks, commodities contracts, and foreign currency. Its application to real assets is a more recent phenomenon, with applications in venture capital investment (Li, 2008), outsourcing (Jiang, Yao, and Feng, 2008), employee incentives (Wang and Lim, 2008), and project management (Miller and Clarke, 2008). A number of highly technical methods exist to value real options, but the basic procedure can be visualized quite easily. To analyze a project in a real-options framework, you must specify the outcomes in various scenarios and use a decision tree to consider the flexible nature of the decision process.

The consideration of real options in project analysis is important for several reasons. First, traditional computations of the net present value and internal rate of return of a project do not capture any inherent flexibility or the dynamic nature of managerial decision making. Second, traditional NPV analysis underestimates the value of a project, which may cause firms to avoid investments that have strategic and (perhaps hidden) financial value. In addition, real-options analysis permits a more complex approach to model the interactions between multiple projects in a firm.

Several types of real options exist. Examples include the option to expand production when things go well, the option to wait and invest after observing the actions of a competitor. The next section presents an example of the option to abandon a project when things go poorly.

Example of real-options analysis

A firm is considering an investment in a project with a cost of \$50,000. It expects after-tax cash flows to be \$16,000 per year for 4 years. At a discount rate of 12%, the net present value of the project is

$$\text{NPV} = (\$50{,}000) + \frac{\$16{,}000}{1.12} + \frac{\$16{,}000}{1.12^2} + \frac{\$16{,}000}{1.12^3} + \frac{\$16{,}000}{1.12^4} = (\$1{,}402)$$

Since the net present value is *negative*, the firm should not make the investment.

Now assume that demand for the product is uncertain, perhaps due to doubts about its technological capabilities. Assume further that the uncertainty will be resolved at the end of the first year. After this "trial" phase, the cash flow will be either \$26,000 or \$6,000, each with 50% probability. In addition, consider an option to abandon the project after the first year, which provides an opportunity for the firm to sell the equipment for \$25,000. The application of real-options analysis using a decision tree permits a re-evaluation of the project.

Consider the alternatives. If demand is high, the firm will realize a cash flow of $26,000 at the end of the first year. In addition, they will continue to operate the project, since it is more beneficial to earn an additional $26,000 per year for 3 years than to abandon the project and receive a single payment of $25,000. On the other hand, if demand is low, the firm will realize a cash flow of only $6,000 at the end of the first year. They will then abandon the project, since receiving $25,000 for the sale of the equipment is better than continuing and receiving $6,000 per year for the next 3 years.

The following table presents the cash flows each year:

Year 0	*Year 1*	*Year 2*	*Year 3*	*Year 4*
High demand	$26,000	$26,000	$26,000	$26,000
−$50,000				
Low demand	$6,000	$6,000	$6,000	$6,000

We can then compute present values as of Year 1, which is when the firm would make the decision about whether to proceed. The following table reports the cash flows in Year 1 in each demand scenario, along with the present values of the cash flows in Years 2 through 4:

Year 0	*Year 1*
High demand	$26,000 + $62,448
−$50,000	
Low demand	$6,000 + $14,411

Since the project will be abandoned in Year 1 in the low-demand scenario, the following table reports the equivalent lump sum cash flows in Year 1 in each demand scenario:

Year 0	*Year 1*
High demand	$88,448
−$50,000	
Low demand	$31,000

The $31,000 in the low-demand scenario is the $6,000 earned by the project plus the proceeds of the sale of the equipment for $25,000.

The net present value of the project considering the abandonment option is

$$\text{NPV} = (\$50{,}000) + 0.50 \times \left[\frac{\$26{,}000}{1.12} + \frac{\$26{,}000}{1.12^2} + \frac{\$26{,}000}{1.12^3} + \frac{\$26{,}000}{1.12^4}\right] + 0.50 \times \left[\frac{\$6{,}000 + \$25{,}000}{1.12}\right] = \$3{,}325$$

The net present value is now *positive* $3,325, so the firm should proceed with the investment. The value of the abandonment option increased the value of the project, and the amount of the increase was sufficient to justify investment.

Conclusion

In summary, real-options analysis is a powerful technique that permits a more detailed analysis of corporate investments. It is most useful in situations involving a high degree of uncertainty and a high degree of managerial flexibility. A meaningful application of real options requires the identification of plausible scenarios, and an estimate of the probability of occurrence and the associated outcome in each scenario. Although the estimates involve some level of subjectivity, the results facilitate the quantification of the value of strategic alternatives.

References

Bin Jiang, Tao Yao, and Baichun Feng. (2008). "Valuate outsourcing contracts from vendors' perspective: A real options approach." *Decision Sciences*, **39**(3), 383–406.

Yong Li. (2008). "Duration analysis of venture capital staging: A real options perspective." *Journal of Business Venturing*, **23**(5), 497–512.

Miller, B., and Clarke, J-P. (2008). "Strategic guidance in the development of new aircraft programs: A practical real options approach." *IEEE Transactions on Engineering Management*, **55**(4), 566–578.

Heli Wang, and Sonya Seongyeon Lim. (2008). "Real options and real value: The role of employee incentives to make specific knowledge investments." *Strategic Management Journal*, **29**(7), 701–722.

9

Technology and Innovation Management: Financing Technology

Patricia Robak

Drexel University

Introduction

Innovation is the key element to economic progress. The basic foundations of economic theory lie in the ability to utilize scarce resources. Without innovation, resources are exhausted rapidly. At the corporate level, the goal of shareholder wealth maximization relies on the efficient use of resources and thereby the prosperous development of technology and innovation. Paradoxically, innovation itself necessitates a significant amount of resources. The ability to acquire capital to fund technology and innovation is the primary challenge facing innovators at both the entrepreneurial level and the corporate level. The main difficulty in obtaining capital lies in the uncertainty surrounding the value of the innovation, coupled with an inherent moral hazard and adverse selection problem. This entry will commence with the issues surrounding capital acquisition and then delineate the many alternatives available for obtaining funding.

Problems in financing innovation: adverse selection and moral hazard

The main issues inherent in funding technology are the asymmetry of information between financier and technology project manager and the high degree of uncertainty and risk surrounding technology development. This explicates the classic moral hazard and adverse selection problems.

Moral hazard suggests that a technology developer who must procure funds to support the development of his project will take greater risks than he otherwise would because he knows that the risk of failure and potential loss of funds is transferred to the financier. Thus, the technology development is paid for by an agent who lacks the innate knowledge of the developer. In this classic principal–agent relationship, the agent has incentive to work in his own best interest rather than that of the principal. Conversely, the financier may be unwilling to offer the appropriate level of funds due to the lack of knowledge. The asymmetry of information is the root of the moral hazard problem.

Adverse selection refers to a type of market failure, where the market fails to allot resources in the most efficient manner. Adverse selection occurs when you engage in a transaction with a counterparty that is sub-optimal and typically arises out of asymmetric information, that is, if complete information was discernible by both parties, the transaction would be valueless. In the case of financing technology, the technology developer possesses information that the financier lacks, hence asymmetric information. In addition, the inherent risk and uncertainty of technological innovations motivates the developer to seek outside sources of funds. He recognizes that the risks are substantial enough to warrant the use of someone else's funds. The notion of adverse selection can appear from the financier's perspective as well. Given the prospect of high-risk developments, the financier rations credit in a disproportionate manner. That is, the market fails to correctly weigh the risk-and-return relationship inherent in the development. Because of the fear of loss and the limited ability to assess risk, the financier will deny funding that might be appropriately priced.

The problem of financing technology is so pervasive as to question the fundamental model of the firm. Firm constituents can and do take actions in obtaining capital that actually place the firm at higher risk. After a particular capital structure has been established, shareholders have incentive to partake in risky innovation as they will bear only a portion of the risk (Scandizzo, 2004). This presents a difficult problem for would-be financiers who must view this risk-taking behavior as a significant threat and begs the question: Why provide funds to highly innovative firms?

The problem of project valuation

The high degree of uncertainty and risk surrounding technology innovation also engenders a significant problem in project valuation. Classic capital-budgeting theory epitomizes the estimation of cash flows arising from an initial investment and ascertains the viability of the investment to procure a sufficient return. A major assumption here is that the cash flows can (a) be ascertained and (b) have a calculable degree of certainty. This contravenes the very nature of research-and-development projects. It has already been established that earnings derived from technological development are highly uncertain and have a renowned risk of failure. This suggests that traditional methods for verifying the viability of technology projects are inapt. The problem of unsuitable or inaccurate methods for valuation exacerbates the financing problem faced by innovators. The level of risk poses a significant problem in ascertaining the viable options for funding technological development.

Option valuation methods may be more appropriate for determining the feasibility of a technology project. At a well-known pharmaceutical company, the financial team has developed and employed option valuation methods to discern the viability of R&D projects. These models enable R&D to be more highly scrutinized and get to the core of the problem of R&D efforts, especially those in early phases: uncertainty of future earnings and high risk of failure. Option valuation allows research projects to be appraised at various stages of the project and typically produces more accurate acceptability estimations (Nichols, 1994).

Financing options

The method of financing largely depends on (1) the amount of capital needed, which in turn depends on the size of the enterprise needing capital and/or (2) the stage of development of the technology. The issues described above plague the attainment of capital at all levels. However, "the primary problem for financing innovative growth companies [is] obtaining equity capital in the range of $250,000 to $5 million" (Shanley, 2004). While this section will focus on procuring capital at all dollar levels, it is informative to note that the most arduous fund raising occurs in this range. The options for funding are presented primarily in order of dollar-size acquisition from smallest to largest.

Bootstrapping

Bootstrapping refers to the technique of obtaining funds for innovation or advancement without resorting to external sources. Bootstrapping is any technique used to obtain resources for financing technology without resorting to traditional debt-and-equity financing (Ebben and Johnson, 2006). The elimination of outside investors and the subsequent relinquishing of control makes bootstrapping a desirable method of financing for small firms and/or early-stage technological development.

Winborg and Landström (2001) expound six categories of financial bootstrapping used by small businesses. First, funds may be derived from resources specific to the owner or manager of the business or technology. Included in this category are the usage of credit cards, personal savings, personal loans, and resources from friends and family to provide mechanisms for securing funds for puerile technology developments. Second, the notion of minimizing accounts receivable warrants another method of bootstrapping. Here, the owners of small businesses, in particular, would take sufficient measures to reduce the level of accounts receivable by rewarding speedy payments and/or ceasing relationship with late-paying customers. Third, "joint utilization" refers to the commingling of resources with other ventures. This could result in the sharing of office space and equipment, for example, or the utilization of one another's employees. The fourth item is categorized as "delaying payment" whereby the venture negotiates contracts with suppliers to "pay late" enabling the use of funds for additional time. An alternate form of payment delay is the use of leased items rather than purchase. Fifth, the minimization of inventory works to alleviate capital stock for technological development. This can be accomplished by developing formalized inventory usage processes or negotiating with suppliers for favorable inventory acquisition. In this way, a portion of technology financing is saddled by the supplier. Finally, the securing of subsidies from government or research-granting programs culminates the methods of bootstrap financing.

Angel investors

Angels are typically private, wealthy investors seeking to invest personal funds in areas outside of their own business framework, yet are involved or interested in. The size of investment is generally up to $1 million, with most under $500,000. Angels are often found through a network of friends and family or other network connections. Officially angels might be defined as "accredited investors" whose characteristics are delineated by federal and state securities laws and regulations. Angel investors typically concentrate their funds in familiar markets or technologies located in their own geographical regions and have long investment horizons. These investors are "adventure investors [who are] prepared to take higher risks and lower rewards when attracted by non-financial aspects of the venture" (Shanley, 2004).

While angel investors are most often found within the friend/family/business acquaintance network, online sources of ANGEL networks such as the ACE-Net (*https://ace-net.sr.unh.edu*) are rapidly emerging. These Internet-based sources connect investors with companies seeking funds. It is estimated that there are more than 2 million angels in the U.S. providing over $20 billion annually (Scandizzo, 2004).

Royalty financing

Royalty financing provides an alternative to regular debt-and-equity financing whereby investment funds are exchanged for future earnings. The basic structure of royalty financing emerges as an alternative to traditional debt-and-equity financing. Businesses engage with investors to provide a specific amount of capital in exchange for a percentage of the company's future revenues. A distinct advantage of royalty financing is that ownership rights are preserved. Investors procure a piece of the revenue stream rather than a piece of the company. An advance is made to the company in return for royalties.

Royalty-financing arrangements can be customized, providing additional flexibility to the user. Tailoring items such as the elapsed time to initial royalty payment and lag time between realization of revenues and royalty payment offer the ability to profit from the capital provided, enabling the business to increase sales prior to repayment.

Royalty financing has some additional benefits. State and federal security laws do not apply, eliminating the need for complex legal and filing fees that erode capital. Investors benefit from garnering profits in early stages rather than waiting for the company to go public or to be sold. Royalty financing allows a business to utilize traditional financing techniques in the future, an option not necessarily available when venture capital funds are employed. Finally royalty financing safeguards the equity position of the founders. However, royalty financing is not beneficial for companies who have low profit margins, do not expect capital influx that will significantly alter revenues, and have inelastic pricing structures (Evanson, 1997).

Private placement

Private placement refers to the obtaining of funds through the sale of equity shares without the regulatory burden of the public market. The issuing firm is not classified as a public company and shares have a limited or non-existent trading sphere. Under U.S. securities laws, advertising and solicitation for investors are generally prohibited and therefore require private-sector investors. Typically, investors must satisfy an "accredited investor" status to be eligible for private placement. Accredited investors are "individuals who, by reason of their financial situation, experience, or knowledge of the company selling the securities, are considered not to require many of the protections afforded by the securities laws" (Shanley, 2004).

While exempt from many of the details of the Securities Act of 1933, private placements must register with the SEC and appropriate state securities commissions. The level of funding is determined under various exemption states. Rule 504, for example, allows up to $1 million in funds during a 12-month period, while Rule 505 allows up to $5 million in a 12-month period (Shanley, 2004).

Ability to utilize private placements depends on each state's Blue Sky Laws, which specify the details of the offering and the number of investors sought. In addition, because private placements contain a considerable risk, the use of an intermediary is usually necessary to develop and position securities.

Venture philanthropy

Venture philanthropy or social venture philanthropy provides a means for capital attainment. Venture philanthropists are non-profit firms willing to contribute capital to a technological effort which coincides with the goals and mission of the philanthropic organization. A partnership arrangement might be formed whereby the philanthropic organization donates capital and then engages with the firm. The control level of a venture philanthropist pales in comparison with that of a venture capitalist, allowing the firm more flexibility in retaining control. The Acumen Fund (*www.acumenfund.org*) is an example of a venture philanthropy whereby funding is provided to social enterprises with innovative approaches to serving underprivileged communities.

In addition, venture philanthropy is also conducted through the advent and use of PRIs (program-related investments) which were initially created by an Act of Congress in 1968. PRIs allow organizations to obtain funds at significantly reduced rates and have been triumphant in promoting economic and social development. The concept is alluring to lenders as a way to encourage the entrepreneurial spirit. In other words, these loans and the like are not grants, but rather repayable loans which in turn promotes profit-seeking activities. Loan fees range from 0% to 4% (Schuerman, 2005).

Research parks

Research parks provide an option for financing technology especially for early-phase R&D projects or start-up technology companies. Research parks are typically aligned with governments or universities for the purpose of transferring intellectual property that arises from technological innovation. While research parks do not necessarily provide funding in the traditional sense, they offer technology starts facilities and services to foster the development process. Universities, in particular, regard research parks as an invaluable source of innovation and ideas. The inimitable exchange of ideas for funds in the form of dollars or research facilities provides a significant advantage to both parties. Research parks proffer services beyond facilities to network affiliations with venture capitalists, lawyers, marketers, and intellectual property rights managers.

Corporations also get in on the action to fund research labs at universities. In 2005, Google, Microsoft, and Sun Microsystems joined forces to underwrite a $7.5 million technology lab. The offerings of such research facilities offer entrepreneurs financial support to enliven their ideas, ideas that will ultimately benefit IT superstars such as Google, Microsoft, and Sun in the form of bringing technological advancements to the market sooner (Markoff, 2005).

Local/State/Federal programs

The requisite to maintain viability in the face of global innovation gives rise to programs at the local, state, and federal levels which provide financial support to innovators. The purpose of such agencies is to promote economic development and competitive positioning. The funding opportunities are delivered in the form of grants or loans which allow for the development of technology.

The Small Business Administration (SBA) is the most notable agency at the federal level. Created in 1953, the SBA offers a range of options to small businesses engaging in development efforts, such as the SBA Loan Guarantee program, the SBA Microloan program, and the SBA Certified Development Company 504 Loan program. The SBA Loan Guarantee program provides financial backing to emerging technologists that generally have minimal credit quality. The SBA Microloan program grants smaller sized loans (up to $35,000) to start-ups or growing small businesses. The SBA Certified Development Company 504 Loan program is designed to provide a significant portion of a development's fixed asset cost, offered by commercial banks and collateralized by the fixed assets themselves (*www.sba.gov*).

The Small Business Innovation Research program and the Small Business Technology Transfer program are additional examples of federal programs and similar agencies exist at the local and state level as well. A major benefit of funding programs is the ability to partner with the agency, yet preserve control of the development.

Asset-based lending

Traditional loans offered to early-stage growth companies base the terms of the contract on the creditworthiness of the company as established by the history of financial statements. Asset-based loans, however, base the terms of the contract on the assets that can back the loan. Lenders evaluate prospective borrowers based on asset value rather than financial statement fundamentals. Such loans extend additional credit to the borrower and typically tender relatively higher interest rates in compensation for the additional risk. Interestingly, however, the actual cost of borrowing tends to be lower given the innate paydown of the principle through lockbox collections that are standard issue for asset-backed offerings. Asset-based lenders provide loans to a wide array of business sales, from less than $250,000 to over $1 billion (CFA, 2008).

Venture capital

Venture capital (VC) is a well-known vehicle for obtaining funding. Venture capital funds are typically set up as limited partnerships with a venture capital firm serving as the general partner. The limited partners provide most of the funding. These include institutional investors, corporations, and wealthy individuals (Shanley, 2004). The critical factor for obtaining VC funds is the matching of the researcher's goals with that of the VC. Venture capitalists maintain criteria for the type of financing they provide; therefore, an entrepreneur must seek VCs whose objectives are aligned. Another key factor for obtaining VC funds is the stage of development. Contrary to popular belief, venture capitalists do not typically fund early-stage development projects. Venture capitalist ascertain viable recipients through an extensive analysis process, including industry, company, and risk analysis. As a result, assessments of newer developments generate incomplete results and thus are unsatisfactory prospects for a VC.

From the venture capitalist's perspective, the success of investing in an R&D endeavor is rendered when the firm goes public or is acquired. From this perspective, venture capital funds are an expensive form of financing. The originators of the development will ultimately lose control of their innovation when the VC takes ownership of the enterprise or technology.

Mergers and acquisitions

While not traditionally a source of financing innovation, mergers and acquisitions (M&As) have grown in popularity for such purpose. Fund-seeking technology developers might gain synergies by combining resources with other institutions in an effort to maximize R&D efforts. Procured by larger firms, the acquisition of development firms creates a continual flow of new technology options for the firm allowing the potential for growth opportunities not probable without. Benefits of M&As flow to both parties. The larger or acquiring firm benefits from the proliferation of new ideas and the smaller or acquired firm benefits from the access to finance and other resources that foster the development process. The downside for the innovator is the loss of control and the relinquishing of proprietary information related to the innovation.

SWORDs and spinouts

Spinouts are separate entities created exclusively for the R&D portion of a business. A SWORD is a specific type of spinout. SWORD stands for stock warrant off-balance-sheet research and development and is generally defined as a set of contracts that define the relationship among parties and the rights accruing to those parties. In a SWORD agreement a separate entity is set up to house an R&D effort. Solt (1993) offers a detailed description of the basics of a SWORD arrangement and describes the process as follows:

> "...(i) during the R&D phase, property rights are transferred to the new venture through the technology license, and funds flow from investors through the new venture and on to the parent via the services and development agreements; (ii) during the commercialization phase, the parent receives a license option on the R&D outcomes and pays royalties to the new venture on successful products; and (iii) the warrant and callable new venture common stock allow both the parent and the new venture shareholders to maintain claims on R&D outcomes."

The SWORD arrangement provides two significant benefits. First, shareholder wealth is preserved by keeping the risks of the technology off the balance sheet of the developing company. Second, SWORDs, as a method of financing, use equity only. While the benefits of debt financing are evident, R&D efforts often have long lead times with payoffs occurring years after development begins. Traditional debt arrangements do not allow for payback at such indistinct intervals. Thus financing R&D efforts with traditional debt/equity combinations often leads to rejection in the project evaluation phase. Rejection under these criteria does not mean that the project is not viable, rather its evaluation method is flawed. Therefore SWORD arrangements allow profitable development efforts to proceed and accrue value to the firm. In addition, it has been shown that the use of SWORDs diminishes the level of information asymmetry inherent in R&D projects (Theodossiou, 2007).

Outsourcing

The relentless rate of change in industries with large technology advancements forces innovation to be outsourced. In industries such as electronics and biotechnology, the necessity to devote substantial resources to innovation causes ample risks and reductions to profit margins. In order to remain competitive, firms are discovering the benefits of outsourcing their R&D. From an economic standpoint, this makes sense. Rather than each firm in the industry hunting for the next advancement, they can outsource to one or a few companies that specialize in a particular development field, gaining huge economies of scale and significantly reducing their R&D expense per dollar of sales. Firms in the tech arena are seeing reductions of several percentage points on the percentage spent on R&D per dollar of sales, this can translate into millions of dollars (Engardio and Einhorn, 2005). The relevance of the outsourcing opportunity is considerable as the productivity levels of R&D are characteristically low.

The stage of development coincides with the use of outsourcing. Typically, firms will outsource their R&D efforts at later stages of development or at re-development phases, this is due to the desire to be the first mover in the industry and to remain proprietary. When describing the point at which outsourced technology is employed, Engardio and Einhorn (2005) state, "You have to draw a line, core intellectual property is above it, and commodity technology is below it." In other words, there exists a point for firms above which they must derive their own innovation, below which the benefits arising from economies of scale for outsourced innovation prevail.

Public markets

The pinnacle moment for an innovator occurs when an initial public offering is rendered. Access to public markets signifies the ability to not only raise large amounts of funds in a single transaction but the potential for additional funds in the future, not only from equity markets but from debt markets as well. The dawning of exchangeable corporate shares provides a payoff in the form of liquidity for the founders or innovators and bestows them with a competitive advantage in their market (Shanley, 2004).

Firms with publicly traded shares have options for raising funds other than straight IPO or seasoned offerings. These forms of public market financing are typically used in down markets or at times of distress. Examples include performance guarantees, directed placements, and private investments in public entities. Performance guarantees offer investors a specified rate of return on their investment. The sale of shares of stock guarantees a particular rate of return over some period of time (usually 1 year). The payoff for the guaranteed rate of return may be compensated by the issuance of additional shares of stock. Directed placements target specific institutional investors offering shares at a discount prior to being publicly traded.

PIPEs (private investment in public entities)

PIPE transactions offer an alternative to publicly held corporations in the acquisition of capital. PIPE transactions tender equity or equity-linked investments to a select group of investors. The major benefits to the issuing corporation are the removal of SEC registration, the flexibility in offering structure and size, and the eradication of the "announcement effect" that typically cause share price decline. PIPE transactions can be tailored in several ways. Registered direct common stock and unregistered direct common stock present shares of registered and unregistered stock, respectively, to a select group of investors. The registered variety offers the benefit of immediate liquidity while the unregistered variety is committed to become liquid under contractual terms (30, 60, 90 days). Another common form of PIPEs is the offering to buy a formula-based quantity of stock at specified intervals at future prices. Typical PIPE transactions range in size from $1 million to $200 million and are commonly used by small-cap to mid-cap companies.

Strategic alliances/partnerships

Partnering or allying with other institutions in order to gain access to capital, markets, and other technologies provides a highly effective capital-raising strategy. Innovators are able to spread the risk of the technology development. Collaboration is the key element to these relationships and the affiliation offers flexibility in terms of design and extent of the association. According to Shanley (2004) the alliance can be defined across three classes. First, the simple contractual agreement is single task–oriented with a short duration. The second is the cooperative contractual relationship whereby technology is appropriated for the purpose of commercial potential. The final arrangement is a joint venture where parties ally for longer term relationships.

Strategic alliances or partnerships develop across all ranges of firm size but usually at later stages of technological development where known success provides a bargaining tool.

Conclusion

Financing technological developments presents a unique problem because the very essence of efficient capital markets is challenged when a firm dedicates resources towards innovation and development (Brown, 1997). While a prolific array of options exist for engendering funds, technological and innovative endeavors must seek out the choices that fit their specific scenario. This, in turn, depends largely on the type of development, the stage of development, and the potential scale and scope of the development.

References

Brown, W. (1997). *R&D Intensity and Finance: Are Innovative Firms Financially Constrained?* (Financial Markets Group Discussion Papers 271). London: London School of Economics.

Commercial Financial Association. (2008). "What is asset-based lending?; *http://www.cfa.com/What_is_ABL/what_is_abl.asp*> (last accessed March 18, 2008).

Ebben, J., and Johnson, A. (2006). "Bootstrapping in small firms: An empirical analysis of change over time." *Journal of Business Venturing*, **21**(6), 851–865.

Engardio, P., and Einhorn, B. (2005). "Outsourcing innovation." *Business Week Online*, March 21; *http://www.businessweek.com/magazine/content/05_12/b3925601.htm*> (last accessed March 20, 2008).

Evanson, D. R. (1997). "Cash is king: Royalty financing keeps investors happy—and keeps your company in your own hands." *Entrepreneur*, December, 62–65.

Houlihan Lokey Howard & Zukin. (2005). "Private investment in public equity (PIPE)"; *http://www.hlhz.com/main.asp?p=CORP_PIPEProductOverview*> (last accessed January 30, 2008).

Markoff, J. (2005). "3 technology companies join to finance research." *The New York Times*, December 15, p. C5.

Nichols, N. (1994). "Scientific management at Merck: An interview with CFO Judy Lewent." *Harvard Business Review*, January/February, 88–99.

Scandizzo, P. L. (2004). "Financing technology: An assessment of theory and practice." *Centre for International Studies on Economic Growth Research Paper Series*, **15**(43).

Schuerman, M. (2005). "Giving due credit." *Worth*, January 1; *http://www.worth.com/Editorial/Money-Meaning/Philanthropy/Best-Practices-Philanthropy-Giving-Due-Credit.asp*> (last accessed March 18, 2008).

Shanley, R. (2004). *Financing Technology's Frontier*. Hoboken, NJ: John Wiley & Sons.

Solt, M. (1993). "SWORD financing of innovation in the biotechnology industry." *Financial Management*, **22**(2), 173.

Theodossiou, A. (2007). "Reasons for financing R&D using the SWORD structure." PhD dissertation, Drexel University.

Winborg, J., and Landström, H. (2001). "Financial bootstrapping in small businesses: Examining small business managers' resource acquisition behaviors." *Journal of Business Venturing*, **16**(3), 235–254.

Part Two

Industry Level

10

Innovation-driven Industry Life Cycles

Susan K. Cohen

University of Pittsburgh

Introduction

Industry life cycle (ILC) refers to a collection of theories and frameworks that explain how the major structural characteristics of industries change, and identify implications for firm strategy, organization, and survival. Our goal is to provide an overview of the ILC concept, which reflects the diverse approaches economists, marketing and strategic management scholars, and organizational theorists have adopted to study it. We first distinguish ILC from the closely related product life cycle (PLC) and technology life cycle (TLC) theories. Next, we introduce the Abernathy–Utterback version of ILC, which has received substantial attention within the management literature. Third, we discuss some of the challenges to the AU ILC, and summarize a predominant alternative to it. Fourth, evidence supporting ILC is offered. We conclude with a discussion of competitive dynamics over the life cycle and their managerial implications.

Industry, product, or technology life cycle?

The terms product life cycle, industry life cycle, and to a lesser extent technology life cycle, are often used interchangeably, which can make distinguishing these theories a chore. However, there are some important differences between them. PLC emerged from the marketing literature, and its unit of analysis is the product. Theodore Levitt introduced the term product life cycle in 1965 to describe the sales cycle of a new product. This cycle involves four discrete stages, through which every new product is expected to pass: introduction, maturity, growth, and decline; product development is sometimes included as a fifth stage (Lambkin and Day, 1989). Stages are distinguished according to shifts in the volume of sales and rate of sales growth, which result from different types of customers (e.g., early adopters, laggards) entering the market. Hence, a major focus of PLC is to identify how marketing managers can reach these different customer groups through their promotion, pricing, and placement decisions.

Theories that pertain to the technology life cycle are a more eclectic group, each explaining different aspects of technological change. Perhaps the earliest was Everett Rogers' (1962) efforts to identify the characteristics of technologies (e.g., complexity, observability) that affect their speed of adoption and diffusion among a population of users. Scholars have also developed typologies of technological innovation (e.g., disruptive versus sustaining, competence enhancing versus destroying) and looked for patterns in their prevalence over time in industries. For example, Tushman and Anderson (1986) describe the evolution of a technology through eras distinguished by the predominance of competence-enhancing and competence-destroying innovation and the variety of designs vying for market dominance. Foster (1986) discusses technology life cycles as S-curves, which depict the progression of a technology from slow to increasing to declining rates of technical performance improvement, as time passes or experience with the technology accumulates (Foster, 1986). Others merge attention to a technology's stage of development and its acceptance in a market to identify distinctive stages (e.g., from bleeding edge, to mainstream, to decline—Popper and Buskirk, 1992). In general, the unit of analysis for TLC is broader than a specific product; it may refer to a component or material technology or a process innovation, which applies to products sold in several different markets.

Insights from PLC and TLC have been incorporated into ILC, and their commonalities are sometimes more clear than their unique contributions (e.g., ILC is often tested at the product market level). Nevertheless, ILC differs in its focus on the relationships between industry structure and aggregate characteristics of organizations—such as their size, age, strategy or business model, as opposed to the revenues or performance of a particular product or technology. ILC has emerged from several research traditions, including scholars seeking to explain the evolution of competitive and management challenges faced by corporations in particular industries (Abernathy and Utterback, 1978), industrial organization economists working to explain observed regularities in patterns of entry and exit across industries (Gort and Klepper, 1982), evolutionary economists interested in how shifting technological opportunities affect competitive dynamics (Winter and Nelson, 1982), and organizational ecologists who focus on population demographics (Hannan and Freeman, 1989). What unites them is a common focus on the changing viability of firms with different characteristics, over the course of an industry's life cycle. Industries are defined as being broader than specific products (e.g., SUVs, trucks, and cars collectively comprise the automobile industry because they can all be manufactured using essentially the same production capabilities). The AU ILC takes *production capability plus product* as its unit of analysis. More generally, the unit of analysis for ILC refers to a dominant business model or strategy, which may encompass multiple products and includes choices such as which activities along the value chain are integrated into the firm.

ILC proposes that industries move through distinctive stages that roughly mimic those of biological organisms (emergence or fragmentation, growth, maturity or shakeout, and decline). The theory explains that these shifts occur as firms learn about customer preferences, and discover the best way to exploit technologies to satisfy those preferences. As these fundamental uncertainties are resolved, firms' strategic choices change. For example, process R&D and outsourcing may appear more attractive at one stage than in another. Moreover, as knowledge about the market and the technology diffuses and firms seek to exploit new opportunities, the predominant focus of competition evolves and the structural character of the industry changes.

Abernathy–Utterback ILC

William Abernathy and James Utterback (1975, 1978) articulated the most studied version of industry life cycle theory.[1] They maintain that industries evolve from a "fluid" to a "specific" state, and that this progression is punctuated by a "transition" period. The three stages are largely distinguished according to the relative emphasis given to product or process innovation. During the *fluid* stage, firms compete to discern user interests and to define core concepts for product design. Fundamental choices, such as whether an automobile ought to have an open or closed body or an airplane ought to have fixed or movable wings, must be made, and these dilemmas dominate the focus of competition. Firms use general purpose labor, techniques, and equipment as long as these core design parameters are in flux, and produce in low volumes. Few entry barriers exist at this stage (e.g., neither brand loyalty nor scale economies are yet possible) and the rate of entry tends to be high.

An industry's *transition* begins when, in response to user feedback and vicarious learning from competitors, the majority of firms adopt common design elements. This marks the emergence of a *dominant design*. A dominant design is most often defined as the one that satisfies the majority of users. It is the creative synthesis of design elements that were introduced in earlier product variations that turns out to have greater appeal to the mass market than others (Abernathy and Utterback, 1978). It is not necessarily the technologically superior or highest performing product design (Tushman and Rosenkopf, 1992). Nor does the emergence of a dominant design mean that competitors' products are identical. Ample room may exist for firms to differentiate their products by introducing novel features through relatively minor changes to a product's architecture and components (Murmann and Frenken, 2006). As market and technological uncertainty decline, firms shift their attention to improving quality and driving costs down through process innovation.

Once the rate of process innovation surpasses that of product innovation, an industry has entered the *specific* phase. Production processes become more specialized, automated, and standardized throughout the industry, and firms seek cost advantages through scale economies as the market expands. As larger firms become more efficient, entry barriers rise and weaker firms exit, leading to consolidation.

Although the life cycle metaphor suggests an ordered progression from birth to death, Abernathy and Utterback (1978) acknowledge that these stages may repeat if "discontinuous" innovations challenge an industry. These are product or process innovations that destroy the value of incumbents' technological and organizational capabilities, overturn established systems of production and marketing, and erode established entry barriers (Utterback, 1994). Discontinuities tend to initiate a new cycle of entry and competition to define a new dominant design. The greater an innovation's potential to disrupt established production systems, the more likely are entrants to be the source of discontinuity (Foster, 1986).

This is not to suggest that incumbents don't innovate. Slowing demand for existing products, changes in customer needs or in the price/performance of complementary technologies, and credible threats from new competitors and technologies may prompt incumbents to innovate. Incumbents engaged in diversifying growth

[1] Sidney Winter (1984) discusses a similar sequence of innovative activity, but focuses on the source of knowledge fueling innovation to distinguish two innovation regimes. AU's "fluid" stage is similar to Winter's "entrepreneurial regime", and their "specific" phase corresponds to Winter's "routinized regime". Tushman and Anderson (1986) discuss three eras—of ferment, substitution, and incremental change—which also closely mirror the AU pattern, but their emphasis is on technology battles more than on industry structural evolution.

initiatives, such as those seeking to leverage a core technology into new product markets, may inject creativity into a stagnating industry. Technologies applied to new contexts must live up to new performance requirements and function with novel sets of complementary resources. As firms adapt their capabilities to meet these demands, they may recognize opportunities to create new market niches or reinvent old products (Levinthal, 1998). In this way, learning about a technology in a new context can reinvigorate product or process innovation in the original application and may shift the life cycle back in time or prolong an existing stage.

Challenges

Despite its intuitive appeal, industry life cycle theory has been challenged on a number of fronts. Better metrics are required for identifying stages and distinguishing ILC transitions from other types of economic change (McGahan, Baum, and Argyres, 2004). "Production–product units" are not easy to identify in a way that is meaningful and easily comparable across contexts. The dominant design concept is similarly difficult to define and measure across industries (see Murmann and Frenken, 2006 for a valuable step toward solving this dilemma). Perhaps most importantly, scholars have argued that the dominant design concept simply does not apply in all industries (Klepper, 1996, 1997). In response, a few authors have developed alternative theories for some widely observed life cycle patterns, such as changes in the rate of entry and exit, and emphasis on process rather than product innovation.

Klepper (1996) offers an alternative explanation for these basic patterns of entry, exit, market growth, and consolidation that does not incorporate the dominant design concept. Instead, he models an industry in which firms are assumed to compete with heterogeneous capabilities and therefore offer unique products. Some of these product variants will be better received by the market than others, and firms that enter earlier and experience greater product success grow faster. Larger firms profit more from process innovation, which reduces unit costs, and hence have an incentive to invest more in it. This departs from the AU model in which product and process innovation are so tightly coupled that firms avoid focusing on production methods until the opportunities to improve a product's design are relatively minor. As incumbents' innovative capabilities improve and costs decline, entry barriers rise and a "shake-out" ensues (entry slows and exit rises), leading to industry consolidation. The rate of product innovation declines, but this is due to there being fewer firms in the industry, rather than to the emergence of a dominant design or exhaustion of technological opportunities. Moreover, as market growth slows, product innovation to attract new customers may appear less profitable.

Adner and Levinthal (2001) explain industry evolution as being driven by firms' responses to heterogeneous customer preferences. Using a computer simulation, they examine the pattern of product and process innovation that emerges as firms try to please customers who accept different levels of performance and vary in their willingness to pay for products that meet their minimum threshold. While the model makes important assumptions that will not hold in some contexts (e.g., firms always incur a trade-off between lowering costs and improving product performance), it is able to reproduce the patterns of product and process innovation ILC predicts, without reliance on the dominant design concept or process innovation advantages of large firms. Three phases are evident as new technologies emerge: firms first emphasize product innovation, switch to process innovation once a sufficient performance level is achieved, and ultimately maintain a balance of the two. In the third phase, price changes little while product performance continues to rise. Firms balance product and process innovation in order to keep prices at the level that maximizes profits, according to the way the aggregate market trades off price and performance. On the one hand, firms seek market share by engaging in process innovation and passing on lower costs to appeal to "low-end" customers. By also engaging in product improvement, firms minimize revenue lost to price reductions, and appeal to "high-end" customers.

Evidence

Evidence in support of ILC

Researchers have sought to verify ILC predictions and understand its boundary conditions. Several patterns predicted by these theories have been observed in industries as diverse as automobiles and penicillin, commercial aircraft and tires, typewriters and television (Klepper, 1997; Malerba, 2007; Utterback, 1994). Dominant designs and industry shake-outs appear to mark a clear transition in the focus of competition and firm survival (Christensen, Suarez, and Utterback, 1998; Dowell and Swaminathan, 2004; Suarez and Utterback, 1995). Several indicators of industry life cycles have been observed. Entry rises or peaks early and declines steadily over time despite continued growth in industry output. Exits increase and industries consolidate following a shake-out or the emergence of a dominant design. Output continues to grow, and price to fall, through the shake-out period, and both eventually reach a constant level. The largest firms' market shares stabilize. The value of size, age, and particular kinds of experience for firm survival depends on an industry's life cycle stage (Agarwal, Sakar, and Echambadi, 2002; Klepper and Simons,

2000; McGahan, Baum, and Argyres, 2004). There is broad agreement that these patterns are driven by a combination of learning by firms, customers, and suppliers; exogenous and endogenous technological change; and the satisfaction of demand.

Evidence contradictory to ILC

However, studies also counter some of the more fine-grained ILC predictions, such as a firm's emphasis on cost reduction versus performance improvement. Two ILC predictions—that the diversity of product offerings and rate of product innovations is greatest early on, and that firms devote increasing effort to process innovation over time—clearly do not hold in all circumstances. Filson (2001) found support for the predicted ILC outcomes only in autos, where the rate of quality improvement was highest early on, and costs improved more rapidly later on. Along the same lines, McGahan and Silverman (2001) found no evidence that innovation shifts from product to process innovation. Nor did they find evidence that overall innovative output declines during industry maturity or that the leaders of mature industries are less innovative than those in emerging industries. Not all industries give rise to a single dominant design, and where competing designs co-exist, product innovation and variety is also sustained (Srinivasan, Lilien, and Rangaswamy, 2006; Windrum, 2005). Moreover, industries tend to engage in persistently different kinds of innovative activity, which makes generalizations about the emphasis on cost reduction versus performance improvement problematic (Klevorick *et al.*, 1995; Pavitt and Patel, 1984).

Reconciling the evidence: boundary conditions

Overall, the evidence suggests that ILC more accurately characterizes the early stages of the life cycle than the mature stages. Its weakest tenet seems to be that product innovation is always highest early on and is eventually surpassed by process innovation. This mixed support for ILC's predictions may reflect important differences across industries (Malerba, 2007). In fact, Abernathy and Utterback (1978) limited their theory to those industries with complex production processes that combine multiple inputs to produce a highly valued product. They felt it was important that a product could be made in multiple ways, permitting firms to differentiate along a number of dimensions but allowing for economies of scale to significantly influence production costs. Suarez and Utterback (1995) expect that products requiring specialized processes, such as synthetic fibers and plastics, will not follow the same pattern. In fact, in many chemical industries, process innovation must occur not only to lower costs sufficiently, but also to achieve the market's minimum acceptable product performance. By contrast, industries that produce customized products in small batches are more likely to exhibit a persistent emphasis on product innovation.

Along the same lines, scholars have suggested that the dominant design concept only applies to markets where the majority of customers have very similar tastes (Windrum, 2005). Tushman and Murmann (1998) argue that a dominant design is unlikely to emerge when market volumes are small, customer preferences change frequently, or when government regulations constrain product variation. St. John, Pouder, and Cannon (2003) more generally argue that a main factor determining whether a dominant design emerges is whether patents, regulations, customer preferences, geography, or competitive behavior create small fragmented markets, which prevent variations from accumulating much volume. Along the same lines, Klepper and Thompson (2003) show that the presence of sub-markets greatly influences how industries evolve. Others propose (and find some empirical support for) the idea that a dominant design only emerges when a new technology is not effectively protected by patents or other intellectual property rights (Srinivasan, Lilien, and Rangaswamy, 2006; Tushman and Anderson, 1990).

Finally, the evidence for the ILC comes largely from industries with rich opportunities for both product and process innovation, which may be a prerequisite for ILC's predictions to hold (Klepper, 1996). However, the nature of these opportunities for innovation seems also to be a potential boundary condition. St. John, Pouder, and Cannon (2003) suggest that the ILC model may not apply to industries in which rapid technological change sustains high levels of uncertainty about market size and requirements, process technologies, and resource availability. Industries that have perennially low-entry barriers and switching costs may never consolidate, as they are more prone to entrants introducing variations that challenge the incumbents' advantages, and may allow for a greater variety of equally viable strategies during any given stage (DeBresson and Lampel, 1985).

Competitive dynamics over the industry life cycle

These theories have a number of managerial implications. Each stage of the life cycle poses a unique set of competitive challenges and opportunities, suggesting that different decisions have greater bearing on firm success in each phase. Critical goals over the life cycle tend to shift from understanding a technology's potential, to assessing market opportunities and defining the product concept for a particular application, to differentiating a firm's product from competing variations, and finally to sustaining operational excellence throughout the value chain (Afuah and Utterback, 1997; Porter, 1980;

Suarez, 2004). Managers must be aware of shifts in the focus of competition and the underlying drivers of these shifts in order to effectively manage critical resource dependencies, allocate attention and resources to the appropriate search/learning activities, and assure the firm acquires the appropriate capabilities in a timely manner.

Fluid stage: product experimentation

New industries tend to arise from technologies that offer novel combinations of user functions and/or require competencies that differ in some important respect from those possessed by the leaders of established markets. Hence, managerial challenges during the fluid stage revolve around *learning*—understanding what the technology can do and what the market wants, and *legitimacy*—influencing and adapting to the evaluative criteria that resource providers use to decide which technologies and organizations to support. In particular, key challenges at this stage include educating and learning from potential users of the new technology in order to refine fundamental product concepts and to create sufficient demand; garnering support for necessary investments in infrastructure and complementary technologies; influencing individuals and organizations who have the capacity to shape the market's selection criteria; and attracting/retaining the requisite scientific/technical talent and financial capital. Firms may influence the criteria that experts and end users use to evaluate their technology by joining standards-setting committees, participating in industry conferences and consortia, forming alliances with providers of complementary technologies, and sharing information with regulatory bodies. When a technology's value is influenced by network externalities, it is particularly important for firms to engage these strategies early.

In addition, managers of new ventures need to develop a firm's capabilities for project management, to assure it stays focused on long-term goals and specific applications and to avoid spreading development efforts too thinly is critical. Also critical to young firms is building endowments, such as a stock of valuable patents and key accounts, to attract funding and talent. Externally, firms need to be able to quickly discern which elements of design to synthesize, as lead users comment on early product versions and the market responds to competitors' product variations. Learning by actively engaging a diverse group of stakeholders, so as not to get swept away in our own vision of the technology, is important. It may also be necessary to work closely with suppliers to adjust general purpose equipment and processes to the new technology and gain an early sense of factors likely to drive its efficacy, reliability, and cost.

Transition: battles for dominance

Eventually, the focus shifts from legitimating the product concept and proving the technology, to influencing users' adoption decisions (Suarez, 2004; Shapiro and Varian, 1999). As the competing technologies stabilize, the performance trade-offs they embody are more easily understood, and firms can focus on crafting marketing messages and pricing policies to build support for their technology. Advertising to differentiate and build brands becomes important, as does proving the organization and technology's credibility, such as by gaining the allegiance of high-status customers or alliance partners. Managers may intensify some of the strategies initiated during emergence to influence selection criteria, and also shift the firm's resources away from innovating around the core design concepts to refining them. Evidence of a substantial market for the technology and faster growth in demand may encourage greater emphasis on cost efficiency and building manufacturing, marketing, and distribution capabilities. Managers must decide carefully how to acquire capabilities (e.g., through internal development, outsourcing, or alliances). Choosing the wrong mode exposes firms to an increased risk of failure during this stage, as compared with the emergence stage; the risk is greater for small firms (Argyres and Bigelow, 2007).

Internally, managers must assure the firm continues to design products that incorporate emerging customer needs, coordinates effectively with complementary technology providers, engages in tactics to build its installed base, and negotiates appropriate contracts for specialized equipment and materials that will be needed in the specific phase.

Specific phase: stabilizing processes

Whether the mature/specific phase of an industry's life cycle comes about through the emergence of a dominant design, the shake-out of weaker competitors, or firms' local adaptations to market needs, this stage tends to be characterized by less uncertainty about how a technology can best satisfy the majority of the market. In light of this, the optimal configuration of the firm's value chain activities and key resource investments may have changed. Successful incumbents may sustain their dominant positions through continuous improvement to product performance, and increasing their emphasis on process innovation (Banbury and Mitchell, 1995). Architectural and modular innovation may also be used to reach niches or untapped market segments, and to bring new customers into the market. Strong customer and supplier relations are essential to supporting efforts. During this stage, firms need to attend to connections between the various activities in their value chains, which enable them to support continuous product and process improvement. For example, assuring that product engineers have the process and customer knowledge they need to design for manufacturability and ease of use, and connecting procurement and logistics with the supply chain to accelerate development cycles and manage inventory costs, may be key drivers of market share and profitability. Opportunities to invest in specialized equipment and processes to drive costs down are likely greater. It may also be important to collaborate with rivals to establish barriers, create boundaries to competition, and to exercise bargaining power over partners. Competitive relationships are especially important toward the end of an industry's life cycle, as firms seek to rationalize and harvest their durable assets.

A special challenge for incumbents during the era of stability is to anticipate the next discontinuity and decide how to prepare for it. In particular, incumbents need to guard against the many sources of incumbent inertia that can obscure the value of emerging technologies and hinder a firm's ability to exploit or adapt to them (Christensen and Rosenbloom, 1995).

Timing investments

If the dominant forces of competition shift in discernible patterns as life cycle theories suggest, then the timing of certain actions should be consequential for a firm's survival and relative success. Two types of decisions seem to have especially important consequences for firms' survival prospects: deciding when to enter the market and choosing which technologies to commit to.

The industry life cycle stage, in which firms enter a market, indicates the challenges it will face. In emergence, the dominant driver of a firm's survival is whether it "picks" the right technology. Early entrants during this phase seem to gain some advantages but their early entry status only favors their survival up to the transition stage (Dowell and Swaminathan, 2006). At that point, firms that misread the market or made the wrong technology bets are likely to fail. Hedging our bets during emergence by investing in multiple technologies can improve a firm's survival prospects (Hatfield, Tegarden, and Echols, 2001).

Firms whose technologies are favored by the market are more likely to survive through the transition stage if they also efficiently configure their value chain (e.g., by vertically integrating into and out of activities to minimize production and transaction costs—Argyres and Bigelow, 2007). Firms that possess superior capabilities for process innovation may also have a greater chance of surviving this stage (Afuah and Utterback, 1997; Klepper, 1996). Accumulating complementary assets, expert endorsement, and high-status customer contracts; engaging a persuasive communication strategy aimed at the mass market; and pricing and licensing to build bandwagons, should further enhance a firm's survival through the transition stages, although there is less evidence on these strategies.

In the specific phase, the key risk for entrants are the reactions of established incumbents and the need to make large, irreversible commitments to enter the industry. Suarez and Utterback (1995) found that the risk of exit or merger was generally smaller prior to emergence of the dominant design. By contrast, Agarwal and Audretsch (2001) found that while smaller entrants generally face lower odds of surviving, their life prospects were not diminished during industry maturity or in technologically intensive products. They suggest that entrants in these contexts focused on filling market niches. In a similar vein, Christensen, Suarez, and Utterback (1998) find that entrants targeting new market segments with architectural innovation—a change in a product's design that offers novel performance attributes—tended to be more successful than those targeting existing market segments or introducing innovative component technology. Entrants include start-ups and diversifying firms. Diversifying entrants have a better chance of survival if their pre-entry experience developed capabilities suitable to the industry life cycle stage in which they enter (Klepper and Simons, 2000). Klepper (2002) found that survivors of industry shake-outs entered earlier and with related pre-entry experience.

While Abernathy and Utterback did not explicitly address when or how firms should expand vertically or horizontally, several authors have examined life cycle implications for firm boundaries. In particular, research on alliances suggests that firms will engage in different kinds of inter-organizational arrangements according to the competitive demands of each stage (Cainarca, Colomba, and Marrioti, 1992; Hagedoorn, 1993; Rice and Galvin, 2006). Firms use alliances to mitigate risk through knowledge acquisition, and in the emergence stage a primary role is to facilitate product innovation. Standards-based alliances may also arise to create design rules, interoperability conventions, and a process for managing paradigmatic uncertainty. During transition, firms increasingly rely on alliances to acquire component technologies and complementary capabilities, and to improve operational efficiency. As the industry moves toward maturity, alliances gain value as a means for firms to recombine their competencies with new technologies and diversify product offerings. Firms may share capital risk and rewards via equity alliances that enable industry consolidation and the attenuation of risks inherent in standalone strategies, as the specific phase progresses.

While the strategic and organizational implications of ILC can appear straightforward, knowing when and how to invoke them is another matter. A particular challenge of applying the theory is that it can be difficult to know what stage you are in—and even harder to anticipate when it will change and how (Henderson, 1995; McGahan, 2000). Although ILC describes the industry structural attributes that may attain at each phase (e.g., dramatically slower growth rates at maturity), to adapt effectively, firms need to understand the causes of these shifts in greater detail than ILC provides. Moreover, adaptation involves a substantial organizational element which the theory does not address.

A few scholars have sought to extend ILC to offer more detailed maps for managers. In particular, McGahan (2000, 2004) develops a typology describing four models of industry evolution (receptive, blockbuster, intermediated, radical organic) in an effort to complement the ILC. Each model identifies the key structural characteristics of an industry, including performance drivers, typical risks, and its historical innovation path. These are meant to help managers recognize what kind of industry they are in, and to appreciate the nature of evolutionary change they are most likely to face. The models also describe whether innovation is likely to alter the predominant mode of organizing productive activity and the distribution of innovation along the chain. Two models of industry evolution are "architectural" in the sense that innovation tends to alter established supplier and customer relationships, and two are non-architectural, meaning that innovation generally occurs within extant customer and supplier relationships. As every industry fits into one of the four models, the framework may help managers to anticipate their industry's evolutionary path, to recognize when it is beginning to shift, and to assess the organizational implications of this change.

In conclusion, "life cycle" is a useful metaphor for thinking about how an industry's structure may evolve. ILC directs managers to attend to the underlying drivers of structural change in their industry and highlights specific aspects of firm strategy and organization that may be affected. Further research on ILC's boundary conditions may help to clarify industry differences that predict variations on the ILC and influence the efficacy of alternative responses to industry structural change.

References

Abernathy, W. J., and Utterback, J. (1975). "Patterns of industrial innovation." *Technology Review*, **50**, 41–47.

Abernathy, W. J., and Utterback, J. M. (1978). "Patterns of industrial innovation." *Technology Review*, **80**(7), June/July, 40–47.

Adner, R., and Levinthal, D. (2001). "Demand heterogeneity and technology evolution: Implications for product and process innovation." *Management Science*, **47**(5), 611–628.

Afuah, A. N., and Utterback, J. M. (1997). "Responding to structural industry changes: A technological evolution perspective." *Industrial and Corporate Change*, **6**(1), 183–202.

Agarwal, R., and Audretsch, D. B. (2001). "Does entry size matter? The impact of the life cycle and technology on firm survival." *Journal of Industrial Economics*, **49**, 21–44.

Agarwal, R., Sarkar, M. B., and Echambadi, R. (2002). "The conditioning effect of time on firm survival: An industry life cycle approach." *Academy of Management Journal*, **45**(5), 971–994.

Anderson, P., and Tushman, M. L. (1990). "Technological discontinuities and dominant designs: A cyclical model of technological change." *Administrative Science Quarterly*, **35**(4), 604–633.

Argyres, N., and Bigelow, L. (2007). "Does transaction misalignment matter for firm survival at all stages of the industry life cycle?" *Management Science*, **53**(8), 1332–1344

Banbury, C., and Mitchell, W. (1995). "The effect of introducing important incremental innovations on market share and business survival." *Strategic Management*, Summer Special Issue, **16**, 161–182.

Burgelman, R. A., and Siegel, R. E. (2007). "Defining the minimum winning game in high-technology ventures." *California Management Review*, **49**(3).

Cainarca, G. C., Colomba, M. G., and Mariotti, S. (1992). "Agreements between firms and the technological life cycle model: Evidence from information technologies." *Research Policy*, **21**, 45–62.

Christensen, C. M., and Rosenbloom, R. S. (1995). "Explaining the attacker's advantage: Technological paradigms, organizational dynamics, and the value network." *Research Policy*, **24**, 233–257.

Christensen, C. M., Suarez, F. F., and Utterback, J. M. (1998). "Strategies for survival in fast changing industries." *Management Science*, **44**(12), S207–S220.

DeBresson, C., and Lampel, J. (1985). "Beyond the life cycle: Organizational and technological design, I: An alternative perspective." *Journal of Product Innovation Management*, **3**, 170–187.

Dowell, G., and Swaminathan, A. (2006). "Entry timing, exploration, and firm survival in the early U.S. bicycle industry." *Strategic Management Journal*, **27**, 1159–1182.

Filson, D. (2001). "The nature and effects of technological change over the industry life cycle." *Review of Economic Dynamics*, **4**, 460–494

Foster, R. (1986). *The Attacker's Advantage*. New York: Summit Books.

Frenken, K., Saviotti, P. P., and Trommetter, M. (1999). "Variety and niche creation in aircraft, helicopters, motorcycles and microcomputers." *Research Policy*, **28**, 469–488.

Garud, R., and Karnøe, P. (2003). "Bricolage versus breakthrough: Distributed and embedded agency in technology entrepreneurship." *Research Policy*, **32**, 277–300.

Gort, M., and Klepper, S. (1982). "Time paths in the diffusion of product innovation." *The Economic Journal*, **92**, 630–653.

Hagedoorn, J. (1993). "Understanding the rationale of strategic technology partnering, interorganizational modes of cooperation and sectoral differences." *Strategic Management Journal*, **14**(5), 371–385.

Hannan, M. T., and Carroll, G. R. (1992). *Dynamics of Organizational Populations: Density, Competition, and Legitimation*. New York: Oxford University Press.

Hannan, M. T., and Freeman, J. (1989). *Organizational Ecology*. Cambridge, MA: Harvard University Press.

Hatfield, D. E., Tegarden, L. F., and Echols, A. E. (2001). "Facing the uncertain environment from technological discontinuities: Hedging as a technology strategy." *Journal of High Technology Management Research*, **12**(1), 63–76.

Henderson, R. (1995). "Of life cycles real and imaginary: The unexpectedly long old age of optical lithography." *Research Policy*, **24**(4), 631–643.

Henderson, R. M., and Clark, K. B. (1990). "Architectural innovation: The reconfiguration of existing product technologies and the failure of established firms." *Administrative Science Quarterly*, **35**, 9–30.

Hsu, D. H. (2007). "Experienced entrepreneurial founders, organizational capital, and venture capital funding." *Research Policy*, **36**, 722–741

Klepper, S. (1996). "Entry, exit, growth, and innovation over the product life cycle." *American Economic Review*, **86**, 560–581.

Klepper, S. (1997). "Industry life cycles." *Industrial and Corporate Change*, **66**(1), 145–181.

Klepper, S. (2002). "Firm survival and the evolution of oligopoly." *Rand Journal of Economics*, **33**(1), 37–61.

Klepper, S., and Simons, K. (2000). "Dominance by birthright: Entry of prior radio producers and competitive ramifications in the US television receiver industry." *Strategic Management Journal*, **21**, 997–1016.

Klepper, S., and Thompson, P. (2003). *Submarkets and the Evolution of Market Structure* (Working Paper). Pittsburgh, PA: Carnegie Mellon University.

Klevorick, A., Levin, R., Nelson, R., and Winter, S. (1995). "On the sources and significance of interindustry differences in technological opportunity." *Research Policy*, **24**, 185–205.

Lambkin, M., and Day, G. S. (1989). "Evolutionary processes in competitive markets: Beyond the product life cycle theory." *Journal of Marketing*, **53**, 4-20.

Levinthal, D. A. (1998). "The slow pace of rapid technological change: Gradualism and punctuation in technological change." *Industrial and Corporate Change*, **7**(2), 217–247.

Levitt, T. (1965). "Exploit the product life cycle." *Harvard Business Review*, **43**, November/December, 81–94.

Malerba, F. (2006). "Innovation and the evolution of industries." *Journal of Evolutionary Economics*, **16**, 3–23.

Malerba, F., and Orsenigo, L. (1996). "The dynamics and evolution of industries." *Industrial and Corporate Change*, **5**, 51–88.

McGahan, A. (2000). "How industries evolve." *Business Strategy Review*, **11**(3), 1–16.

McGahan, A. (2004). *How Industries Evolve: Understanding the Critical Link between Strategy and Innovation* (p. 1). Cambridge, MA: Harvard Business School Press.

McGahan, A., and Silverman, B. (2001). "How does innovative activity change as industries mature?" *International Journal of Industrial Organization*, **19**, 1141–1160

McGahan, A., Baum, J., and Argyres, N. (2004). "Context, technology and strategy: Forging new perspectives on the industry life cycle." *Advances in Strategic Management*, **21**, 1–23.

Murmann, J. P., and Frenken, K. (2006). "Toward a systematic framework for research on dominant designs, technological innovations, and industrial change." *Research Policy*, **35**, 925–952.

Pavitt, K., and Patel, P. (1984). "Sectoral patterns of innovation: Towards a taxonomy and a theory." *Research Policy*, **13**, 343–374.

Popper, E. T., and Buskirk, B. D. (1992). "Technology life cycles in industrial markets." *Industrial Marketing Management*, **21**(1), 23–31.

Porter, M. E. (1980). *Competitive Strategy*. New York: The Free Press.

Rice, J., and Galvin, P. (2006). "Alliance patterns during industry life cycle emergence: The case of Ericsson and Nokia." *Technovation*, **26**, 384–395.

Rogers, E. M. (1962). *Diffusion of Innovations*. New York: The Free Press.

Shapiro, C., and Varian, H. R. (1999). "The art of standards wars." *California Management Review*, 41(2), 8–32.

Srinivasan, R., Lilien, G. L., and Rangaswamy, A. (2006). "The emergence of dominant designs." *Journal of Marketing*, **70**(April), 1–17.

St. John, C. H., Pouder, R. W., and Cannon, A. R. (2003). "Environmental uncertainty and product–process life cycles: A multi-level interpretation of change over time." *Journal of Management Studies*, **40**(2), March.

Suarez, F. F. (2004). "Battles for technological dominance: An integrative framework." *Research Policy*, **33**, 271–286.

Suarez, F. F., and Utterback, J. M. (1995). "Dominant designs and the survival of firms." *Strategic Management Journal*, **16**, 415–430.

Tushman, M. L., and Anderson, P. (1986). "Technological discontinuities and organizational environment." *Administrative Science Quarterly*, **31**, 439–465.

Tushman, M. L., and Murmann, J. P. (1998). "Dominant designs, technology cycles, and organizational outcomes." *Research in Organizational Behavior*, **20**, 231–267.

Tushman, M., and Rosenkopf, L. (1992). "Organizational determinants of technological change: Towards a sociology of technological evolution." *Research in Organizational Behavior*, **14**, 311–347.

Utterback, J. (1994). *Mastering the Dynamics of Innovation*. Cambridge, MA: Harvard Business School Press.

Utterback, J. M., and Abernathy, W. J. (1975). "A dynamic model of process and product innovation." *Omega*, **3**, 639–656.

Utterback, J. M., and Suarez, F. F. (1993). "Technology, competition, and industry structure." *Research Policy*, **22**, 1–21.

Winter, S. (1984). "Schumpeterian competition in alternative technological regimes." *Journal of Economic Behavior and Organization*, **5**, 287–320.

Windrum, P. (2005). "Heterogeneous preferences and new innovation cycles immature industries: The amateur camera industry 1955–1974." *Industrial and Corporate Change*, **14**(6), 1043–1074.

Windrum, P., and Birchenhall, C. (1998). "Is lifecycle theory a special case? Dominant designs and the emergence of market niches through co-evolutionary learning." *Structural Change and Economic Dynamics*, **9**, 109–134.

Winter, R., and Nelson, S. (1982). *An Evolutionary Theory of Economic Change*. Cambridge, MA: Harvard University Press.

Zott, C., and Quy Nguyen Huy. (2007). "How entrepreneurs use symbolic management to acquire resources." *Administrative Science Quarterly*, **52**, 70–105.

11

Technological Characteristics of Industries

Donna Marie De Carolis

Drexel University

Competing in the 21st century mandates that managers understand current technological trends and devise strategies based on future technological trajectories. The increasing use of multiple technologies that are embedded in products and services complicates strategic technology planning. The technology and innovation management literature provides some frameworks and models that facilitate insights into the relationship between industry dimensions and technological characteristics that are relevant to competition and planning.

Several characteristics of industries will influence technological innovation and the strategic decisions that managers pursue: (1) the extent to which there is potential for technological innovation in an industry; (2) the technological dynamism of the industry; (3) industry reliance on technical standards; (4) the extent of collaboration among firms; and (5) government regulation.

Industries differ in the extent to which there is potential for technological innovations. This is due primarily to the types of production processes that are used in various industries. Production processes may be characterized as assembled products, non-assembled products, and assembled systems. Examples of non-assembled products would be steel or aluminum. Furniture is an example of an assembled product and televisions or computers are examples of assembled systems. Generally, technological innovation will vary according to the type of production processes that are utilized in an industry. For example, in non-assembled products, innovation stems from improvements in materials, products, processes, or scale. In simple assembled products, industry technological innovations may occur in input materials. Historical examples of innovation in assembled systems have been mechanization, automation, and product changes.

Another differentiating characteristic of industries in terms of technology is the degree of technological dynamism that characterizes the industry. This is manifested in the timing and speed of industry innovations. In other words, a technologically dynamic industry is characterized by rapid advances in science and technology necessitating that firms move very quickly to sustain a technological edge and bring new products to market. Innovation speed is a term that refers to the time it takes from concept to commercialization. This gap is closing in many industries. Subsequently, in terms of technology strategy, this translates into continuous investment in resources and capabilities and cultivating of human capital. It might also mean collaborations with other firms and organizations in order to develop rapidly evolving competencies needed to compete. Examples of such types of industries would be nanotechnology, biotechnology, and information technology.

The extent to which an industry relies on technological standards is another distinguishing characteristic of industries. Technical standards are a somewhat broader concept than dominant design. They refer to specifications or requirements to make sure that different components of systems can be connected regardless of who the manufacturer of the component might be. These standards are important in many industries such as telecommunications and electronics. Standards allow companies to enter industries at various places on the value chain, producing different components for the same overall product. Established companies may have the resources to produce several components for a system. Technical standards become very important for new ventures. Startups, as part of their technology strategy, may need to begin with one component that will then tie into a system of products.

Industries also differ in the degree to which they rely on collaboration. As a result of the maddening pace of technological advances, developing collaborative strategic alliances has become in many industries a necessary strategy to compete. Horizontal alliances typically involve firms within the same industry who collaborate to achieve knowledge sharing, create synergies, or engage in product/process development. Vertical alliances represent collaborations among firms in different industries in which firms are seeking similarities and complementarities among upstream and downstream partners along the value chain. Two industries that have the most horizontal strategic alliances are the defense and the biotechnology industries. It is virtually impossible for companies in these industries to keep pace with scientific and technological advances without research partners. Alliances have varying types of contract structures which will depend in part on the purpose of the alliance. For example, if an alliance between two or more firms is created for a product development purpose, the contract might address licensing rights, equity stakes, etc. On the other hand, if the alliance is vertical in nature, such as between a manufacturer and distributor, it may involve less governance and equity rights and be more contractual in nature.

Government regulation is another dimension of industries that will impact technological innovation. The financial, transportation, and communication industries are heavily regulated compared with the consumer goods sectors. Certainly there are good reasons for regulation yet government oversight adds a layer of complexity to decision making which will slow innovative advances. Everything from procurement practices to liability concerns can seriously erode a firm's ability to innovate.

Frameworks for examining technological characteristics of industries

There exist several frameworks that can be applied to the analysis of the technological characteristics of industries. Perhaps one of the most useful was proposed by David Teece (1986). Teece discussed how three elements—(1) the state of intellectual property in an industry (or firm); (2) the phase of dominant design (pre-paradigmatic or paradigmatic) and (3) the use of complementary assets—affects competition and innovation for firms. Specifically, industries can be characterized as having strong or weak appropriability regimes. The appropriability regime is the ability of firms to retain profits from their inventions and innovations. In a "tight" appropriability regime, firms within an industry have strong intellectual property protection either from a patent, copyright, or trade secret. For example, in the pharmaceutical and chemical industries, tight appropriability regimes are the norm due to the underlying nature of science and technology in this industry. It is difficult for competitors to duplicate innovations as patents on molecules are very strong. This allows for a period of monopoly profits for a technological innovation, provided of course the commercialization strategy is sound.

Teece (1986) also points to the dominant design dimension of technological advances. Dominant designs are defined as those product or process configurations of components and underlying core concepts that will not vary substantially from one model to another. Dominant designs hold a great deal of market share in a particular market. If a dominant design does not exist in an industry, or if the technological and scientific states of an industry are in flux, there is room for a firm to establish a dominant design (Afuah, 2003). On the other hand, if a dominant design does exist, then an incumbent or new entrant's firm technological strategy may be to improve upon that dominant design, and competition will move toward pricing issues.

Complementary assets are the final element in Teece's framework, defined as those activities that are required to successfully commercialize an innovation. Teece suggests that when appropriability regimes are weak (which is the case in a majority of industries) and there already exists a dominant design, then successful innovation will depend on strategy execution through complementary assets. Firms can either own or outsource complementary assets such as manufacturing facilities, marketing and distributing activities, etc. For startup companies, the decision to own or outsource can be critical to technology strategy.

Another framework for understanding strategies in technology-based industries is provided by Christenson, Verlinden and Raynor (2001) who have argued that in the early stages of a technology it may be critical for firms to engage in all aspects of the value chain as the "connections" between components of a technology are still immature and are not yet standardized. As the component portions of a technology become more standardized, there is room for companies to supply those interlocking parts. Once this happens, industry value chains begin to disintegrate. The computer industry provides an example. In the early 1980s IBM was a vertically integrated company and engaged in all aspects of the value chain in the production of personal computers. As components such as disk drives and software became standardized, it was easy for smaller companies to competitively provide "pieces" of that value chain. The value chain of the computer industry "disintegrated" as technologies matured. The industry as a whole became more diverse as markets emerged and were centered on "pieces" of the personal computer itself. In contrast, the pharmaceutical industry provides an example of a vertically integrated industry value chain where a large majority of the firms still engage in several value-creating activities from research and development, through manufacturing, marketing, and distribution. In other industries, it is not as strategically important for firms to be vertically integrated.

Industry life cycle models provide another method of analyzing industry characteristics and implications for technology strategy and competition. The Utterback–

Abernathy evolutionary model depicts technology as moving from a fluid phase to a dominant design phase to a specific phase. The fluid phase describes the early stages of a technology's birth and development; firms are competing on design not price. When a dominant design emerges, the technology stabilizes and firms compete on incremental innovation and price. This is the specific stage.

The Utterback–Abernathy model has implications for firm performance as it relates to new versus established firms in technology-intensive industries. The model suggests that the fluid stage favors new firms while the specific phase favors established firms. This is due to several factors. Prior to the emergence of a dominant design, new firms can operate on smaller scales, have less bureaucracy, and experiment with new designs. Once a dominant design is established, established firms have an advantage in that they can operate on larger scales as competition shifts to efficiency and economies of scale. Bureaucracy and hierarchy are organizational attributes that enable economies of scale. Similarly, prior to a dominant design, learning curves are weak and a new firm can enter the competitive space without a disadvantage. This disappears, however, after the emergence of a dominant design and learning curves become important to production and competition.

There are limitations to this model. Most importantly, the model does not work well in service industries or in industries characterized by non-assembled components. It works best in industries with assembled products and in which customer tastes are homogeneous.

A note on the technological characteristics of service industries

The underpinning of service industries such as financial, insurance, telecommunications, and health care is technological innovation in both product and processes. Information technologies, in particular, are the backbone of a majority of service industries. Despite the label of "service", these industries require strong in-house technical and/or research personnel to sustain the product and process innovations for competitive advantage.

In service industries Barras (1986) has proposed the "reverse product life cycle". In the Utterback–Abernathy model, the technology life cycle starts with radical innovations; in the Barras reverse product life cycle model, the technology life cycle ends with radical innovations. The Barras framework describes service industries as adopting new technologies that are first developed in manufacturing industries. This adoption allows for greater efficiencies in the service industry facilitating incremental innovations in the service. In the final stage, the service industry is able to focus on radical product innovations that are based on the technology innovations adopted earlier.

Conclusion

Competing in any industry requires an understanding of the nature of the technologies that underlie the products and services of that industry and, equally as important, the impact of industry characteristics on technology innovation and development. Outlined above are several industry characteristics that will play a role in technological innovation and progress, which include (1) the potential for technological innovation in an industry; (2) technology dynamism; (3) technical standards; (4) the extent of collaboration among firms; and (5) government regulation. An understanding of these basic technological characteristics of industries should assist managers and entrepreneurs in technology strategy planning and implementation. Moreover, the frameworks presented can be applied in tandem with the knowledge of technological characteristics of industries. For example, technology life cycles will differ between dynamic and stable industries. The implication of this dynamic has consequences for the planning and implementation of technology strategies.

References

Abernathy, W.J., and Utterback, J.M. (1978). "Patterns of innovation in technology." *Technology Review*, **80**(7), 40–78.

Afuah, A. (2003). *Innovation Management: Strategies, Implementation and Profits*. New York: Oxford University Press.

Barras, R. (1986). "Towards a theory of innovation in services." *Research Policy*, **15**, 161–173.

Christensen, C. M., Raynor, M. E., and Verlinden, M. C. (2001). "Skate to where the money will be." *Harvard Business Review*, November.

Teece, D. J. (1986)."Profiting from technological innovation: Implications for collaboration, licensing and public policy." *Research Policy*, June.

12

Competitive Dynamics in High-technology Industries

Ram Mudambi and Tim Swift*[†]

*Temple University and [†]St. Joseph's University

12.1 Introduction

Competition may be the most important determinant of firm performance. Industrial economics research has shown that firms not only compete with similar firms, but also with their buyers, sellers, and firms from other industries, and that those competitive forces explain a large portion of the variation in firm performance (Porter, 1980). The purpose of this chapter is to review the types of competition that exist in high-technology industries.

Literature in technology and innovation management has identified important ways in which high-tech firms compete. Some of the most well-recognized dimensions along which technology-based competition is waged include the choice of timing of entry, the nature of innovativeness (i.e., radical, incremental, or architectural), and the degree to which firms strive to establish the dominant design within their industries.

"First-mover" firms seek to enter markets first while "fast-followers" let other firms expend resources to build new markets and then conduct lower cost market entry later. Literature on first-mover advantage has yielded some surprising findings. Intuitively, one might expect that the firm that is first to exploit a new opportunity reaps the greatest rewards. Such first-mover firms have privileged access to new customers, and have the opportunity to be perceived by customers as a trend-setting firm. However, there is evidence that in some circumstances fast-follower firms outperform the first firms that enter a market (Mitchell, 1991).

A common perspective is that innovation can be characterized as either breakthrough or incremental. Incremental innovation has been defined as refinement of an established design in a way that yields price or performance improvements (Banbury and Mitchell, 1995; Dosi, 1982). Breakthrough innovation generally breaks paradigms, is based on new product designs, and is generally incompatible with existing dominant products (Dosi, 1982; Sheremata, 2004). Henderson and Clark (1990) re-directed this discussion by pointing out that some of the most disruptive innovations in industrial history were based on existing technologies. However, the innovating firms re-combined existing technologies in new ways. This ability to combine or recombine existing components in new ways was called architectural innovation. Thus, research began to focus on the difference between incremental innovation and architectural innovation.

It is not necessary to be on the cutting edge of technology to produce disruptive innovation. While some firms constantly strive to introduce the most sophisticated technologies, other firms rely on existing competencies and upon the components made by other firms to assemble larger solutions using "architectural" innovation. Several examples can be used to illustrate the point that the standard or product that the market adopts *en masse* is not necessarily the most technologically advanced one, and that the firms reaping the greatest benefit are the ones that produce the "dominant design" adopted by the market. In the 1880s, the direct current standard was initially adopted over the competing alternating current standard, though that latter was eventually proved to be superior (McNichol, 2006). Many observe that the IBM version of the personal computer, JVC's VHS video cassette format (Anderson & Tushman, 1990), or Microsoft's Windows operating system (Welch, 2007) were not the best versions of the new technology available at the time of their acceptance by the market. Standards and dominant designs arise in an industry as a critical mass of stakeholders is persuaded to adopt one out of several competing novel prototypes.

From a macro perspective, these forms of competition are related. Architectural innovations often result in the adoption of a new dominant design, while incremental innovations generally result in modest improvements to a market's existing design. First-movers are unlikely to create new dominant designs because initial entrants usually make pioneering mistakes. Fast-followers offer refinements to initial innovations that are more likely to be widely adopted.

The nature of ensuing competition is sensitive to the extent of network externalities and the consequent importance of standards. These externalities depend on the importance of complementary products and services (i.e., the extent to which the value of the innovation is enhanced by being placed within the context of a network). For example, the value of the Windows operating system is greatly enhanced by the enormous number of software applications that are written for it. Once this operating system emerged as the *de facto* standard for personal computers, its position was reinforced by the incentives to develop a widening range of complementary products and services.

In this chapter we explore in detail the current views on each of these important areas of technologically based competitive dynamics.

12.2 The timing of competitive entry

A broad body of research in competitive dynamics shows that early-movers generally outperform late-movers. In product markets, firms that are the first to introduce new products generate superior profit (Lieberman and Montgomery, 1988; Nelson and Winter, 1982; Porter, 1980; Schumpeter, 1934, 1950; Lee *et al.*, 2000) and superior stock returns (Chaney, Devinney, and Winter, 1991; Eddy and Saunders, 1980). Resource-based theorists predict that first-movers are better able to create proprietary resources that will promote sustainable competitive advantage (Barney, 1986; Conner, 1991; Makadok, 2001; Wernerfelt, 1984). Vanderwerf and Mahon (1997) find that first-movers and early-movers generate superior market share across a broad range of situations and industries. Late-movers are less likely to be successful because of the existence of well-established competitors and less growth opportunities (Carpenter and Nakamoto, 1989; Lilien and Yoon, 1990; Makadok, 2001; Robinson, Kalyanaram, and Urban, 1994; Shaw and Shaw, 1984; Teplensky *et al.*, 1993). Studies on firm decision-making speed find that fast decision making is positively associated with firm performance across a number of contexts (Baum and Wally, 2003; Eisenhardt, 1989; Judge and Miller, 1991).

However, first-mover firms make pioneering mistakes due to a lack of experience and precedents to guide their actions. First-movers may invest in creating public awareness of the benefits of a new technology through marketing expenditures. Second-movers can be free-riders, benefiting from the awareness created by the first-movers. In some industries, large first-movers that introduce a new technology can invest heavily to create scale economies, which can be exploited by smaller late entrants (Mitchell, 1991). Fast-follower firms may generate superior performance by taking the extra time to develop a comprehensive plan (Grinyer, Mayes, and McKiernan, 1988; Miller and Friesen, 1984; Virany, Tushman, and Romanelli, 1992). In addition, previous research has identified several contingencies under which late-movers generate superior performance (Shamsie, Phelps, and Kuperman, 2004). Finally, fast-followers can defeat first-movers by developing a better design that improves upon earlier prototypes offered by first-movers. This form of competition is explored in the next section.

12.3 Incremental, breakthrough, and architectural innovation

Breakthrough innovations require higher levels of tacit knowledge (Nelson and Winter, 1982; Zucker, Darby, and Brewer, 1998). Tacit knowledge is not transmitted easily; when it is transferable, it is transmitted most effectively in face-to-face settings (Cantwell and Santangelo, 2000; Sorenson, Rivkin, and Fleming, 2006). There is a large body of literature establishing the highly local nature of scientific knowledge flows (Almeida and Kogut, 1999; Henderson, Jaffe, and Trajtenberg, 1998; Jaffe, Trajtenberg, and Henderson, 1993; Zucker, Darby, and Brewer, 1998), underlining the importance of clusters in technology-intensive industries. Thus, firms that utilize high levels of tacit knowledge in their innovation efforts compete by accessing regional knowledge that resides in geographic centers of excellence. These firms create a presence within knowledge-intensive regions so that they can interact with locally embedded R&D subject matter experts (Cantwell and Janne, 1999; Cantwell and Santangelo, 1999). Such regions are relatively abundant in the resources that support R&D that is relevant to the firm's activities, such as an available supply of specialized knowledge workers.

This line of reasoning underlines the importance of geographic areas known as "technological clusters", where R&D work that involves highly tacit knowledge can be performed by co-located R&D scientists (Mudambi, 2008b). Clusters often represent worldwide centers of excellence in particular industries or technologies. Silicon Valley outside of San Francisco is a well-known technological cluster of software development; Boston's Route 128 is a well-known biotech cluster, and southern Germany has a renowned technological cluster for high-precision machinery (Saxenian, 1994; Storper, 1995). Literature suggests many reasons for technological clusters, such as the existence of large, incumbent firms in certain regions (Agrawal and Cockburn, 2003), and the co-location of firms that are participating in different but complementary technological fields (Robinson, Rip, and Mangematin,

2007). The most competitive firms seek to place R&D units inside technological clusters. Firms locate R&D resources in technological clusters in order to gain access to "locally embedded sectoral specialists" (Cantwell and Santangelo, 1999, p. 120). By interacting with R&D workers that are on the cutting edge of innovation, firms may gain a competitive advantage by applying this new, tacit knowledge in less competitive markets elsewhere.

While much R&D is aimed at creating technology with superior performance, many major innovations are created by combining existing components already created by other firms. In fact, research suggests that while incremental innovations often involve cutting edge technology, major architectural innovations are often low-tech: "Producers and customers accept a package of relatively well-known innovations and forego the best technical performance in order to reduce technological uncertainty. State-of-the-art designs typically achieve superior performance through experimental, risky advances that may be too unreliable and expensive for the majority of adopters" (Anderson and Tushman, 1990, p. 617).

In general, research relies upon two different criteria to distinguish between incremental, architectural, and breakthrough innovation. While breakthrough or radical innovation introduces a significant new dominant design, incremental innovation "refines and extends" existing designs (Henderson and Clark, 1990, p. 11). Innovation can also be characterized as "modular" or "architectural" (Henderson and Clark, 1990). Modular innovation introduces new standalone components that replace existing components (Mudambi, 2008a). Architectural innovation changes the way that components link together to create a larger solution.

Emerging research suggests that organizational form plays a role in the firm's decision to compete on the basis of either architectural or incremental innovation (Mudambi, 2008a). Vertically integrated firms rely on their ability to coordinate the activities of upstream component manufacturers and downstream solution providers. Thus, it is likely that vertically integrated firms have the ability to coordinate multiple processes within the value chain. Small advances anywhere in the value chain are communicated effectively within the firm. For example, new marketing intelligence can be crisply passed on to design, and a small design advance is effectively communicated to manufacturing. Vertically integrated firms should be more adept at incremental innovation (Mudambi, 2008a).

Since firms that are not vertically integrated have not created the "linkage economies" that enable integrated firms to pass information efficiently along the value chain, such unintegrated firms are more likely to compete on the basis of architectural innovation (Mudambi, 2008a). Such firms are likely to be particularly adept at recombining diverse bodies of knowledge, likely sourced from multiple external organizations. They are skilled at identifying and controlling the creative heart of the value proposition, while outsourcing all other activities. Taken together, firms that successfully implement such specialization strategies have "orchestration" competencies.

12.4 Standard-setting and dominant designs

An important aspect of technology evolution over the last few decades is the move from stand-alone to networked systems and products. This evolution has dramatically increased the importance of standards and dominant designs. The importance of interconnection and interoperability means that many markets coalesce around one type of technology while shunning others. For example, in the 1980s Microsoft's Windows operating system became the *de facto* standard operating system for most personal computers, beating out IBM OS/2 and the Apple MAC operating systems. Today, Windows is the most widely used personal computing operating system worldwide. Accordingly, many application developers have built their software to be compatible with the Windows environment, and major PC manufacturers such as Lenovo and HP invest significant R&D in concert with Microsoft to ensure that their devices are optimized to run Windows. IBM discontinued support for OS/2 in 2006.

Dominant designs are persistent architectures that are widely adopted by the industry (Anderson and Tushman, 1990). This concept has its roots within the population ecology literature, which has studied how some forms survive while others fail (Hannan and Freeman, 1977). Previous research has asserted that the emergence of new dominant designs is the primary driver of industry evolution (Clark, 1985; Henderson and Clark, 1990; Utterback and Abernathy, 1975). Standards are often important elements of dominant designs. The two concepts are closely related. However, standards have much stronger relationships with the markets for complementary goods and services. This may be why a firm that establishes a dominant design does not often appear to reap competitive advantages from it, though one that establishes a standard may (Gallagher, 2007).

Kuhn (1962) was among the first to suggest that ordinary technological progress occurs along a normal trajectory based on previous innovations, but extraordinary innovations overthrow the paradigm. In a similar way, new dominant designs can rely on existing competencies within an industry, or can destroy competencies in order to introduce new ones (Tushman and Anderson, 1986). Previous scholars argue that competency-supporting innovations have defensive value; market incumbents can maintain higher market share by introducing more incremental innovations (Banbury and Mitchell, 1995).

However, radical innovation has disruptive power, in that it helps challengers make inroads against established firms (Reinganum, 1983). This has immediate implications for technology-based competition. Incumbent firms are defending established market positions that entrant firms seek to undermine.

Incumbents succeed by providing the best possible response to the demands of an extant client base. Since existing customers have day-to-day operating requirements, incumbent firms are naturally inclined towards incremental innovation. In a number of segments of the computer industry (supercomputers, vector processors, mini-supercomputers, and massively parallel processors) it has been shown that incumbents possess advantages over new entrants when implementing incremental technologies (Afuah, 2004). Entrants succeed by pursuing technological trajectories that are not immediately demanded by customers. Such an approach has the lowest opportunity cost for an entrant that has no current customers to lose. There is evidence that entrants are often firms started by former employees of an incumbent firm whose knowledge went unused (Klepper, 2002; Klepper and Sleeper, 2005). This is precisely because the new knowledge threatened the incumbent's existing competencies or because the cost structure of the leading firm was incompatible with the new revenue potential.

Thus, entrants are more likely to undertake radical innovation that introduces turbulence, depletes the value of the incumbents' stock of dedicated resources, and selects out those that are unable to adapt (Mudambi, 2008a; Schmalensee, 2000). For example, numerous instances of technology "leapfrogging" have been documented for the U.S. video game console industry, wherein entrants introduced radical new technological standards that displaced existing industry leaders (Schilling, 2003).

Some firms rely more on marketing and sociological forces than on technological ones. Once a breakthrough innovation disrupts the usual ways of doing things, firms' strategies are highly sensitive to the extent of network economies. When network economies are high, as in the case of mobile telecommunications, firms usually form consortia within which they discuss, negotiate, and align positions on technical features with their peers (Leiponen, 2008), a process known as "standard setting". Often, multiple consortia arise, each promoting its own standard. On the other hand, when network economies are not as high, as in the case of the personal computer, firms compete to have their version of the innovation adopted by the market in order to become the "dominant design" (Anderson and Tushman, 1990).

How are such relatively inferior forms adopted while superior forms existed? Firms can create the industry standard encouraging the development of complementary goods and services. Microsoft accomplished this by licensing their software to numerous computer manufacturers while other firms sought to keep their operating systems proprietary. Today, a new personal computer user is very likely to select a computer that runs on Microsoft Windows not because of Windows' performance, but because of the broad availability of software, equipment, peripheral devices, and professional services that all work well with Windows. Thus, firms that control standards can create and maintain competitive advantage on the basis of the depth and breadth of complementary goods and services.

Another example that helps to illustrate the importance of the availability of complementary goods in winning a standards war is the adoption of Dolby noise reduction technology in the music-recording industry. In the late 1960s and early 1970s, Dolby noise reduction was among several emerging technologies that claimed to improve the quality of audio recordings. In an effort to encourage adoption of Dolby technology, Thomas Dolby provided his solution to recording studios at minimal cost in order to ensure that a wide variety of music records and tapes were made with Dolby technology. Once it became clear that the majority of music recordings were being produced with Dolby noise reduction technology, manufacturers of stereo systems adopted Dolby *en masse*, making Dolby the dominant design (Hill and Jones, 2007, p. 237).

Significant advantages accrue to the makers of a dominant design (as noted, these need not be the originators). Learning curve benefits and economies of scale benefit those producing the dominant design (Anderson and Tushman, 1990; Arrow, 1962; Rosenberg, 1982). Downstream support providers such as maintenance and consulting services providers become more efficient at servicing the dominant design over time, thus driving down the total cost of ownership of this product versus other less dominant ones (Anderson and Tushman, 1990).

12.5 Concluding remarks

We have identified several dimensions along which high-technology firms compete. By being first-movers, firms seek to lock up significant market share before other competitors follow. Alternatively, fast-followers allow other firms to incur the expense of building up new markets, observe their pioneering mistakes, and enter later with a less costly, hopefully superior offering. While some firms rely on cutting edge technological skills to incrementally improve existing products and services, other firms combine the components made by other firms in order to create significant architectural innovations. These firms also compete with each other, individually and in consortia, to influence early-adopters and lead-users in order to establish new standards and dominant designs.

These types of competition are exceptionally important when observing the competition between incumbents and entrants. The motivational and strategic predispositions of incumbents towards incremental innovation and entrants towards radical innovation are clear. However, a number of factors affect the likelihood of success of these two strategies. High overhead costs, restrictions on flexibility (both externally and internally imposed) and simplistic historical pricing are some factors that are likely to reduce the effectiveness of incumbents' incremental innovation strategies (Clemons, Croson, and Weber, 1996). A lack of complementary technologies and assets, multiple competing technological standards, and high switching costs militate against the success of an entrant's radical innovation (Schilling, 2003).

In summary, technology-based competition is based on more than technology. A number of economic and sociological factors have an important bearing on the nature and success factors in this arena. For example, both switching costs and user conservatism play a role in the continuing dominance of the QWERTY standard that has migrated from typewriters to modern keyboards (David, 1985). These factors are likely to be reinforced by the rapid pace of technological advances as well as the increasing importance of networked systems. Firms that succeed in such competition are those that master the commercial rather than the technological aspects of innovation.

References

Afuah, A. (2004). "Does a focal firm's technology entry timing depend on the impact of the technology on co-opetitors?" *Research Policy*, **33**(8), 1231–1246.

Agrawal, A., and Cockburn, I. (2003). "The anchor tenant hypothesis: Exploring the role of large, local, R&D-intensive firms in regional innovation systems." *International Journal of Industrial Organization*, **21**(9), 1227–1254.

Agrawal, A., and Knoeber, C. R. (1996). "Firm performance and mechanisms to control agency problems between managers and shareholders." *Journal of Financial and Quantitative Analysis*, **31**(3), 377–398.

Almeida, P., and Kogut, B. (1999). "Localization of knowledge and the mobility of engineers in regional networks." *Management Science*, **45**(7), 905–917.

Anderson, P., and Tushman, M. L. (1990). "Technological discontinuities and dominant designs: A cyclical model of technological change." *Administrative Science Quarterly*, **35**, 604–633

Arrow, K. J. (1962). "Welfare and the allocation of resources for invention." In: R. Nelson (Ed.), *The Rate and Direction of Economic Activity*. Princeton, NJ: NBER/Princeton University Press.

Banbury, C. M., and Mitchell, W. (1995). "The effect of introducing important innovations on market share and business survival." *Strategic Management Journal*, **16**, 161–182.

Barney, J. (1986). "Strategic factor markets: Expectations, luck and business strategy." *Management Science*, **32**, 1231–1241.

Baum, R. J., and Wally, S. (2003). "Strategic decision speed and firm performance." *Strategic Management Journal*, **24**(11), 1107–1129.

Cantwell, J., and Janne, O. (1999). "Technological globalisation and innovative centres: The role of corporate technological leadership and locational hierarchy." *Research Policy*, **28**, 119–144.

Cantwell, J., and Santangelo, G. D. (1999). "The frontier of international technology networks: Sourcing abroad the most highly tacit capabilities." *Information Economics and Policy*, **11**, 101–123.

Cantwell, J., and Santangelo, G. D. (2000). "Capitalism, profits, and innovation in the new techno-economic paradigm." *Journal of Evolutionary Economics*, **10**, 131–158.

Carpenter, G. S., and Nakamoto, K. (1989). "Consumer preference formation and pioneering advantage." *Journal of Marketing Research*, **26**, 285–298.

Chaney, P. K., Devinney, T. M., and Winter, R. S. (1991). "The impact of new product introductions on the market value of firms." *Journal of Business*, **64**, 573–610.

Clark, K. B. (1985). "The interaction of design hierarchies and market concepts in technological evolution." *Research Policy*, **14**, 235–251.

Clemons, E. K., Croson, D. C., and Weber, B. W. (1996). "Market dominance as a precursor of a firm's failure: Emerging technologies and the competitive advantage of new entrants." *Journal of Management Information Systems*, **13**(2), 59–75.

Cohen, W. M., and Levinthal, D. A. (1990). "Absorptive capacity: A new perspective on learning and innovation." *Administrative Science Quarterly*, **35**, 128–152.

Conner, K. (1991). "A historical comparison of resource-based theory and five schools of thought within industrial organization economics: Do we have a new theory of the firm?" *Journal of Management*, **17**, 121–154.

David, P.A. (1985). "Clio and the economics of QWERTY." *American Economic Review*, **75**(2), 332–337.

Dosi, G. (1982). "Technological paradigms and technological trajectories: A suggested interpretation of the determinants and directions of technological change." *Research Policy*, **11**, 147–162.

Eddy, A. A., and Saunders, G. B. (1980). "New product announcements and stock prices." *Decision Sciences*, **11**, 90–97.

Eisenhardt, K. M. (1989). "Making fast strategic decisions in high-velocity environments." *Academy of Management Journal*, **27**, 299–343.

Gallagher, S. (2007). "The complementary role of dominant designs and industry standards." *IEEE Transactions on Engineering Management*, **54**(2), 371–379.

Grinyer, P. H., Mayes, D. G., and McKiernan, P. (1988). *Sharpbenders: The Secrets of Unleashing Corporate Potential*. Oxford: Basil Blackwell.

Hannan, M. T., and Freeman, J. (1977). "The population ecology of organizations." *American Journal of Sociology*, **82**(5), 929–964.

Henderson, R. M., and Clark, K. B. (1990). "Architectural innovation: The reconfiguration of existing product technologies and the failure of established firms." *Administrative Science Quarterly*, **35**, 9–30.

Henderson, R., Jaffe, A. B., and Trajtenberg, M. (1998). "Universities as a source of commercial technology: A detailed analysis of university patenting." *Review of Economics and Statistics*, **80**(1), 119–128.

Hill, C. W. L., and Jones, G. R. (2007). *Strategic Management: An Integrated Approach* (Seventh Edition). Boston, MA: Houghton-Mifflin

Jaffe, A. B., Trajtenberg, M., and Henderson, R. (1993). "Geographic localization of knowledge spillovers as evidenced by patent citations." *Quarterly Journal of Economics*, **108**(3), 577–98.

Judge, W. Q., and Miller, A. (1991). "Antecedents and outcomes of decision speed in different environmental contexts." *Academy of Management Journal*, **34**, 449–463.

Klepper, S. (2002). "The capabilities of new firms and the evolution of the US automobile industry." *Industrial & Corporate Change*, **11**(4), 645–666.

Klepper, S., and Sleeper, S. (2005). "Entry by spinoffs." *Management Science*, **51**(8), 1291–1306.

Kuhn, T. S. (1962). *The Structure of Scientific Revolutions*. Chicago, IL: University of Chicago Press.

Lee, H., Smith, K. G., Grimm, C. G., and Schomburg, A. (2000). "Timing, order and durability of new product advantages with imitation." *Strategic Management Journal*, **21**(1), 23–30.

Leiponen, A. E. (2008). "Competing through cooperation: The organization of standard setting in wireless telecommunications." *Management Science*, **54**(11), 1904–1919.

Lieberman, M., and Montgomery, D. (1988). "First-mover advantages." *Strategic Management Journal*, Summer Special Issue **9**, 41–58.

Lilien, G., and Yoon, E. (1990). "The timing of competitive market entry: An exploratory study of new industrial products." *Management Science*, **36**, 568–585.

Makadok, R. (2001). "Toward a synthesis of the resource based and dynamic-capability views of rent creation." *Strategic Management Journal*, **22**(5), 387–402.

McNichol, T. (2006). *AC/DC: The Savage Tale of the First Standards War*. San Francisco, CA: Jossey-Bass.

Miller, D., and Friesen, P. (1984). *Organizations: A Quantum View*. Englewood Cliffs, NJ: Prentice-Hall.

Mitchell, W. (1991). "Dual clocks: Entry order influences on incumbent and newcomer market share and survival when specialized assets retain their value." *Strategic Management Journal*, **12**(2), 85–100.

Mudambi, R. (2008a). "Location, control and innovation in knowledge-intensive industries." *Journal of Economic Geography*, **8**, 699–724.

Mudambi, R. (2008b). "Spikes, blocs and the 'death of distance'." *Journal of International Business Studies*, **39**(6), 1091.

Nelson, R. R. (1959). "The simple economics of basic scientific research." *Journal of Political Economy*, **67**(3), 297–306.

Nelson, R. R., and Winter, S. (1982). *An Evolutionary Theory of Economic Change*. Cambridge, MA: Belknap Harvard.

Polanyi, M. (1962). *Personal Knowledge: Towards a Post-critical Philosophy*. Chicago, IL: University of Chicago Press.

Porter, M. (1980). *Competitive Strategy*. New York: Free Press.

Reinganum, J. F. (1983). "Uncertain innovation and the persistence of monopoly." *American Economic Review*, **73**, 741–748.

Robinson, D. K. R., Rip, A., and Mangematin, V. (2007). "Technological agglomeration and the emergence of clusters and networks in nanotechnology." *Research Policy*, **36**, 871–879.

Robinson, W. T., Kalyanaram, G., and Urban, G. L. (1994). "Firstmover advantages from pioneering new markets: A survey of empirical evidence." *Review of Industrial Organization*, **22**, 1–23.

Rosenberg, N. (1982). *Inside the Black Box: Technology and Economics*. New York: Cambridge University Press.

Saxenian, A. (1994). *Regional Advantage: Culture and Competition in Silicon Valley and Route 128*. Cambridge, MA: Harvard University Press.

Schilling, M. A. (2003). "Technological leapfrogging: Lessons from the U.S. video game console industry." *California Management Review*, **45**(3), 6–32.

Schmalensee, R. (2000). "Antitrust issues in Schumpeterian industries." *American Economic Review*, **90**(2), 192–196.

Schumpeter, J. A. (1934). *The Theory of Economic Development*. Cambridge, MA: Harvard University Press.

Schumpeter, J. A. (1950). *Capitalism, Socialism and Democracy*. New York: Harper.

Shamsie, J., Phelps, C., and Kuperman, J. (2004). "Better late than never: A study of late entrants in household electrical equipment." *Strategic Management Journal*, **25**, 69–84.

Shaw, R., and Shaw, S. (1984). "Late entry, market shares and competitive survival: The case of synthetic fibers." *Managerial & Decision Economics*, **5**, 72–79.

Sheremata, W. A. (2004). "Competing through innovation in network markets: Strategies for challengers." *Academy of Management Review*, **29**(3), 359–377.

Sorenson, O., Rivkin, J. W., and Fleming, L. (2006). "Complexity, networks and knowledge flow." *Research Policy*, **35**, 994–1017.

Storper, M. (1995). "Regional technology coalitions an essential dimension of national technology policy." *Research Policy*, **24**(6), 895–911.

Teplensky, J. D., Kimberly, J. R., Hillman, A. L., and Schwartz, J. S. (1993). "Scope, timing and strategic adjustment in emerging markets: Manufacturer strategies and the case of MRI." *Strategic Management Journal*, **14**(7), 505–527.

Tushman, M. L., and Anderson, P. (1986). "Technological discontinuities and organizational environments." *Administrative Science Quarterly*, **31**, 439–465.

Utterback, J., and Abernathy, W. J. (1975). "A dynamic model of product and process innovation." *Omega*, **33**, 639–656.

Vanderwerf, P. A., and Mahon, J. F. (1997). "Meta-analysis of the impact of research methods on findings of first-mover advantages." *Management Science*, **43**, 1510–1519.

Virany, B., Tushman, M. L., and Romanelli, E. (1992). "Executive succession and organization outcomes in turbulent environments." *Organization Science*, **3**(1), 72–91.

Welch, J. C. (2007). "Review: Mac OS X shines in comparison with Windows Vista"; *www.informationweek.com*

Wernerfelt, B. (1984). "A resource-based view of the firm." *Strategic Management Journal*, **5**(2), 171–180.

Zucker, L., Darby, M., and Brewer, M. (1998). "Intellectual human capital and the birth of the U.S. biotechnology enterprises." *American Economic Review*, **88**(1), 290–306.

Zucker, L., Darby, M., and Armstrong, J. (2002). "Commercializing knowledge: University science, knowledge capture, and firm performance in biotechnology." *Management Science*, **48**(1), 148–153.

Part Three

Innovation

13

Types of Innovation

Rosanna Garcia

Northeastern University

It is the goal of this chapter to offer an organizing framework, or typology, of types of innovations that a firm may create or experience, and to briefly describe the differences in management processes required by firms for executing these various innovation types. Because of the plethora of terms used to label innovations (see Table 13.1 and Garcia and Calantone, 2002 for details), the focus here is on the most widely referenced types of innovations: product/service versus process; radical versus incremental; technological versus administrative; architectural versus modular; and disruptive versus sustaining. There are many ways of categorizing innovations (Table 13.1), with the greater number of categories adding greater and greater refinement for understanding the differences in innovation typology. However, we use the bi-level categorization because of its simplicity and frequency of use when discussing innovations.

First, we provide a definition for the general conceptualization of innovations. A classic dictionary definition of innovation is the embodiment, combination, or synthesis of knowledge into a new idea, method, or device. Peter Drucker, famed management consultant, defined innovation as a change that creates a new dimension of performance. We use this broader description to define innovation as a new idea, method, process, or device that creates a higher level of performance for the adopting user. From an economics perspective, the change resulting from the implementation of the innovation typically increases customer or producer value. However, it should be noted that some innovations may be destructive to one party while beneficial to another (Christensen, 1997).

Innovation in not synonymous with invention. Inventions are the first conceptualization of an innovation and do not result in economic contributions to the originating entity. An invention that moves through the innovation process, which includes the acts of research, development, and commercialization, becomes an innovation. Thus, an innovation provides economic value and diffuses to other parties beyond the inventor(s). Economic contribution is not limited to revenues but may also be realized as cost savings, such as with administrative innovations, or time savings, such as with process innovations, both of which we discuss below.

Another useful definition to understand before moving forward is that of "dominant design". Dominant designs are the benchmark features with which subsequent designs are compared. They are the industry standard. The classic example is the QWERTY keyboard, which became the dominant design moving from the typewriter to the computer keyboard. Dominant designs are characterized by (1) a set of core design concepts that correspond to the major functions performed by the product and (2) a basic architecture that defines the ways in which the components of an innovation are integrated (Henderson and Clark, 1990).

Dominant designs emerge after a technological breakthrough, which initiates an era of intense technical variation and selection in an industry, when a basic architecture of product or process becomes the accepted market standard (Utterback and Abernathy, 1975). Once a dominant design is established, architectural knowledge becomes stable and embedded in the practices and procedures of the organization. Other examples of dominant designs include Microsoft Windows, Ford's Model T automobile, and gas combustion engines for automobiles.

With this foundation terminology set, we now provide definitions for the different types of innovations.

13.1 Product/Service versus process innovations

A commonly used classification for innovations is the product/service/process typology. Product innovations are tangible objects that deliver a new level of perform-

Table 13.1. Categories of innovations with increasing refinement.[a]

Bi-level categorization

- discontinuous/continuous innovation
- instrumental/ultimate innovation
- variations/reorientations
- true innovation/adopted innovation
- original/reformulated innovation
- innovations/reinnovations
- radical/routine innovation
- evolutionary/revolutionary innovation
- sustaining/disruptive innovation
- breakthrough/incremental innovation
- business model/radical

Tri-level categorization

- low innovativeness/moderate innovativeness/high innovativeness
- incremental/new generation/radically new innovation
- platform/design/component innovation
- radical/really new/incremental innovation

Quad-level categorization

- incremental/evolutionary market/evolutionary technical/radical innovation
- incremental/market breakthrough/technological breakthrough/radical innovation
- incremental/architectural/fusion/breakthrough innovation
- incremental/modular/architectural/radical innovation
- niche creation/architectural/regular/revolutionary innovation

Higher level categorization

- systematic/major/minor/incremental/unrecorded innovation
- reformulated innovation/new parts/remerchandising/new improvements/new products/new user/new market/new customers
- improvements/new product lines/additions to existing products/new-to-the-world products/ cost reduction–process development/repositionings
- research/breakthrough/platform or generational/derivative or incremental/step-out or break-out

[a] See the Glossary for definitions and Garcia and Calantone (2002) for details regarding these classifications.

ance to adopting users. Examples of product innovations include Apple's iPod, video/camera mobile phones, Procter & Gamble's Febreze odor eliminator and automatic teller machines (ATMs). Service innovations are intangible methods of serving users with a new level of performance. They can be new service concepts (iTunes), a new way to interact with customers (Dell Direct online computer stores), or a new way of service delivery (Peapod grocery delivery). Process innovations deliver a new level of performance to the method by which a company operates. Process innovations can increase bottom-line profitability, reduce costs, improve efficiency, improve productivity, and/or increase employee job satisfaction. Examples of process innovations include ISO 9000 quality management, enterprise resource planning (ERP), Henry Ford's assembly manufacturing, and more recently nano-manufacturing. Process innovation can increase the relative value of a product or service by improving quality and reducing manufacturing or development costs for the innovating organization.

Traditionally, product and service innovations have been combined together in a single category because both focus on performance enhancements benefiting end-user customers, whereas process innovations' performance enhancements benefit the company implementing the process. As the importance of the service industry has grown in the U.S. market place (U.S. Census Bureau reports that service industries represent 55% of the economic activity in the U.S. during 2007), there has been a renewed interest in how product and service innovations

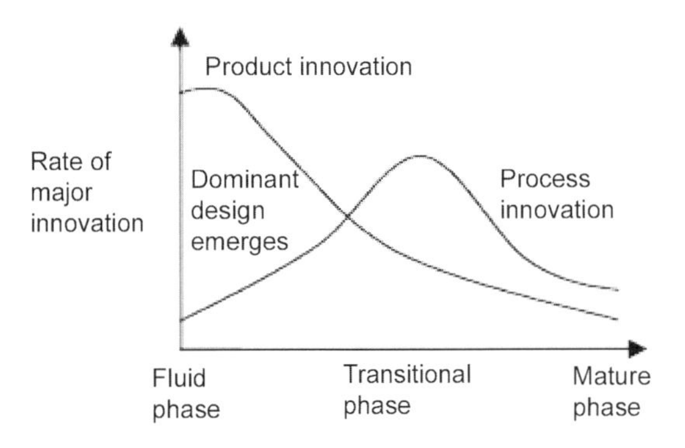

Figure 13.1. Patterns of innovation.
Utterback (1996, p. 130).

differ. For the present discussion, we will continue to include them in one category.

Simultaneous engagement in both product/service and process innovations is important to a company's new-product development (NPD) effectiveness. This dual focus is required because, as industries and markets mature, innovation efforts tend to shift from creating products and services to cost-reducing process innovations (Utterback, 1996). As graphically depicted in Figure 13.1, during the early emergence period of an innovation within an industry, the rate of product innovation exceeds the rate of process innovation. After a dominant design emerges, companies focus on process improvements to increase productivity and operational reliability in order to improve industry competitiveness. The two innovation trajectories become distinct in that each type accomplishes different goals for the company. Product and service innovations shift the demand curve whereas process innovation reduces costs and initially shifts the supply curve through lowering the marginal cost of production.

Process innovations should not be confused with the innovation process. The innovation process encompasses: (1) the invention or research phase, in which ideas are generated; (2) the development phase, in which the best ideas are selected and developed further; and (3) the market penetration/adoption phase, in which ideas are exploited for their new level of performance generated through the innovation. Thus, the innovation process should not be confused with the outcome of that process, the actual innovation, which may or may not be a process innovation.

What is not evident and can lead to confusion in differentiating process innovations from the innovation process is that process innovations have been known to foster technological breakthroughs resulting in radically new innovations.[1] An example would be the process innovations coming from nanotechnologies that are resulting in radically new medical treatments for cancer.

[1] Radical innovations are defined in the next section.

13.2 Radical versus incremental innovations

"Radical" and "incremental" are common terms used to describe the innovativeness of a product/service or process. We define incremental innovation as the refinement, improvement, and exploitation of existing innovations. Incremental innovations build on and reinforce the applicability of existing knowledge, and subsequently strengthen the dominance and capabilities of incumbent firms and the dominant design. Incremental innovations are characterized by reliability, predictability, and low risk. Incremental innovations are also referred to as sustaining innovations, continuous innovations, derivative innovations, evolutionary innovations, improvements, low innovativeness, minor innovations, and variations (see Table 13.1 and the Glossary). Examples of incremental innovations include the video iPod, whitening toothpaste, and Microsoft's Window Vista operating system.

We define radical innovations as innovations with features offering dramatic improvements in performance or cost, which result in transformation of existing markets or creation of new ones. They involve fundamental technological discoveries for the firm, and thus are new to the firm and/or industry, and offer substantially new benefits and higher performance to customers. Radical innovations are rare in occurrence. It has been suggested that only 10% of all new innovations fall into the category of radical innovations. Examples include magnetic resonance imaging (MRI), personal computers, the Internet, and cellphones.

Radical innovations are differentiated from other types of innovations as they cause discontinuities in the *status quo* on more than one strategic level, thus requiring the innovating firm to develop new situation-specific competencies in more than one domain. The domains most frequently impacted are technology, market, organization, and social. For example, the development of personal computers required (1) technological developments by the innovating organization, (2) new market developments as the end-user moved from corporations to individual consumers, (3) organizational changes in manufacturing and marketing, and (4) social changes by consumers who had no realization that they "needed" a computer.

Developing radical innovation requires understanding how to strategically plan for discontinuities on one or more of the levels noted above. Most firms are unable to alter the inertial forces driving the firm down a particular path, thus to plan for major strategic changes based on radical innovations is rare. This is the reason that so few radical innovations are introduced to the market place annually. Below we summarize the four levels that may be impacted:

- *Technology*. Radical innovations often rely on completely new technological principles, new architectures,

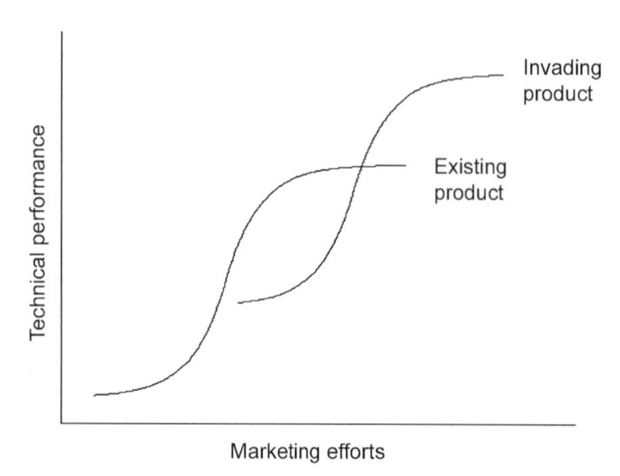

Figure 13.2. Technology/Marketing S-curve phenomena. Adapted from Garcia and Calantone (2002).

or new materials. Radical innovations replace old technology and initiate a new technological trajectory (see Figure 13.2).

- *Market.* From a market perspective, radical innovations act as the catalyst for the emergence of new markets and/or new industries. New markets evolve that support the new technological innovation, new competitors enter the market, and new partners and distribution channels emerge to exploit the new technology. Radical innovations often satisfy unmet customer needs for the first time, resulting in a quantum leap in customer value.
- *Organizational.* Organizational changes by the innovating organization are usually required in strategy, structure, processes, competences, incentive systems, and/or culture when developing radical innovations. Radical innovations are often disruptive to producers because they require new skills sets and competences not required when developing incremental innovations.
- *Social.* Radical innovations often require end-users to undergo considerable attitudinal and behavioral changes. Consumers are often called upon to "destroy their existing knowledge base" such as when home computers replaced typewriters. Despite their substantial benefit, radical innovations are disruptive to consumers because they introduce products and value propositions that disturb prevailing consumer habits and behaviors in a major way.

Discontinuities in any combination of more than one of these levels identifies a radical innovation. Innovations that only impact the *status quo* on one level are not considered radical innovations. For example, Figure 13.2 shows the interactions between technology advances and marketing efforts for radical innovation where technological product performance moves along an S-shaped curve until technical limitations cause research effort, time, and/or resource inefficiencies that then result in diminishing returns.

A final note regarding radical versus incremental innovation is that the degree of innovativeness is relative to "whom" (firm, customer, competitor, industry). What one firm identifies as a radical innovation can be labeled as an incremental innovation by another firm. One firm may be called upon to significantly alter their organizational competences, whereas another firm may operate in the *status quo*—even though they are both developing the same innovation. For example, the Swiss Watch Company, known for its trendy Swatch watches, partnered with Daimler-Benz AG to develop the smart® car. Developing small, fuel-efficient automobiles would be considered radical for the Swatch manufacturers, but incremental for Daimler-Benz. Attesting to the difficulties in developing radical innovations shortly after launching the smart® car, the Swiss Watch Company left the partnership due to issues with market place acceptance of the initial offering.

13.3 Technological versus administrative innovations

Technological innovations are innovations reflecting the application of science and/or engineering to develop technical applications or to accomplish a specific technical task. For example, desktop computers represent advanced electronic technology. Although traditionally, new technology has been concerned with the construction of machines, structures, and tools on a relatively large scale more recent technological innovations are focused on micro-level sciences. Nano-based products and microelectronics typify this new technological trend, as does the blossoming of genetic/DNA engineering. Technological innovations may be product (optoelectronics communications systems), services (the World Wide Web), or process innovations (nano-manufacturing). They may be incremental innovations (online banking utilizing the World Wide Web) or radical (the World Wide Web itself).

In contrast, administrative innovations refer to those innovations that change an organization's structure or its administrative processes. They are indirectly related to the basic work activity of the organization and are more directly related to management activities (Kimberly and Evanisko, 1981). Administrative innovations promise to further managerial efficiency or effectiveness as they involve organizational structure and administrative processes. Administrative innovations are process-oriented as opposed to product-oriented. Examples of administrative innovations are management by objectives, six-sigma processes, job rotation, staff incentive systems, and telecommuting.

Some researchers believe that the technological versus

administrative innovation distinction captures the foremost, fundamental dichotomy in the innovation typology (Damanpour, 1987). Developing technological innovations is the production focus of the organization, whereas developing administrative innovations is required to bring new performance levels to the organizational structure set for fulfilling the production focus.

13.4 Disruptive versus sustaining innovations

Disruptive innovation, a term popularized by Christensen (1997), refers to a technological innovation, product, service, or process with a different set of features and performance attributes, relative to existing products, which broadens or develops new markets by providing functionality that undermines existing market strategies. Disruptive innovations may replace the existing dominant design or technology in the market (e.g., desk-top computers replacing mainframes), or may provide an opportunity for building new markets (online university degree programs).

In contrast, sustaining innovations improve performance levels of established products (video-capable cellphones) and provide incumbent firms an opportunity to reinforce their core competences. Companies that focus on sustaining innovations excel at knowing the market, listening to the voice of the customer, and designing incremental improvements into existing technology to meet the needs of their core customers. However, a focus on sustaining innovations tends to dismiss the value of disruptive innovations, which do not reinforce current company goals or strategies (e.g., Polaroid's inability to compete against digital photography). Christensen (1997) labeled this phenomena the "innovator's dilemma"; the tendency for incumbent firms to ignore disruptive innovations and, thus, miss out on new market opportunities.

Past research (Christensen, 1997 among others) suggests that disruptive innovations have five characteristics:

(1) Performance: relative to existing products, disruptive innovations introduce a different set of features and performance attributes that are cheaper, simpler, smaller, and/or more convenient. This combination of features is unattractive to mainstream customers due to inferior performance on the dominant design attributes these customers value. However, a new customer segment or a price-sensitive mainstream customer sees value in the innovations' new attributes and the lower price (e.g., point-to-point low-frill airlines such as Southwest and easyJet).
(2) Niche market: disruptive innovations' power lies in their ability to meet the needs of a niche market that is unaddressed or undervalued by the current leading companies. The new-customer segment sees value in the innovation's new attributes and lower price. However, these niche customers are seen as insignificant by industry leaders.
(3) Performance improvement: the new disruptive technology steadily improves in performance until it meets the standards of performance demanded by the mainstream market. Performance improvements may surpass the prevailing performance level, thus new functionality disrupts existing market structures.
(4) Gradually erode: disruptive innovations do not destroy the value of established technology quickly, but instead gradually erode its value. If the new (disruptive) technology does not eventually displace the dominant one outright, niche markets initially created by the disruptive innovation will grow to be formidable competitors to incumbents.
(5) Innovator's dilemma: incumbent firms frequently fail to respond to disruptive innovations because addressing this new market requires building competencies they are unable to acquire easily. By focusing on the needs and wants of mainstream customers instead of the emerging niche market, dominant incumbents are in time significantly challenged, if not replaced, by younger more agile companies.

Disruptive innovations generally tend to be broken into two categories: low-end (incremental) innovations and high-end (radical) innovations. These two types pose significantly different challenges for incumbent firms and have different strategic implications for managers such as those issues referenced in the earlier section on incremental versus radical innovations. One type of low-end disruptive innovation is business model innovation, which is the design of a fundamentally different business model in an existing mature industry (Markides, 2006). Examples of business model innovations include Dell Direct, Amazon Online book seller, and the University of Phoenix's online higher education programs. Business model innovators do not discover new products or services; they simply redefine how an existing product or service is delivered to the customer. They may use strategic alliances, structural changes and/or emergent technologies to create and deliver new and differentiating value. The new business model increases the existing market size, either by attracting new customers or by encouraging existing customers to consume more.

Many companies see business model innovation as an important differentiator because imitation or commoditization is difficult or unattractive for incumbent firms. Given that business model innovations target niche markets that have been ignored by established firms and require different value chains that conflict with the *status quo* market structure, incumbent firms rarely adopt the new strategy or alter their business strategy in response to this new competition. The large retail bookstore, Barnes and Noble, had many missteps in their effort to compete with Amazon. Many stock brokerage firms have decided not to follow the low-end services of Schwab's discount offering. Because the economic pie grows, thereby increasing the market size, business model innovations can be a competitive advantage for emergent firms.

13.5 Architectural versus modular innovations

A product's architecture refers to the ways in which components are integrated and linked together to form the product. Architectural knowledge focuses on the linkages between components as opposed to the components themselves. Modularity refers to how the components are integrated into the product using sub-elements, subassemblies, sub-systems or "modules" that independently perform distinctive functions. Module knowledge (also called component knowledge) focuses on these modules (components) themselves as opposed to the linkages between components.

Accordingly, architectural innovation reconfigures the linkages between the components of established products in new ways while leaving the core design elements untouched (Henderson and Clark, 1990). The Sony Walkman is an example of an architectural innovation, where miniaturization of radio technology allowed portability, thereby significantly changing how music was listened to. The Walkman fueled an entire industry of portable music players. Conversely, modular innovation involves the introduction of new technology to specific modules of a product that displaces the core design concepts while leaving the established linkages between components relatively untouched. The antilock-braking system, now available in most automobiles, is an example of a modular innovation.

Differentiating between architectural and modular innovations is important for managers because it provides guidelines for the firm's focus on NPD strategies. Henderson and Clark (1990) suggest that firms focus on architectural innovations until a dominant design is in place. The emergence of a dominant design signals the industry-wide acceptance of an architecture, after which the NPD strategic focus benefits by shifting to modular innovations. New component knowledge becomes more valuable to a firm than new architectural knowledge as competition begins to revolve around component refinements and cost savings. Modular innovations require firms to focus on standardization of modules to allow for greater substitutability of processes or components across product families. The advantages of modularity are twofold: to increase product line variations without adding excessive complexity to the manufacturing system, and, secondly, to enable mixing-and-matching of components facilitating the mass customization of products. Customers' demands for customization have been increasingly making product-specific elements more essential, thus increasing the value of modular innovations.

Once a dominant design has been set, architectural innovation can present established organizations with challenges that may have significant competitive implications. Since architectural knowledge becomes embedded in the structure and information-processing procedures of established organizations through the dominant design, the potential destruction of the dominant architectural design is difficult for firms to recognize and address. Many organizations encounter difficulties in their attempts to make the transition to a new architecture. More agile new entrants, whose learning is less entrenched, may find it easier to build the organizational flexibility that abandoning old architectural knowledge and building new ones requires. Thus, architectural innovations introduced after the acceptance of a dominant design is a disruptive innovation.

13.6 Putting it all together: the product portfolio

Managing the types of products that a firm decides to develop requires evaluating the product portfolio. Cooper (1993) defines new-product portfolio management as the ongoing decision process, where the mix of a business's active new-product projects is constantly reviewed and revised. In the review process, new projects are evaluated, selected, and prioritized; existing projects are accelerated, killed, or de-prioritized; and resources are allocated and reallocated to active projects. Organizational performance may depend more on synergies and compatibility between different types of innovations than on each type alone. A firm's myopia on one type of innovation may be profit-maximizing in the short term but can be detrimental to the longevity of the firm. For example, few firms can focus only on developing long-term and highly risky radical innovations; incremental innovations are also required to bring in revenues that can cover the R&D expenses of radical innovations.

Included in portfolio management is product life cycle management. Determining the life expectancy of each innovation in the product portfolio can help firms recognize a need to reallocate resources into R&D efforts. Life cycle management is also useful for helping firms acknowledge the emergence of a dominant design and the need to focus on modular innovations for cost savings through commoditization or to focus on process innovations for manufacturing efficiencies. Additionally, as previously discussed, the interplay between product innovations and process innovations can be monitored to determine how the proper innovation focus evolves as the product's life cycle evolves.

A variety of portfolio tools are available to assist in the reviewing of projects. Examples include financial return metrics (NPV, ROI, IRR), prioritized scored lists of projects, Allied Signal–Honeywell's strategic buckets, Procter & Gamble's three-dimensional risk–reward bubbles, and portfolio mapping (for more details, see Cooper, Edgett, and Kleinschmidt, 2001). The fundamental concept behind portfolio management is akin to portfolio diversification in investment portfolios. The proper mix of products helps to mitigate the business risks of new-product development with a reported 35% to 90% of all projects initiated never making it to commercialization. Portfolio management is also important because otherwise

- strategic criteria are missing in project selection;
- less risky low-value projects—such as product extensions, modifications, enhancements, and short-term projects—are undertaken;
- no new-product strategic focus evolves; and
- wrong projects are selected based on politics, opinions, or emotions.

Portfolio management processes can be successfully developed to help executives in their attempts to obtain better results from scarce R&D dollars, achieve the balance needed between short-term pressures and the future, longer term needs of the organization and to ensure that R&D efforts are being directed towards helping the organization achieve its strategic objectives.

References

Christensen, C. M. (1997). *The Innovator's Dilemma: When New Technologies Cause Great Firms to Fail.* Boston, MA: Harvard Business School Press.

Cooper, R. G (1993). *Winning at New Products* (Third Edition). Cambridge, MA: Perseus Publishing.

Cooper, R., Edgett, S., and Kleinschmidt, E. (2001). "Portfolio management for new product development: Results of an industry practices atudy." *R&D Management*, **31**(4), 361–380.

Damanpour, F. (1987). "The adoption of technological, administrative, and ancillary innovations: Impact of organizational factors." *Journal of Management*, **13**(4), 675–688.

Drucker, P. F. (2006). *Innovation and Entrepreneurship*. Harper Collins.

Garcia, R., and Calantone, R. (2002). "A critical look at technological innovation typology and innovativeness typology: A literature review." *Journal of Product Innovation Management*, **19**, 110–132.

Henderson, R. M., and Clark, K. B. (1990). "Architectural innovation: The reconfiguration of existing product technologies and the failure of established firms." *Administrative Science Quarterly*, **35**, 9–30.

Kimberly, J. R., and Evanisko, M. J. (1981). "Organizational innovation: The influence of individual, organizational, and contextual factors on hospital adoption of technological and administrative innovation." *Academy of Management Journal*, **24**, 689–713.

Markides, C. (2006). "Disruptive innovation: In need of better theory." *Journal of Product Innovation Management*, **23**(1), 19–25.

Utterback, J. M. (1996). *Mastering the Dynamics of Innovation.* Boston, MA: Harvard Business School Press.

Utterback, J. M., and Abernathy, W. J. (1975). "A dynamic model of process and product innovation." *Omega*, **33**, 639–656.

14

Sources of Innovation

Jenny Darroch and Morgan P. Miles*[†]

*Claremont Graduate University and [†]Georgia Southern University

The need to reconfigure and innovate in the face of change is one of the dominant issues that underlies business strategy making today (Covin and Slevin, 2002). Firms are constantly attempting to leverage innovation in order to gain a competitive advantage or simply to survive. Hamel (2000, p. 11) suggests that:

> "Somewhere out there is a bullet with your company's name on it. Somewhere out there is a competitor, unborn and unknown, that will render your strategy obsolete. You can't dodge the bullet—you're going to have to shoot first. You're going to have to out-innovate the innovators. Those who live by the sword will be shot by those who don't."

The tremendous changes in technology, strategy, culture, and business models have greatly increased competitive pressures on firms. Accordingly, resources, routines, behaviors, and practices are frequently examined as firms strive to become more innovative. The first step in the innovation process is to determine where to begin; that is, to identify a source of innovation (see von Hippel's 1988 comprehensive work in this area). However, therein lies the problem as "spotting" or recognizing attractive economic opportunities is seldom an easy task. Hamel (1998a, p. 12) offers the following story to illustrate how pig on the hoof became roast pork:

> "One day, a wild pig wandered into a hut; lightning struck the hut; the hut burned down; a human poked through the charred remains, touched the pig, sucked on a finger, and voila! Yummy . . ."

Hamel (1998b) notes that the real question for all firms attempting to be innovative is: "is there a pig spotting guide that can help firms find innovations?" This question lies at the core of work on innovation—where do ideas for innovations come from? We focus on this question by proposing a dynamic innovation framework that integrates Drucker's (1985) seminal work on the seven sources of innovation. This chapter is divided into three sections: first, we outline the demand-side and supply-side sources of innovation. Here, we argue that innovations can originate from the supply side, where the focus is on the products or services firms are willing to offer (Sarasvathy and Dew, 2004), or from the demand side, where the focus is on revealed preferences for a certain combination of attributes (Lancaster, 1971) that appear either as existing or "yet-to-be-invented" products. Second, we introduce Drucker's (1985) seven sources of innovation and identify them as either demand-side or supply-side approaches to innovation. Finally, we provide a dynamic innovation framework that brings together demand-side and supply-side sources into an integrated system.

Demand-side sources of innovation

For demand-side sources of innovation, managers identify emerging tastes and preferences that typically arise due to social, technological, or regulatory environmental changes. These tastes and preferences manifest themselves as unmet needs and wants for which managers develop new products (see, e.g., Allen and Marquis' 1964 work on government requests for proposals). Consumers can state tastes and preferences but may not be able to articulate their needs and wants. In fact, it is quite likely that consumers harbor latent, but detectable, unmet needs and so can describe a problem they have with an existing product without offering a solution (e.g., the digital picture frame as a solution to displaying digital photographs). Importantly though, a sufficient number of consumers with a homogeneous set of tastes and preferences is assumed to exist, and this provides the incentive for managers to develop new products to satisfy the needs of this new market. Thus, one task of marketing management is to detect consumer preferences in order to identify unmet latent needs (Kotler, 1973). Kotler (1973, p. 44) notes that:

> "*latent demand* exists when a substantial number of people share a strong need for something which does not exist in the form of an actual product. The *latent demand* represents an opportunity for the marketing innovator to develop the product that people have been wanting."

The more latent the need, the more sophisticated managers' market-sensing and opportunity recognition capabilities must be, and the more entrepreneurial the manager must behave, in order to make the linkages between unmet market needs and innovative solutions.

Marketing, as a discipline, has provided managers with a plethora of tools and techniques aimed at keeping current customers at the center of the business, involving customers as part of the new-product development process, managing customer relationships and accessing customers in order to measure attitudes and opinions. Marketing managers generally have superior market-sensing capabilities (Day, 1994) and increasingly look to alternate, non survey–based methodologies such as demographic trend analysis (Drucker, 1985) or anthropological studies (see, e.g., Arnould and Wallendorf, 1994) in order to uncover latent needs. This demand-side approach to sources of innovation is embedded in the value creation approach to marketing, described in Kotler and Keller's *Marketing Management* textbook and offered as the most effective approach to marketing (Kotler and Keller, 2006, p. 36). By following the value creation approach, managers first do their homework by conducting marketing research in order to identify market segments that exist, thereby identifying consumers with homogeneous tastes and preferences. Next, managers choose a segment or segments upon which to focus and then set about creating a value proposition for each segment. Managers then develop a product that will provide value to consumers. Once launched, marketers embark upon a campaign to communicate the product's value to consumers. We contend that this demand-side approach to innovation is reflected in the past American Marketing Association (2004) definition of marketing, which emphasizes value creation:

> "Marketing is an organizational function and a set of processes for creating, communicating and delivering value to customers and for managing customer relationships in ways that benefit the organization and its stakeholders."

One of the criticisms of the demand-side approach to innovation, however, is that managers may not be able to "read" the market in order to accurately identify gaps or opportunities. In addition, and as Kaldor (1971) noted, consumers often do not really know or acknowledge their needs. Similarly, Houston (1986, p. 86) suggested that marketers use their capabilities to create future markets:

> "(c)ustomers are not necessarily good sources of information about their needs a decade from now ... Anticipating future needs and wants are consistent with the marketing concept".

Once gaps are identified, however, it is assumed that managers can leverage or acquire those resources necessary to support the selected value-creating strategy (Barney, 1991; Penrose, 1959; Wernerfelt, 1984), question routines and long-held assumptions of the market in order to facilitate quick adaptation to changing conditions (Teece, Pisano, and Shuen, 1997), and leverage tacit knowledge and adjust routines in order to accommodate those changes (Nelson and Winter, 1982).

Supply-side sources of innovation

Alternately, an innovation can be created by first developing a new product and then leading consumers to that product; we call this a supply-side approach to innovation. Here the focus is on leveraging innovations around existing products, processes, strategies, domains, or business opportunities (see, e.g., Morris, Kuratko and Covin, 2008). For many managers, this internally driven option is often more certain, manageable, and economically attractive (see, e.g., Burgelman and Doz, 2001; Campbell and Park, 2004) because these entrepreneurial initiatives are operationally or strategically derived from core business capabilities. In some situations, however, new-product development may not be coupled with strong market-sensing capabilities and so the risk is that consumers may not adopt the product because managers have misread the market.

If the firm does succeed in creating primary demand for the new product, a market segment of consumers with homogeneous needs and wants emerges and a new market is created. An example of a supply-side approach to innovation is provided by Akio Morita, the founder of Sony, who pursued his idea for a portable cassette player (the Sony Walkman) on the basis that "Sony does not serve markets, it creates them" (Kotler and Keller, 2006, p. 353). With the Walkman, Sony gained an early position of market leadership in a newly created market and, in so doing, influenced emerging industry standards and enjoyed strategically significant cost advantages.

A slightly different take on supply-side sources of innovation comes from the theory of effectuation

(Sarasvathy, 2007). With effectuation, the starting point for an entrepreneurial new business is based on the answers to three questions: (1) who I am—my traits, tastes, and abilities; (2) what I know—my education, training, expertise, and knowledge; and (3) who I know—my work and social networks. An entrepreneur will start small, working with that which is close at hand, before growing the business through effective execution. Planning is not a precursor to execution; rather the entrepreneur simply makes a start and awaits the outcome of his or her actions.

Just as marketers use tools and techniques to identify sources of innovation from consumers, those working in technology-related roles have provided managers with a range of techniques to facilitate supply-side sources of innovation. For example, Ladewig (2007, p. 3) describes the TRIZ process, a Russian acronym for The Theory of Inventive Problem Solving, a technique developed to support engineering creativity. The TRIZ framework provides a stepwise process to stimulate innovative technology-based solutions by specifying: (1) the product's core function, (2) the constraints on product performance, (3) the "ideal product", and (4) potential solutions to overcome the constraints on the product and bridge the gap between the current product and an ideal one.

One of the early criticisms of supply-side sources of innovation is that inventions were often developed without taking into account the context of end-users. Much progress has been made, however, and supply-side techniques have been altered to better integrate the "voice of the customer" (Griffin and Hauser, 1993; Katz, 2004). For example, quality function deployment (Katz, 2007) breaks the new-product development process down into a number of steps, with each step focusing on transforming user demands into design quality (Akao, 1994); therefore, each step in a demand-side process includes the voice of the customer. Similarly, Orban and Miller (2007) introduce the notion of a "prosumer," an individual who plays a dual role as both a new-product development specialist and consumer of the product. The prosumer must represent the needs of the customer throughout the NPD process.

Drucker's sources of innovation

In Drucker's (1985, p. 95) paper titled "The discipline of innovation", he argued that innovation "can and should be managed like any other corporate function." Drucker notes that while some innovations "spring from a flash of genius," most are simply the result of a "conscious, purposeful search" for exploitable opportunities. He offers seven sources of innovation: (1) unexpected occurrences, (2) incongruities, (3) process needs, (4) industry and market changes, (5) demographic changes, (6) changes in perception, and (7) discovery of new knowledge. The remainder of this section explains each of Drucker's sources of innovation.

Drucker's (1985) first source of innovation, that of "unexpected occurrences", mandates that firms seeking to innovate must reconsider social, political, cultural, economic, and macro-environmental problems as potentially attractive and exploitable entrepreneurial opportunities that might result in commercially successful innovations.

An example includes Pfizer's well-known drug Viagra, which was initially developed to provide relief from angina and found to be largely ineffective. However, as Langreth (1998, p. B1) reports:

> "The program was about to be shelved permanently in 1993, when Pfizer's researchers noticed something quite unexpected: Several men who had received higher than usual doses in a small study told doctors they had achieved improved and more frequent erections than before. At that time, it seemed like a side effect rather than a remedy. But the Pfizer scientists, trying to salvage a drug they had worked on for years, believed the erection effect might represent a significant advance . . . Some patients were so enamored of Viagra that they refused to give the pills back when their tests had ended."

This example illustrates how Viagra went from being an ineffective cardiovascular drug to a blockbuster solution for a condition that was emerging as a social problem: male erectile dysfunction. In this case a new and highly profitable market was created by exploiting the completely unintentional outcomes of a new product.

Incongruities exist where there are opportunities to better integrate the actions of producers and consumers, something Drucker refers to as "expectations and results" (Drucker, 1985, p. 97). For example, the creation of Apple's iTunes came about because the MP3 player created incongruities by allowing music that had been "pirated" off the Internet to be stored, classified, and physically moved from a PC to a very small, inexpensive, and portable device. Apple recognized that many consumers wanted to legally purchase specific tracks without purchasing an entire CD. Thus, the iTunes/iPod business model was created (Kessler, 2003).

Innovations can sometimes arise due to "process needs"; that is, the inability of current market offerings to meet the functional needs of the market. An example is the e-book reader, which allows a consumer to download PDF files to a portable device that has a relatively large screen and high storage capability. As more and more information is marketed in PDF format, the e-book reader will, for example, allow students to download a textbook or article anywhere and at anytime. Since Amazon sells the Kindle reader and 90,000 e-book titles, which can be downloaded directly to an e-reader (i.e., without a PC) at an average price of $9.99/e-book (Mossberg, 2007), the e-reader will also alleviate the problem of high book costs that consumers, especially students, face. Another example of a process need comes from the recent energy shortage, which is driving innovation in new forms of energy such as ethanol, bio-fuels, and hydrogen and more efficient technologies for human transport (see PBS, 2004; Miles, Darroch, and Munilla, forthcoming).

Changes in the regulatory, industry, technology, political, economic, cultural, and/or market environments often stimulate innovation as well. For example, the recent green revolution is forcing innovation along the food supply chain (Mangu-Ward, 2006). Many food producers are offering organic alternatives, which in many cases offers the opportunity for product and process innovations (see Miles, Darroch, and Munilla, forthcoming). Consumer interest in organics and other types of healthier functional foods has dramatically increased and created many opportunities for innovation across products, strategies, processes, and business models.

Drucker (1985) also identifies demographic changes as a source of innovation. Changes in the age distribution, average income levels, and racial composition of a region are all demographic factors that can create opportunities for innovation. A noteworthy example is the recent emergence of high-performance, three-wheel, human-powered tricycles that offers innovative answers to the problems faced by an aging, but fitness-seeking, older generation. Likewise, with regard to income distribution, Hart and Christensen (2002) suggest that the rapidly growing population of less developed nations is providing many opportunities for socially useful innovations that meet basic human needs. The creation of the $100 laptop to provide computer-assisted education in the less developed world by the NGO One Laptop Per Child is a result of innovation directed at meeting these very critical needs (Anon., 2008). Even the financial service industry has seen disruptive innovation with micro-lending being offered by institutions such as the Grameen Bank and the establishment of financial person-to-person small business loan organizations such as Kiva.org (Anon., 2008).

In addition, changes in perceptions and social cognitive structures can provide sources of innovation. By this, Drucker means changes in social attitudes that may impact the desirability of a product. For example, prior to the 1960s, a deep dark suntan was socially desirable. Then research suggested that there may be health concerns associated with excessive sun tanning and so the sun block industry was born. Likewise, an alliance between Ericsson, Skandia, and the Huddinge University Hospital in Sweden formed to develop an innovative health care product that leveraged telecom technology and remote sensing to create a medical monitoring system that allows older people to remain at home rather than move to hospital (Covin and Miles, 2007). However, for this innovation to work, caregivers, patients, and the medical service industry in Sweden had to change their perceptions and agree that home health care was in fact as good as, if not superior to, institutional health care.

The discovery of new knowledge can also provide a source of innovation. Examples include the creation of food products from genetically modified organisms (GMOs) and the cloning of livestock due to the dramatic advances in biotechnology. Monsanto has completely reconfigured its industry by the effective application of GMO technology to agricultural seeds and other production inputs (see Magretta, 1997). This newest green revolution has allowed more efficient and profitable farming practices to be developed enabling food production to increase while providing farmers with alternate ways to manage rising costs.

Five of Drucker's sources of innovations are largely driven by consumer demand and include exploiting (1) market incongruities, (2) process needs, (3) industry and market changes, (4) demographic changes, and (5) changes in perception. The remaining two sources of innovation, exploiting unexpected occurrences and discovering and leveraging new knowledge, are predominately driven by the producers' ability to supply the market with innovations. Table 14.1 classifies Drucker's sources of innovation into demand-side and supply-side approaches to innovation.

An integrated dynamic model of innovation

Since managers should not emphasize one source of innovation over another, the many sources of innovation become an integrated whole. Drucker's framework suggests that innovation is more of a business process and less of a chance occurrence. In the spirit of Drucker's systems approach to innovation, we contend that managers need to develop a comprehensive process that allows the firm to cycle between the demand and supply side so that once an opportunity is discovered or created, supporting processes assess the innovation for strategic and operational fit and enable the firm to assess and then exploit the entrepreneurial opportunity (see Burgelman, 1984; Shane and Venkataraman, 2000).

In this section, we bring the demand-side and supply side perspectives together into one framework and allow for dynamic interactions between the two (Robertson and Yu, 2001). Our approach is outlined in Figure 14.1.

Identifies and exploits opportunities

Initially, the market is in disequilibrium, and demand exceeds supply because consumers have needs that are

Table 14.1. The interrelationship between Drucker's sources of innovation and supply-side and demand-side sources innovation.

Drucker's (1985) sources of innovation	*Selected example*	*Demand-side and supply-side sources of innovation*
Unexpected occurrences	Viagra	Supply side
Incongruities	i-Tunes	Demand side
Process needs	e-book readers	Demand side
Industry and market changes	Organic foods	Demand side
Demographic changes	$100 laptops and Kiva.org	Demand side
Changes in perceptions	Tele-home-health-care	Demand side
Creation/Discovery of new knowledge	Cloned livestock	Supply side

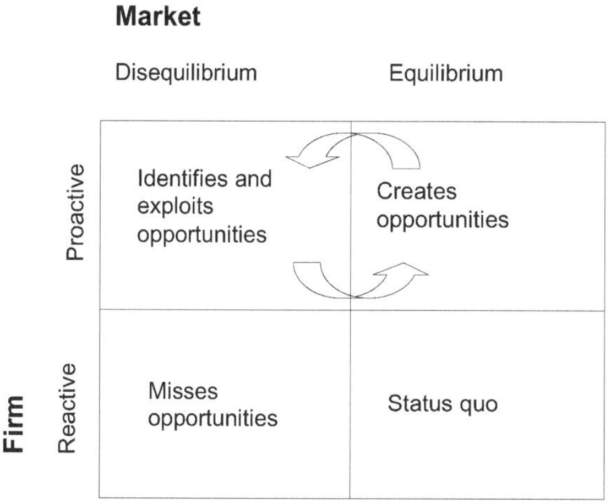

Figure 14.1. A dynamic model of market creation.

not currently being met by existing products (Kotler, 1973, p. 44). Consumers might be able to articulate their unmet needs or they might be able to articulate problems they have with current product offerings. However, because existing products do not satisfy those unmet needs, consumers are unlikely to offer a solution. Therefore, managers must possess superior market-sensing and opportunity recognition capabilities (Day, 1994; Hayek, 1948; Kirzner, 1997) because traditional marketing research methods might not successfully uncover latent demand. However, those within the firm are likely to be immersed in the market and be very connected to the players in it. Thus, the task of marketing management is to actively identify and then exploit new opportunities by developing new products. Once a product is developed, firms need to adopt innovative marketing practices to persuade and educate consumers about new-product/market opportunities and develop additional demand for that product (Houston, 1986). Thus, the firm generates supply to satisfy latent demand, builds more demand, and in so doing moves the market back to equilibrium.

Creates opportunities

Here, the market starts out in equilibrium and managers actively seek to create new opportunities through innovation (Schumpeter, 1934), the result of which is that firms generate supply that pushes the market out of equilibrium. In order to develop an innovation, managers make use of existing resources by, for example, leveraging R&D to produce technology-driven innovations. Kotler's (1973) demand framework suggests that there are eight states of demand possible for any product including (1) negative, (2) nonexistent, (3) latent, (4) declining, (5) irregular, (6) full, (7) overfull, and (8) unwholesome demand. In this situation we suggest there is nonexistent demand for the innovation: consumers are either disinterested or indifferent to the innovation. Once the product is developed, the task of marketing management is to create or stimulate demand by making consumers aware of the innovation and demonstrating the value the innovation has over current offerings. When full demand has been created the market moves back to equilibrium.

Missing opportunities

Under this scenario, the market is in disequilibrium because, as before, there are unmet consumer needs (i.e., demand exceeds supply). However, those within the firm sometimes fail to identify market opportunities. Firms operating in this quadrant are in danger of losing ground to competitors because they are neither satisfying existing consumers (who have unmet needs) nor creating new consumer groups. In addition, the firm does not possess the market-sensing capabilities that will allow it to discover, assess, and exploit these opportunities. Such a firm is very tied to the security offered by maintaining the *status quo*. The firm does not want to cause conflict in the market by confusing or alienating its consumers and so those within the firm tend to listen to only a small group of customers. For example, firms in the computer industry, such as the Digital Equipment Company (DEC), were developing mini-computers with the attributes most desired by their best customers. However, DEC and many other firms in the mini-computer industry simply did not understand the competitive threats that emerged out of the development of the micro personal computer such as the Apple II or Commodore (Christensen, 2000).

Serves customers

The market is in equilibrium and supply equals demand. Because the firm is not proactive, however, it will not endeavor to alter the supply curve. The firm will stick to its knitting, preferring to serve current customers well. The task of marketing management, in this context, is to maintain full demand or revitalize declining demand (Kotler, 1973). Such a stance might result in making incremental adjustments to existing products or revitalizing current offerings in response to feedback from customers to avoid a situation of faltering demand (Kotler, 1973). This strategy is highly effective in a market for which preferences are relatively stable but have been disrupted by an environmental shock. For example, the beer-brewing industry was negatively impacted by the success of the Atkin's diet and its low-carbohydrate mantra. However, brewers quickly responded by either reformulating beers as low-carbohydrate (such as the development of Michelob Ultra) or better communicating that some existing beers, such as Miller Light, were already low in carbohydrates. The danger, however, is that a competitor might engage in Schumpeterian-type innovation (Schumpeter, 1934) and upset the *status quo* by altering consumer preferences and creating demand; once again the firm runs the risk of losing ground and becoming uncompetitive.

Conclusions

The central thesis of this chapter is that innovation is a systematic process requiring managers to cycle between demand-side and supply-side sources of innovation—sometimes beginning with the product/technology and sometimes beginning with the consumer. Ultimately, however, and in order for the innovation to succeed, both the demand and supply side need to be brought back into equilibrium. We argue that adopting a blended approach to innovation, and moving comfortably between demand-side and supply-side sources, is important to the eventual success of innovation in the market place.

Implicit in the integrated framework shown in Figure 14.1 is a recommendation that successful new-product development requires cross-functional integration as the firm cycles between the talents of those in marketing (demand-side innovation), who are likely to excel in sourcing ideas from the market, and the talents of those in technology, R&D, and operations, who are skilled at leveraging internal core competencies (supply-side innovations).

References

Akao, Y. (1994). "Development history of quality function deployment." In: *The Customer Driven Approach to Quality Planning and Deployment*. Minato-ku, Tokyo: Asian Productivity Organization.

Allen, T. J., and Marquis, D. G. (1964). "Positive and negative biasing sets: The effects of prior experience on research performance." *IEEE Transaction on Engineering Management*, **11**(4), 158–161.

AMA (2007). *http://www.marketingpower.com/Community/ARC/Pages/Additional/Definition/default.aspx#*

Anon. (2008). "Online banking report publishes 'Person-to-Person' lending 2.0: Disruptive service or market niche?" *Business Wire*, New York, January 21.

Arnould, E. J., and Wallendorf, M. (1994). "Market-oriented ethnography: Interpretation building and marketing strategy formulation." *Journal of Marketing Research*, **31**(4), 484–504.

Barney, J. (1991). "Firm resources and sustained competitive advantage." *Journal of Management*, **17**, 139–157.

Burgelman, R. A. (1984). "Designs for corporate entrepreneurship in established firms." *California Management Review*, **26**(3), 154–167.

Burgelman, R. A., and Doz Y. L. (2001). "The power of strategic integration." *Sloan Management Review*, **42**(3), 28–38.

Campbell, A., and Park, R. (2004). "Stop kissing frogs." *Harvard Business Review*, **82**(7/8), 27–28.

Christensen, C. (2000). *Strategies for the Competitive Edge* (Stanford Executive Briefing). Mill Valley, CA: Kantola Productions.

Covin, J. G., and Miles, M. P. (2007). "The strategic use of corporate venturing." *Entrepreneurship Theory and Practice*, **31**(2), 183–207.

Covin, J. G., and Slevin, D. P. (2002). "The entrepreneurial imperatives of strategic leadership." In: M. A. Hitt, R. D. Ireland, S. M. Camp, and D. L. Sexton (Eds.), *Strategic Entrepreneurship: Creating a New Mindset* (pp. 309–327). Oxford: Blackwell Publishing.

Day, G. (1994). "The capabilities of market-driven organizations." *Journal of Marketing*, **58**, October, 37–52.

Drucker, P. F. (1985). "The discipline of innovation." *Harvard Business Review*, **63**(3), 67–73.

Griffin, A., and Hauser, J. (1993). "The voice of the customer." *Marketing Science*, **12**(1), 1–27.

Hamel, G. (1998a). "Strategy innovation and the quest for value." *Sloan Management Review*, **39**(2), 7–14.

Hamel, G. (1998b). *Creating the Future* (Stanford Executive Briefing). Mill Valley, CA: Kantola Productions.

Hamel, G. (2000). *Leading the Revolution*. Boston, MA: Harvard Business School Press.

Hart, S. L., and Christensen, C. M. (2002). "The great leap: Driving innovation from the base of the pyramid." *MIT Sloan Management Review*, **44**(1), 51–56.

Hayek, F. A. (1948). *Individualism and the Economic Order*. Chicago, IL: University of Chicago Press.

Houston, F. S. (1986). "The marketing concept: What it is and what it is not." *Journal of Marketing*, **50**(2), 81–87.

Kaldor, A. G. (1971). "Imbricative marketing." *Journal of Marketing*, **35**, April, 19–25.

Katz, G. M. (2004). "The voice of the customer." In: *The PDMA Toolbook 2 for New Product Development*. Hoboken, NJ: John Wiley & Sons.

Katz, G. M. (2007). "Quality Functional Deployment and the house of quality." In: A. Griffin and S. M. Somermeyer (Eds.), *The PDMA Handbook for New Product Development* (pp. 41–69). Hoboken, NJ: John Wiley & Sons.

Kessler, A. (2003). "In the fray: The music industry needs hackers, not lawyers." *The Wall Street Journal*, September 9, p. D6.

Kirzner, I. M. (1997). "Entrepreneurial discovery and the competitive market process: An Austrian approach." *Journal of Economic Literature*, **35**(1), 60–85.

Kotler, P. (1973). "The major tasks of marketing management." *Journal of Marketing*, **37**, October, 42–49.

Kotler, P., and Keller, K. L. (2006). *Marketing Management* (12th Edition). Englewood Cliffs, NJ: Prentice-Hall.

Ladewig, G. R. (2007). "TRIZ: The theory of inventive problem solving." In: A. Griffin and S. M. Somermeyer (Eds.), *The PDMA Handbook for New Product Development* (pp. 3–40). Hoboken, NJ: John Wiley & Sons.

Lancaster, K. (1971). *Consumer Demand: A New Approach*. New York: Columbia University Press.

Langreth, R. (1998). "Pfizer pins multibillion-dollar hopes on impotence pill: Aimed at heart, research leads to a surprise." *The Wall Street Journal*, March 19, p. B1.

Magretta, J. (1997). "Growth through global sustainability: An interview with Monsanto's CEO, Robert B. Shapiro." *Harvard Business Review*, January/February, 78–90.

Mangu-Ward, K. (2006). "Taste: Food fight." *The Wall Street Journal*, June 9, p. W13.

Miles, M. P., Darroch, J., and Munilla, L. S. (forthcoming). "Sustainable corporate entrepreneurship." *International Entrepreneurship and Management Journal*.

Morris, M. H., Kuratko, D. K., and Covin, J. G. (2008). *Corporate Entrepreneurship and Innovation*. Mason, OH: Thomson South-Western.

Mossberg, W. S. (2007). "Amazon's Kindle makes buying e-books easy, reading them hard." *The Wall Street Journal*, November 29, p. B1.

Nelson, R. R., and Winter, S. G. (1982). *An Evolutionary Theory of Economic Change*. Cambridge, MA: Belknap.

Orban, A., and Miller, C. W. (2007). "The slingshot: A group process for generating breakthrough ideas." In: A. Griffin and S. M. Somermeyer (Eds.), *The PDMA Handbook for New Product Development* (pp. 107–140). Hoboken, NJ: John Wiley & Sons.

PBS. (2004). "Alan Alda in future car." *Scientific American Frontiers* (Public Broadcasting Service).

Penrose, E. T. (1959). *The Theory of the Growth of the Firm*. New York: John Wiley & Sons.

Robertson, P. L., and Yu, T. F. (2001). "Firm strategy, innovation, and consumer demand: A market process approach." *Managerial and Decision Economics*, **22**(4/5), 183–199.

Sarasvathy, S. D. (2007). "What makes entrepreneurs entrepreneurial?"; *http://www.effectuation.org/ftp/What%20makes%20entrs%20entl%20note.pdf*

Sarasvathy, S. D., and Dew, N. (2004). *When Markets Are Grue* (Darden Business School Working Paper No. 04-06). Charlottesville, VA: University of Virginia.

Schumpeter, J. A. (1934). *The Theory of Economic Development*. Cambridge, MA: Harvard University Press.

Shane, S., and Venkataraman, S. (2000). "The promise of entrepreneurship as a field of research." *Academy of Management Review*, **25**(1), 217–226.

Teece, D., Pisano, G., and Shuen, A. (1997). "Dynamic capabilities and strategic management." *Strategic Management Journal*, **18**(7), 509–533.

von Hippel, E. (1988). *The Sources of Innovation*. New York: Oxford University Press.

Wernerfelt, B. (1984). "A resource based view of the firm." *Strategic Management Journal*, **5**(2), 171–180.

15

Innovation Models

Allan N. Afuah and Kwaku O. Prakah Asante*[†]

*University of Michigan and [†]Ross School of Business

The rapid pace of technological innovation, business model innovation, and globalization that firms face raises some interesting questions. Why do some firms innovate more than others? Why are some firms more likely to profit from innovation than others? What is innovation and how does a firm manage the creative process for profitability? Understanding the critical success factors and inhibitors that underpin successful innovation—be it technological or business model—is critical to gaining and sustaining a competitive advantage. This chapter presents a summary of the models and concepts which have been developed to understand innovation phenomena and contribute towards answering these questions.

What are innovation models?

Innovation models evaluated in this entry contribute to the understanding of firms likely to introduce, exploit, and sustain profits from innovation. Static and dynamic models are presented that capture several internal and external factors of the firm-impacting innovation. These factors include the size of the firm, technological and market capabilities, technology imitation, complementary assets, and technology evolution. Static models capture the cross-sectional perspective of a firm's capabilities and knowledge, in addition to the firm's incentive to invest at specific instances in time. The static models do not characterize how innovations evolve over time. Conversely, dynamic models incorporate the effect of time in the innovation process. Dynamic models take a longitudinal view of innovation and explore its evolution following introduction. Technological evolution is characterized as involving different phases. Each phase of evolution presents different challenges and may take a different type of firm to succeed. The static and dynamic innovation models are described in the following sections.

Static and dynamic innovation models

This section presents several static and dynamic innovation models for understanding the success factors and inhibitors that underpin successful innovation. A summary of the background of the models, main contributions, and why they are important are presented (Afuah, 2003). In addition, the key features of the models, how they differ from each other, and value-added contributions are presented in Tables 15.1 and 15.2.

Static models

Schumpeter

Schumpeterian models aim to discern the impact of the size and age of firms on their likelihood of innovating (Schumpeter, 1934, 1950). Schumpeter's first assertion was that small, entrepreneurial firms are better positioned to innovate due to their often nimble organization. He later postulated that large firms with various degrees of monopoly power facilitate innovation. Schumpeter later supported the notion that large firms are more likely to innovate because they tend to have the complementary assets to readily take an idea generated to concept realization and to production. Furthermore, larger firms have more opportunities for raising capital, exploiting scale economies along the value chain, and better positioned to protect their innovations from competitors. Consequently, larger firms are more likely to invest in innovation as they can manage more of the risk involved. It is important to note that empirical studies of innovative activity (Cohen, 1995; Kamien and Schwartz, 1975) have not confirmed a clear correlation between firm size, market power, and innovation.

Other factors which are important in determining whether established large firms or new entrants are able to introduce an innovation include how new the idea and the product are, and whether an innovation is incremental or radical. The discussion of the Abernathy-Clark and the

Table 15.1. Relationships between static models.

Model	*Key features*	*Value added*
Schumpeter I Schumpeter II	Entrepreneurs are the most likely to innovate. Large firms with some degree of monopoly power are the most likely to innovate.	Attempt to answer the question: Who is the most likely to innovate? The type of firm is what matters.
Abernathy–Clark	Unbundles technological and market knowledge. Highlights the importance of market capabilities.	Explains why incumbents may do well at radical technological innovations.
Henderson–Clark	Unbundles technological knowledge into component and architectural. Defines innovation as: incremental if both architectural and component knowledge are enhanced; architectural if component knowledge is enhanced but architectural knowledge is destroyed.	Explains why incumbents fail at what appears to be incremental innovations. These are actually architectural innovatioons.

From Afuah (2003).

Table 15.2. Relationships between dynamic models, as well as value added.

Model	*Key features*	*Value added*
Utterback–Abernathy	Three phases in an innovation's life cycle—fluid, transitional, and specific.	Introduces dynamism.
	Dominant design defines a critical point in the life of an innovation.	Concept of dominant design.
	From radical product innovation to dominant design to incremental innovation.	Industries evolve relatively predictably from one phase to the other.
	From major product innovation to major process innovation.	
	From many small firms offering unique products to few firms offering similar products. From profitable firms to less profitable ones.	
Tushman–Rosenkopf	Similar in features to the Utterback–Abernathy model: technological discontinuity, era of ferment, emergence of a dominant design, and era of incremental change.	Technological progress depends on factors other than those internal to the technology.
	The more complex an innovation the more intrusion from sociopolitical factors during evolution of the technology.	The more complex the technology, the more it is underdetermined by factors internal to it.
Foster's S-curve	The returns on the effort put into a technology fall off as the limits to the technology are approached.	How to predict the end of an existing technology and the arrival of a technological discontinuity.
	The limits of a technology can be predicted from knowledge of its physical limits.	

From Afuah (2003).

Henderson–Clark models which follow present additional factors to consider in determining who is likely to introduce and exploit an innovation.

Abernathy–Clark

The Abernathy and Clark model supports the notion that incumbents may do well at radical technological innovations (Abernathy and Clark, 1985) compared with new entrants. The model dis-integrates technological and market knowledge and stresses the importance of market capabilities. A firm may not have the resources or capabilities for technological innovation; however, it may have established market capabilities. If the market capabilities are critical to the appropriation of profits, and not easily obtained, an incumbent can capitalize on its market capabilities at the expense of a new entrant.

The Abernathy and Clark model classifies innovations according to their impact on the existing technological and market knowledge of the manufacturer. Innovation phenomena are characterized as regular, niche, revolutionary, or architectural. An innovation is considered "regular" if the manufacturer's existing technological and market capabilities are preserved. A "niche innovation" is described as one that preserves technological capabilities but renders market capabilities obsolete. "Revolutionary innovation" renders technological capabilities obsolete but enhances market capabilities. If both technological and market capabilities become obsolete Abernathy and Clark classify the innovation as "architectural". A key contribution of this model is that market knowledge can be just as important as technological knowledge for successful innovation. Incumbent firms tend to possess marketing capabilities as key complementary assets, and where such capabilities are critical to profitability so incumbents are more likely to surpass new entrants.

Henderson–Clark

Henderson and Clark's model postulates why incumbents fail at what might seem to be incremental innovations but are actually architectural innovations (Henderson and Clark, 1990). They suggest that products are made of components connected together, and building them requires technological knowledge of not only the components, but the linkages between the components. Architectural knowledge is characterized as knowledge of the linkages between the components. Henderson and Clark contend that innovation can impact component knowledge and/or architectural knowledge with significant consequences for the firm.

The Henderson and Clark model classifies innovation phenomena into four categories. If innovation enhances both component and architectural knowledge it is incremental. Radical innovation destroys both component and architectural knowledge. Innovation is considered architectural if component knowledge is enhanced but architectural knowledge is destroyed. It is called modular innovation if component knowledge is destroyed but architectural knowledge is enhanced. These definitions assist in explaining why firms had issues with what appeared to be incremental innovation. Architectural innovation may have been confused with incremental innovation. While the component knowledge required to exploit the innovation had not undergone modification the architectural knowledge had changed. Architectural knowledge tends to be tacit and embedded in the routines and procedures of an organization. Consequently, recognizing architectural knowledge and effectively responding are often extremely challenging.

Teece model

The Teece model characterizes how value is appropriated from the imitability of technology and complementary assets (Teece, 1986). Imitability refers to the extent to which technology can be imitated. Protection of technology from imitators may come from intellectual property protection. Another barrier to imitation could be that potential imitators might not have the competences to imitate the given technology. Complementary assets are all the other capabilities apart from those that constitute the technology which the firm requires to exploit the technology. Manufacturing, distribution channels, service, reputation, brand name, and complementary technologies constitute complementary assets.

The Teece model helps to explain whether an innovator is likely to profit from an innovation. If imitability is high and the technology can be easily imitated, it is difficult for the innovator to make money if complementary assets are easily available or unimportant. If the complementary assets are tightly held and important the owner of such assets makes money. If imitability is low and is difficult to imitate the technology the innovator stands to profit if complementary assets are freely available or unimportant. If imitability is low and complementary assets are important and difficult to acquire, whoever has both or the more important of the imitability or complementary regimes wins. The firm needs to appropriately position itself to obtain or negotiate additional assets required to introduce technology to the market place.

Dynamic models

Utterback–Abernathy

The Utterback–Abernathy model (1978) captures the dynamic processes within an industry and its firms during the evolution of a technology. The model suggests that technologies evolve from one well-defined phase to another. Phases in an innovation's life cycle include fluid, transitional, and specific. In the fluid phase, there are considerable technological and market uncertainties. Firms have no clear idea whether, when, or where to invest in research and development. Custom designs are common with the new-product technology and are often expensive and unreliable. However, custom designs are able to meet the requirements of market niches. There is minimal process innovation in the fluid phase.

Evolution enters the transitional phase when producers learn more about how to meet customer needs through producer–customer interaction, and standardization leads to a dominant design. The dominant design signals a substantial reduction in uncertainty, experimentation, and major design changes. A dominant design describes a design with the constituent components and underlying core concepts which do not vary substantially from product model introductions. In addition, the design commands a high percentage of market share. Competitive emphasis shifts to meeting the needs of specific customers with increased understanding of system requirements. The rate of product innovation in the transitional phase decreases and emphasis shifts to process innovation. During the specific phase, products built around the dominant design proliferate and there is more of an emphasis on process innovation. Product innovation is primarily incremental during the specific phase. Materials and equipment are very specialized during this dynamic phase, products are highly defined, and the basis for competition is low cost. The pattern described repeats itself when a new technology with the potential to render the old one obsolete is introduced. This results in a discontinuity, and the innovation cycle starts again with the fluid phase with another wave of entering firms.

It is important to note that in the fluid phase firms with product innovation competences that allow them to differentiate their products are more likely to perform better than those that do not. In the specific case, low-cost competences are particularly important. Since control of a standard can be an asset, measures to win such a standard can also be instrumental in determining who succeeds in exploiting an innovation. The significance of this model is that as technology evolves through different phases, firms need different kinds of capabilities in order to profit from the technology.

Tushman–Rosenkopf

To what extent can a firm influence the evolution of an innovation? Moreover, to what extent can a firm design to an industry standard? The Tushman–Rosenkopf (1992) model argues that this depends on the amount of technological uncertainty. Technological uncertainty depends on the complexity of the technology and the state of evolution. Tushman and Rosenkopf describe complexity as a function of the innovation's dimensions of merit, which corresponds to its attributes perceived by its local environment, and the number of interfaces between the innovation and complementary innovations. In addition, the number of components and linkages that constitute the innovation and the number of organizations in the innovation's local environment add to complexity.

Tushman and Rosenkopf define technological evolution phases to consist of a technological discontinuity, an era of ferment, emergence of a dominant design, and an era of incremental change. Technological discontinuities are unpredictable innovations which advance a relevant technological frontier substantially. Product or process design emerges with a significant cost, performance, or quality advantage over prior product forms before the discontinuity. The discontinuity can lead to the enhancement of a firm's competence or rather be competence destroying. In the era of ferment, there is a significant amount of technological and market uncertainty. Competition is prevalent for acceptance between designs using the new technology—each of which may have a different technical approach—as well as between the old and new technologies. A dominant design, a design that establishes dominance in a product class, emerges from the competing technologies. The emergence of a dominant design substantially reduces technological uncertainty and ushers in the beginning of the era of incremental change. Product features are established and attention is turned to incremental innovation until replaced by a technological discontinuity.

Similar to the Utterback–Abernathy model, the Tushman–Rosenkopf model indicates that a firm may need different capabilities for success at each phase of the life cycle. How effective these competences are in influencing the evolution of the technology is a function of the complexity of the product. A firm's innovation strategy and technological resource allocations should address the stages of its technology product life cycle. The more complex an innovation the more intrusion can be expected from sociopolitical factors during evolution of the technology. Both the Utterback–Abernathy and the Tushman–Rosenkopf models have shortcomings. Predicting the emergence of a dominant design is challenging and not all products have dominant designs. In addition, it is difficult to infer when each technology evolutionary phase begins and ends.

Foster's S-curve

The Utterback–Abernathy and Tushman–Rosenkopf models indicate that an era of incremental change ends with the arrival of a technological discontinuity. However, as discussed above, predicting when this discontinuity occurs is challenging. The S-curve provides a model which facilitates how to predict the end of an existing technology, and the arrival of a technological discontinuity, based on knowledge of the technology's physical limits (Foster, 1986). Foster postulates that the returns on the efforts put into a technology diminish as the limits to the physics of the technology are approached. Consequently the limits of a technology's commercial life can be predicted from knowledge of its physical limits. Foster contends that the rate of technological advancement is a function of the amount of effort put into the technology and follows the S-curve characteristic. Technological progress starts slowly, increases rapidly, and finally diminishes asymptotically as the physical limits of the technology are reached. Correspondingly, the returns on efforts diminish, promoting investigations into new technology with physical properties which overcome the physical limit of the old technology. The introduction of a new technology initiates a new S-curve.

It is important to note that technology S-curves can be computationally applied to the forecasting of technology trends. As data on the performance of a technology are available the rate of change can be predicted from (Ettlie, 2000):

$$Y = \frac{L}{1 + ae^{-bt}}$$

where Y is the rate of change in technological progress;
L is the value of the curve at the upper limit of the growth value;
t is time.

From the expected shape of the curve (S-curve), the coefficients a and b can be determined to fit the data. Scenario analysis of the shape and location of the curve can be performed to provide insight about how quickly a technology may emerge and plateau. It is also important to emphasize that such projections should be taken in the light that they are empirical and might not capture emerging competition in the market place. Contributions from other innovation models as discussed in this chapter should be explored in innovation strategy formulation.

Integrated model and framework

In the discussion of innovation models it is apparent that one model alone does not fully capture the complexities involved in innovation. Each of the models explored makes a contribution towards understanding who is likely to introduce or exploit an innovation. To profit from innovation, effective integration of aspects of these models in the development of an innovation strategy is constructive. The contributions made from the models described can be summarized as information that underlines the introduction and exploitation of an innovation—the *how*, *who*, *what*, and *when* questions of innovation (Afuah, 2003). Exploring these related questions about innovation leads to the integrated profit chain model and framework of Figure 15.1. The integrated framework incorporates the models discussed, and embodies the technology, capability, and business model imperatives required to profit from innovation.

A firm profits from an innovation by using new knowledge to offer new products and services at a lower cost than its competitors (Figure 15.1). How different is the new knowledge required to offer the new product from a firm's existing knowledge, and how different is the new product from existing products? In the presence of an innovation the difference between the new knowledge and the old is critical in determining how well the firm can recognize and exploit the innovation. Who does the innovation impact and to what extent? If, for example, an innovation is characterized from Henderson–Clark's model as being incremental, architectural, or radical, who along the firm's innovation value-added chain does the innovation impact? The firm's value-chain could consist of suppliers, customers, and complementary innovators, as well as the global context within which the firm is operating. Is the innovation radical to R&D, manufacturing, marketing, or another function?

In order to offer new products a firm must implement superior activities based on its capabilities relative to competitors. What is it about some firms that allows them to innovate better than their competitors? For example, does the firm size, unique skills, or complementary assets provide strategic advantages over imitators or new entrants (Teece, Schumpeter models)? Underlining the firm's capabilities are technological and market knowledge (Abernathy–Clark). In addition, a firm's capabilities are a function of its overall strategy, structure, systems, people that make up the firm, and its local environment. Competitors may introduce radical technological changes, causing a firm's existing capabilities and underlying knowledge to become obsolete. When in the life of the innovation (Utterback–Abernathy, Tushman–Rosenkopf, and Foster models) are the *how*, *who*, and *what* questions being answered? The extent to which an innovation is radical, incremental, or architectural and how much of the new knowledge the firm must deal with are functions of the innovation life cycle. Difference capabilities are required during different phases along the life cycle to profit from innovation.

Summary

Firms are driven to innovate to create value, appropriate profits, and enhance shareholder value. However, the innovation process from new-idea generation, to realization, and successful commercialization is challenging. A systematic, structured, and disciplined approach to the innovation process along the value chain of a company is imperative. The models and concepts presented in this chapter are valuable tools for understanding innovation success factors and inhibitors. The models make contributions towards who is likely to introduce and exploit

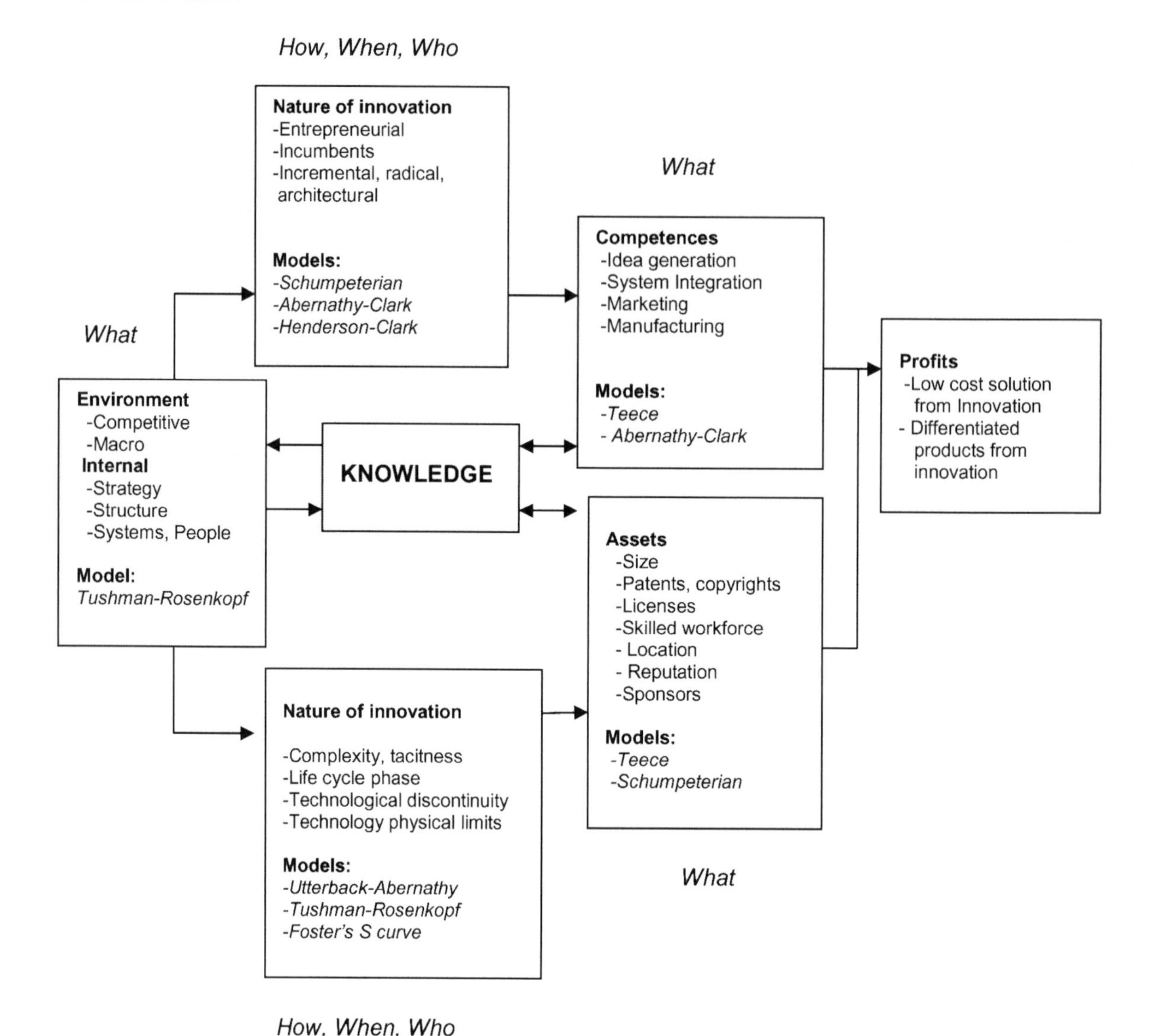

Figure 15.1. Integrative framework incorporating key innovation models for exploring how to profit from an innovation—the *how*, *who*, *what*, and *when* of innovation.

innovation and assist firms in innovation strategy formulation to create, capture value, and sustain profits.

References

Abernathy,W., and Clark, K. B. (1985). "Mapping the winds of creative destruction." *Research Policy*. **14**, 3–22.

Abernathy, W. J., and Utterback J. M. (1978). "Patterns of innovation in technology." *Technology Review*, **80**(7), 40–47.

Afuah, A. (2003). *Innovation Management: Strategies, Implementation, and Profits*. Oxford: Oxford University Press.

Cohen, W. (1995). "Empirical studies of innovative activity." In: P. Stoneman (Ed.), *Handbook of Economics of Innovation and Technological Change*. Oxford: Blackwell..

Ettlie, J. E. (2000). *Managing Technological Innovation* (p. 72). New York: John Wiley & Sons.

Foster, R. (1986). *Innovation: The Attacker's Advantage*. New York: Summit Books.

Henderson, R., and K. B. Clark. (1990). "Architectural innovation: The reconfiguration of existing product technologies and the failure of established firms." *Administrative Science Quarterly*, **35**, 9–30.

Kamien, M. I., and Schwartz, N. L. (1975). "Market structure and innovation: A survey." *Journal of Economic Literature*, **13**(1), 1–37.

Roberts, E. B., and C. A. Berry. (1985). "Entering new businesses: Selecting strategies for success." *Sloan Management Review*, **26**(3), 3–17.

Sauber, T., and Tschirky, H. (2006). *Structured Creativity: Formulating an Innovation Strategy*. New York: Palgrave Macmillan.

Schumpeter, J. A. (1934). *The Theory of Economic Development*. Boston, MA: Harvard University Press.

Schumpeter, J. A. (1950). *Capitalism, Socialism and Democracy* (Third Edition). New York: Harper.

Teece, D. J. (1986). "Profiting from technological innovation: Implications for integration, collaboration, licensing and public policy." *Research Policy*, **15**, 285–306.

Tushman, M. L., and Rosenkopf, L. (1992). "Organizational determinants of technological change: Towards a sociology of technological evolution." *Research in Organizational Behavior*, **14**, 311–347.

16

Diffusion of Innovation

C. Anthony Di Benedetto

Temple University

Diffusion of innovation has been defined as "the process by which an innovation is spread within a market, over time and over categories of adopters" (Crawford and Di Benedetto, 2008, p. 241). This definition indicates that diffusion refers to the spread of an innovative product category through a market or a society. (Note that the focus is on product categories, not competitive brands within a single product category.) This definition distinguishes diffusion from a related concept, adoption, which has more to do with processes at the individual level. One can think of diffusion, then, as an aggregate of the adoption process over the whole market.

The Rogers' diffusion model

The groundbreaking work in diffusion of innovations was done by Everett Rogers and first published in 1962. According to Rogers (1962), several product characteristics will influence its rate of diffusion (these are sometimes called the barriers to diffusion). The most important of these are:

1. Relative advantage of the new product, as compared with other products with which it is competing. For example, Google was immediately adopted in the Internet community once it was launched, as it was perceived to offer superior search capabilities relative to the available alternatives.
2. Compatibility of the new product with the end-user's experiences and values. The more innovative the product is, for example, the more incompatible with prior experiences and the more learning will be required. Microwave ovens didn't sell well at first as they prepared food in such a different way compared with traditional cooking.
3. Complexity, or how difficult the new product will be for the end-user to understand, or how confusing it will be to use. People found Apple's Newton message pad (and in particular its handwriting recognition function) hard to use, and it didn't sell; by contrast, Apple's iPod, with all functions controlled by a single button, could not be easier to use.
4. Divisibility, which here refers to how easily it is for the new product to be purchased and used in smaller quantities. A new food product is easily divisible, but a new car or a new satellite TV service would not be.
5. Observability of results, or the extent to which the benefits of using the new product are easily seen. For example, it would be hard to observe the benefits of using a new decay-preventing toothpaste without many years of constant use.

In addition, one must also consider the end-users that first adopt the new product, as their positive experiences and good word-of-mouth will influence later users also to adopt the product and it will diffuse successfully through the market. Rogers (1962) envisioned that the diffusion of a product into the market could be defined by an S-shaped curve, where initial adoption would be slow, then ramp up through a growth phase, and eventually slow down again once most of the market had adopted the product.

To understand the way the product diffuses through the market, Rogers envisioned a series of adopter categories, or market segments defined in terms of how quickly they purchase the new product. The earliest users (the first 5% or so of the market) are referred to as the innovators, and they are followed by the early-adopters who make up about the next 10% or 15% of the market. If successfully adopted by these early-users, the product diffuses to the early majority, the late majority, and finally to the last 15% to 20% known as the laggards. The adopter categories are important for marketers to understand, because of the influence of the innovators and early-adopters. Individuals in these categories are often viewed as opinion leaders, who can have a very strong influence on whether the rest of the market ever adopts the product. It is critical, then, to get the opinion leaders to adopt the product so that, through word-of-mouth communication, later segments will also follow their lead. The opinion leaders (in North America at least) never adopted Sony's Minidisc recordable CD player, and consequently this product never took off in the North American market (though it was quite popular for some time in Europe and Asia). Since, by definition, the innovators are a very small percentage of the market, marketers will often attempt to reach both innovators and early-adopters to stimulate more positive communication and to get more complete adoption by the majority in the market.

Since it is so important to get adoption among the innovators and early-adopters, the logical next issue would be to try to identify them so that the majority of marketing efforts can be focused on them in the early launch stage. While it is not always possible to target them precisely, they often share certain identifiable traits. In a study of business-to-business firms, innovators were characterized as having the following traits: venturesomeness (a desire to do something new and different), social integration (extensive contact with others, on the job or socially), cosmopolitanism (interest in world affairs and travel, and interests outside the immediate community), social mobility (being in a position or occupation where one travels in many social circles, such as successful executives and professionals), and priviligedness (being better off financially) (Gauvin and Sinha, 1993).

Rogers' model of diffusion of innovations is related conceptually to the product life cycle (PLC), perhaps one of the most enduring models used in marketing management today. Like the diffusion model, the PLC models the sales (and also the profits) of a product category or product form through time. It also assumes S-shaped sales growth through time: a slow introductory period, followed by a period of rapid growth, then a slowing down as the market matures. But whereas the diffusion model is more concerned with the adopter categories and the barriers to diffusion (relative advantage, compatibility, and so on), the PLC is more concerned with the appropriate strategies at each of the stages (introduction, growth, maturity, and decline). It also makes predictions about industry profit through time (profit is maximized during the growth stage, and begins to decline as firms battle each other during the maturity stage).

The Moore diffusion model

Probably the most significant extension to the original diffusion model is that of Geoffrey Moore, as popularized in his book *Crossing the Chasm* (Moore, 1991). Moore's model is most concerned about the diffusion of discontinuous, disruptive high-tech products and is of particular interest to entrepreneurs involved in high-tech innovation. Moore's model is distinct from Rogers' in that he views innovators and early-adopters as visionaries and the early majority as pragmatists—and suggests that these two groups have different expectations about the new product. The chasm refers to this difference in expectations between the two groups, and the fact that pragmatists are not likely to use visionaries as reference points. The visionaries may readily adopt the product, but if the chasm is not successfully crossed, the product never will cross over into the mainstream and be adopted by the majority in the market place. This is a key difference between Moore's and Rogers' models: whereas Rogers assumed that the majority would probably follow the example of the opinion leaders, Moore proposes that the majority's expectations are quite different from those of the opinion leaders, and therefore they may not necessarily follow the opinion leaders' example.

Moore suggests techniques by which managers can cross the chasm. These techniques are all rooted in the recognition that the chasm exists, and that the expectations and desires of the two groups may be very different. For example, visionaries may be quick to adopt the product and drive up its reputation as being the "latest thing", but pragmatists are not likely to be impressed by this and therefore their biggest influence is not the word-of-mouth from visionaries. Rather, pragmatists tend to care about what affects their businesses, what technologies their competitors have adopted, what the new technology can specifically offer their industry, what the reputation is of the manufacturing firm, what is recommended in the publications that they read (not what the visionaries read!), and so forth. More specific recommendations from Moore (1991) include planning the point of market attack: the firm must carefully choose a value proposition that will show the pragmatist what the value of the new product is, in terms of how it solves customers' problems. Then it must devise a positioning strategy, define the competition in meaningful terms for the pragmatist, and develop communications, distribution, and pricing strategies that will effectively reach the pragmatists.

Some criticisms have been aimed at these diffusion models. Both models (even Moore's comparatively recent one) are now sometimes criticized for being somewhat outdated, as the rise of the Internet has changed many of the strategies for promotion and distribution. Further, it has been noted that both models are concerned with the rate of adoption, and not with profitability. It is not guaranteed that a firm that successfully crosses the chasm will ever make a profit. Nevertheless, the insights derived from these models provide managers with much useful insight on the mechanisms by which a new product diffuses through the market place.

Using diffusion models for sales prediction

Diffusion models are often used for prediction of the diffusion of a new product into the market place. The best known of these is the Bass model (Bass, 1969), which is designed to estimate the diffusion of a durable good into the marketplace through time, to predict how long it will take for the product to reach maximum sales per year (or whatever time measure is used), and what the maximum level of sales will be that year. Mathematically, the Bass model is simple. In the following equation, $s(t)$ is the level of sales at a future time t, and it is predicted as:

$$s(t) = pm + (q - p)Y(t) - (q/m)[Y(t)]^2$$

where p and q are innovation and imitation rates, m is the size of the market, and $Y(t)$ is the cumulative number of adoptions by time t. If the size of the market is known, then, the entire model is driven by two parameters: the proportion of innovators in the population (p) and how quickly the rest of the market imitates the innovators (q). Conceptually, the first term shows the sales in the first time period as simply the total number of innovators (the innovation rate times the size of the market). The second term in the model captures sales growth as diffusion passes from the innovators to the early-adopters and early majority (the larger the imitation rate and the larger the number of cumulative adoptions, the more sales). Inevitably, however, as the product diffuses through the market, the number of individuals that have not yet adopted the product dwindles, and additional sales become harder to achieve. This corresponds to progression to the late majority and laggards, and the effect is captured by the negative third term in the Bass model (once the cumulative sales gets to a certain point, additional sales per period begin to decrease). The Bass model, thus, returns an S-shaped curve of cumulative sales through time, which is consistent with Rogers' conception of diffusion and also with the PLC.

A characteristic of the Bass model is that the time required to reach the sales peak, t^*, and the peak level of sales at that time, s^*, can be predicted. These equations are:

$$t^* = [1/(p + q)] \ln(q/p)$$

$$s^* = (m)(p + q)^2/4q$$

In the original article, Bass (1969) showed how one could predict sales peaks and the required time to peak sales for durable products such as color televisions, using preliminary sales figures. More recent articles have found that the average value for p is about 0.03, and for q it is about 0.38, though these values will vary by type of product (see Sultan, Farley, and Lehmann, 1990 and Mahajan, Muller, and Bass, 1995 for discussions of parameter estimation).

The Bass model has been criticized for its limiting assumptions (e.g., no marketing variables are included, and the size of the market is taken as a constant). Later studies, such as those of Norton and Bass (1987) and Bass, Krishnan, and Jain (1994) have attempted to relax some of these assumptions. For excellent reviews of this literature, please see Mahajan, Muller, and Bass (1990) or Bass *et al.* (2000).

The diffusion literature has a comparatively long history, with its roots in the groundbreaking work by Rogers (1962) and the earliest attempts to quantify the diffusion process by Bass (1969). As noted above, a criticism of the work of both Rogers and Moore is that their theories had not been empirically tested. Nevertheless, academic researchers seem to have reached agreement on the Rogers' model of diffusion (both his list of barriers to diffusion, and the naming of the adopter groups), since these categories are now listed in virtually every marketing and product management textbook.

The Bass model in particular has stood the test of time as, many years after its original publication, it has been commonly used by managers to predict new-product diffusion. According to van den Bulte (2002), RCA used it in the early 1980s to predict sales growth in music CDs, DirecTV used it in 1992 to predict satellite TV diffusion through the latter part of the 1990s, and even today a form of it is used by movie exhibitors in Europe to predict box office revenues and to determine the number of screens on which a movie should be shown. Given the rate of new product introductions, both high-tech and otherwise, it seems that the possible applications of the Bass model are greater now than ever.

Perhaps the most intriguing application of the Bass model is in predicting the size and growth of Internet communities (Firth, Lawrence, and Clouse, 2006). Membership in an Internet community is much like owning a durable good: individuals either have joined or have not joined the community (in the durable goods case, individuals either have or have not adopted the good, and repeat purchase is not really an issue). Further, some individuals will be innovators, or quick to join Internet communities such as MySpace virtually from the moment they are available, as they immediately see the membership benefits. Others see membership as perhaps less pressing, and will take longer to join. Though it is out of the scope of this chapter, one might also need to factor in the network externality effects on the rate of diffusion. For a social community like MySpace, for a new communication category like instant messaging, or for a new communication technology like the iPhone, the more individuals that have joined, the greater the value to all members (how much value would there be in MySpace if there were only 100 members worldwide?). Thus, if the innovators are particularly influential, the imitation rates may be much higher than would have been otherwise predicted and the product can quickly become the standard.

Final thoughts: global product diffusion

Finally, as the market place becomes increasingly global, there will be a need for greater attention to global diffusion of products. Each country or culture is likely to have its own particular barriers to adoption, and these will vary for different categories of products. Also, the imitation rate is likely to differ from country to country, for infrastructure reasons (physical distances, lower use of the Internet or instant messaging, even lower penetration of telephones) or for cultural reasons. Still, companies will seek to get their product into multiple markets worldwide to exploit scale and scope economies. From a Bass model point of view, the size of the potential market will be much larger and there will be many more unknowns affecting the innovation and imitation rates. This will increase the risks of making predictions about diffusion based on this model.

Bibliography

Bass, F. M. (1969). "A new product growth model of consumer durables." *Management Science*, **15**(1), 215–227.

Bass, F. M., Krishnan, T. V., and Jain, D. C. (1994). "Why the Bass model fits without decision variables." *Marketing Science*, **13**, 203–223.

Bass, F. M., Jain, D. C., Krishnan, T. V., Mahajan, V., Muller, E., and Wind, Y. (2000). "Modeling the marketing–mix influence in new-product diffusion." In: *New Product Diffusion Models* (pp. 99–122). New York: Springer-Verlag.

Crawford, M., and Di Benedetto, A. (2008). *New Products Management* (Ninth Edition). Burr Ridge, IL: Irwin/McGraw-Hill.

Firth, D. R., Lawrence, C., and Clouse, S. F. (2006). "Predicting Internet-based online community size and time to peak membership using the Bass model of new product growth." *Interdisciplinary Journal of Information, Knowledge, and Management*, **1**, 1–12.

Gatignon, H., and Robinson, T. S. (1985). "A propositional inventory for new diffusion research." *Journal of Consumer Research*, March, 849–867.

Gauvin, S., and Sinha, R. K. (1993). "Innovativeness in industrial organizations: A two-stage model of adoption." *International Journal of Research in Marketing*, **10**, 165–183.

Hahn, M., Park, S., and Zoltners, A. A. (1994). "Analysis of new product diffusion using a four-segment trial–repeat model." *Marketing Science*, **13**(3), 224–247.

Mahajan, V., Muller, E., and Bass, F. M. (1990). "New product diffusion models in marketing: A review and directions for research." *Journal of Marketing*, **54**(1), 1–26.

Mahajan, V., Muller, E., and Bass, F. M. (1995). "Diffusion of new products: Empirical generalizations and managerial uses." *Marketing Science*, **14**(3), G79–G89.

Moore, G. A. (1991). *Crossing the Chasm.* New York: Harper Business Essentials.

Norton, J. A., and Bass, F. M. (1987). "A diffusion theory model of adoption and substitution for successive generations of high-technology products." *Management Science*, **33**(9), 1069–1086.

Rogers, E. M. (1962). *Diffusion of Innovations*. New York: Free Press.

Sultan, F., Farley, J. U., and Lehmann, D. R. (1990). "A meta-analysis of applications of diffusion models." *Journal of Marketing Research*, **27**(1), 70–78.

Sultan, F., Farley, J. U., and Lehmann, D. R. (1996). "Reflection on 'A meta-analysis of applications of diffusion models'." *Journal of Marketing Research*, **33**(2), 247–249.

van den Bulte, C. (2002). "Technical report: Want to know how diffusion speed varies across countries and product? Try using a Bass model." *Visions*, October.

17

Consumer Adoption of Technological Innovations

Paschalina (Lilia) Ziamou

The City University of New York

The consumer adoption process refers to the process consumers use to determine whether or not to adopt an innovation. This process is influenced by consumer characteristics, such as personality traits and demographic or socioeconomic factors, the characteristics of the new product, such as its relative advantage and complexity, and social influences, such as opinion leaders (Gatignon and Robertson, 1991).

In the context of technological innovations, the adoption process is also influenced by one or several new technologies that are incorporated in the new product. New technologies are likely to significantly affect the innovation's functionality or interface. Functionality refers to the set of potential benefits that a product can provide the consumer. Interface refers here to the specific means by which a consumer interacts with a product to obtain a particular functionality (see, e.g., Ziamou and Veryzer, 2005). Specifically, new technologies suggest four types of innovations (see Figure 17.1) with unique characteristics that are likely to affect the adoption process.

Functionality \ Interface	Existing	New
Existing	E.g., Smart phones in multiple colors	E.g., Voice recognition software
New	E.g., Multi-mode cell phones	E.g., Car GPS navigation systems

Figure 17.1. Typology of technological innovations.

Existing functionality and existing interface

These innovations are incremental in nature since they offer an existing functionality and an existing interface; however, they are usually characterized by esthetic changes that affect the product's appearance. Smart phones, for example, are usually black or silver when first introduced into the market but are available in multiple colors several months later.

Existing functionality and new interface

These innovations provide benefits available by existing products but result in a new set of actions for the consumer. Voice recognition software is one example of this type of innovation. Consumers create documents or emails, for example, by dictating (instead of typing) to a computer.

New functionality and existing interface

These innovations do not change consumer interaction with a device; they offer, however, a new functionality. Multi-mode cellphones, for example, operate in more than one frequency and enable roaming between different countries.

New functionality and new interface

Car GPS navigation systems, for example, fall under this category. These products provide the consumer with novel functionality, such as door-to-door navigation and real-time traffic information. The novel interface implies a new set of actions for using the device, such as using a touchscreen and voice recognition interfaces.

Innovations that incorporate a novel interface require significant learning cost from the consumer since they imply learning a new set of tasks. High learning cost is likely to hinder the adoption of such innovations (Hill,

Smith, and Mann, 1987), unless the functionality provided is new and provides significant benefits to the consumer (Ziamou and Ratneshwar, 2003). Furthermore, innovations incorporating a novel interface often result in fear of technological complexity leading to feelings of ineptitude and frustration (Mick and Fournier, 1998). Conversely, innovations that provide the consumer with a new functionality are characterized by a high relative advantage, which is likely to facilitate adoption (Rogers, 1983).

References

Gatignon, H. and Robertson, T. S. (1991). "Innovative decision processes." In: T. S. Robertson and H. H. Kassarjian (Eds.), *Handbook of Consumer Behavior* (pp. 316–348). Englewood Cliffs, NJ: Prentice-Hall.

Hill, T., Smith, N. D., and Mann, M. F. (1987). "Role of efficacy expectations in predicting the decision to use advanced technologies: The case of computers." *Journal of Applied Psychology*, **72**(2), May, 307–313.

Mick, D. G., and Fournier, S. (1998). "Paradoxes of technology: Consumer cognizance, emotions, and coping strategies." *Journal of Consumer Research*, **25**(2), September, 123–143.

Rogers, E. M. (1983). *The Diffusion of Innovations*. New York: Free Press.

Ziamou, P. L. and Ratneshwar, S. (2003). "Innovations in product functionality: When and why are explicit comparisons effective." *Journal of Marketing*, **67**(2), April, 49–61.

Ziamou, P. L., and Veryzer, R. (2005). "The influence of temporal distance on consumer preferences for technology-based innovations." *The Journal of Product Innovation Management*, **22**(4), July, 336–346.

Part Four

Firm Level

18

Open Innovation

Mariann Jelinek

College of William and Mary

Introduction

"Open innovation", the term popularized recently by Henry Chesbrough (Chesbrough, 2003), is the contemporary label for a much older idea—that the majority of the knowledge in the world exists outside the walls of any given firm, and thus that drawing on resources beyond the firm's own R&D, scientists, engineers, and technicians—or its own marketing or manufacturing experts—is desirable. Contract research, consulting, and design firms, from Arthur D. Little (founded in 1886) and Battelle Institute (founded in the 1930s) to IDEO and Frogdesign have long offered outside assistance to firms seeking to innovate. In the U.S. universities have also been active research partners to U.S. industry since the mid–19th century, responding rapidly to new industry needs with new curriculum designs. Faculty often formed alliances with firms, doing critical research and placing many graduates with their industrial partners, thus fueling the growth of new industries (Mowery and Rosenberg, 1998).

Factors encouraging open innovation

Open innovation has gained additional prominence and momentum in the context of three specific factors: financial pressures for efficiency, burgeoning globalization, and increased technical complexity. First, financial pressures drove firms to eliminate expenditures that did not lead promptly to financial returns. As a result, longer horizon, centralized corporate R&D was eliminated in many major U.S. firms during the 1980s. These activities were dismantled and distributed to corporate divisions (where their focus was tightened, their budgetary support much reduced, and their temporal horizon shortened). Corporate research centers at numerous firms, and even such centers of excellence as AT&T's Bell Labs, are no more. By 2000, many had been downsized or eliminated in favor of some form of outsourcing. Icons like Procter & Gamble, Lucent (formerly AT&T), Intel, and IBM have begun to source substantial research from outside the firm (Chesbrough, 2003). Outsourcing has achieved a sort of apotheosis with online systems such as InnoCentive, "a Web-based community matching scientists to R&D challenges presented by companies worldwide."[1]

Globalization is a second big factor. Genuine commercial success is attributed to reconfiguration or reapplication of existing solution elements into new forms or applications (Hargadon, 2003; Hargadon and Sutton, 1997), while regional economic growth is attributed to enhanced interconnections and mobility of personnel and ideas among firms (Saxenian, 1994, 1999)—both forms of open innovation. Yet the scope for accessing new ideas and circulating personnel has become increasingly global (Doz and Hamel, 1998; Doz, Santos, and Williamson, 2001; Saxenian, 2000). Some researchers have argued that a vastly enhanced knowledge landscape mandates a shift to "open innovation", drawing upon external resources at virtually every stage of the discovery-to-commercialization process, and every stage of the firm's value chain.

Against this global environment, industry–university consortia and collaborative research efforts have proliferated in the U.S., and industry investment in sponsored university research has soared, surpassing Federal support, just as friction around issues of intellectual property between industry and university participants reached a crescendo. Today, industrial funding accounts for roughly 70% of all U.S. R&D on an annual basis, including sponsored university research, which is the locus of most fundamental research. But U.S. universities are by no means the only source of research, ideas, and innovation

[1] *http://www.innocentive.com/* (consulted May 2, 2007). Nor is InnoCentive the only such site: Nine Sigma promises to "accelerate the innovation cycle" via requests for proposals in a similar fashion (*http://www.ninesigma.com/mx/newsletter.html?gclid=CIirlfX774sCFQ0eSgodJSHkOA*; consulted May 2, 2007), and there are numerous others.

possibilities: firms have seen dramatic increases in alliances, strategic partnerships, and other cooperative arrangements, including (but not limited to) externally purchased research, contract research, sponsored university research carried out abroad to avoid U.S. intellectual property restrictions, joint ventures, and offshore research subsidiaries in places like China and India. These latter efforts have been substantially supported by foreign governments, and in some cases required as a condition for market access. By 2005, multinationals—that is, most large U.S. firms—have as much as 20% of their assets engaged in alliances, a trend growing for some time (Barney, 1991; Cook, Halevy, and Hastie, 2003; Culpan, 2002; Doz and Hamel, 1998; Dyer and Singh, 1998).

Increasing complexity

The third major factor encouraging open innovation practices is the increasing complexity of many contemporary processes and products, which incorporate more science and broader technology in a given product. For example, contemporary automobiles contain more computer power than went to the Moon on the Apollo space shots. Then too, where once telecommunications and computers seemed entirely separate industries, products like Apple's iPhone mean that "phone" and "computer" expertise must come together for successful product design. The iPhone offers an especially interesting example because it incorporated design elements from Apple's earlier (and wildly successful) iPod, including the disk drive and display screen, along with a variety of software, like that for iTunes access. The iPod's success resulted from collaboration between partners including (among others) PortalPlayer, Philips, IDEO, General Magic, Connectix, WebTV, Texas Instruments, and Toshiba. The rapid-fire development of Apple's iPod, a breakthrough product, occurred in about 6 months (Slowinski and Tao, 2007)—solutions and relationships that Apple made excellent re-use of for the iPhone. Both iPod and iPhone benefitted enormously from in-depth expertise that only innovation partners could provide in anything remotely resembling such a time frame. Such collaborations can achieve ambitious goals, like developing flat-screen technology simply too complex for any single company, or *country*, to undertake alone (Murtha, Lenway, and Hart, 2001).

The emerging alliance imperative extends beyond technical complexity to entrepreneurship: in particular, high-technology entrepreneurial start-ups are often founded by teams of entrepreneurs, who are more successful when their networks have broad access to a range of resources (Boeker, 1989; Schoonhoven, Eisenhardt, and Lyman, 1990). Networks extend the access of firms to important resources (Gulati, Nohria, and Zaheer, 2000), assist in identifying opportunities (Cooper, 2001), as well as providing formal and informal relationships, information, insight, and legitimacy (Kogut, 2000), among other benefits: in short, whether the complexity at issue is technical or organizational, open relationships contribute to innovative efforts. Some firms building their business model on innovation are open by design, and "multinational from birth", like Logitech, founded by Swiss principals with major operations in California and Asia. Entrepreneurship has expanded exponentially around the globe, fueled in part by the ready access to markets the Internet provides.

Emerging issues in open innovation

Globalization, complexity, and financial pressures encourage further open innovation because many more capable potential partners are available, interested, and feasibly accessible through virtual links. Good ideas are by no means limited to North America, or Europe (or, indeed, to any firm's home country or region) (Doz and Hamel, 1998; Doz, Santos, and Williamson, 2001). Formerly isolated economies, like China and India, and the old Iron Curtain countries are rejoining the global economy. Many of these countries are actively seeking foreign collaboration as well as direct investment to bolster their innovation capabilities. While "developed country" partners gain cut-rate technical talent and inexpensive research, "developing country" partners gain much more than foreign revenues: they also benefit from priceless exposure to state-of-the-art problems and procedures, no less than to research mentoring in the very questions that are posed. In addition, as U.S. firms have moved offshore, the price has often been technology: the price of entry into China for General Motors was "our latest manufacturing technology", according to Larry Burns, Vice President of R&D and Strategic Planning (August 14, 2006).[2]

What are the consequences for strategy and for innovation? Alliances carry both benefits and risks, and must be managed carefully. Describing an alliance as *any limited-scope, flexible, deliberately delineated collaboration*, Slowinski and Sagal (2003) identify three fundamental requirements for success. First, each firm must clearly understand their own business strategic plan and the ties between that plan and the potential alliance. Second, both (or all) firms must clearly and honestly reveal that plan to the other party. Finally, both (or all) must agree on alliance intentions, commitments, rights, and limitations that satisfy the strategic plans of both firms (Slowinski and Sagal 2003, p. 6). Yet where governments are players, the parties may not be equal, or equally forthcoming. Further, while sharing "our latest manufacturing technology" in return for entry to the rapidly expanding Chinese market seems pragmatic at the level of the firm, larger implications may reside at the level of the industry or the country. Without substantial internal innovation resources neither

[2] Presentation at the NSF R&D Industry Expert Panel, August 14, 2006.

firms nor industries will remain effective contributors to the state of the art; they risk being surpassed by their partners. Without continued interplay between a country's academic research and its industry problems, universities may become simply irrelevant, incapable of advancing industry capabilities or producing employable graduates. Finally, like Japan, Korea, and other "Asian Tigers" before it, China is clearly building its industries' competitive capabilities for the world market. China is becoming a formidable competitor in many markets.

Contemporary thinking about open innovation recognizes potential and benefits, but also acknowledges risks. Traditionally feared outcomes—loss of proprietary secrets inadvertently leaked to the partner; theft of technology, or poaching of skilled employees; asymmetric benefits favoring one party to a deal—have given way to broader considerations. How shall a firm balance what is outsourced against the need to maintain and foster internal capabilities and "absorptive capacity" (Cohen and Levinthal, 1990)? How shall the firm's dependency upon outside partners be managed? How should such partnerships be managed long term, beyond any single transaction? Can (or should) the partners' fundamental independence be preserved? The questions are important,[3] but should not preclude careful strategic thinking about the potential and benefits of externally sourced skills, ideas, and other innovation assistance. If products and processes are much easier to copy than methods of organizing work and innovation (Kogut, 2000), how shall firms develop sustainable competitive advantages at collaborative innovation *per se*? These larger concerns will be the focus of open innovation challenges to come.

References

Barney, J. B. (1991). "Firm resources and sustained competitive advantage." *Journal of Management*, **17**(1), 99–120.

Boeker, W. (1989). "Strategic change: The effects of founding and history." *Academy of Management Journal*, **32**(3), September, 489–515.

Chesbrough, H. W. (2003). *Open Innovation: The New Imperative for Creating and Profiting from Technology*. Boston, MA: Harvard Business School Press.

Cohen, W. M., and Levinthal, D. A. (1990). "Absorptive capacity: A new perspective on learning and innovation." *Administrative Science Quarterly*, **35**(1), 128–152.

Cook, J., Halevy, T., and Hastie, B. (2003). "Alliances in consumer packaged goods." *McKinsey on Finance*, 16–20.

Cooper, A. C. (2001). "Networks, alliances and entrepreneurship." In: M. A. Hitt, R. D. Ireland, S. M. Camp, and D. L. Sexton (Eds.), *Strategic Entrepreneurship: Creating a New Integrated Mindset*. London: Basil Blackwell.

Culpan, R. (2002). *Global Business Alliances: Theory and Practice*. Westport, CT: Quorum Books.

Doz, Y. L., and Hamel, G. (1998). *The Alliance Advantage*. Boston, MA: Harvard Business School Press.

Doz, Y., Santos, J., and Williamson, P. (2001). *From Global to Metanational: How Companies Win in the Knowledge Economy*. Boston, MA: Harvard Business School Press.

Dyer, J. H., and Singh, H. (1998). "The relational view: Cooperative strategy and sources of interorganizational competitive advantage." *Academy of Management Review*, **23**(4), 660–679.

Gulati, R., Nohria, N., and Zaheer, A. (2000). "Strategic networks." *Strategic Management Journal*, **21**(3), Special Issue, 203–213.

Hargadon, A. (2003). *How Breakthroughs Happen: The Surprising Truth about How Companies Innovate*. Boston, MA: Harvard Business School Press.

Hargadon, A., and Sutton, R. I. (1997). "Technology brokering and innovation in a product development firm." *Administrative Science Quarterly*, **42**(4): 716-749.

Kogut, B. (2000). "The network as knowledge: Generative rules and the emergence of structure." *Strategic Management Journal*, **21**(3), Special Issue: Strategic Networks, 405–425.

Mowery, D. C., and Rosenberg, N. (1998). *Paths of Innovation: Technological Change in 20th-Century America*. Cambridge, U.K.: Cambridge University Press.

Murtha, T. P., Lenway, S. A., and Hart, J. A. (2001). *Managing New Industry Creation: Global Knowledge Formation and Entrepreneurship in High Technology*. Stanford, CA: Stanford University Press.

Saxenian, A. (1994). *Regional Advantage: Culture and Competition in Silicon Valley and Route 128*. Cambridge, MA: Harvard University Press.

Saxenian, A. (1999). *Silicon Valley's New Immigrant Entrepreneurs*. San Francisco, CA: Public Policy Institute of California.

Saxenian, A. (2000). "Silicon Valley as a regional system of innovation: Its international linkages. *China–U.S. Joint Conference on Technological Innovation Management*. Beijing, PRC: Reseach Center for Technological Innovation, Tsinghua University.

Schoonhoven, C. B., Eisenhardt, K. M., and Lyman, K. (1990). "Speeding innovation to market: The impact of technology-based innovation on waiting times to first product introduction in new semiconductor ventures." *Administrative Science Quarterly*, **35**(1), March, 177–207.

Slowinski, G., and Sagal, M. W. (2003). *The Strongest Link: Forging a Profitable and Enduring Corporate Alliance*. New York: AMACOM.

Slowinski, G., and Tao, J. (2007). "Implementing open innovation: Twenty years of ROR thought leadership." *Industrial Research Institute 2007 Member Summit: Building the Future of R&D*. Lincolnshire, IL: Industrial Research Institute.

[3] And are discussed in detail in Slowinski and Sagal (2003).

19

Technology Strategy

Steven W. Floyd and Carola Wolf

University of St. Gallen

Technology strategy represents managers' efforts to think systematically about the role of technology in decisions affecting the long-term success of the organization. This involves deciding issues such as what technologies are strategically important, how to position the firm relative to technology development, and how much to spend in research and development, among others. Table 19.1 groups these and other questions into those related to formulating technology strategy and those associated with implementing it. Though they oversimplify the complexity of any particular company's technology strategy, these questions provide a useful framework for organizing the discussion in this chapter.

Table 19.1. Key questions in technology strategy.

Formulating technology strategy

- What are the organization's core technologies?
- How should the firm position itself relative to technology development?
- Should the firm seek to establish a technology standard?
- Why do technological discontinuities arise and how should the firm respond?

Organizing for technology strategy

- What are the risks and rewards of strategic alliances as a means of developing new technology?
- How much should the firm invest in research and development relative to competition?
- How can human capital be managed to produce superior technology?

Formulating technology strategy

Core technologies

To a strategist, "technology" is usually meant in the broadest possible terms—knowledge of how to do things and how to accomplish human goals (Simon, 1973). The first issue that arises in formulating a technology strategy therefore is narrowing the scope. Of all the technologies that are relevant to an organization, which technologies should be the focus of strategy? Classical organizational theory distinguishes between technologies deployed in the "technical core"—the units where the product or service is produced—and technologies used in support tasks (Jelinek, 1977). To increase efficiency, intervening technologies and elaborated structures are designed to protect the technical core from too much uncertainty, to coordinate among organization elements, and to mediate the fit between the organization and its environment (Thompson, 1967). Within this broad organizing framework, technologies are also likely to differ in terms of whether they are simply necessary to perform a particular task (basic technologies), or whether they are strategically important (critical technologies) (Ford, 1988). According to contemporary theory, the most important strategic issue is whether technology provides the basis for competitive advantage.

The word "core" is also used to describe a class of technologies that are based on new science and that spawn subsequent research, discovery, and commercialization of products (Kim and Kogut, 1996). Sometimes, core technologies may evolve over long periods of time and lead to many different derivatives. Core technologies like the automobile or microprocessor, for example, are the technical ancestors of a vast array of products. Other non-core technologies may produce very few "children", becoming essentially dead-ends. A technology's pattern of development is called its trajectory. Firms may attempt to capitalize on such trajectories by developing proprietary products based on a core technology. This often takes the form of technology platforms, such as Intel's X-86 microprocessor.

Platforms are deliberately designed to spawn multiple product derivatives.

Technological pioneering strategy

One of the key roles for technology in competitive strategy is its effect on the timing of a firm's entry into product markets. Generally, three alternative competitive positions are identified with respect to the timing of market entry: market pioneers who are first to market with new products, quick-followers who introduce close approximations of the new product without incurring the pioneer's development expenses, and late-entrants who enter only when most technology and market uncertainties are already resolved (Robinson and Chiang, 2002). When a firm seeks to offer the latest technology in its new products or services, it is said to be a technology leader, and technology leadership is one of the principle means by which firms pursue market pioneering strategies (also known as first-mover strategies) (Lieberman and Montgomery, 1988).

Pioneering based on technology leadership carries a number of risks and benefits. Potential benefits include gaining pre-emptive access to scarce assets such as distribution channel relationships and physical locations (Lieberman and Montgomery, 1988), establishing the product/service as an industry standard and building reputation and customer loyalty that is difficult for rivals to imitate. Any one or a combination of these benefits may produce first-mover advantage as evidenced by the ability of the pioneer to earn profits above the industry average (Lieberman and Montgomery, 1988).

The primary risk to the pioneer is that the developmental costs associated with achieving technology leadership are not recovered in the profit margins associated with the sale of the product or service. This risk is increased whenever fast-followers or late-entrants are able to leverage the investments of the pioneer in the development and marketing of their own products. This, in turn, becomes more likely when there are few barriers to imitation and rapid technology diffusion among competitors. Property rights in the form of patents, copyrights, and trademarks may help protect the first-mover's advantage, but market uncertainties may induce additional risks. If the users' needs are dynamic, for example, late-movers may succeed with less innovative technology by investing resources otherwise spent on new technology into tailoring the product's appeal in ways that are more responsive to customer needs. The complex mix of technical and market uncertainties surrounding the pioneering strategy are reflected in research findings that show an inconsistent empirical relationship between pioneering strategies and economic performance.

Technology standards

One of the benefits of a pioneering strategy may be the opportunity to establish a technological standard. The focus of standardization may be at the component level (e.g., Sharp Corporation's liquid-crystal displays), end-product level (e.g., Adobe's document reader software), or system level (e.g., Apple's iTunes/iPod music-downloading service). Standards tend to emerge when there are substantial benefits from using devices with compatible technical elements in a particular domain. These compatibilities provide the basis for developing complementary products so that the value of using one increases the value of the other. This makes the original product more valuable to users and increases the motivation of manufacturers, distributors, and customers to adopt it, thereby increasing the installed base and motivating others to extend the range of complementary products and services (Anderson and Tushman, 1990).

The combination of Intel's microprocessor and Microsoft's Windows operating software is an example of a proprietary standard at the system level in the personal computer industry. It arose in part because users benefit from the large installed based of other users using the same hardware and software combination. The potential effects of technology standards include locking in users by encouraging them to invest in and depend on the standard and locking out potential competitors by making their products technologically incompatible. In combination, a well-established standard is often the source of durable and uncontested streams of revenues and profit.

While industry conditions and the nature of the technology play a big role in determining whether a standard emerges, firm strategy is also important. In particular, strategies that lead to the development of a large installed base increase the benefits that accrue to users from complementary products (Schilling, 2002). Aggressive pricing strategies are one way to establish a large installed base relatively quickly. Adobe Systems Inc., for example, chose to make part of its electronic document software (the document reader) available for free and established the .pdf file as the *de facto* industry standard. Adobe realizes economic benefits by selling complementary products.

Because user behavior plays such a large role in determining which technology becomes the standard, the standard is not always the most technically advanced or the best technology (Lee *et al.*, 1995). The layout of keys on the keyboard of most computers provides a famous example. From the perspective of keyboard efficiency, the best layout would be to locate more frequently used keys toward the centre so that they are struck by one of the larger fingers. Instead, the most common layout (the QWERTY keyboard) puts two of the most frequently used vowels ("a" and "o") where they are struck by the smallest fingers. This configuration was established in the era of the mechanical typewriter when keys were positioned to create separation so that they would not stick together. QWERTY became the basis for developing the typing skills of users, and this complementarity led to its adoption as the industry standard. Thus, the migration of users' typing skills to the computer keyboard explains why a technically suboptimal keypad layout developed in the 19th century continues as the standard for computers being built today.

Technological discontinuities

Although the QWERTY standard has endured, the trajectory of many more technologies is less continuous. Technologies emerge and sometimes go on to become industry standards. While they are dominant, the pattern of technological change is one of incremental improvement in the core technology together with the development of additional complementary technologies (Tushman and Anderson, 1986). These evolutionary change processes preserve the technical competencies of incumbent firms and, because they make the standard more attractive, these evolutionary changes help to ward off challenges to the standard by alternative technologies. Incumbents therefore typically develop a vested economic interest in preserving the technological *status quo*.

Because they are perpetuated for economic reasons, however, standards and dominant designs tend to lag behind technical frontiers. Incumbent firms therefore face the possibility that a superior technology will be developed and gain a foothold in the market. When this happens, the standard or dominant design may be unseated, creating a shift in the viable basis of competitive advantage. Typically, such competence-destroying change is so fundamentally different from the dominant technology that the skills and knowledge base required to operate the core technology shift dramatically. This is usually accompanied by major changes in skills, distinctive competences, and production processes, all of which are accompanied by shifts in the distribution of power and control within organizations and industries (Chandler, 1977). In sum, research has described the process of technological change as triggered by a breakthrough, or discontinuity, followed by a relatively long period of intense technical variation and selection, culminating in a single dominant design that is subject to further discontinuity in the future (Anderson and Tushman, 1990).

The organizational consequences of this pattern help to explain why incumbents often seem unresponsive to dramatic technological changes. Why, for example, have the major record companies failed to respond successfully to the Internet and music downloading? Part of the problem is that managers in these organizations have neglected investment in the technological resources necessary to a successful Internet strategy. Influenced by culture and mindsets more in tune with history than contemporary realities, resource allocation decisions have favored the CD format. What were once core capabilities have become what one author describes as "core rigidities" (Leonard-Barton, 1992). Moreover, the dilemma is that the more they invest in the Internet revolution, the more they cannibalize their core business—selling recorded music. Thus, despite what seems an obvious need for radical change, these incumbents persist with an obsolete technology and fail to migrate successfully to a new dominant design (Christensen, 1997).

Some discontinuities can be caused by firms who deliberately choose to use new technologies to drive market change. Distinct from strategies that are "market-driven", such *market-driving* strategy means using technology to create change in business models and to reshape the market in fundamental ways. Companies like Amazon, Dell, Starbucks, and Southwest Airlines provide examples of market-driving strategies. In each case, rather than take the markets as a given, these companies fundamentally changed the competitive equilibrium in ways that shifted share and profitability in their favor.

Organizing for technology strategy

Technology alliances

Increasingly, the financial stakes involved in major technology investments have motivated firms to seek out strategic alliances as a way to share the risks. In addition to risk-sharing, learning and knowledge transfer are primary motives for seeking an alliance partner as part of a technology strategy. Research suggests that equity-based joint ventures promote greater knowledge transfer between partners than non-equity alliances. The absorptive capacity of a partner defined as "the ability of a firm to recognize the value of new, external information, assimilate it, and apply it to commercial ends" (Cohen and Levinthal, 1990, p. 128) has been found to explain the extent of capability learning. Research results do not support the naive view that organizations can "acquire a capability" in an alliance. For example, among several impediments to realizing the potential of an alliance, cultural differences between a large, integrated firm and a biotechnology startup have been shown to frustrate the success of alliances in the pharmaceutical industry.

One problem that plagues technology-based alliances more than other forms of partnership is *knowledge spill-over*. Partners must be concerned about whether proprietary information unrelated to the alliance leaks out inadvertently as a part of deliberate knowledge-sharing efforts. Jet engine–manufacturing units from General Electric (GE) and United Technologies (UT), for example, formed an alliance to develop a new-generation jet engine. The rationale for the alliance was risk-sharing, and the relationship paired UT's ability for world-class engineering with GE's ability to manage costs and quality. Since GE's strength is also seen as a UT weakness (and *vice versa*), however, both firms are motivated to learn the other's strength. While this suits the goals of collaboration, GE and UT are also fierce competitors, and knowledge spill-over may threaten their respective competitive positions—

particularly if one learns better or more quickly than the other. Ironically, the principal means used to control spillover (e.g., dividing work in a way that limits interaction across areas of expertise) makes learning and knowledge exchange more difficult and therefore may defeat the alliance's purpose.

R&D intensity

Alliances are not the only way that managers seek to learn strategically important technologies. The portion of firm revenue/income invested in research and development, conceptualized as R&D intensity, is also a primary means of learning. In technologically intensive industries, such as pharmaceuticals, firms may invest as much as 15% of annual revenues in R&D. R&D intensity is usually measured as annual R&D expenses divided by annual sales. Research suggests that R&D intensity is closely related to the success of a technology strategy in many firms. For example, one study found evidence of a positive relationship between a new venture's R&D intensity, late-stage technical capabilities, and wealth creation (Deeds, 2001). In addition to being a predictor of technology strategy success, R&D intensity has been assessed as a substitute for diversification strategy in the pursuit of corporate growth. In fact, firms that reduce their level of diversification as a part of a restructuring strategy tend to increase their R&D intensity, while firms that increase their diversification tend to reduce R&D intensity.

The allocation of R&D investments across projects directed at basic research versus those targeting product development is contingent on technology strategy. Typically, technology leaders, for example, would be expected to invest relatively greater amounts on basic research than others because being a pioneer relies on applying the latest scientific knowledge. For multi-business firms, another contingency that arises with respect to R&D spending is the proportion spent in centralized activities versus what is allocated to support divisional efforts. Where the corporate strategy relies on developing and leveraging core competencies, the tendency is to invest more heavily in centralized activities for basic research and divisional activities for product development. Since United Technology's corporate strategy is based partly on a core competence in materials (e.g., carbon fiber applications), for example, resources are allocated within the corporate R&D function to study materials sciences. This knowledge is then deployed within UT's subsidiaries, including Sikorski's helicopter fuselages, Pratt & Whitney's jet engine housings and Otis Elevator's car bodies.

Managing scientific human resources

Human capital constitutes the lion's share of expenditures on research and development in most organizations. Leveraging these scientific human resources is challenging, and a significant body of research exists on managing R&D personnel (for a review see Farris and Codero, 2008).

Research shows, for example, that intrinsic rewards are more important in the scientific context. However, the use of extrinsic and team rewards is increasing compared with the past. Studies also show that multiple performance metrics and multiple appraisers are often used to evaluate the performance of scientists.

Until recently, much research has assumed that scientists are not motivated by private interests and that they are stimulated by recognition and working in the public domain. Research on "star scientists" in biotechnology—those who are responsible for major breakthroughs—paints a different picture, however. Here, scientists are seen to take an entrepreneurial interest in their subject and to value the prestige and resources that accompany their accomplishment (Zucker and Darby, 1996) This personality profile makes it more likely that scientists will seek to protect the rewards from their discovery by limiting the diffusion of knowledge. One of the implications of this work is that firms seeking to obtain world-leading positions in a scientific domain need direct access to star scientists rather than being able to rely on public channels for the diffusion of scientific knowledge.

In summary, strategic thinking about the role of technology in long-term firm success is especially important in rapidly changing environments characterized by intensified global competition and technological dynamism. Strategically appropriate investments in technology have been shown to positively affect business performance in a range of different contexts (e.g., Khan and Manopichetwattana, 1989). Managers should therefore consider technology strategy as central to their leadership responsibility.

References

Anderson, P., and Tushman, M. L. (1990). "Technological discontinuities and dominant designs: A cyclical model of technological change." *Administrative Science Quarterly*, **35**, 604–633.

Chandler, A. D. (1977). *The Visible Hand*. Cambridge, MA: Harvard University Press.

Christensen, C. M. (1997). "Making strategy: Learning by doing." *Harvard Business Review*, November/December, 141–156.

Cohen, W. M., and Levinthal, D. A. (1990). "Absorptive capacity: A new perspective on learning and innovation." *Administrative Science Quarterly*, **35**, 128–152.

Deeds, D. L. (2001). "The role of R&D intensity, technical development and absorptive capacity in creating entrepreneurial wealth in high technology start-ups." *Journal of Engineering and Technology Management*, **18**, 29–47.

Farris, G. F., and Codero, R. (2008). "What we know about managing scientists and engineers: A review of the recent literature"; *http://cims.ncsu.edu/documents/managingscientist.pdf*

Ford, D. (1988). "Develop your technology strategy." *Long Range Planning*, **21**(5), 85–95.

Jelinek, M. (1977). "Technology, organizations, and contingency." *Academy of Management Review*, **2**, 17–26.

Khan, A. M., and Manopichetwattana, V. (1989). "Innovative and noninnovative small firms: Types and characteristics." *Management Science*, **35**, 597–606.

Kim, D., and Kogut, B. (1996). "Technological platforms and diversification." *Organization Science*, **7**(3), 283–301.

Lee, J. R., O'Neal, D. E., Pruett, M. W., and Thomas, H. (1995). "Planning for dominance: A strategic perspective on the emergence of dominant design." *R&D Management*, **25**, 3–15.

Leonard-Barton, D. (1992). "Core capabilities and core rigidities: A paradox in managing new product development." *Strategic Management Journal*, **13**, 111–125.

Lieberman, M. B., and Montgomery, D. B. (1988). "First-mover advantages." *Strategic Management Journal*, **9**, 41–58.

Robinson, W. T., and Chiang, J. (2002). "Product development strategies for established market pioneers, early followers, and late entrants." *Strategic Management Journal*, **23**, 855–866.

Schilling, M. (2002). "Technological success and failure in winner-take-all markets: The impact of learning orientation, timing, and network externalities." *Academy of Management Journal*, **45**(2), 387–398.

Simon, H. A. (1973). "The structure of ill-structured problems." *Artificial Intelligence*, **4**, 181–201.

Thompson, J. D. (1967). *Organizations in Action*. New York: McGraw-Hill.

Tushman, M. L., and Anderson, P. (1986). "Technological discontinuities and organizational environments." *Administrative Science Quarterly*, **31**, 439–465.

Zucker, L. G., and Darby, M. R. (1996). "Star scientists and institutional transformation: Patterns of invention and innovation in the formation of the biotechnology industry." *Proceedings of the National Academy of Sciences*, **93**(12), 709–716.

20

Absorptive Capacity and Technological Innovation

Shaker A. Zahra, Bárbara Larrañeta,† and J. Luis Galán‡*

*University of Minnesota, †Pablo Olavide University, and ‡University of Seville

Technological innovation is the foundation of competitive distinctiveness that gives the firm an advantage over its rivals, enabling it to achieve superior performance. Successful technological innovation requires the integration of multiple capabilities. These capabilities are usually grounded in knowledge-based routines (Helfat and Peteraf, 2003; Nelson and Winter, 1982). The knowledge used to develop these routines could be internally developed or acquired from external sources. The dynamism and complexity of today's competitive landscape often makes it essential for companies to use knowledge generated by other firms and institutions. Determining which types of knowledge to bring to the organization, how to best assimilate this knowledge, and how to exploit it for competitive advantage are important decisions that are shaped by recipient companies' absorptive capacity (Lane, Koka, and Pathak, 2006). Companies that do not have this capacity may not benefit from the rich and varied information that exists in their industry and markets.

The concept and its dimensions

Absorptive capacity refers to the ability of a firm to "recognize the value of new external information, assimilate it and apply it to commercial ends" (Cohen and Levinthal, 1990). In recent years, definitions of absorptive capacity have highlighted its power in converting the knowledge gained from external sources into usable ideas, products, goods, services, and models (Zahra and George, 2002; Zahra, van deVelde, and Larrañeta, 2007). This process of knowledge conversion—the translation of abstract knowledge into more concrete prototypes, designs, etc.—make it possible to exploit externally generated knowledge (Zahra, van deVelde, and Larrañeta, 2007). As such, new conceptualizations of absorptive capacity highlight effective exploitation of knowledge gained externally or by integrating them with the firm's own knowledge base (Zahra and George, 2002).

The focus on knowledge conversion and subsequent commercial exploitation adds richness to the literature that has traditionally focused on a firm's knowledge base. In one of the earliest discussions of the concept, Cohen and Levinthal (1990, 1994) equated absorptive capacity with the firm's R&D intensity, treating it as a static resource, not a capability. By definition, a capability embodies and integrates multiple skills and resources that enable the timely and efficient completion of a task (Helfat and Peteraf, 2003; Miller, 2003). A capability-based view implies learning and evolving. Learning means acquiring new knowledge and gaining new insights from its creation and use (Helfat and Peteraf, 2003). The more the organization and its managers learn, the more likely they will gain new insights about what they are doing and how to do it better. This learning is crucial because it helps managers conceive of different strategic options, redefine the market arena, and reconceptualize cause–effect relationships differently. Equally important, this allows the firm to discover and pursue different market opportunities (Zahra, 2009). These activities drive the firm's evolution. Thus, absorptive capacity is intimately connected to organizational learning, capability building, and organizational evolution (Zahra, Sapienza, and Davidsson, 2006).

There are several conceptualizations of absorptive capacity (Cohen and Levinthal, 1990; Lane, Koka, and Pathak, 2006; Todorov and Durisin, 2007). Zahra and George (2002) propose that absorptive capacity has four dimensions: acquisition, assimilation, transformation, and exploitation. Each of these dimensions serves a unique purpose and can enrich the firm's technological innovation. A deficiency in one of these dimensions, however, can weaken the firm's overall innovation activities and hamper its ability to develop and commercialize innovative technologies.

As a key component of the firm's absorptive capacity, *acquisition* refers to the firm's ability to identify value and acquire externally generated knowledge that is critical to its operations. Identification of potentially valuable knowledge is usually based on a thorough understanding of the firm's opportunity set, strategy, and current product portfolios. It depends also on the nature of the firm's appreciation of the evolutionary forces that govern the markets and potential technological trajectories. Knowing who controls which types of knowledge and how to gain access to them is another important consideration, which enables the firm to connect with these sources, developing beneficial relationships that facilitate knowledge transfer. Of course, the firm can rely on traditional market mechanisms to acquire this knowledge (e.g., through purchase or licensing).

Once knowledge is gained, the firm has to work hard at assimilating it. *Assimilation* refers to the mechanisms and routines a firm can use to process, interpret, and understand the information obtained from external sources (Kim, 1998). Assimilation makes it possible for the firm to proceed with *transformation*, which refers to a firm's ability to integrate, develop, and refine the routines that generate combinative or new knowledge (Garud and Nayyar, 1994). Finally, *exploitation* denotes a firm's ability to refine, extend, and leverage existing competencies by incorporating acquired and transformed knowledge into its operations by developing strategic initiatives such as embarking on radical technological innovation (Zahra and George, 2002).

Absorptive capacity and technological innovation

Researchers have noted the importance of absorptive capacity for promoting and sustaining technological innovation (Zahra, 2009; Zahra and George, 2002). They note that incoming knowledge flows replenish the firm's knowledge base, infusing new ideas and processes that stimulate technological innovation. One weakness of prior research is failing to delineate when absorptive capacity can lead to radical versus incremental technological innovation. Incremental innovations are extensions, refinements, and upgrades of the firm's technology-based products, processes, and services. Radical innovations represent major advancements on what is known and embodies a variety of options that include truly new-to-the world innovations and major technological shifts that qualitatively exceed what currently exists in the market.

A convenient way of conceptualizing the effect of absorptive capacity on technological innovation is to consider its breadth versus depth. Breadth refers to the extent to which the knowledge contained in the firm's absorptive capacity is multifaceted and comprehensive in its coverage of a multitude of fields. As a result, absorptive capacity could be narrow (covering only a few fields) versus broad (covering a wide range of fields). A broad absorptive capacity can give the firm a wider range of options when it comes to technological innovation. In contrast, depth refers to the extent to which the firm has developed expert-type mastery of a particular technological domain. Depth could form a continuum ranging from shallow (where the firm has some or even a superficial level of skill in a given field) to deep (where the firm has great expertise in a given field). Plotting the orthogonal dimensions of depth and breadth provides some insights into the strategic value of absorptive capacity *vis-à-vis* technological innovation, as shown in Figure 20.1.

Quadrant 1 in Figure 20.1 shows the situation where the firm's absorptive capacity is narrow and shallow. This might occur from overspecialization, absence of scanning systems that gather information about the competitive terrain, or the lack of sustained investments in R&D and other innovative activities. As a result, the firm's product lines are apt to be limited, old, and even decaying. As a result, the firm's ability to engage in technological innovation is constrained. Infusion of knowledge may help to some extent but the full benefits associated with knowledge inflows are not likely to materialize because of the company's limited absorptive capacity.

Quadrant 2 depicts the situation where the firm has knowledge in different areas but lacks sufficient expert knowledge. The firm might benefit from the depth of its absorptive capacity by being prolific in upgrading its products. Of course, a series of incremental innovations lead to radical technological innovations.

In quadrant 3 the firm has deep knowledge in a few fields. This combination is conducive to pioneering technological innovations as well as building and protecting a viable niche. The depth of the firm's knowledge, however, could become a strategic handicap if the external environment changes drastically and a new technological paradigm emerges.

In quadrant 4 the combination of broad and deep knowledge could enrich the firm's technological innovation by providing opportunities for integration, fusing technologies (Zahra, 2009), creating new product platforms, or simply adding more lines to existing products through upgrades. However, the firm's success in cultivating its absorptive capacity depends on its integrative capability. This capability refers to the firm's skill in managing and harvesting different sources of knowledge, assimilating them, and envisioning strategic uses for them. In addition, organizational processing of knowledge could be cumber-

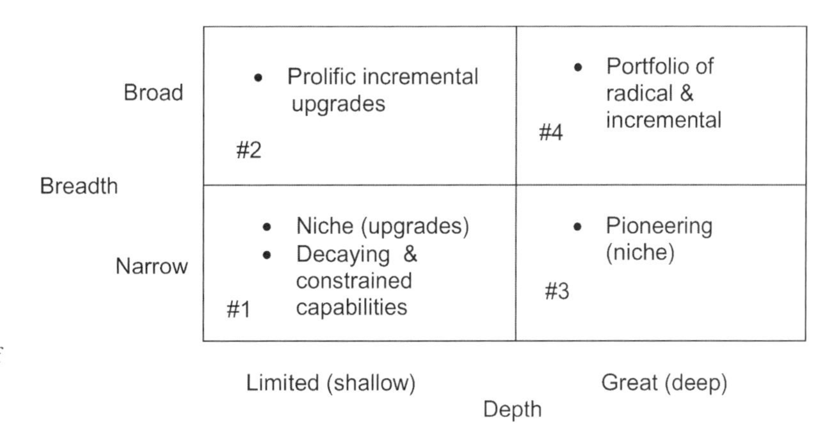

Figure 20.1. Depth versus breadth of absorptive capacity and technological innovation.

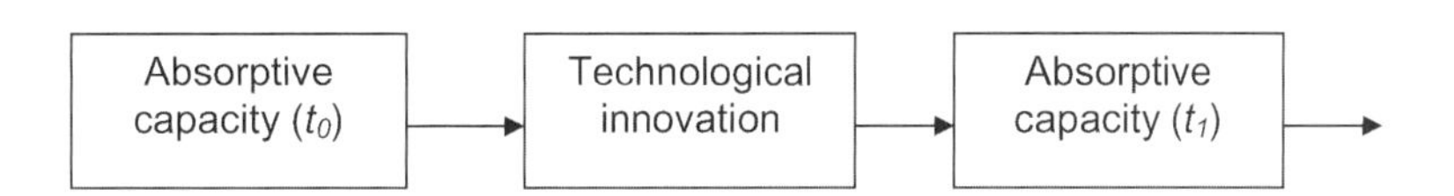

Figure 20.2. Temporal dialog between absorptive capacity and technological innovation.

some because of the tension that could arise from the need to assimilate and connect different technological fields and master the knowledge that prevails.

Absorptive capacity, learning, and building innovative capability

Thus far, we have suggested that absorptive capacity provides a key foundation for technological innovation. Of course, these innovations enrich the firm's skill base by promoting learning by doing and learning from failures. This knowledge could be used to upgrade the firm's absorptive capacity. This learning could guide the selection of potential innovative capabilities that can be built and the products the firm develops. It also enhances managers' awareness of the linkages among different technological fields and how they might converge or fragment. This awareness can serve to guide managers' decisions about how to develop, acquire, accumulate, and integrate different types of knowledge to build the firm's technological portfolio. Consequently, managers interested in building new capabilities can learn a great deal from analyzing their company's experiences with technological innovations, capturing this learning and integrating it with the firm's absorptive capacity. Figure 20.2, therefore, highlights the dynamism that characterizes the relationship between absorptive capacity and technological innovation. While some of these benefits might occur serendipitously, the firm is likely to gain more from deliberately cultivating the ongoing relationship between its absorptive capacity and technological innovation.

Conclusion

Absorptive capacity is important for enriching organizational learning, acquiring diverse knowledge from multiple sources, and using this knowledge to build new capabilities. Absorptive capacity is especially useful in stimulating technological innovation (Figure 20.1), a key foundation of competitive advantage. The knowledge gained and accumulated from technological innovation could enhance absorptive capacity (Figure 20.2), thus increasing the firm's strategic repertoire and its ability to develop new ways of competing and prospering in dynamic markets. Therefore, managing the ongoing relationship (dialog) between absorptive capacity and technological innovation is an onerous but rewarding managerial challenge.

References

Cohen, W. M., and Levinthal, M. D. A. (1990). "Absorptive capacity: A new perspective on learning and innovation." *Administrative Science Quarterly*, **35**(1), 128–152.

Cohen, W. M., and Levinthal, M. D. A. (1994). "Fortune favours the prepared firm." *Management Science*, **40**(2), 227–251.

Garud, R., and Nayyar, P. R. (1994). "Transformative capacity: Continual structuring by intertemporal technology transfer." *Strategic Management Journal*, **15**(5), 365–386.

Helfat, C. E., and Peteraf, M. A. (2003). "The dynamic resource-based view: Capability lifecycles." *Strategic Management Journal*, **24**(10), 997–1010.

Kim, L. (1998). "Crisis construction and organizational learning: Capability building in catching-up at Hyundai Motor." *Organization Science*, **9**(4), 506–521.

Lane, P. J., Koka, B., and Pathak, S. (2006). "The reification of absorptive capacity: A critical review and rejuvenation of the construct." *Academy of Management Review*, **31**(4), 833–863.

Lenox, M., and King, A. A. (2004). "Prospects for developing absorptive capacity through internal information provision." *Strategic Management Journal*, **25**(4), 331–345.

Miller, D. (2003). "An asymmetry-based view of advantage: Towards an attainable sustainability." *Strategic Management Journal*, **24**(10), 961–976.

Nelson, R., and Winter, S. (1982). *An Evolutionary Theory of Economic Change*. Boston, MA: The Belknap Press of Harvard University Press,

Todorova, G., and Durisin, B. (2007). "The concept and reconceptualization of absorptive capacity: Recognizing the value." *Academy of Management Review*, **32**(3), 774–786.

Zahra, S. (2009). "The virtuous cycle of discovery and creation of entrepreneurial opportunities." *Strategic Entrepreneurship Journal*, **2**(1), September, 243–257.

Zahra, S. A., and George, G. (2002). "Absorptive capacity: A review, reconceptualization, and extension." *Academy of Management Review*, **27**(2), 185–203.

Zahra, S., Sapienza, H., and Davidsson, P. (2006). "Entrepreneurship and dynamic capabilities: A review, model and research agenda." *Journal of Management Studies*, **43**(4), 917–955.

Zahra, S., van deVelde, E., and Larrañeta, B. (2007). "Knowledge conversion capability and the performance of corporate and university spin-offs." *Industrial & Corporate Change*, **16**, 569–608.

21

The Generations of R&D and Innovation Management

William L. Miller

4G Innovation LLC

The understanding of R&D and innovation can be greatly improved by reviewing human history, which has evolved through four ages: Hunter/Gatherer, Agriculture, Industrial, and Information (Grove, 1997). A fifth age appears to be emerging now as the Age of Innovation that will be dominant the 21st Century. Each age was enabled and driven by the evolution of a new radical core capability which was centered on a new integrated collection of knowledge, tools, technology, and processes. The Agricultural Age was enabled by the new radical capability for farming that began to emerge about 7000 BC without the benefit of "modern" R&D organizations and labs—"modern" R&D only appeared in the latter part of the 19th century. The Industrial Age was enabled by the new radical capability for manufacturing that began about 1770 in England—again without the benefit of "modern" R&D. However, the acceleration of the rate of innovation since 1900 has been largely due to the existence of "modern" R&D organizations and methods that have accelerated the development of technology. As an example, the Information Age was enabled by a radically different and improved capability for information processing that began with the first computer in 1946 and has rapidly evolved in economic value driven by the capability of information technology such as microprocessors and other electronic integrated circuits. The value (performance/cost) of these integrated circuits was driven by the exponential improvement described by Moore's Law (Moore, 1965).[1]

However, technology development only partly describes the rate of innovation. The variance in the rate of diffusion and adoption of innovation has been described by Everett Rogers (1962) and Jared Diamond (1997). In addition, Chris Argyris (1992) and Edgar Schein (1993) have described how the complexity of organizational learning affects innovation.

The recent beginning of the Age of Innovation has been recognized by various groups and documented in their reports. A good example is the report by the Council on Competitiveness under the National Innovation Initiative (NII) entitled *Innovation America* (2005) which says that

> "Innovation will be the single most important factor in determining America's success through the 21st Century."

In the 21st century, the scope of R&D has broadened and become more integrated with innovation management (Miller, 1995). An effective R&D organization now needs to consider how technology and the principles of knowledge management apply to organizations and how knowledge is linked to technology management and business processes (Miller, 1995). As an example, the need for service innovation has broadened the agenda for R&D. This broader scope is important for R&D to support a market "pull" rather than just provide a technology "push".

Research and development (R&D) within an organization is defined as the discovery of new knowledge which is generally focused differently during the phases of research and then development. During research, the discovery is focused on resolving the scientific and technical uncertainty to supply the new knowledge in the improved capability needed for competitive products, processes, services, markets, business models, and industry structures such as supply chains and distribution channels. During development, the discovery is focused on combining the new knowledge with tools, technology, and processes to build a capability to create new competitive value that serves market needs. The new value is generally created with combinations of improved products, processes, services, business models, and industry structures. Patents are generated as part of R&D to protect intellectual property

[1] See also *http://www.computerhistory.org/semiconductor/timeline/1965-Moore.html*

(IP). The capability produced by R&D has an associated architecture that determines the structure of product and business applications and how the components of systems fit together. An example of an innovation architecture is the Apple™ business with iTunes™, iPods™, and iPhones™.

Managing R&D properly requires the management of knowledge, technology, and innovation including strategy and investment, which includes resource allocation. In summary, the output of R&D to an organization is an acquired valuable new internal capability and architecture with intellectual property (IP) protection that enables the delivery of a new external capability and architecture to markets.

The result of R&D, when integrated into innovation management, becomes "technical progress" according to neoclassical growth theory, which was described in the 1950s by Robert Solow, winner of the 1987 Nobel Prize for Economics (Gordon and Wilcox, 1998). Solow calculated that about seven-eighths of the growth in output per worker in the United States over the period 1909 to 1957 was attributable to "technical progress". Earlier work by Joseph Schumpeter (1950) helped the emergence of new economic growth theory when he described innovation as waves of "creative destruction" or radical innovation assisted by entrepreneurs. Growth theory has progressed since Solow's work with a new understanding of capability that includes both technology and knowledge. Paul Romer (Warsh, 2006) has been the main author of insights focused on the economics of knowledge that have formed the core of the new theory, called endogenous growth theory. Recently, William Baumol (2002) has also contributed to economic growth theory and, like Schumpeter, has emphasized the importance of entrepreneurship to drive radical innovation. Robert Atkinson (2004) has also written about how radical technological innovation has been the direct cause of the major economic periods in American history. Both of these authors, however, leave gaps in the analysis of how the management of R&D and innovation has evolved.

Since 1900, multiple generations of R&D and innovation management have evolved and coexisted as widely accepted theory and best practice. Competitiveness and economic development have been determined by the application of these generations. These generations can be identified and classified by their superior dominant capabilities and dominant architectures that become nearly universally accepted because they provide a competitive advantage during a period of time—the generation.

A capability in a generation becomes dominant when it evolves from the ability to produce only largely random outcomes to the ability to produce largely predictable, efficient outcomes with superior economic advantages. The evolution of a capability inside a generation follows a pattern of four stages that begins with the creation of a new candidate capability and a corresponding architecture (structural design rules for application of the capability). That is followed in the next stage by the creation and application of a new means to measure performance of the new capability and architecture. In the third stage, new methods are created and applied for the efficient production/delivery of competitive advantage and value using the new capability. When the capability can predictably deliver economic results it rapidly becomes a dominant capability at the end of the third stage. What comes last is the creation of different types and levels of management that yield incremental improvement of the dominant capability and architecture and produce applications that form new value chains, value networks, and innovation ecosystems.

R&D evolves with incremental changes inside a generation and then jumps to a new generation with radical enhancements. The need for a new generation becomes apparent when problems persist and can't be solved with the current generation. As an example, the "dry R&D pipeline" in the pharmaceutical industry that has persisted for more than the last decade is an example of the failure to improve the "how" of R&D.

Each new generation of innovation management has coexisted with previous generations. These generations have been characterized by the evolution of management capability such as strategic planning; R&D including technology management; marketing, management of information systems, operations including manufacturing, sales, and logistics/distribution; human resource management; and finance.

The four generations have also been characterized by the management of architecture including organizational structure, product/service architecture, business models with partnerships, and industry structure. Industry structure is understood as more than a static structure of value chains or networks, but as a collection of dynamically interacting enterprises in an innovation ecosystem.

First-generation R&D (1G)

1G began about 1900 and evolved from 1900 to 1940 with the creation of the first industrial R&D labs, and the creation of new capabilities in other separate functional departments inside the emerging modern industrial corporation as defined by Alfred Sloan (1990). A core new capability in 1G was the capability to do low-cost "mass production" in manufacturing with a new architecture in the factory—interchangeable parts combined with a capability for measurement; the micrometer was developed by Joseph Whitworth and patented by Jean Laurant Palmer in 1848[2]—and the moving assembly line. Com-

[2] See *http://www.shef.ac.uk/hawley/project/research/micrometers*

bined, the new functional departments formed the dominant architecture for the modern industrial corporation. The economy in the United States from 1900 to 1940 was driven by technological innovation and emerged as a "factory-based", industrial economy (Atkinson, 2004). Major new infrastructures such as electricity and telephone communications were introduced.

The main objective of 1G R&D was to create an innovation capability inside then emerging modern industrial corporations that could be driven by science and technology. The core functional departments that enable and drive innovation such as R&D and marketing were created in 1G. The first industrial R&D laboratories were also created at companies such as GE. Thomas Edison's Menlo Park R&D lab near Edison, New Jersey (created in 1876) is the best example of a first-generation R&D lab with long laboratory desks or benches and shelves stocked with chemicals, materials, and test instruments for conducting experiments. Menlo Park produced many important technological innovations such as the electric light bulb, the microphone for the telephone, the phonograph, and a system based on direct current (DC) rather than alternating current (AC) for producing and distributing electricity.

The strengths of 1G included the introduction of essential functional disciplines—such as R&D, marketing, and operations in the modern model of the corporation—and that entrepreneurial activity inside the corporation was led by higher level corporate executives. Innovation leadership in 1G was not just focused on the management of R&D labs as science and technology "push". Entrepreneurs such as Edison and Ford provided the leadership for innovation and R&D with market "pull". They acted as the innovation managers who overcame internal and external political opposition and defined the important problems that guided R&D. Entrepreneurial activity gave 1G a capability to create radical innovation, although 1G could not manage radical innovation with predicable rather than random success.

The main challenge in 1G was that the innovation outcome, even for incremental innovation, was still largely random since R&D labs were being driven by the random, experimental discovery of new technologies. Other challenges were that the speed and efficiency of new-product development was relatively slow and poor, since no structured process existed that supported projects and teams. And since the development of technology was largely dependent on internal R&D, the scope of problems that could be solved was limited by internal capability and budgets. There were few options to acquire external technology.

The management of 1G innovation was not very effective, since there was limited methodology for coordinating the separate departments of R&D, marketing, manufacturing, and finance and no well-defined stage-gate process for managing new-product development (NPD) projects staffed by multidisciplinary teams. Product development was inefficient as activity slowly flowed from department to department. As a result of poor coordination, product designs could be created that couldn't be efficiently manufactured, readily sold, or serviced.

1G was replaced by 2G to substantially improve the predictability, speed, cost, and scope of innovation capability within a corporation.

Second-generation R&D and innovation management (2G)

2G began about 1940 and evolved from 1940 to 1975. The efficacy of 2G was demonstrated in World War II by the Manhattan Project which only required 3 years to develop the first atomic bomb and many other rapid product development projects. As the chief science advisor to the U.S. federal government in WWII, Vannevar Bush had great success with a new model for innovation that connected university research with industrial development. That success led to the creation of the National Science Foundation after WWII and the adoption of the new dominant capability in 2G R&D as the management of innovation projects with teams supported by the extension of the internal industrial R&D lab to include collaboration with university basic and applied research. The economy in the United States from 1940 to 1975 was also driven by both market and technological innovation and emerged as a "corporate-based, mass production economy" (Atkinson, 2004).

The primary objective of the second generation or "2G" was to improve the predictability, speed, cost, and scope of innovation capability within a corporation. A secondary objective of 2G was to improve the capability of R&D by doing collaborative research with external organizations such as universities.

The new dominant capability in 2G was the ability to do rapid, complex project management for product development with multidisciplinary teams. A stage-gate process for new-product development (NPD) forms the core of 2G capability. A good explanation of the NPD process is described in a popular book by Robert Cooper (2001). However, 2G is focused on the development of individual products and has limited ability to develop and manage product families. The new model for 2G was tested and proven by the effectiveness of the Manhattan Project and many other rapid product development projects in WWII.

The management of 2G innovation focused on project management for new-product development (NPD) using teams staffed with both internal and external resources. NPD projects were managed with a stage-gate process with NPD milestones. In addition, R&D had evolved from the

1G R&D lab that required direction by a strong entrepreneur to multidisciplinary, distributed projects in R&D directed by complex requirements and also managed with a stage-gate process with technology milestones. Projects were staffed by teams from government, industry, and academia organizations and the team performed collaborative R&D. The understanding and management of market behavior was enhanced in 2G, and market segmentation became routine practice in 2G. Brand management also emerged with hierarchical levels that included the corporate brand at the top and product brands at the bottom.

The strengths of 2G included the speed and predictability of new-product development and the enhanced capability to leverage resources from external organizations including universities to supplement internal R&D.

Clearly, 2G had many new enhancements over 1G while it adopted most of 1G's best practices. But 2G also had serious limitations in innovation management.

The challenges of 2G were focused on the lack of integration with strategic planning and the inefficiency and high cost of NPD required for multiple products in a family. Also, a lack of senior executive (CEO or Chief Innovation Officer) leadership for innovation including entrepreneurship existed in 2G and was demonstrated by the delegation of innovation management to steering committees that govern projects. The formal discipline to manage NPD projects, which is a strength in 2G, also became a challenge and barrier to radical innovation that included business innovation from internal capability development. This focus in 2G on planning and executing projects with predicable results, with projects managed by stage-gate processes with predictable milestones that don't require either a complex learning to achieve or a re-planning (causing a spiral process) to incorporate the learning, created a nearly impenetrable barrier to radical innovation (Miller, 1995).

However, even with a focus in 2G on incremental innovation as NPD, the probability of success for new products and services in 2G was only about 20% and about 3,000 ideas were required to generate one successful new product (Womack, Jones, and Roos, 1991).

The second generation of innovation management evolved concurrently with the Information Age, which began with the first computer in 1946. As the new technology of electronics evolved in the Information Age, R&D had to keep pace and manage the applications of new technology in NPD and internal operations. Products were no longer mainly mechanical. Products and chemical or manufacturing processes had new control systems enabled by electronics.

The development of capabilities that will be adopted in future generations frequently occurred during an earlier generation. Even though Toyota had developed the Toyota Production System (TPS) in the time period of 2G, the rest of the world didn't understand its advantages as a new internal dominant capability until MIT in 1990 published the results of its study of the TPS in a book entitled *The Machine that Changed the World* (Stevens and Burley, 1997).

2G was replaced by 3G mainly to improve planning for innovation and R&D and improve execution with overlapped development of product and technology lifecycles, portfolio management, and cost reduction with platforms. The oil crisis in the 1970s triggered the need for an improvement in planning, and the proliferation of market segments caused the need for a reduction in innovation cost since each new segment to be served required new, specific products. Increased competitiveness in markets caused a need for more efficient and faster R&D.

Third-generation R&D and innovation management (3G)

3G began about 1975 and evolved to 2000. The beginning of 3G was centered on the capability to develop new products or services much faster and at much lower development cost than 2G. Eventually, the new capability in development was combined with more efficient operations supported by supply chain management and lean manufacturing.

From 1975 to 2000, the economy in the United States grew with new services in the service sector faster than it did with new products in the manufacturing sector. Progressive organizations such as GE targeted the same opportunity in services and adopted hybrid business models that combined manufacturing with services. Another example is GM with OnStar™. Therefore, 3G managed business innovation in addition to product and service innovation.

The primary objective of the third generation or 3G was to improve the overall financial performance and competitiveness of the corporation and improvement was directed to the strategic planning of innovation and R&D with enhancements such as roadmaps, portfolios, scenarios, new marketing methodologies including product lifecycle planning. A secondary objective was to improve the cost of innovation with product platforms and new digital development tools. Another secondary objective was to improve the performance of operations with new radical capabilities such as lean manufacturing.

The dominant new capability in 3G was faster and lower cost product development combined with lean operations , all driven by technology and product lifecycle planning based on multi-generational, platform-based, product families. Markets had fragmented into many segments and a broadened family of products was now required to compete. A new 3G stage-gate process for product/

service platform development was developed to feed the 2G NPD process and was a primary enhancement in 3G innovation management. By basing 3G product development on product/service platforms, both the time and cost of development of a global product family were reduced by sharing platforms across all of the products/services in the family. To better enable rapid change at lower risk, 3G strategic planning was enhanced with scenarios, the impact of dominant designs in products/services or processes as described by Utterback (1994), and entrepreneurship supported by corporate and external venture capital. 3G, however, was very limited in its ability to effectively manage radical innovation that included business innovation. Only a few established companies such as GE and GM successfully performed business innovation that combined products with services.

3G R&D witnessed the adoption of new digital design and engineering tools such as CAD/CAE systems and the use of these tools to support system integration, portfolio management of technology, technology roadmaps, and technology platforms (Roussel, Saad, and Erickson, 1991). The capability for technology development in R&D was extended by technology acquisition from external sources including technology licensing, consortia, and acquisition of small companies in order to supplement the development of technology already coming from internal R&D laboratories and university-funded research.

William Baumol (2004) recognized that during the period of 3G, nearly three-quarters of the R&D money in the U.S. was spent by large corporations, but most of the economic growth came from small companies including startups that can't afford to spend much at all on R&D. Changes in R&D and innovation management should enable large companies to grow faster and small companies to grow even faster by adopting new methods that include being able to access R&D at an affordable cost.

In 3G marketing, brand management evolved with the industry restructuring to permit embedded brands such as "Intel Inside" to be visible to the consumer. Also in 3G marketing, Geoffrey Moore (1991) extended the diffusion model of Everett Rogers (1962) into a strategy to guide the repetitive application of the 2G NPD process to manage the rapid evolution of the new products driven by a new technology to ensure adoption by different types of customers. The implementation of this product evolution resulted in changes in 3G R&D and innovation management.

The strengths of 3G innovation management were in the new planning and execution methods that improved

(a) quality measured by new standards such as ISO9000 that were driven by customer satisfaction;
(b) financial performance driven by methods such as the "balanced scorecard" and measured by economic value added (EVA);
(c) core business processes with new methods such as reengineering.

Business processes included planning, new-product development (NPD), platform development, and operations that included sales. Selling was improved with problem-based methods that created a customer dialog on their situations–problems–implications–needs (SPIN). Lean manufacturing in operations required internal capability development and new architectures for supply chain management and quality management with six-sigma methods. NPD improved the understanding of customer requirements with new methods such as the "voice of the customer" and "quality function deployment (QFD)".

The scenarios for multi-generational product lifecycle planning based on platforms that were introduced in strategic planning were supported with new capabilities in 3G R&D such as technology road mapping and portfolio management of technologies (Roussel, Saad, and Erickson, 1991). The goal in 3G was to align corporate strategy, the planning for new products/services (NPD), and the planning for new technology coming from either internal R&D or external sources.

As competitiveness increased in the period 1975 to 2000, in response, corporations and industries introduced the development of platform-based 3G R&D and 3G product development beginning in the mid-1980s to late-1980s. For example, the computer industry adopted platform-based architectures developed by Microsoft and Intel as dominant designs. That adoption led to the restructuring of the industry from vertical to horizontal with most of the profits being realized no longer at the top of the value chain for the assembly of computers.

One of the most significant enhancements in 3G innovation management was the introduction of support by venture capitalists for external entrepreneurship to generate radical innovation. Most corporations practicing 3G focus their investments on incremental innovation.

3G discovered and adopted many other new management concepts and practices such as focusing innovation as incremental innovation on core competencies. But 3G didn't provide a solution to the evolution of core competencies. For example, Kodak had a core competency in the chemistry of photography and had problems effectively managing the transition to digital photography.

The primary challenge of 3G innovation and R&D management is centered on the inability in 3G to create sustainable, profitable corporate business growth that is driven by radical innovation (Leifer *et al.*, 2000). There are several reasons for this inability and a solution to this problem was the main reason for the introduction of 4G. The lack of effective senior executive (CEO, etc.) leadership for innovation is still a major limiting factor in

3G, except when innovation exists as part of external entrepreneurship.

Here are some examples of the failure of 3G as practiced by senior management to manage radical innovation:

- GM R&D had developed a minivan in 1978, 6 years before Chrysler commercialized its minivan, but the GM Board of Directors killed it as a radical innovation without a tangible market demand.
- AT&T invented the cellphone and hired McKinsey in 1985 to analyze the market opportunity. The resulting low forecast killed the cellphone business within AT&T until 1994 when AT&T acquired McCaw Communications, the largest cellular carrier at the time, for $11.5 billion.

Even with 3G capability, the probability of success for new products and services is no better than about 20% as it is with 2G. What makes the difference in performance between 2G and 3G is mainly the cost effectiveness of platform management in 3G. Platforms can save up to 90% of the cost of product development (Meyer and Lehnerd, 1997).

There are several core reasons why 3G is being replaced by 4G.

(1) The weak correlation between R&D and innovation or R&D and economic development

The correlation is weak between 3G R&D spending and successful innovation, which includes the commercialization of new products and services that produce economic value added (EVA) and business growth (Baumol, 2004). Established companies have had major problems with managing innovation for sustained organic growth, especially since radical innovation is required.. Startup companies with CEOs as entrepreneurs generally perform better with radical innovation than established companies. This performance can be partly explained by the leadership activity of a CEO in a startup, who also frequently acts as a Chief Innovation Officer and does not delegate the leadership of innovation to project teams governed by steering committees. The CEO, as an entrepreneur, is more likely to learn and revise plans that result in a spiral innovation process similar to a 4G process rather than executing a 2G or 3G stage-gate process according to frozen plans with milestones governed by steering committees.

(2) New market/business opportunity assessment

The political, emotional, and complex nature of economic rational decision making for both customers and suppliers is difficult in the context of innovation. Barriers to innovation include the risk, uncertainty, and politics associated with the lack of user experience with "new to the world" products/services, business models, and industry structures.

The 3G practice that listens to the "voice of the customer", as collected in surveys or in focus groups, only works when the customer has had adequate experience using similar products or services. An example of the issue is the difficulty to adequately estimate and validate the value of an innovation in a future market or market segment that has not yet experienced such a solution to its unmet needs. 4G practices a new methodology as part of its innovation process that includes experimentation by customers and analysis of customers with new analytical methods fed by new observational methods such as video ethnography to identify hidden unmet needs. The experimentation enables customers to experience new innovations by using them in actual situations. With a process for innovation management that enables experimentation in new capabilities combined with analytical observation, 4G manages this core issue of opportunity assessment and overcomes the limitations of traditional market research.

3G financial practice relies on discounted cash flow analysis and net present value which, when combined with false assumptions about the longevity of a competitive advantage, creates illusionary advantages for doing incremental innovation in existing product/service lines rather than radical innovation in new-product markets. These illusions create barriers to innovation that can be overcome with a new spiral innovation process as defined in the next generation of R&D and innovation management, which is the fourth generation (Miller and Morris, 1998).

(3) Management of radical innovation

Strategy has the goal of guiding resource allocation, but 3G strategy did not adequately specify what process, organization, and leadership should be used to manage R&D and innovation. Strategy needs to allocate and guide the investment for the next generation of internal competitive capabilities and the next generation of competitive (candidates for dominant) capabilities and architectures (designs) for external customers. 4G (Miller and Morris, 1998) offers a new strategic planning methodology and a new business process for innovation—including both incremental and radical innovation—that closes the gap between business strategy, public policy, and the implementation of radical innovation for economic and social development.

Fourth-generation R&D and innovation management (4G)

4G began about 1990, but its evolution and adoption has accelerated since 2000 based on the need for effective management of radical innovation to sustain the economic growth of companies and improve both the competitiveness of nations, local regions, and companies. 4G accelerates the growth in both large and small companies and improves the probability of a successful commercialization.

4G has a unique, new spiral process for managing innovation, which has a larger scope than either the 2G process used for new-product development (NPD) or the 3G process used for platform development. The spiral process integrates knowledge management to accelerate the rate of mutual learning between customers and suppliers to mutually discover what's needed and what's possible as a new capability. The 4G innovation process, which discovers unmet needs and feasible capabilities, feeds both the 2G NPD process and the 3G process of platform development.

Whereas 1G, 2G, and 3G capabilities for innovation management were focused on incremental improvements in products, services, and processes, 4G adds the capability to radically improve not only products, services, and processes but also business models, industry structures, and internal capabilities such as R&D, marketing, and manufacturing. At the core of its process for managing innovation, 4G R&D has a new business process for capability and architecture development that also discovers and targets new unmet needs.

In 4G, capability is defined as people with knowledge, tools, technology, and process(es). And in 4G, architecture is defined as the structure of the capability.

In 4G, value is defined as either internally applied or externally delivered new capability and architecture that meet competitive economic and emotive needs. These needs typically fall into several categories of value with specific attributes. The categories are

- health, safety and comfort protection;
- productivity improvements;
- lifecycle cost advantages;
- technology integration or compatibility to "fit";
- emotional design appeal.

"Value innovation" is simply the creation of a unique profile of specific 4G value attributes for a new capability.

The fourth generation (4G) of R&D and innovation management was introduced beginning in 1990 with the primary objective to effectively manage both incremental and radical innovation for sustained, organic business growth. Secondary objectives were to use capital much more efficiently to produce economic value added (EVA) by improving the probability for the success of new products, services, new business models, and new industry structures. Specifically, the objective for 4G was to enable radical innovation faster and with much leaner resources at a much higher yield than 3G.

The management of 4G operates with guidance from 12 new principles and practices (Miller, 2001). As an example of one of these principles, 4G leadership comes from new types of people with both broad and deep skills who have become "innovators" to lead innovation projects rather than people who might be specialists from either marketing, R&D, finance, or engineering and who become project managers. A 4G innovation manager is focused on identifying and solving core problems with competitive innovation as "technology pull" rather than relying on R&D to produce "technology push".

In 4G management practice, problems as opportunities are defined by the "jobs" (such as use cases) that need to be done by people in the context of a work process with their tools. The people can be either at a customer's or inside the corporation. Jobs are then defined in terms of the "as-is" capability and the "to-be" required new capability. Capability is then analyzed in terms of a value proposition to be delivered to either internal or external customers with certain key value attributes and levels. Another example of 4G is that management uses a new innovation process and new innovation tools to discover and understand unmet needs and develop candidate solutions. The degree of integration of R&D with innovation management and the type of leadership of both R&D and innovation are major factors that differentiate 4G from earlier generations. Another example of 4G leadership is that a chief innovation officer is appointed within a corporation.

Compared with 3G, the strengths of 4G are centered in the effective management of radical innovation which is guided by 12 principles and practices that enable faster identification of unmet needs and faster development of new capabilities and architectures with sustainable competitive advantages to serve those needs. The management of radical innovation using 4G methods requires fewer resources and results in a much higher probability of commercial success than 3G methods.

The 12 principles and practices of 4G are described as follows:

(1) First principle

A broader definition of innovation with a larger scope is required to be competitive in a globally interconnected economy and to effectively guide organizations that include industrial corporations, governments, foundations, and universities. Radical innovation is required in all companies in addition to incremental innovation. Innovation needs to occur in the internal capabilities within a business, not just in new products/services or new businesses.

Business organizations now appear to be threatened with "death" (such as bankruptcy or forced acquisition) unless they renew themselves with radical innovation within 15 years (Foster and Kaplan, 2001).

The 5:1 salary gap between the U.S. and China/India has caused the rapid offshoring of many high-paying professional jobs including those in engineering.[3] Universities in the U.S. have permitted undergraduate and graduate education to become a global commodity even in technical fields such as engineering. New upgraded, competitive capabilities are required in U.S. engineering practices, or engineering jobs may disappear in the U.S. within 15 years.

(2) Second principle
Innovation management that creates new sustainable value depends on lifecycle management of basic and competitive capabilities and the structures of these capabilities as architectures.

Value is created when a new competitive capability is externally delivered to customers or users by a supplier using an internal capability with a competitive advantage.

(3) Third principle
Competitive capabilities are built from layered "capability stacks" that form the building blocks of value chains, value networks, and other industry structures. Dominant architectures (designs) that form the structure of capability have three parts:

- capability "stacks" that include people with knowledge, tools, technology, and processes;
- business models with partners;
- industry/market structures.

(4) Fourth principle
Capability and architecture are the core building blocks of economic value and competitiveness and can be mapped onto enhanced financial-accounting models that include both tangible and intangible assets (Miller and Morris, 1998).

(5) Fifth principle
Markets have a new architecture with dual distribution channels—one channel for sales transactions that delivers a combination of products and/or services, and another channel for knowledge that must flow bi-directionally between the user of a product and/or service at the point of use back to the suppliers.

(6) Sixth principle
A new spiral business process is required for competitive capability and architecture development (Miller, 2001, 2006; Miller and Morris, 1998).

[3] *National Science Foundation Workshop: 5XME*, held May 10–11, 2007 and organized by Professor Galip Ulsoy, University of Michigan.

Professional education offered by universities and consultants needs to support the teaching of this process in actual projects with customers and suppliers. This process drives innovation as business development that precedes the traditional processes for product or service development or platform development. Analysis of stakeholder needs is performed to identify problems and root causes. The process guides the creation of scenarios for solutions to be acquired, developed, and tested. Solutions are developed in "innovation labs" with technology partners such as universities and then tested in "application labs" with lead users who are customers. Intellectual property protection is obtained in "innovation labs" before customer testing begins. The process guides and measures experience-based learning as rigorous analytical experiments. In these experiments, R&D and innovation management, which includes both knowledge and technology management, operate to guide the work of innovation groups at suppliers and customers. New methods for collecting data by observation such as video ethnography are used to feed new analytical tools to help accelerate the valid discovery of unmet user needs and test solutions. These methods supplement traditional market research such as the "voice of the customer" and other market tools to rank the priority of attributes in a value proposition such as conjoint analysis.

An example of the application of the 4G process is shown in Figure 21.1 (Miller, 2006). This process was applied at ETAS, a supplier of vehicle development tools to the global automotive industry, to create a new capability called immune system engineering (ImSE), which improves the competitiveness of corporations in the automotive industry by reducing vehicle development time while lowering the cost of development and reducing the cost of warranty repair. Vehicles with an immune system have the capability for self-diagnosis and self-repair. As part of 4G, ImSE enables the creation of the next level of competitive capability in engineering, manufacturing, and service operations that is similar to what lean manufacturing accomplished as part of 3G.

(7) Seventh principle
Strategic planning needs to manage several related dimensions of technology as part of capability planning including product/service development, tools/processes, lifecycles, innovation roadmaps, technology portfolios, internal R&D and external acquisitions (open innovation), and the discovery of unmet needs for radical innovation in existing and emerging markets. New capabilities and architectures (designs) need to coexist in a value proposition with the current capabilities that exist in a market or inside an organization and need to enable a migration from the current state to a future, more competitive state. In 4G an innovation roadmap describes the pathways for the migration and shows the contributions of

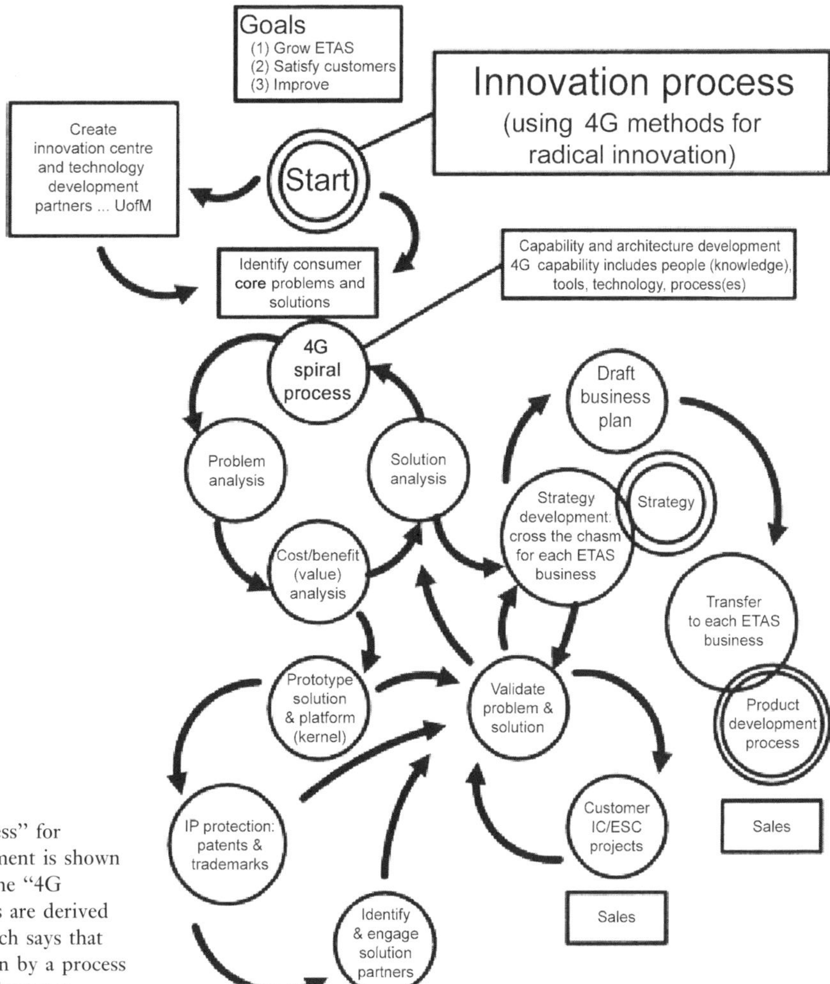

Figure 21.1. The "4G spiral process" for capability and architecture development is shown to be embedded in an example of the "4G innovation process". Both processes are derived from the sixth principle in 4G, which says that innovation is a process that is driven by a process for capability and architecture development.

education and research to learning which makes progress on the pathways.

(8) Eighth principle

New value propositions need to be defined to target solutions for problems as opportunities with the value of the solutions defined with several value attributes for multiple stakeholders.

(9) Ninth principle

New types of "T-shaped" innovation leaders who have greater breath and depth of capability need to be created and given career paths to senior management.

(10) Tenth principle

New types of R&D labs as "innovation labs" and "application labs" need to be created to further accelerate the rate of innovation and supplement incubators and accelerators.

These labs would operate in conjunction with a new professional education model and organization. These labs create more effective environments to support the spiral innovation process. As an example, Intel applied 4G to revise its management of R&D and the types of university research projects that receive funding. Intel created "lablets" which are a form of 4G's innovation labs, which are labs focused on solving a specific problem per "lablet". The "lablets" are located next to a university campus and staffed by a combination of industrial and university subject matter experts. Application labs are then also formed to test prototypes with customers in support of the spiral process.

In 1997, *Pasteur's Quadrant* (Stokes, 1997) was published as a study of limitations in 3G R&D. The author, Donald E. Stokes, concluded that the linear R&D process

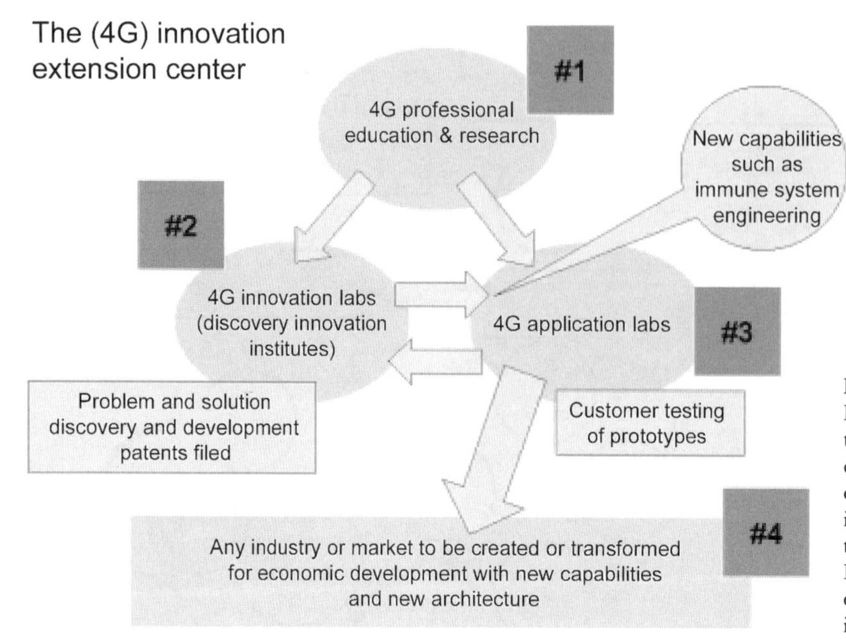

Figure 21.2. The 4G Innovation Extension Center (IEC) is part of the twelfth principle of 4G which creates a new professional educational model to support innovation that supplements the traditional university model. 4G IECs should be created either inside or outside corporations but not inside universities.

practiced in 2G and 3G R&D was best supplemented with a spiral process directed by "use-inspired R&D". The new 4G spiral process, operating in innovation and application labs, implements Stokes' recommendations.

(11) Eleventh principle

A new corporate organization is needed with a chief innovation officer (CINO) who is the process manager for the spiral innovation process that develops new strategic capability and architecture.

The chief technology officer would report to the CINO as would the head of business development and exploratory marketing. The CINO manages all radical innovation with help from an innovation group.

(12) Twelfth principle

Innovation operates in an organizational ecology with groups of partners who compete with different capabilities and architectures. The ecology must be managed.

See Figure 21.2 which describes the four parts of an innovation ecology supported by a 4G innovation extension center (IEC), which is similar to agriculture extension centers and manufacturing extension centers. The 4G IEC can supplement and guide innovation incubators and accelerators.

Developing a new industry or market such as alternative energy requires a coordinated approach to developing an "innovation ecology". The ecology is an extended enterprise with partners (including educational organizations, industrial companies, and government) who need to manage innovation as an ecosystem.

The challenges of 4G are centered on the classic problem of innovation itself—getting skeptics to adopt 4G as an innovation. And the solution follows the patterns of innovation in human history. Those that are willing to experiment and see results will adopt 4G much faster than those who are unwilling to experiment.

Summary

Figure 21.3 shows a comparison of the four generations of R&D. Entrepreneurship had been introduced in 1G, but

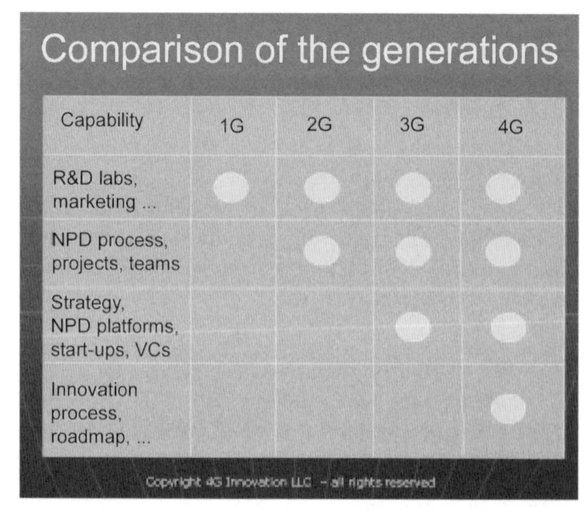
Comparison of the generations

Capability	1G	2G	3G	4G
R&D labs, marketing ...	●	●	●	●
NPD process, projects, teams		●	●	●
Strategy, NPD platforms, start-ups, VCs			●	●
Innovation process, roadmap, ...				●

Figure 21.3. The four generations of R&D and innovation management are compared using some of the important attributes as capabilities.

had been suppressed in 2G with the focus on making improvements in current product and services. One of the most significant was the 3G introduction of formal support for external and internal entrepreneurship such as sources of investment from venture capitalists (VCs). But entrepreneurships were not sufficient even in 3G to enable the effective management of radical innovation. 4G uniquely can effectively manage either incremental or radical innovation with its new innovation process and planning roadmap. Each newer generation builds on the important capabilities of earlier generations and deletes the practices that cause limitations and poor performance.

References

Argyris, C. (1992). *On Organizational Learning*. Oxford, U.K.: Blackwell.

Atkinson, R. (2004). *The Past and Future of America's Economy: Long Waves of Innovation that Power Cycles of Growth*. Northampton, MA: Edward Elgar Publishing.

Baumol, W. (2002). *The Free-Market Innovation Machine*. Princeton, NJ: Princeton Univeristy Press.

Baumol, W. (2004). *Education for Innovation: Entrepreneurial Breakthroughs vs. Corporate Incremental Improvements* (Working Paper 10578). Cambridge, MA: National Bureau of Economic Research (NBER); *http://www.nber.org/papers/w10578*

Cooper, R. (2001). *Winning at New Products*. New York: Basic Books.

Council on Competitiveness. (2005). "Innovate America" (a report); *http://www.innovateamerica.org/webscr/report.asp*

Diamond, J. (1997). *Guns, Germs, and Steel*. New York: W. W. Norton.

Foster, R., and Kaplan, S. (2001). *Creative Destruction: Why Companies that Are Built to Last Underperform the Market—and How to Successfully Transform Them*. New York: Doubleday.

Gordon, G., and Wilcox, J. (1998). *Macroeconomics* (Seventh Edition). Reading, MA: Addison-Wesley.

Grove, N. (1997). *Atlas of World History*. Washington, DC: National Geographic Society.

Leifer, R., McDermott, C., O'Connor, G. , Peters, L., Rice, M., and Veryzer, R. (2000). *Radical Innovation: How Mature Companies Can Outsmart Upstarts*. Boston, MA: Harvard Business School Press.

Meyer, M., and Lehnerd, A. (1997). *The Power of Product Platforms*. New York: The Free Press.

Miller, W. (1995). "A broader mission for R&D." *Research Technology Management*, November/December, 21–36.

Miller, W. (2001). "Innovation for business growth." *Research Technology Management*, September/October, 26–41.

Miller, W. (2006). "Innovation rules!" *Research Technology Management*, March/April, 8-13.

Miller, W. and Morris, L. (1998). *Fourth Generation R&D: Managing Knowledge, Technology and Innovation*. New York: John Wiley & Sons.

Moore, G. (1965). "Cramming more components onto integrated circuits." *Electronics*, **38**(8), April 19.

Moore, G. (1991). *Crossing The Chasm*. New York: HarperCollins.

Rogers, E. (1962). *The Diffusion of Innovations*. New York: The Free Press.

Roussel, P., Saad, K., and Erickson, T. (1991). *Third Generation R&D*. Boston, MA: Arthur D. Little.

Schein, E. (1993). "How can organizations learn faster? The challenge of entering the green room." *Sloan Management Review*, Winter.

Schumpeter, J. (1950). *Capitalism, Socialism, and Democracy* (Third Edition). New York: Harper & Row.

Sloan, A. (1990). *My Years with General Motors*. New York: Doubleday.

Stevens, G. A., and Burley, J. (1997). "3000 raw ideas = 1 commercial success." *Research Technology Management*, May/June, 16–27.

Stokes, D. (1997). *Pasteur's Quadrant: Basic Science and Technological Innovation*. Washington, DC: Brookings Institution Press.

Utterback, J. (1994). *Mastering the Dynamics of Innovation*. Boston, MA: Harvard Business School Press.

Warsh, D. (2006). *Knowledge and the Wealth of Nations: A Story of Economic Discovery*. New York: W. W. Norton.

Womack, J., Jones, D., and Roos, D. (1991). *The Machine that Changed the World*. New York: HarperCollins.

22

Dual Career Ladders in Organizations

Christy H. Weer and Jeffrey H. Greenhaus*[†]

*Radford University and [†]Drexel University

Organizations with dual career ladders provide technical employees with alternative opportunities for organizational advancement. Technical professionals may choose to advance through one of two ladders—the management ladder or the technical ladder. Employees who wish to advance into management positions shift their focus from technical application to administration and leadership responsibilities. Those who choose to advance up the technical ladder continue to focus their efforts on increasingly difficult and important technical tasks. Ideally, both ladders provide similar career advancement opportunities and organizational rewards.

In essence, the goal of the dual-ladder system is to provide technical professionals with options for career development without limiting their career growth or income potential. The system is based on the notion that not all technical professionals have the desire (or the skills) to move into management, but want the status and rewards typically associated with management positions.

Dual-ladder career alternatives were designed in the 1970s in response to career development concerns among technical professionals. Employee attitude surveys indicated that many technical professionals were disappointed by a lack of career guidance and development. With little alternative for advancement, top technical performers often chose to be promoted out of technical jobs and into management positions where they were neither well suited for the positions nor satisfied with their jobs.

Raelin (1987) provides an example of a typical dual-ladder promotion system for a general engineering career path. Each rung up the ladder, whether on the managerial side (from department head to R&D manager to vice president of R&D) or the technical side (from engineering associate to senior engineering associate to principal engineering associate), requires an increasing degree of expertise and responsibility (Raelin, 1987). It has been suggested that specific sets of skills, personality traits, and interests are necessary for success within each of these ladders (Toto, 1988). Employees seeking to climb the managerial ladder should enjoy working with people and be skillful in motivating, directing, and leading others. Moreover, those seeking managerial jobs should be comfortable being removed from the day-to-day technical work, thus having an impact on work only through others. On the other hand, employees seeking to successfully climb the technical ladder should enjoy working with materials rather than with people, and gain gratification from working on their own unique and creative projects (Toto, 1988).

It has been suggested that providing technical professionals with career advancement alternatives can have beneficial consequences for employees as well as organizations. Individuals who work for organizations with dual-ladder systems should experience increased job satisfaction and motivation as they are able to pursue the career path that best suits their skills, talents, and desires. Moreover, organizations that successfully implement dual-ladder systems should be better able to attract and retain key technical professionals (Goldstein, 1988; Mainiero and Upham, 1986).

Despite the potential benefits, dual-ladder systems have undergone significant criticism. It has been said that in some organizations dual ladders may indeed be available; however, they are often less than equal in status and rewards (Allen and Katz, 1986). Management positions tend to incur greater power, prestige, and salary. Moreover, advancing on the technical side often takes significantly longer than advancing through management, and may indeed be seen as a "loyalty prize" rather than true career advancement (Allen and Katz, 1992; Goldstein, 1988). Thus, many technically devoted employees ultimately shift to the management side or leave the organization altogether. Proponents suggest that in order for the dual-ladder system to be successful, organizations must be committed to maintaining consistent career advancement opportunities for both ladders.

Despite the criticisms, companies that have successfully implemented the dual-ladder career development system have been considered among the best companies to work for by technical professionals. It is these companies that are able to recruit and retain top technical contributors.

Bibliography

Allen, T. J., and Katz, R. (1988). "The dual ladder: Motivational solution or managerial delusion?" In R. Katz (Ed.), *Managing Professionals in Innovative Organizations: A collection of Readings* (pp. 516–529). New York: Ballinger Publishing/Harper & Row.

Goldstein, M. L. (1988). "Dual-career ladders still shaky but getting better." *Industry Week*, **236**, January, 57–60.

Katz, R., Tushman, M., and Allen, T. J. (1995). "The influence of supervisory promotion and network location on subordinate careers in a dual ladder RD&E setting." *Management Science*, **41**, 848–863.

Mainiero, L. A., and Upham, P. (1986). "Repairing a dual-ladder CD program." *Training and Development Journal*, **40**, 100–104.

Raelin, J. A. (1987). "Two-track plans for one-track careers." *Personnel Journal*, **66**, 96–101.

Toto, J. V. (1988). "Consultant know theyself." *Journal of Management Engineering—ASCE*, 167–168.

23

Human Resources in R&D

George F. Farris and Yi-Yu Chen*[†]

*Rutgers University and [†]New Jersey City University

Effective management of technical professionals has been studied for many years, yet it remains a challenging topic for managers and scholars alike. The annual survey of research and development trends by the Industrial Research Institute (Cosner, 2009) identified "attracting and retaining talent" as the third "biggest problem" facing technology leaders, behind "growing the business through innovation" and "accelerating innovation", each of which also has a component involving the management of technical professionals.

The fundamental nature of work in scientific fields is changing with more complex knowledge-based work, more dependence on social capital and technological competence of technical professionals, more team-based collaborative projects, and more time pressure. Scientists and engineers are required to cultivate new sets of skills to cope with flatter organizational structures and less job security, to develop inter-disciplinary knowledge bases through social networks, and to be able to leverage Web-based collaboration tools and cope with work diversity to perform effectively in cross-functional, cross-cultural, or virtual teams.

Two important review articles, Farris and Cordero (2002) and Badawy (1988) identified 10 key topics in the management of scientists and engineers: (1) human resources planning, (2) rewarding, (3) appraising performance, (4) career management, (5) cross-functional teams, (6) leading scientists and engineers, (7) knowledge management, (8) demographic diversity, (9) leveraging electronic technology, and (10) outsourcing. In this chapter we will review each area, emphasizing research done since 2002. Among the important trends we found were an increasing emphasis on scientists and engineers as participants in social networks, both domestic and global. Collaboration in diverse social contexts (e.g., different cultural settings and ethical mixes) is becoming increasingly common, and a great deal of technical work now occurs in virtual (non-collocated) teams. Key findings in each of 10 areas are summarized in Table 23.1.

Human resources planning

The scope of R&D is broadening to involve multiple disciplines and interaction with different functional groups in the organization. Thus, in addition to technical skills, scientists and engineers are expected to possess diversified skills to perform better (Farris and Cordero, 2002). They are being asked to play several critical roles, including champion, gatekeeper, idea generator/key innovator/rainmaker, and project/team leader. A champion role is an informal leadership role for projects and is especially critical when a formal leadership role and support from senior management is lacking (McDonough, 2000). Gatekeepers are those that span between organizations and external environments as well as among functional areas of organizations to communicate and detect changes that need to be integrated within organizations. Idea generators, key innovators, and rainmakers are those who constantly come up with innovative ideas to inspire the team to achieve higher performance and improved team cohesiveness. Finally, a project leader is a formal leadership role and can influence the outcome of projects through leadership styles and expertise in dealing with crisis management.

With complex and diversified information required to perform tasks in a timely manner, external organizational linkages of R&D units are very important factors to increase performance. The role of gatekeepers is especially important in complex R&D projects. R&D gatekeepers are not only first-line supervisors, but can include people in middle and upper R&D management and non-supervisory R&D staff members as well.

In addition, an agile workforce becomes attractive due to its cost-saving advantage and flexible working capacity. An agile workforce includes the use of cross-trained workers to broaden the skill sets and depth of workers, the use of contingent and temporary workers, and the use of flexible working hours in practice. Workforce agility not only

Table 23.1. Changes since 2002.

Topic area	*Highlights*
Human resources planning	More diverse skills are required; the gatekeeper role has evolved; use of agile workforces and flexible work practices, such as flexible work schedule and pay incentives, has increased.
Rewards	Intrinsic rewards are still more appealing than extrinsic ones; different social contexts present different preferences for reward systems; a community-like environment reduces volunteer turnover.
Appraising performance	Multiple sources and measures are used to enhance objectivity; social contexts need to be analyzed to interpret performance outcomes from different appraised groups; ethical behavior should also be evaluated.
Career management	Dual-ladder systems are being embellished due to environmental changes; social contexts help shape the career orientations of scientists and engineers; organizations should provide a variety of career development programs.
Cross-functional teams and global virtual teams	Much technical work is done in cross-functional teams and global virtual teams; many teams are cross-cultural and cross-national; team diversity has significantly increased and needs to be carefully managed to enhance productivity.
Leading networked scientists and engineers	Scientists and engineers are involved in professional and social networks; technical leaders should create networking opportunities when possible and manage networks well to achieve favorable outcomes.
Knowledge and creativity management	Technical leaders should shape the interdependencies among scientists and engineers to promote creativity; scientists and engineers with weak network ties can promote creativity.
Demographic diversity	The workforce is increasingly diversified; technical leaders need to manage demographic diversity to reduce conflict and to achieve desired objectives.
Electronic technology	The emergence of Web-based collaboration tools facilitates R&D teamwork across geographical boundaries; effective use of different forms of digital media depends on task characteristics; new leadership skills are required to better leverage electronic technology to promote collaboration.
Outsourcing	Outsourcing of R&D projects is increasing; new leadership skill requirements, organizational structures, and workforce compositions are the consequences.

represents flexible employee capacity via cross-trained skills, but it also provides benefits beyond pure efficiency improvements to include quality improvement, learning curve acceleration, and economy of scope and depth (Hoop and van Oyen, 2006).

Rewards

As consistently supported in the literature, intrinsic rewards are more effective than extrinsic rewards in motivating engineers and scientists to perform well (Farris & Cordero, 2002). Intrinsic rewards are those that satisfy psychological needs rather than material needs. Thus, intrinsic implies motivation to perform the task to fulfill self-actualized purposes. The satisfaction is reached from within oneself. Examples of effective intrinsic rewards include technically challenging work assignments, opportunities to pursue personal research interests, and interesting project work.

In contrast to intrinsic rewards, extrinsic rewards are those that compensate high performers with material rewards. Monetary compensation is the most common type. Extrinsic rewards can be effective as well, especially when procedural justice is in place. (Procedural justice refers to the fairness of organizational procedures used to determine the outcomes an employee receives, and it can positively influence an employee's satisfaction on pay and incentive plans.) As a result, if an organization structures long-term incentives such as stock options into its group incentive plans, scientists and engineers may be motivated to commercialize technical ideas into marketable products.

Corporations compensate technical professionals with a combination of intrinsic and extrinsic rewards. Finding the right mix of rewards and incentives to motivate technical professionals effectively is critical to the organization's success. The relationship between the right mix of intrinsic and extrinsic rewards should consider factors such as the research environment and technological strategy of the firm and diversified national contexts (Manolopoulos, 2006). For example, if the technological strategy of an organization is to focus primarily on breakthroughs in basic science disciplines, the organization will include intrinsic rewards to motivate its R&D professionals more than an organization that emphasizes incremental innovations. In addition, when competition escalates to a global scale, it is necessary to evaluate theories in a global context. Manolopoulos (2006) found that R&D professionals in a European Union peripheral economy, Greece, are motivated by extrinsic rewards and mainly by economic compensation. As more cross-disciplinary, cross-national, and cross-cultural teams are used, reward systems should consider team-based performance evaluation mechanisms and different perceptions from diversified social contexts to promote collaborative behavior.

Effective reward systems also keep the turnover of technical professionals low. Non-monetary recognition and opportunities for competency development have a negative impact on turnover intentions while procedural justice and organizational commitment partially mediate the turnover intentions of technical professionals (Pare and Tremblay, 2007). In addition to the reward system, organizations can also create a community-like environment to promote the socialization of employees to reduce volunteer turnover (Lee *et al.*, 2004) and to enhance knowledge sharing.

Appraising performance

Performance measurement and appraisal are essential parts of any reward system. Since R&D work is so complex, different individuals may evaluate the same work differently. Multiple sources and measures can improve the validity and accuracy of appraisal outcomes. Thus, instead of relying on one single source of performance evaluation, such as the immediate supervisor, the 360-degree appraisal system is a common practice in which supervisors, peers, subordinates, and customers provide performance feedback.

However, the 360-degree feedback does not guarantee an objective outcome if results are not interpreted accurately due to consistent biases among different groups of evaluators. Recently, researchers have argued that the study of performance appraisal should shift its focus from format research (i.e., the construction of rating scales) to an emphasis on the social environment that affects performance evaluation (see Levy and Williams, 2004 for a complete review). Aspects of the social environment (e.g., organizational culture, HR policies) influence the effectiveness of a performance appraisal system indirectly. For example, an organization that implements a learning culture will structure its performance appraisal system differently than one organization without such a culture. Thus, it is critical to identify, measure, and define the social context of an organization where performance appraisal takes place to truly understand the effectiveness of performance appraisal mechanisms.

Career management

Traditional wisdom differentiates two divergent career orientations among scientists and engineers: "cosmopolitan" or "local" (Gouldner, 1957). A person with a cosmopolitan orientation demonstrates a tendency to strive for achieving technical performance, cultivating technical expertise, and building reputation in his/her professional reference groups. In contrast, a person with a local orientation demonstrates a tendency to strive for achieving commercial success, transferring technical knowledge to commercial success, and building a reputation within the organization.

These differences in career orientation are the basis for the dual-ladder system. Cosmopolitans can be promoted through the technical ladder while locals can be promoted through the management ladder. Several studies have been conducted to identify which technical professions will choose each ladder. For example, Allen and Katz (1992) argued that age and educational backgrounds make a difference for people advancing through the technical ladder. Changes in the external environment have complicated the effective use of the dual ladder. As new tensions between science and commercialization arose (Turpin and Deville, 1995), a third career path—commercial management—has emerged in addition to the traditional double paths.

In addition to the organizational context, scientists and engineers are embedded in social contexts such as professional reference groups and families, and they have different personal perceptions of jobs and preferences at different stages of life. Organizations need to take these factors into account in providing different career paths which align organizational needs with individual needs. Mallon, Duberley, and Cohen (2005) developed a categorization scheme for career management of technical people, which accounts for both available organizational opportunities and individual technical expertise and life preferences. Four orientations were identified: the impassionate scientist, the strategic opportunist, the balanced scientist, and the organizational careerist. Impassionate scientists are those who view their work as work but nothing else while the strategic opportunists are those who actively make opportunities for self-benefits. Impassionate scientists stick to their disciplines, but strategic opportunists proactively manage to further their sciences and careers to seek for the most fruitful opportunities. Balanced scientists seek balance among life, family, and leisure in conjunction with their careers, while organizational careerists focus on advancing through the hierarchical systems within organizations.

In general, organizations need to provide diversified career development programs to accommodate the needs of scientists and engineers to further their professional knowledge, techniques, and expertise. Chen, Chang, and Yeh (2003) present a framework on the effective coordination of career needs of technical professionals (i.e., the demand side) and organizational career development programs (the supply side). They argue that R&D personnel have diversified career needs at different stages of their careers, which include exploration, establishment, maintenance, and disengagement.

Thus, effective management systems are those that take into account the different career orientations among scientists and engineers and try to align organizational needs with their individual needs to motivate technical professionals. Furthermore, organizations need to provide diversified career growth programs to increase job satisfaction, to motivate, and to reduce voluntary turnover of technical professionals.

Cross-functional teams and global virtual teams

The increasing adoption of cross-functional teams is critical for bringing new products to the market in today's environment, characterized by shortened product life cycles and rapid technological change. The success of cross-functional teams depends on how well team members can integrate different skill sets from functional areas such as marketing, R&D, manufacturing, and others. The benefits of cross-functional teams can include shorter product development time and reduced costs.

In addition, globalization inevitably involves the utilization of teams across cultural, graphical, and ethnical boundaries. Global virtual teams offer similar benefits to organizations as cross-functional teams but face different challenges due to the high degree of team diversity. Team diversity, if not managed well, will lead to team conflict to cause the project to fail. Several methods can be used to improve the performance of global teams (Kankanhalli, Tan, and Wei, 2006). First, diversified backgrounds of team members (e.g., cultural, linguistic, and functional backgrounds) should be made known to team members to promote shared understanding and reduce team conflict. Second, team members should be carefully selected to minimize the impact from cultural diversity or to impose one dominant organizational culture on teams. Third, cultivating trust among team members should reduce their opportunistic behavior. Fourth, though communication technology facilitates work coordination, the lack of immediate feedback or face-to-face communication may lead to conflict. Thus, periodic communication channels (e.g., conference calls) will help reduce team conflict.

Leading to networked scientists and engineers

Farris and Cordero (2002) pointed out that the role of technical managers has shifted from the traditional command-and-control role to the leadership role. Technical managers assumed two important roles: the catalyst role and the captain role (Cordero, Farris, and DiTomaso, 2004). In the catalyst role, technical managers focus on providing a stimulating working environment and resources to facilitate the performance of scientists and engineers. In the captain role, they supervise scientists and engineers directly to achieve performance outcomes and job satisfaction. Cordero, Farris, and DiTomaso (2004) further pointed out that technical managers can assume both catalyst and captain roles by effectively utilizing technical, people, and administrative skills.

Leading scientists and engineers involves facilitating knowledge sharing, motivating knowledge creation, and retaining high-quality scientists and engineers. Better utilization of the capabilities of scientists and engineers depends on how well technical leaders understand their knowledge bases, skill sets, and capability potential so that different teams can be assembled to perform diversified tasks effectively (Reagans, Zuckerman, and McEvily, 2004).

Scientists and engineers actively engage in cultivating their networks to get various advantages; that is, they take advantage of their social capital or relational assets. For example, professionals use their networks to learn new and diverse technical knowledge to help with present and future projects (Katz, 2005). Networked relationships of scientists and engineers also create opportunities for them to leave their organizations and join others. Effective management of these relationships should foster job embeddedness, "the combined forces that keep a person from leaving his or her job" (Yao *et al.*, 2004, p. 159), and reduce voluntary turnover. Finally, it is critical to manage the social capital of team members carefully because it predicts the group's performance level. Social capital can improve the performance of functionally and cognitively diversified groups by enhancing the group processes of communication, social integration, and coordination.

Knowledge and creativity management

Scientists and engineers are knowledge workers, and organizational performance depends on how well organizations can manage the creation, diffusion, and sharing of knowledge. Farris and Cordero (2002) pointed out that culture and structure are two primary factors which enable knowledge management in organizations. They identified several areas that foster the success of knowledge management, including a culture that fosters learning and knowledge sharing, senior management support of knowledge management efforts, and a performance appraisal system that rewards knowledge sharing.

Early work on knowledge management focused on explicit knowledge, but now the focus is shifting to tacit knowledge sharing among people. It is important to understand how knowledge is created and shared among scientists and engineers. The exploration of the social networks of these technical professionals becomes relevant and useful because these formal and informal connections of relationships are where knowledge sharing occurs. Creativity in organizations is very much a social process. Technical leaders need to shape and reshape the interdependencies among their technical professionals to create an environment that promotes creativity.

As a result, organizations need to be carefully structured to increase interactions among their technical professionals to encourage creativity. To promote a process that stimulates creativity, technical leaders need to stimulate information flows rather than simply to direct the process of knowledge sharing and acquisition (Assimakopoulos and Yan, 2006). To appoint scientists and engineers with weak ties as information gatekeepers or opinion leaders also helps promote creativity (Burt, 2001).

Demographic diversity

Technical work groups have become increasingly diverse with more and more women and minorities, especially Asian, joining the workforce. Some studies have demonstrated consistent advantages enjoyed by different gender, age, and racial groups. Studies on the relationships between stress levels of technical professionals and outcomes of working practices, turnover rate, morale, and performance also have highlighted the differences in stress levels and sources of stress based on gender.

Demographic diversity brings several benefits to technical organizations (see Cordero and Farris, 2002 for a review). Inconsistency has been found in the literature between demographic diversity and team performance (Reagans, Zuckerman, and McEvily, 2004). Reagans *et al.* offer a social network perspective to suggest that the inconsistencies are caused by internal network density (i.e., the intensity level of interactions among team members) and external network range (i.e., the breadth of contacts outside the team). Demographic diversity decreases internal network density because team members with different backgrounds are assumed to maintain relatively few interactions. On the other hand, demographic diversity increases the external network range because team members with different backgrounds are assumed to connect with wider ranges of contacts outside the team. And both network variables (i.e., internal network density and external network range) have positive effects on team performance. Clearly, it is important to consider both internal and external networks in managing diversity.

Electronic technology

The Internet has changed the way scientists and engineers work, especially when computer-mediated tools are utilized to promote communication and when collaboration tools are adopted to facilitate all types of virtual meetings. Web-based collaboration enhances knowledge sharing and information flow without the geographical limitations. A variety of tools are available to enhance the collaboration of the workforce. These tools include Google docs and spreadsheets, Web conferencing, Wikis (group-editable webpages), desktop sharing/sharing workspaces, discussion forums/bulletin boards, blogs, instant messaging, podcasts, group decision support system, and social networking tools. However, how and in what situations these tools can enhance knowledge sharing should be further studied to fully take advantage of these tools. Farris *et al.* (2003) have proposed a web-of-innovation model to outline how organizations can evaluate the effects of Web-based tools on facilitating and enhancing knowledge flows in the innovation process. Kock and Davison (2003) also suggested that even lean media (i.e., email conferencing) would foster knowledge sharing when accompanied by carefully designed social processes.

Organizations are actively promoting the use of advanced electronic technologies for project collaboration and increasingly investing in e-learning and technology-delivered instruction to cultivate the skills of their workers. New skill sets and strategies are thus required to effectively leverage electronic technology for technical professionals, especially those who work together from different locations. However, the adoption of electronic technology is not without risks. Kratzer, Leenders, and van Engelen (2005) argue that it is critical to analyze the characteristics of tasks assigned to R&D teams to better utilize different forms of electronic technologies. Furthermore, companies should have clear policies on information sharing to protect proprietary information.

Outsourcing

There is a growing trend to outsource more R&D projects, especially in the information technology industry. To manage outsourcing effectively, different leadership skills are required for different projects. Karlsen and Gottschalk (2006) proposed six different roles as leader (i.e., one that is responsible for supervising, hiring, training, organizing, coordinating, and motivating team members), resource allocator (i.e., one that is responsible for allocating human, financial, and information resources), spokesman (i.e., one that is responsible for getting acceptance of the project within the organization), entrepreneur (i.e., one that is responsible for developing solutions for constantly changing technical opportunities), liaison (i.e., one that is responsible for communicating with the external environment), and monitor (i.e., one that is responsible for scanning the external environment to keep up with technical changes and competition) to manage different projects of client outsourcing, vendor outsourcing, client termination, and vendor termination. Empirical results show that the appropriate leadership role depends on the outsourcing perspective, time, and situation.

Another impact brought by outsourcing R&D projects is the change in the workforce for scientists and engineers discussed previously. Certain job categories (e.g., call centers, manufacturing functions, customer service departments, etc.) have been scattered in regions where organizations can either reduce costs or increase efficiency. Contract and temporary workers are being used more. These changes create new challenges in many regions of the globe.

Where to go in the field of managing scientists and engineers

The review by Farris and Cordero (2002) identified 10 interrelated areas that cover the management of scientists and engineers. This chapter updates previously identified areas to evaluate how organizations adapt to the changing R&D environment.

The field of managing technical professionals today encompasses the management of not only scientists and engineers but also the professional and social relationships they have established. Technical leaders need to not only manage the intellectual capital of scientists and engineers, but also to manage social relationships and networks in which they are embedded. The interaction of social capital with organizational contextual factors creates a new page for the study of how to manage scientists and engineers effectively in the 21st century.

References

Allen, T., and Katz, R. (1992). "Age, education and the technical ladder." *IEEE Transactions on Engineering Management*, **29**(3), 237–245.

Assimakopoulos, D., and Yan J. (2006)."Sources of knowledge acquisition for Chinese software engineers." *R&D Management*, **26**(1), 97–106.

Badawy, M. K. (1988). "What we have learned about managing human resources." *Research-Technology Management*, September/October, 19–35.

Burt, R. S. (2001). "Structural holes versus network closure as social capital." In: N. Lin, K. S. Cook, and R. S. Burt (Eds.), *Social Capital*. Chicago, IL: Aldine/Gruyter.

Chen, T.-Y., Chang, P.-L., and Yeh, C.-W. (2003)."The study of career needs, career development programmes and job satisfaction levels of R&D personnel: The case of Taiwan." *International Journal of Human Resource Management*, **14**(6), 1001–1026.

Cordero, R., Farris, G. F., and DiTomaso, N. (2004). "Supervisors in R&D laboratories: Using technical, people, and administrative skills effectively." *IEEE Transactions on Engineering Management*, **51**(1), 19–30.

Cosner, R. (2009). "Industrial Research Institute's R&D trends forecast for 2009." *Research-Technology Management*, January/February, 19–26.

Ettlie, J. E., and Elsenbach, J. M. (2007). "The changing role of R&D gatekeepers." *Research-Technology Management*, September/October, 59–66.

Farris, G. F., and Cordero, R. (2002). "Leading your scientists and engineers 2002." *Research-Technology Management*, November/December, 1–13.

Farris, G. F., Hartz, C. A., Krishnamurthy, K., McIlvaine, B., Postle, S. R., Taylor, R. P., and Whitwell, G. E. (2003). "Web-enabled innovation in new product development." *Research-Technology Management*, **46**(6), 24–35.

Gouldner, A. W. (1957). "Cosmopolitans and locals: Toward an analysis of latent social roles." *Administrative Science Quarterly*, **2**(3), 281–306.

Hoop, W. J., and van Oyen, M. P. (2006). "Agile workforce evaluation: A framework for cross-training and coordination." *IIE Transactions*, **36**, 919–940

Kankanhalli, A., Tan, B. C. Y., and Wei, K.-K. (2006). "Conflict and performance in global virtual teams." *Journal of Management Information Systems*, **23**(3), 237–274.

Karlsen, J. T., and Gottschalk, P. (2006). "Project manager roles in IT outsourcing." *Engineering Management Journal*, **18**(1), 3–9.

Katz, R. (2005). "Motivating technical professionals today." *Research-Technology Management*, November/December, 19-27.

Kock, N., and Davison, R. (2003). "Can lean media support knowledge sharing? Investigating a hidden advantage of process improvement." *IEEE Transactions on Engineering Management*, **50**(2), 151–163.

Kratzer, J., Leenders, R. Th. A. J., and van Engelen, J. M. L. (2005). "Keeping virtual R&D teams creative." *Research Technology Management*, 13–16.

Lee, T. W., Mitchell, T. R., Sablynski, C. J., Burton, J. P., and Holtom, B. C. (2004). "The effects of job embeddedness on organizational citizenships, job performance, volitional absences, and voluntary turnover." *Academy of Management Journal*, **47**, 711–722.

Levy, P. E., and Williams, J. R. (2004). "The social context of performance appraisal: A review and framework for the future." *Journal of Management*, **30**(6), 881–905.

Mallon, M., Duberley, J., and Cohen, L. (2005). "Careers in public sector science: Orientations and implications." *R&D Management*, **35**(4), 395–407.

Manolopoulos, D. (2006). "What motivates R&D professionals? Evidence from decentralized laboratories in Greece." *International Journal of Human Resource Management*, **14**(4), 616–647.

McDonough III, E. F. (2000). "Investigating the factors contributing to the success of cross-functional teams." *Journal of Product Innovation Management*, **17**, 221–235.

Pare, G., and Tremblay, M. (2007). "The influence of high-involvement human resources practices, procedural justice, organizational commitment, and citizenship behaviors on information technology professionals' turnover intentions." *Group & Organization Management*, **32**(3), 326–357.

Reagans, R., Zuckerman, E., and McEvily, B. (2004). "How to make the team: Social networks vs. demography as criteria for designing effective teams." *Administrative Science Quarterly*, **49**, 101–133.

Turpin, T., and Deville, A. (1995). "Occupational roles and expectations of research scientists and fesearch managers in scientific research institutions." *R&D Management*, **25**(2), 141–157.

Yao, X., Lee, T. W., Mitchell, T. R., Burton, J. P., and Sablynski, C. S. (2004). "Job embeddedness: Current research and future directions." In: R. Griffeth and P. Hom (Eds.), *Understanding Employee Retention and Turnover* (pp. 153–187). Greenwich, CT: Information Age.

24

The Stage-Gate® Product Innovation System: From Idea to Launch

Robert G. Cooper
McMaster University

What is a Stage-Gate® system?

A Stage-Gate® system[1] is a conceptual and operational map for moving a new-product project from idea though to launch and beyond—a blueprint for managing the product innovation process to improve effectiveness and efficiency.[2,3] According to studies done by the Product Development and Management Association (PDMA) and the American Productivity and Quality Center (APQC), approximately 75% of U.S. product developers now use some form of Stage-Gate as the model to drive new-product projects to market.[4,5]

Stage-Gate breaks the product innovation process into a predetermined set of stages preceded by gates, as in Figure 24.1. Typically there are four, five, or six stages depending on the company and nature of the project. Each stage is comprised of a set of prescribed best practice activities designed to progress the project and to gather the essential information needed to make the next "go/kill" or investment decision. Following each stage, there is a go/kill decision point called a "gate", which is staffed by the senior management team, the owners of the resources required to move into the next stage. This incremental investment, stage-and-gate format led to the name "Stage-Gate system".

[1] Stage-Gate® is a trademark of Product Development Institute Inc. in the U.S.A. and Australia; of R. G. Cooper & Associates Consultants Inc. in Canada; and of R. G. Cooper and Jens Arleth (Innovation Management-U3) in the European Union.
[2] For a detailed description of Stage-Gate®, see Cooper (2001).
[3] Parts of this chapter are based on material in Cooper (2000, 2003, 2005).
[4] Studies have pointed to the extensive use of Stage-Gate in industry; for example, see Adams and Boike (2004), Griffin (1997), and PDMA Foundation (2004).
[5] See APQC best practices study (Cooper, Edgett, and Kleinschmidt, 2002a) and in shortened form, as a three-part series (Cooper, Edgett, and Kleinschmidt, 2004a, b, 2005).

The need for a formal new-product process

New products have an alarming failure rate. Of every nine new-product concepts, only one becomes a commercial success, according to PDMA studies (Adams and Boike, 2004; Griffin, 1997; PDMA Foundation, 2004). And a review of many investigations suggests that about 40% of new products fail at launch, even after all the product tests, customer trials, and even test markets.[6] Another study reveals that 46% of company's resources spent on new product development (NPD) go to unsuccessful ventures (as cited in Cooper, 2001). Finally, in only 21.3% of companies do their total new-product efforts meet their annual profit objectives (see APQC studies: Cooper, Edgett, and Kleinschmidt, 2002a, 2004a, b, 2005).

Why do so many new products fail and why do the majority of businesses underperform in NPD? The reasons for failure or poor performance have been widely studied; some of the more important ones include (Cooper, 2001):

- A lack of understanding of the market and customer—market potential, customer needs and wants, and the competitive situation.
- A failure to commit the necessary resources to product development (or to the project team)—people assigned to the project are simply stretched too thinly, so they cut corners and execute in haste.
- The lack of solid upfront homework before development begins—a "ready, fire, aim" approach to NPD.
- A lack of discipline in the new-product process, and paying lip service to quality of execution—things simply

[6] The PDMA best practices study reports a 59% success rate after launch; see Adams and Boike (2004), Griffin (1997), and PDMA Foundation (2004).

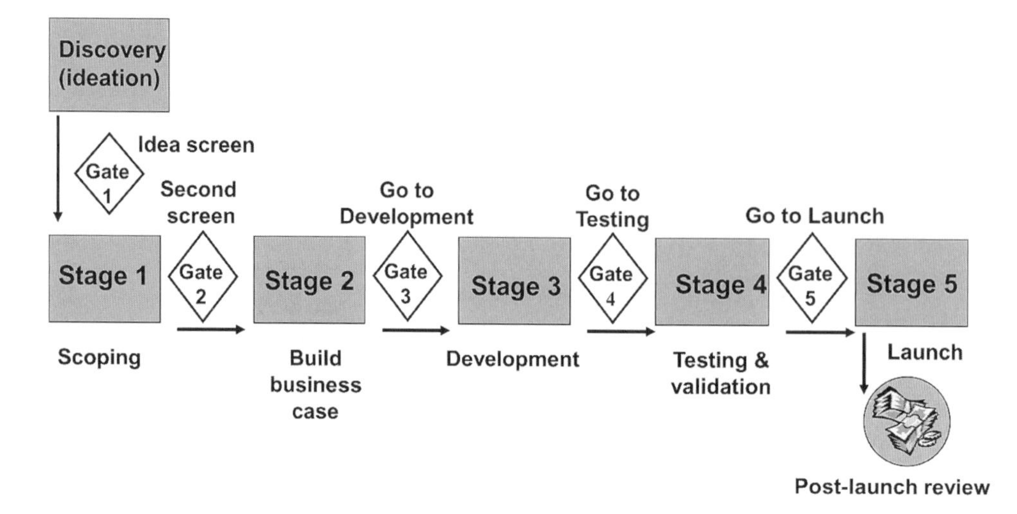

Figure 24.1. An overview of the Stage-Gate® system.
Source: Cooper (2001).

don't happen as they should, when they should, or as well as they should in NPD projects.

- A lack of good new-product ideas as feedstock to the process.
- Moving targets—unstable product specifications and project scope creep.
- A lack of senior management commitment to and engagement in NPD.

The existence of these and other recurring reasons for new-product failure is cause for rethinking the way companies go about conceiving, developing, and launching new products. The fact is that many of these failures could have been prevented by employing different or better methodologies and practices. What was missing was a systematic process to guide the project team from idea through to launch and beyond, much like a playbook guides a football team down the field to a touchdown in North American football.

A little history: the roots of Stage-Gate

Where did Stage-Gate come from? A little history will help to lead to a better understanding of the role and design of the model. There had been attempts to craft product development processes that date back to the 1960s. For example, Procter & Gamble had a new-product process model, dated 1964—a fairly crude process by today's standards, but at least an attempt to capture the best practices of the day and design an idea-to-launch system (P&G's current idea-to-launch system is a third-generation Stage-Gate system, quite different from the early model: Cooper and Mills, 2005).

Another early system, the Phase-Review process, was developed by the military and NASA in the 1960s and was popular through to the 1980s (and is still used by some organizations even today). Indeed, many higher technology and engineering companies that dealt with government and the military adopted the approach (Cooper, 1994). Here, the project was broken into discrete phases, with review points at the end of each phase: Government funding for the next phase was conditional on certain tasks being completed. While the system worked well for government contractors, the approach was very technically oriented—essentially an engineering or design-and-develop process; and it dealt largely with technical issues, not business issues. For example, the review points did not ask the question: "should we continue to invest in this project?" but rather: "is the project ready for the next stage of funding?" The notion of TRLs or technology readiness levels as the criteria for moving to the next phase is still used by NASA. Some pundits criticized the process as being very cumbersome and bureaucratic, with long delays awaiting decisions. On the other hand, it couldn't have been all bad, as that generation of managers and engineers did manage to put a man on the moon in less than a decade using the system!

From the beginning, the Stage-Gate process was envisioned as a model based on observing and measuring what successful product development project teams do—an attempt to understand and replicate the behaviors of these winners. Stage-Gate did not magically appear overnight from some consultant's toolbox! Rather, the Stage-Gate model was originally conceived in the early 1980s by the author, the result of several studies of successful project teams, as well as large-sample investigations into new-product success factors.[7] The analogy is that of watching many video replays of football games, analyzing the videos, and then trying to understand how winning teams win. Unless one is blind, patterns begin to emerge—I called them "critical success factors" or factors that distinguish the successful teams.

Next, these patterns of success can be integrated into a playbook, game-plan, or unified model to drive new products to market. That's how Stage-Gate was born.[8] Indeed this notion of constructing a best practices idea-to-launch model was almost an afterthought, and the model was tacked on as a final chapter in an early university research publication back in 1976! A series of large-sample success/failure studies then followed this initial research, and identified yet other factors, practices, and behaviors that lead to success. These best practices in turn were built into early versions of Stage-Gate. The first companies to implement Stage-Gate (Exxon Chemical, DuPont, Procter & Gamble, and Swarovski) provided excellent proving grounds, and we learned a lot about how to make Stage-Gate work, and more importantly, what not to do.

The name "Stage-Gate" first appeared in print in 1988 (Cooper, 1988, 1990). Since then—and after many research studies, in-depth investigations, articles, books, and company implementations—Stage-Gate has become the most pervasive tool for driving new products to market in industry.

The point I want to make (and it's an important one) is that we did not begin with the goal of designing a process.[9] Rather, we undertook academic research and asked the fundamental question: "Why do winning teams win the game?" We tried to model what they did and, in so doing, crafted Stage-Gate. And as more was learned about winners and what they did, these new insights were built into the model. It is thus a system that was created not because of a desire to design a process, but rather a system based on observing what winners do. When you think about Stage-Gate, don't lose sight of the original intent—a system to guide a team down the field to the goal line and to victory, quickly and effectively!

Research that led to Stage-Gate

Before getting into the details of a Stage-Gate process, consider some of the success factors—behaviors, practices, common denominators—that have been found to separate winning project teams from losers, and businesses that do well at product development from the poorer performers. Much research has been undertaken to uncover these success factors—research that typically quantitatively compares a large sample of successful projects or businesses versus poor performers and seeks those factors that discriminate between the winners and losers (a good review of success factors is provided in Cooper, 1996 and Montoya-Weiss and Calantone, 1994). Here is a sample of some of these best practices and success factors; Figure 24.2 shows typical summary correlations with two measures of new-product performance, namely profitability of the new product and timeliness (time efficiency and on-time performance) from one study (for the source of correlations see Cooper, 1995):

1. *Differentiated, superior products*: the number-one driver of profitability is delivering a differentiated product with unique customer benefits and superior value for the user. Such superior products with compelling value propositions
   - meet customer needs better than competitive products (but often customers do not know what they need!);
   - are higher quality products, however the customer or user defines quality;
   - solve a customer problem with existing products;
   - save the customer or user money—better value-in-use.
2. *Voice of the customer*: new-product projects that feature high-quality VoC market research are blessed with more than double the success rates and 70% higher market shares than those projects with poor marketing information. Sadly, a strong market orientation and customer focus is noticeably lacking in many businesses' new-product projects and is consistently rated one of the weakest areas in projects. VoC methods are designed to probe the customer's unmet or unarticulated needs, and not just to focus on desired product features and specifications. On the basis of such customer or user insights, the project team is in a much better position to design a truly superior product—item 1 above.
3. *Sharp, stable, and early product definition*: a failure to define the product and project before development begins is a major cause of both new-product failure and serious delays in time-to-market. In spite of the fact that early and stable product definition is consistently cited as a key to success, firms continue to perform poorly here. Terms such as unstable product specs and project scope creep describe far too many new-product projects. It is essential that the project team achieve

[7] Early versions of Stage-Gate are in Cooper (1988, 1990).
[8] The first article to describe Stage-Gate was Cooper (1983).
[9] "We" refers to the author, as well as academic colleagues who undertook this pioneering research with the author, namely Professor Elko Kleinschmidt, Dr. Scott Edgett, and Professor Ulrike de Brentani.

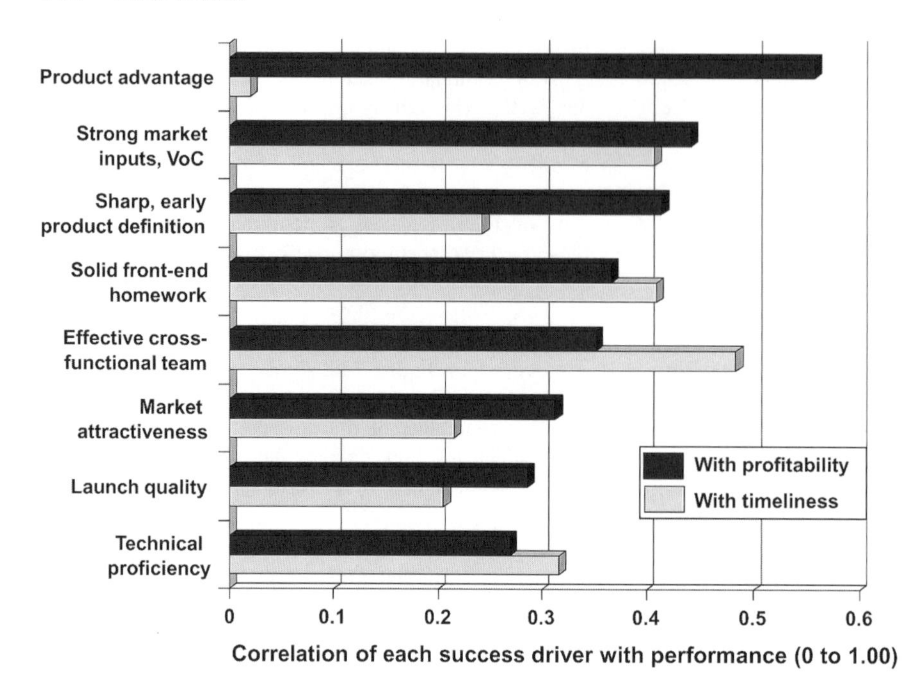

Figure 24.2. Impact of drivers on profitability and timeliness.
Adapted from Cooper (1995).

fact-based product and project definition before moving into Development (prior to Stage 3 in Figure 24.1).

4. *Upfront homework*: too many new-product projects move from the idea stage right into Development with little or no upfront homework. The results of this "ready, fire, aim" approach are usually disastrous. Solid pre-Development homework drives up new-product success rates significantly and is strongly correlated with financial performance. Sadly, firms devote on average only 7% of a project's funding and 16% of person-days to these critical upfront homework activities.
5. *True cross-functional project teams*: good organizational design is strongly linked to both success rates and shorter times-to-market. This means that projects are organized by a cross-functional team, led by a strong project leader, accountable for the entire project from beginning to end, and dedicated and focused (as opposed to spread over many projects). While the ingredients of a "good team" should be familiar ones, surprisingly many projects are found lacking here.
6. *Quality of the market launch*: not surprisingly, a strong market launch underlies successful new products. For example, new-product winners devote more than twice as many person-days and dollars to the launch as do teams that fail. Similarly, the quality of execution of the market launch is significantly higher for winners. The need for a quality launch—well planned, properly resourced and well executed—should be obvious. But in some businesses, the launch is an afterthought—something to worry about after the product is fully developed.
7. *Tough go/kill decision points in the NPD process—a funnel, not a tunnel*: too many projects move far into Development without serious scrutiny: once a project begins, there is very little chance that it will ever be killed.[10] The result is that many marginal development projects are approved, with the improper allocation of scarce resources. The solution is effective portfolio management, where each new-product project is viewed as an investment. Here, tough go/kill decision points are built into the new-product process in the form of gates, that successively cull mediocre projects. The result is a funnel with fewer and fewer projects proceeding, but those that remain are high-value projects (by contrast, "tunnels" occur when a set of projects is initially approved and none is subsequently killed, even though the project's prospects have turned negative—the result is too many projects and often weak projects).
8. *Quality of execution*: a quality crisis exists in the new-product process. Simply stated, key activities and tasks, from initial screening to market research and even executing the launch, are poorly executed, or in some cases, not even done at all! Figure 24.3 shows quality-of-execution results across many companies for some activities in the new-product process (these quality-of-execution ratings are data from the APQC studies: Cooper, Edgett, and Kleinschmidt, 2002a, 2004a, b, 2005). For example, only 18% of companies

[10] Cooper, Edgett, and Kleinschmidt (2001). See also APQC studies (Cooper, Edgett, and Kleinschmidt, 2004a, 2005) and especially RTM—Part II (Cooper, Edgett, and Kleinschmidt, 2004b).

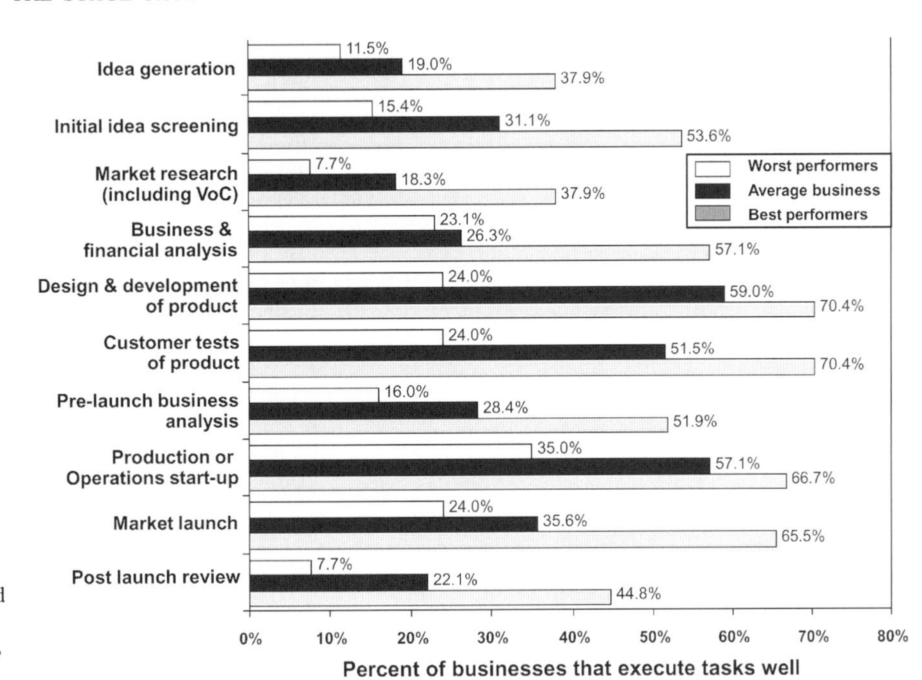

Figure 24.3. Quality of execution is strongly linked to success.
Adapted from Cooper, Edgett, and Kleinschmidt (2002).

carry out market research proficiently; 26% undertake an adequate business and financial analysis; and only 36% execute the launch well. Note how much better top-performing businesses execute these activities in Figure 24.3.

9. *An international orientation*: new products aimed at international markets (as opposed to domestic) and with international requirements built in from the outset are more profitable. By contrast, products that are developed for domestic markets and sold locally yield lower profits. And the strategy of "design for local needs, and adjust for export later" also does not work well; the product is usually compromised.
10. *Speed and reducing time to market*: speed yields competitive advantage, the notion that "first to market wins"; it means less likelihood that the market or competitive situation has changed by the time one launches; and it means a quicker realization of profits. So the goal of reducing the development cycle time is admirable (taken from Cooper, 2005). Additionally, many managers complain about projects taking too long: the APQC study found, for example, a slip rate of 35.4%, while 32% of executives rate their time efficiency in NPD as "very poor".[11] Many firms have redoubled their efforts to reduce product development cycle times over the last 5 years, with the average reduction being about one-third (see PDMA best practices study: Griffin, 1997).
11. *The role of top management*: top-management support is a necessary ingredient for product innovation. Top management's main role is to set the stage, to be a behind-the-scenes facilitator, and much less an actor, front-and-center. Common behaviors of effective senior management teams include making a long-term commitment to product development as a source of growth; developing a vision, objectives, and strategy for product innovation for the business; and making available the necessary resources to product development, ensuring that they aren't diverted to more immediate needs in times of shortage. Most important, senior management must be engaged in the new-product process, reviewing projects, making timely and firm go/kill decisions, and if "go", making resource commitments to project teams. Further, management must empower project teams and support committed champions by acting as mentors, facilitators, "godfathers," or sponsors of project leaders and teams.

These and other success factors and best practices have been uncovered in countless studies of successful projects, project teams, and businesses. When I first crafted the original Stage-Gate process, I tried to build in some of these success factors into a model, process, or methodology to guide project leaders and teams from idea through to launch and beyond. Now let's see how the

[11] Slip rate is the difference bdetween actual time to market and projected time in the business case, as a percent of projected time. See APQC studies (Cooper, Edgett, and Kleinschmidt, 2002a, 2004b, 2005) and especially Part I (Cooper, Edgett, and Kleinschmidt, 2004a).

process works, and how these and other success factors are built in.

How Stage-Gate works

The stages

A typical five-stage, five-gate Stage-Gate system is shown in Figure 24.1. Each stage consists of a set of prescribed activities—that is, tasks within a stage are done concurrently and in parallel. These tasks are a list of best practices, such as voice-of-customer research, building a business case, or conducting a customer product test, and are deliberately defined and built into these stages as either mandatory or highly recommended activities for the project team to execute.

Each stage is cross-functional: there is no "R&D Stage" or "Marketing Stage". Rather, the project is undertaken by people from different functional areas within the firm—a cross-functional project team; thus every stage is marketing, R&D, production, engineering, and so on.

Stage-Gate is also designed as a risk management model. The activities within each stage are crafted to gather critical information and to reduce the project's unknowns and uncertainties. Each stage also involves more resources than the preceding one: the process is an incremental commitment one, much like buying options on an investment. But with each step increase in project cost, the unknowns and uncertainties are correspondingly driven down, so that risk is effectively managed.

The typical stages in Stage-Gate, as shown in Figure 24.1, are:

Discovery: this first stage is very early work designed to uncover opportunities and generate new product ideas.[12] Discovery, sometimes called Ideation, includes everything from major efforts such as strategic analysis, scenario generation, and ethnography through to fostering ideation and submission by a company's own employees and by welcoming outsiders to participate (open innovation). The deliverable is an idea for a new product.

Scoping: this is a quick, preliminary investigation and scoping of the project—largely desk research. This stage includes a preliminary market, technical, source-of-supply and business assessment of the proposed idea. The deliverable at the end of Stage 1 is a preliminary business case based on rough estimates, and a tentative product and project definition.

Build business case: this is a make-or-break stage—a much more detailed investigation than Scoping. It involves primary and potentially extensive research, both market and technical. Activities such as in-depth voice-of-customer research, competitive analysis, market research, detailed technical assessment and financial analysis are undertaken here. The deliverable here is a full business case, including a detailed product and project definition, a project justification, and a project plan in the form of a timeline.

Development: this potentially expensive stage entails the actual detailed design and full development of the new product, as well as in-house, lab, or alpha-testing of the product. Additionally, the project team develops the operations or production process on paper, as well as test plans for Stage 4 and full commercialization plans for Stage 5. The financial and business analyses are updated.

Testing and validation: this stage involves tests or trials in the marketplace, and operations trials to verify and validate the proposed new product and its marketing and production. Tasks such as trial production, in-home consumer testing, beta-tests and test markets are carried out here. The deliverable is a validated product and project, along with full commercialization and rollout plans for the next stage.

Launch: this is the commercialization stage—the implementation of full operations or production, marketing, and selling. Implementation of the post-launch plan also begins.

Following the Launch Stage is a post-launch review which terminates the new-product project. Here, performance is assessed by comparing the new product's revenues, costs, profits, and timing to projections made at key go/kill decision points; and a post-audit—an assessment of the project's strengths and weaknesses, what can be learned, and how to undertake the next project better—is carried out. In this way, continuous improvement via closed-loop feedback—preventing recurrence of deficiencies—is built into the system. As well, the product life cycle plan is also approved and implemented.

The gates or go/kill decision points

Following each stage is a gate or decision point. Gates serve as quality control checkpoints, as go/kill and prioritization decisions points, and as points where the go-forward plan and resources for the next stage of the project are agreed to and committed.

The structure of each gate is similar:

Deliverables are what the project leader and team bring to

---

[12] This Discovery Stage is not numbered (or is often designated Stage 0), because it was added some years after Stage-Gate became popular. New research revealed that companies had a shortage of solid new-product ideas, and needed to systematize idea generation, capture, and handling, thus the Discovery Stage.

the decision point (e.g., the results of a set of completed activities). These deliverables are visible and are based on a standard menu for each gate. Thus project teams clearly understand what they must deliver to each gate, and as a result, what work is required of them.

Criteria are questions on which the project is judged. These criteria often include "must meet" or knock-out questions (a checklist) designed to weed out misfit or bad projects quickly; and also "should meet" criteria or desirable factors, which are scored (using a scorecard or a point count system). Project attractiveness scores from the scorecard (the weighted or unweighted addition of the criteria scores) are used to make the go/kill decisions and to prioritize projects.

Outputs of the gate include a decision (go/kill/hold/recycle); an approved action plan for the next stage (complete with people required, money and person-days committed, and an agreed timeline); and a list of deliverables and date for the next gate.

Built-in success factors

The logic of a well-designed product innovation system, such as Stage-Gate in Fiigure 24.1, is appealing because it incorporates many of the critical success factors—the drivers of success and speed—that have been found to impact on new-product success in countless research studies. For example:

1. *Team-based*: the process is cross-functional, built around a project team with authority. Stage-Gate is not a technical or marketing process, but very much a business process. Because each stage consists of technical, marketing, sales, operations, and even financial activities, the active engagement and commitment of people from all of these areas is essential. Teams have authority (control over their resources) but are also accountable for the end-result of the project.
2. *Focus*: the process features a governance process in the form of gates, with visible go/kill criteria. Gates weed out poor projects early and help focus scarce resources on the higher value projects. Gates are also cross-functional: gates are staffed by gatekeepers from different functions or departments in the firm, the senior managers who own the resources needed for the next stage. And the gates provide the transfer of power to the project team from functional bosses: a gate is an irrevocable decision to commit resources to a project.
3. *Differentiated products, unique benefits, superior value for the customer*: this is one key to new-product success, yet all too often, when redesigning their new-product processes, firms fail to build in any attempt to seek truly superior products. Stage-Gate drives the quest for product advantage by
   - ensuring that some of the criteria at every gate focus on product superiority;
   - requiring that key customer actions designed to deliver product superiority be included in each stage of the process (examples are given in the "customer-focused" item, next);
   - demanding that project teams deliver evidence of product superiority to gate reviews.
4. *Customer-focused and sharp product definition*: market insights and voice-of-customer are a part of the system: market inputs begin in the Discovery Stage, and are evident in every stage until the end of project. This extensive VoC and market emphasis is designed to uncover unmet and unarticulated needs, and thus lead to a product with a compelling value proposition for the user, and also to validate the product with the customer. Additionally, a product definition step is built into the process in Stage 2, Build Business Case, so that the project scope and product specs remain relatively stable from Gate 3 onward.
5. *Front-end loaded*: Stage-Gate emphasizes doing the due diligence or pre-development activities early in a project. Stages 1 and 2, the Scoping and Build Business Case stages, are the essential homework steps before the door to Development is opened at Gate 3. Thus, Stages 1 and 2 in Figure 24.1 typically have a list of best practice tasks, investigations, and analyses, complete with "how-to guides" and worksheets, that the project team must undertake or employ when building their business case.
6. *Agile*: agility is built in through spiral processing in order to speed the process. Spirals are iterations with the customer, namely a series of checks of the product with the customer as it takes shape. That is, a series of "build–test–feedback–and–revise" iterations or spirals are built in, beginning with a concept test in Stage 2, moving to virtual prototype, protocept, and/or rapid prototype tests in Stage 3, and ending with a beta-test or a customer field trial in Stage 4. In this way, the project team moves rapidly towards a final and fact-based product design in the early stages, and has a proven product by the time the Launch Stage is reached.
7. *Quality of execution*: the stages define the recommended best practice activities for execution by the project team. They map out what needs to be done and how, and thus expectations are clearly defined for teams. At the same time, the gates provide the critical quality control checks in the process: unless the project meets certain quality standards, it fails to pass the gate.
8. *A well-conceived, properly resourced market launch*: marketing planning is an integral part of Stage-Gate. Indeed, a preliminary launch plan is required as part of the business case delivered to Gate 3: that is, the project team must think about how they plan to sell the new product before they even develop it! And at each gate

thereafter, the launch plan and required launch resources is a key deliverable and vital discussion item.

9. *Fast-paced via parallel processing*: parallel processing is one solution to the need for a complete and quality process, yet one that meets the time pressures of today's fast-paced business world. Traditionally, new-product projects have been managed via a series approach: one task strung out after another, in sequence, much like a relay race. In marked contrast, with parallel processing many activities are undertaken concurrently rather than in series. Thus the process is far more intense and more work gets done in a given time period. Also, there is less chance of an activity or task being cut out, simply because of lack of time: as a result of activities being done in parallel, not in series, no one activity extends the total elapsed project time.
10. *An international orientation*: Stage-Gate builds in a strong international component for companies that operate globally, and indeed it becomes a transnational process. First, in every stage, there are several key international activities, such as market research or concept testing undertaken in multiple international markets concurrently. The option to design a global product (one product for the world) or a local product (one product concept, common technology, but the product is locally tailored to suit different national markets) is a feature of Stage-Gate. Finally, the organizational structure—cross-functional teams with members drawn from several continents, and global gatekeeping executive groups as the governance structure—is the way Stage-Gate works internationally.

The impact of installing a Stage-Gate process

Most best practice firms have a Stage-Gate new-product process, or one like it, in place, according to the PDMA best practices study noted above. And properly implemented, Stage-Gate processes really work, according an in-depth study of their firms' new-product processes—see Figure 24.4 (Cooper and Kleinschmidt, 1991):

- improved teamwork between functional areas;
- less recycling and rework due to built-in quality-of-execution checks;
- improved success rates, the result of better project evaluations at the gates and more attention to key success activities;
- earlier detection of failures by employing tough, rigorous gates with clear go/kill criteria;
- better launch because marketing, planning, and other market-oriented activities are integral to the process;
- shorter elapsed time due to better front-end homework, more multi-functional inputs, better market and product definition, and less recycle work (Griffin, 1993).

The APQC study (cited earlier) found that "all the elements of this process [Stage-Gate] are very evident in top performing businesses" (see APQC studies: Cooper, Edgett, and Kleinschmidt, 2002a, 2004a, b, 2005). A sub-sample of best practice businesses were identified in this study for more in-depth investigation, and each company indicated that "a solid, well-defined process with clearly defined activities in each stage and a well-defined decision framework for the gates (decision points) was a critical best practice for them."

Advancements in the system: NexGen Stage Gate®

Stage-Gate is an evergreen process, constantly evolving. Not surprisingly, a number of companies, with years of experience with Stage-Gate and a desire for productivity increases, have moved to a next-generation process. Four themes emerge in these newer approaches:[13]

1. *Flexible, scalable, and adaptable*: Stage-Gate systems in smart companies are flexible and scalable, designed to suit the needs and risk levels of different types of projects. For example, they use the full five-stage process in Figure 24.1 for large, high-risk product developments, but a truncated three-stage and even two-stage process for smaller, well-defined ones, such as line extensions, product improvements, or customer requests (as shown in Figure 24.5). There is even a TD version of Stage-Gate for technology development, fundamental research, or the development of new technology platforms (Cooper, 2006b). NexGen Stage-Gate is also flexible: key activities and even entire stages can be overlapped (start the next stage or activity before the last one is completed—the principle of simultaneous execution). Finally, the process is adaptable and dynamic: it adapts to the changing situation and fluid circumstances of the project as the project evolves, and as new or more reliable information becomes available.
2. *An open system*: progressive companies build into their Stage-Gate system the necessary flexibility, capability, and systems to work with outside organizations in product innovation. Companies such as P&G, Kimberly Clark, and Air Products have moved to open innovation and have modified their Stage-Gate process in order to build in a network of partners, alliances, customers, and outsourced vendors from idea generation right through to launch.[14]
3. *Effective portfolio management*: most businesses have

[13] For more on NexGen Stage-Gate, see Cooper (2006a) and Cooper and Edgett (2005).

[14] For P&G's approach to open innovation, see Huston and Sakkab (2006). Open innovation is described in Chesbrough (2003), Docherty (2007), and Cooper and Edgett (2007).

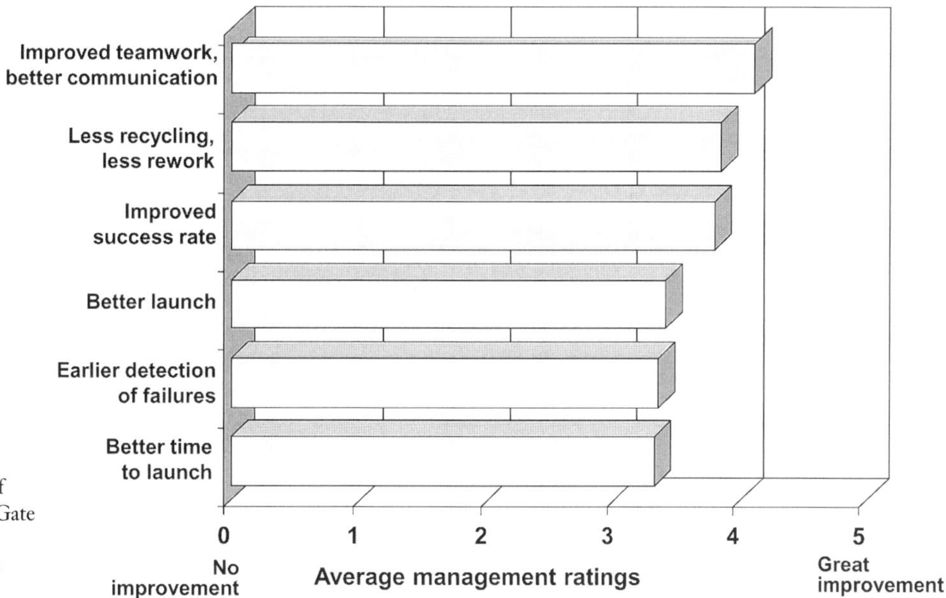

Figure 24.4. Impact of implementing a Stage-Gate NPD process.
Adapted from Cooper and Kleinschmidt (1991).

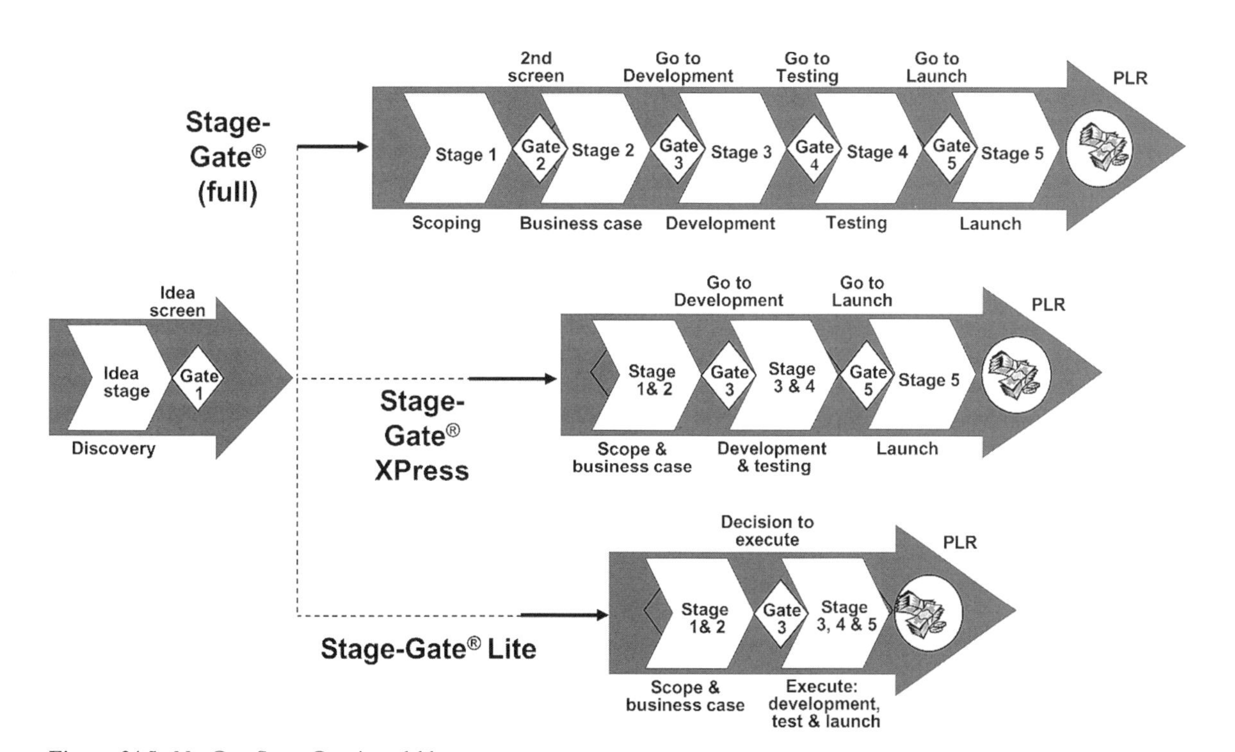

Figure 24.5. NexGen Stage-Gate is scalable.
Use the full five-stage model (top) for major NPD projects; the three-stage XPress model for product extensions, modification, and improvements (sustaining innovation); and the two-stage Lite version for minor customer requests involving small changes and minimal expenditures.

too many development projects underway, and often the wrong ones. Making the right go/kill decisions and effectively allocating development resources is fundamental to productivity improvement in product innovation. The tools that best performers utilize to make better portfolio decisions, and are often built into NexGen Stage-Gate, include:[15,16]

- *Strategic buckets*: setting up "buckets of resources" to ensure the right mix and balance of projects (by project type, and across market segments and technologies).
- *Product and technology roadmaps*: to map out the major development initiatives (major projects, technologies, platforms) strategically required over the next 5 to 7 years.
- *Scorecards*: rigorous qualitative methods employed by gatekeepers at gate meetings to help select and prioritize the best development projects.
- *The productivity index*: a financial tool that attempts to maximize the economic value of the portfolio, subject to personnel or financial resource constraints.

4. *A lean system*: many businesses' idea-to-launch processes contain much bureaucracy, time-wasters, and make-work activities. Even worse, they have rigid procedures, demanding too much paperwork, forms, meetings, and committees, regardless of the project. By contrast, smart companies have borrowed concepts from lean manufacturing and have streamlined their idea-to-launch process, removing waste and inefficiency at every opportunity. For example, they utilize value stream analysis in order to remove all non-value-added activities and have created a "lean Stage-Gate".

Conclusion

Product innovation is one of the most important endeavors of the modern corporation.[17] Without a systematic new-product process, however, often the product innovation effort is a chaotic, hit-and-miss affair. The Stage-Gate® system is an enabler or guide, building in best practices and ensuring that key activities and decisions are done better and faster. Many leading companies, however, have taken the necessary step, and have designed and implemented a world-class idea-to-launch system such as Stage-Gate, and the results have been positive: better, faster, and more profitable new-product developments!

[15] These portfolio tools can be found in Cooper and Edgett (2006) and Cooper, Edgett, and Kleinschmidt (2002b).
[16] This section adapted from Cooper and Edgett (2008).
[17] Paragraph adapted from Cooper (2003).

References

Adams, M., and Boike, D. (2004). "PDMA Foundation CPAS Study reveals new trends." *PDMA Visions*, **XXVIII**(3), July, 26–29.

Chesbrough, H. (2003). *Open Innovation: The New Imperative for Creating and Profiting from Technology*. Boston, MA: Harvard Business School Press, 2003.

Cooper, R. G. (1983). "A process model for industrial new product development." *IEEE Trans. on Engineering Management*, **EM-30**(1), February, 2–11.

Cooper, R. G. (1988). "The new product process: A decision guide for managers." *Journal of Marketing Management*, **3**(3), 238–255.

Cooper, R. G. (1990). "Stage-Gate systems: A new tool for managing new products." *Business Horizons*, **33**(3), May/June.

Cooper, R. G. (1994). "Third-generation new product processes." *Journal of Product Innovation Management*, **11**(1), 3–14.

Cooper, R. G. (1995). "Developing new products on time, in time." *Research-Technology Management*, **38**(5), September/October, 49–57.

Cooper, R. G. (1996). "New products: What separates the winners from the losers." In: M. D Rosenau Jr. (Ed.), *PDMA Handbook for New Product Development*. New York: John Wiley & Sons.

Cooper, R. G. (2000). "Doing it right: Winning with new products." *Ivey Business Journal*, July/August, 54–60.

Cooper, R. G. (2001). *Winning at New Products: Accelerating the Process from Idea to Launch* (Third Edition). New York: Perseus Books; *www.stage-gate.com*

Cooper, R. G. (2003). "Stage-Gate new product development processes: A game plan from idea to launch." In: E. Verzuh (Ed.), *The Portable MBA in Project Management* (pp. 309–346). Hoboken, NJ: John Wiley & Sons.

Cooper, R. G. (2005a). *Product Leadership: Pathways to Profitable Innovation* (Second Edition). New York: Perseus Books.

Cooper, R. G. (2005b). "A Stage-Gate® idea-to-launch framework for driving new products to market." In: H. Levine (Ed.), *Project Portfolio Management: A Practical Guide to Selecting Projects, Managing Portfolios, and Maximizing Benefits*. San Francisco, CA: Jossey-Bass Business & Management (John Wiley & Sons Imprint).

Cooper, R. G. (2006a). "Formula for success." *Marketing Management Magazine* (American Marketing Association), March/April, 21–24.

Cooper, R. G. (2006b). "Managing technology development projects: Different than traditional development projects." *Research-Technology Management*, November/December, 23–31.

Cooper, R. G., and Edgett, S. J. (2005). *Lean, Rapid and Profitable New Product Development*. Ancaster, ON: Product Development Institute; *www.stage-gate.com*

Cooper, R. G., and Edgett, S. J. (2006). "Ten ways to make better portfolio and project selection decisions." *PDMA Visions*, **XXX**(3), June, 11–15.

Cooper, R. G., and Edgett, S. J. (2007). *Creating Breakthrough New Product Ideas: Feeding the Innovation Funnel* (Chapter 5). Ancaster, ON: Product Development Institute; *www.stage-gate. com*

Cooper R. G. and Edgett, S. J. (2008). "Maximizing productivity in product innovation." *Research-Technology Management*, January (forthcoming).

Cooper, R. G., Edgett, S. J., and Kleinschmidt, E. J. (2001). "Portfolio management for new product development: Results of an industry practices study." *R&D Management*, **31**(4), October, 361–380.

Cooper, R. G., Edgett, S. J., and Kleinschmidt, E. J. (2002a). *New Product Development Best Practices Study: What Distinguishes the Top Performers*. Houston, TX: APQC (American Productivity & Quality Center).

Cooper, R. G., Edgett, S. J., and Kleinschmidt, E. J. (2002b). *Portfolio Management for New Products* (Second Edition). New York: Perseus Publishing.

Cooper, R. G., Edgett, S. J., and Kleinschmidt, E. J. (2004a). "Benchmarking best NPD practices, Part I: Culture, climate, teams and the senior management role." *Research-Technology Management*, **47**(1), January/February, 31–43.

Cooper, R. G., Edgett, S. J., and Kleinschmidt, E. J. (2004b). "Benchmarking best NPD practices, Part II: Strategy, resources and portfolio management practices." *Research-Technology Management*, **47**(3), May/June, 50–60.

Cooper, R. G., Edgett, S. J., and Kleinschmidt, E. J. (2005). "Benchmarking best NPD practices, Part III: The NPD process and decisive idea-to-launch activities." *Research-Technology Management*, **47**(6), January/February, 43–55.

Cooper, R. G., and Kleinschmidt, E. J. (1991). "New product processes at leading industrial firms." *Industrial Marketing Management*, **10**(2), May, 137–147.

Cooper, R. G., and Mills, M. (2005). "Succeeding at new products the P&G way: A key element is using the 'Innovation Diamond'." *PDMA Visions*, **XXIX**(4), October, 9–13.

Docherty, M. (2007). "Primer on 'Open Innovation': Principles and practice." *PDMA Visions*, April, 13–15.

Griffin, A. (1993). "Metrics for measuring product development cycle time." *Journal of Product Innovation Management*, **10**(2), March, 112–125.

Griffin, A. (1997). *Drivers of NPD Success: The 1997 PDMA Report*. Chicago, IL: Product Development & Management Association.

Huston, L., and Sakkab, N. (2006). "Connect and develop: Inside Procter & Gamble's new model for Innovation." *Harvard Business Review*, **84**(3), March.

Montoya-Weiss, M. M., and Calantone, R.J. (1994). "Determinants of new product performance: A review and meta analysis." *Journal of Product Innovation Management*, **11**(5), November, 397–417.

PDMA. (2004). *The PDMA Foundation's 2004 Comparative Performance Assessment Study (CPAS)*. Chicago, IL: Product Development and Management Association.

25

Learning and Experience

Paul Olk and Robert McGowan

University of Denver

Introduction

Effective management of knowledge and how an organization learns from its experiences have become increasingly relevant to most technology and innovation management issues. Reflecting an overall trend towards a greater percentage of work involving individuals sharing knowledge (Beardsley, Johnson, and Manyika, 2006), effective organizational learning has been identified in technology and innovation management research as critical to a firm achieving and sustaining a competitive advantage. We address this development by focusing on what is meant by knowledge, how individual knowledge leads to organizational learning, and the various ways firms learn.

What is knowledge?

Knowledge is defined variously as (i) expertise and skills acquired through experience and education, the theoretical or practical understanding of a subject; (ii) what is known in a particular field or in total, facts and information; or (iii) awareness or familiarity gained by experience of a fact or situation.[1] Most start with Plato's formulation of knowledge as "justified true belief". Plato aside, the focus in this discussion will be on "organizational knowledge" and how it is acquired. Naturally, skills, familiarity, and experience all play a pivotal role.

An organization's ability to perform an activity rests on its knowledge of that activity; that is, the competencies rest on both technological and market knowledge (Tidd, Bessant, and Pavitt, 2005). Technological knowledge is knowledge of components, linkages between components, methods, processes, and techniques that go into a product or service. Market knowledge is knowledge of distribution channels, product or service applications, and customer expectations, preferences, needs, and wants. It is important to note that the two are not necessarily separate and discrete: technological knowledge can help identify new market opportunities in terms of features and differentiation; market knowledge can assist in determining those features and attributes that should be addressed. Abernathy and Clark (1985) refined this distinction. They state that a firm's technological capabilities could become obsolete while its market capabilities remain intact. If such market capabilities are important and difficult to acquire, an incumbent whose technological capabilities have been destroyed can use market ones to its advantage over a new entrant.

Three properties of technical knowledge determine how well a firm performs the activities that rest on knowledge: newness, quantity, and tacitness.

Newness: one critical property of the knowledge that underpins an activity is how new it is to the function or organization performing the activity. If it is very different from existing knowledge, it is said to be radical or competence-destroying. For example, the transition from the traditional travel agent to online or Internet booking has transformed the entire role of travel agents today. Travel agents are now able to utilize the Internet to develop customized travel packages for particular market segments. If newness builds on existing knowledge, it is said to be incremental or competence-enhancing. The newer the knowledge, the more difficult it is for firms to perform these activities.

Quantity: a second factor of technical knowledge is the quantity of the new knowledge. The move from a three-piece steel can to a two-piece, thin-drawn aluminum can, or the move from a Boeing 747 to a Boeing 777 each entailed new knowledge. But the amount of knowledge needed for a plane is a lot greater than that needed for a

[1] *Oxford English Dictionary*.

can. This amount is a function of the complexity of the activities that go into the product, which may or may not result in a complex product. For example, the activities that go into the discovery and development of pharmaceutical products are complex and knowledge-intensive. Yet the final product is relatively simple. The activities that go into making a plane are also complex and knowledge-intensive and the final product is also complex.

Tacitness: in examining the nature of organizational knowledge, most researchers draw from Polanyi's (1966) distinction between articulated (or explicit) versus tacit knowledge. The former consists of information that can readily be codified, and therefore transferred. It is explicit if it is spelled out in writing, verbalized, or codified in drawings, software programs, or other products. The latter is information that is not codifiable, often developed and transmitted through experience, and represented in the routines and standard operating procedures of an organization. Tacit knowledge may not even be verbalized or articulated (Hedlund, 1994). It is also the core of the firm's prior knowledge base. The firm may have some proprietary explicit knowledge such as firm-specific blueprints and standard operating procedures. However, they are useful only when tacit knowledge enables its members to utilize them. Much of the knowledge that underlies the effective performance of an organization is the tacit knowledge embodied in its members (Howells, 1996; Nelson and Winter, 1982).

Nonaka and Takeuchi (1995) used these two types of knowledge to identify four modes of organizational knowledge creation. Depending on whether the knowledge is initially explicit or tacit and on whether it becomes explicit or tacit, the knowledge creation process may consist of socialization (tacit to tacit), externalization (tacit to explicit), internalization (explicit to tacit), or combination (explicit to explicit). Socialization occurs through a process of shared experiences which lead to the individuals having common mental models and similar technical skills. Externalization represents the process of converting tacit knowledge into explicit concepts. Nonaka, Takeuchi, and Umemoto (1996) suggest that knowledge can be converted from tacit to explicit via metaphors, analogies, and models. Metaphors allow one to understand something by seeing it in terms of something else. An analogy helps us understand the unknown through the known and bridges the gap between an image and a logical model. As these concepts become explicit, the concept can then be modeled. According to Nonaka and Takeuchi (1995), this process is quintessentially knowledge creation since as the concept moves from tacit knowledge through to a logical model, inherent contradictions and insufficient (or inadequate) conceptualizations are addressed and the outcome reflects a new concept. Internalization is the opposite process, converting explicit knowledge into tacit knowledge. It is related to "learning by doing", which will be discussed below; the individual takes the documented or verbalized knowledge and converts it through experience to tacit knowledge. Combination involves converting explicit knowledge into another type of explicit knowledge. This may occur through recombination or reconfiguration of existing knowledge into new knowledge.

Converting individual knowledge into organizational learning

Central to a learning perspective is that these different four modes of knowledge conversions are embedded in the organizing principles of an organization. Organizational learning, defined as the process of improving actions through better knowledge and understanding (Fiol and Lyles, 1985), involves transferring this individual knowledge into organizational routines. While individuals play a critical role in creating the knowledge structure of an organization, the knowledge resides at the organizational level.

Nonaka and Takeuchi (1995) modeled this through a spiral of organizational knowledge creation. Knowledge moves from the individual to higher levels through repeated transformations between articulated and tacit forms of knowledge. They distinguished between four levels of carriers or agents of the knowledge (individual, group, organizational, and inter-organizational). As knowledge is continually transformed between tacit and articulated states, it encompasses more individuals in and around the organization, moving from knowledge shared by individuals to becoming group-level knowledge, then to an organizational level of understanding and eventually becoming inter-organizational knowledge.

Different ways of learning

Having addressed what is knowledge and how it becomes organizational knowledge, we now turn our focus to the different ways in which companies learn. Specifically, we focus on the concepts of exploitation and exploration, learning by doing, trial and error, improvisation, learning by diffusion, and learning by scaling.

Explore and exploit: the most commonly discussed type of organizational learning involves exploration or exploitation. Exploration focuses on the development of new knowledge the organization does not possess—either through internal development by such actions as experimentation, discovery or play, or from external acquisition.

It is often more expensive, and the returns are often uncertain, possibly negative, and relatively far off into the future. Exploitation addresses how a company can make use of its existing knowledge or routines through refinement or extension. Its returns are typically more certain, positive, and faster. March (1991, p. 71) explained the concepts and the central tension for organizational learning:

> "Exploration includes things captured by terms such as search, variation, risk taking, experimentation, play, flexibility, discovery, innovation. Exploitation includes such things as refinement, choice, production, efficiency, selection, implementation, execution. Adaptive systems that engage in exploration to the exclusion of exploitation are likely to find that they suffer the costs of experimentation without gaining many of its benefits. They exhibit too many undeveloped new ideas and too little distinctive competence. Conversely, systems that engage in exploitation to the exclusion of exploration are likely to find themselves trapped in suboptimal stable equilibria."

Organizations must balance the trade-off between the short-term, immediate benefits from exploitation with the longer term, less certain benefits but ones that will more likely lead to continuing future performance benefits from exploitation.

The benefits from balancing exploration and exploitation have led a focus on the ambidextrous organization. Tushman and O'Reilly (1996) argued that firms must be good at both short-term incremental adaptation as well as long-term change to accommodate discontinuous change. For the short-term orientation, companies will organize to gain efficiency and quickly make use of existing knowledge. This may include good alignment with product markets, strong relationships with existing customers, strong development divisions, and strong internal communication efforts. For long-term orientation, these companies may set up separate divisions (sometimes referred to as skunkworks) that focus on different activities and can develop the specific knowledge without interference from other knowledge practices. The challenge for this structure is to not only manage the separation between these two efforts but then to integrate the new knowledge with the existing knowledge when it has sufficiently developed.

Learning by doing: learning by doing describes situations in which experience leads to an increasing understanding of a process, which tends to result in increased efficiency (Argote, 1999). The emphasis is often on the development of strategic capabilities through a process of building routines. That is, as a company proceeds down the "learning curve" it acquires, develops, and refines its routines as a consequence of success in responding to challenges and opportunities. Successful responses are repeated and become encoded into an organization's knowledge base as part of its standard operating procedures. Because the emphasis is on enhancing efficiency, learning by doing generally reduces the range of acceptable behaviors and is less likely to lead to novel solutions. While there is an intuitive understanding of this relationship, the limited, fine-grained, empirical evidence supports the underlying assertion that deliberate actions by management—particularly aimed at solving problems and transferring these practices throughout the organization—accelerates learning and improves performance (Macher and Mowery, 2003).

Trial and error: a trial-and-error process, sometimes referred to as adaptive learning, begins when a firm undertakes a course of action. After this action, there is some outcome response from the environment, which the firm interprets and evaluates the response, and then adapts its course of action to enhance the likelihood of the desired response (Van de Ven and Polley, 1992). Throughout multiple iterations of this process, the firm continually adjusts to new information about the relevant environment as the information becomes available. As part of an intentional learning process, trial and error can produce novel responses and the development of new routines. It is more likely to be successful when the goals are clear, when the response from the environment is sufficiently strong, and when the connection between the novel elements of actions and external feedback is strong.

Improvisation: improvisation differs from both learning by doing and trial by error in that it occurs in situations in which neither prior practices nor external feedback are the primary source of the learning. Rather, during improvisation, firms engage in real-time, short-term learning that may lead to the creation of a new capability (Miner, Bassoff, and Moorman, 2001). Rather than continuing with prior practices, a firm may deliberately break from earlier behaviors and try a novel approach in an effort to learn. While this may appear to be similar to a formal experimental effort—sometimes labeled selectionism (Sommer and Loch, 2004)—where the firm pursues multiple solutions independently of one another and picks the best one *ex post*, improvisation is generally focused on a deliberate effort to try a single, novel approach. Learning is more improvisational when the design and execution of a solution occur simultaneously, and more experimental when it involves planned variations in underlying conditions (Miner, Bassoff, and Moorman, 2001). The effect of improvisation may be ephemeral if the effort, even though successful in creating new knowledge, is not incorporated into a routine and becomes part of a long-term learning outcome. However, if a particular improvisation act is combined with other improvisations, perhaps as part of an experimentation or learning-by-doing effort, it may help produce long-term learning by the firm. Similar to the discussion of exploration and exploitation, a challenge for managers is to find the proper use of improvisation. If done too frequently, too independently, or on too large of a scale, improvisation may lead to a fragmentation of learning, to distraction from more formal experimental efforts, or to poorer organizational performance.

Learning by scaling: knowledge and learning can also be gained through scaling (Amidon, 1997; Nonaka and Takeuchi, 1995; Senge, 1990). The scale of a firm is the extent of its activity described by its size. The scale of a firm's activity can be described by its revenues, units sold, or some other measure of size. Economies of scale are based on the concept that larger quantities of units sold will result in reduced per-unit costs. Economies of scale are generally achieved by distributing fixed costs such as rent, general and administrative expenses, and other overheads over a larger quantity of units sold. When significant economies of scale exist in manufacturing, distribution, service, or other functions of a business, larger firms (up to some point) have a cost advantage over small firms. Thus, smaller, new-entrant firms need to differentiate their product on qualities other than price. As the smaller, new-entrant firm grows in size, it can also learn to reduce its costs per unit and price competitively with larger firms. Scaling over time allows the firm to take advantage of learning curve effects in which quality or efficiencies increase.

Another issue related to scale is the concept of scalability (Brafman and Beckstrom, 2006; Dyer and Ericksen, 2007). Scalability refers to how big a firm can grow in various dimensions to provide more service. There are several measures of scalability. They include volume or quantity sold per year, revenues, and number of customers. These measures are not independent, as scaling up the size of a firm in one dimension can affect the other dimensions. Easily scalable ventures are attractive, while ventures that are difficult to grow are less so. In terms of knowledge and learning, scalable ventures spend a great deal of time and energy transferring best practices as the enterprise grows. This may entail the rotation of individuals or briefing sessions in which best practices are discussed and disseminated.

Learning by diffusion: a great deal of research has been conducted to try and identify what factors affect the rate and extent of adoption of an innovation or new knowledge (Rogers, 2003). As a new technology or application becomes more widespread and accepted, firms will need to be able to learn quickly how it can be effectively integrated into existing (as well as future) product or service offerings. Naturally, there can be significant first-mover advantages for such firms—particularly if the new technology displaces existing technology.

There are a number of characteristics that have been found to affect diffusion:

—relative advantage that can be gained from the new technology;
—compatibility with other products/services and how possible synergies can be achieved;
—complexity (the more complex the feature or attribute, the less likely individuals will be willing to switch) or the lack of willingness to re-learn;
—trialability (learning by doing);
—observability (the more visible the attributes, the more likely others will adopt) which will also ease the process of learning new uses or applications.

In addition to the characteristics of adoption, there are key processes of diffusion. The well-known technology S-curve provides one example of diffusion patterns of new technology. There are also extrapolative or normative diffusion patterns. Extrapolative approaches entail using statistical models to predict trends (stock prices, energy consumption, agricultural production, etc.). Normative models start with a desired end-state and try to determine the optimal path (space exploration, modes of transportation).

Firms can be characterized as early-adopters, fast-followers, as well as imitators. The key to which mode to adopt depends on the newness of the new technology, the degree to which other firms have already entered into this market space, as well as the level of investment required. Timing of adoption is important since firms may have significant switching costs that will need to be addressed. For example, the switch from film print media to digital technology had a profound effect on a number of industries.

Finally, forecasting patterns of adoption is central to learning by diffusion. Knowledge from one application or product can serve as a useful proxy in determining patterns of growth and time required for adoption. For example, in determining the likely acceptance of high-definition television sets, experts looked at the adoption pattern of transitioning from black-and-white sets to color. They looked at which customers tended to be early-adopters as well as the importance of relative price differentials. Forecasting can also help to identify what might be required in the future, and to estimate how many are likely to be required in a given time period. However, in the case of new products or services, forecasting is difficult, as the products and markets may not be well defined.

Learning how to learn

Each of the above types of learning represents alternatives for how a firm can develop a new routine or a new process. The learning that occurs is related to a specific behavior and how it is enhanced by the type of learning. Another dimension to learning and experience is whether the firm engages in the appropriate type of learning or is particularly adept at this type of learning. That is, companies have to learn how to learn. Sometimes also referred to as higher order or second-order learning, this learning involves a firm's managers acquiring the knowledge to select the appropriate type of learning for the context or to improve its skill with a particular learning type. Similar to the specific types of learning, the ability to learn how to learn is also likely to be heterogeneous across firms and a likely source of sustained competitive advantage for those firms with this capability.

References

Abernathy, W., and Clark, K. B. (1985). "Mapping the winds of creative destruction." *Research Policy*, **14**, 3–22.

Amidon, D. M. (1997). *Innovation Strategy for the Knowledge Economy*. Newton, MA: Butterworth-Heinemann.

Argote, L. (1999). *Organizational Learning: Creating, Retaining and Transferring Knowledge*. Norwell, MA: Kluwer Academic.

Beardsley, S., Johnson, B. C., and Manyika, J. M. (2006). "Competitive advantage from better interactions." *McKinsey Quarterly*, **2**.

Brafman, O., and Beckstrom, R. A. (2006). *The Starfish and the Spider*. London: Penguin Books.

Dyer, L., and Ericksen, J. (2007). "Dynamic organizations: Achieving marketplace agility through workforce scalability." In J. Storey (Ed.), *Human Resource Management: A Critical Text* (Second Edition). London: Thompson Learning.

Fiol, C. M., and Lyles, M. A. (1985). "Organizational learning." *Academy of Management Review*, **10**, 803–813.

Hedlund, G. (1994). "A model of knowledge management and the n-form corporation." *Strategic Management Journal*, **15**, 73–90.

Howells, J. (1996). "Tacit knowledge, innovation and technology transfer." *Technology Analysis and Strategic Management*, **8**(2), 91–106.

Macher, J. T., and Mowery, D. C. (2003). " 'Managing' learning by doing: An empirical study in semiconductor manufacturing." *Journal of Product Innovation Management*, **20**, 91–410.

March, J. G. (1991). "Exploration and exploitation in organizational learning." *Organization Science*, **2**, 71–87.

Miner, A. S., Bassoff, P., and Moorman, C. (2001). "Organizational improvisation and learning: A field study." *Administrative Science Quarterly*, **46**, 304–337.

Nelson, R. R. and Winter, S. G. (1982). *An Evolutionary Theory of Economic Change* (p. 134). Cambridge, MA: Harvard University Press.

Nonaka, I., and Takeuchi, H. (1995). *The Knowledge Creating Company*. New York: Oxford University Press.

Nonaka, I., Takeuchi, H., and Umemoto, K. (1996). "A theory of organizational knowledge creation." *International Journal of Technology Management*, **11**, 833–845.

Polanyi, M. (1996). *The Tacit Dimension*. London: Routledge & Kegan Paul.

Rogers, E. (2003). Diffusion of Innovations. Free Press: New York.

Senge, P. M. (1990). *The Fifth Discipline*. New York: Doubleday.

Sommer, S. C., and Loch, C. H. (2004). "Selectionism and learning in projects with complexity and unforeseeable uncertainty." *Management Science*, **50**, 1334–1347.

Tidd, J., Bessant, J., and Pavitt, K. (2005). *Managing Innovation: Integrating Technological, Market and Organizational Change*. New York: John Wiley & Sons.

Tushman, M., and O'Reilly, C. A. (1996). "Ambidextrous organizations: Managing evolutionary and revolutionary change." *California Management Review*, **38**, 8–30.

Van de Ven, A. H., and Polley, D. (1992). "Learning while innovating." *Organization Science*, **3**, 92–116.

26

New-product Development Innovation and Commercialization Processes

Abbie Griffin
University of Utah

History of product development processes

Much new-product development (NPD) research over the last two decades has focused on improving the "process" of new-product development. It has taken the perspective that NPD could be managed like any other (complex) process. The underlying assumption is that standard methods and protocols could be put into place, and individuals and teams could follow a process to repeatedly commercialize a stream of successful new products. That is, the field has worked to change the "art" of product development to the "science" or process of product development.

The *PDMA Handbook of New Product Development* defines a product development process as:

> "A disciplined and defined set of tasks, steps and phases that describe the normal means by which a company repetitively converts embryonic ideas into salable products or services" (Kahn, Castellion, and Griffin, 2005, p. 601).

Formal product development processes were first developed by NASA in the 1960s (Cooper, 1996). Their phased project planning (PPP) process was an elaborate and detailed scheme for working with contractors and suppliers on very complex space projects. The PPP broke development into discrete phases. The formal review points at the end of each phase ensured that all of the tasks in the phase had been satisfactorily completed prior to committing funding for the next phase of the program. However, the process was very engineering-driven, applying strictly to the product's physical design and development. No marketing, manufacturing, finance considerations, or people representing those functions were included in the process.

Eventually, use of the process migrated to the Department of Defense to aid in developing complex planes and equipment. Ultimately, PPP was adopted by a number of firms, starting with government contractor firms, but eventually moving into more general use. While the system did bring discipline to a previously chaotic and frequently *ad hoc* set of tasks, it was cumbersome and too narrow in scope, dealing only with the physical development phase of the innovation process.

The first mention of a "product development process" in the academic literature is in 1966. Sherman (1966) quotes a study conducted by the consulting firm Booz, Allen & Hamilton (BAH) on new-product management. In this study, BAH reported that internal development required more than just focusing on R&D, stating that "every step in the entire process of new-product evolution must be carefully planned" (Sherman, 1966, p. 42). They then outline a six-stage process that they suggest firms should follow in developing new products: exploration, screening, business analysis, development, testing, and commercialization. Additionally, they indicate that a go/no-go decision must be made by management at the end of each stage.

Since 1966 a significant amount of research on NPD processes has been done. One of the earliest scholars on this topic is Robert G. Cooper, inventor of the Stage-Gate® process. In his first article on NPD processes, he concluded, based on three case studies, that effective processes need to (a) consist of a sequence of discrete stages; (b) proactively integrate marketing and technical activities; (c) allow for activities to be conducted in sequence at times and in parallel at other times; and (d) provide for making incremental commitments to projects over time (Cooper, 1976). Over the next decade, he (and co-authors) further developed and refined what effective processes include, how well various steps are carried out, and what impact each step had on new-product outcomes. Ultimately, Cooper's research culminated in the well-publicized generalized Stage-Gate® process (Cooper, 1996, and see Chapter 24 in this encyclopedia).

The overall innovation process

The NPD process typically is represented by some number of generic stages or phases (see Figure 24.1 in the Stage-Gate® chapter). One of the underlying premises of Stage-Gate® types of processes is their seemingly linear nature. Indeed, one of the objectives of developing formal NPD processes was to eliminate iterations back into the earlier phases of the process necessitated by infeasible or un-manufacturable concepts having proceeded far along the development path toward commercialization.

In actuality, however, Stage-Gate® types of processes and the majority of the research and literature on NPD processes focuses on tasks to be completed *after* new technology capabilities have been developed and the idea concept is generated, when it is ready to go into initial screening. Two other aspects of NPD processes that generally fall outside the scope (and most frequently precede) the more formalized NPD process stages represented by Stage-Gate® types of processes also must be considered in managing the overall innovation and new-product development process: the front end of innovation, or the "fuzzy front end" (Smith and Reinertsen 1991), and gaining product acceptance into the formalized process development structure. Figure 26.1 presents a simplified, linear illustration of this total process of innovation and commercialization. Figure 26.2 presents a more expanded illustration of the complexities of innovation and new-product development, including feedback loops from later stages of the overall process back to earlier ones, when initial plans could not be realized due to technical or commercial issues. In Figure 26.2 the successful outcome of the process is "Routine Product Shipments". The Stage-Gate® process fits into the three-box group of tasks in the lower central portion of the figure.

The fuzzy front end of innovation: creating the concept

It is in the fuzzy front end (FFE) that the concepts that will go into development are generated and all of the technical unknowns are eliminated. Understanding the upfront sub-processes necessary to develop high-potential concepts are key to overall NPD success (Smith and Reinertsen, 1991). Success in this step requires both technical and business knowledge, as well as a creative ability to see potential across boundaries and synthesize across fields. Study of the FFE suggests that it, in and of itself, consists of a complex set of sub-tasks,

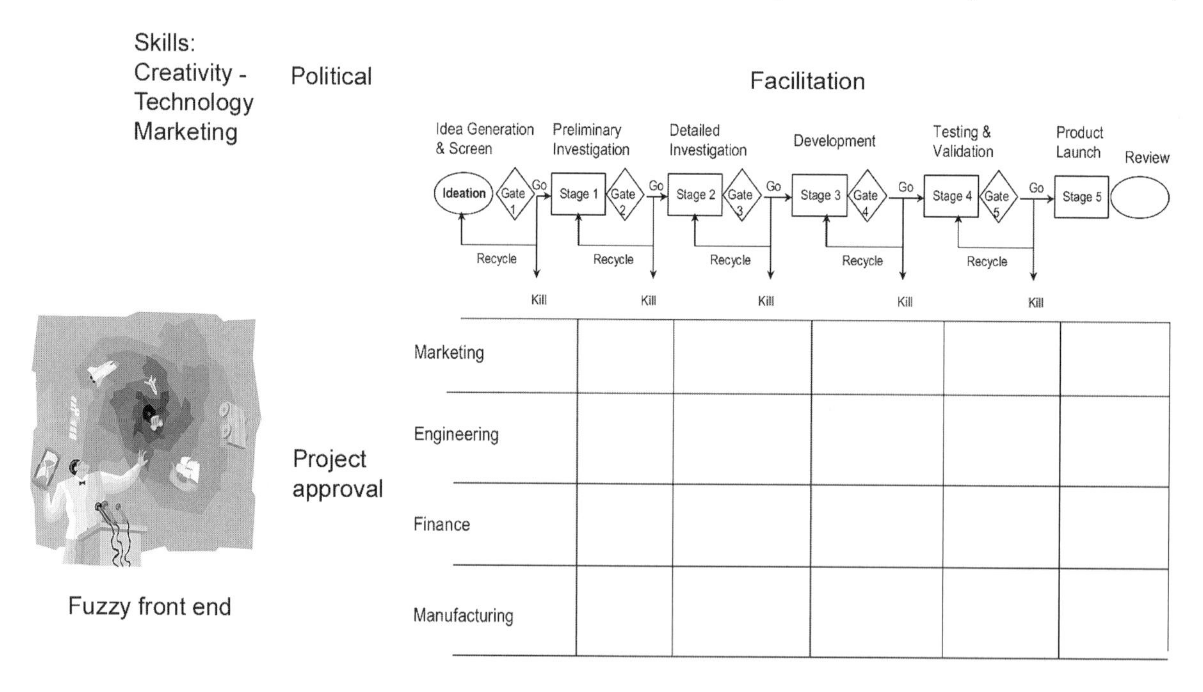

Figure 26.1. The innovation and commercialization process.

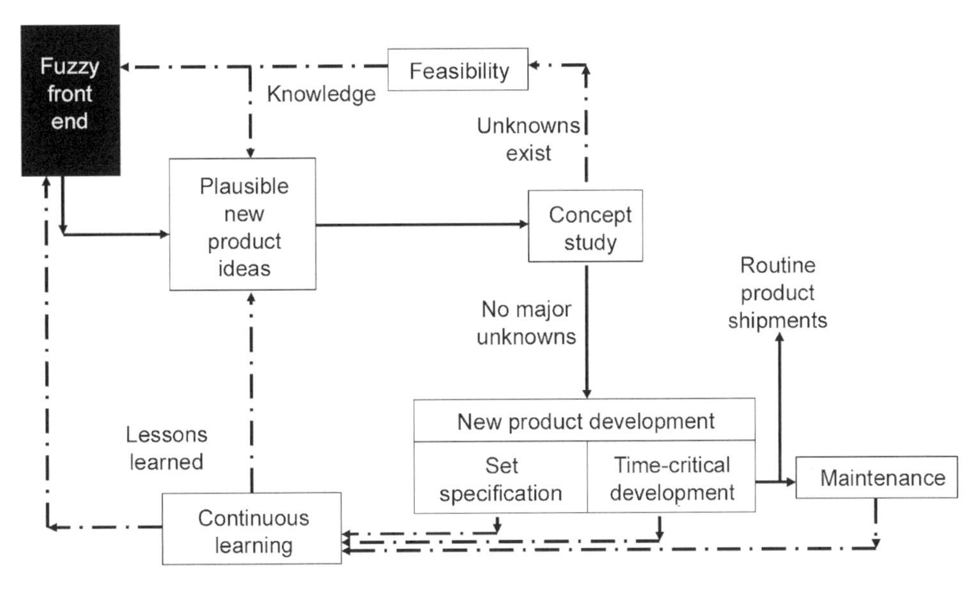

Figure 26.2. An expanded view of the innovation and commercialization process.

which are more iterative and flexible than the formal aspects of later processes.

In the earliest study of the FFE, Khurana and Rosenthal (1997, 1998) found that U.S. firms predominantly used a rather ordered process for the front end, with product and portfolio strategy and the organizational structure feeding a pre-phase 0 stage where preliminary opportunity identification and market and technology analysis occurred (Figure 26.3). This in turn fed into a phase 0, product concept definition phase, which in turn fed into phase 1, the product definition and project planning phase. After that, the formal project acceptance decision was made, and projects typically entered into the firm's formal product development process. The pre-phase 0 stage starts when someone at the firm first recognizes in a semi-formal way an opportunity. How this occurs is unspecified, and no detail is provided for how to systematize opportunity recognition. The key insight obtained from their FFE study is that the greatest success comes to organizations that take an holistic approach to the front end, which arises as a derivative of a "company-wide culture" focused on business vision, technical feasibility, customer focus, schedule, resources, and coordination. However, they also find that such holistic approaches are rare. Only 1 of the 18 com-

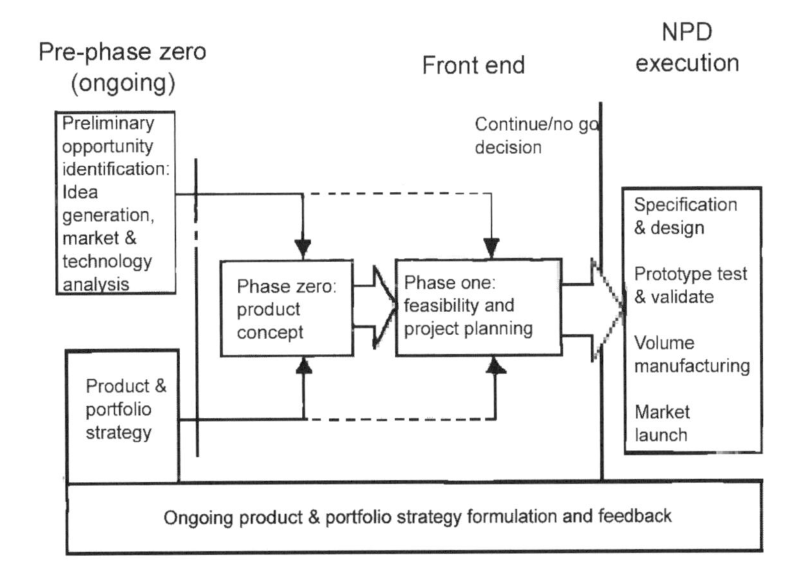

Figure 26.3. A stylized view of the FFE. *Source*: Khurana and Rosenthal (1998).

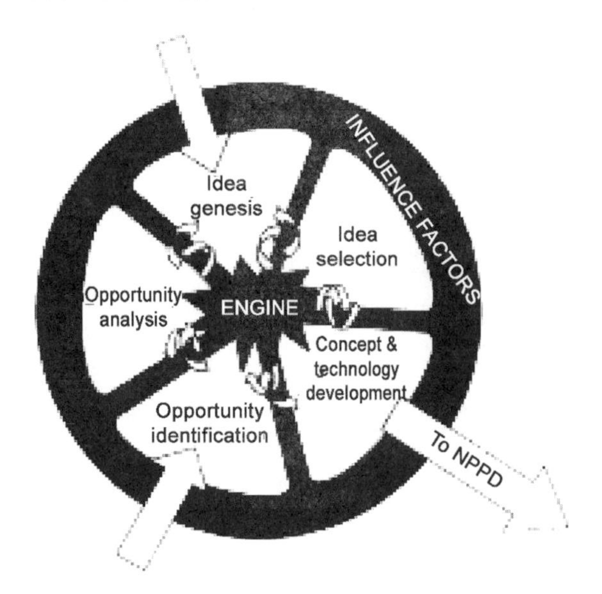

Figure 26.4. New Concept Development Model for the front end of innovation.
Source: Koen *et al.* (2001).

panies in their study was identified as having developed a coherent set of approaches covering all four front-end elements: product strategy, product definition, project definition, and organizational roles (Khurana and Rosenthal, 1998).

Since this initial research into the FFE several teams of researchers have proposed various models for how the FFE operates and suggestions for how best to manage across it. Perhaps the most comprehensive model of the FFE was developed collectively by a team of eight companies, led by an academic researcher, ultimately developing the New Concept Development Model depicted in Figure 26.4 (Koen *et al.* 2001). The key elements in the front end—opportunity identification, opportunity analysis, idea genesis, idea selection and concept, and technology development—are driven by an "engine" representing senior management support. Teams iterate across the elements, using each as needed in whatever order necessary, until a sufficiently interesting concept has been developed to the point that it can enter the formal NPD process at the firm. Koen *et al.* (2001) specifically avoid the word "process" for this front end of innovation because they believe the term implies a structure that may not be applicable. The FFE process resides within a circle of factors that influence its operation, including the organizational capabilities of the firm, the business strategy, the outside world, and the enabling science that will be used. Of all the FFE models in the literature, this model most stresses the iterative nature of the task.

Gaining project acceptance: from concept to formal project

Even once plausible concepts have been developed and their major (technical and commercial) unknowns have been eliminated, they still must be accepted into the formal product development processes. A proposed opportunity must cross the initiation gap and gain corporate support as a formal new-product development project (Markham and Kingon, 2004). Gaining project acceptance requires developing a business plan for the proposed product or service and shepherding that plan through the firm's funding and staffing process. Although several processes and structures for increasing the rationality of decision making have been developed, this task fundamentally remains one of managing the politics of gaining project acceptance in the firm. Frequently, managing the politics of this task is led by a product champion who understands the technology and connects it to a market need, recognizing its potential.

Project implementation: from project to product launch

After the initial concept has been defined and has management support, a formal project is launched to implement the concept and bring it to market, usually following some sort of formal Stage-Gate® type of product development process (Griffin, 1997). At this point in the NPD process, a project manager generally is assigned formally to the project. His responsibility is to organize the execution of the project and ensure that each task and milestone is completed on time and within an allocated budget.

The whole innovation process: more than the sum of the parts

Invention in the FFE, project definition and acceptance, and process execution each require different competencies (Schumpeter, 1934). The FFE requires technical competency, business knowledge, and creativity. Project acceptance requires business, market, and some technical knowledge, coupled with a driving political capability. Project implementation requires project management and facilitation skills. In a traditional NPD setting, different individuals, with different types of skills, undertake the different roles involved with moving an innovation opportunity through the laboratory, gaining political acceptance for it as a project, and managing the formal commercialization process. In the classical view of NPD, these stages are driven or managed by inventor technologists, champions, and implementers, respectively.

In summary, the formal NPD "process" is presented in the literature as a fairly linear, stylized, and rather rational process, which is fed by ideas and concepts that are generated in the fuzzy front end. However, when one considers the entire innovation and commercialization process, more than just a formalized implementation process is required for success over the long term. Capabilities to manage the political aspects of gaining product acceptance into the organization's formal structures must be put into place. Additionally, capabilities for developing new technologies and for linking those new technologies to potentially interesting applications must also be put into place: mechanisms for innovating in the fuzzy front end are needed. Finally, the most successful firms are those that also put into place feedback mechanisms such that new lessons learned and even new knowledge generated in one generation of innovation and commercialization can be utilized to work more effectively in subsequent generations of both incremental and more innovative development.

References

Cooper, R. G. (1976). "Introducing successful new industrial products." *European Journal of Marketing*, **10**(6), 300–329.

Cooper, R. G. (1996). *Winning at New Products*. Reading, MA: Addison-Wesley.

Griffin, A. (1997). "PDMA research on new product development Practices: Updating trends and benchmarking best practices." *Journal of Product Innovation Management*, **14**(6), November, 429–458.

Kahn, K., Castellion, G., and Griffin, A. (Eds.). (2005). *The PDMA Handbook of New Product Development* (Second Edition). Hoboken, NJ: John Wiley & Sons.

Koen, P., Ajamian, G., Burkart, R., Clamen, A., Davidson, J., D'Amore, R., Elkins, C., Herald, K., Incorvia, M., Johnson, A. *et al.* (2001). "Providing clarity and a common language to the 'Fuzzy Front End'." *Research-Technology Management*, March/April, 46–55.

Khurana, A., and Rosenthal, S. R. (1997). "Integrating the Fuzzy Front End of new product development." *Sloan Management Review*, Winter, 103–120.

Khurana, A., and Rosenthal, S. R. (1998). "Towards holistic 'front ends' in new product development." *Journal of Product Innovation Management*, **15**(1), 57–74.

Markham, S. K., and Kingon, A. (2004). "Turning technical advantage into product advantage." In: P. Belliveau, A. Griffin, and S. M. Somermeyer (Eds.), *The PDMA ToolBook2 for New Product Development*. Hoboken, NJ: John Wiley & Sons.

Schumpeter, J. A. (1934). *The Theory of Economic Development*. Cambridge, MA: Harvard University Press.

Sherman, R. F. (1966). "New-product development: Get the whole company into the act." *Management Review*, **55**(9), 41–44.

Smith, P. G., and Reinertsen, D. G. (1991). *Developing Products in Half the Time*. New York: John Wiley & Sons.

27

Service Innovation

John E. Ettlie, Iliana Payano Mejia, and Jessica M. Walker

Rochester Institute of Technology

It is a sublime irony that the economic dominance of the service sector in developed economies has generated little or no applied research on service innovation. Many even consider the distinction between a service and a service innovation at best uninteresting and at worst an annoying waste of time. In response, the intention here is to be at least provocative and at best provide a documentation of the headwaters of a rich and fruitful new research stream in the field of innovation management and operations.

Numerous typologies of services have been published (see Berry *et al.*, 2006) but they all fall short of being very useful to service innovators and their systematic observers. Most likely, this is a consequence of the fact that the service sector spends little of its retained earnings on R&D. Not surprisingly, much of the academic research on services appears in the marketing and operations literature—how to package, position, sell, and deliver a service offering. While this literature provides a solid context for the current research, it does not give us much information on the whys and wherefores of the service innovation phenomena. In this report, it is argued that service innovation is different—different from just a clear understanding of R&D, manufactured new products, and services offered through any channel. Up until now, service innovators went on intuition and the crude, approximate application of what we know to be state of the art in product innovation. Although not irrelevant, this approximation is much farther from the reality when we make detailed, microscopic observations on the new-service development process.

The idea of service innovation—some new, non-manufacturing, construction, or extraction offering that departs in an important way from precedent—can be divided into three very broad categories: new-service offerings by incumbent service providers (e.g., Bank of America's high-technology branches in Atlanta), recent startup service innovations (e.g., Google), and service innovations offered by manufacturing firms (e.g., GM's OnStar). Although none of these examples is part of this study, the latter category is the subject of this chapter.

Most developed countries rely heavily on services to sustain their economies—ranging from 70–80% and more of GDP. Ironically, we know very little about how new services come about and add commercial advantage at the firm, industry, or country level of aggregation. This state of affairs is about to change. IBM, among others, has launched a campaign to add to our knowledge base of service sciences at the strategic (as opposed to the tactical) level of firm analysis. Several new programs in service sciences build on service sector concentrations at universities and a rich tradition of service sector research in operations (e.g., Chase and Apte, 2007; Heineke and Davis, 2007) and marketing, especially in North America. These initiatives promise a new, vigorous, systematic re-evaluation of the service sector. However, with rare exceptions, which are reviewed below, little or no systematic, rigorous applied research has appeared on service innovation.

There are exceptions to the generally agreed-upon observation that little or no applied research on service innovation has appeared in the literature. For example, Susman, Warren, and Ding (2006) review the literature on service innovation with attention to manufacturing and find a number of European scholars writing and observing cases in this arena. There are others (Sawhney, Balasubramanian, and Krishnan, 2004; Tidd and Hull, 2003), including the study that was the progenitor of the current research (Ettlie, Rosenthal, and Hall, 2007) but little else has appeared in the literature on service innovation. This seems especially remarkable when compared with the apparent unabated burgeoning of the literature on new-product innovation processes.

It could be that the very nature of the service innovation process contributes to the paucity of literature on the subject: it is less formalized generally than most other critical functions of organizations. Some definitions of service even highlight the absence of metrics associated with the subject: "series of activities" and "more or less intangible" are typical of the basics for the sector (Gustafsson, Nilsson, and Johnson, 2003, p. 4). Even potentially more challenging is the addition of innovation to this definition—lack of precedence and the absence of concentration of technical specialists to take responsibility for this process. It may not surprise most readers that since formalized R&D is totally absent in the service sector, general managers are often the source of successful new-service ideas—just the opposite as findings on idea sourcing for new products (Ettlie and Elsenbach, 2007). The latter result was replicated in the case studies discussed below of this project, although the dependent variable in this investigation was the persistence and sustaining of new-service ideas, concepts, or new offerings, not economic success by traditional accounting measures *per se*.

Although systematic theory-driven empirical research may be severely lacking in this field, this has not stopped publication of descriptive studies and case histories of service innovation. In fact, many of these services, including those of particular interest to this project—service innovation offerings by incumbent manufacturing firms— have been celebrated in the trade and business literature. Many of these cases are discussed below. The Bank of America teaching case (Thomke, 2003) is widely used in business schools and tells the story of how the Atlanta branches of this bank were significantly modernized by an erstwhile quality group within BoA in response to a competitor's initiative.

What is a significant new-service offering by a manufacturing firm? For example, in manufacturing one important category for the source of service innovation and for our purposes here is that an innovation cannot be "just" post-sales service. Some authors argue even this type of service is rare among manufacturers, preferring to leave service to the retail or wholesale sector. These are services integrated into their core product offerings, and it would seem a natural transition to this activity, but apparently it is not (Oliva and Kallenberg, 2003; Wise and Baumgartner, 1999). Again taking manufacturing as the example, this represents a shift in strategy for most firms from cost control to profit growth, often being pushed by customers to introduce new services (Auguste, Harmon, and Pandit, 2006; Reinartz and Ulaga, 2006). Others have observed how different truly radical innovation and the processes needed to offer breakthroughs are (Hargadon, 2003; Rothaermel and Hill, 2005). An example of a truly innovative offering by an incumbent manufacturing firm is the General Motors Corporation OnStar system. This case is included in the summary examples of published focal cases of interest below.

Published examples of service innovation

We divide service innovations into three broad categories and give published examples of each below. First, published examples of significant, new services offered by manufacturing firms. Then incumbent services and their innovations of note are featured. Finally, startup service firms offering significant, innovative, new offerings. Two notes of caution are in order here. These cases were not selected for any representation of concepts, nor a random-sampling method. They were convenient illustrations of their category of service innovation. Second, the reader should keep in mind that innovation is rare in any context, and seeing more than a dozen examples of service innovation might give the impression that this phenomenon is common—just the opposite is true. Significant departure from the past is never commonplace—it is always rare. These are examples of the exceptions, not the typical practice of offering something new—with a label that it is innovative—most "new" offerings, are just copies with little improvement over extant offerings, repackaged and sold as the latest fad.

Manufacturing firms offering significant new-service innovations

In Table 27.1 published examples of service innovations by incumbent manufacturing firms are summarized. These are included to provide clarity, context, and breadth to the category of inquiry in this chapter. None is part of the project primary data collection, but these cases are similar in their nature to the cases compiled in this research, and they are useful for setting the stage for this nascent research stream.

GM OnStar

General Motors Corporation's OnStar is an on-board information technology service that has been offered for 10 years as of this writing, and is, perhaps, the best known of any service innovation offered by an incumbent manufacturing firm. The range of services is large, but in an average month (in the third-quarter of 2005) there were 12.6 million hands-free calls, 340,000 route support calls, 44,000 remote door unlocks, 32,000 remote diagnostics, and 25,000 roadside assistance calls, 15,000 emergency calls, 5,500 Good Samaritan calls, 900 automatic airbag notifications, and 500 stolen vehicle location assists (see Parekh, 1998). Although most of the technology is off-the-shelf, OnStar has applied for 300 patents. The secret of the success of this initiative, if there is one, is that GM allowed OnStar management to adopt the time horizon, pacing, and decision processes that the consumer

Table 27.1. Published cases of service innovations by manufacturing firms.

Case	*Description*	*Citation*
1. GM OnStar	GM's road service delivered via cellphone	Vasilash (2006)
2. John Deere	Remote machine-monitoring system	Press release[a]
3. Stryker	Physiotherapy Associates	Press release[b]
4. CAT	Remanufacturing	Arndt (2005)
5. UT	Pratt & Whitney parts services	Lunsford (2006)
6. UT	Otis Elevator's OTISLINE	Parekh (1998)

[a] *http://www.qualcomm.com/press/releases/2006/060818_john_deere_construction.html*
[b] *http://www.myphysio.com/index.cfm?fuseaction=goHome*

electronics industry uses, like their partner Motorola. Six generations of technology in 9 years resulted.

John Deere remote machine–monitoring system

John Deere and Company has partnered with Qualcomm to develop a remote machine-monitoring system on forestry and construction equipment. JDLink automatically will collect and transmit information on how equipment is being used in order to ultimately allow better equipment utilization. Four levels of service are offered, from the most basic on location and use to more advanced levels on fuel consumption, engine performance, gear selection, temperatures, and diagnostics. This service extends the "heritage of integrity, quality and ... innovation around the globe" (see Hoovers.com). Qualcomm is a leader in providing wireless data platforms and has shipped (2006) over 711,000 mobile systems to business customers in 39 countries.

Stryker Physiotherapy Associates

Stryker is a well-known manufacturer of alloy implants like artificial knee and hip joints. The subsidiary company started Physiotherapy Associates in 1985 as many patients entered into physical therapy after surgery. With 475 clinics in 31 states, including 35 in Florida alone, the company specializes in orthopedics, spinal care, and neurological rehabilitation. Patients include injured workers and athletes as well as diabetic wound care. Federal health care programs like Medicare and Medicaid account for about 15% of Physiotherapy revenues (Hoovers.com). In June of 2007, Physiotherapy was purchased by Water Street Healthcare Partners located in Chicago, a private equity firm, for about $150 million in cash and committed additional equity capital. Water Street intends to continue pursuing acquisitions of small and mid-sized physical therapy companies.

Caterpillar tractor remanufacturing services

The CAT plant in Corinth, MS, employs 600 people rebuilding two diesel engines per shift. CAT has additional remanufacturing plants and opened its 14th specialized plant in Shanghai in 2006. Annual revenue for this part of CAT's business exceeds $1 billion and is growing at a rate of 20% a year. "Services are the key" to solving CAT's cyclical business. CAT has three service divisions: financial, logistics, and remanufacturing, and services are expected to contribute 20% to total revenues by 2010 (about $10 billion).

CAT got into remanufacturing by accident: originally asked by Ford to rebuild diesel engines to save money in 1973, which cost half the price of a new purchase, CAT opened a repair shop in Bettendorf, IA. By 1982 the Corinth plant was opened to handle demand and reclaim crankshafts. But it was not until 2000 that management at CAT understood the true potential of this service as a business, not just to support the purchase of new engines. One of the hidden advantages of the business is that remanufacturing discourages knock-off parts from entering the aftermarket, which is a very profitable segment (see Thomke, 2003). Even though this strategic move by a large manufacturer seems obvious now, it took 35 years for this company to move into a service business which is a logical extension of a core competence.

United Technologies' Pratt & Whitney parts services

Using a similar logic to that which drove CAT into remanufacturing diesel engines, Pratt & Whitney recently announced they were going to enter the jet engine spare parts business, until now dominated by GE. Airline customers continue to complain about the high cost of spare parts and want an alternative, lower cost supplier.

The significance of this move is that Pratt & Whitney will produce parts for GE jet engines as well as their own, and said that United Airlines would be their first customer, supplying parts for the 15,000 CFM56 jet engines currently in service. This would save millions of dollars annually in maintenance costs, according to UAL, parent of the airline. Pratt & Whitney expects to achieve $500 million in annual revenue from this service venture in 5 years. Pratt & Whitney originally decided to forego this engine segment in the 1980s for the Boeing 737, allowing GE to pass Pratt & Whitney in global jet engine production. Supply of spare parts would help Pratt & Whitney rebuild relationships (see HSBC, 2007).

United Technologies' Otis Elevator's OTISLINE

The case well known to business school students because of the Harvard Business School Case, originally published in 1986 on OTISLINE (*9-186-304*, updated in 1990), involved an elevator service repair and emergency line, which continues to flourish (see MSNBC, 2007). The Harvard case focuses on John Miller, director of information services for the Otis Elevator division of United Technologies, and on improving the responsiveness of OTISLINE to customers including adding more services.

According to the Harvard case, the idea for a centralized service capability, which became known as OTISLINE, originated in late 1981 in its North American operations and was successfully pilot-tested in August 1982 as a way to improve responsiveness to customers' calls for service. Local, commercial answering services simply did not have the same commitment to responding that Otis wanted to achieve. The ultimate goal was to reduce callbacks for each installed elevator, which, at the time, would save Otis $5 million a year. Success of the project at the time was attributed to top-management support. Nearly all aspects of Otis' business were impacted by this service including information services, service mechanic dispatching and control, customer service, service marketing and engineering, especially with the addition of remote diagnostic technology.

More recently (Dann, 2003), OTISLINE supported 105,000 elevator passengers in distress in the U.S., Canada, Guam, and the Caribbean islands. But the phone rings in Connecticut where OTISLINE operators are located in a call center. Otis is now the largest manufacturer of elevators and escalators, but owners must contract for OTISLINE. Most of the calls are minor problems like elevator door closing and noise like rattles, but operators can respond in at least three languages. Even pushing the stop button will activate a call.

These six cases (Table 27.1) vary in the degree to which the new-service offering was truly innovative (e.g., OnStar had no precedent versus CAT remanufacturing, which was common even at the time among many firms like machine tool builders), the degree to which this was a calculated, strategic intent, based on general manager leadership and vision (e.g., Pratt & Whitney spare parts service was calculated versus CAT remanufacturing which started by responding to one customer and no plan), the degree to which the new service was part of a much larger strategic service strategy (e.g., John Deere required a strategic technology agreement with Qualcomm versus GM OnStar, which operates as a standalone business), and lastly, but surely not finally, the degree to which this was a natural extension of existing core strengths of the manufacturing business or a real stretch in terms of unaligned diversification (e.g, CAT remanufacturing aligned almost perfectly with manufacturing core strength versus Physiotherapy Associates, and John Deere remote sensing, which clearly go beyond core capabilities).

Incumbent service with innovative offerings

When one thinks of service innovations, the typical examples that leap to mind are usually from incumbent service firms: banks, hospitals, etc. In Table 27.2 a sample of significant new offerings by incumbent firms is summarized.

Starwood Hotels and six-sigma innovation

Westin, owned by Starwood Hotels & Resorts Worldwide Inc., adopted the six-sigma management process in 2006 to increase efficiency and promote innovation within its services. The Westin Chicago River North Hotel piloted Unwind, a result of six-sigma. This brought guests together for nightly events, ultimately leading to increased brand loyalty and more spending. Teams of trained six-sigma employees, headed by Brian Mayer, have worked with Westin and Starwood management to look at the various areas of improvement. Massage services were introduced to cater to guests' needs, while new work processes emerged to provide employees with a healthier work environment. In 2006, the six-sigma service programs generated over $100 million in profits for Starwood Hotels. The creative and design-oriented culture helped to make the service innovations a reality for the business, raising customer satisfaction and revenues without sacrificing quality.

Bank of America and new-branch banking

One of the most difficult issues facing service innovation firms is the ability to test changes in a real-world environment. Bank of America originally renovated and altered 20 branches in the Atlanta region to accomplish three separate types of service experiments. To both plan and carry out the efforts, an Innovation & Development (I&D) Team was formed. The ID team wanted to first strengthen relationships with customers but not lose sight of branch efficiency in the process. The first service innovation was to transform five branches into "express

Table 27.2. Summary of incumbent service firm innovative new offerings.

Case	*Description*	*Citation*
1. Starwood Hotels	Six-sigma conflicts with innovation	Ante (2007)
2. Bank of America	New-branch banking	Thomke (2003)
3. HSBC	Green banking	HSBC (2007)
4. Parkland Hospital	Emergency room kiosks	MSNBC (2007)
5. Levi Strauss	Can khakis really be disruptive?	Dann (2003)

centers" focused around ease of access for customers performing simple transactions, such as deposits and withdrawals. Second, the I&D Team turned five different branches into "financial centers", where sophisticated services, such as stock trading and portfolio management, could be supported by knowledgeable staff and advanced technologies. Finally, 10 Atlanta branches were named "traditional centers" with a focus on conventional banking services, better technologies, and more efficient processes. Bank of America's customers were carefully observed in their new environments, with the I&D Team constantly examining behaviors and reactions. Service experiments were introduced for a mandatory period of 90 days to gather positive and negative feedback. In some cases, such as a mortgage loan experiment, the time period was reduced when it was clear that there were aspects that required modifications. The diverse base of new services and the carefully designed experimentation process helped Bank of America increase its customer base and begin "an unprecedented surge of creative thinking about branch banking."

HSBC and green banking

Sustainability has become a major feature of today's business world, and HSBC has brought this idea into one of their bank branches in Greece, NY. To achieve a triple bottom line of economic, environmental, and social benefits, various features were implemented to achieve a "zero carbon branch". Geothermal ground source heat pumps and 36 solar panels provide year-round energy. HSBC has also implemented an indoor lighting system and air quality–monitoring system to effectively and efficiently utilize energy resources. As a result, energy consumption is down 52%, as compared with traditional bank branches. A second major innovation has been a 15,000-liter rainwater collection system for use in restrooms and landscaping. The Greece HSBC branch has also witnessed a reduction in water usage of 71%. Not only have these environmental benefits led to reduced costs, but customers are given an opportunity to do business with a bank that is located in a green building. Creating an innovative building structure and management program has allowed HSBC to reduce its impact on the environment while focusing more time and energy on providing higher quality banking services.

Parkland Hospital and emergency room kiosks

The emergency room at Parkland Hospital has served approximately 300 people per day. Not only has there been an extended waiting time to see a doctor, but patients have also waited up to 2 hours simply to check in. To combat this problem, the hospital installed electronic kiosks in the emergency room designed to assess symptoms and prioritize medical problems. This innovative program was introduced in 2007, taking its inspiration from the airline industry and illustrating how an innovative idea can be successfully implemented across several business sectors. ER patients now experience an almost instantaneous check-in process, while also benefiting from a much shorter overall wait time. Added to the increased efficiency for the hospital staff are the benefits from raising customer satisfaction and allowing patients to be more in control.

Levi Strauss and disruptive khakis

The Dockers brand, produced by Levi Strauss & Co., has found a way to target "less capable [and] less skilled" (Dann, 2003, p. 3) market segments with its new innovations. Specifically, Dockers khakis have been outfitted with stain-resistant material, hidden pockets, and expandable waistbands. Not only do these new technologies offer convenience to their users, but these additions have also helped men adapt to a more business-casual work environment. Andrea Corso, brand spokesperson, explained how "We analyzed this core customer group and found ways to address their needs as a consumer." Although the technological innovations have been vital in expanding the Dockers brand, there is an ultimate focus on the customer.

These cutting edge ideas were introduced during "an explosion in the business casual clothing market segment." Indeed, businesses have allowed employees to dress more informally since 1998, according to the Society of Human Resource Management. It is important to notice how the product was created to respond to both customer needs and workplace conditions. Fashion-challenged men are the

Table 27.3. Summary of new-entrant service innovations.

Case	*Description*	*Citation*
1. RockBottomGolf.com	Golf ball diving leads to riches	Chao (2006)
2. CB2	Spin-off retailer from Crate & Barrel	Napolitano (2003)
3. Zipcar	Influencing customer behavior	Frei (2005)
4. Askmen.com	Men's magazines online	Ettlie (2006)
5. Trader Joe's	Boutique grocery	Abraham (2002)

target market for many of Dockers' innovations, and it appears that this marketing strategy has proven to be successful in the world of highly competitive fashion companies.

New-entrant service innovations

The final category of service innovation and case examples is new-entrants—startups, spin-offs, and new ventures that have shown promise and distinction in the literature. The summary of these cases appears in Table 27.3, and they are discussed next.

RockBottomGolf.com and online retailing

Locating an advantageous niche market is one of the most lucrative aspects of the business world today. Online retailers are often faced with risky choices in order to get ahead. RockBottomGolf was established in 2002 as an eBay retailer of used golf equipment. After spending some time collecting and re-selling golf balls from ponds, it was made clear that there was a market for other types of used golf equipment and accessories. Starting off as an eBay retailer allowed the company to gain exposure in an increasingly competitive environment, although the sales focus has shifted more towards its own website. As JoBeth Rath, customer service manager, stated, "The traffic isn't there on eBay anymore" (Chao, 2006, p. 12D).

RockBottomGolf has been so successful due to its low prices and ability to provide its customers with a wide selection of used golf equipment. Although its 1st year experienced a period of 4 months with no income, 2002 ended with $3.2 million in sales. As of 2006, the company was expected to realize sales of $24 million. It is important to note that RockBottomGolf.com is not simply a product sales website: the focus on meeting customer needs allows the business to be a part of the service industry. Although it is a relatively new business, the leap to becoming self-reliant in the online retailing market has allowed Rock-BottomGolf to put itself in a strong position.

CB2: new retail market segments

CB2 is a small chain under Crate & Barrel, owned by Gordon Segal. It is similar to Crate & Barrel in a broad sense, but it targets a younger audience. Rather than designing the flagship store in Chicago around the C&B model, there is a more whimsical feel to the environment. Products such as eye masks for hangovers and drinking glasses decorated with cocktail recipes point to a very different vision. Indeed, "CB2 strives to be hipper and more urban than Crate & Barrel, to reach younger shoppers who may be bored or intimidated by the original chain." Although there have been opportunities for rapid expansion, Segal has taken a cautious position to evaluate the successes and setbacks of various features. Debbie Kushnir, manager of the Chicago CB2 store, has pointed out how involved she has been in staying on top of every aspect of the business, down to the lighting and music. Convenience, quality, and affordability have become strong elements of the CB2 model, and continue to attract younger customers searching for "inexpensive, fun home-decorating" items.

Zipcar and car-sharing services

In an age of increasing importance on environmental consciousness and a decreased need for individual transportation in cities, Zipcar has developed a very successful service innovation. In essence, its car-sharing program has allowed people to enjoy the benefits of personal transportation only when necessary. For many city inhabitants who work close to home, owning a vehicle often means absorbing excessive and unnecessary prices. Zipcar estimates that it can cost up to $843 per month to own a car, while using one of its shared vehicles can cost as little as $34 per month for 4 hours of use. The company has also been innovative in offering its services to businesses, where cars can be shared between employees.

Zipcar members typically pay a $25 application fee and either an annual or monthly fee, in addition to any hourly/daily usage fees. For people who only require a car for short trips several times a month, the savings are tremendous. Zipcar is also very convenient, allowing its members to reserve a car online or through the phone anywhere from a few minutes to 1 year before it will be needed. This revolutionary business idea allowed the company to expand to 21 cities with a fleet of over 400 cars by 2004. Recognizing the needs of the residents in urban centers has created a robust opportunity for Zipcar to continue offering its car-sharing services.

Askmen.com and pacing growth

The dot.com bubble burst forced online businesses to search for innovative ways to reach their audiences, while also somehow finding ways to be profitable. Askmen.com was one such business, although its situation is quite unique. With only a handful of employees and a lifestyle website specifically targeted towards men, it was difficult to fathom how to rapidly expand in 2002. Sales had been recorded at $2.5 million, which includes such sources as advertising and e-commerce, while profits before taxes rang in at $600,000. Despite this positive financial position, the owners of Askmen.com knew that it would be imperative to offer new services to its customers. Its target market of 18 to 49-year-old men now represented approximately 90% of all site visitors. One of the greatest challenges would be to control the growth of the business, which ran somewhat against the current market conditions that demanded growth for sustainable websites. Askmen.com came up with a variety of projects to explore, including a video game section, movie section, reality show, and affiliate program initiative. These possible new services were designed to put the company in a strong market position while carefully and continuously targeting the correct audience.

Trader Joe's and store branding

Creating a unique in-store experience has evolved into an important differentiation strategy for many areas of the economy. Trader Joe's is a western U.S. grocery chain with a unique strategy focused on offering customers a memorable experience. Rather than developing a narrow profit-driven vision, the company is built upon seven values that range from integrity to treating the store as a brand in itself. The goals are focused on finding unique products and on cultivating loyal customers. To this end, Trader Joe's has made customer satisfaction a key component of every store. Management, which works in a one-level organizational structure, strives to keep stores very small and intimate at about 10,000 square feet. In comparison, Safeway, another western grocery chain, averages 55,000 square feet per store. The benefits come in the way of store loyalty and letters from customers expressing their dedication to a grocery store that is like a small community of shoppers unto themselves. In essence, Trader Joe's has combined the positive results of unique product offerings with a neighborhood-style store that appeals to many consumers.

An evolving model of service innovation

Most Americans work in the service sector and yet we know precious little about how innovation comes about in this vast sector of the economy. Perhaps this is due, in part, to the dominance of the manufacturing sector with respect to formal R&D: "United States' manufacturing contributes $1.5 trillion to GDP, employs 20 million workers, accounts for more than 70% of industrial R&D, and constitutes the main source of technology for the larger service sector" (Tassey, 2004, p. 153).

The service component of the economy continues to grow in significance, especially through business opportunities associated with the application of new information technology (Gallouj and Weinstein, 1997; Gustafasson and Johnson, 2003; Quinn, 1992). Lack of innovation in services has resulted in the general regional decline of manufacturing sectors when people are forced to migrate to lower paying jobs in alternate sectors of the economy (Beeson and Tannery, 2004). Despite the huge scale, breadth, and scope of the U.S. service sector,[1] innovation in this sector remains a mysterious process. Research aimed at improving understanding of service innovation could offer significant contributions to theory and practice.

Service sector innovations have been studied primarily in isolation of mainstream innovation research. Recent examples include those in financial services (Adams, 2004; Vermeulen, 2004), banking (Thomke, 2003), software (Capaldo *et al.*, 2003), and the public sector (Hinnant and O'Looney, 2003). There have also been a few studies in other industrialized countries; for example: business services in Singapore (Kam and Sing, 2004); hotels in Taiwan (Lin and Su, 2003); the application of software in the Italian service sector, which has enhanced productivity (Cainelli, Evangelista, and Savona, 2004); and public sector service innovation and telecom in the U.K. (Miozzo and Ramirez, 2003; Walker, 2003) and Switzerland (Hollenstein, 2003). In addition, a few comparative studies like that of Ramirez (2003) and survey and case data from marketing researchers (Martin and Horne, 1993; Martin, Horne, and Chan, 2001; Martin, Horne, and Schultz, 1999) have been conducted.

Even more importantly, the most recent professional and applied research literature reveals service innovation as conspicuous by its absence. For example, the Radnor and Probert (2004) compilation, in *Research-Technology Management*, of 11 articles on technology road mapping contains only one paper from the service sector (postal service). And, in the recent special issue of *Decision Sciences*, **35**(3), Summer 2004, devoted to service management, of 10 articles only 1 considers innovation and that is

[1] Approximately 8 out of 10 people working in the U.S. are employed in services, in Canada, Japan, and France, the percentage is over 70% (Bretthauer, 2004).

focused on network-based customer service systems, hardly a radical or disruptive technology.

Although limitations of data availability and industry classification cloud the overall phenomena, it also seems clear that R&D and innovation in services continue to grow, often in ways quite unlike the precedents of manufacturing and other sectors (see Gallaher, 2003).[2] From the perspectives of service innovation and service delivery, perhaps the most important difference between the two is that services are always co-produced by the provider and customer or client. For us, this is the defining difference that has implications for the innovation process. In particular, it suggests that the study of service innovation should include, at minimum, the co-development role of key customer relationships in helping define the new-service concept, designing and testing early versions of the service before a full commercial launch is attempted.

As we explore the nature of service innovation, we question whether or not this sector is truly poised for significant new growth through a unique innovation process at the divide or if its innovation is due to ongoing incremental changes—what we call perpetual beta-testing (see Ettlie, Mejia, and Walker, 2004). The latter has been identified as being typical of the history of this sector of the economy unless managed in novel ways (Bullinger, Fahnrich, and Meiren, 2003; Tidd and Hull, 2003).

References

Abraham, S. (2002). "Dan Bane, CEO of Trader Joe's." *Strategy & Leadership*, **30**(6): 30–32.

Adams, M. (2004). "Real options and customer management in the financial services sector." *Journal of Strategic Marketing*, **12**(1), 3.

Ante, S. E. (2007). "Starwood Hotels: Rubbing customers the right way." *Business Week*, October 8.

Arndt, M. (2005). "CAT sinks its claws into services." *Business Week*, December 5, No. 3692, 56–59.

Auguste, B., Harmon, E. P., and Pandit, V. (2006). "The right service strategies for product companies." *McKinsey Quarterly*, 40

Beeson, P., and Tannery, F. (2004). "The impact of industrial restructuring on earnings inequality: The decline of steel and earnings in Pittsburg." *Growth and Change*, **35**(1), Winter, 21.

Berry, L., Shankar, V., Parish, J. T., Cadwallader, S., and Dotzel, T. (2006). "Creating new markets through service innovation." *Sloan Management Review*, Winter, 56–63.

Bretthauer, K. (2004). "Service management." *Decision Sciences*, Special Issue, **35**(3), Summer, 325).

Bullinger, K.-J., Fahnrich, K.-P., and Meiren, T. M. (2003). "Service engineering: Methodical development of new service products." *International Journal of Production Economics*, **85**(3), September 11, 275.

Cainelli, G., Evangelista, R., and Savona, M. (2004). "The impact of innovation on economic performance in services." *Service Industries Journal*, **24**(1), January, 116.

Capaldo, G., Iandoli, L., Raffa, M., and Zollo, G. (2003). "The evaluation of innovation capabilities in small software firms: A methodological approach." *Small Business Economics*, **21**(4), December, 343.

Chao, M. (2006). "Golf-ball diving leads to riches." *Democrat and Chronicle*, Monday, September 25, 11D, 12D.

Chase, R. B., and Apte, U. M. (2007). "A history of research in service operations: What's the big idea?" *Journal of Operations Management*, **25**(2), March, 375,

Dann, J. B. (2003). "Can khakis really be disruptive?" *Strategy & Innovation*, September 1.

Ettlie, J. E. (2006). *Managing Innovation* (pp. 41–46). New York: Elsevier.

Ettlie, J. E., and Elsenbach, J. (2007). "The changing role of R&D gatekeepers." *Research-Technology Management*, **50**(5), September/October, 59–66.

Ettlie, J. E., Mejia, I. P., and Walker, J. M. (2004). "Is service innovation at the divide or the perpetual beta machine?" Paper presented at *The Meeting of the Academy of Management, New Orleans, LA, August 9.*

Ettlie, J. E., Rosenthal, S. R., and Hall, M. (2007). "Formalizing the new service development process." Paper presented (and awarded best-in-conference) at *Decision Sciences Institute Mini-conference on Service Science, Carnegie-Mellon University, May 23–25.*

Frei, F. X. (2005). *Zipcar: Influencing Customer Behavior* (June 30, pp. 1–8). Harvard Business School.

Gallaher, M. P. (2003). "Measurement issues in a changing environment." Paper presented to the Panel to Review Research and Development Statistics at *The National Science Foundation Workshop, July 24*. Research Triangle Park, NC: RTI International.

Gallouj, F., and Weinstein, O. (1997). "Innovation in services." *Research Policy*, **26**(4/5), 537–557.

Gustafsson, A., and Johnson, M. D. (2003). *Competing in a Service Economy*. San Francisco, CA: Jossey-Bass, a Wiley imprint.

Gustafsson, A., Nilsson, L., and Johnson, M. D. (2003). " The role of quality practices in service organizations." *International Journal of Service Industry Management*, **14**(2), 232–245.

Hargadon, A. (2003). "Organizaiton in action: Social science bases of administrative theory." *Administrative Science Quarterly*, **3**, September, 498.

Heineke, J., and Davis, M. M. (2007). "The emergence of service operations management as an academic discipline." *Journal of Operations Management*, **25**(2), March, 364.

Hinnant, C., and O'Looney, J.A. (2003). "Examining pre-adoption interest in online innovations: An exploratory study of e-service personalization in the public sector." *IEEE Transactions on Engineering Management*, **50**(4), 436.

Hollenstein, H. (2003). "Innovation modes in the Swiss service sector: A cluster analysis based on firm-level data." *Research Policy*, **32**(5), 845.

HSBC. (2007). "Zero carbon branch." *HSBC Global Environmental Efficiency Programme* (June); *www.hsbc.com/committochange*

[2] U.S. service sector R&D had grown to 31% in 1999, but there were still issues of misclassification and measurement.

Kam, W. P., and Sing, A. (2004). "The pattern of innovation in the knowledge-intensive business services sector of Singapore." *Singapore Management Review*, **26**(1), 21–44.

Lin, Y., and Su, H.-Y. (2003). "Strategic analysis of customer relationship management: A field study on hotel enterprises." *Total Quality Management & Business Excellence*, **14**(6), August, 715.

Lunsford, J. L. (2006). "Pratt & Whitney will produce parts for rival GE jet engine." *Wall Street Journal*, February 16.

Martin, C. R. Jr., & Horne, D. (1993). "Services innovation: Successful versus unsuccessful firms." *International Journal of Service Industry Management*, **4**(1), 49–63.

Martin, C. R., Horne, D. A., and Chan, W. S. (2001). "A perspective on client productivity in business-to-business consulting services." *International Journal of Service Industry Management*, **12**(2), 137–152.

Martin, C. R., Horne, D. A., and Schultz, A. M. (1999). "The business-to-business customer in the service innovation process." *European Journal of Innovation Management*, **2**(2), 55–62.

Miozzo, M., and Ramirez, M. (2003). "Services innovation and transformation of work: The case of UK telecommunications." *New Technology, Work, and Employment*, **18**(1), 62.

MSNBC. (2007). "Hospital installs electronic check-in kiosks." *Nightly News* (narrated by Brian Williams). MSNBC, July 16.

Napolitano, J. (2003). "Crate & Barrel handles its offshoot with care." *The New York Times*, June 22.

Oliva, R., and Kallenberg, R. (2003). "Managing the transition from products to services." *International Journal of Service Industry Management*, **14**(2), 160–173.

Parekh, S. (1998). "Elevator have you in a jam? Help may be just hundreds of miles away"; "Services: From the lawyer recommending mirrors to the woman venting about teen"; "24-hour telephone workers hear myriad woes." *Los Angeles Times*, April 27, p. 7.

Quinn, J. B. (1992). "The intelligent enterprise: A new paradigm." *The Academy of Management Executive*, **6**(4), 48–64.

Radnor, M., and Probert, D. R. (2004). "Viewing the future." *Research Technology Management*, **47**(2), March/April, 25–27.

Ramirez, M. (2003). "Economic performance in the Americas: The role of the service sector in Brazil, Mexico and the USA." *Service Industries Journal*, **23**(5), November, 163.

Reinartz, W., and Ulaga, W. (2006). "Growth beyond the core." *Financial Times* (UK), March 31, p. 10

Rothaermel, F. T., and Hill, C. W. (2005). "Technological discontinuities and complementary assets: A longitudinal study of industry and firm performance." *Organization Science*, **16**(1), January/February, 52–71.

Sawhney, M., Balasubramanian, S., and Krishnan, V. V. (2004). "Creating growth with services." *Sloan Management Review*, **45**(2), Winter, 34.

Susman, G., Warren, A., and Ding, M. (2006). *Product and Service Innovation in Small and Medium-Sized Enterprises* (Contract No. SB1341-03-Z- 0015/65332, 70 pp.). Gaithersburg, MD: Department of Commerce, NIST, Manufacturing Extension Partnership.

Tassey, G. (2004). "Policy issues for R&D investment in a knowledge-based economy." *Journal of Technology Transfer*, **29**(2), April, 153.

Thomke, S. (2003). "R&D comes to services." *Harvard Business Review*, **81**(4), 70–79.

Tidd, J., and Hull, F. (2003). *Service Innovation: Organizational Responses to Technological Opportunities and Market Imperatives*. London: Imperial College Press.

Vasilash, G. S. (2006). "OnStar: 10 years later." *Automotive Design and Production*, February, 60–61.

Vermeulen, P. (2004). "Managing product innovation in financial services firm." *European Management Journal*, **22**(1), February, 43.

Walker, R. M. (2003). "Evidence on the management of public services innovation." *Public Money & Management*, **23**(2), April, 93–102.

Wise, R., and Baumgartner, P. (1999). "Go downstream: The new profit imperative in manufacturing." *Harvard Business Review*, **77**(5), September/October, 133–141.

28

Process Innovation in Operations

John E. Ettlie and Shalini Khazanchi

Rochester Institute of Technology

In order to improve efficiency when producing a product or providing value to customers, organizations routinely innovate their service and production operations by introducing new tools and technologies, devices, and knowledge (e.g., Damanpour and Gopalakrishnan, 2001; Utterback and Abernathy, 1975). Service innovation is taken up in Chapter 27 so we restrict ourselves here to process innovation primarily in manufacturing where the accumulated literature has matured.

In the 40 years since Charles R. Walker's book *Technology, Industry, and Man: The Age of Acceleration* (1968) was published, writers such as David F. Noble (*Forces of Production: A Social History of Industrial Automation*, 1984) and many others have been prophetic in predicting the folly of context-free thinking which dominated much of the philosophy of managers of that era. In a similar vein, Wickham Skinner approached the issue of organizing production from a different perspective, asking the simple question: "What is the purpose of corporate manufacturing and strategies?" This question ultimately resulted in the concept of the focused factory—still independent of the R&D context which was about to dominate the scene and deflect attention once again from the context, since the focused factory materials did not deal with innovation. Therefore, in this chapter we will discuss in detail innovation in operations with the hope that we have not excluded any important imbedding context.

We begin with Jay Jaikumar's seminal work on flexible manufacturing and move on to the theory of scope versus scale which began modern treatments of this subject and which have changed our views forever on organizing operations. Specifically, following flexible manufacturing, we discuss mass customization followed by synchronous innovations that emphasize the organizational innovations needed to realize the benefits of process innovations, such as adoption of advanced manufacturing technologies (AMT). We then discuss various organizational innovations such as changes in organizational strategy, structure, culture, and human resource management (HRM) practices critical to realizing the benefits of advanced manufacturing technologies.

Flexible operations

Jaikumar's (1986) seminal work comparing Japanese and U.S applications of flexible manufacturing began a sea of change in how manufacturing organizations strategically manage their operations. The reason was simple. The Japanese were earlier to adopt and achieve substantially higher levels of flexibility with the same equipment when compared with their American competitors and counterparts. It turns out that it was not a matter of which technology was adopted and how early it was installed but *how* it was applied. Scope is the flexibility of a production line (number of products that can be made) and scale the reduction in unit cost with volume.

As subsequent applied research revealed, the science behind flexibility, that is, scope versus scale in manufacturing (Milgrom and Roberts, 1990), revealed a much more complicated scenario than Jaikumar captured with his extensive case studies. One of the first indications of the nuances in this research stream came with the publication of Mansfield's (1993) article on diffusion of flexible manufacturing in the U.S. Mansfield studied the rate of diffusion of flexible manufacturing systems among 175 firms in the U.S., Western Europe, and Japan to understand why the U.S. had been relatively slow to introduce flexible manufacturing. By the time Mansfield's article appeared in 1993, there was little doubt in a convergent literature about the economic advantages of flexible (scope-based) manufacturing: Increased utilization of equipment, more scheduling and prospects for new product choices, improved product quality, reduced (primarily in-process) inventories, reduced lead times, labor costs, and floor space.

Mansfield (1993) found that the reason U.S. firms adopted flexible manufacturing systems (FMSs) at a lower rate than Japanese firms was that Japanese firms achieved higher profitability with FMS. If we control for average rate of return on investment and number of years since first introduction, there is actually no statistically significant difference between Japanese and U.S. firms in the rate of imitation. U.S. firms return less profit with FMS as opposed to other investments (e.g., robots). Additionally, the first major firm in the U.S. began using FMS later than in Japan. Users of FMS were larger companies than non-users, and non-adopters in the U.S. estimated profitability return to be less (26%) than required (36%) according to Mansfield.

Mansfield also found inter-firm diffusion rates (an imperfect measure of utilization) to favor Japanese firms (5.59 systems per 10,000 employees), and surprisingly, European firms (2.61 systems per 10,000 employees) ahead of the U.S (0.38 systems per 10,000 employees). Even taking industry differences into account (e.g., electrical equipment was highest with 4.14 systems per 10,000 employees and automobiles lowest at 0.49 systems per 10,000 employees), the country differences stand out as rather remarkable. Benefits such as quality, intergenerational product capability, and inventory savings apparently are not accounted for directly in these firm estimates of returns. Further (see Small, 2006) strategic versus economic justification is quite different in firms considering complex versus less complex portfolio adoption (many technologies at once) of advanced design and manufacturing technology. Given that at least one study found the impact of IT on performance was higher when firms de-emphasize quality innovation strategy (Theodorou and Florou, 2008), it is clear that more research on this topic is needed.

Improvements in methodology have accelerated our understanding of benefits captured from the adoption of operations technology. Large-sample studies, often focusing on global manufacturing firms (Ettlie, 1998; Sohal *et al.*, 2006), SMEs (small and medium-sized enterprises, Marri, Irani, and Gunasekaran, 2007), and occasionally using longitudinal data (Boyer, 1999) have contributed significantly to our knowledge base in this field. For example, Boyer (1999, p. 824) used data from 112 plants in 1994 and 1996 and found, significantly, that investing incrementally in technology rather than using an "all-or-nothing approach" paid greater dividends.

Further, the more strategically placed the manufacturing function is in a firm (e.g., participation in business strategy making) the greater will be the resources devoted to new manufacturing technology. Testimony to the longitudinal approach comes with Boyer's finding that there is a lag in the impact of these investments on performance, but they are significant and positive in the end. The greater the investment made, he found, the greater the impact on profits and growth.

Ettlie (1998) studied 600 global firms and found that market share improvements were significantly related to R&D investments (R&D as a percentage of sales) and agility (scope promoted by flexible manufacturing systems). Unclear and remaining to be documented with global data is the relationship between technology strategies and investments in both product and process technologies and the streams and timing of these investments.

Sohal *et al.* (2006, p. 5225) studied 224 Australian companies and found that there was a "dynamic accumulation of both technology and human assets," which is consistent with the emerging view of dynamic capabilities of the firm (Teece, Pisano, and Shuen, 1997) and the synchronous innovation model of the firm (Ettlie, 1988) which is taken up next.

Mass customization

It is probably safe to say that we have wrestled with the concept of mass customization (Davis, 1987; Pine 1993; Ward, Milligan, and Berry, 2000) and attendant enabling technologies (e.g., flexible manufacturing) and production systems (e.g., modular manufacturing) for at least 20 years, and still not seen a truly widespread, satisfactory instantiation of the concept (Piller, Moeslein, and Stotko, 2004). Actually, the term "mass customization" is an oxymoron because there is no such thing as customization for the masses (Welborn, 2007), it refers instead to the way flexible demand is met. It has been recently defined as ". . . the ability to provide individually designed products and services to every customer through high process flexibility and integration" (Zerenler and Özihan, 2007).

But the promise of radically flexible production technology has not lived up to the original projections (Ettlie, 2000). And it is not as if we have not tried. For example, Ward, Milligan, and Berry (2000) review 126 cases of mass customization in their review article. The best likely explanation is that most customer markets or only some customer segments simply cannot sort out the options possible in most products at the cost they can be produced profitably (Dellaert and Stremersch, 2005; Squire *et al.*, 2004; Kaplan, Schoder, and Haenlein, 2007; Zhang and Tseng, 2007). Further, there is still the nagging issue of cost since suppliers often have to carry more inventory, (Aigbedo, 2007) and cost of waiting, and the complexity of the process goes up (Blecker and Abdelkafi, 2006)

Applications of these related concepts in the automotive industry can be used to illustrate both the promise and progress of this concept as well as the unfulfilled aspiration of mass customization. When VW opened their first modular manufacturing plant to assemble trucks (Schemo, 1998), the world followed soon after trying to emulate and improve this early example installed primarily as a radical new way of reducing the cost through co-location and co-production with suppliers. We have come full circle with announcements like those of Ford Motor Company which intends to make a push anew for simplification of options and production systems in order to become more competitive (Bailey, 2008). Other U.S. manufacturers have done the same, including Chrysler and General Motors Corporation.

The concept of mass customization is quite simple: if customization of products (and services) is what customers want and report higher satisfaction when they think products have been tailored to their unique needs, then give them what they want if production methods allow this to be done economically. With the advent of flexible manufacturing systems in discrete parts production during the 1960s and 1970s, it was not long before mass customization systems, often using modular design concepts, followed. It was not a matter of lot sizes of one, with infinite variety, but significant extension of product options, at no increased cost for the variety. However, it seems clear now that even with the best plans for manufacturing scope rather than scale and modular flexible manufacturing production technology, systems that have very limited variety are more than adequate to meet customers' demands and are significantly less costly to implement.

Honda and Toyota have demonstrated the advantages of restricting flexibility quite well and enjoy stable or increasing sales and profits while their competitors, for the most part, are struggling with cost and quality issues in automobiles, and there are serious theoretical and practical concerns about modular manufacturing that seem to limit the concept because of appropriation issues (e.g., technology and benefit spill-overs to supplier and in the supply chain) as well as customers' inability to take advantage in timely fashion (e.g., they have to wait for order fulfillment if they exploit options offered by manufacturers: Bailey, 2008).

There appear to be many successful applications of mass customization in other industries like computer assembly (Chen and Wang, 2007), apparel and truck manufacturing (Arndt, 2006), and car manufacturing (smart car) and bicycles (Kotha, 1996). But wholesale adoption of this philosophy has not been forthcoming. It appears to be restricted in its application to just a few, high-variety contexts where design and production are more closely coupled and in less complex production situations. Some manufacturers (e.g., Zara) appear to be approximating the concept without much fanfare and mature technologies where styles change quickly and close links between design and manufacturing can be maintained with organizational integrating mechanisms and flat organizational structures. The quality of products continues to dominate in many industries, as measured by the absence of things gone wrong rather than variety (things gone right and customization). Both types of quality seem to maximize customer satisfaction in a more restricted range of technological intensity (medium ranges) than previously thought (Ettlie and Johnson, 2001), as in the auto industry, but not in their high-tech or low-tech suppliers.

Postponement of production is a big part of implementing many mass production systems from HP's original concept (Feitzinger and Lee, 1997) to more recent applications (Shao and Ji, 2008; Piller, Moeslein, and Stotko, 2004) which show it is not beneficial to delay production of high value-added process in order to achieve differentiation. More recently, mass customization is seen as an alternative to off-shore outsourcing (Bock, 2008), and the component matrix has been used to bring mass customization to small and medium-sized firms (Ismail *et al.*, 2007). Work design practices have also been shown to influence the success of implantation of mass customization (Liu, Shah, and Schroeder, 2006).

Synchronous innovation

As mentioned earlier, synchronous innovation is the simultaneous adoption of process innovation technology and organizational innovations designed specifically to successfully capture the benefits of this new technology, which is typically available for competitors and precipitates a weak appropriation condition since these technologies typically have no intellectual property rights to adopters. This comes about as a result of the absence of intellectual capital protection for adopted technologies—the rights to these hardware–software systems are usually retained by the supplier of the adopted systems, and rarely developed by the adopting organization. The theory of synchronous innovation was originally advanced in a conference paper (Ettlie, 1985) and then later initially documented in a book (Ettlie, 1988) with case illustrations (see Figure 28.1).

The first full empirical support of synchronous innovation appeared in Ettlie and Reza (1992) and has since been replicated in several large-sample studies (e.g., Brandyberry, Rai, and White, 1999; Carroll, 1999; Gittleman, Horrigan, and Joyce, 1998; Pagell, Handfield, and Barber, 2000; Schroeder and Congden, 2000; Snell *et al.*, 2000) and extended theoretically as well (e.g., Swink and Nair, 2007). In the latter study of 224 manufacturing plants, the authors found that design–manufacturing integration plays a complementary role to advanced manufacturing usage, for quality, flexibility, and delivery, but not for cost nor new-product flexibility.

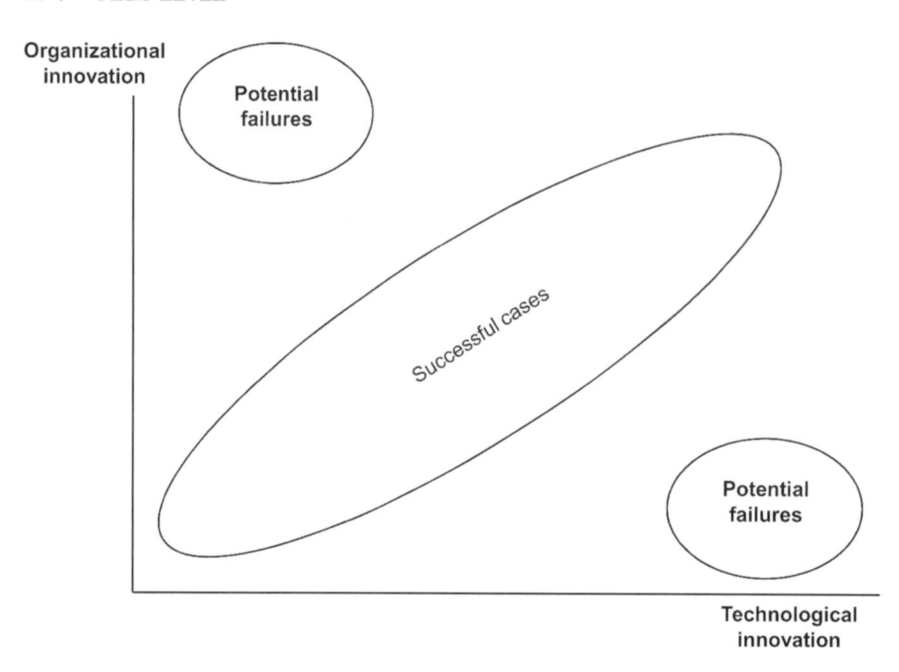

Figure 28.1. Successful management of the New Adopted Technology®.

Assessing the benefits of advanced manufacturing technologies: the role of context

Organizations adopt various programmable machinery called advanced manufacturing technologies (AMT) such as flexible manufacturing systems, computer-aided manufacturing, and computer numerically controlled machines to organize production or manufacturing process for improved production quality, efficiency, and flexibility (Hayes, Wheelwright, and Clark, 1988; Lewis and Boyer, 2002). AMT offers two seemingly paradoxical benefits: flexible product customization and cost efficiency (Ettlie and Reza, 1992; Zammuto and O'Connor, 1992). AMT can improve work flow, information sharing across functions, and managerial control which can, in turn, enhance employees' abilities to diagnose and solve problems effectively and efficiently (McDermott and Stock, 1999). Yet, research shows mixed results regarding the outcomes of AMT; while some organizations report gains in flexibility or productivity or both, many report AMT failure. Between 50% and 75% percent of implementations fail in terms of quality, flexibility, and reliability (Chung, 1996).

Small (2006) found that the more complex the AMT portfolio, the more departments were involved in the justification and installation stage of adoption. Regardless of portfolio, more departments tend to be involved early (justification) rather than later (installation). Finance and general management tend to drop involvement when installation begins. Marketing involvement tended to be low regardless of portfolio complexity. MIS/IT involvement was higher for the more complex portfolios of AMT adoption, as would be expected. Further, Small (2006) focused on the justification process for advanced manufacturing and design technologies with a survey of 82 plants. He found that larger firms (sales and employees) and firms using a combined justification approach (strategic and economic) adopted the more complex technology portfolios. The most popular measures of performance of these new systems was payback (58.7%) and ROI (53.3%) with cost–benefit analysis used third most often by 44% of respondents. Small (2006) found no statistically significant relationship between type of performance measure and complexity of the adopted portfolio. However, complex portfolio adopters were significantly more likely to use DFC (discounted cash flow) justification and significantly more likely to use multiple justification techniques. One implication of these results is that complex justification techniques probably need to be simplified so non-financial managers can use them routinely.

Thus, while there are many factors that determine the success of AMT implementation, amongst the ones often sighted by researchers and, perhaps, ignored by managers are issues related to organization strategy, design or structure, culture, and HRM practices. Researchers have consistently documented the need for changes in organization strategy, structure, and HRM practices to complement AMT, and not doing so is listed among the reasons for AMT failure in a number of studies (e.g., Dean, Yoon, and Susman, 1992; Small, 2007; Snell and Dean, 1992). Similarly, there are significant differences in organization culture and practices between high- and low-performing AMT plants (Lewis and Boyer, 2002). Indeed, Jaikumar and Bohn (1992) stated early on that the problem is not with the technology itself but *how* it is implemented. Below, we discuss in turn the findings and implications related to organization strategy, organization structure, organization culture, and HRM practices.

Organization strategy

Small (2007) found that the more complex the AMT portfolio, the more likely strategic benefits were measured. The majority of the 82 plants he studied (82.3%) reported evaluation intentions for operations and technical performance. But only 38% said they were going to evaluate strategic benefits. This seems rather unusual since 75% of these plants use both strategic and operational justification in general, but for AMT this number drops to 33% and 62%, respectively.

Operational improvements, under the synchronous innovation approach, have been reported to be rather impressive and significant. For example, throughput time reduction of 50% with AMT is typical (Ettlie, 2006, p. 344). For large integrated system adoption, like ERP (enterprise resource planning) systems, general management appears to have a much bigger role in directly influencing outcomes. For example, in a study of 80 ERP systems, Ettlie *et al.* (2005) found that general managers needed to demonstrate support for ERP adoption by actually modeling the behavior they wanted others in the firm to emulate. They had to actually use the ERP system rather than just command its use.

Organization structure

While it is clear that AMT implementation inherently requires changes in organizational structure, there has been some debate over the type of organizational structure best suited to achieve both flexibility and productivity benefits. Some researchers have argued that AMT results in a more mechanistic organization structure emphasizing control and coordination while others have argued for a more organic structure emphasizing flexibility and empowerment. In an attempt to resolve this controversy, Dean, Yoon, and Susman (1992), in a seminal study, found that AMT implementation requires aspects of seemingly contradictory organic and mechanistic forms of organization design. Specifically, they found support for decentralization and formalization with the former being a characteristic of an organic form and the latter of a mechanistic form. They explain the findings in view of key AMT characteristics—their integrative capacity and capital intensive nature. Integrative capacity allows for better work flow, information-sharing, and communication across functions enabling employees at the lower levels to make decisions independently. Thus, centralized decision making will prevent organizations from achieving flexibility benefits.

Furthermore, AMT's special ability to integrate across functions can also increase interdependence and, therefore, the need for control and coordination (Dean, Yoon, and Susman, 1992). Formalization can provide guidance on acceptable form and boundaries for employee behavior and actions. It can also help minimize the risk that may come with decentralized decision making, given that AMT are often expensive requiring large capital investments. Taken together, it seems that AMT characteristics such as their integrative capacity and capital intensive nature pose challenges that may be managed if paradoxical organization structural elements—formalization and decentralization—coexist.

Organization culture

A debate somewhat similar to the one about the nature of the relationship between organization structure and AMT has gone on about the relationship between organization culture and AMT benefits. Organization culture is often defined broadly as a collection of values, beliefs, and norms shared by its members and reflected in organizational practices and goals (Hofstede *et al.*, 1990). Not surprisingly, given that organization structure and culture are intertwined with one shaping the other, research suggests a paradoxical culture embracing seemingly competing values, flexibility- and control-oriented, as best suited for achieving AMT benefits. In an earlier study, McDermott and Stock (1999) found that flexibility-oriented values emphasizing creativity, spontaneity, and empowerment and internal focus on maintenance and improvement of existing organization were significantly related to managers' satisfaction with AMT implementation. Thus, organization culture indirectly influences AMT plant performance as managerial satisfaction can be an important factor in encouraging managers to use AMT in a way that would help realize organization goals (McDermott & Stock, 1999). It is also important to note that managerial satisfaction was found to be significantly correlated with more tangible AMT benefits such as productivity and flexibility. They also found control values emphasizing stability and order along with external focus on market, competition, and adaptation with external environment were significantly related to increase in growth and market share following AMT implementation.

Recently, Khazanchi, Lewis, and Boyer (2007) in an attempt to further clarify the relationship between culture and plant performance following AMT implementation found flexibility-oriented values as more proximal and control-oriented values as distant predictors of AMT plant performance. They suggest that control values with supporting practices can help alleviate fear of losing control associated with flexibility-oriented values and practices. For example, established routines and clearly specified AMT objectives can facilitate trust in employees' spontaneous decision making and creative problem solving necessary to realize AMT's flexibility benefits. Taken together, an organization culture that embodies paradoxical values and beliefs of flexibility and control reinforced through complementary practices that provide clearly established goals and objectives as well as discretion to employees are critical in achieving AMT benefits.

HRM practices

As discussed above, AMT adoption changes the work environment by streamlining the work flow, distributing information across functions resulting in job recompostion and employee empowerment (Snell and Dean, 1992). In essence, AMT adoption alters the nature of factory jobs from being *job-based* where employees perform a fixed set of tasks, duties, and functions to *skill-based* where employees engage in a range of activities with greater discretion and empowerment. AMT adoption requires employees to utilize all the available information and engage in independent decision making and creative problem solving. Employees' knowledge about their own job or function may not be sufficient; knowledge about other functions may be equally, if not more, important to handle problems related to other functions. HRM practices should be designed to support the nature of jobs as well as accommodate new orientation towards employees where they are selected, trained, evaluated, and compensated for skills as well as cross-functional experience and knowledge.

Earlier research (e.g., Snell and Dean, 1992) as well as a more recent study (Snell *et al.*, 2000) examined the relationship between AMT implementation and HRM practices including staffing, training, performance appraisal, and compensation. Their research suggests that HRM practices, training in particular, should be changed to meet new-job requirement and human capital needs. Specifically, Snell and Dean (1990) and Snell *et al.* (2000) found that organizations should invest particularly in their training programs that would help employees gain conceptual knowledge about other functions as well as learn and acquire new problem-solving skills critical in the changed work environment following AMT implementation. Both job rotation and on-the-job training can serve this purpose. Similarly, performance appraisal and compensation systems should be used to encourage and support learning and empowerment by evaluating and rewarding behaviors that demonstrate creative problem solving and practices.

In summary, AMT implementation requires changes in the work environment in terms of the organizational strategy, structure, culture, and HRM practices—all of which are intertwined—in order to realize AMT benefits such as productivity and flexibility. AMT plants often require competing values and practices, emphasizing both flexibility and control, to co-exist which increases the possibility of mixed messages and resulting conflict. It is therefore very important that structure, culture, and HRM practices are aligned and fit one another. Any mis-match between the three is likely to result in mixed messages to employees and dysfunctional conflict. In other words, organization structure, culture, and HRM practices should not only complement AMT implementation but also be congruent with one another. This is the essence of the synchronous innovation model. However, more work is needed in this area, and what remains is to develop a theory of synchronous innovation that applies to all situations or at least can be programmed in a contingent fashion—factory, office (i.e., service), farm, and construction.

Conclusion

What emerges in this chapter on innovation and operations is no less than a general paradigm for successful capture of benefits from both adoption of process-enhancing innovation and weak appropriation conditions of all types. Although the details may vary by technology type and degree of departure from existing practice, the key is found in understanding the nuances of administrative innovations needed to support unique absorption of the full potential of the adopted changes.

The most typical mistake most adopting units make is in underestimating how much change is actually represented by the adopted process improvement—usually a radical technology is at hand, and it often takes time to appreciate how much change is precipitated by the new requirements. In the learning process, the typical reaction is to back off on the extent of adoption or simplify the actual new process technology installed in order to find a more reasonable pace of change. This may or may not lead directly to discovery of the organizational innovations that will be required to capture the benefits of the adopted innovation. For example, unionized production plants and operations like the dock workers in Long Beach, CA (Ettlie, 2006), often take years to understand the process of establishing a union–management technology agreement or modifying an existing technology agreement in order to go forward in the implementation process and achieve a successful transition to the new-process technology platform.

This tendency of finding a better way to learn and accelerate change is not limited to production operations, nor is it restricted to just hourly employees. In fact, it is often middle and top management, as well as technical staff, that will be ultimately required to make the greatest changes and adapt to absorb the new-technology potential. The adoption of computer-aided drafting, design, engineering, and virtual development networks typically requires a complete change in the strategy and structure of the organizations in questions (e.g., Ettlie and Pavlou, 2006).

Organizations that anticipate the need for change in both process technology and supporting innovations in procedures like business processes are well ahead of the game—even if the details are not known in advance—and are much better prepared to learn faster than competitors. The learning process often requires abandonment of the original ideals for the project, but, in the end, benefits will be uniquely captured that are difficult for competitors to emulate because they cannot easily duplicate the conditions that led to this learning process. Therein lies the secret to the successful adoption of any challenging new process-enhancing technology.

References

Aigbedo, H. (2007). "An assessment of the effect of mass customization on suppliers' inventory levels in a JIT supply chain." *European Journal of Operational Research*, **181**(2), 704–715.

Arndt, M. (2006). "Paccar: Built for the long haul." *Business Week*, January 30, 66.

Bailey, D. (2008). "U.S. carmakers are adopting simplicity as a path to recovery." *International Herald Tribune*, p. 11.

Blecker, T., and Abdelkafi, N. (2006). "Complexity and variety in mass customization systems: Analysis and recommendations." *Management Decision*, **44**(7), 908–929.

Bock, S. (2008). "Supporting off-shoring and near-shoring decisions for mass customization manufacturing processes." *European Journal of Operational Research*, **184**(2), 490–508.

Boyer, K. (1999). "Evolutionary patterns of flexible automation and performance: A longitudinal study." *Management Science*, **45**(6), 824–842.

Brandyberry, A., Rai, A., and White, G. P. (1999). "Intermediate performance impacts of advanced manufacturing technology systems: An empirical investigation." *Decision Sciences*, **30**(4), 993–1020.

Carroll, B. (1999). "Motorola makes the most of teamworking." *Human Resource Management International Digest*, **7**(5), 9-11.

Chen, Z., and Wang, L. (2007). "A generic activity-dictionary-based method for product costing in mass customization." *Journal of Manufacturing Technology Management*, **18**(6), 678–700.

Chung, C. A. (1996). "Human issues influencing the successful implementation of advanced manufacturing technology." *Journal of Engineering and Technology Management*, **13**(3/4), 283–299.

Damanpour, F., and Gopalakrishnan, S. (2001). "The dynamics of adoption of product and process innovations in organizations." *Journal of Management Studies*, **38**(1), 45–65.

Davis, S. M. (1987). *Future Perfect*. Reading, MA: Addison-Wesley.

Dean, J. W., Yoon, S. J., and Susman, G.I. (1992). "Advanced manufacturing technology and organization structure: Empowerment or subordination?" *Organization Science*, 3(2), 203–229.

Dellaert, B. G. C., and Stremersch, S. (2005). "Marketing mass-customized products: Striking a balance between utility and complexity." *Journal of Marketing Research*, **42**(2), 219–227.

Ettlie, J. E. (1985). "The impact of interorganizational ianpower flows on the innovation process." *Management Science*, **31** (9), 1055–1071.

Ettlie, J. E. (1988). *Taking Charge of Manufacturing*. San Francisco: Jossey-Bass.

Ettlie, J. E. (1998). "R&D and global manufacturing performance." *Management Science*, **44**(1), 1–11.

Ettlie, J. E. (2000). *Managing Technological Innovation*. New York: John Wiley & Sons.

Ettlie, J. E. (2006). *Managing Innovation*. New York: Elsevier.

Ettlie, J. E. and Johnson, M. D. (2001). "Technology, customization, and reliability." *Journal of Quality Management*, **6**, 193–210.

Ettlie, J. E., and Pavlou, P. A. (2006). "Technology-based new product development partnerships." *Decision Sciences*, **37**(2), May, 117–144.

Ettlie, J. E. and Reza, E. M. (1992). "Organizational integration and process innovation." *Academy of Management Journal*, **35**(4), 795–827.

Ettlie, J. E., Perotti, V., Cotteleer, M., and Joseph, D. (2005). "Strategic predictors of successful enterprise system deployment." *International Journal of Operations and Production Management*, **25**, 953–1153.

Feitzinger, E., and Lee, H. (1997). "Mass customization at Hewlett-Packard: The power of postponement." *Harvard Business Review*, **75**(1), 116–121.

Gittleman, M., Horrigan, M., and Joyce, M. (1998). "'Flexible' workplace practices: Evidence from a nationally representative survey." *Industrial & Labor Relations Review*, **52**(1), 99–115.

Hayes, R. H., Wheelwright, S. C., and Clark, K. B. (1988). *Dynamic Manufacturing: Creating the Learning Organization*. New York: Free Press.

Hofstede, G., Neuijen, B., Ohayv, D. D., Sangers, B. (1990). "Measuring organizational cultures: A qualitative and quantitative study across twenty cases." *Administrative Science Quarterly*, **35**(2), 286–316.

Ismail, H., Reid, I. R., Mooney, J., Poolton, J., and Arokiam, I. (2007). "How small and medium enterprises effectively participate in the mass customization game." *IEEE Transactions on Engineering Management*, **54**(1), 86–97.

Jaikumar, R. (1986). "Postindustrial manufacturing." *Harvard Business Review*, **64**(6), November/December, 69–77.

Jaikumar, R., and Bohn, R. E. (1992). "A dynamic approach to operations management: An alternative to static optimization." *International Journal of Production Economics*, **27**(3), 265–282.

Kaplan, A. M., and Haenlein, M. (2006). "Toward a parsimonious definition of traditional and electronic mass customization." *Journal of Product Innovation Management*, **23**(2), 168–182.

Kaplan, A. M., Schoder, D., and Haenlein, M. (2007). "Factors influencing the adoption of mass customization: The impact of base category consumption frequency and need satisfaction." *Journal of Product Innovation Management*, **24**(2), 101–116.

Khazanchi, S., Lewis, M. W, and Boyer, K. K. (2007). "Innovation supportive culture: The impact of organizational values on process innovation." *Journal of Operations Management*, **25**, 871–884.

Kotha, S. (1996). "From mass production to mass customization: The case of the national industrial bicycle company of Japan." *European Management Journal*, **14**(5), 442–450.

Lewis, M. W., and Boyer, K. K. (2002). "Factors impacting AMT implementation: An integrative and controlled study." *Journal of Engineering Technology Management*, **19**(2), 111–130.

Liu, G., Shah, R., and Schroeder, R. G. (2006). "Linking work design to mass customization: A sociotechnical systems perspective." *Decision Sciences*, **37**(4), 519–545.

Mansfield, E. (1993). "The diffusion of flexible manufacturing systems in Japan, Europe, and the United States." *Management Science*, **39**(2), February, 149–159.

Marri, H. B., Irani, Z., and Gunasekaran, A. (2007). "Advance manufacturing technology implementation in SMEs: A framework of justification criteria." *International Journal of Electronic Business*, **5**(2), 124–140.

McDermott, C. M., and Stock, G. N. (1999). "Organizational culture and advanced manufacturing technology implementation." *Journal of Operations Management*, **17**(5), 521–533.

Milgrom, P., and Roberts, J. (1990). "The economics of modern manufacturing: Technology, strategy, and organization." *American Economic Review*, **80**, 511–528.

Noble, D. F. (1984). *Forces of Production: A Social History of Industrial Automation*. New York: Oxford University Press.

Pagell, M., Handfield, R. B., and Barber, A. E. (2000). "Effects of operational employee skills on advanced manufacturing technology performance." *Production and Operations Management*, **9**(3), 222–238.

Piller, F. T., Moeslein, K., and Stotko, C. M. (2004). "Mass customization: Reflections on the state of the concept." *International Journal of Flexible Manufacturing Systems*,**16**(4), 313–334.

Piller, F. T, Moeslein, K., and Stotko, C. M. (2004). "Does mass customization pay? An economic approach to evaluate customer integration." *Production Planning & Control*, **15**(4), 435–444.

Pine, B. J. (1993). *Mass Customization: The New Frontier in Business Competition*. Boston, MA: Harvard Business School Press.

Schemo, D. J. (1998). "Is VW's new plant lean or just mean?" *New York Times*, Sunday, January 19, pp. D1, D6.

Schroeder, D. M., and Congden, S. W. (2000). "Aligning competitive strategies, manufacturing technology, and shop floor skills." *Production & Inventory Management Journal*, **41**(4), 40–47.

Shao, X-F., and Ji, J-H. (2008). "Evaluation of postponement strategies in mass customization with service guarantees." *International Journal of Production Research*, **46**(1), 153–171.

Small, M. H. (2006). "Justifying investment in advanced manufacturing technology: A portfolio analysis." *Industrial Management and Data Systems*, **106**(4), 485–508.

Small, M. H. (2007). "Planning, justifying and installing advanced manufacturing technology: A managerial framework." *Journal of Manufacturing Technology Management*, **18**(5), 513–537.

Snell, S. A., and Dean, J. W. (1992). "Integrated manufacturing and human resource management: A human capital perspective." *Academy of Management Journal*, **35**(3), 467–504.

Snell, S. A., Lepak, P. D., Dean, J.W., and Yodunt, M. A. (2000). "Selection and training for integrated manufacturing: The moderating effects of job characteristics." *Academy of Management Journal*, **37**(3), 446–466.

Sohal, A. S., Sarros, J., Schroder, R., and O'Neill, F. (2006). "Adoption framework for advanced manufacturing technologies." *International Journal of Production Research*, **44**(24), 5225–5246.

Squire, B., Readman, J., Brown, S., and Bessant, J. (2004). "Mass customization: The key to customer value?" *Production Planning & Control*, **15**(4), 459–471.

Swink, M., and Nair, A. (2007). "Capturing the competitive advantages of AMT: Design-manufacturing integration as a complementary asset." *Journal of Operations Management*, **25**(3), 736–754.

Teece, D. J., Pisano, G., and Shuen, A. (1997). "Dynamic capabilities and strategic management." *Strategic Management Journal*, **18**(7), 509.

Theodorou, P., and Flourou, G. (2008). "Manufacturing strategies and financial performance: The effect of advanced information technology." *Omega*, **36**(1), 107–121.

Utterback, J. M., and Abernathy, W. J. (1975). "A dynamic model of process and product innovation." *Omega*, **3**, 639–656.

Walker, C. R. (1968). *Technology, Industry, and Man: The Age of Acceleration*. New York: McGraw-Hill.

Ward, P. T., Milligan, G. W., and Berry, W. L. (2000). "Approaches to mass customization: Configurations and empirical validation." *Journal of Operations Management*, **18**(6), 605–625.

Welborn, C. (2007). "Mass customization." *OR-MS Today*, **34**(6), 38.

Zammuto, R. F., and O'Connor, E. J. (1992). "Gaining advanced manufacturing technologies benefits: The role of organizational design and culture." *Academy of Management Review*, **17**(4), 701–728.

Zerenler, M., and Özilhan, D. (2007). "Mass customization manufacturing (MCM): The drivers and concepts." *Journal of American Academy of Business*, **12**(1), 262–269.

Zhang, M., and Tseng, M. M. (2007). "A product and process modeling based approach to study cost implications of product variety in mass customization." *IEEE Transactions on Engineering Management*, **54**(1), 130–144.

29

Organizing for Innovation

Mariann Jelinek

Mason School of Business, College of William and Mary

Structure has been a central focus of organization theory from its beginnings in the 1890s when Max Weber defined managers' choice as "bureaucracy or dilettantism". Bureaucracy centered on structural means—assignment of responsibility, reporting relationships, and decision authority, for instance—to control variance in behavior and information, assuring appropriate assignment of attention and responsibility for clearly identified tasks. Yet another view suggests that, while rules and routines help for some tasks, they offer little to address unpredictable circumstances or to facilitate innovation. Both have become increasingly central to business, as indicated in IBM's surveys of global executives, who repeatedly point to internally generated innovation as their key concern. Meanwhile, new products, services, and technologies, new global competitors and methods roil markets and demand fresh responses.[1] Moreover, products and more recently services have become increasingly complex, intertwined with customers' (or customers' customers') needs and organizations. So, how can firms organize for innovation? Many do so by emphasizing networks of relationships beyond traditional organizational boundaries, as we shall show.

This chapter will begin with an account of how organizing for innovation has moved from classic emphasis on structures to systems to teams (increasingly, to virtual teams) and now to networks as the focal point for innovation. We will briefly turn to examples of highly innovative contemporary firms, noting the importance of entrepreneurship and innovation, as well as their interesting variants on classic structure: these variations address what traditional structural options do not. *En route*, we will briefly note insights from psychology about innovation, relating these to the changing demands organizations face, including the increasingly important roles of information technology, communication, and collaboration. We will close with the recent rise of outsourcing and of "open innovation" including such public but initially anonymous Internet sites as InnoCentive,[2] and their implications.

A brief history of organizing for innovation: from structure to systems to teams

A first question to ask is, "Why organize at all?" The short answer is, "Because only by organizing can larger or more complex tasks be accomplished." Traditionally, the theory of organization (OT) emphasized structure as a means of control: "The design of an organization refers, of course, to its structural characteristics ... that influence or constrain important aspects of the organization's total behavior (Haberstroh, 1965, p. 1171), via the central configuration or arrangement of tasks and resources, information, and authority. Among the earliest forms of organization are "agency" organizations in which someone in charge tells others what to do. The leader (typically the owner, manager, foreman, or chief craftsperson) directs the help, with each worker acting as the agent or extension of the leader. Coordination and direction were achieved through direct supervision. Examples include royal appointees such as generals (or tax farmers), whose authority depends on their personal relationship with the monarch. Similarly, medieval bankers typically

[1] IBM's (2006) survey, provocatively titled *Expanding the Innovation Horizon*, is typical in depicting CEOs as embracing change and innovation, drawing widely on global resources to do so.

[2] A Web-based community matching scientists to R&D challenges presented by companies worldwide; *http://www.innocentive.com/* (consulted May 2, 2007). Nor is InnoCentive the only such site: Nine Sigma promises to "accelerate the innovation cycle" via requests for proposals in a similar fashion (*http://www.ninesigma.com/mx/newsletter.html?gclid=CIirlfX774sCFQ0eSgodJSHkOA*; consulted May 2, 2007), and there are others.

relied on family members, whose authority derived from their familial relationship.

Agency might suggest appointing a creative, innovative person to lead innovation—but creativity or innovative capability in one area is no guarantee of capability in multiple disciplines typically needed for commercial innovation. Moreover, even a highly innovative leader will be limited to his or her own creativity, restricting the size and potential reach of an agency-based organization. Contemporary products and services are often vastly more complex than the knowledge base of any single person, however brilliant. Thus, direct personal authority has limits, although it was typical of virtually all organizations until its displacement in the 19th century by bureaucratic, rules-based organizations.

Systematic task assignment

Assigning the work in systematic fashion to facilitate development of specialist expertise gave rise to advantages recognized as early as 1776 in Adam Smith's observations of pin making, and widely utilized in many organizational arrangements that group like activities together. Specialization facilitates development of expertise because workers do the same tasks repeatedly. Formal work assignments substitute structural arrangements for the preferences of workers or the vagaries of chance availability—and practice, development of better methods, and focus all contribute to better productivity and greater output. As the potential danger and complexity of routine business operations increased—in steel mills, for example—attention to systematic rules became more important. One especially dangerous and visible activity, the long-distance railroad, quickly gave rise to the first "modern" organization, coordinated by that era's "high technology" for communication, the telegraph, and distant electric signals—all innovative for their time, as was the coordinated organization they supported (Chandler, 1965). This coordination avoided costly and often fatal accidents of trains colliding.

The birth of the bureaucratic organization

Max Weber described such innovative systems of his time as "bureaucracy", and noted their capabilities. While bureaucracy has acquired a bad name in our times, it was a significant advance in efficiency and fairness, because it was based on objective rules, formally identified (and limited) responsibilities, it made use of precedent so that decisions could be made more quickly and consistently, and because records enabled rapid training of new staff, among other benefits. Yet bureaucracy's focus on precedent and fixed rules underline its limitations for innovation: it tends to look backwards, rather than forward, and most innovations will necessarily fall outside the rules.

Early organization theorists

When Chester Barnard, an influential early management thinker, defined organization as "the consciously-coordinated activities of three or more persons" (Barnard, 1938, 1956), he pointed to communication and incentives as critical to enlisting willing participation. Such willing participation, engaged interest, and motivation have been the focus of much research since Weber and Barnard, along with specialties arising around motivation, incentives, and performance assessment (among other focal points). Like Barnard, Mary Parker Follett, another early management theorist, also pointed to the importance of employees' potential for creativity and innovative problem solving. Including these factors pointed to critical issues for organizations and management—and began to focus attention beyond structure *per se* to structure's effects on organization members. Yet structure remained an important focus, centered on a few major configurations. Functional structures group similar activities together—design, manufacturing, shipping, or finance into separate departments, for example. Within a hospital the departments may center on medical specialties (radiology, oncology, or pediatrics, for example), plus business functions (billing, insurance, payroll). Clustering such activities together can foster specialization and deeper expertise in each specialty—but, at the same time, such an arrangement often hinders innovation because the interfaces between departments and activities may resist change. Crossing functional and departmental boundaries to collaborate around innovation is notoriously difficult. Nor is the problem lessened when divisional or geographic structures create multiple functional structures. One response is to designate a structural "home" for innovation, in a research and development (R&D) function.

Organization and strategy

Early corporate efforts at innovation in large-scale organizations, most notably at DuPont, identified links between organizing arrangements and strategy. DuPont and Corning were among the earliest corporations to set up R&D labs, with resources specifically devoted to innovation. These early innovation efforts embraced a wide array of topics that remain challenges for innovation in our own time, among them the place and focus of R&D, the need for additional resources to implement innovations in practice, and the need for administrative structures and information flows to support innovation. DuPont and General Motors were among the first practitioners of divisional structure—initially incorporating all the business functions within each distinct car's division—to separate corporate governance from the divisions, reserving financial and strategic controls to the corporation, and enabling the corporate level to direct innovation efforts while divisions focused on separate markets. Corporate resources underwrote more fundamental R&D, and articulated distinctions among and between divisions' product portfolios to create a coordinated corporate strategy, focusing the innovative efforts of each division to a particular target market.

What was so special about linking organization to strategy? For the first time, firms could envision proactive innovation to drive market demand as well as responding to it. For the first time, too, these firms were able to contemplate deliberate innovation for specified targets—whole new product lines like cellophane at DuPont, or a car priced between Chevrolet and Cadillac at GM. The top officers of the firm—Alfred P. Sloan and Pierre S. DuPont, for example—could envision arrays of products to suit different needs, and direct their subordinates' efforts to innovation. Organizational innovation could be generated via research directed for particular targets, and structured into R&D departments; the rest of the organization could then execute the plans. Employees would function like well-oiled machines, turning out innovations on demand.

"Organic" and "mechanistic" organization

Some of the first empirical studies on organization seemed to confirm a dichotomy between innovation and order. Joanne Woodward (1958, 1965) studied different types of technology and observed variations in formality, the degree to which orders were directed from a central authority, whether activities were governed by protocols or routines and set responses, and levels of authority. Where authority was centralized, rules were primary, and levels many, the organization could be said to be "bureaucratic" or "mechanistic"—because it is expected to operate with machine-like precision (Burns and Stalker, 1962). Reserving decision making to a few senior executives, insisting on rigid adherence to rules or procedures, and seeking to specify every contingency in advance makes sense only where it can be assumed that the task, the environment, and the wider circumstances around them will remain unchanged—a dubious and increasingly untenable assumption in our globally linked economy.

By contrast, researchers found that the most successfully innovative organizations were "organic": they focused on the task at hand, rather than on rules or formal hierarchy. Where decision making was delegated to those nearest the task, initiative was permitted and encouraged, and informal arrangements could vary the procedures in the interest of task accomplishment, organizations were less mechanistic and thus more organic. Work environments that demanded rapid, creative response to periodic disasters tended toward looser, more task-focused, and less hierarchical organizing methods. Technology-based innovation more often succeeded where "organic" structures encouraged initiative and hybrid thinking in place of rigid routines.

Organizing for innovation

While much earlier organization focused on operations, World War II gave rise to urgent efforts to innovate, among the most famous of which may be the efforts to develop working radar, decode German encryption, and produce the atomic bomb. Each of these efforts involved the clash of then-dominant ideas about order, on the one hand (principally military hierarchy and command), with empowerment, innovation, and creativity, on the other (principally the unpredictable and unscripted nature of creative thinking, especially among the array of scientists and technical experts whose willing participation was essential, and whose life choices were anything but military). Indeed, these experiences highlight what seems to be a central conundrum in organizing for innovation: genuine innovations are unpredictable, while "organizing" immediately calls up images of prediction, coordination, and regularity.

Innovation in teams

These early innovation teams were often "co-located" in today's language—sequestered in secure locations where they could interact intensively in pursuit of their goal. These teams were typically focused on particular innovation targets, like radar or creating an atomic bomb. In industrial situations, the Lockheed "Skunkworks", a secret group separated from ordinary activities to work on innovation, is the closest analog. The benefits of separation are often said to include freedom from the demands of daily routines and supervision, and thus the ability to focus more intently on the innovation task. The hazards often include difficulties in reintegrating the innovation to implement it in routine practice. Resistance by those responsible for "ordinary business" may stem from resentment of the innovators' privileges, as well as the lack of important tacit understanding about the innovation.

The traditional R&D labs of Corning, DuPont, and General Motors, or AT&T's Bell Labs, gathered technical experts together, allocated funding for them, and grouped them around projects: innovation teams, in essence. These labs allowed greater (Bell Labs) or lesser (GM) amounts of researcher autonomy and choice of project to work on. Discoveries that did not fit the corporation's business model might be spun off (as Gore-tex was, to its discoverer, when DuPont declined to commercialize it), or simply sat on the shelf. In the 1980s, under pressure of downsizing and in search of more immediate gains from R&D expenses, many U.S. corporations reduced their central laboratories or disbanded them altogether (Corning is one notable exception, but there are others), distributing research to product divisions. One result was significant foreshortening of research time horizons: divisions' work was more applications-oriented, so that today most "research" in the industrial sector is best classified as "development", according to the National Science Foundation.[3]

Networks of innovation

Co-located teams constitute a face-to-face network, yet often—particularly for scientists who are part of larger, external communities—the network of innovation far transcends the immediate team: "cosmopolitan" scientists may consult with colleagues around the world, for example. We shall return to networks of innovation in contemporary organizations momentarily. Here, we merely note that teams can and do extend beyond organizational boundaries.

The rise of the matrix organization

So-called matrix organizations sought to blend task focus on innovative project activities in temporary teams with more stable formal arrangements for administrative purposes. However, such organizations were unduly complex, resulting in high overhead, and they simply failed to facilitate effective innovation. Thoughtful observers have ascribed strategic and operational difficulties at Hewlett-Packard to the firm being "imprisoned in its structure" (Loomis, 2005)—it was structure itself that eroded performance. In response, emphasis has moved away from structure as the means for organizing and toward less hierarchy and more cooperation. In short, task-based relationships have begun to take precedence over formal structures, and networks of innovation increasingly extend beyond firm boundaries.

From innovation structures to innovation networks

New ways of organizing and new modes of coordination arise to address the limits of older approaches. Today's challenge is innovation that often must draw on resources that reside well beyond traditional organizational boundaries. Because the pace of technology development is so rapid, an organization focused on exploiting established knowledge and technology—that is, one deploying most classical organizational configurations—is likely to be less able to explore and discover new knowledge. Moreover, a structure well-suited to one purpose, product, and market may be ill-suited to recognizing need or opportunity for new options, let alone generating innovation to meet the changed conditions. The North America downsizing of corporate R&D mentioned previously withdrew resources from innovation efforts just at the point when innovation became vastly more important. As a result, firms that have stripped out what seemed "extraneous" capabilities were unable to autonomously innovate. They have been compelled to seek resources beyond their boundaries for help, via strategic alliances, outsourced research, and other extra-organizational relationships to foster innovation.

The more firms engage in broader networks of relationship, the more difficult it is to determine where "the organization" begins and ends, and the more important the links among and between partners become. According to some, close inspection "reveals most organizations to be mere legal fictions with no 'inside' or 'outside' analogous to borders—they are simply dense spots in networks of contracts . . ." (Davis and Marquis, 2005). Some researchers now see organizations as protean social and enacted constructs, in which structure, understanding, and culture constantly co-evolve. Others say organizations are action streams or action nets (Czarniawska, 2004). Observers agree that organizations are (much) more than structure; for innovation, the critical addenda center on influencing human cognition, social cognition and team work, and network relationships (both at individual and company levels). Thus, from "organization" structure for innovation, we must now turn to the role of multi-location and often multi-organization teams and networks in innovation.

As strategic environments and the technologies and information used to exploit them become more complex, the needed information for effective innovation rapidly exceeds the capabilities of any single individual. Add to this observations about the importance of ancillary technologies (Mowery and Rosenberg, 1998), knowledge brokering, and technology reconfiguration (Hargadon, 2003) in innovation efforts, and the growing importance of teams, networks, and alliances becomes clear. The pharmaceutical industry offers one model: existing major pharmaceutical companies, built on a business model of expertise in clinical trials, regulatory matters, advertising ("Ask your doctor!"), and distribution, generally lacked in-depth capacity in such new and rapidly evolving fields as genomics, proteomics, and molecular biology—the feedstock for contemporary pharmaceutical discovery. The established firms quickly developed strategic alliances with so-called "boutique" pharma companies, often university spin-offs like Genentech, Biogen, and Amgen that specialized in the new areas. While some of the start-ups were acquired, many others remained independent participants in ongoing alliances. Both parties benefited: the large companies leveraged their existing (and expensive) human resources and infrastructure for legal, regulatory, and clinical trials management to test the growing array of genetics-based therapies; while the smaller firms retained their agility and focus on innovation, leaving the "downstream" activities of bringing discoveries to market to their partners.

A very different model can be seen in the globally distributed R&D activities of Nokia: designing a cellphone

[3] Data regarding the focus and classification of R&D expenditures can be tracked through the semi-annual *NSF Engineering and Science Indicators* reports.

for use in Japan, the company quickly concluded that Japanese characters required more space, so a larger screen and scrolling capability were developed by Japanese researchers working for the company—and rapidly adopted by Nokia's customers elsewhere. Nokia was an early-adopter of globally distributed R&D, fueling its constant innovation from a wealth of good ideas drawn from many markets. Creative businesses generally, including music and other cultural creatives, also draw on global resources for innovation. Yet another approach can be seen in the entry of outside firms into China. The Chinese government initially required foreign direct investment and joint ventures as the price of entry, facilitating technology transfer along with modern management methods, training, and first-hand experience for Chinese in the mysteries of capitalist competition. (Similar relationships were required for entry into the Japanese market after World War II, and were supported by U.S. foreign policy.)

Whether for reasons of burgeoning, differentiated specialist expertise, design ideas, market insight, or political reality, risk, uncertainty, and complexity render standalone efforts impossible: that is, *organization* structure for innovation has been superseded by *networks* structured for innovation. Yet, despite this focus on alliances between organizations, the real work ultimately happens when people collaborate, whether multidisciplinary teams within a single organization or "trans-border" teams that come from two or more firms, government entities or other organization types, or with input from non-affiliated persons like individual customers, listserv participants, or bloggers. Traditional views of innovation assume a single moment of invention, by an identifiable inventor. When economists "opened the black box of technology", realizing that innovation was a driver of economic development, they distinguished between the (still opaque) invention stage and the subsequent product development stage (Rosenberg, 1994). By contrast, Tuomi (2002, pp. 9–10) notes, "Science and technology studies . . . provide ample historical evidence that this fundamental assumption [of a single moment of invention and a single inventor] is not valid in general," and proposes an alternative view in which "Innovation happens when social practice changes. If new technology is not used by anyone, it may be a promising idea, but strictly speaking, it is not technology. Similarly, if new knowledge has no impact on anyone's way of doing things—in other words, if it doesn't make any difference—it is not knowledge. Only when the way things are done changes, an innovation emerges." In other words, innovation happens beyond firms in networked social relationships that include non-firm members.

Internet connections around the globe have kept pace with globalizing business, so that "virtual teams" are widely (if loosely) spoken of. Real teams, virtual or otherwise, are relatively stable, with defined membership and an established, shared working process to achieve a goal that can only be achieved together. Here, in real teams, is where the work of innovation actually gets done, transcending the limitations of individuals to create something truly new. Yet teams are fragile social structures, often vulnerable to hostile surrounding cultures, norms, and broader organizational systems, measures, habits, and intentions (for further discussion of teams see Earley and Gibson, 2002; Gibson and Cohen, 2003; Hinds and Hiesler, 2002; and Wageman *et al.*, 2008). If firms are fragile inside organizations, they are still more problematic when separated by geographic and cultural distance, as when team members reside in different countries—Russian or Chinese or Indian researchers interacting with others from North America or Europe. The notion of "cultural distance" has been expanded to include other factors creating cognitive differences: cultural, administrative, geographic, or economic distance (the so-called CAGE framework) (Ghemawat, 2001) seeks to guide considerations of how those distances might be bridged and what resources are necessary.

Alternative approaches to innovation: the challenges of constraints

So, how then can people be organized for innovation? More particularly, in an era of global business, globally distributed and multinational work teams, how can managers foster innovation?

Innovation for fun and profit

IDEO, an exemplar as an innovative firm, has thrived on challenging assignments to fuse design and engineering, technical know-how, and art (Kelley and Littman, 2005). A central pattern in IDEO's success is its design teams' ability to reconfigure and re-purpose technologies and ideas that proved successful in other applications. Hargadon (2003) goes so far as to identify this as a central aspect of breakthrough innovations, and IDEO fosters this kind of idea reuse by a variety of administrative methods, including varying teams from project to project so that designers don't get too used to one another; sharing "best" ideas in a toolkit of elegant solutions; and adding challenges (like taking on time-constrained projects for the fun of it). Flat hierarchy, open information exchange, diverse employee backgrounds (a range of technical and artistic expertise and cultural backgrounds) facilitate "mixing and matching" ideas by fostering an open culture of sharing and discussion where reputational credibility, fun, and elegant design achievement are more central than sheer remuneration (for a description of practices see Tomke and Nimgade, 2000).[4]

A process/cultural approach to innovation

The architectural firm of Frank Gehry Associates encourages innovation by keeping ideas fluid for as long as possible, "brainstorming" dramatically different concepts of buildings through "shreck models"—from a Yiddish word meaning "monster"—that are deliberately so rough and unfinished that neither clients nor designers become fixated on the design too early (Gehry, 2004). Repeated iterations on seemingly "finished" projects contribute to getting past compromises that seemed unavoidable to more integral designs. Here, too, open communication, "social innovation" coupled with technological innovation—the very early use of computer-aided design programs drawn from the aerospace industry to assure the precise manufacture of components for Gehry's notoriously "not square" buildings, for example—produce a rich mix of ideas and variants among design team members (including the client) who are encouraged to stretch their ideas and evolve toward genuine innovation.

Innovation due to serendipitous resource couplings

Both Gehry Associates and IDEO teams feature carefully selected members of broad experience and deep expertise. But innovation in other firms may rely more on chance resources. Mergers and acquisitions and strategic alliances have boosted reliance on cross-disciplinary and often cross-organizational teams, many of them virtual. Because innovation is increasingly important, researchers have generated a host of relevant findings about how foundational elements such as shared understanding, trust, and integration might be created among individuals. Researchers agree that trust is critical; they note the benefits of face-to-face introductions as an important starting point for successful virtual teams, and of very real barriers to trust and effective communication that arise as teams expand.

Innovation and institutional infrastructure

While there are many studies of factors concerning innovation at the individual level, and numerous studies of creativity, innovation has its impact at the firm level, where it is also visible as the locus of entrepreneurship and technological innovation. Firm-level results aggregate to significant national and international impact. We know, for example, that high-technology entrepreneurship in startup firms has been a central driver of economic growth and development in the U.S., and an array of economists from Schumpeter in the 1930s to the present point to innovation as the primary factor. Moreover, entrepreneurship has surged around the world since 1975, flourishing in a wide array of countries, including China, the Czech Republic, India, Ireland, Korea, Peru, and Turkey in addition to the U.S., according to the *Global Entrepreneurship Monitor* (GEM), a 42-country, 5-continent study of entrepreneurial propensities of countries.

Entrepreneurs innovate by identifying underserved niches in existing markets, or generating whole new markets (and industries). They may also generate wholly new business models, innovating in managerial terms—one form of "social technology". The firms that have persistently made the greatest innovative contributions in the U.S. have been high-technology startups, which have higher rates of patenting, higher rates of job and wealth creation. Small firms (10–200 employees) account for 45% of the total non-farm payroll, and just over half of the private, non-farm employment, creating 60% to 80% of *the net new jobs annually* for the last decade. These small firms are *more innovative* than larger firms, producing 13 to 14 times as many patents per employee, and their new-product development accounts for up to 80% of sales of new innovative products in their first years after launch. Patents filed by innovative small businesses are twice as likely as those filed by large firms to be among the *top 1% of patents* in subsequent citations, according to the U.S. Small Business Administration.

Yet, not all startups survive; these nascent firms face a "liability of newness" because they are dependent upon the social structure in which they are embedded, lacking the network of established relations larger firms enjoy, along with a variety of other established and trusted relationships, perhaps even among the founding members. Similar patterns for young firms in a wide array of industries suggest that those firms most likely to survive are those that are founded by more than a single individual (i.e., by a *team of entrepreneurs*), because they have larger and more diverse networks to access for resources, information, expertise, perspectives, and other benefits. These relational resources translate into higher survival and success rates. The best survival is enjoyed by firms with a "strong" founding team of three or more members that spans the necessary, varied areas of expertise and experience, a mix of experienced veterans plus more recently trained technical specialists; these firms have the highest revenue growth rate in their first 4 years, a higher probability of reaching $20 million in revenues, and a higher probability of going public (Schoonhoven *et al.*, 1990). High-technology startups are explicitly organized for innovation, their critical lifeblood.

[4] Interested readers are also referred to the IDEO website, which provides abundant additional information about the company's practices; *http://www.ideo.com/*

What infrastructure supports this innovation? By the logic of our argument, critical infrastructure supporting innovation includes the institutional arrangements that facilitate creation of entrepreneurship, whether inside an existing firm, or as a startup, and increasingly this "infrastructure" consists of a network of knowledgeable partners. The creation of new knowledge, acquisition and accumulation of the range of expertise, connections, and other resources—financial support, expert legal or accounting advice, and the like—are all needed to nurture innovations through the "Valley of Death" between scientific discovery and market launch (Jelinek, 2005; Jelinek and Markham, 2007). Silicon Valley's notable success at innovation during more than half a century is arguably rooted in the plethora of supporting resources readily available there—the raw materials of constantly forming new networks of innovation.

In the U.S. at large, a long-standing national innovation system has relied heavily on university research as the wellspring for fundamental research and scientific discovery—long before the Bayh–Dole legislation often portrayed as its start[5] (Mowery *et al.* 2004). Close industry–university relationships have been central both to initial discovery and subsequent development, on the one hand, and to the rapid training of expert employees needed for such diverse industries as chemicals, petroleum and aerospace in earlier times (Mowery and Rosenberg, 1998), and biotechnology more recently, on the other hand. Startups that retain close links to universities are more likely to survive and prosper. So, universities have been important network players contributing to innovation.

The open innovation model

Open innovation (Chesbrough, 2003) has been offered as a contemporary and more effective alternative to older, proprietary innovation efforts inside a single firm's boundaries. Open innovation can be said to be both the cause and effect of the increasing strategic alliances, collaborative innovation, and partnerships noted earlier. At the extreme, these outreach efforts may include "seekers" posting problems that "solvers" answer on websites like InnoCentive, which open the innovation process to all-comers. Open innovation is a contemporary name for an approach long practiced, to some degree at least: Pierre S. duPont and his cousins were MIT graduates, as were many of the chemists they hired to do research for them; the petrochemical and aeronautics industries grew up around collaborative university research that firms commercialized. What is different today is that the openness and collaboration are explicitly recognized, studied, and actively managed.

Organizing for innovation is an enormous topic accessible from innumerable perspectives. This chapter has not sought to survey this range, but rather to offer a brief overview of traditional responses to the challenge of organizing for innovation, contemporary issues, and practices, and point to emerging issues.

[5] Bayh-Dole legislation, named for its U.S. Senate sponsors, permitted universities to patent discoveries made on their campuses with Federal research support, thereby clarifying ownership and subsequent licensure. Bayh-Dole has been widely praised for releasing the innovative, entrepreneurial spirit of university researchers (by its supporters) or excoriated as having commercialized higher education (by its critics). Mowery *et al.* (2004) demonstrate that both extremes of opinion are unwarranted.

References

Barnard, C. I. (1938; 1956). *The Functions of the Executive*. Cambridge, MA: Harvard University Press.

Burns, T., and Stalker, G. M. (1962). *The Management of Innovation*. Chicago, IL: Quadrangle Books.

Chandler, A. D. J. (1965). "The railroads: Pioneers in modern corporate management." *Business History Review*, **XXXIX**, Spring, 16–40.

Chesbrough, H. W. (2003). *Open Innovation: The New Imperative for Creating and Profiting from Technology*. Boston, MA: Harvard Business School Press.

Czarniawska, B. (2004). "Management as the designing of an action net." In: R. J. Boland and F. Collopy (Eds.), *Managing As Designing* (pp.102–105). Stanford, CA: Stanford University Press.

Davis, G. F., and C. Marquis (2005). "Prospects for organization theory in the early twenty-first century: Institutional fields and mechanisms." *Organization Science*, **16**, 332–343.

Earley, P. C., and C. B. Gibson (2002). *Multinational Work Teams: A New Perspective*. Mahwah, NJ: Lawrence Erlbaum Associates.

Gehry, F. O. (2004). "Reflections on designing and architectural practice." In: R J. Boland and F. Collopy (Eds.), *Managing As Designing*. Stanford, CA: Stanford Business Books.

Ghemawat, P. (2001). "Distance still matters: The hard reality of global expansion." *Harvard Business Review*, **79**(8), 137–147.

Gibson, C., and Cohen, S. (2003). *Virtual Teams that Work*. San Francisco, CA: Jossey-Bass.

Global Entrepreneurship Monitor; *http://www.gemconsortium.org*

Haberstroh, C. J. (1965). "Organization design and systems analysis." In: J. G. March (Ed.), *Handbook of Organizations* (1247 pp.). Chicago, IL: Rand McNally.

Hargadon, A. (2003). *How Breakthroughs Happen: The Surprising Truth about How Companies Innovate*. Boston, MA: Harvard Business School Press.

Hinds, P. J., and Hiesler, S. (Eds.). (2002). *Distributed Work*. Cambridge, MA: MIT Press.

IBM. (2006). *Expanding the Innovation Horizon*; *http://www-935.ibm.com/services/uk/bcs/html/bcs_landing_ceostudy.html* (consulted November 3, 2007).

Jelinek, M. (2005). "Crossing Death Valley together: Cultural dynamics of industry–university IP." Paper presented at *The SMA Conference, November 9–12*. Charleston, SC: Southern Management Association.

Jelinek, M., and S. Markham (2007). "Industry–university IP relations: Integrating perspectives and policy solutions." *IEEE Transactions in Engineering Management*, **54**(2).

Kelley, T., and Littman, J. (2005). *The Ten Faces of Innovation: IDEO's Strategies for Beating the Devil's Advocate and Driving Creativity throughout Your Organization*. New York: Doubleday.

Mowery, D. C., and N. Rosenberg (1998). *Paths of Innovation: Technological Change in 20th-century America*. Cambridge, U.K.: Cambridge University Press.

Mowery, D. C., Nelson, R. R., Sampat, B. N., and Ziedonis, A. A. (2004). *Ivory Tower and Industrial Innovation: University–Industry Technology Transfer before and after the Bayh–Dole Act*. Stanford, CA: Stanford University Press.

NSF Report; *http://www.nsf.gov/statistics/seind08/*

Rosenberg, N. (1994). *Exploring the Black Box: Technology, Economics, and History*. New York: Cambridge University Press.

Schoonhoven, C. B., Eisenhardt, K. M. *et al.* (1990). "Speeding innovation to market: The impact of technology-based innovation on waiting times to first product introduction in new semiconductor ventures." *Administrative Science Quarterly*, **35**(1), March, 177–207.

Tomke, S., and Nimgade, A. (2000). *IDEO Product Development* (Harvard Business School Case 9-600-143). Boston, MA: Harvard Business School Press.

Tuomi, I. (2002). *Networks of Innovation*. Oxford, U.K.: Oxford University Press.

Wageman, R., Nunes, D. A. *et al.* (2008). *Senior Leadership Teams: What It Takes to Make Them Great*. Boston, MA: Harvard Business School Press.

Woodward, J. (1958). *Management and Technology*. London: Her Majesty's Stationery Office.

Woodward, J. (1965). *Industrial Organization: Theory and Practice*. London: Oxford University Press.

30

The Concept of Corporate Entrepreneurship

Jeffrey G. Covin and Donald F. Kuratko

Indiana University

Introduction

Many have observed that corporate entrepreneurship (hereafter CE) is a common and/or inevitable byproduct of organizational activity and, therefore, has existed as long as organizations themselves (e.g., Burgelman and Sayles, 1986; Covin and Miles, 1999; Drucker, 1985; Kuratko, Ireland, and Hornsby, 2001). However, CE *as a concept* has surfaced within the academic literature largely over the last four decades. There is no universally agreed-upon meaning attached to the term CE. Moreover, depictions of the CE concept have varied considerably over time.

Early scholars tended to use the label CE solely in reference to the pursuit of new business venturing opportunities within established firms (e.g., Hill and Hlavacek, 1972; Peterson and Berger, 1971). The terms CE and corporate venturing were often used interchangeably and in reference to this same "new business" phenomenon. More recent conceptualizations of CE suggest that entrepreneurship also exists in other forms within established firms. For example, Guth and Ginsberg (1990) and Zahra (1991) have argued that CE may occur as strategic renewal, whereby an established firm instigates significant changes to its strategy and/or structure in pursuit of greater organizational efficiency or effectiveness. In their review of the CE literature, Sharma and Chrisman (1999, p. 19) proposed that CE may exist as corporate venturing, strategic renewal, or innovations "of the Schumpeterian (1934) variety"—that is, innovations involving "the introduction of an original invention or idea into a commercially usable form that is new to the marketplace and has the potential to transform the competitive environment as well as the organization."

There is no objectively "correct" meaning to the term CE. Nonetheless, there are some ways of viewing CE that have greater utility than others. In particular, a valuable CE conceptualization scheme will be one that captures the essence of the CE construct as generally discussed within the literature. That is, CE scholars and practitioners must be able to implicitly agree that the scheme accurately represents conventional understandings and interpretations of the concept. Most writings on the CE concept seem to embrace the notion that CE involves new-business entry by established firms (i.e., corporate venturing). Beyond this conclusion, little agreement exists within the literature on the matter of CE's appropriate definition and domain.

With this proviso in mind, the purpose of this chapter is to propose a CE conceptualization scheme that meaningfully reflects the complexity and heterogeneity of current thought on the nature of CE. Of critical concern in this chapter is the identification and discussion of specific entrepreneurial phenomena that can and have been associated with the term CE. Two empirical phenomena are herein proposed as constituting the domain of CE—namely, corporate venturing and strategic entrepreneurship (see Figure 30.1). Corporate venturing approaches have as their commonality the adding of new businesses (or portions of new businesses via equity investments) to the corporation. This can be accomplished through three implementation modes—internal corporate venturing, cooperative corporate venturing, and external corporate venturing. By contrast, strategic entrepreneurship approaches have as their commonality the exhibition of large-scale or otherwise highly consequential innovations that are adopted in the firm's pursuit of competitive advantage. These innovations may or may not result in new businesses for the corporation. With strategic entrepreneurship approaches, innovation can be in any of five areas—the firm's strategy, product offerings, served markets, internal organization (i.e., structure, processes, and capabilities), or business model. Each of these categories of corporate entrepreneurship is further reviewed below. For additional discussion of the CE conceptualization scheme proposed here, the reader is

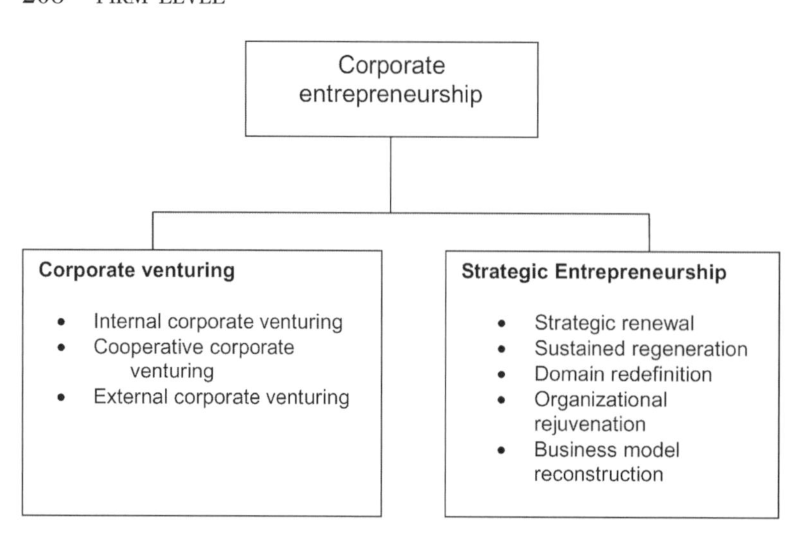

Figure 30.1. Defining corporate entrepreneurship.
Source: Morris, Kuratko, and Covin (2008, p. 81).

referred to the authors' text on this topic (see Morris, Kuratko, and Covin, 2008).

The concept of corporate venturing

Corporate venturing refers to a set of entrepreneurial phenomena through which new businesses are created by, added to, or invested in by an existing corporation. With internal corporate venturing, new businesses are created and owned by the corporation. These businesses typically reside within the corporate structure, but, occasionally, may be located outside the firm and operate as semi-autonomous entities. Among internal corporate ventures that reside within the firm's organizational boundaries, some may be formed and exist as part of a pre-existing internal organization structure and others may be housed in newly formed organizational entities within the corporate structure. Cooperative corporate venturing (a.k.a. joint corporate venturing, collaborative corporate venturing) refers to entrepreneurial activity in which new businesses are created and owned by the corporation together with one or more external development partners. Cooperative ventures typically exist as external entities that operate beyond the organizational boundaries of the founding partners. External corporate venturing refers to entrepreneurial activity in which new businesses are created by parties outside the corporation and subsequently invested in (via the assumption of equity positions) or acquired by the corporation. These external businesses are typically very young ventures or early growth–stage firms. In practice, new businesses might be developed through a single venturing mode, any two venturing modes, or all three venturing modes. A firm's total venturing activity is equal to the sum of the ventures enacted through the internal, cooperative, and external modes. With corporate venturing, an established firm's entry into an entirely new business is the key entrepreneurial act. Figure 30.2 depicts the meaning of the term "new business."

Motives for corporate venturing

Corporations create new businesses for multiple reasons. Having a clear understanding of the motives for corporate venturing is critical for effective venture management. In the final analysis, it is impossible to evaluate the success or failure of corporate-venturing initiatives unless it is clear what management's goals were in the first place. Companies must create venture evaluation and control systems that assess venture performance on criteria that follow from the venture's founding motive.

Tidd and Taurins (1999) concluded that there are two sets of motives that drive the practice of internal corporate venturing: leveraging (to exploit existing corporate competencies in new-product or market arenas) and learning (to acquire new knowledge and skills that may be useful in existing product or market arenas). When the overall motive is leveraging, some of the specific reasons that firms engage in corporate venturing include:

- To exploit underutilized resource—build a new business around internal capabilities that remain idle for prolonged periods; the new business becomes the vehicle for outsourcing those capabilities to others.
- To extract further value from existing resources—build a new business around corporate knowledge, capabilities, or other resources that have value in product–market arenas not currently being served by the firm.
- To introduce competitive pressure onto internal suppliers—build a new business that becomes an alternative supplier to existing internal supply sources.
- To spread the risk and cost of product development—build a new business whose target market promises to be

Market focus of the venture	Current product of the corp.	Extension of current product	New product for the corp. in current industry	New product for the corp. in new industry (i.e., diversification)
Market creation (new to "World")	**New business**	**New business**	**New business**	**New business**
New market for the corp.	Major market development	**New business**	**New business**	**New business**
Extension of current market	Minor market development	Minor product-market development	**New business**	**New business**
Current market of the corp.	Market penetration	Minor product development	Major product development	**New business**

Product focus of the venture

Figure 30.2. The concept of a "new business".

larger than that for which the core product to be offered by the business was initially developed.
- To divest non-core activities—build a new business to pursue business opportunities that the firm is in a favorable position to exploit and that the firm has no strategic interest in.

The learning motives can also be broken down further as well. Three major types of organizational learning tend to receive the greatest emphasis:

- To learn about the process of venturing—build a new business as a laboratory in which the innovation process can be studied.
- To develop new competencies—build a new business as a basis for acquiring new knowledge and skills pertaining to technologies, products, or markets of potential strategic importance.
- To develop managers—build a new business as a training ground for the development of individuals with general management potential.

In another study of corporate-venturing practice (one that included firms engaged in both internal and external corporate venturing) Miles and Covin (2002) reported that the firms pursued venturing for three primary reasons: (1) to build an innovative capability as the basis for making the overall firm more entrepreneurial and accepting of change; (2) to appropriate greater value from current organizational competencies or to expand the firm's scope of operations and knowledge into areas of possible strategic importance; and (3) to generate quick financial returns. Where the motivation is to generate quick financial returns, firms often concentrate on the external mode of venturing. Specifically, many corporations invest in new, externally founded businesses in hopes of realizing significant financial gains, returns beyond those easily obtainable within the firm's current scope of operations. This type of venturing is often pursued through the use of corporate venture capital funds.

Corporate venture capital

The term corporate venturing is sometimes interpreted to mean corporate venture capital investment activity. Many corporations have internal venture capital funds that are used to invest in external new ventures deemed strategically important or financially attractive to the corporation. Corporations also sometimes operate as investment partners in external venture funds that are owned and controlled by multiple parties. The external venture funds in which corporations invest often target new businesses in specific technology or product–market arenas of interest to the fund-owners and in which those owners have, ideally, developed some investment competence.

Over the past few decades, the popularity of directly investing corporate funds into external business startups has varied with the strength of the market for new public offerings (IPOs). In general, when the stock market has been strong, corporate venture capital investments have been popular. In times when the stock market has dropped, investment in external startups has fallen off. Historically, corporate venture capital investments have occurred in three waves or boom-and-bust cycles (see Gompers, 2002).

A useful framework for linking corporate venture capital investments with a company's larger strategic agenda has been proposed by Chesbrough (2002). He observes that corporate venture capital investments can be sorted into categories according to (1) their objectives (strategic or financial) and (2) the degree to which the new business being invested in (typically a startup) has operational capabilities (i.e., resources and processes) that are linked to those of the investing corporation (tight or loose linkages). Firms with strategic investment objectives invest in external startups whose success promises to increase the sales and profits of the firm's existing businesses. Firms with financial investment objectives, by contrast, are principally seeking attractive financial returns through their investments. Tight operational linkages portend the possibility of leveraging the corporation's competencies in ways that contribute to the success of the venture (e.g., the venture might use the existing distribution channels of another business of the corporation); loose operational linkages imply that the corporation and the startup have minimal resource or process overlap and that the startup operates in a partially or wholly autonomous manner *vis-à-vis* the investing corporation. Based on these two considerations, Chesbrough (2002) discusses four pure types of corporate venture capital investment:

- Driving investments (strategic rationale for investment and tight operational links between the startup and the investing company)—these investments extend the corporation's presence in product–market or technological arenas regarded as strategic to the corporation.
- Enabling investments (strategic rationale for investment and loose operational links between the startup and the investing company)—these investments complement the strategy of the corporation by stimulating demand for the corporation's current products through the development of the larger ecosystem within which the corporation or its businesses operate.
- Emergent investments (financial rationale for investment and tight operational links between the startup and the investing company)—these investments are targeted toward startups whose success may be of strategic relevance to the corporation or toward startups for whom the corporation's resources or processes provide needed and critical value within the startup's overall business model.
- Passive investments (financial rationale for investment and loose operational links between the startup and the investing company)—these investments are diversification actions in which the corporation operates as a money manager or investment intermediary for its shareholders.

According to Chesbrough (2002), passive investments are seldom if ever justifiable inasmuch as the corporate shareholders can diversify their own portfolios—they do not need the corporation to do this for them. Driving, enabling, and emergent investments, on the other hand, can foster growth in the corporation's current business or lead the firm into desirable new-business arenas. Perhaps most significantly, poor short-term financial returns should not discourage corporations from making investments in external startups deemed to be of strategic importance within the scope of the corporation's current or planned business operations.

The concept of strategic entrepreneurship

Strategic entrepreneurship constitutes a second major category of approaches to corporate entrepreneurship. While corporate venturing involves company involvement in the creation of new businesses, strategic entrepreneurship corresponds to a broader array of entrepreneurial initiatives which do not necessarily involve new businesses being added to the firm. All forms of strategic entrepreneurship have one thing in common: they all involve the exhibition of organizationally consequential innovations that are adopted in the pursuit of competitive advantage.

Strategic entrepreneurship involves simultaneous opportunity-seeking and advantage-seeking behaviors (Ireland, Hitt, and Sirmon, 2003). The innovations that are the focal points of strategic entrepreneurship initiatives represent the means through which opportunity is capitalized upon. These are innovations that can happen anywhere and everywhere in the company. By emphasizing an opportunity-driven mindset, management seeks to achieve and maintain a competitively advantageous position for the firm.

These innovations can represent fundamental *changes from the firm's past* strategies, products, markets, organization structures, processes, capabilities, or business models. Or, these innovations can represent fundamental bases on which the firm is fundamentally *differentiated from its industry rivals*. Hence, there are two possible reference points that can be considered when a firm exhibits strategic entrepreneurship: (1) how much the firm is transforming itself relative to where it was before (e.g., transforming its products, markets, internal processes, etc.) and (2) how much the firm is transforming itself relative to industry conventions or standards (again, in terms of product offerings, market definitions, internal processes, and so forth).

With regard to differentiation relative to one's industry, it should be noted that some firms consistently exhibit high levels of innovativeness from the time of their founding and, as such, they have always been tagged with the label of *entrepreneurial firm*. Moreover, some industries invite continuous entrepreneurial behavior (e.g., fashion-related industries, technology-based industries). Therefore, innovativeness *per se* may not be a basis on which firms in those industries are differentiated from their industry rivals. Rather, it is the products, services, and processes that result from this innovativeness that determine how well they are differentiated.

Strategic entrepreneurship can take one of five forms—strategic renewal, sustained regeneration, domain redefinition, organizational rejuvenation, and business model reconstruction (Covin and Miles, 1999). Some defining attributes of these different forms or approaches are presented in Table 30.1.

Table 30.1. Forms of strategic entrepreneurship.

Form of strategic entrepreneurship	*Focus of the entrepreneurial initiative*[a]	*The entrepreneurial event*	*Typical frequency of the entrepreneurial event*
Strategic renewal	*Strategy* of the firm	Adoption of a new strategy	Low
Sustained regeneration	*Products* offered by the firm or *markets* served by the firm	Introduction of a new product into a pre-existing product category or introduction of an existing product into a new (to the firm) but pre-existing market	High
Domain redefinition	*New competitive space*	Creation of new or reconfiguration of existing product categories or market space	Low
Organizational rejuvenation	*Organization structure, processes, and/or capabilities* of the firm	Enactment of a major, internally focused innovation aimed at improving strategy implementation	Low to moderate
Business model reconstruction	*Business model* of the firm	Design of a new or redesign of an existing business model	Low

[a] The focus of the entrepreneurial event can be the entire firm or, in the case of multi-business firms, one or more of its businesses.
Source: Morris, Kuratko, and Covin (2008, p. 89).

The forms of strategic entrepreneurship

Strategic renewal is a type of entrepreneurship whereby the firm "seeks to redefine its relationship with its markets or industry competitors by fundamentally altering how it competes" (Covin and Miles, 1999, p. 52). As originally defined by Guth and Ginsberg (1990, p. 5), the label "strategic renewal" referred to "the transformation of organizations through renewal of the key ideas on which they are built." Yet, strategic renewal has a more specific meaning and focus. As shown in Table 30.1, with strategic renewal the focus of the entrepreneurial initiative is the firm's strategy. However, not all firms that adopt new strategies are pursuing strategic renewal. Rather, new strategies constitute strategic renewal when they represent fundamental repositioning efforts by the firm within its competitive space. Additionally, firms that are founded based on unique value propositions that deviate from accepted industry strategic recipes would be considered entrepreneurial firms that practice the strategic renewal form of strategic entrepreneurship. These firms are playing new strategic games designed to place the firms in more favorable industry positions. Within the popular business press and practitioner-oriented journals, the phenomenon of strategic renewal has also been called strategic innovation (e.g., Hamel and Prahalad, 1995) or value innovation (e.g., Kim and Mauborgne, 1997, 1999).

Sustained regeneration refers to the entrepreneurial phenomenon whereby the firm "regularly and continuously introduces new products and services or enters new markets" (Covin and Miles, 1999, p. 51). With this form of strategic entrepreneurship the firm is in constant pursuit of entrepreneurial opportunities. Most of these opportunities will result in incremental innovations as represented by the offering of product extensions or movement into adjacent market arenas. On occasion, employment of the sustained regeneration form of strategic entrepreneurship will result in new-business creation. Sustained regeneration is most commonly employed as a basis for attaining or sustaining competitive advantage under conditions of short product life cycles, changing technological standards, or segmenting product categories and market arenas. Arguably, sustained generation is the most recognized and common form of strategic entrepreneurship. Firms that successfully practice sustained regeneration have reputations as "innovation machines". Unlike the other forms of strategic entrepreneurship, sustained innovation cannot be represented by a single, discrete event. Rather, sustained regeneration exists when firms exhibit an ongoing pattern of new-product introductions and/or new-market entries.

Domain redefinition refers to the entrepreneurial phenomenon whereby the firm "proactively creates a new product–market arena that others have not recognized or actively sought to exploit" (Covin and Miles, 1999,

p. 54). Through domain redefinition, firms move into uncontested markets, or what Kim and Mauborgne (2005) have called "blue oceans". These are product–market arenas in which new product categories are represented. A product category refers to a group of products that consumers view as substitutable for one another yet distinct from those in another product category. These new product categories can either give rise to completely new industries or redefine the boundaries of existing industries. Domain redefinition renders a firm's current competition moot, at least temporarily, inasmuch as this entrepreneurial activity takes place in unoccupied competitive space. The entrepreneurial firm's hope is that its first-mover status will create a basis for sustainable competitive advantage when and if competitors follow. The domain redefinition phenomenon is discussed within the business literature under a variety of labels including bypass strategy (Fahey, 1989), market pioneering (Golder and Tellis, 1993), whitespace marketing (Maletz and Nohria, 2001), and blue ocean strategy (Kim and Mauborgne, 2004, 2005). Unlike the other forms of strategic entrepreneurship, domain definition necessarily results in new-business creation.

Organizational rejuvenation refers to the entrepreneurial phenomenon whereby the firm "seeks to sustain or improve its competitive standing by altering its internal processes, structures, and/or capabilities" (Covin and Miles, 1999, p. 52). With organizational rejuvenation, the focus of the innovation effort is a core attribute or set of attributes associated with the firm's internal operations. The objective of these efforts is to create a superior organizational vehicle through which the firm's strategy can be implemented. When pursued successfully, organizational rejuvenation enables a firm to achieve competitive advantage without changing its strategy, product offerings, or served markets. Sometimes organizational rejuvenation will entail a fundamental redesign of the entire organization, such as might result from major business process reengineering projects that reconfigure the firm's internal value chain. Organizational rejuvenation can also involve single innovations that have sweeping implications for the firm (e.g., major restructuring efforts) or multiple smaller innovations that collectively contribute to significantly increased organizational efficiency or effectiveness at strategy implementation (e.g., administrative innovations designed to facilitate inter-unit communications or the transference of core competencies). In order to constitute organizational rejuvenation, the innovation(s) in question cannot simply imitate initiatives that are common to the firm's industry. Rather, the innovation(s) must, at least temporarily, distinguish the firm from its industry rivals.

Business model reconstruction refers to the entrepreneurial phenomenon whereby the firm designs or redesigns its core business model(s) in order to improve operational efficiencies or otherwise differentiate itself from industry competitors in ways valued by the market. Business models have been described as "stories that explain how enterprises work" (Magretta, 2002, p. 87). According to Magretta (2002), these stories address four basic questions:

1. Who is the customer?
2. What does the customer value?
3. How do we make money in this business?
4. What is the underlying economic logic that explains how we can deliver value to customers at an appropriate cost?

Firms may be founded on the basis of novel business models or they may adopt new business models in their pursuit of competitive advantage. Common activities within business model reconstruction include outsourcing (i.e., relying on external suppliers for activities previously performed internal to the firm) and, to a lesser extent, vertical integration (i.e., bringing elements of the supplier or distributor functions within the ownership or control of the firm).

Summary

In summary, we have reviewed the two categories of corporate entrepreneurship. These categories are corporate venturing and strategic entrepreneurship. Corporate venturing refers to a set of entrepreneurial phenomena through which new businesses are created by, added to, or invested in by an existing corporation. Corporate venturing occurs in three modes: internal corporate venturing (new businesses are created and owned by the corporation), cooperative corporate venturing (new businesses are created and owned by the corporation and one or more external development partners), and external corporate venturing (new businesses are created by parties outside the corporation and subsequently invested in—via the assumption of equity positions—or acquired by the corporation). External corporate venturing is often pursued through corporate venture capital investments. Strategic entrepreneurship refers to a broad array of entrepreneurial phenomena, which may or may not result in new businesses being added to the corporation, in which large-scale or otherwise organizationally consequential innovations are adopted in the firm's pursuit of competitive advantage. The forms of strategic entrepreneurship include strategic renewal, sustained regeneration, domain redefinition, organizational rejuvenation, and business model reconstruction.

References

Burgelman, R. A., and Sayles, L. R. (1986). *Inside Corporate Innovation Strategy, Structure and Managerial Skills*. New York: The Free Press.

Chesbrough, H. W. (2002). "Making sense of corporate venture capital." *Harvard Business Review*, **80**(3), 90–99.

Covin, J. G., and Miles, M. P. (1999). "Corporate entrepreneurship and the pursuit of competitive advantage." *Entrepreneurship Theory and Practice*, **23**(3), 47-63.

Drucker, P. (1985). *Innovations and Entrepreneurship*. New York: Harper & Row.

Fahey, L. (1989). "Bypass strategy: Attacking by surpassing competitors." In: L. Fahey (Ed.), *The Strategic Planning Management Reader* (pp. 189–193). Englewood Cliffs, NJ: Prentice Hall.

Golder, P. N., and Tellis, G. J. (1993). "Pioneer advantage: Marketing logic or marketing legend." *Journal of Marketing Research*, **30**, 158–170.

Gompers, P. A. (2002). "Corporations and the financing of innovation: The corporate venturing experience." *Federal Reserve Bank of Atlanta Economic Review*, Fourth Quarter, 1–17.

Guth,W. D., and Ginsberg A. (1990). "Guest editors' introduction: Corporate entrepreneurship." *Strategic Management Journal*, **11**, Summer Special Issue, 5–16.

Hamel, G., and Prahalad, C. K. (1995). "Thinking differently." *Business Quarterly*, **59**(4), 22–35.

Hill, R. M., and Hlavacek, J. D. (1972). "The venture team: A new concept in marketing organizations." *Journal of Marketing*, **36**, 44–50.

Ireland, R. D., Hitt, M. A., and Sirmon, D. G. (2003). "A model of strategic entrepreneurship: The construct and its dimensions." *Journal of Management*, **29**(6), 963–989.

Kim, W. C., and Mauborgne, R. (1997). "Value innovation: The strategic logic of high growth." *Harvard Business Review*, **75**(1), 103-112.

Kim, W. C., and Mauborgne, R. (1999). "Creating new market space." *Harvard Business Review*, **77**(1), 83–93.

Kim, W. C., and Mauborgne, R. (2004). "Blue ocean strategy." *Harvard Business Review*, **82**(10), 76–84.

Kim, W. C., and Mauborgne, R. (2005). "Blue ocean strategy: From theory to practice." *California Management Review*, **47**(3), 105–121.

Kuratko, D. F, Ireland, R. D., and Hornsby, J. S. (2001). "The power of entrepreneurial outcomes: Insights from Acordia, Inc." *Academy of Management Executive*, **15**(4), 60–71.

Maletz, M. C., and Nohria, N. (2001). "Managing in the whitespace." *Harvard Business Review*, **79**(1), 102–111.

Magretta, J. (2002). "Why business models matter." *Harvard Business Review*, **80**(5), 86-92.

Miles, M. P., and Covin, J. G. (2002). "Exploring the practice of corporate venturing: Some common forms and their organizational implications." *Entrepreneurship Theory and Practice*, **26**(3), 21-40.

Morris, M. H., Kuratko, D. F., and Covin, J. G., (2008). *Corporate Entrepreneurship and Innovation*. Mason, OH: Thomson/South-Western Publishers.

Peterson, R., and Berger D. (1971). "Entrepreneurship in organizations." *Administrative Science Quarterly*, **16**(1), 97–106.

Schollhammer, H. (1982). "Internal corporate entrepreneurship." In C. Kent, D. Sexton, and K. Vesper (Eds.), *Encyclopedia of Entrepreneurship*. Englewood Cliffs, NJ: Prentice Hall.

Schumpeter, J. A. (1934). *The Theory of Economic Development*. New Brunswick, NJ: Transaction Publishers.

Sharma, P., and Chrisman, J. J. (1999). "Toward a reconciliation of the definitional issues in the field of corporate entrepreneurship." *Entrepreneurship Theory and Practice*, **23**(3), 11–28.

Tidd, J., and Taurins, S. (1999). "Learn or leverage? Strategic diversification and organizational learning through corporate ventures." *Creativity and Innovation Management*, 8(2), 122–129.

Zahra, S. A. (1991). "Predictors and financial outcomes of corporate entrepreneurship: An exploratory study." *Journal of Business Venturing*, **6**(4), 259–285.

31

Intellectual Property Strategy at the Firm Level

Mariann Jelinek

College of William and Mary

Intellectual property—legal ownership rights to creative works that are "novel, non-obvious, and useful"—have long been associated with physical works like inventions, written work like books or articles, and even artistic works. As the world economy increasingly emphasizes "knowledge work", it turns to "bits and bytes"—electronic "clicks" rather than physical "bricks". Knowledge, often captured in digital content, the expertise on which it is based, and the products and services to which it gives rise, are all considered "intellectual capital" to be protected, nurtured, and husbanded equally as much as financial capital. Moreover, since design specifications or chemical formulas can be readily digitized, the "recipes" for physical items enters the realm of "clicks", easily copied, easily transferred, and much more difficult to control.

Intellectual property eights (IPR) have risen to prominence in innovation discussions, fueled by numerous assertions and assumptions about the nature and quality of IP, what it can do and what it can't. This chapter seeks to sketch a basic framework for considering IP strategy at the firm level, clarify some of the contested issues (or at least some appropriate grounds for contest), and suggest approaches for considering IP strategy options. The chapter proceeds as follows: the first section defines intellectual capital, intellectual property, and IPR, distinguishing patents from copyrights. Next, the industry effect on IP is discussed, offering some observations of what industries appear to benefit most from patents, and what dimensions of industry dynamics highlight patents as an appropriate strategy. Observations about strategy options for firms come next, reflecting different options and uses for IP. The U.S. IP system's discontents, especially as these relate to IPR in an increasingly global economy, is the final topic—of interest because the U.S. innovation system has been so successful, and many advocate emulating its IP system in hopes of duplicating its entrepreneurial success.

Intellectual capital: types and options

A variety of intangible assets—brand names, logos, business processes, knowledge, databases, software, and more—may be termed "intellectual capital" insofar as they are knowledge-based forms of value. Loosely speaking, the intellectual base that creates intellectual property—close to the usage in information technology circles—may be generalized to include other items to recognize that certain types of knowledge can create value, despite enormous difficulty in placing any rigorous value on them. But what, then, is "intellectual property"? Intellectual property (IP) refers to formally recognized ownership claims for devices (patents) or names, written or recorded media (copyright), as formally recognized by a government. Patent regulations vary by jurisdiction, and in a global economy these differences (and efforts to harmonize across them) are an important issue (for a detailed discussion, see Chapter 7 on intellectual property in this volume.)

Much of the remainder of this chapter will focus on patents, rather than copyright. However, as knowledge becomes increasingly specifiable in digital terms (e.g., the molecular structure of an alloy or a drug; the digitized specification for an instrument) the distinction between patent and copyright may be less meaningful in the future.

Patent strategies for the firm

Assuming a firm has a discovery in hand, an early decision will be whether to patent (and thus disclose) the new knowledge, or sequester the knowledge as a trade secret (risking deliberate or inadvertent disclosure by knowledgeable employees, on the one hand, or reverse-engineering by rivals, on the other). The trade-offs surrounding this question include considerations of the cost of patenting; the likelihood of success with the patent (since an unsuccessful patent application will nevertheless reveal the discovery); the value of the patent *per se*; and the role of other constraints on simple duplication by rivals. For example, the formula for Coca-Cola has been kept as an unpatented trade secret, said to be known only to a handful of senior officers, for decades. Near-duplicate sodas exist; but Coke's enormous brand franchise, advertising, and global reach, even into China where savvy marketing and adaptation of practices to Chinese sensibilities—including becoming an official sponsor of the Chinese soccer team in the Olympics—have assured Coke's ascendancy.[1] The major form of IP discernible here is, of course, Coke's enormous brand equity. The strength of this perennial brand is an effective barrier to rivals, even beyond any patent (which would, in any case, long since have expired).

Patent protection for some discoveries is enormously valuable, as in the pharmaceutical industry. Brand name products like Pfizer's Viagra®, Merck's FOSAMAX®, or Genentech's Avastatin® can be sufficiently unique as to be the treatment of choice for their target conditions or diseases. Because these medications may sometimes be the only remedy effective for a particular condition, or the only example of a new treatment modality—genetically tuned anti-cancer drugs, for example—their greater effectiveness or lessened side-effects can garner effective monopolies that can endure for many years, advantages bolstered by advertising, clinical trial extensions of the drug to new conditions, and the like, as well as established brand recognition.

Patent protection in markets where the patent holder chooses to manufacture and sell serves the purpose of preventing rivals from entry. Until and unless an alternative is developed, or unless misfortunate, unanticipated side-effects like those discovered for VIOXX®[2] arise, market dominance and profits continue while the patent is in force. Once a drug goes off-patent, it may be offered as an over-the-counter (non-prescription) remedy in some cases, perhaps with a weaker reformulation to make it safer for self treatment. In these instances, while patent protection has waned, the maker's advertising and distribution can create enduring brand equity that continues to protect market share.

Patents on business processes and methods are an important and relatively new category of patent with profound implications. Amazon's "one-click" ordering, Priceline.com's "name your own price" auctions, and Dell's business model of integrating customer IT systems and online order interfaces with Dell's production systems are all protected (Dell's system with more than 42 issued and pending patents, as of 2000). These business model patents are especially interesting because they prohibit rivals from simply duplicating procedures that have proven so successful for Amazon, Priceline.com, and Dell. Since rivals cannot duplicate the methods, the patent holders can enjoy superior performance and returns on the basis of their patented proprietary methods. The resulting protected stream of profits has been successfully collateralized by patent holders in license exchanges and other transactions, translating to enormous value (Rivette and Kline, 2000). Business process patents continue to be hotly contested, and wealthy patent holders like Dell and Amazon are tempting targets for suits.

However, not all industries or situations are appropriate for standalone monopoly operations and, often, patents serve other purposes. Because most patent disputes are settled by licensure and the payment of royalties, we can conclude that patents frequently serve as negotiating chips. The semiconductor industry is a case in point, due in part to historical accident. In the early days of the U.S. semiconductor industry, the government was a primary customer, seeking advanced capabilities for military equipment. The government first, and many major corporations thereafter, frequently demanded "second sources" for critical semiconductor components—particularly in the early days of the industry, when semiconductor firms were young, and their manufacturing capabilities not well established. A "second source" firm was a rival to whom the first firm had to deliver the complete recipe for its discovery, so that the second source could also provide the product to the customer. Intel as a first source for microprocessors provided just such information to AMD, a firm that continues to compete with Intel decades later. Intel and AMD have sued and counter-sued about AMD's rights to use various Intel technologies. Using patents to negotiate access to another firm's IP is a routine matter, and the terms of exchange can be simply access for access, or can involve license fees in some fashion (per unit or fractional share of profits or sales).

Some firms choose to discover, patent, and license, seeking a stream of license revenues, rather than sales of product. The increasing prevalence of strategic alliances and outsourcing agreements—including outsourced R&D—favors such arrangements. So, too, does a strong brand or market position, or a strong technology position: either can enable the discovering firm to also harvest license revenues. For example, Canon's laser engine was a significant advance on previously available technology and, as the *de facto* standard for laser printer engines, was widely licensed into other companies' printers, generating significant revenues. Further, any large firm devoting substantial revenues to R&D may from time to time discover patentable knowledge not within its chosen realm of

[1] See, for example, "Coca-Cola's re-entry and growth strategies in China," ICMR case BSTR140 (reproduced in Carpenter and Sanders, 2007, and also available at *http://www.icmrindia.org/casestudies/catalogue/Business%20Strategy2/BSTR140.htm*

[2] Some years after its launch, VIOXX was discovered to increase the likelihood of heart attack and stroke, and was consequently withdrawn from the market. Numerous lawsuits ensued, along with the largest product liability settlement in U.S. history, for $4.8 billion.

interest. Since most large corporations routinely patent discoveries, such patents can languish, unexploited. A more productive response is to spin these products off for others to develop. One famous example is Gore-tex®, discovered as Teflon by a DuPont scientist, Willard Gore. Because the initial applications of the product were in consumer goods, DuPont declined to commercialize it, and allowed Gore and his wife to do so themselves instead. While much further work was required to develop the product, its manufacturing processes and uses, Gore-tex® is a highly valuable brand with more than $2 billion in annual revenues, in areas as diverse as sporting goods and outerwear, biologically non-reactive devices for insertion in the human body, low-friction mechanical fittings, and cookwear surfaces, among others.

Beyond the issues of whether to patent or retain knowledge as a trade secret, whether to aim at a standalone monopoly, use the patent for negotiating, or to seek license revenues as the key aim, further strategic considerations arise over IP as commercial intelligence. Tight relationships among a firm's patent holdings can create a strong area into which rivals cannot penetrate without infringing. Interlocking or adjacent patents in related fields can fence off substantial areas of technology. Just as the Japanese had earlier filed numerous patents to foreclose various areas of technology, Hewlett-Packard's ink-jet printing technology was protected with a veritable thicket of closely linked patents that excluded rivals from the most effective technology. HP declined to license, electing instead to further develop ink-jet technology, extending its dominance and enjoying substantial revenues and profits from a truly superior product. Failure to consider the strength of a rival's patents can be disastrous, as Kodak discovered when its decades-old thrust into instant photography ran afoul of Polaroid's patents. Not only did Kodak have to pay $925 million in fines, but its legal costs, the entire instant photography R&D effort, and its factory were all chalked up as losses, along with additional costs to purchase back all of the millions of instant cameras it had sold in over a decade of contested business operations, for a total cost of at least $3 billion, plus strategic foreclosure of a promising new arena of business (Rivette and Kline, 2000).

Another interesting example is Amgen,[3] a biotechnology company and the discoverer of a viable process for manufacturing EPO, human erythropoietin, a hormone regulator of red blood cell production. While EPO had been predicted in medical literature in the 1920s, only with the advent of recombinant DNA technology was commercial manufacture made feasible. Amgen devoted years and billions to developing its process, but initially failed to receive patent protection, because a rival firm (which was unable to produce the hormone in commercial quantities) had filed first. While Amgen eventually prevailed in the courts, the company has also experienced repeated challenges to its enormously lucrative monopoly. The company's patent holdings are especially interesting: Amgen holds multiple patents for different aspects of EPO production processes, and also patents related to the activity of EPO, such as methods to activate cell receptors using antibodies, DNA sequences encoding EPO, novel methods of administration of the hormone (e.g., pulmonary administration), and EPO purification. Amgen has filed a long series of related patents extending its claims and knowledge base. These closely related patents serve to fence in EPO for Amgen, making rivals' participation in the EPO market difficult, if not impossible.

The patent position of another biotech-related company, Affymetrix, is equally interesting. Affymetrix makes the equipment that permits molecular analysis, the chips and data arrays to support research by others, and products that encode known genetic knowledge so that others can explore further. Affymetrix's holdings fall into eight families: DNA sequencing, mask or probe design, reagents, software, arrays, arrays and analysis, mechanical devices and uses, and synthesis. These varied categories protect different aspects of Affymetrix's "core technologies, including DNA and protein arrays, scanner and detector technology, and microfluidics, as well as a broadened portfolio of patents related to the use of beads to measure nucleic acids or peptide binding for genomic analysis" (Li and Yu, 2007). Affymetrix's patents fall all along its value chain, from analytical materials through multiple aspects of the molecular analysis process, thereby creating a robust protective scheme.

Firms do not perform all of their own research, however, and understanding both what is wanted by way of discovery and, more importantly, what is needed in terms of patent access and ownership is crucial. A first response may be that ownership of a patent is essential, but this is by no means always the case. It is especially important to realize this in dealing with others' patents: access may be attainable where exclusivity or ownership is not—and access may do. Sometimes, where it is possible to exploit a monopoly position, a firm can choose to do its own research, patent, and retain exclusivity. At other times, perhaps because fundamental knowledge is potentially far more broadly applicable or beyond the firm's own scope of knowledge, a firm may join a university research consortium with non-competing firms. Here, early access to potentially usable knowledge comes along with substantial financial leverage: each partner pays a small fraction of the total R&D cost. However, in the U.S. at least, such consortia rarely yield ownership possibilities, although exclusive field-of-use licenses may be available. Sometimes, a firm may wish simply to avoid the possibility

[3] This section draws on a very interesting Web article on *Patent Strategies for Biotechnology Enterprises*, by Wei Li and Xiang Yu of BioPharm International; visit *http://www.biopharminternational.com/biopharm/Business/Patents-Patent-Strategies-for-Biotechnology-Enterp/ArticleStandard/Article/detail/444984* (consulted November 14, 2007).

of others patenting knowledge: in this case, rapid publication to disclose the knowledge early is a desirable strategy, putting the information into the public domain and foreclosing others' efforts to control the knowledge. This may be especially true where a firm can gain access to early-phase, pre-competitive information. Where exclusivity is not available, a firm may still be able to develop its own further proprietary knowledge to gain exclusive patent protection for its intended business.

U.S. universities typically claim ownership of discoveries made on their campuses or by faculty, students, or staff, and typically will not sell the patents nor exclusively license them. These considerations have led some firms to outsource university research to non-U.S. universities: those in China, Russia, and India are frequently mentioned as having inexpensive, well-trained researchers eager for work. Such outsourcing is undeniably cheaper, as typical overhead rates for in-house research, contract research by U.S. firms, or U.S. university research can easily reach 60% to 90% of the cost of the research itself. Furthermore, non-U.S. universities will often sell IP rights, or give them outright to the research sponsor. Yet strategic considerations of outsourced R&D are even more interesting.

So-called "open innovation" advocates note that there are always more minds outside a firm than inside it, so that reaching out for new knowledge broadens the possible range of discovery (Chesbrough, 2003). Outsourcing enables a firm to access much infrastructure for research that it may not itself choose to support because it is rarely needed: expensive research equipment falls into this category. In addition, talking to others outside the firm brings fresh perspectives and habits of mind to bear. But there are other strategic considerations, too. Outsourcing to domestic firms puts the new knowledge into play to some degree. A research firm may use the knowledge it develops for one client in other applications (IDEO's famous Inventor's Box is an illustration: ideas are recycled, albeit in perhaps wholly different contexts). Another consideration is whether outsourcing the difficult problems results in a decline in the outsourcing firm's capability. If others are doing the research and making the discoveries, important tacit knowledge may be foregone, as the focal firm's absorptive capacity erodes. That is, without having done the research, a firm may not understand new discoveries sufficiently well to evaluate them, or to make effective use of them (see Chapter 18, Open Innovation).

Finally, outsourcing to offshore research partners, especially in India and China, comes with special hazards due to well-known shortcomings in patent and copyright enforcement in those areas. General Motors arranged for contract manufacture of a small car in China, only to find an identical vehicle (with a new badge) being sold in China by its partner in competition with GM's own offering. While the Chinese partner claimed prior knowledge had been acquired from Daiwoo engineers before that company was purchased by GM, it is by no means clear that the supposed sellers had the right to sell. Even if direct pirating is not at issue, joint ventures or contract research can have an important effect in enhancing the partner's knowledge and capabilities at the edge of the state of the art. In short, placing leading edge research abroad may create the competitors of tomorrow, trained with the outsourcing firm's research dollars.

Patents as strategic levers

Beyond the issues of patents as discussed thus far, patents constitute a potential treasure trove of useful strategic intelligence. In the first instance, a firm's own patents—what they cover, and what they don't; the relationship among areas of strong coverage; how strong or fundamental the firm's knowledge base is, and whether it can be strengthened in one direction or another—should be the basis for considerations of R&D planning, product development, and market strategy, among other uses. A strong patent position, covering fundamental knowledge in a particular area can be the basis of a preemptive claim to a market: whole new industries can rest on proprietary knowledge, as with Chester Carlson's original patent on xerography. Similarly, rivals can be warned off, reserving a new market for the patent holder, as Avery thwarted Dow's entry into a new film label market segment, by careful deployment of strong patent protection. As a result, Avery enjoyed uncontested dominance of a profitable new line of business.

Patent mapping—laying out the patent coverage and state of the art in a given arena—is essential business intelligence for enforcing patents (as with Polaroid and Kodak), but also for due diligence and valuation of merger and acquisition (M&A) candidates. M&A specifically targeting patents can be powerful strategic moves, as when S3 anonymously outbid Intel for the patents of bankrupt chipmaker Exponential Technologies—obtaining a patent that pre-dated Intel's Merced chip patents. S3 thus held Intel's next-generation chip business hostage, and forced Intel to cross-license its patents, leaving S3 free to pursue its own high-performance graphic chip business (Rivette and Kline, 2000). Moreover, such considerations should also guide strategic planning for new-product development and R&D targets, lest the firm attempt markets already foreclosed by others' patents, or miss the opportunity to protect arenas of potential dominance. Cross-licensing can make customers out of competitors, as when IBM and Dell cross-licensed, assuring IBM a market for its components and Dell access to low-cost components paid for by license access.

The industry effect

Patents are not of equal importance in every industry, and not all patents are equally valuable. This fact may be discerned from examination of the sources and revenues of university patents in the U.S.: while the overwhelming majority of patent disclosures come from engineering disciplines, the overwhelming majority of license revenues arise from a very few patents—five patents or fewer generate 90% or more of license revenue at virtually every U.S. university (Mowery *et al.* 2004), including giants like the University of California system, with a patent portfolio of over 5,000 active patents (Ku, 2001). Virtually every one of the high-revenue patents concerns biotechnology. Agricultural and biotechnology patents were among the very earliest patents filed by universities in the U.S.: the University of Wisconsin's 1925 patent on vitamin D was among the early and most lucrative, widely licensed university patents. But not all patents are lucrative; indeed, the vast majority of all university patents are never licensed and, of those that are licensed, most have only one bidder. In short, many university patents never yield license revenues, although university patents are of high quality, cover fundamental insights, and are frequently cited. Patenting in the U.S. might easily cost $10,000 or more, while worldwide protection requires additional tens of thousands of dollars for preparation, filing, and maintenance fees.

The value of a patent to its holder will be affected by how readily its benefits can be captured, which in turn reflects industry conditions. Where technology *per se* can be brought to bear readily—where no dominant rival controls access to the market, for example—a patent may be highly valuable. Examples of industries where patents have yielded substantial benefits include, most prominently, pharmaceuticals, medical instruments, and computers. Computers offer a particularly interesting example in that the move toward specialization of components and away from integrated value chains creates many more options for entry, and thus more potential value to patents (Nelson, 2007) as the once-monolithic computer industry reconfigures. The pharmaceuticals industry, which remains dominated by large firms with skills at clinical trials, marketing, and sales and distribution, has also seen significant entry by standalone firms like Amgen, as well as strategic alliances between "big pharma" and smaller, highly specialized "boutique biotech" firms with specialty expertise in the newer biotech, genomics, and proteomics areas where larger firms are void.

In part because the market for therapeutic drugs or medical equipment is easy to estimate—potentially, all those with diabetes or a particular form of cancer, or all the candidates for a particular surgical procedure—valuations for patents on these items appear more solid. Moreover, third-party payer systems (in the U.S., in particular) and the value to the patient of health have historically supported substantial margins and growth in the pharmaceutical and medical instrumentation domains. Those margins and that growth translate to potential advantages that enrich patent valuations.

By contrast, industries like commodity chemicals have razor-thin margins and far more restrained growth. Catalysts, for example, or additives to facilitate plastics forming, may offer little additional benefit, at the same time that process use of IP is extremely hard to police. As a result, patents in these areas have far less value. Thoroughly mature markets and technologies can also be difficult venues for patent value because "upstream" changes can require "downstream" customers to change their production processes, making change complex, expensive, and easy to derail. In these circumstances, patents may be of lesser value. In general, engineering patents are more numerous but less lucrative; the most valuable patents are almost invariably biotech-oriented, whether as agricultural biotech or as pharmaceuticals.

The U.S. IPR system and its discontents in a global economy

The U.S. IPR system is especially important because the U.S. has been the wellspring of so many fundamental discoveries and resulting new industries, including petrochemicals, plastics, aerospace, computers, semiconductor electronics, pharmaceuticals, software, and more. Because such discoveries fuel whole new industries, they create a vibrant economy that other countries seek to duplicate. One aspect of the U.S. innovation system that has attracted much attention is the Bayh–Dole legislation that permits U.S. universities to claim ownership of discoveries resulting from Federal funding. In practice, this has meant claimed ownership to all discoveries by faculty, students, or staff on university premises. By clarifying ownership, the logic goes, deals can be arranged with industry to facilitate the substantial investments required to commercialize early-stage research discoveries. *Economist* magazine asserted that Bayh-Dole was "innovation's golden goose" and had generated much of the U.S. success in innovation (2002). But it did so just as the latest lucrative patents (the Cohen–Boyer patents for recombinant DNA and Columbia's co-transformation patents) gained notoriety, fueling even greater interest among other countries (and their universities) in emulating Bayh–Dole. Others raised concerns about commercializing universities, undercutting academic freedom, and stifling fundamental research.

These disputes matter enormously, in light of efforts to harmonize varying IP laws across jurisdictions, especially under the World Trade Organization's 1994 Agreement on

Trade-related Aspects of Intellectual Property Rights (TRIPs). TRIPs sets forth standards on such matters as the rights of performers, producers of recorded media, authors of works of fiction, and creators of films and photographs (among others). Notably, and reflecting the growth and development of technology, not only such obviously creative works as fiction and film, but also integrated circuit designs, software, new drugs, and newly created plant varieties are all covered under TRIPs. Such extended protections immediately raise further issues of constraint of innovation, particularly where they prohibit the use of foundational techniques—in software, for example, or the "business practice" patents described earlier. The more fundamental the patent or copyright protection, the more likely it is to be valuable, on the one hand, and the more likely it is to prohibit others' further invention, on the other.

Despite efforts to harmonize so far, differences remain between the U.S. system and those of much of the rest of the world, as well as between developed-country systems (like those of Europe, Japan, and the U.S.) versus those of India, China, and African countries, for example. In seeking to resolve the issues, understanding what ails the U.S. system (as well as what works well within it) may help to avoid preserving the flaws while undermining innovation. The U.S. system can be defined as consisting of the U.S. Patent and Trademark Office (USPTO), its judicial counterpart, the Court of Appeals for the Federal Circuit (CAFC), and the Justice Department, which pursues antitrust and infringement actions with greater or lesser rigor, depending on the political philosophy of the administration in power. The interaction of these elements creates a general atmosphere of enforcement, patent strength (and thus value). The U.S. system relies on "first to discover" as the basis for award, in contrast to other jurisdictions" "first to file" bases. This translates into important shortcomings in the system's operation, including potential for awarded patents to be undercut by claimants who wait until the worth of an idea is proven before stepping forth to undercut the successful innovator firm, a form of free riding sometimes termed "submarine patents". Moreover, the chronic and systemic underfunding of the USPTO—particularly during recent decades' explosive growth of science, technology, and commercial activities based upon them—has produced a dearth of knowledgeable, well-trained patent officers, a huge increase in workload, and thus less capable examinations, resulting in weaker patents that invite litigation (Jaffe and Lerner, 2004). The problem is exacerbated when attorneys or scientists with a bit of Patent Office experience are regularly hired away by litigation specialists at much higher salaries. The litigious character of U.S. business culture undoubtedly adds to the difficulty, but underfunding of government IP functions is unlikely to be solely a U.S. problem: other countries without a tradition of strong IP enforcement will lack needed resources as well, if for other reasons. In any event, the more underfunded the IP enforcement system, the weaker the patent protection it is likely to afford.

"Weak" patents come in several forms. One type of weak patent is insufficiently specific to exclude other, related phenomena. Another form of weak patent simply should not have been granted in the first place, either because the patent covers a natural phenomenon or because the patent conflicts with a prior patent that should have taken precedence. Still other kinds of weakness in the patent-granting operation include unwarranted extensions of prior patents. Such weaknesses testify in part to the naturally complex, overlapping, and difficult-to-distinguish nature of applied knowledge: if a particular insight can be applied in multiple areas, then patent claims should specify which ones the applicant claims primacy in, and seeks protection for. However, such specificity may be both difficult to articulate, in the first place, and difficult to distinguish, in the second. In any event, new applications areas may be described as "different" by new applicants, yet seen as "the same" by the first claimant, whose patent would rise in value thereby. All of these "weaknesses" and others invite litigation, potentially undermining the security of the patent regime (although in fact most patent disputes are settled by negotiation).

As government acknowledgements of ownership rights, patents are also subject to special circumstances, exceptions, and exclusions. In the U.S., for example, the government reserves so-called "march-in" rights to use any discovery based on Federally funded research or development. While march-in rights generally don't exclude the patentee from continuing to exploit the patent, national security concerns may prevent production for export, thus potentially limiting the value of a patent. The U.S. also provides exceptional treatment for "orphan" drugs—those treating diseases or conditions affecting only a small number of patients, said to be too small a market to attract ordinary commercial interest. Here, the exceptional treatment favors the patent holder, by granting exclusivity or extending favorable tax treatment to encourage development.

In other countries, different exclusions or exceptions apply—some of those in developing countries in sharp contrast to claims by U.S. or other developed-country patentees. For example, Indian pharmaceutical manufacturers enjoy government permission to reverse-engineer and improve the manufacturing processes of drugs developed in the U.S. or Europe, and sell them in India as well as in other (mostly developing) countries that excluded some drug patents on humanitarian grounds. AIDS remedies are among the exemplars in such actions: large fractions of the populations of many African countries could never afford North American prices for the "AIDS cocktails" that transform HIV into a serious chronic disease, rather than an immediate death sentence. But Viagra is also widely sold in Africa, at a small fraction of the U.S. price, garnering substantial revenues for the copycat, rather than the patent holder.

Agricultural biotechnology is another hotly contested arena: while hybrid seeds have long been a feature of the agricultural scene, Monsanto's development of "Round-Up Ready™" seeds—hybrids immune to Monsanto's herbicide "Round-Up" and thus easier to plant for more productive yields—met with huge protest and resistance, when Monsanto demanded that harvested seed from the crop not be saved for the following year. This request was depicted by protesters as an especially exploitative act on Monsanto's part, although hybrid seeds often don't breed true, as Gregor Mendel documented more than 100 years ago. Opponents also protested that the seeds were genetically modified, and thus constituted "Franken foods"—a play on the name of Frankenstein, the fictional monster created by a mad scientist—despite a dearth of any credible evidence that "genetically modified" food carries any risk different from foods modified by humans for millennia.

Indigenous "intellectual property" is another contested area: while not officially recognized by governments, indigenous peoples have used a wide array of natural substances for medicinal purposes for millennia—and developed-country pharmaceutical companies seek to patent such uses. Similarly, cultural products of indigenous peoples have been appropriated (and copyrighted) by outsiders, who then seek to prevent the original owners from commercializing their own creations (Lessig, 2002).

Exclusions of drug patents on humanitarian grounds, protests against genetically modified foods and agricultural biotechnology more generally, and indigenous outrage at outsiders' appropriation of their knowledge point to limits of patent protection and of external constraints that may affect the value of a patent. We may note that these factors are also both political and cultural, and thus not necessarily logical, rational, or science-based. Nevertheless, they must clearly be part of considerations for firms seeking to craft a patent strategy, since they affect the degree to which a patent can, in fact, be exploited.

The U.S. IPR system and the issues it gives rise to are also interesting ultimately for one other reason: they point to the future of IPR. Many of the difficulties visible in the U.S. may well be implicit in other countries' systems, as well. Simply because the U.S. has traveled so far along the IP road, issues of relevance to the hoped-for worldwide system may be more visible in the U.S. Thus rather than simply emulating the U.S. system in hopes of reproducing its entrepreneurial success, a much more judicious consideration of the system's strengths and weaknesses is desirable. Attacks on U.S. practices regarding critical pharmaceuticals, indigenous ownership of long-used natural medicines, and cultural properties (among others) point to issues that any worldwide system must address. Finally, it is worth noting that the foundations of U.S. patent law reside in the U.S. Constitution, conceived in a dramatically different scientific, technical, and cultural milieu. It is not obvious that a global IPR system should necessarily enshrine the U.S. system's emphasis on individual innovation, nor strike the same balance on who enforces IP rights, nor embrace the economic rationality arguments that appear increasingly suspect.

Conclusions

Contemporary good patent practice is far more aggressive in enforcement (including, importantly, business method patents like those of Amazon and Priceline.com), in patent valuation and auditing, and in licensing patents out. IBM's annual free cash flow from patenting in excess of $1 billion, Texas Instruments's aggressive licensing, and the practices of numerous other firms in pruning unneeded patents for donation (and tax credits) all point to more careful consideration of patents and their potential strategic uses. Yet patents are differently construed and differently enforced in other countries—a matter of no small concern in an increasingly global economy.

Intellectual property covers an array of non-tangible assets, including brands, logos, trade- and service-marks as well as patents and copyrighted materials. Each category of item has its own complexities, but strategies for acquiring, maintaining, and using the protection of intellectual property rights should be a consideration for every firm. A place to begin is by understanding the nature of the intellectual capital to be protected, and the range of possible means to do so. Another early consideration should be the conditions of the industry of competition—is it fragmented? Dominated by a single rival or a few firms that control access to key distribution channels? Where entry is not foreclosed by such dominance, patents can effectively enable a firm's positioning. In other markets, where technology is moving very rapidly, patenting may not be an especially useful strategy—technology may well move on before a patent can be obtained. Yet fundamental discoveries, even in a rapidly moving technology arena, can be enormously valuable, especially where these are carefully positioned to fence, surround, or interlock with other firms' IP. Finally, patents are critical industrial intelligence—for positioning R&D and product planning, and for assessing M&A. In short, strategy is essential for effective IP positioning, and IP positioning is essential for effective strategy.

Finally, while significant differences in IPR across jurisdictions remain, increasing multinational or transnational economic activity argues for harmonization of IPR. Any new system will face issues of patent strength, enforcement mechanisms, the basis for patents (e.g., first to file or first to discover), and their limits (critical medications for life-threatening diseases, epidemics, indigenous knowledge, or cultural products). The underlying philosophical issues of how to encourage innovation without unduly rewarding the innovators at the expense of society are at the heart of the matter.

References

Anon. (2002). "Innovation's golden goose." *Economist*, **365**, 3.

Carpenter, M. A., and Sanders, W. G. (2007). *Strategic Management: A Dynamic Perspective* (pp. 534–541). Upper Saddle River, NJ: Pearson/Prentice Hall.

Chesbrough, H. W. (2003). *Open Innovation: The New Imperative for Creating and Profiting from Technology*. Boston, MA: Harvard Business School Press.

Jaffe, A. B. and Lerner, J. (2004). *Innovation and Its Discontents: How Our Broken Patent System Is Endangering Innovation and Progress, and What to Do About It*. Princeton, NJ: Princeton University Press.

Ku, K. (2001). *Effects of Patenting and Technology Transfer on Commercialization*. Washington, D.C.: National Academies Board on Science, Technology and Economic Policy Committee on Intellectual Property Rights in the Knowledge-Based Economy.

Lessig, L. (2002). *The Future of Ideas: The Fate of the Commons in a Connected World*. New York: Vintage Press/Random House.

Li, W., and Yu, X. (2007). *Patent Strategies for Biotechnology Enterprises*; *http://www.biopharminternational.com/biopharm/Business/Patents-Patent-Strategies-for-Biotechnology-Enterp/ArticleStandard/Article/detail/444984* (last accessed November 14).

Mowery, D. C., Nelson, R. R., Sampat, B. N., and Ziedonis, A. A. (2004). *Ivory Tower and Industrial Innovation: University–Industry Technology Transfer before and after the Bayh–Dole Act*. Stanford, CA: Stanford University Press.

Nelson, R. (2007). "Understanding economic growth as the central task of economic analysis." In: F. Malerba and S. Brusoni (Eds.), *Perspectives on Innovation* (pp. 27-41). Cambridge, U.K.: Cambridge University Press.

Rivette, K. G. and Kline, D. (2000). "Discovering new value in intellectual property." *Harvard Business Review*, **78**(1), 54–66.

Part Five

Project Level Concepts

32

Innovation Teams

Gloria Barczak

Northeastern University

This chapter focuses on the teams who do the work of innovation and new-product development. It begins with a description of two different types of innovation teams and the advantages and disadvantages of each. Next, team size, diversity, and proximity are discussed. This is followed by a section that highlights team process issues such as communication and cooperation. The chapter ends with a discussion about the effective leadership of innovation teams.

Definition of innovation teams

Innovation teams are cross-functional groups of individuals who are charged with creating and developing new products and services. Members typically come from a variety of functional disciplines including marketing, engineering, product design, and manufacturing. Innovation teams are temporary in that they are together for the life of the project from idea conception to launch. Team members are likely to work on multiple projects simultaneously.

In the recent American Productivity Quality Center (APQC) benchmarking study (Cooper, Edgett, and Kleinschmidt, 2002), 72% (75/105) of survey participants indicated that they used cross-functional teams for new-product development. These respondents were representatives of mid to large U.S. business units from a number of different industries. Nearly 74% of the respondent firms used clearly identified team leaders who were responsible for driving projects through their entirety. This study also found that the best performing businesses provide adequate resources to their innovation teams and ensure that team members are not working on too many projects simultaneously. However, not all innovation teams are the same. It is important to distinguish between two types of innovation teams—heavyweight and lightweight teams.

Heavyweight versus lightweight teams

Heavyweight teams are those in which team members report directly to the project leader and are dedicated to the project for its duration. As a result, team members are likely to feel ownership of and responsibility for the project and be more likely to commit to its effective completion. In these teams, members' first allegiance is to the project and secondarily to their functional group. Functional managers provide the human resources for the team and act in an advisory role, if needed; however, power and responsibility for the project lie with the team leader (Wheelwright and Clark, 1992).

An advantage of heavyweight teams is that the team is responsible for how particular tasks are delegated, organized, and accomplished. As well, the project leader is accountable for the success or failure of the project and for management of the team and its members. A disadvantage of such teams is they can create tension with the functional organization over who has more control with regard to the project (Wheelwright and Clark, 1992).

Lightweight teams, in comparison, are those in which functional members act as representatives of their functional area on a particular project. Team members are assigned by their functional leaders and remain under their control. Thus, the allegiance of lightweight team members is, first, to their functional group and, second, to the project. Leaders of such teams are also "lightweight" in that they are responsible for overall coordination of the project and oversight of the project budget and schedule. They do not, however, have control over the human resources (Wheelwright and Clark, 1992).

The advantage of lightweight teams is that one person, the project leader, ensures that tasks are done in a timely manner and that all functional areas are informed about project issues leading to improved coordination and communication. A disadvantage of such teams is that power resides in the functional group managers who can influence their representatives to delay and even sabotage a particular development effort, thereby rendering the lightweight project leader superfluous (Wheelwright and Clark, 1992).

Team formation and size

Innovation teams can be formed in several ways. The most common approach is for a functional manager to assign a subordinate to a given project by default; that is, because he or she is available. Alternatively, members may be assigned to a project due to their expertise and experience. In rare situations, members may volunteer or be asked to join a project (Barczak and Wilemon, 2003). The method of team formation depends on the importance and innovativeness of the project as well as the practices of the firm. For example, teams destined to work on a radical project which is of greater risk and importance to the firm are more likely to be formed by asking particular individuals to be part of the team and/or by asking for volunteers. In this way, management controls team membership to include individuals with the appropriate functional and interpersonal skills thereby enhancing the probability of project success.

The size of the team can influence its effectiveness. As team size increases, trust, participation, and productivity all decrease. To keep team size reasonable, core teams consisting of members from functional disciplines critical to the project are created. Core team members are dedicated to the project throughout its life cycle. Core team size can vary depending on the scope and scale of the project or program and the innovativeness of the project.

In addition to the core team, other functional representatives may be solicited to help on the project as needed. Such team members are part of the extended or peripheral team and are brought in for specific tasks and activities.

Diversity of teams

Team diversity has two dimensions: tenure diversity and functional diversity (Ancona and Caldwell, 1992a). Tenure diversity refers to the multiplicity of members' length of time working in the organization. Functional diversity relates to the assortment of functional disciplines represented by the core team. Diverse groups are more creative and able to solve problems; however, they seem to be less effective at social integration and cohesion because they lack capabilities for teamwork (Ancona and Caldwell, 1992a).

Tenure diversity bestows the team with a range of skills, experiences, perspectives, and networks because members came into the organization at varied points in time (Ancona and Caldwell, 1992a). High levels of tenure diversity enable the team to better define project goals, develop plans, and prioritize work (Ancona and Caldwell, 1992a). On the negative side, innovation teams comprised of individuals who arrived in an organization at different times may have different perspectives and experiences thereby reducing their attraction to each other and limiting the opportunity for interaction.

Teams that are highly functionally diverse may find it difficult to develop a shared purpose and collaborate effectively due to the different perspectives of the members. This view is supported by Ancona and Caldwell (1992a) who found that functional diversity was not related to a team's internal processes (i.e., ability to define and prioritize goals). However, creating a team climate in which team members are strongly identified with the team rather than their functional disciplines can reduce the adverse effects of diverse functional identities.

On the positive side, functional diversity provides the team with access and information that they would not have otherwise (Ancona and Caldwell, 1992a) and brings diverse input into new-product decision making. High levels of functional diversity are associated with high levels of communication with individuals outside the team because team members feel more comfortable with their external networks than their team members (Ancona and Caldwell, 1992a). This lack of team camaraderie motivates individual team members to seek outsiders (their networks) with whom to communicate.

Functional diversity is related to faster development time. Heterogeneous teams of representatives of various functional disciplines enable forward-thinking and integration of marketing, technical, and manufacturing activities thereby shortening the time to market. Both tenure and functional diversity, however, have a negative, direct effect on performance. Specifically, group tenure is negatively related to management ratings of adherence to budgets and schedules while functional diversity is negatively associated with both management ratings of innovation and team-rated performance (Ancona and Caldwell, 1992a). In other words, functionally diverse teams seem to have a lower capacity for teamwork, are more open to political and goal conflicts amongst functional disciplines, and are less likely to achieve agreement on decisions (Ancona and Caldwell, 1992a).

Team proximity

Distance between team members creates communication difficulties in many cross-functional teams; however, these difficulties are exacerbated in teams that are dispersed. According to O'Leary and Cummings (2007), dispersion of team members should be viewed as a multi-dimensional construct consisting of three elements: spatial, temporal, and configurational dispersion. Spatial dispersion refers to the geographic distance among team members. Temporal dispersion relates to the extent to which team members' normal work hours overlap. Configurational dispersion pertains to the location and arrangement of team members across sites regardless of the spatial and temporal distances among them. Specifically, configurational dispersion is assessed by ascertaining the number of sites where members are located and the number of members at each site. With regard to the latter, it is necessary to examine the degree of isolation of team members (i.e., how many members are alone at their site) and the balance of the configuration (i.e., is there a similar number of team members at each site).

As spatial distance increases, spontaneous communication decreases because it decreases the likelihood of face-to-face interaction. Likewise, temporal dispersion results in a reduction in real-time interactivity and problem solving (O'Leary and Cummings, 2007). As a team's configuration becomes more complex, effective team functioning becomes more difficult to achieve, although information technology (IT) can assist in this effort as it helps to bridge the dispersion gap (O'Leary and Cummings, 2007). Technology, however, can never replace the benefits of face-to-face interaction. Low proximity increases the importance of teamwork in achieving high team performance.

Innovation teams that are more geographically and culturally distant (i.e., global teams) face more behavioral (e.g., trust, communication) and managerial (e.g., budget, schedule) challenges than co-located teams. Moreover, more dispersed teams have lower levels of performance than teams that are closer together geographically, in part because teamwork quality decreases with increased proximity. However, dispersed teams can be more effective and efficient than co-located teams if they achieve high levels of teamwork (i.e., communication, coordination, cohesion, effort, mutual support, balanced contributions) (Hoegl, Ernst, and Proserpio, 2007). Thus, teamwork quality is more important to team performance as team dispersion increases (Hoegl, Ernst, and Proserpio, 2007).

Team process

Characteristics of effective team members and teams

The performance of innovation teams is influenced by the caliber of people on the team (Barczak and Wilemon, 2003). According to Holahan and Markham (1996) effective team members possess functional expertise, strong interpersonal abilities, and robust communication skills. Barczak and Wilemon (2003) found that important interpersonal skills include being cooperative and getting along with other team members. As well, effectual team members have valuable personal qualities such as a strong work ethic, discipline, self-motivation, and determination. Overall, successful team members require a combination of skills with interpersonal skills being particularly valued by other team members (Barczak and Wilemon, 2003; Holahan and Markham, 1996).

Not surprisingly, effective innovation teams are cooperative, committed to a common objective, and get along well which result from having the right mix of team members with the right skills (Barczak and Wilemon, 2003). Such teams also have an effective leader, high levels of teamwork, and empowerment (Holahan and Markham, 1996).

Teamwork

In the NPD literature, the words integration, coordination, collaboration, and teamwork are often used interchangeably (Pinto and Pinto, 1990) contributing to confusion in the field. Kahn (1996), for example, labels integration as a combination of interaction and collaboration. Interaction emphasizes communication and information exchange while collaboration focuses on an effective process where departments work together with shared resources and goals (Kahn, 1996). Song, Montoya-Weiss, and Schmidt (1997) define cooperation as a high degree of interdependency and information sharing between various organizational units, which are akin to Kahn's (1996) definition of interaction. Jassawalla and Sashittal (1998) argue that integration is a subset of collaboration, rather than collaboration being a subset of integration as Kahn (1996) suggests. They define collaboration as being cross-functional, involving high levels of integration, coordination, and cooperation *as well as* high levels of at-stakeness (i.e., equitable interest in implementing objectives and outcomes), transparency, mindfulness, and synergies from their interactions. Hoegl, Ernst, and Proserpio (2007) maintain that teamwork includes all of these concepts—communication, coordination, cohesion, effort, mutual support, and balanced contributions.

In spite of these conflicting definitions, there is evidence that each of these team processes is linked to performance. Frequency of communication amongst innovation team members is an important contributor to team performance. However, high frequency of communication decreases the creative performance of the team (Kratzer, Leenders, and van Engelen, 2004). High-performing teams communicate more and interact more frequently with people outside the team than low-performing teams. But, high-cooperation teams use more informal methods of communication than low-cooperation teams. There are no significant differences between the two types of teams with regard to formal communication (Pinto and Pinto, 1990). High-cooperation innovation teams are more likely to communicate to share project-related information, to review the progress of the project, and to receive feedback on performance (Pinto and Pinto, 1990). Such teams also spend less time resolving interpersonal conflicts amongst team members and thus are able to devote more time and energy to project-related endeavors. Internal mechanisms such as evaluation criteria, rewards, and senior management support can be used to foster cooperation amongst innovation team members. High levels of cooperation and collaboration lead to higher new-product success (Pinto and Pinto, 1990) while cooperation and integration positively impact team performance (Kratzer, Leenders, and van Engelen, 2004). However, moderate levels of both lead to the highest team performance; very low levels and very high levels of team cooperation and integration impede team performance (Kratzer, Leenders, and van Engelen, 2004).

Rewards

Cooper, Edgett, and Kleinschmidt (2002) find, among a sample of 105 U.S. business units from various industries that 30% of respondents indicated that NPD teams are rewarded and/or recognized when they do a good job meeting commercialization objectives, such as launching the product on time or meeting sales targets. Rewards tend to be non-financial in nature and include project completion celebrations, words of praise, getting to work on a bigger project next time, and plaques/pins. Yet, project team members state that most often rewards are individually based, rather than team-based. In fact, individual performance is the primary driver of salary and promotions (Barczak and Wilemon, 2003).

Little research has explored the relationship between rewards and performance. However, it has been shown that when it is easy to evaluate individual performances in the team, differential rewards based on position/status in the firm result in higher member satisfaction. When ease of individual evaluation is low, then position-based and equal rewards exert a negative effect on performance. Thus, firms need to find effective and unbiased methods of evaluating individual team members.

Leadership of innovation teams

The increased usage of cross-functional innovation teams for NPD work has, in turn, raised the importance of the project leader's role. Effective project leaders need to perform five roles that enable the team's performance: technical expert, strategic planner, team builder, gatekeeper, and champion/boundary spanner (Barczak and Wilemon, 1989; Elkins and Keller, 2004).

Though project leaders need some level of technical skill in order to be respected by team members, it is not essential that the project leader be the technical expert on the team. More technically skilled project leaders can facilitate more team innovativeness; however, less technically skilled leaders can also enhance team innovativeness if they give the team greater autonomy to make decisions regarding the new product (Elkins and Keller, 2004).

In their strategic planning role, team leaders are responsible for developing the project strategy and setting goals and a project schedule. A clearly articulated strategy and goals are critical to building and maintaining focus for the team. Without it, members can resort to narrow viewpoints based on their functional affiliations and/or personal goals (Elkins and Keller, 2004). However, flexibility in achieving the strategy and the goals is necessary as shifts in technology, customer needs, and organizational constraints occur (Barczak and Wilemon, 1989).

The team builder role involves selecting team members, resolving conflicts, and creating a climate in which team members feel comfortable and secure with their team members. As part of this climate, the leader needs to keep members challenged and motivated and instill a positive attitude. Leaders tend to mediate when conflicts are ongoing and have the potential to disrupt the project (Barczak and Wilemon, 1989).

In the gatekeeper role, the project leader is responsible for disseminating information and knowledge within the team and with relevant entities outside the team. As well, the leader needs to ensure and facilitate clear and continued communication amongst team members. This can be done through physical work arrangements, informal gatherings, and facilitating communication between individuals who have common interests or need particular information (Barczak and Wilemon, 1989). Although frequency of communication with individuals/groups outside the team impacts performance, it is the quality of the content of the communication with external groups which is most important to performance (Ancona and Caldwell, 1992b).

Finally, as a champion for the project, the leader needs to secure resources, promote and protect the project, and act as a liaison between the team and other groups (Barczak and Wilemon, 1989; Elkins and Keller, 2004). However, the effect of championing has no effect on project performance. Champions encourage their targets through their relationships, not through their influence tactics, but they have no direct impact on the project.

Project team leaders are overwhelmingly appointed by management rather than selected by team members; however, this does not guarantee their success. Jassawalla and Sashittal (2000) argue that team leaders who are from R&D are less effective, in part because they tend to maintain their allegiance to their functional discipline. As well, such leaders promote the notion that new-product development is an R&D activity thereby limiting the commitment of other functional areas. By contrast, project leaders carefully selected by senior management for both their intrapersonal and interpersonal skills are more committed to the team and the project (Jassawalla and Sashittal, 2000). As a result, they view functional areas as equal in power and facilitate team interaction so that information is freely shared.

With regard to leadership style, the literature has advocated two different approaches—emergent and planned (Lewis *et al.*, 2002). An emergent style reflects a flexible approach that encourages and facilitates improvisation as teams learn through doing and gain knowledge as they experiment with trying new approaches to problems

(Lewis *et al.*, 2002). Such a style of leadership is based on the notion that innovation is ambiguous and uncertain. By contrast, a planned style of leadership assumes that product development is a predictable and rational process requiring structure and discipline. As a result, leaders utilizing this style will set milestones and establish a schedule, track the project, undertake formal reviews, and adjust elements of the project as needed (Lewis *et al.*, 2002; Wheelwright and Clark, 1992). According to Lewis *et al.* (2002), both leadership styles—emergent and planned—are needed throughout the life of a project, sometimes concurrently. The challenge for the project leader, therefore, is to use "subtle control" whereby he or she keeps the team focused and on schedule, yet empowers and motivates the team to be creative and innovative (Lewis *et al.*, 2002; Wheelwright and Clark, 1992).

Conclusion

Teams are the foundation of successful innovation efforts. However, putting together a group of cross-functional employees does not equate to an effective team. Effective innovation teams require careful deliberation about the type of team to use for the task as well as the size, diversity, and proximity of the team. The composition of the team, beyond just considering the functional skills of team members, is also critical to facilitate communication and cooperation. Finally, effective teams necessitate leaders who enable the team to be its best and to work through difficult challenges.

References

Ancona, D., and Caldwell, D. (1992a). "Demography and design: Predictors of new product team performance." *Organization Science*, **3**(3), 321–341.

Ancona, D., and Caldwell, D. (1992b). "Bridging the boundary: External activity and performance in organizational teams." *Administrative Science Quarterly*, **37**, December, 634–665.

Barczak, G., and Wilemon, D. (1989). "Leadership differences in new product development teams." *Journal of Product Innovation Management*, **6**, 259–267.

Barczak, G., and Wilemon, D. (2003). "Team member experiences in new product development: Views from the trenches." *R&D Management*, **33**(5), 463–479.

Cooper, R. G., Edgett, S. J., and Kleinschmidt, E. J. (2002). *Improving New Product Development Performance and Practices* (Benchmarking Study). Houston, TX: American Productivity & Quality Center.

Elkins, T., and Keller, R. (2004). "Best practices for R&D project leaders: Lessons from thirty years of leadership research." *International Journal of Innovation and Technology Management*, **1**(1), 3–16.

Hoegl, M., Ernst, H., and Proserpio, L. (2007). "How teamwork matters more as team member dispersion increases." *Journal of Product Innovation Management*, **24**(2), 156–165.

Holahan, P., and Markham, S. (1996). "Factors affecting multi-functional team effectiveness." In: M. Rosenau, A. Griffin, G. Castellion, and N. Anscheutz (Eds.), *The PDMA Handbook of New Product Development* (Product Development & Management Association). New York: John Wiley & Sons.

Jassawalla, A., and Sashittal, H. (1998). "An examination of collaboration in high-technology new product development processes." *Journal of Product Innovation Management*, **15**, 237–254.

Jassawalla, A., and Sashittal, H. (2000). "Cross-functional dynamics in new product development." *Research-Technology Management*, **43**(1), 46–50.

Kahn, K. (1996). "A definition of interdepartmental integration with implications for product development performance." *Journal of Product Innovation Management*, **13**, 137–151.

Kratzer, J., Leenders, R., and van Engelen, J. (2004). "A delicate managerial challenge: How cooperation and integration affect the performance of NPD teams." *Team Performance Management*, **10**(1/2), 20–25.

Lewis, M., Welsh, M. A., Dehler, G., and Green, S. (2002). "Product development tensions: Exploring contrasting styles of project management." *Academy of Management Journal*, **45**(3), 546–564.

O'Leary, M. B., and Cummings, J. (2007). "The spatial, temporal, and configurational characteristics of geographic dispersion in teams." *MIS Quarterly*, **31**(3), 1–19.

Pinto, M. B., and Pinto, J. (1990). "Project team communication and cross-functional cooperation in new program development." *Journal of Product Innovation Management*, **7**, 200–212.

Song, X. M., Montoya-Weiss, M. M., and Schmidt, J. B. (1997). "Antecedents and consequences of cross-functional cooperation: A comparison of R&D, manufacturing, and marketing perspectives." *Journal of Production Innovation Management*, **14**(1), 35–47.

Wheelwright, S., and Clark, K. (1992). *Revolutionizing Product Development*. New York: The Free Press.

33

Understanding Customer Needs

Abbie Griffin

University of Utah

The most successful product development efforts match a cost-competitive solution to a set of fully understood customer problems. Customer needs are the problems that a product or service solves and the functions it performs. They describe what products let you do, not how they let you do it. For example, many businesspeople have a need to "be able to do any work I want, wherever I am." Products and features deliver the solutions to people's problems. Features are the ways in which products function—a portable PC delivers a partial solution to being able to work wherever I want. So does taking a secretary and all our paper files on a trip, but although this was a preferred solution for some in past millennia, this is not a feasible solution today. Solutions and features change more rapidly than needs and problems to be solved.

Customers have general problems for which they need a solution and which relate to the overall product function. General needs and problems are fairly stable. They change only slowly, if at all, over time. For example, there is a general need to "protect my feet from the environment," which shoes, as a product, solve.

However, as they say, "the devil is in the details." Customers also have very specific needs or aspects of the overall function that a successful product must also solve. These more detailed needs, which may be related to a particular feature or technical solution, often are specific to the particular contexts in which the product is used. For instance, shoes "keep my feet dry," "keep my feet warm," "protect my feet from sharp objects," and even "help keep me from tripping."

Customer problems generally are very complex, and frequently different needs conflict. At the same time I want to be able to "protect my feet from the environment," I also want to "look stylish," and "keep my feet from overheating." Sandals may look stylish and keep our feet cool, but they do not protect our feet from the environment very well. The development team thus needs to have a good understanding of the relative importance of all the contexts and ways in which the product will be used, misused, and abused to select the most appropriate feature set for their product. It is first uncovering and understanding and then delivering against these detailed needs that differentiates between product successes and failures.

The four Cs of good statements of customer needs and problems:

- Customer language: they are not the words of the team and do not contain company-specific or technical jargon.
- Clear: they are easily understandable by all, over time. Some teams even create dictionaries with detailed definitions of terms and phrases.
- Concise: they are not too wordy. They contain only the words necessary to describe the need.
- Contextually specific: they include all contextual references and provide situational details. For example, "Protects my feet when it is raining outside."

Only the people involved with the details of how a problem affects the day-to-day way they perform their job or live their lives can provide you with their needs. And only the people who interact with, use, or are affected by the operation of a particular product can provide you with the details of how that product excels at or fails to solve their problems. A purchasing agent cannot identify the logistical and physical problems that a grocery clerk has operating a point-of-sale scanner system. Nor can they help a company understand the procedure the general manager of the store must go through to produce a daily income statement or rectify the store's inventory position at the end of the month using the software associated with the scanner system. The details of customer needs and problems must be gathered directly from the people who have them.

It is critical to ask customers about functions (what they want to do), not features (how it is done), because only by

Table 33.1. Summary of techniques for obtaining needs.

Needs uncovering techniques	*Information obtained*	*Major benefits*	*Major drawbacks*
Be a user	• Tacit knowledge • Feature trade-off impacts on product functionality	• Knowledge depth • Generates irrefutable belief in identified needs	• Hard to transfer knowledge to others • Time and expense
Critically observe users	• Process knowledge • Tacit knowledge	• Learn customer language • Find unarticulated needs	• Time and expense • Must translate observations into words
Interview users for needs	• Large volumes of details • Context-specific needs	• Speed of information collection • Breadth of information	• Ability to elicit reliable, tacit, and process needs • Viewed as marketing's job

understanding functional needs can teams make the appropriate trade-offs in technologies and features as they become feasible in the future. It is continual probing as to "why" something is wanted or works well that gets underlying needs.

It is also key to talk to customers only about situations that they have actually experienced. If someone has never been on a romantic picnic, they cannot be asked about what they would like in this situation because they do not know. What they would relate is fantasy.

Finally, needs are best uncovered by asking detailed questions about specific use instances. General questions produce general needs. General needs are not as useful in designing products as are the details of problems. Customers are very capable of providing an excruciating level of detail when they are asked to relate the story of specific situations that have occurred in the last year.

One pitfall to avoid is not talking to enough customers to obtain a complete set of needs. Only observing one firm's business processes or only talking to your firm's people as surrogates for actual customers is almost more dangerous than not interacting with customers at all. No one customer provides a full set of customer needs for any product area. Interaction with about 20 customers is required to obtain about 90% of customer needs (Griffin and Hauser, 1993).

Techniques for deeply understanding customer needs

Firms can obtain detailed understanding of customer needs through at least three market research techniques:

- Be an involved customer who has those needs and problems.
- Critically observe and live with customers who have those needs, also called customer visits (McQuarrie, 2008).
- Talk to customers with those needs, also called voice of the customer (Griffin and Hauser, 1993).

Table 33.1 summarizes the main aspects of these techniques.

References

Griffin, A., and Hauser, J. R. (1993). "The voice of the customer." *Marketing Science*, **12**(1), 1–27.

McQuarrie, E. F. (2008). *Customer Visits: Building a Better Market Focus* (Third Edition). Armonk, NY: Sage Publications

34

The Front End of Innovation in Large Established Firms

Peter A. Koen and Heidi M. J. Bertels

Stevens Institute of Technology

Introduction

Innovation in large established firms may be broadly divided into three areas: the front end of innovation (a.k.a. fuzzy front end or predevelopment), the new-product development process, and commercialization. The majority of scholarly research has been focused on the development process (see Brown and Eisenhardt, 1995; Montoya-Weiss and Calantone, 1994 for reviews), which requires a well-practiced routine and has become regimented through the widespread adoption of total quality management practices and six-sigma programs. Managing the front end is crucial since choices made during this part determine the company's options of which products will be ultimately developed and commercialized.

Research streams in the front end may be divided into two broad categories depending on whether the learning behavior that is required is exploitative or explorative (March, 1991). Exploitative behavior focuses on incremental and competency-enhancing product development efforts. Exploration refers to more radical (or breakthrough) and competency-destroying product development efforts.

In contrast to a relative consensus on how to manage exploitation processes and practices, our understanding of similar exploration processes and practices is just emerging with a lack of consensus as to the definitions, the required organizational structure, or even the need to have large corporations simultaneously pursue both exploration and exploitation (see Gupta *et al.*, 2006; Markides, 2006 for a discussion of the controversies). The front-end processes and practices for exploration projects are dependent on whether the innovation is aligned with the company's strategy, structure, culture, operational processes, and customers or the context in which the firm competes in and solves customers' problems. Christensen and Rosenbloom (1995) defined this industry structure as a value network. Using this definition our discussion of the front end for exploration projects is divided into those that are in alignment with the company's value network and those that require a new value network.

Based on the exploration/exploitation and value network frameworks the remaining part of this chapter is divided into three parts: (1) exploitation projects, (2) exploration projects that reinforce the organization's current value network, and (3) exploration projects that require a new value network, For all three, the organizational processes and practices used in the front end are discussed as well as project initiation (i.e., where the good ideas come from) and selection.

Front end of innovation for exploitation projects

Process and practices

Approximately 80% (Adams-Bigelow, 2005) of U.S. companies have adopted a formal Stage-Gate® process (Cooper, 2001) for exploitation projects (see also Chapter 24 of this encyclopedia). The first three stages/gates of the process, shown in Figure 34.1, are focused on the front end. They divide the front end into three distinct sequential stages separated by management decision gates. The remaining three stages, not shown in the figure, concern the development process. In the discovery stage, pre-work is done to discover and uncover opportunities and generate ideas. In the subsequent stage 1, scoping, a quick and inexpensive assessment of the technical merits of the project and its market prospects is carried out. The final stage, stage 2, focuses on building a more detailed investigation leading to a business case.

Cooper (2001) determined best practice factors based on studies of large samples of successful and unsuccessful exploitation products. While the research of Cooper and

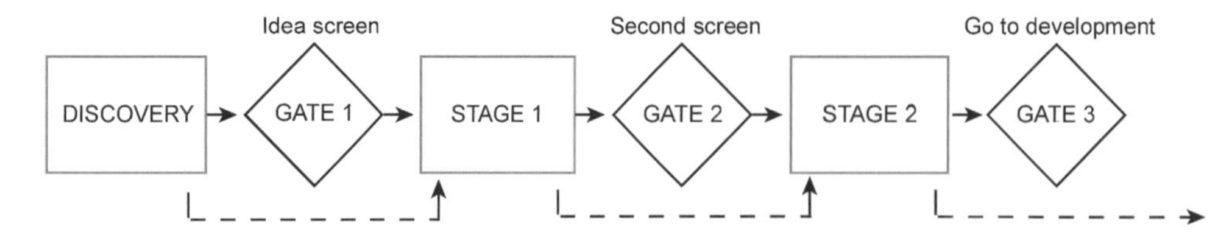

Figure 34.1. Stage-Gate process for the front end of innovation. *Source*: from Cooper (2001).

colleagues embraces the entire product development process, we will discuss only three of nine factors they found that are applicable to the success of the front end. The first, "product superiority", was measured as the relative product performance in terms of benefits, quality, reduced costs, superiority, and problem-solving ability compared with competitive products. The second factor, "product definition", reflects the "whats" and "for whoms" of the product concept prior to product development. The third factor, "upfront activities", refers to the initial screening, preliminary market assessment, and financial analysis of the project. Similar results were found by Song and Parry (1996) who looked at new-product introductions in Japanese high-technology firms.

There have been four empirical studies that specially focused on the front end, and for the most part, studied exploitation projects. Bacon *et al.* (1994) indicated that the front end will lead to a well-understood product definition which in turn will be linked to overall product development success. Moenaert *et al.* (1995) investigated the integration of marketing and research and development (R&D) activities and how information exchange between marketing and R&D affects success. Khurana and Rosenthal (1997, 1998) published the first comprehensive study of front-end activities based on case studies of ten incremental and two radical projects. They indicated that successful organizations follow a holistic approach. Furthermore, Khurana and Rosenthal (1997, 1998) classified front-end success factors into foundation and project-specific elements. Foundation elements refer to having a clear product strategy, well-planned portfolio of new products, and an organization structure that facilitates product development via ongoing communication and cross-functional sharing of responsibilities. Project-specific elements help clarify the product concept, evaluate it through executive reviews, and develop plans, schedules, and estimates of the project's resource requirements. This holistic approach addresses one of the major shortcomings of the serial model, shown in Figure 34.1, by addressing the front-end process within its context: the organization as well as the broader environment.

Koen *et al.* (2002) extended the work of Khurana and Rosenthal and created a holistic process model (see Figure 34.2) for the front end. This model divides the front end into three distinct areas. The first is the engine, or center of the model, which accounts for the vision, strategy, and climate and which drives the five key activities of the front end. The second or inner part of the model defines the five activity elements of the front end. These include opportunity identification, opportunity analysis, idea generation, idea analysis, and concept definition. The third element consists of the external environmental factors that influence the engine and the five activity elements.

In this model (Figure 34.2) "opportunity" refers to the business or technology gap, which a company realizes by design or accident, that exists between the current situation and the envisioned future. Opportunity recognition has received considerable scholarly attention (e.g., Christensen and Peterson, 1990; Hills and Shrader, 1998; Kaish and Gilad, 1991; O'Connor and Rice, 2001; Zietsma, 1999) in the entrepreneurship literature. A product "idea" stands for the most embryonic form of a new product or service, while a "product concept" refers to having a well-defined form which includes its primary features and customer

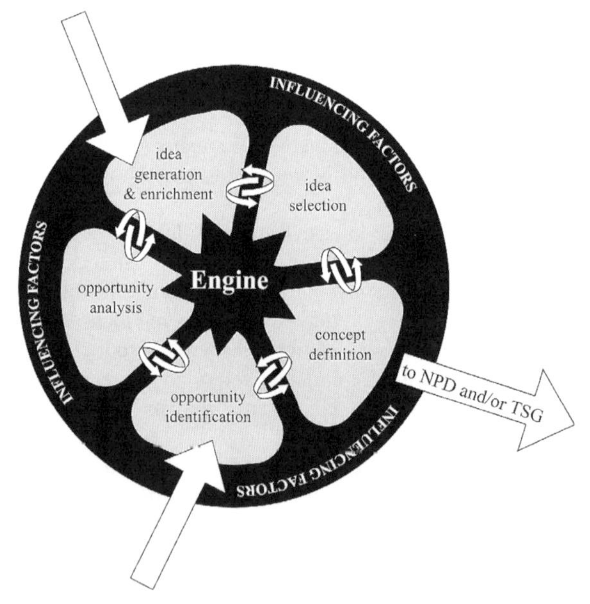

Figure 34.2. Holistic model of the front end. The engine accounts for vision, strategy, and climate and drives the five activity elements. The model rests on the outer elements, which are the influencing factors such as the environment and regulatory factors that the company cannot control.

P. A. Koen *et al.*, "Fuzzy front end: Effective methods, tools, and techniques." In: P. Belliveau, A. Griffin, and S. M. Somermeyer (Eds.), *The PDMA Toolbook 1 for New Product Development* (pp. 5–35). New York: John Wiley & Sons, Inc. Reproduced with permission of John Wiley & Sons, Inc.

benefits. These expanded definitions more precisely explain the activities that occur in the front end and prevent confusion between opportunity recognition where the company identifies an unmet customer need and idea generation which produces specific ideas that seek to provide solutions to the problems indicated in the opportunity space. The model is circular in shape, in contrast to the previous serial models, in order to indicate that ideas are expected to flow, circulate, and iterate between and among all the five elements (Koen *et al.*, 2002, p. 9).

Project initiation and selection

The process for identifying new ideas for exploitation projects in large established companies follows a well-honed process, which is extensively discussed by Cooper (2001). Exploitation projects in large companies may be grouped into one of two areas: new-platform projects (Meyer and Lehnerd, 1997) or incremental extensions to the existing platforms. An example of a platform would be Apple's first iPod—the 5 GB or 10 GB version. The first iPod established the overall technology and manufacturing architecture. Incremental extensions include other generations of iPods (e.g., 20 GB, 40 GB, Photo, and Video). New platforms are established when they are consistent with the overall strategy and value network of the company, when significant gaps are identified in the opportunity space and the expected economic value exceeds the company's financial hurdle rate. The establishment of a new platform is based on extensive market, competitive, and financial assessments. Typically, platform teams under the sponsorship of senior management are established to develop business assessments for these new initiatives (Meyer, Anzani, and Walsh, 2005).

The remaining part of the exploitation portfolio contains incremental extensions and improvements to the existing platforms. The development of a new platform is usually a top-down initiative. In contrast, the recommendation of incremental extensions is bottom-up, driven by suggestions from the sales force and customer recommendations made to improve the offering or by operations to lower the cost or improve the manufacturability of the product.

In most companies the effort required to support the proposed new platforms and incremental extensions exceeds the development resources available. The ultimate development decision is based on extensive portfolio analysis (Cooper, Edgett, and Kleinschmidt, 2001) which evaluates the overall value of the projects to the company with respect to each other. Approval for exploitation projects in most public U.S. companies tends to be driven by projects with both the earliest and largest expected financial returns.

The search for new ideas in many companies is no longer exclusively focused within the confines of the business. Companies have begun to embrace open innovation which focuses on obtaining ideas external to the company (Chesbrough, 2003). In 2000, Procter & Gamble's CEO A. G. Lafley made it a company-wide "... goal to acquire 50% of our (i.e., P&G's) innovations outside the company ..." (Huston and Sakkab, 2006, p. 61). The model is working at P&G as well as many other companies throughout the world. Huston and Sakkab (2006) indicated that 45% of the new initiatives at P&G have key elements that were discovered externally. They reported that "... R&D productivity has increased by nearly 60% ..." (Huston and Sakkab, 2006, p. 61). Since 2000, R&D investment as a percentage of sales has decreased from 4.8% to 3%, which attests to the value of the open innovation. To sustain this momentum, P&G has a network of 70 technology entrepreneurs based around the world to create external connections with universities, industry researchers, and suppliers.

The standard approach for understanding unmet customer needs through traditional focus groups, affinity diagrams, and quality function deployment processes has been found to fall short of delivering the deep insights needed. Traditional market research is being augmented with ethnographic techniques. Ethnography is a descriptive methodology for studying the customer in relation to his or her environment (Kelley, 2005; Perry, Woodland, and Miller, 2004). Ethnographic techniques involve methods for gaining intimate knowledge of the customer by becoming part of their habitat. P&G has its

> "... employees spend hours with women, watching them do laundry, clean floors, apply makeup and diaper the children. They look for nuisances that a new product might solve. They return to the labs determined to address the feature women care about most" (*Wall Street Journal*, June 1, 2005).

For example, P&G changed the packaging of individual tampons, as a result of ethnographic research, to non-crinkle plastic, which allows teenage girls to be able to open them in the bathroom without anyone knowing. In its first year (2004) of introduction the product obtained a 17% market share.

Front end of innovation for exploration projects aligned with the value network

Process and practices

The invention of stents is a good example of an innovation that required the development of new competencies, but was aligned with the value network (e.g., interventional medicine) which the company was already serving. Stents are used in a procedure for non-invasively expanding restricted cardiac arteries called angioplasty, which allowed companies to rapidly expand their market by solving one of the major problems associated with this procedure. Before the discovery of stents, angioplasty procedures to open obstructed cardiac arteries often could not prevent the spontaneous closure of the arteries resulting in repeat procedures or even open-heart surgery. This limited the usefulness of the procedure. Stents essentially solved this problem by preventing the artery from closing. Stent front-end efforts would require considerable research in order to perfect the technology and would utilize a Technology Stage-Gate (TSG) process to reduce technical uncertainty.

The objective of the TSG process (Ajamin and Koen, 2002; Schwartz, Yu, and Modlin, 2004) is to build confidence in a technology to a point where feasibility is demonstrated and a product development program can be initiated. The TSG lies within the front end of innovation since traditional development efforts (i.e., past gate 3) should not be started when there is significant technical uncertainty. The TSG process holds the team accountable only to the next gate since the outcomes of the technology efforts cannot be predicted *a priori*. To a certain extent the TSG embraces the concepts embodied in real-options theory (McGrath and MacMillan, 2000) where a limited investment is made at each stage. A go/no-go decision is made after each stage dependent on overall risk and value assessment based on the information collected during the previous stage.

Project initiation and selection

Initiatives for exploration projects that are aligned within the company's value network and solve a significant customer need or dramatically decrease costs also follow a well-honed process in large companies. In many cases these improvements lead to entirely new platform products. For example, the development of stents discussed previously solved one of the major issues preventing the use of angioplasty in more complicated cardiac vessel disease. In this case senior management challenged the research and development group to develop ways to solve the artery closure problem. In a similar way Intel challenged its research and development group to solve the highly complex and seemingly intractable microprocessor designs as the chips get smaller and more powerful.

The selection process for projects that are well aligned with the strategy and structure of the company follows a portfolio analysis (Cooper, Edgett, and Kleinschmidt, 2001), as discussed previously in the exploitation part of this entry. However, many of the measures (e.g., technology risk, research expenditure, ultimate cost, time to market, etc), which can be well defined for exploitation projects, are difficult if not impossible to determine for exploration projects thus making the decision process difficult. Further, the ultimate research expenditure and extended time to develop a solution often frustrates senior management.

Individual initiatives are often key to sustaining development. O'Connor and Rice (2001) discuss a number of organizational competencies needed to sustain the process:

1. Articulating a call to action: senior management needs to establish a context to encourage idea generation by continually articulating the future direction of the company and the need to develop new innovations,
2. Investing in organizational enablers: these can be broadly thought of as a combination of the right culture and communication mechanisms that stimulate idea generation.
3. Project oversight board: strategic priorities constantly wax and wane—especially during a poor financial quarter. Promising exploration ideas often need a longer term strategic view and need to be protected during their initial growth stage by a project oversight board,
4. Promoting and nurturing networks: ideas often require the knowledge of multiple disciplines and individuals. Companies need to actively promote a culture that encourages both formal and informal communication networks.

Front end of innovation for exploration projects *not* aligned with the value network

Process and practices

Corning's optical fiber program, General Electric's development of Computerized Axial Tomography (CAT), Motorola's development of cellular phones, and Searle's development of NutraSweet (Lynn, Morone, and Paulson, 1996), created entirely new markets which required the companies to develop not only new technical competencies, but also an understanding of how the product would fit within a entirely new value network. These new products require different operational competencies, different vendor relationships, and filled unmet customer needs which were unarticulated by the market.

Lynn, Morone, and Paulson (1996) found that such products, which create new value networks relative to the incumbent's existing value network, did not follow the traditional serial Stage-Gate approach employed for exploitation products, but followed a learning strategy which they called the "probe-and-learn process". These products required an iterative process of development where initial versions of the product were tested in the market in order to gain insight into what markets to pursue and what features and benefits provided value to the customer. The process can be schematized (see Figure 34.3) into four parts: discovery, incubation, acceleration, and institutionalization.

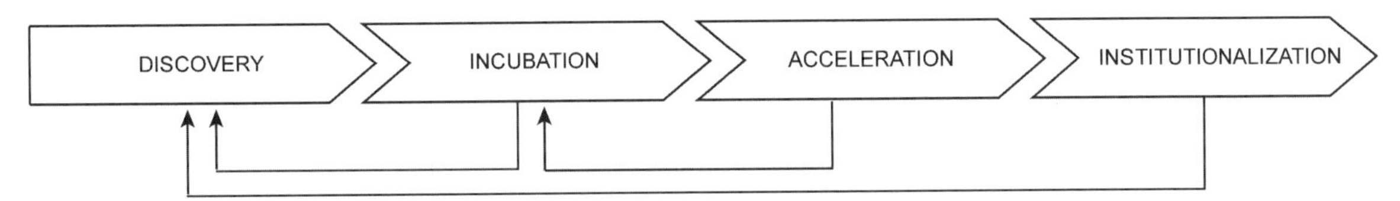

Figure 34.3. Front end for exploration projects that follow a probe-and-learn process.
Source: adapted from O'Connor *et al.* (2008).

The first three parts build on work of both Lynn, Morone, and Paulson (1996) as well as O'Connor and DeMartino (2006). In the discovery stage, ideas are formulated and screened against development costs, market potential, and overall strategic value to the company. In the next stage, incubation, initial prototypes are created which are suitable for sales and testing. The third, or acceleration stage, requires the new product to begin to scale up and create an actual business. The final stage, institutionalize, requires the new business to become integrated with the existing infrastructure of the corporation. This last stage, institutionalization, is the integration of the new opportunity with the existing business units or the development of a new business unit reporting to the traditional hierarchy of the company. The first three stages, while described in a sequential nature, are *not* sequential in practice. There is considerable learning from the 2nd stage and 3rd stage back into the discovery or 1st stage. In essence the 2nd and 3rd stages should also be considered as part of the front end since the learning causes new ideas to be discovered and tried out in the marketplace.

New technology advances, which may initially be aligned with the strategy and structure of the company, may often have even more value in markets that are external to the value network of the company. A classic example of this is Xerox and its Palo Alto Research Center (Smith and Alexander, 1988). Researchers developed numerous computer hardware (e.g., Ethernet) and software technologies (e.g., Graphical User Interface) that proved extremely valuable to companies active in other value networks. Xerox never leveraged these breakthrough innovations since their value network was focused on large high-speed copiers and printers.

Garvin and Levesque (2005, p. 9) describe an emerging business opportunity (EBO) structure and process being used at IBM to develop opportunities in new value networks as the "... white spaces between established businesses." The EBO unit is separated from the main stream with its initial efforts focused on understanding the basic unmet customer needs in the marketplace emphasizing the use of ethnographic methodologies. In the beginning the strategy for a new EBO is an iterative process. "Sometimes it would take a year to a year and half to get the strategy we were happy with. It would change three or four times" (Garvin and Levesque, 2005, p. 10). In contrast to the more structured TSG process, review meetings for the EBO, which are held at monthly intervals, are focused on strategic clarity and understanding significant unmet customer needs. Financial *pro formas* while required are used more to understand the key assumptions of the project, rather than to establish a rigid hurdle. Project initiation and selection in this structure follows the probe-and-learn approach where ultimate project approval is the result of many iterations guided by monthly senior management dialogs. The EBO system, which was started in September 2000, has proven to be quite successful. As of spring 2003 the EBO Group has created two new business units that are generating over $1 billion in annual sales and a number of others with yearly sales which exceed $100 million.

Project initiation and selection

von Hippel (1986), who has pioneered the lead-user approach, has indicated that many of the classic market research techniques do not lead to breakthrough ideas. Lead-users, or the first inventors of a product, across seven different industries, were found to be responsible for the development of a minimum 11% (electronic assembly) to a maximum of 100% (scientific instruments and semiconductor equipment) of breakthrough inventions (von Hippel, 1986). Lead-users are different from other users because they are at the leading edge of an identified trend in terms of new-product and process needs and because they expect to obtain a relatively high net benefit from solutions to their own needs (von Hippel, 1986). Similarly, Lilien *et al.* (2002) found that projects, in a study at 3M which drew on lead-users and users of "advanced analog" markets, possess significantly more novelty and annual sales than the average "traditional" project.[1]

In order for companies working with lead-users to be successful, they must find a way to bring together the user's tacit knowledge about needs and information and the manufacturer's solution technologies at one locus to create a solution in the form of a product. The difficulty resides in the observation that user's and manufacturer's information

[1] Analog users are people who are innovating in areas significantly outside the industry, but whose innovations may have direct applicability to providing new insights. For example, the search for new skin creams which would lead to a smooth skin surface without any fissures lead the team to look at earthquake specialists who are expert at measuring and predicting fissures.

is sticky (i.e. costly to acquire, transfer, and use). For example, the manufacturer's information can be highly specialized technical information usually not known by its product's users, while the nature of user information might be difficult to express in explicit terms. In order to solve this problem von Hippel and Katz (2002) recommend that companies develop toolkits, which customers can use, that are integrated with the company's product and design tools.

Kim and Mauborgne (2004) suggest that companies look for breakthrough innovations in the wrong places. They posit that breakthrough innovations are more easily achieved if companies look for opportunities in blue oceans rather than red ones. Blue oceans represent uncontested, unknown market space where large uncontested opportunities may be found. These are in contrast to red oceans, which are known markets where companies typically focus their innovation efforts. Red-ocean markets are fiercely defended by competitors who attack each other with the cruelness of sharks—creating a bloody mess. New markets are born in blue oceans and evolve to red through competition, commoditization, and oversupply. In their study of 150 companies they found that blue-ocean opportunities contributed to 14% of new-product launches, were responsible for 38% of the revenue, and an amazing 61% of the profits. They use *Cirque de Soleil* as an example of a new blue ocean. While it draws heavily from the circus tradition, it emphasizes the use of humans as performers instead of animals. The show combines the elements of street performances combined with the pageantry of ballet, theater, circus entertainment, opera, and music into an exquisitely choreographed entertainment. When compared with traditional circuses, *Cirque de Soleil* eliminated star performers, animal shows. and multiple show arenas. However, what they added was a refined entertainment environment combined with artistry, music, and dance.

Summary and conclusions

The front end is a crucial and path-dependent part of the innovation process where new products are born. The front end for exploitation projects has been extensively studied and follows a well-honed process driven by the strategy of the company. Significant improvements in the process have come about from the use of ethnographic techniques where companies are better able to understand unarticulated customer needs. Companies have also embraced open-innovation concepts so they no longer are dependent on finding all of the ideas within the company's boundaries. The front end for exploration within the company's current value network, while less studied and understood, also follows a well-orchestrated process.

In contrast, the front end for exploration in new value networks is much less understood. Process and practices, which have worked for exploitation projects and exploration projects within the same value network such as Stage-Gate, have been found to be inappropriate for this domain. Companies, who have had success in new value networks, have adopted a more *ad hoc* process that embraces a learning strategy.

References

Adams-Bigelow, M. (2005). "First results from the 2003 Comparative Performance Assessment Study (CPAS)." In: K. B. Kahn (Ed.), *The PDMA Handbook of New Product Development* (Second Edition). Hoboken, NJ: John Wiley & Sons.

Ajamin, G. M., and Koen, P. A. (2002). "Technology Stage-Gate™: A structured process for managing high-risk new technology projects." In: P. Belliveau, A. Griffin, and S. M. Somermeyer (Eds.), *The PDMA Toolbook One for New Product Development*. New York: John Wiley & Sons.

Bacon, G., Beckman, S., Mowery, D., and Wilson, E. (1994). "Managing product definition in high-technology industries: A pilot study." *California Management Review*, **36**, 32–56.

Brown, S. L., and Eisenhardt, K. M. (1995). "Product development: Past research, present findings, and future directions." *Academy of Management Review*, **20**, 343–378.

Chesbrough, H. (2003). *Open Innovation*. Boston, MA: Harvard Business School.

Christensen, P. S., and Peterson, R. (1990). "Opportunity identification: Mapping the sources of new venture ideas." In: N. E. Churchill, W. D. Bygrave, J. A. Hornaday, D. F. Muzyka, K. H. Vesper, and W. E. J. Wetzel (Eds.), *Frontiers of Entrepreneurship Research*. Wellesley, MA: Babson College.

Christensen, C. M., and Rosenbloom, R. S. (1995). "Explaining the attacker's advantage: Technological paradigms, organizational dynamics, and the value network. *Research Policy*, **24**, 233–257.

Cooper, R. G. (2001). *Winning at New Products: Accelerating the Process from Idea to Launch*. Cambridge, MA: Perseus Publishing.

Cooper, R. G., Edgett, S. J., and Kleinschmidt, E. J. (2001). *Portfolio Management for New Products*. Cambridge, MA: Perseus Publishing.

Garvin, D. A., and Levesque, L. C. (2005). *Emerging Business Opportunities at IBM* (Harvard Business School Cases). Boston, MA: Harvard Business School.

Gupta, A. K., Smith, K. G., and Shalley, C. E. (2006). "The interplay between exploration and exploitation." *Academy of Management Journal*, **49**, 693–706.

Hills, G. E., and Shrader, R. C. (1998). "Successful entrepreneurs' insights into opportunity recognition." *Frontiers of Entrepreneurship Research*. Wellesley, MA: Boston College.

Huston, L., and Sakkab, N. (2006). "Connect and develop." *Harvard Business Review*, **84**, 58–66.

Kaish, S., and Gilad, B. (1991). "Characteristics of opportunities search of entrepreneurs versus executives: Sources, interests, general alertness." *Journal of Business Venturing*, **6**, 45–61.

Kelley, T. (2005). *The Ten Faces of Innovation: IDEO's Strategies for Beating the Devil's Advocate and Driving Creativity throughout Your Organization*. New York: Doubleday.

Khurana, A., and Rosenthal, S. R. (1997). "Integrating the fuzzy front end of new product development." *Sloan Management Review*, **38**, 103–120.

Khurana, A., and Rosenthal, S. R. (1998). "Towards holistic 'front ends' in new product development." *Journal of Product Innovation Management*, **15**, 57–75.

Kim, C. W., and Mauborgne, R. (2004). "Blue ocean strategy." *Harvard Business Review*, **82**(10), 76–84.

Koen, P. A., Ajamian, G. M., Boyce, S., Clamen, A., Fisher, E., Fountoulakis, S., Johnson, A., Puri, P., and Seibert, R. (2002). "Fuzzy front end: Effective methods, tools, and techniques." In: P. Belliveau, A. Griffin, and S. M. Somermeyer (Eds.), *The PDMA Toolbook for New Product Development*. New York: John Wiley & Sons.

Lilien, G. L., Morrison, P. D., Searls, K., Sonnack, M., and von Hippel, E. (2002). "Performance assessment of the lead user idea-generation process for new product development." *Management Science*, **48**, 1042–1059.

Lynn, G. S., Morone, J. G., and Paulson, A. S. (1996) "Marketing and discontinuous innovation: The probe and learn process." *California Management Review*, **38**, 8–37.

March, J. G. (1991). "Exploration and exploitation in organizational learning." *Organization Science*, **2**, 71–87.

Markides, C. (2006). "Disruptive innovation: In need of a better theory?" *Journal of Product Innovation Management*, **23**, 19–25.

McGrath, R. G., and MacMillan, I. C. (2000). "Assessing technology projects using real options reasoning." *Research-Technology Management*, **43**, 35–49.

Meyer, M. H., Anzani, M., and Walsh, G. (2005). "Organizational change for enterprise growth." *Research-Technology Management*, **48**(6), 48–56.

Meyer, M. H., and Lehnerd, L. (1997). *The Power of Product Platforms*. New York: The Free Press.

Moenaert, R. K., De Meyer, A., Souder, W. E., and Deschoolmeester, D. (1995). "R&D/Marketing communication during the fuzzy front-end." *IEEE Transactions on Engineering Management*, **42**, 243–258.

Montoya-Weiss, M. M., and Calantone, R. (1994). "Determinants of new product performance: A review and meta-analysis." *Journal of Product Innovation Management*, **11**, 397–417.

O'Connor, G. C., and DeMartino, R. (2006). "Organizing for radical innovation: An exploratory study of the structural aspects of RI management systems in large established firms." *Journal of Product Innovation Management*, **23**, 475–497.

O'Connor, G. C., and Rice, M. P. (2001). "Opportunity recognition and breakthrough innovation in large established firms." *California Management Review*, **43**, 95–116.

O'Connor, G. C., Leifer, R., Poulson, A. S., and Peters, L. S. (2008). *Grabbing Lightning: Building a Capability for Breakthrough Innovation* (First Edition). San Francisco, CA: Jossey-Bass, a Wiley imprint.

Perry, B., Woodland, C. L., and Miller, C. W. (2004). "Creating the customer connection: Anthropological/ethnographic needs discovery." In: P. Belliveau, A. Griffin, and S. M. Somermeyer (Eds.), *The PDMA Toolbook 2 for New Product Development*. Hoboken, NJ: John Wiley & Sons.

Schwartz, K. J., Yu, E. K., and Modlin, D. N. (2004). "Decision support tools for effective technology commercialization." In: P. Belliveau, A. Griffin, and S. M. Somermeyer (Eds.), *The PDMA Toolbook 2 for New Product Development*. Hoboken, NJ: John Wiley & Sons.

Smith, D. K., and Alexander, R. C. (1988) *Fumbling the Future: How Xerox Invented Then Ignored the First Personal Computer*. New York: William Morrow.

Song, M. X., and Parry, M. E. (1996). "What separates Japanese new product winners from losers?" *Journal of Product Innovation Management*, **13**, 422–439.

von Hippel, E. (1986). "Lead users: A source of novel product concepts." *Management Science*, **32**, 791–805.

von Hippel, E., and Katz, R. (2002). "Shifting innovation to users via toolkits." *Management Science*, **48**, 821–833.

Zietsma, C. (1999). "Opportunity knocks—or does it hide? An examination of the role of opportunity recognition in entrepreneurship." *Frontiers of Entrepreneurship Research*. Wellesley, MA: Babson College.

35

Project Management under High Uncertainty

*Svenja C. Sommer** *and Christoph H. Loch*[†]

*HEC, Paris and [†]INSEAD, Fontainebleau

A project can be defined as a sequence of activities undertaken to accomplish a temporary endeavor (with a defined completion date) to create a unique product or service (Meredith and Mantel, 2003, p. 8). Projects are temporary structures (comprising reporting lines, goals and incentives, budgets, and team memberships), use flexible methods, and are dismantled after the work is done. Each project is unique in some respect and is thus managed slightly differently from other projects or ongoing repetitive processes. The ability of an organization to use temporary structures to flexibly respond to risks is particularly relevant in innovation contexts but has gained in relevance in many functional areas and business contexts. Projects are now seen as an important strategy to manage change in organizations.

Because projects are used to manage novel issues, they fundamentally must deal with uncertainty: resources are not available or are less productive than planned, tasks do not have the planned-for outputs, the context changes, the customer needs change, or any of myriad other influences change. The project management discipline has developed, in addition to careful planning, a number of methods to respond to uncertainty. In this chapter we give a brief overview of widely used methods, and then explain the challenges of managing highly novel projects that face extreme uncertainty. We summarize existing methods for dealing with such extreme uncertainty.

Project management methodologies and tools

This section briefly describes widely used methods for managing project uncertainty.

Variation and buffers

All non-trivial projects face possible (uncertain) variations in task durations, costs, and outcome quality. Firms commonly use project buffers to respond to variation. Project buffers take three main forms. First, a schedule buffer leaves a safety margin between promised completion time and "expected" completion time. Historically, firms placed schedule buffers at the level of important activities (e.g., a new feature in software development; see Cusumano and Selby 1995). Such buffers can help to plan for uncertainty in some activities, but they have the disadvantage of making project workers focused on meeting the deadlines of "their" own milestones, possibly undermining collaboration in the face of interactions. Goldratt (1997) proposed the "critical chain" methodology, which locates one central buffer at the end of the project, and each delay of any activity (on the critical path) reduces that buffer. This is not mainly a calculation device, but a tool to change attitudes: Project workers no longer need to protect their own schedule, rather the entire team "sits in one boat". Thus, the centralized buffer creates mutual commitment; moreover, a combined buffer is more efficient (Herroelen and Leus, 2001). These two advantages have made buffer management popular. Centralized schedule buffers have been incorporated as add-ons to commercial scheduling software packages such as Microsoft Project or Primavera (see an overview in Herroelen, 2005).

Second, budget contingencies serve as buffers for errors in cost estimates. A budget reserve of 5% to 10% of the estimated cost is a common rule of thumb, rarely exceeding 15% even for highly uncertain projects. The size of the budget contingency may be refined when a firm possesses systematic data of similar novel endeavors in the past (Kezsbom, Schilling, and Edward, 1989, p. 63), or when the firm can estimate the most likely and pessimistic estimates for the specific project (Meredith and Mantel, 2003, Ch. 8). Budget contingencies can also serve as an alternative to schedule buffers: if the project completion time starts to slip, the duration of activities may be shortened by reverting to project crashing (e.g., working overtime or outsourcing activities; see Kendrick, 2003, p. 150). The time cost trade-off has been extensively studied (Berman, 1964; Kelly and Walker, 1959), more recently also considering the possibility of overlapping activities (Roemer and Ahmadi, 2004).

Third, specification compromises can also serve as buffers. Cusumano and Selby (1995, 1996) describe the use of features as buffers in the development process at Microsoft. By assigning priorities to product features and by developing and testing feature by feature, Microsoft was able to relax some performance goals or scale back features to keep the project on schedule.

Project risk management and contingency planning

Project buffers are typically used when there are *many* uncertain events, but each individual event has only a *small* impact (e.g., the completion of a task may be delayed by a few days because of weather conditions, availability of materials, sickness of personnel, a regulatory delay, waiting for coordinating information from another task, etc.). In this case, none of the individual events is worth being managed by itself, rather it is effective to "statistically" manage their collection with a buffer. This does not mean that these myriad small uncertain events are not important; if not managed, they may well have a devastating effect on the project.

The buffer approach is appropriate for many projects that are threatened by an accumulation of small deviations, but not by major failures. However, many face just that: major uncertain events whose occurrence would have a large impact. Project risk management is a formally developed methodology for managing such risk factors. It consists of four phases: (a) risk identification, (b) risk assessment and prioritization, (c) risk response planning, and (d) documentation and learning (Chapman and Ward, 1997; Loch, De Meyer, and Pich, 2006; Wideman, 1992).

During risk identification the nature and consequence of each possible risk is summarized in a risk list. To ensure that no risks are overlooked, many firms use risk categories or "generic" templates that group risks that have occurred in the past (for an example see Loch, De Meyer, and Pich, 2006). During risk assessment and prioritization the team assesses the likelihood of an event occurring as well as its potential financial impact on the project. The risk status, defined as (event amount at stake) × (event probability), is often used to prioritize the risks. Many scholars propose to complement this assessment with scenario planning (Huss, 1988; Schoemaker, 1991), the consideration of extreme situations that the project may evolve into, to explore the joint impact of various important risk factors and to stimulate thinking about further possible changes (Schoemaker, 1995). Prioritization of risks, in order to focus the scarce response capacity of the project management office on the highest exposure, typically uses the criteria of likelihood of the risk, and impact of the risk. High-probability, high-impact risks are closely monitored, while low-likelihood, low-impact risks are "delegated" into a buffer.

Risk response planning (management) can take two forms: risk prevention for risks at least partially under the control of the project team, and contingency planning for uncontrollable risks. Risk prevention includes risk avoidance (completely remove the cause of a risk by changing the plan), risk mitigation (reduce the probability or the impact), and risk transfer to outside parties (e.g., insurance or incentive contracts, see Kendrick, 2003, Ch. 8; von Branconi and Loch, 2004).

Contingency planning identifies in the planning phase an alternative course of action that will be triggered if and when the event occurs. The two most common methods for contingency planning are decision trees and extended risk lists. Decision trees are logical and graphical representations of sequential decisions over time, explicitly recognizing that both an earlier decision and the outcome of an earlier uncertain event influence the decision options later.

Figure 35.1 provides an example of a decision tree for part of a drug research project (Loch and Bode-Greuel,

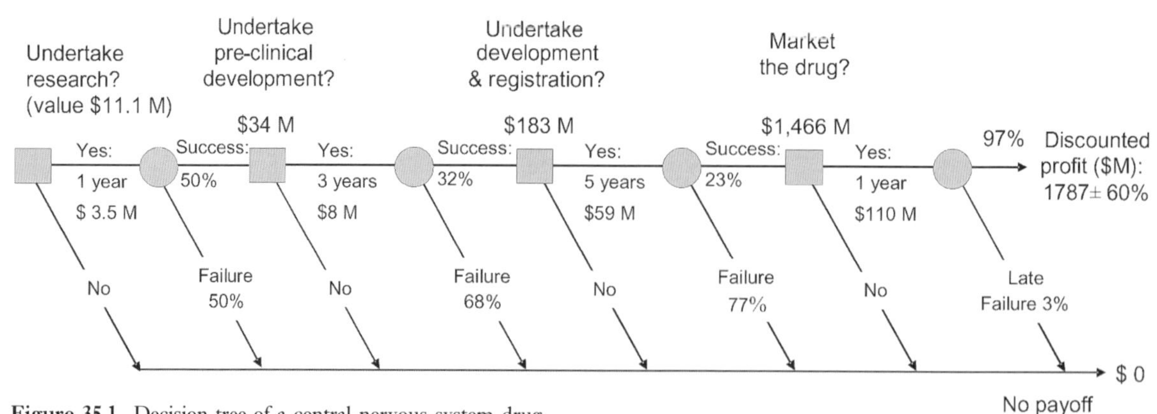

Figure 35.1. Decision tree of a central-nervous-system drug.
Source: Loch and Bode-Greuel (2001).

2001). The squares denote decision nodes (here: whether or not to continue the project at various stages). Under each branch, the time and cost of continuing are indicated; for example, if the first decision in the tree in Figure 35.1 (undertake research?) is "yes", the subsequent research work over the next year will cost of $3.5 million. The circle following the decision node represents a "chance node", which represents an uncertain event: Did the research provide promising results (success) or not? Here, success and failure have each a probability of 50%. Failure (negative results of research) forces termination of the project. Success allows the next stage decision (here: undertake pre-clinical work?).

By representing the dependence and ordering of risks, decision trees help establish a natural priority of attention for the project manager. In addition, they allow management to estimate the value of managerial flexibility and of preventive or mitigating actions. Formally, a decision tree is a graphical application of dynamic programming, a mathematical modeling technique that allows capturing sequences of uncertain decisions with complex outcomes.

An important drawback of decision trees is that their complexity explodes exponentially with the number of risks and decisions considered (for each decision and risk with n branches, the number of subsequent subtrees is multiplied by a factor of n). Therefore, decision trees are commonly only used to *focus* on a handful of the most important risks, incorporating smaller risks through extended risk lists containing the risk's probability, and preventive, mitigating, or contingent actions (Table 35.1).

While information is lost for risks that interact, risk lists are sufficient for most project risks. For example, a risk list for a drug research project contains risks from multiple sources, as shown in Table 35.1. The risks listed are aggregated to categories, as a detailed risk list for a pharmacological project easily fills 30 pages.

The value of managerial flexibility is often referred to as "real option value" (Dixit and Pindyck, 1994). However, the unavailability of market-traded assets to replicate project-specific risks limits the application of option theory (Smith and McCardle, 1999).

The last step, documentation and learning, provides a post-project assessment and an update of the historical database. Documenting the experience from past projects extends and improves available risk lists, including the estimation of risk probabilities and impact, and helps project teams to better understand the range of possible mitigating or contingent actions. Thus, annotated risk lists can become important elements of an organization's knowledge management—codified expertise that allows project teams to quickly access the whole repertoire of actions that the organization has used in previous similar situations.

Residual risk management

In many complex and novel projects, good risk planning can establish a solid general approach but cannot possibly identify all risks. No matter how well the project team does its homework, a few risks will inevitably be overlooked. Three widely used approaches exist to manage such "residual risk": robust planning, improvisation, and control-and-fast-response.

Robust planning leaves enough margin in the project plan for small changes. For example, "information gap decision theory" (Ben-Haim, 2001; Regev, Shtub, and Ben-Haim, 2006) develops a mathematical model of the situation where the variation of a project influence parameter around its believed value is unknown. The method proposes to choose actions that maximize the project's robustness or immunity to the unknown variation. The method formalizes the intuition of many project managers in doing things "just in case": for example, wider tolerances than ideally necessary for technical solutions, choosing performance curves that trade off a higher performance peak against a wider range over which performance is acceptable (robust satisficing), or choosing a supplier that does not offer the lowest cost but has experience and a track record of collaborative problem solving.

The second approach consists of leaving resources aside to "improvise" around the basic plan in response to residual risks. In some projects, a "risk management office" holds resources to develop *ad hoc* solutions to problems that creep up, and then share these solutions with all parties involved (Loch, De Meyer, and Pich, 2006, Ch. 1). In artificial intelligence, a similar approach is called "execution monitoring" or "dynamic replanning" (Ambros-Ingerson and Steel, 1988; Weiss, 2000). For example, the risk management office in a large IT integration project held experts on call (at considerable cost) from PC, software, network, and server vendors. With this problem-solving resource, they helped the decentralized migration teams to solve small unexpected problems, which were inevitable to emerge, as it was impossible to fully foresee and plan for all possible events in a 40,000 PC migration involving 1,200 project workers across international operating divisions. The risk management office also spread the learnings as standards, updating and extending the project knowledge, plans, and risk lists as they went along (Loch, 2005).

The third approach is relevant in complex projects in which the causal connections among myriad, partially unknown, influence parameters are not understood, and the team must establish a "green area" of a "safe mode of operation": for example, operators in a nuclear power plant know that certain system parameters must be kept within certain bounds (e.g., pressure in certain pipes, temperature in certain vessels). No one knows all causal connections, but this "green range" has been established as safe. Moreover, no one knows exactly what will happen if one deviates on any one of them; it is indeed possible that catastrophic failure may result from deviation of the system outside

Table 35.1. Generic risk list (template) of a pharmaceutical development project.

Risk category	*Detailed sub-categories*
Substance and production	
Ingredients	Risk from suppliers (dependency, stability, transfer, contracts), cost of production, availability of drug substance, process (reproducibility, scale-up, impurities), stability (shelf life)
Final product	As above, plus dosage changes, formulation changes
Analytical methods	Specificity, transfer of license or to a different site
Regulatory issues	Ingredient status, toxicity documentation, mixtures, impurity limits
Preclinical	
Safety pharmacology	Findings in core battery/supplemental studies, toxicity in cell cultures
Primary pharmacology	Choice of endpoints and species, target selectivity, and specificity
Bioanalytics	Detection of parent compound and metabolites, toxicity or metabolism in test species different from humans, drug accumulation, oral bioactivity, *in vivo* tests, body penetration
Toxicology	Availability of test substance, pharmacodynamic side-effects, high mortality rate in long-term studies, drug-specific side-effects
Clinical	
Phase I	Pharmacokinetics (e.g., different in sub-populations, interactions with other compounds or foods), pharmacodynamics (e.g., subject tolerance different from patient tolerance)
Phase II	Appropriate dosage, exposure duration, relevance of placebo control
Phase III	Study delay, patient recruitment, negative outcome (not significant), new regulatory requirements
General regulatory risks	Status of comparator, toxants in environment, availability of guidelines, interaction with agencies (e.g., process time, contradictions among different agencies), requirement differences across countries
General risks	
Licenses	Dependence on licensing partners
Patents	Disclosure of new patents
Trademarks	Viability/Acceptance of trademark at submission
Costs	Currencies, inflation, additional patients or studies needed
Market risks	New competitors, new therapies, patient acceptance, target profile, political risks (e.g., pricing, prevention versus therapy)

Source: Loch, De Meyer, and Pich (2006, Ch. 3).

the green area. Control-and-fast-response refers to the approach of preventing deviations if possible and, if one occurs, containing it immediately (Loch, De Meyer, and Pich, 2006, Ch. 3; Weick and Sutcliffe, 2001). Control takes the form of ever-paranoid and pervasive monitoring. For example, aircraft carriers conduct foreign-object-damage walk-downs on deck several times a day to prevent small objects (such as bolts or trash) from being sucked into airplane engines. In the constant chatter of simultaneous loops of conversation among project participants, "seasoned personnel do not 'listen' so much as they monitor for deviations, reacting instantly to anything that does not fit their expectations of the correct routine" (Weick and Sutcliffe, 2001, p. 32). Deviations are not conveniently explained away but investigated until their root cause is found, even if it is disruptive or painful, no matter where they are in the hierarchy.

When a slight deviation is discovered, even if it seems inconsequential, corrective and if necessary drastic action is taken. For example, during a naval exercise in the 1980s, a seaman on the U.S. nuclear carrier *Carl Vinston* reported the loss of a tool on the deck. All aircraft aloft were redirected to land bases until the tool was found, and the seaman was commended for his action—recognizing a potential danger—the next day at a formal ceremony (Weick and Sutcliffe, 2001, p. 59).

Types of uncertainty and unforeseeable uncertainty

The classic methods of project management are most suitable for the simplest form of uncertainty, or "risk". Risk refers to the possibility of several potential outcomes of a situation, each with a probability of occurrence that can be measured (e.g., from experience or experiments). For risks we can estimate "optimal" project buffers, and we can use dynamic programming to perform optimal risk management and contingency planning.

However, Knight (1921) pointed out that often the probabilities are not known. As Keynes (1937) put it later, "there is no scientific basis on which to form any calculable probability whatever. We simply do not know." This more challenging situation is referred to as "Knightian uncertainty", and sometimes as "ambiguity".

Formal methods exist to deal with ambiguity. A mathematical treatment is still possible when people "guess" their own "subjective" probabilities (Savage, 1954). Second, ambiguity can be represented as a probability distribution over a multitude of possible probability distributions (e.g., Camerer and Weber, 1992), making possible the mathematical treatment of this extended concept of uncertainty. Project risk management, working with *ranges* of outcomes, and robust planning and replanning all can deal with ambiguity as long as all important factors are in principle known (e.g., Amit, Brander, and Zott, 1998; Chapman and Ward, 1997; Smith and Merritt, 2002).

However, these concepts do not fully capture the uncertainty faced by many novel projects, in which management often faces serious gaps in knowledge: important influence factors or their relationships are not known at all at the outset (Schrader, Riggs, and Smith, 1993). Economists have called this "unawareness" or "unforeseen contingencies" (Kreps, 1992; Modica and Rustichini, 1994), scholars in public policy term it "wicked problems" (as opposed to "tame problems" that we know how to analyze, see Rittel and Webber, 1973), and engineering and project management professionals have used the term "unknown unknowns" (expanding the "known unknowns" of Knightian uncertainty), or "unk unks" (Wideman, 1992).[1] When a project develops a new technology or tackles a new market, unknown unknowns are rampant (for details see Pich, Loch, and De Meyer, 2002).

[1] The term is, in fact, "folklore": it has been widely used in aerospace, electrical machinery, and nuclear power project management for decades.

Approaches to unforeseeable uncertainty

The described approaches work well if the project goal and fundamental project approach remain unchanged. However, if unk unks are so fundamental that the project goal and path are, themselves, fundamentally unknown, then local flexibility, residual replanning, and robustness are insufficient. Any plan will run into major surprises, and the project team will be required to abandon assumptions and look for solutions in non-anticipated places (Brokaw, 1991; Loch, De Meyer, and Pich, 2006; Miller and Lessard, 2000). As an illustration, consider the following quote from Drucker (1985, p. 189):

> "When a new venture does succeed, more often than not it is in a market other than the one it was originally intended to serve, with products and services not quite those with which it had set out, bought in large part by customers it did not even think of when it started, and used for a host of purposes besides the ones for which the products were first designed."

Of course, the project team should attempt to convert unknown unknowns into risk, by performing careful upfront feasibility and risk analyses—this is precisely the risk identification step described earlier. However, in novel or breakthrough projects, this is simply not possible, no matter how hard one tries.

A problem of buffer and risk management (or similar planning approaches) is that an organization cannot list or even anticipate all possible events. In addition, determining contingent actions upfront might result in overlooking unexpected events and less improvisation (Weick and Sutcliffe, 2001). Adner and Levinthal, (2004, p. 77) made a similar observation for the real-options approach; namely, new opportunities might be foregone when rigid abandonment criteria are used.

Thus, the ability to effectively respond to unforeseeable uncertainty requires a *mindset* of realizing that planning and initially identified risks and contingencies are insufficient. The necessity for adopting a flexible mindset can indeed be identified at the outset of the project: while the unk unks themselves cannot be foreseen, the *situation that bears their presence may well be diagnosable*. For example, discovery-driven planning (McGrath and MacMillan, 1995, 2000) proposes to explicitly acknowledge that unknown unknowns exist and to uncover them with analyses such as assumptions checklists. Similarly. Loch, de Meyer, and Pich (2006) illustrate, using a startup venture project as an example, how the presence of unknown unknowns can be diagnosed by systematically probing what one knows about the project and where one has the intuition of being on unsafe ground.

Two fundamental approaches exist for this level of unforeseeable uncertainty: trial-and-error learning and

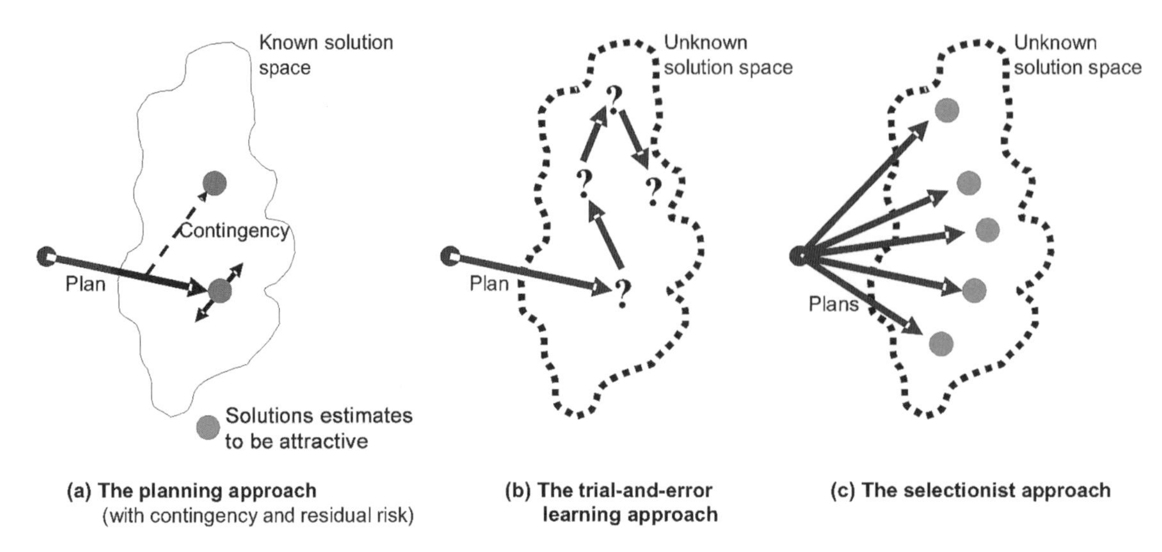

Figure 35.2. Three approaches to unknown unknowns.

selectionism (Leonard-Barton, 1995; Pich, Loch, and De Meyer, 2002), summarized in Figure 35.2 (Loch, De Meyer, and Pich, 2006).

The left panel of the figure summarizes the normal risk management mindset: a most attractive decision (in the known "solution space" of decision variables) can be identified, subject to some variation and buffer "wiggling". In case a large risk strikes, a contingent solution (Plan B) can be identified.

However, in the presence of unk unks, the solution space of decision variables is not known. Under trial-and-error learning (center panel), the team starts moving toward one outcome (the best it can identify), but is prepared to repeatedly and fundamentally change both the outcome and the course of action as it proceeds and as new information becomes available. Exploratory experiments, aimed at gaining information without contributing any progress to the current version of the plan, are an important part of this approach, and organizations need to view failure of such experiments as a source of learning rather than a mistake (Schoemaker and Gunther, 2006). It is therefore important to track the learning and reduction in knowledge gaps rather than tracking only the progress towards a target. The approach has been described under various names by numerous authors in technology transfer, new-product development, engineering projects, and new ventures (Chesbrough and Rosenbloom, 2002; Chew, Leonard-Barton, and Bohn, 1991; Drucker, 1985; Leonard-Barton, 1995; Loch, Solt, and Bailey, 2008; Lynn, Morone, and Paulson, 1996; Mullins and Sutherland, 1998; O'Connor and Veryzer, 2001; Pitt and Kannemeyer, 2000; Thomke, 2003; van de Ven *et al.*, 1999, Ch. 2). Thomke (1997, p. 105) calls the approach "flexibility", describing a design that can be modified at relatively low cost to account for unexpected changes in the environment (see also Angelmar, 1994; Bhattacharya, Krishnan, and Mahajan, 1998; Eisenhardt and Tabrizi, 1995; Iansiti and MacCormack, 1997; MacCormack, Verganti, and Iansiti, 2001).

As an example of trial-and-error learning, consider British Telecom's development of the 1471 "call to identify the last caller" service. The initial idea was to develop a digital display phone, which would show the caller identity. When testing the prototypes in a small town, BT provided the remaining households with a service number to obtain the caller ID retroactively. The test revealed that households with the service number were equally satisfied and BT decided to offer a service instead of the hardware product. BT changed the business model a second time from a service priced at 2c per call to a free service, once they realized that people would usually call back the last caller, and the additional call volume alone made the service profitable. Both product features and market acceptance were initially unclear, and BT changed the business model twice (private conversation with BT manager, 1997).

Alternatively (right-hand panel of Figure 35.2), the team might choose to "hedge" and opt for selectionism, or pursuing multiple approaches in parallel, observing what works and what doesn't (without necessarily having a full explanation why) and choosing the best approach *ex post*. Examples of this approach abound, including Microsoft's pursuit of several operating systems during the 1980s (Beinhocker, 1999), Toyota's "set-based engineering" (Sobek, Ward, and Liker, 1999), and "product churning" by Japanese consumer electronics companies in the early 1990s (Stalk and Webber, 1993). Or, consider Option International, a Belgian startup that became a leader in modem and mobile phone cards in the 1990s. While following a trial-and-error learning approach, on the technology side, they pursued several business models in parallel in a selectionist approach, on the business side. They developed the FirstFone subscriber identity card as a product under their own brand name, served as a development contractor for telecom original equipment manufacturers (OEMs), and co-developed with Lucent a high-speed data solution for Universal Mobile Telecommunications System (UMTS) networks. In 2003 this co-development resulted in a breakthrough contract with a large mobile operator—the third of the trials proved to be the winner that made the company successful (Loch, De Meyer, and Pich, 2006, Ch. 6).

The relative attractiveness of trial-and-error learning versus selectionism depends on its relative costs (including cost of delay for trial-and error learning) as well as its relative benefits. The cost trade-off is well understood (Loch, Terwiesch, and Thomke, 2001; Sommer and Loch, 2004). Often, however, the costs are unknown, or their difference is dominated by differences in value creation. The benefit comparison depends both on the presence of unknown unknowns and the complexity of the project (the number of relevant influence variables and their interactions, see Loch *et al.*, 2006; Simon, 1969, p. 195; Sommer, Loch, and Dong, 2009).

If neither unforeseeable uncertainty nor complexity are high, the standard risk management methods described earlier are sufficient. High complexity increases the benefit of selectionism, a result well known in the search literature (e.g., Fox, 1993; Loch, Terwiesch, and Thomke, 2001). If major unforeseeable uncertainty is present, but complexity is not very high, trial-and-error learning offers the largest benefit: by identifying previously unknown influence variables and adjusting the course of action to account for them the team can determine the optimal course of action that maximizes the performance with respect to the full state space.

The most difficult situation arises when unforeseeable uncertainty and complexity combine. Due to the complexity, learning can no longer find the optimal course of action even after all unk unks are revealed, while selectionism is of little help, if the "best" parallel trial must be chosen before the unforeseeable uncertainty is resolved. If the parallel trials can all be kept alive until they can be tested under fully realistic circumstances (e.g., by market-testing fully functioning prototypes, including product mix and promotion mix), selectionism can do much better, since at least the truly best of the trials can be chosen. Empirical evidence suggests that selectionism with full market feedback significantly improves the success of projects, while learning does not seem to do so (Loch *et al.*, 2006; Sommer, Loch, and Dong, 2009).

It is critical for project management teams to realize that the management systems of a project need to be adjusted, in comparison with standard planning projects, when trial-and-error learning or selectionism are used: incentives need to change (e.g., experimental outcomes need to be rewarded even when the outcome is negative), milestones must be planned differently (according to learning outcomes rather than "progress" towards a goal that may prove to be a mirage), and the type of information exchanged must change (from deliverables to learning). Otherwise, the project facing unknown unknowns is likely to fail (see Loch, De Meyer, and Pich, 2006, Ch. 8).

Project management phases

Projects are well known to go from initial high uncertainty (as the project is defined) to lower uncertainty as delivery or market introduction approaches. Indeed, product development can be viewed as an "uncertainty reduction process" (Browning *et al.*, 2002). This reduction of uncertainty associated with project management phases has been known since the end of the 1960s. In their study of several hundred industrial innovations over the previous decades, Meyers and Marquis (1969, p. 3) described "a series of stages in the process of innovation": (1) recognition of demand and technical feasibility, (2) idea generation (identification and formulation of problems worth working on), (3) problem-solving activity (unanticipated problems usually arise), (4) solution (may be a solution to a modified problem with somewhat different objectives), (5) implementation and use (see also Paap and Katz, 2004). Thus, the classic Stage-Gate process of product development was originally conceived as risk reduction.

Today, many associate product development "stages" make use of the Stage-Gate process described by Cooper (1983). The Stage-Gate process assumes that the basic problem solution (including product definition, target market, and business model) are known and thus expands the implementation phase of Meyers and Marquis' (1969) model (Paap and Katz, 2004). Cooper later incorporated additional features such as flexibility, overlapping stages, and fuzzy gates (Cooper, 1994). This model is applicable to most "next generation development" projects with moderate uncertainty: When a new car model or a new generation of a consumer good is being developed, market needs, technical challenges, and the business model are well known; mainly cost and budget uncertainties remain.

However, truly new technology development projects and startups entering new markets still need to go through all the uncertainty reduction stages identified by Meyers and Marquis (1969). Recent descriptions of project stages recognize the fact that new technology and market development projects are fundamentally different and expand and elaborate on the early phases to include activities to collect information and reduce the risk over the course of the project (Cooper, 2006, p. 25). In summary, the impor-

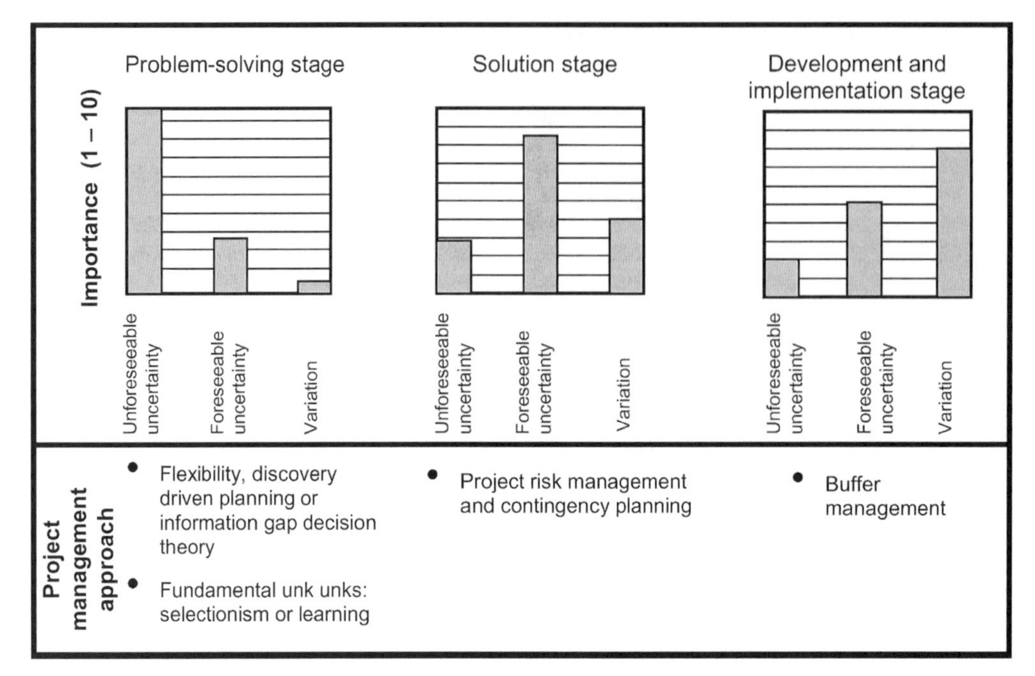

Figure 35.3. Uncertainty profiles over the phases of a novel project.
From De Meyer, Loch, and Pich (2002) and Loch, Solt, and Bailey (2008).

tance of the various types of uncertainty changes over the course of a novel project (Figure 35.3).

While the early phase tends to have a high potential for unforeseeable uncertainty (unanticipated problems or opportunities), the solution phase mainly faces foreseeable or known uncertainty or risks, and the implementation phase is dominated by schedule and budget variations. The importance of each of these uncertainty types (variation, foreseeable uncertainty, and unforeseeable uncertainty) can be represented in uncertainty profiles (De Meyer, Loch, and Pich, 2002). As the importance shifts from unforeseeable uncertainty to variation, an adjustment of the management techniques from selectionism and learning toward planning and control is required.

References

Adner, R., and Levinthal, D. A. (2004). "Real options and real tradeoffs." *Academy of Management Review*, **29**(1), 120–126.

Ambros-Ingerson, J., and Steel, S. (1988). "Integrating planning, execution and monitoring." *Proceedings Seventh National Conference on Artificial Intelligence* (AAAI-88 735-740).

Amit, R., Brander, J., and Zott, C. (1998). "Why do venture capital firms exist? Theory and Canadian evidence." *Journal of Business Venturing*, **13**, 441–466.

Angelmar, R. (1994). *Capital A* (INSEAD Case Study 05/94-4310). Fontainebleau, France: INSEAD.

Beinhocker, E. D. (1999). "Robust adaptive strategies." *Sloan Management Review*, **40**(3), 95–106.

Ben-Haim, Y. (2001). *Information Gap Decision Theory: Decisions under Severe Uncertainty*. London: Academic Press.

Berman, E. B. (1964). "Resource allocation in a PERT network under continuous activity time-cost functions." *Management Science*, **10**(4), 734–735.

Bhattacharya, S., Krishnan, V., and Mahajan, V. (1998). "Managing new product definition in highly dynamic environments." *Management Science*, **44**, S50–S64.

Brokaw, L. (1991). "The truth about start-ups." *Inc*, April, 52–67.

Browning, T. R., Deyst, J. J., Eppinger, S. D., and Whitney, D. E. (2002). "Adding value in product development by creating information and reducing risk." *IEEE Transactions on Engineering Management*, **49**(4), 443–458.

Camerer, C., and Weber, M. (1992). "Recent developments in modeling preferences: Uncertainty and ambiguity." *Journal of Risk and Uncertainty*, **5**, 325–370.

Chapman, C., and Ward, S. (1997). *Project Risk Management: Processes, Techniques and Insights*. Chichester, U.K.: John Wiley & Sons.

Chesbrough, H., and Rosenbloom, R. S. (2002). "The role of the business model in capturing value from innovation: Evidence from Xerox Corporation's technology spinoff companies." *Industrial and Corporate Change*, **11**(3), 529–555.

Chew, W. B., Leonard-Barton, D., and Bohn. R. E. (1991). "Beating Murphy's Law." *Sloan Management Review*, Spring, 5–16.

Cooper, R. G. (1983). "A process model for industrial new product development." *IEEE Transactions on Engineering Management*, **EM-30**, 2–11.

Cooper, R. G. (1994). "Third-generation new product processes." *Journal of Product Innovation Management*, **11**, 3–14.

Cooper, R. G. (2006). "Managing technology development projects." *Research-Technology Management*, **49**(6), 23–31.

Cusumano, M. A., and Selby, R. W. (1995). *Microsoft Secrets: How the World's Most Powerful Software Company Creates Technology, Shapes Markets, and Manages People*. New York: Free Press.

Cusumano, M.A., and Selby, R. W. (1996). "How Microsoft competes." *Research-Technology Management*, **39**(1), 26–30.

De Meyer, A., Loch, C. H., and Pich, M. T. (2002). "Managing project uncertainty: From variation to chaos." *Sloan Management Review*, **43**(2), 60–67.

Dixit, A. K., and Pindyck, R. S. (1994). *Investment under Uncertainty*. Princeton, NJ: Princeton University Press.

Drucker, P. (1985). *Innovation and Entrepreneurship: Practice and Principles*. New York: Harper & Row.

Eisenhardt, K. M., and Tabrizi, B. N. (1995). "Accelerating adaptive processes: Product innovation in the global computer industry." *Admin. Sci. Quart.*, **40**, 84–110.

Fox, B. L. (1993). "Random restarting versus simulated annealing." *Comput. Math. Appl.*, **27**, 33–35.

Goldratt, E. M. (1997). *Critical Chain*. Great Barrington, MA: North River Press.

Herroelen, W. (2005). "Project scheduling: Theory and practice." *Production and Operations Management*, **14**(4), 413–432.

Herroelen, W., and Leus, R. (2001). "On the merits and pitfalls of critical chain scheduling." *Journal of Operations Management*, **19**, 559–577.

Huss, W. R. (1988). "A move towards scenarios." *International Journal of Forecasting*, **4**, 377–388.

Iansiti, M., and MacCormack, A. (1997). "Developing products on internet time." *Harvard Business Review*, September/October, 108–117.

Kelly, J. E., and Walker, M. R. (1959). "Critical-path planning and scheduling." *Proceedings of Eastern Joint Computer Conference*, pp. 160–173.

Kendrick, T. (2003). *Identifying and Managing Project Risk*. New York: American Management Association.

Keynes, J. M. (1937). "The general theory of employment." *Quarterly Journal of Economics*, **51**, 209–223.

Kezsbom, D. S., Schilling, D. L., and Edward, K. A. (1989). *Dynamic Project Management: A Practical Guide for Managers and Engineers*. New York: John Wiley & Sons.

Knight, F. H. (1921). *Risk, Uncertainty and Profit*. Boston, MA: Houghton Mifflin.

Kreps, D. (1992). "Static choice and unforeseen contingencies." In: P. Dasgupta, D. Gale, O. Hart, and E. Maskin (Eds.), *Economic Analysis of Markets and Games: Essays in Honor of Frank Hahn* (pp. 259–281). Cambridge, MA: MIT Press.

Leonard-Barton, D. (1995). *Wellsprings of Knowledge*. Boston, MA: Harvard Business School Press.

Loch, C. H. (2005). *PCNet* (INSEAD Case Study). Fontainebleau, France: INSEAD.

Loch, C. H., and Bode-Greuel, K. (2001). "Evaluating growth options as sources of value for pharmaceutical research projects." *R&D Management*, **31**(2), 231–248.

Loch, C. H., De Meyer, A., and Pich, M. T. (2006). *Managing the Unknown*. Hoboken, NJ: John Wiley & Sons.

Loch, C. H., Solt, M. E., and Bailey, E. (2008). "Diagnosing unforeseeable uncertainty in a new venture." *Journal of Product Innovation Management*, **25**(1), 28–46.

Loch, C. H., Terwiesch, C., and Thomke, S. (2001). "Parallel and sequential testing of design alternatives." *Management Science*, **47**(5), 663–678.

Loch, C. H., Sommer, S. C., Dong, J., and Pich, M. T. (2006). "Step into the unknown." *Financial Times*, Mastering Risk, March 24, 2006, pp. 4–5.

Lynn, G. S., Morone, J. G., and Paulson, A. S. (1996). "Marketing and discontinuous innovation: The probe and learn process." *California Management Review*, **38**(3), 8–37.

MacCormack, A., Verganti, R., and Iansiti, M. (2001). "Developing products on 'Internet time': The anatomy of a flexible development process." *Management Science*, **47**(1), 133–150.

McGrath, R. G., and MacMillan, I. C. (1995). "Discovery driven planning." *Harvard Business Review*, July/August, 44–54.

McGrath, R. G., and MacMillan, I. C. (2000). *The Entrepreneurial Mindset*. Boston, MA: Harvard Business School Press.

Meredith, J. R., and Mantel, Jr., S. J. (2003). *Project Management: A Managerial Approach* (Fifth Edition). New York: John Wiley & Sons.

Meyers, S., and Marquis, D. G. (1969). *Successful Industrial Innovation: A Study of Factors Underlying Innovation in Selected Firms* (NSF 69-17). Washington, D.C.: National Science Foundation.

Miller, R., and Lessard, D. R. (2000). *The Strategic Management of Large Scale Engineering Projects*. Cambridge, MA: MIT.

Modica, S., and Rustichini, A. (1994). "Awareness and partial information structure." *Theory and Decision*, **37**, 107–124.

Mullins, J. W., and Sutherland, D. J. (1998). "New product development in rapidly changing markets." *Journal of Product Innovation Management*, **15**, 224–236.

O'Connor, G.C., and Veryzer, R. (2001). "The nature of market visioning for technology based radical innovation." *Journal of Product Innovation Management*, **18**, 231–246.

Paap, J., and Katz, R. (2004). "Anticipating disruptive innovation." *Research-Technology Management*, **47**(5), 13–22.

Pich, M. T., Loch, C. H., and De Meyer, A. (2002). "On uncertainty, ambiguity and complexity in project management." *Management Science*, **48**(8), 1008–1023.

Pitt, L. F., and Kannemeyer, R. (2000). "The role of adaptation in microenterprise development: A marketing perspective." *Journal of Developmental Entrepreneurship*, **5**(2), 137–155.

Regev, S., Shtub, A., and Ben-Haim, Y. (2006). "Managing project risks as knowledge gaps." *Project Management Journal*, **37**(5), 17–25.

Rittel, H. W. J., and Webber, M. M. (1973). "Dilemmas in a general theory of planning." *Policy Sciences*, **4**, 155–169.

Roemer, T. A., and Ahmadi, R. (2004). "Concurrent crashing and overlapping in product development." *Operations Research*, **52**(4), 606–622.

Savage, L. J. (1954). *The Foundations of Statistics*. New York: John Wiley & Sons.

Schrader, S., Riggs, W. M., and Smith, R. P. (1993). "Choice over uncertainty and ambiguity in technical problem solving." *Journal of Engineering and Technology Management*, **10**. 73–99.

Schoemaker, P. J. H. (1991). "When and how to use scenario planning: A heuristic approach with illustrations." *Journal of Forecasting*, **10**, 449–564.

Schoemaker, P. J. H. (1995). "Scenario planning: A tool for strategic thinking." *Sloan Management Review*, **36**(2), 25–40.

Schoemaker, P. J. H., and Gunther, R. E. (2006). "The wisdom of deliberate mistakes." *Harvard Business Review*, June, 109–115.

Simon, H. A. (1969). *The Sciences of the Artificial* (Second Edition). Boston, MA: MIT Press.

Smith, J. E., and McCardle, K. F. (1999). "Options in the real world: Lessons learned in evaluating oil and gas investments." *Operations Research*, **47**(1), 1–15.

Smith, P. G., and Merritt, G. M. (2002). *Proactive Risk Management*. New York: Productivity Press.

Sobek, D.K. II, Ward, A.C., and Liker, J. K. (1999). "Toyota's principles of set-based concurrent engineering." *Sloan Management Review*, **40**, 67–83.

Sommer, S. C., and Loch, C. H. (2004). "Selectionism and learning in projects with complexity and unforeseeable uncertainty." *Management Science*, **50**(10), 1334–1347.

Sommer, S. C., Loch, C. H., and Dong, J. (2008). "Managing complexity and unforeseeable uncertainty in startup companies: An empirical study." *Organization Science*, **20**(1), 118–133.

Stalk, G. Jr., and Webber, A. M. (1993). "Japan's dark side of time." *Harvard Business Review*, **71**(4), 93–102.

Thomke, S. H. (1997). "The role of flexibility in the development of new products: An empirical study." *Research Policy*, **26**, 105–119.

Thomke, S. H. (2003). *Experimentation Matters*. Cambridge, MA: Harvard Business School Press.

van de Ven, A. H., Polley, D. E. Garud, R., and Venkataraman, S. (1999). *The Innovation Journey*. Oxford, U.K.: Oxford University Press.

von Branconi, C., and Loch, C. H. (2004). "Contracting for major projects: Eight business levers for top management." *International Journal of Project Management*, **22**(2), 119–130.

Weick, K. E, and Sutcliffe, K. M. (2001). *Managing the Unexpected*. San Francisco, CA: Jossey Bass.

Weiss, G. (2000). "Planning and learning together." *Proceedings of the International Conference on Autonomous Agents, Barcelona, Spain* (pp. 102–113).

Wideman, R. M. (1992). *Project and Program Risk Management*. Newton Square, PA: Project Management Institute.

36

Evaluating Innovation Projects

Mark P. Rice

Babson College

All enterprises face a common challenge: creating the future (the domain of innovation and entrepreneurship) while simultaneously exploiting the past (the domain of ongoing operations). Given the differences along multiple dimensions between projects that exploit the past versus those that create the future, particularly with respect to the extent and magnitude of uncertainties that must be resolved, it is necessary to deploy different evaluation processes. Delineating these differences is the primary focus of this chapter.

Project evaluation can and should be viewed as an opportunity to compare and contrast all projects across the firm—those that are part of ongoing operations and those that are part of the portfolio of innovation projects. It is possible—even likely—that some of the projects supporting ongoing operations are less important to the short and long-term health of the firm than innovation projects that result in a future that is different from the past. Hence, evaluation that leads to reductions in the portfolio of projects supporting ongoing operations is a form of innovation, as it results in a reshaping of the firm. Decisions made at the conclusion of an evaluation process determine which projects will be pursued and which will be killed or shelved. Performance of the portfolio is determined as much by evaluation outcomes leading to the cancellation of poor performers early in the project life cycle as by evaluation outcomes leading to persistence (Block and MacMillan 1993).

Adopting different evaluation approaches for different projects

Variation in the severity of uncertainties that must be confronted across the spectrum of a firm's projects is embodied in the aggregate project-planning framework originally developed by Wheelwright and Clark, and extended by Christensen (Christensen, 2000; Wheelwright and Clark, 2003). This framework identifies categories of projects—ranging from those supporting current products/services and the entire spectrum of innovation activities (derivative innovation, platform innovation and breakthrough innovation), all of which vary as a function of technical and market uncertainty. Though innovation is not a discrete variable, throughout the rest of this chapter projects are segmented into two categories:

- Category 1 projects: those that support or extend ongoing operations and that are marked by relatively low uncertainty (product/brand support, and derivative or incremental innovation).
- Category 2 projects: those involving more than incremental change with respect to ongoing operations and that are marked by relatively high uncertainty on multiple dimensions (platform, radical, breakthrough, discontinuous, or game-changing innovation).

The questions about project evaluation that will be addressed in the remainder of this chapter include:

1. Why is project evaluation important?
2. How should project evaluation be conducted?
3. When should project evaluation occur?
4. What criteria should be used for project evaluation?
5. Who should engage in project evaluation?

"Firm" is used as the general term for the organizational unit in which the project will be evaluated. It is recognized that for large firms the relevant organizational unit may be the firm itself, a business unit of the firm, the R&D organization, corporate new-business development unit or some other sub-unit of the firm.

Why is project evaluation important?

Naturally a firm conducts project evaluation exercises seeking to make the best possible decisions about optimizing allocation of resources in order to achieve its strategic and tactical objectives. This applies not only to the initial investment but also to ongoing evaluation that determines whether successive investments will be made. Sub-optimal evaluation processes will lead to sub-optimal allocation of resources, underperformance of the firm's portfolio of projects and by extension underperformance of the firm overall—leading to a weakening of the firm's competitive position. Hence, through effective project evaluation processes the firm seeks to maximize return on investment within the context of strategic relevance and competitive positioning. A further discussion of the strategic and tactical objectives that should drive project evaluation is presented in the section entitled "What criteria should be used for project evaluation?"

The relatively straightforward and commonly understood evaluation processes are most applicable to projects with relatively low uncertainty—those that fall within the domains of ongoing operations and incremental innovation. As the number of categories of uncertainty and the criticality of uncertainties increase across a portfolio of innovation projects, simplicity and clarity in the evaluation process diminish and the demand for creativity and sophistication of judgment increases.

For radical innovation projects, allocation of resources may become less important and acquisition of resources—both internal (bootlegging) and external—may become more critical to project success (Leifer *et al.*, 2000). Hence, assessing the capacity and appetite of the firm for enabling the informal flow of resources internally (via bootlegging) and for engaging with external partners who can provide key resources (financial, competencies, access to potential customers, and so forth) may be an important part of project evaluation (Rice *et al.*, 2000).

Rather than relying on a pre-established and fixed process and set of criteria, the firm can use the project evaluation process to uncover latent uncertainties, clarify assumptions about those uncertainties,determine tests that can be used to test those assumptions, and prioritize project activities to be addressed during the next chapter in the life of the project (Rice, O'Connor, and Pierantozzi, 2008).

How should project evaluation be conducted?

The project evaluation process that the firm will employ needs to be communicated broadly, frequently, and effectively throughout the organization. Articulation of strategic intent by senior management with respect to: (1) expectations about developing and managing a dynamic portfolio of projects; and (2) communication about organizational systems, processes, and structures designed to support innovators—will heighten the awareness of employees. Effectiveness in pre-project activities will result from better preparation for the evaluation process, thereby enabling the firm to gain a better yield from pre-project activities.

Given that informal project evaluation often occurs before the firm engages in a formal initial evaluation, it is important that those involved in pre-project development know where to go when progress is sufficient to warrant triggering a formal evaluation. O'Connor and Rice (2001) elucidated the concept of hunters/gatherers to address the issue of specifying a project proposal–receiving function. The gatherer of project proposals may be an individual designated by the firm. For example, the firm may expect employees engaged in informal pre-project activity to approach a supervisor for Category 1 projects (those supporting ongoing operations), whereas for Category 2 projects (those that may challenge ongoing operations) a new business development organization may serve as the gatherer. In some firms a suggestion box or website is designated as the gatherer of new-project proposals. These are then logged into a tracking system and submitted to a review process. Some firms also designate an individual or individuals to act as hunters—proactively seeking out promising project ideas within the pre-project activity milieu.

The firm needs to set expectations regarding what presenters of project proposals should prepare for evaluation and in what form—presumably both written and oral. Many useful guides for preparing business proposals or plans are readily available—in print and on the Web (see, e.g., Mullins, 2003; Sahlman, 1997). The firm may choose to provide a simple Web-based template that will provide the opportunity for a rapid, cursory, informal assessment through which the reviewer can provide guidance to the proposal team, thereby improving the quality of the final proposal submitted into the formal evaluation process. Particularly for radical innovation projects involving high uncertainty on multiple dimensions, the learning plan methodology can be useful in generating questions to be asked and an approach to resolving them that can be incorporated into a useful proposal (Rice, O'Connor, and Pierantozzi, 2008). The evaluator(s) will need to assess what is known and the credibility of the plan for resolving uncertainties to determine whether the magnitude and probability of the projected outcome is sufficient to warrant initiating a project or, for an ongoing project under review, to determine whether and how to sustain the project.

The evaluation team should be clear about possible outcomes of the evaluation process, particularly so because innovation projects should have a more expansive set of decision outcomes than the go/no-go outcomes typically perceived as the appropriate outcomes for incremental innovation projects being managed through a Stage-Gate® process. In addition to (1) go and (2) no-go (kill), potential outcomes for innovation projects may include (3) redirect based on learning that has occurred through cumulative development; (4) delay or discontinue business development aspects of the project and reassign the activity to R&D for further technical development; and (5) discontinue but seek to license the technology in order to gain licensing revenue as a form of return on investment.

It is noted that the Stage-Gate model specifies four decision outcomes. Beyond those that are typically cited and applied—"go" and "kill"—Cooper (1990) also specified "hold" and "recycle" as possible decision outcomes. The fourth decision option is particularly important for innovation projects that are more than incremental, as learning that accumulates through the process of testing assumptions about critical uncertainties may be sufficiently promising to warrant project continuation, but often the learning reveals the need to take the project in a new direction. However, in practice, this option is often overlooked and, when project outcomes don't conform to plan, the project is killed.

The Stage-Gate framework (Cooper, 1990) is complemented by the milestone-planning framework developed by Block and MacMillan (1985) and by the discovery-driven planning framework developed by McGrath and MacMillan (1995). All three are useful frameworks for managing initial and subsequent evaluations of an innovation project.

When should project evaluation occur?

Evaluation needs to occur at project initiation and at key decision points along the path to commercialization. The decision gates may be aligned with the budget review schedule (appropriate for Category 1 projects that are aligned with and in support of ongoing operations) and/or may take place on an *ad hoc* basis as project milestones are achieved or deadlines established within the project plan are reached (appropriate for Category 2 projects—i.e., for radical innovation projects). However, informal evaluation also typically occurs before a formal evaluation is triggered that leads to the formation of an officially sanctioned project. This "pre-project evaluation" may be conducted by an individual or small group of individuals who decide to invest discretionary resources in exploring the potential of an opportunity. The decision to explore may occur within a strategic context or may simply reflect the interests of an inquisitive scientist or engineer. A project that is ultimately adopted by the firm and extends the firm's strategic framework illustrates the concept of "retroactive rationalization" of unplanned and successful innovations that are incorporated into the firm's strategy (Burgelman, 1986).

At some point during informal pre-project development, the participants determine that it is time to request a formal evaluation—because sufficient development progress has been achieved and/or because continued progress requires more resources than can be accessed informally. If the project is expected to be straightforward and of limited duration, then an appropriate manager may be approached with the expectation that a singular decision may be made to fund the project with a one-time investment from start to finish. However, for projects involving moderate to high uncertainty or complexity that are projected to require a more substantial commitment of time and resources, the evaluator or evaluation team may opt to establish a staged approach to evaluation and decision-making regarding commitment of resources. This approach is commonly used for entrepreneurial ventures by professional venture capitalists and has been adopted by corporate venture capital organizations (Rice *et al.*, 2000).

What criteria should be used for project evaluation?

For Category 1 projects supporting ongoing operations, projected performance as a function of financial metrics is often the primary objective criteria used to determine whether a project should be funded. Because uncertainty is relatively low, a rigorous quantitative analysis is possible and appropriate. A project may be approved if its projected financial performance exceeds an established threshold and if its projected performance exceeds that of enough competing projects that it makes the cut for funding within the budget allocated for new projects. Strategic fit is assumed and complementary resources (skills, production facilities, distribution channels, and so forth) are typically available or can be readily accessed, since the project is related to ongoing operations.

Evaluation of Category 2 projects involving radical innovation is more challenging, since many of the conventional evaluation criteria are neither applicable nor useful—particularly early in the project life cycle. Hence a different set of criteria may need to be developed and deployed. Rather than relying on net present value, return on investment, and other traditional financial criteria for which defensible, data-driven market forecasts and cost models are needed—evaluation teams may consider inclusion of some or all of the following:

- Alignment with the strategic intent of the company. Will the project enable the firm to explore and develop new technology and market domains? Will the project engage the firm in developing competencies that will enable it to succeed in these new domains? A firm may choose to stimulate growth through expansion of its strategic framework or by finding new businesses within the white spaces between current lines of business.

- Progress against goals. Particularly for projects involving uncertainty, careful setting of goals is important. The Learning Plan Methodology enables project teams to establish goals related to rapid and inexpensive learning related to critical uncertainties. Progress can be measured as a function of the degree to which critical uncertainties are resolved and/or new, unexpected, and value-producing insights are generated.
- Defense against competition. Even though a firm may have little interest in pursuing a project that may cannibalize current products or services, it may choose to proceed in order to deter or discourage competitors who are considering encroaching on the firm's markets by developing substitutes.
- Cultural transformation. A firm may perceive that—though it has in the past enjoyed a relatively stable competitive environment in which focusing on operational excellence was sufficient—the competitive environment is now undergoing a change toward an industry profile requiring more innovation. Hence, the firm may pursue a highly visible project or set of projects in order to stimulate awareness of and openness to innovation across the entire firm.
- Talent development. A firm may recognize that it needs to increase its innovation productivity and that it has insufficient human resources to do so. Hence it may choose to launch a project or set of projects specifically to grow the cadre of employees at various levels who are skilled and sophisticated with respect to innovation management.
- Development of systems, processes, and structures for supporting a portfolio of innovation projects. Firms that make a strategic commitment to catalyze growth through innovation may recognize that a comprehensive system needs to be established in order to support sufficient flow of a portfolio of projects through the innovation pipeline. Leifer *et al.* (2001) described this organizational capacity as a radical innovation hub.
- Degree of innovativeness. Within the concept of an aggregate project plan (Christensen, 2000; Wheelwright and Clark, 2003), those involved in the evaluation process may wish to consider degree of innovation in terms of risk and uncertainty in order to achieve some desired distribution of projects along this dimension.
- Business model considerations. It is likely that innovation projects will challenge the firm to project which parts of the value chain it will attempt to retain for itself and which parts it will choose to outsource to partners. This in turn will determine what resources (and competencies) the firm will need in order to participate in the value chain as envisioned. If the firm doesn't currently have some or all of those resources and competencies, it will want to consider whether this gap can be closed, and if so, how (Rice, 1996).

Other criteria may be identified that are important in a particular industry, to a particular firm, or to a particular project.

Who should conduct project evaluation?

Presumably, the firm's leadership makes resource allocation decisions on a periodic basis that establish budgets for two categories of projects across the spectrum of innovation, arbitrarily segmented earlier in this article into Category 1 projects and Category 2 projects. These allocation decisions represent the involvement of senior management in strategic evaluation of the portfolio of the firm's projects. This allocation can be adjusted as necessary during routine periodic strategic reviews or on an *ad hoc* basis when circumstances warrant an interim adjustment to the strategic plan.

For projects in Category 1, uncertainty is generally low, and thus individual managers can be given authority to commit project funds at or below a threshold level of investment appropriate to the level of the manager. It will fall to the manager to decide whether to involve others in the decision to initiate or sustain a project and, if so, how.

For most established firms, evaluation of projects in Category 2 is problematic. Managers typically have spent the majority of their professional lives working in ongoing operations and have limited experience in the radical innovation arena. Hence the sophistication of judgment that comes with experience is limited.

For firms in which the entrepreneurial founder is still active, it is often the entrepreneurial founder who overcomes organizational resistance to dealing with risk and uncertainty, who conducts evaluation of Category 2 projects, and who assumes the role of project champion.

For firms that no longer enjoy the involvement of the entrepreneurial founder and that seek effective evaluation of Category 2 projects, developing a cadre of decision makers with experience in evaluating projects involving high uncertainty on multiple dimensions is the key. These individuals can be either internal or external. Developing internal expertise can be accomplished over time by assigning senior managers to internal venture capital boards or on advisory boards for innovation organizations. Accessing external expertise has occurred in some firms by retaining outsiders with significant experience in evaluating early stage venture proposals (e.g., retired venture capitalists, intellectual property officers at technological universities, and managers of technology incubators), or by partnering with venture capital firms (Rice *et al.*, 2000). An ideal

Table 36.1. A summary of contrasts.

Dimension	*Category 1 projects (supporting ongoing operations)*	*Category 2 projects (creating the future)*
Level of uncertainty	Low	High
Planning approach	Operating or business plan	Learning plan
Evaluation criteria	Traditional financial metrics; strategic positioning; progress against plan	Strategic alignment with respect to opportunity to explore new technologies/markets; value of learning per unit of time and per dollar spent; variety of non-financial metrics
Investment approach	Tied to budget allocation on established budget cycles.	Staged investing; also bootstrapping and resource acquisition from informal sources will be important, particularly in the fuzzy front end
Timing of evaluation	Total commitment before project initiation—with periodic review	Thorough reviews conducted before each staged investment on an *ad hoc* basis
Who should conduct the evaluation?	Operating managers at a level appropriate for the project	Those who are skilled in judging high-uncertainty projects—veteran innovation project leaders and potentially outsiders (e.g., retired venture capitalists)

evaluation team will include one or more senior managers with the willingness and capacity to serve as project champions or sponsors and who may or may not be sophisticated in evaluation of radical innovation projects, complemented by evaluators with extensive experience and a track record of successful judgment in the evaluation of Category 2 projects.

Project evaluation: an opportunity for creating competitive advantage

A firm that aspires to implement a comprehensive and effective process of project evaluation needs to address the five questions presented in this article—in a way that is appropriate for the firm's context and that supports the optimal performance of the firm's portfolio of projects. In contrast with the Category 1 projects that directly support the ongoing operations of a firm, the firm's Category 2 innovation initiatives are confronted by a broad range of uncertainties—some of which can be severe. Established enterprises have highly developed and broadly implemented practices for evaluation of projects supporting ongoing operations. In contrast, managerial practices for evaluation of innovation projects that embody high uncertainty are often limited and less mature. This presents an opportunity for firms to gain a competitive edge by developing and implementing robust project evaluation processes across the entire innovation spectrum.

References

Block, Z., and MacMillan, I. C. (1985). "Milestones for successful venture planning." *Harvard Business Review*, September 1.

Block, Z., and MacMillan, I. C. (1993). *Corporate Venturing: Creating New Businesses within the Firm*. Boston, MA: Harvard Business School Press.

Burgelman, R. A. (1986). *Inside Corporate Innovation*. New York: The Free Press.

Christensen, C. (2000). *Using Aggregate Project Planning to Link Strategy, Innovation, and the Resource Allocation Process* (Note No. 9-301-041). Boston, MA: Harvard Business School Publishing.

Cooper, R. G. (1990). "Stage-gate systems: A new tool for managing new products." *Business Horizons*, May/June, 44–54.

Leifer, R., O'Connor, G., and Rice, M. (2001). "Implementing radical innovation in mature firms: The role of hubs." *Academy of Management Executive*, August.

Leifer, R., McDermott, C., O'Connor, G., Peters, L., Rice, M., and Veryzer, R. (2000). *Radical Innovation: How Mature Companies Can Outsmart Upstarts*. Boston, MA: Harvard Business School Press.

McGrath, R. G., and MacMillan, I. C. (1995). "Discovery-driven planning." *Harvard Business Review*, July 1.

Mullins, J. (2003). *The Nw Business Road Test: What Entrepreneurs and Executives Should Do before Writing a Business Plan* (Second Edition). London: FT/Prentice Hall.

O'Connor, G., and Rice, M. (2001). "Opportunity recognition and breakthrough innovation." *California Management Review*, Winter, 95–116.

Rice, M. (1996). "Virtuality and uncertainty in the domain of discontinuous innovation." *Proceedings of the 1996 International Conference on Engineering and Technology Management (IEEE). Vancouver, BC, August 18–20* (pp. 528–532). Red Hook, NY: Curran Associates/IEEE Publishing.

Rice, M., O'Connor, G., and Pierantozzi, R. (2008). "Driving breakthrough innovation: Implementing a learning plan to counter project uncertainty." *Sloan Management Review*, Winter.

Rice, M., O'Connor, G., Leifer, R., McDermott, C., and Standish-Kuon, T. (2000). "Corporate venture capital models for promoting radical innovation." *Journal of Marketing Theory and Practice*, Summer, 1–9.

Sahlman, W. A. (1997). "How to write a great business plan." *Harvard Business Review*, July 1.

Wheelwright, S., and Clark, K. (2003). "Creating project plans to focus product development." *Harvard Business Review*, September.

37

Managing Project–Organization Coupling in Breakthrough Innovation

Trudy Heller

Executive Education for the Environment

Still crazy after all these years

(Paul Simon)

Managers of disruptive innovations face the thorny dilemma of granting enough autonomy to a project that thinking outside the box may take place, yet connecting the project securely enough to the resources and strategies of the organization that it may receive an intelligent vetting and, if deemed fitting, survive. Heller (1999) suggested loosely coupled systems as a conceptual framework for increasing the focus on this critical aspect of innovation management. She argued that framing project and organization as loosely coupled systems had advantages: escaping oversimplified strategic forcing (bottom-up) and top-down models, emphasizing the mutual influence of project and organization, and enriching the conversation about this aspect of breakthrough innovation among researchers and practitioners.

The significance of breakthrough innovation has been heightened in recent years, as awareness of the urgent need to develop discontinuous, leapfrog technologies to address environmental sustainability challenges has increased (Christensen, 2000; Foster and Kaplan, 2001; Hart, 2005; Ottman, 1999; Pernick and Wilder, 2007). Hart (2005) asserts that capitalism is at a crossroads. Companies that cannot make the radical transformation to cleaner technologies and sustainability strategies will be losers in a Schumpeterian process of creative destruction. As Jeffrey Immelt, CEO of General Electric, noted about their "Eco-imagination" strategy, "The whole world is moving in a new direction. We've got to try to keep pace" (Kranhold, 2007).

This chapter will re-present the project–organization relationship through a loosely coupled systems lens. It will explore the concept of loosely coupled systems as an image representing the interface between innovation projects (emerging organizations) and their host corporations (existing organizations). The model calls attention to ways in which management's actions create more or less independence between innovation and host organization systems. It suggests everyday actions that loosen and couple innovation and management systems. The model offers new ways of conceptualizing and managing this important interface. Finally, more recent tools for managing the interface as a field of mutual influence will be noted.

Models of the innovation project/host organization interface

Entrepreneurial organizations have been described as "emerging organizations", and have been contrasted with "existing organizations" (Gartner, Bird, and Starr, 1992). In the case of corporate entrepreneurship, the emerging organizations (innovation projects) are embedded in existing host organizations so that the interface of the two organizational systems must be managed. Recent reviews of product development research concur that more research-based knowledge has been generated on innovation systems, or projects, than on the interface of projects and host organizations (Brown and Eisenhardt, 1995; Leonard-Barton, 1992). Existent images of the project/host organization interface tend to focus on the project or the organization, rather than the relationship between the two. Yet this interface is an important aspect of product development because it is the arena wherein the marrying of innovation to established organizational capabilities and strategies is consummated.

The dominant model of how synergies between innovation projects and host organizations are created is the bottom-up or strategic forcing model. In this process, the innovation is, forcibly, fitted into the resisting organization and strategy. The "champion" (Schon, 1963) or "heavyweight" (Clark and Fujimoto, 1990) project leader uses a mix of political power, organizational savvy, and technical prowess to get the innovation past organizational roadblocks to implementation. This model has focused research attention on political processes, on championing activity, and on corporate entrepreneurship as an impetus for total organizational renewal. The strategic forcing, or championing, model has generated a great deal of research (cf. Chakrabarti, 1974; Clark and Fujimoto, 1990; Frost and Egri, 1991; Day, 1994; Schon, 1963; Shane, 1994). The questions underlying this research are: "What is the nature of a successful innovation system?" and "How does the innovation fare as it makes its way through the (resistant) organization?"

An alternative model focuses attention on the organization side of the interface. Here, the creation of project/organization synergies is conceptualized as "induction", a process through which innovators are imbued with the values and strategies of the firm and thereby channel their creations to be consistent with established capabilities (van de Ven, 1986). In this top-down model, senior managers oversee and encourage the creative output of the innovation process. This model is associated with Japanese management and advocates the use of metaphor and other means of gentle persuasion to align the innovators with the strategic thrusts of the corporation (cf. Nonaka, 1991). This model has focused research attention on senior managers as purveyors of vision, on the constraints of strong organizational cultures (Jelinek and Schoonhoven, 1990), and on the nature of host organizations that are hospitable to innovation (Adler and Borys, 1996; Dougherty and Hardy, 1996). The top-down model of induction is less well developed, as Brown and Eisenhardt's (1995) review of product development research notes: "... the management-related concepts ... such as vision, subtle control, and even support are vague." The question underlying this research is: "What is the nature of a host organization that supports, inspires, and controls entrepreneurial activity?"

Aim of this chapter

This chapter operationalizes and explores loosely coupled systems as an alternative to forcing and induction images to represent the relationship that takes place at the interface of the innovation project and its host organization. The loosely coupled systems model focuses attention on: (1) the dialectical quality of the relationship between the two systems (at once loosened from and coupled to), (2) the independence *and* mutual influence of the emerging and existing organizations, (3) both tacit and explicit loosenings and couplings, and (4) both innovators and managers as co-creators of synergy between project concept and organization. Thus, unlike the inducing and forcing models, it promises to provide a map of the terrain that captures a relationship of mutual influence between innovation and host organizational systems.

The exploration will proceed as follows: The concept of loosely coupled systems and its application to product development will be introduced. Specific mechanisms that serve to loosen and/or couple project and organization are then described, as well as the dynamics of the loosely coupled systems model in this application. Finally, contributions that this construct can make to the task of imagining and managing the project/organization interface are discussed.

Loosely coupled systems for corporate entrepreneurship

One theoretical statement of the notion that innovations belong in organizational structures that are somewhat independent of the operating organization comes from the research stream generated by Weick's (1976) notion of loosely coupled systems. This stream of research carries both methodological and substantive implications for the study of corporate entrepreneurship.

Methodological implications rest on the conceptualization of loosely coupled systems. The dialectical nature of the concept suggests that innovation systems are at once loosened from and coupled to the host organizational system. Also, since "tight couplings in one place may imply loose couplings elsewhere," the challenge to the researcher of loosely coupled systems is to discern the patterns of couplings that produce a particular outcome (Weick, 1976). This research task is particularly challenging, according to Weick, because tightly coupled regions of the organization are more explicit and easier to see and talk about. Hence the loosely coupled aspects of organization may only be discerned through comparative studies, where the researcher may see that what is there and what is talked about in one setting may not be there and may not be talked about in another setting.

Substantive findings from the empirical work generated by this research stream include the following three that are salient to the case of new-product development: (1) Loosely coupled systems may foster creativity: "researchers have suggested that loosely coupled systems create a haven of psychological safety in which deviance and experimentation are protected" (Meyerson and Martin, 1987, p. 636), and where creativity may "flourish unencumbered by the demands of the ongoing operation" (Kazanjian and Drazin, 1986). (2) Loosely coupled systems appear to be "an organizational form which obscures subunit activities from the monitoring of top management," giving system members more behavioral discretion to act illegally (Vaughan, 1982), or at least illegitimately (Dougherty and Heller, 1994) in the pursuit of new-product development. (3) Loosely coupled systems may serve to seal off or buffer the spread of problems (Weick, 1976), or, in the case of new-product development, failure.

The discovery of loosening and coupling mechanisms

How are innovation projects coupled to and loosened from their host organizations? The literature reports eight specific mechanisms that link innovation and routine organizational systems, creating more or less independence for innovators (Heller, 1993, 1999). Table 37.1 lists the mechanisms clustered into two categories. The first cluster of three mechanisms connects innovators to the strategic identity of the firm. The second cluster of five mechanisms connects innovators through the organization and provision of resources for the innovation project.

Table 37.1. Categories of loosening and coupling mechanisms.

Project–Strategy connections
—Widely held understandings about what the organization does: strategy, mandate for innovation, risk climate
—Technological compatibility
—Established markets

Project–Organization connections
—Funding
—Senior management attention
—Structural location
—Standard operating procedures
—Human resource deployments

The three mechanisms of the first, strategic cluster—(1) strategy, (2) existent technology, and (3) established markets—link projects to sponsoring organizations through widely held understandings of what it is that the organization does: Is the innovation aligned with the firm's ongoing strategy? Does it utilize technologies in which the company has established competencies? Does the innovation target markets that the company is already geared up to serve?

For each of these mechanisms, conforming to the understanding couples project and organization, while deviating loosens project and organization. Thus, while innovators may or may not conform their new-product concept to the *status quo*, they still use their understanding of the norms as a standard, and a single set of categories describes both loosening and coupling mechanisms.

The five mechanisms of the second, organization cluster, listed in Table 37.1, link projects through practices concerning (4) funding, (5) senior management attention, (6) structural location of new-product development activities, (7) standard operating procedures, and (8) human resource deployments. A "big budget" project that represents a larger percentage of a firm's investment in innovation is likely to be more tightly coupled to its sponsor and receive greater scrutiny from senior management. Similarly, a project that is developed in an established innovation center will have less independence, or be more tightly coupled, than a project developed as an informal skunkworks[1] operation outside any formal organizational structure. When innovators follow standard operating procedures their project becomes more tightly coupled, as the innovation is then limited to what can be developed under pre-existing rules. Finally, assigning inexperienced, junior personnel to an innovation project loosens its connection to the authority structure of the organization, providing both greater freedom to create more radical innovation and diminishing the chances that the innovation will be connected enough to the sponsoring organization to be implemented.

Thus daily decisions concerning funding, human resource deployment, where to locate the innovation activity, etc. create more or less independence for innovation and host organizational systems. Each of these linking mechanisms is a two-edged sword, bringing needed resources and legitimacy to the projects and, at the same time, exposing the activities of innovators to scrutiny, intervention, and possible sanctions.

A project may be given free rein to develop outside legitimate development structures and standard operating procedures and be staffed by a team that is loosened from functional ties, or it may operate within an official development unit, be closely scrutinized by top management, and have its funding tied to a business unit with specific commercialization interests. Only the former project will, theoretically, be endowed with the advantages noted in the literature on loosely coupled systems for innovation—that is, (1) providing a psychological haven for creativity, (2) providing obscurity from senior management scrutiny, and (3) buffering the organization from innovators' failures.

Dynamics of loosely coupled systems: managing independence

In addition to the categories of coupling/loosening mechanisms described above, three dynamics of project/organization independence are suggested. Each is discussed below:

(1) Loosenings (not couplings) are easier to observe, discuss, and manage. Innovators who are interviewed regarding coupling are more likely to talk about how they are "not" coupled (Heller, 1993, 1999) than how

[1] SkunkWorks® is a registered service mark of Lockheed Martin.

they are coupled. For example, one project manager is reported as telling his technical workers not to follow standard operating procedures for documentation on his project.

Thus the discovery of coupling and loosening mechanisms does not conform to Weick's (1976) suggestion that the tightly coupled elements of an organization are more explicit and easier to see and talk about. On the contrary, Dougherty and Heller (1994) found that when innovators conformed to organizational norms this information was implicit in the interviews and in the telling of the tale of the project. Their conformity was taken for granted. When innovators deviated from the usual, however, they were aware of the deviation and would speak explicitly about this (e.g., that people assigned to the project team had been freed of the usual constraints by their functional bosses).

(2) Loosening on one level of the organization may exist with coupling on another level (pseudolooseness): Many of the loosening mechanisms described in the literature on corporate entrepreneurship (e.g., the use of *ad hoc*, flexible structures, and teaming) may loosen the project at the local level. Discussions of these mechanisms ignore the question of whether these same projects may be tightly connected at a higher level of the organization through senior management attention or through the new product's compatibility with the technical core of the organization. When this occurs, we may get a false, or incomplete, impression of the project's independence.

For example, Heller (1993) reports on a project team that was given free rein to work with minimal supervision from team members' functional supervisors. However, the project was closely watched by senior management. Weick (1982, p. 395) uses the term "pseudolooseness" to describe loose coupling at the local level with tight coupling at a higher level of the organization "because members are tightly coupled to a limited set of decision premises determined by top management and implemented during socialization experiences."

(3) Motivational aspects of loosening and coupling. The motivational surge described by proponents of the skunkworks operation (cf. Peters, 1983) for innovation may actually be the outcome of the direction of change, rather than a reflection on the overall separateness of the project. Heller (1993), for example, reports about an innovation team that described exhilaration when placed on a project that loosened them from their usual functional ties. On the other hand, innovators in another company grumbled when they became more closely tied to a business unit's commercialization interests and their mandate for innovation narrowed. Earlier researchers who attributed the surge in motivation to the independence skunkworks operations afford their members may be only partly right. Innovators who are set loose in a skunkworks operation may actually be motivated by the *increase* in their independence. Presumably, if the skunkworks context provided less independence than the innovators' previous work situation, they would experience a decrease in motivation.

Discussion

This chapter began with the suggestion that loosely coupled systems was an image that would be useful as a map of the interface of emerging and existing organizations. It identifies some of the mechanisms that actually function to loosen or couple elements of entrepreneurial and existing organizational systems as reported in the literature and explores some of the dynamics of these mechanisms in practice.

What does this exploration offer to corporate entrepreneurship theory? Fiol (1995) notes the richness of the corporation as a context for entrepreneurship because of the existence of contradictory thought worlds—the right stuff for creative innovation. The loosely coupled systems image preserves independence and creative tension among thought worlds, while allowing managers to think consciously about and manage the relative independence of project and organization.

The project–organization interface has been underdeveloped as a focus of study and practice because of a lack of constructs and language for understanding this boundary. This chapter's identification of specific loosening/coupling mechanisms and their dynamics permits a richer and more precise discussion of the project/organization interface in the following ways:

1. The mechanisms described above detail how, through ordinary daily decisions, projects and host organizations become more or less independent. Hence it is not just through the extraordinary actions of champions or the establishment of project teams that the interface is managed. This exploration suggests other ways in which innovators may become independent of their host organization (e.g., through managerial neglect, through the pursuit of new markets, through the deployment of inexperienced personnel). Similarly, this suggests many ways that projects become coupled to their organizational hosts (e.g., through development in formal entrepreneurial units, through senior management attention, through the use of established technologies).
2. The image of loosely coupled systems provides a language to describe the relative independence of project and organization that preserves the dialectical nature of the relationship. Configurations of loosening and

coupling mechanisms can capture the ways in which projects are at once loosened from and coupled to their corporate hosts. Other models focus attention on either separation or integration.
3. Identification of specific loosening/coupling mechanisms opens discussion regarding the level of organization at which the mechanism operates. The construct of pseudolooseness invites a more holistic, multi-level view of the question of independence of innovation projects. Thus, an innovation project team may be loosened from its host organization at the local level but tightly connected at a higher level of the organization through senior management attention or through the new product's compatibility with the technical core of the organization.

Future studies on this topic may investigate ways in which these mechanisms are configured in different organizational contexts and whether different mechanisms are operational in other industries and cultures. Future work may also investigate the three variables that emerged from this preliminary study as the dynamics of loosely coupled systems: (1) loosenings are more easily observed; (2) all levels of the organization must be considered in assessing the independence of the project and organization (pseudolooseness); and (3) motivational effects come from the direction of change (more or less independence) rather than just the extent of independence. Finally, other images of the project–organization interface may also contribute to the development of a more imaginable and manageable interface of emerging and existing organizational systems.

Epilogue: tools for managing the project–organization linkage

Getting the (in)dependence of innovation project and organization right remains a difficult management challenge. Nike, for example, set up the "World Shoe Project" to develop athletic footwear for the Chinese market. While conceived as a separate pilot project, close ties to the ROI requirements of established product lines led to its demise (McDonald and London, 2002). Established organizations may also take the acquisition route to gaining discontinuous capabilities, as when Clorox acquired Burt's Bees to expand its market with environmentally conscious consumers (Storey, 2008). The acquisition bypassed the challenges of innovating breakthrough products in-house.

Useful tools, both conceptual and practical, are nonetheless addressing the challenge. A few are briefly noted. Strategic conversations (Dougherty and Hardy, 1996; Manning, 2002) may enable the process of weaving the innovation and the organization's strategy together. Once viewed as strategic, the innovation is in a more powerful position to attract sustained resources, senior management attention, and successful implementation.

Interaction between innovators and others is also recommended (Heller, 2000) to facilitate the development of interpretive knowledge—an essential addition to technical and marketing knowledge. Innovators must not only know what organizational changes are required to implement a particular innovation, but what it means for the strategic identity of the business to carry out those changes. How will Clorox, a chlorine bleach company, recreate a strategic identity that integrates the natural skin care products of Burt's Bees?

Learning plans have demonstrated their usefulness to managers of breakthrough innovation (Rice, O'Connor, and Pierantozzi, 2008). This tool overcomes the limitations of other methods of standardizing the innovation process (e.g., Stage-Gates, milestones). By creating "iterative learning loops" managers can more quickly and wisely determine whether a project is worth continuing.

Gone are the days when the skunkworks was the ideal method for established firms to create breakthrough innovations. (Just create it in a separate unit and then throw it over the wall.) The conceptual frameworks and practical tools described here are enabling better management and understanding of the mutual influence of project and organization.

References

Adler, P. S., and Borys, B. (1996). "Two types of bureaucracy: Enabling and coercive." *Administrative Science Quarterly*, **41**, 61–89.

Brown, S. L., and Eisenhardt, K. M. (1995). "Product development: Past research, present findings, and future directions." *Academy of Management Review*, **20**(2), 343–378.

Clark, K. B., and Fujimoto, T. (1990). "The power of product integrity." *Harvard Business Review*, **68**(6), 107–118.

Chakrabarti, A. K. (1974). "The role of champions in product innovation." *California Management Review*, **17**(2), 58–62.

Christensen, C. M. (2000). *The Innovator's Dilemma*. New York: Harper Business.

Day, D. L. (1994). "Raising radicals: Different processes for championing innovative corporate ventures." *Organization Science*, **5**(2), 148–173.

Dougherty, D., and Hardy, C. (1996). "Sustained product development in large, mature firms: Overcoming innovation-to-organization problems." *Academy of Management Journal*, **39**(5), 1120–1153.

Dougherty, D., and Heller, T. (1994). "The illegitimacy of product innovation in large firms." *Organization Science*, **5**(2), 200–218.

Fiol, C. M. (1995). "Thought worlds colliding: The role of contradiction in corporate innovation processes." *Entrepreneurship Theory and Practice*, **3**, 71–90.

Foster, R., and Kaplan, S. (2001). *Creative Destruction: Why Companies that Are Built to Last Underperform the Market—and How to Successfully Transform Them.* New York: Doubleday.

Frost, P. J., and Egri, C. P. (1991). "The political process of innovation." In: L. L. Cummings and B. M. Staw (Eds.), *Research in Organizational Behavior* (Vol. 13, pp. 229–295). Greenwich, CT: JAI Press.

Gartner, W. B., Bird, B. J., and Starr, J. A. (1992). "Acting as if: Differentiating entrepreneurial from organizational behavior." *Entrepreneurship Theory & Practice*, **3**, 13-31.

Hart, S. L. (2005). *Capitalism at the Crossroads.* Philadelphia, PA: Wharton School Publishing.

Heller, T. (2000). " 'If only we'd known sooner' ": Developing knowledge of organizational changes earlier in the product development process." *IEEE Transactions on Engineering Management*, **47**(3), 335–344.

Heller, T. (1999). "Loosely coupled systems for corporate entrepreneurship: Imagining and managing the innovation project/host organization interface." *Entrepreneurship: Theory and Practice*, Winter, 25–31.

Heller, T. (1993). "Organizing for innovation: The mutual influence of project and organization." Ph.D. dissertation, University of Pennsylvania.

Jelinek, M., and Schoonhoven, C. B. (1990). *The Innovation Marathon: Lessons from High Technology Firms.* Oxford, U.K.: Basil Blackwell.

Kazanjian, R. K., and Drazin, R. (1986). "Implementing manufacturing innovations: Critical choices of structure and staffing roles." *Human Resource Management*, **25**, 385–403.

Kranhold, K. (2007). "Greener pastures: GE's environment push hits business realities"; "CEO's quest to reduce emissions irks clients"; "The battle of the bulbs." *Wall Street Journal* (Eastern edition), September 14, p. A1.

Leonard-Barton, D. (1992). "Core capabilities and core rigidities: A paradox in managing new product development." *Strategic Management Journal*, **13**, 111–125.

Manning, T. (2002). "Strategic conversation as a tool for change." *Strategy & Leadership*, **30**(5), 35–37.

McDonald, H., and London, T. (2002). *Expanding the Playing Field: Nike's World Shoe Project.* Washington, D.C: World Resources Institute.

Meyerson, D., and Martin, J. (1987). "Cultural change: An integration of three different views." *Journal of Management Studies*, **24**, 623–647.

Nonaka, I. (1991). "The knowledge-creating company." *Harvard Business Review*, November/December, 96–104.

Ottman, J. (1999). "Five strategies for stimulating out-of-the- box thinking regarding environmentally preferable products and services." In: M. Charter and J. Polonsky (Eds.), *Greener Marketing.* Sheffield, U.K.: Greenleaf Publishing.

Pernick, R., and Wilder, C. (2007). *The Clean Tech Revolution.* New York: Harper Collins.

Peters, T. J. (1983). "The mythology of innovation, or a skunkworks tale, part II." Reprinted in: M. L. Tushman and W. L. Moore (Eds.), *Readings in the Management of Innovation* (pp. 138–147). Cambridge, MA: Ballinger.

Rice, M. P., O'Connor, G. C., and Pierantozzi, R. (2008). "Implementing a learning plan to counter project uncertainty." *MIT Sloan Management Review*, **49**(2), 54–62.

Schon, D. A. (1963). "Champions for radical new innovations." *Harvard Business Review*, **40**(2), 77–86.

Shane, S. (1994). "Cultural values and the championing process." *Entrepreneurship Theory & Practice*, **4**, 25–41.

Storey, L. (2008). "Can Burt's Bees turn Clorox green?" *New York Times*, January 6.

Strauss, A. L. (1987). *Qualitative Analysis for Social Scientists.* Cambridge, U.K.: Cambridge University Press.

van de Ven, A. (1986). "Central problems in the management of innovation." *Management Science*, **32**(5), 590–607.

Vaughan, D. (1982). "Toward understanding unlawful organizational behavior." *Michigan Law Review*, **80**, 1377–1402.

Weick, K. E. (1976). "Educational organizations as loosely coupled systems." *Administrative Science Quarterly*, **21**, 1–19.

Weick, K. E. (1982). "Management of organizational change among loosely coupled elements." In: P. S. Goodman (Ed.), *Change in Organizations* (pp. 375–408). San Francisco, CA: Jossey-Bass.

38

Promotors and Champions of Innovation: Barriers to Innovation and Innovator Roles

Søren Salomo and Hans Georg Gemünden*[†]

*Danish Technical University and [†]Technical University Berlin

A lot of research and practical effort has been devoted to devising effective formal structures of supporting the innovation activities of firms. Nevertheless, innovation management research has continuously identified more informal mechanisms, which also play an important role in achieving and securing innovation success. Among those informal means dedicated individuals driving the innovation effort of firms have for a long time been recognized as an important factor for innovation success: "There is plenty of reason to suppose that individual talents count for a good deal more than the firm as an organization". The basic idea is that the management of innovation requires persons who commit themselves with enthusiasm and self-motivation to the new-product or new-process idea (Chakrabarti, 1974; Howell and Shea, 2001; Markham, 1998; Rothwell *et al.* 1974; Witte, 1973, 1977). These persons may or may not have been officially assigned to the innovation process. They do, however, show a high personal involvement in the innovative project and nurture the project often in addition to their official organizational position. Such behavior has been summarized in the concept of champion or innovation promotor roles.

The champion concept is, in its simplest form, a monopersonal concept in which the success or failure of the innovation process is attributed to one single person who makes the decisive contributions. Schon (1963, p. 82) describes him as "... a man willing to put himself on the line for an idea of doubtful success. He is willing to fail. But he is capable of using any and every means of informal sales and pressure in order to succeed." Howell, Shea, and Higgins (2005) identify three typical behavioral traits—that is, a champion (1) expresses enthusiasm and confidence, (2) persists under adversity, and (3) gets the right people involved. These traits are correlated with transformational leadership and with the success of new-product development.

More elaborate concepts are based on the assumption that innovations are faced with different kinds of barriers, and that different kinds of power sources and different kinds of contributions are required to overcome these barriers. Specialized innovator roles are defined by specific barriers, specific power sources, and specific value-creating contributions or functions. A two-power source typology was derived and empirically tested by Witte (1973), who advocated a distinction between expert and power promotors. The word "promotor" is derived from the Latin word and should therefore not be confused with the term "promoter" typically used in the sports environment. Witte posits that expert knowledge is needed to overcome the barrier of not knowing and innovation, and that hierarchical power and power over resources is needed to overcome barriers of not wanting an innovation. In a similar vein Rothwell *et al.* (1974) found that a "technical genious" and an executive champion are critical. Based on an extensive literature analysis, Hauschildt and Chakrabarti (1989) further suggest that organizational barriers hinder communication between both role players. They identify internal networkers, labeled "process promotors", who are needed for bringing the technical genius and the executive champion together. The positive impact of such process promotors, when forming a troika structure of champions with experts and senior management champions, was shown by Hauschildt and Kirchmann (2001). Barriers external to the individual organization may also hinder innovations—an issue particularly relevant in times of open innovation. To overcome technology-related barriers of know-what, know-whom, and know-how, Allen (1970) introduced the concept of a gatekeeper. Technology gatekeepters build on extensive internal and external networks, which they use to gather, evaluate, and internally disseminate technology-related information. Market-related barriers are addressed in the relationship promotor concept of Walter and Gemünden (2000).

Promotors and champions overcoming barriers

Five innovator roles

From the perspective of barriers and corresponding "energies" to overcome these, five different promotor roles can be identified:

First is the *power promotor*, who has the necessary hierarchical power to drive the project, to provide needed resources, and to help to overcome obstacles of unwillingness that might arise during the course of the project. The hierarchical position does not only lend organizational power to enforce innovative action and block powerful organizational opposition, but also goes along with strategic oversight. Hence, the power promotor may be able to provide a more long-term perspective to innovation and can prioritize innovation projects against other competing projects. Finally, as an important internally and externally visible representative of the organization and proponent of innovation, the power promotor lends otherwise scarce legitimacy to corporate innovative endeavors.

Second, the *expert promotor*, who carries authority from a functional perspective. As a task-related expert this promotor role can contribute with specific, typically technical knowledge, which is relevant to the innovation task. The ability of the expert promotor to gather and evaluate innovation-related information, to generate and test alternative solutions, and to build expertise in innovation-related domains, allows the barriers of incompetence to be reduced.

Third, the *process promotor*, who derives influence primarily from organizational know-how. The process promotor establishes the connection between the power and the expert promotor and has the necessary diplomatic skills to bring together the needed people for the innovation process (Howell, Shea, and Higgins, 2005). Compared with the two previous innovator roles, the process promotor has a very distinct power base from which to draw when supporting innovation processes. It is neither the formal hierarchical position nor the functional expert authority but a strong social network within the firm on which the process promotor relies. Based on case study evidence, Griffin *et al.* (2009, p. 222) show that this role behavior relies on sound organizational knowledge, which enables those exceptional innovators to "both understand and participate in the politics necessary to gain acceptance of and resources for their project." This also corresponds to the contribution of successful R&D project managers identified by Katz and Allen (1985). They show that project managers concerned with gaining recognition for the innovation project and linking it to other parts of the organization realize higher project performance.

Four different types of characteristics mainly support such process promotor role behavior. (1) Social networks only build over time and hence require the process promotor to be part of the organization for a longer period of time. (2) In order to direct innovative efforts and guide the innovation process, leadership qualities are necessary. Charisma, ability to inspire others, and offering intellectual stimuli are relevant competencies of the process promotor in this context (Shane, 1994). (3) Beyond these characteristics, process promotors exert more informal influence on the people involved and affected by innovation (Howell and Higgins, 1990). They use influence tactics, which include the ability to mobilize cooperative coercion means, coalition building, rational justification, and continuous communication of enthusiasm and an attractive vision of the innovation (Dean, 1987). (4) In terms of personal characteristics, risk affinity, need for achievement, and self-confidence combined with an almost unconditional devotion to innovation seem to be an underlying basis for showing this socially focused role behavior.

Fourth, the *relationship promotor*, who has strong personal ties not only inside but in particular outside the organization (i.e., to customers, suppliers, and research partners). In times of "open innovation" the relationship promotor acts as an ambassador of the organization facilitating cooperation and effective exchange of information with partners beyond the organizational boundaries.

Fifth, the *technological gatekeeper* who primarily acts in the context of R&D tasks (Allen, 1970; Domsch, Gerpott, and Gerpott, 1989; Tushman and Katz, 1980). The technological gatekeeper is mainly active in the research and development area establishing an information and communication exchange network, filtering the information needed, assembling information from internal and external sources, and providing it to relevant members and working groups of the organization itself.

The main difference between these last two roles lies in the fact that technological gatekeepers mainly import technological knowledge and distribute it within their organization. Relationship promotors import market-related knowledge (i.e., customer requirements, exploitation opportunities, and information about the competitive status in the markets).

While relationship role behavior to a large extent mirrors the competences and characteristics of the process promotor, it is also different as it focuses on solving inter-organizational conflicts and aims at overcoming physical and in particular mental distances between partner firms. Thus, the gatekeeper and the relationship promotor also know external partners that engage in cooperative projects, and they have good personal relationships with them. They know how to define, establish, and run cooperative R&D projects, how to create and secure trust, and how to get third-party money for such projects, particularly from public research institutions (Gemünden, Salomo, and Hölzle, 2007). In order to fulfill these tasks this boundary-spanning role (Aldrich and Herker, 1977) builds upon primarily three sources of power (Allen, 1970; Walter and Gemünden, 2000). (1) Social skills (i.e., the ability to perceive and adapt to the complex social situation of interorganizational cooperation). (2) Network knowledge (i.e., information about the formal and informal relationships of the relevant interaction partners of the organization). (3) Portfolio of relationships (i.e., individually well connected both inside and outside the organization in terms of resilient personal relationships to relevant actors).

Whereas the role of top management is stressed in the entrepreneurship and strategic management literature and technology expert roles are recognized in particular in the context of R&D, the empirical research on key innovators stresses much more the role of internal and external networkers. These networking roles are typically rooted in middle management levels of larger organizations, supporting both external and internal lower-level experts in order to help those overcoming bureaucratic barriers and resource-related challenges. Additionally, these networkers may also serve as important advocates of innovation-oriented customers, who are willing to take risks and to fund an innovation in the early stages of its diffusion. Thus, it is not only the single role that matters but the cooperation between different kinds of promotors, champions, and gatekeepers.

Cooperation between promotors

Distinguishing between these innovator roles builds upon the idea of specialization or division of labor. The promotor or champion functions just described correspond to critical barriers in the innovation process. Each of the innovator roles is "specialized" in overcoming these barriers through building upon specific sources of power and competences. While barriers typically emerge as combinations of forces restricting innovation processes, interaction between promotor roles is required to secure efficient and effective innovation (Burgelman, 1983). When innovator roles join forces and cooperate to bring along the innovation task, they increase the probability of successful innovation. Hence, successful innovator role behavior includes role-specific contributions to overcoming barriers and establishing relationships with other innovator roles in order to leverage individual contributions (Kimberly and Evanisko, 1981).

Role combination and role plurality

The idea of individual role behavior suggests that one person obtains one role throughout the innovation process. From the perspective of positive effects through specialization, such "role exclusivity" is appealing. However, empirical evidence points towards multiple "role combinations" and "role plurality" in many innovation projects. In fact, it is inherent in the understanding of behavior from a role perspective that individuals involved in innovation projects may take more than one role. A committed chief technology officer (CTO) with access to resources and hierarchical power may act both as a senior management champion and as an expert promotor. The latter role behavior is possible if the CTO has, for example, expert knowledge in a technical domain relevant to the innovation problem. Rost, Hölzle, and Gemünden (2007) found that an accumulation of roles is helpful when innovativeness is very high, so that learning can take place in a smaller group, particularly in the early stages of the innovation process. This is also supported by the case study findings of Griffin *et al.* (2009), who show that exceptional innovators accumulate role behavior, spanning from technology expert roles to organizational boundary spanners.

"Role plurality" is another issue relevant in many innovation projects. In particular, when confronted with more complex innovation projects, characterized by multiple stakeholders, increased demand for cross-organizational cooperation, and high technological uncertainty (e.g., in highly innovative ventures), many individuals may take upon themselves role-specific behavior. Highly complex technological problems, which require input from diverse technical domains, often require input from different technical experts, which then may exhibit expert promotor role behavior. Hence, many projects see more highly committed people involved, who show role-specific behavior, resulting in multiple and parallel enacted promotor or champion roles.

Dynamics of promotor roles

Additionally, promotor role constellations in innovation projects are not a static issue. Role-taking by individuals may not be stable over a project's duration. In many instances, individuals who show role-specific behavior in one project stage may not be able to provide the same input in later stages. Typically, as projects progress, different competencies, networks, and resources are required. As promotors offer a very specific contribution to innovation projects based on distinct and above-average competencies and power bases, they also stand a good chance that these competencies are only fit for solving specific problems. Alterations of problem structures resulting from project progress will encourage other committed individuals to engage in promotor role behavior, supplementing or even making existing promotor roles obsolete.

Limitations of the promotor concept

Although understanding the critical input of dedicated individuals in innovation projects from a role behavior perspective is conceptually appealing and meta-analytic research lends strong empirical support to the suggested positive performance effects of these individuals, some limitations have to be noted.

The promotor or champion role concept builds on an underlying assumption that every innovation is necessary, useful, and better than the *status quo*, and, hence merits to be implemented against opposition. As with many other advocated innovation management approaches this is a biased perspective. Promotors may also support and drive ultimately wrong projects (Markham, 2000). Contrary to formal means of supporting innovation projects, like formalized process models, steering groups, etc., informal promotor activity lacks aspects of institutionalized rational decision making and thus runs a higher risk of sustaining projects that warrant termination (Gemünden and Lechler, 1996; Gemünden, Salomo, and Hölzle, 2007). Power promotor role behavior appears to be most prone to these negative effects. On the product portfolio level Ernst (2001) showed an inverted U-shaped relationship between top management support and profitability, because in case of increased resource commitment the additional projects do not cover the cost of capital. Scholl, Hoffmann, and Gierschner (2004) document further defects of power promotor involvement, which result from de-motivation, if some preferred people get support and others do not. The often used naive statement that top management support is a "success factor" thus needs to be qualified (Markham and Griffin, 1998). In fact, recent research shows top management commitment to be a "two-sided sword". Organizations exhibiting senior management involvement in innovation in terms of active power promotors are able to establish a favorable behavioral environment for innovation activity as innovation receives legitimacy. At the same time, interference from top management in day-to-day business and the details of new-product development seems to impact innovation performance negatively (Kleinschmidt, de Brentani, and Salomo, 2007).

Performance limitations are adequately attributed to the role behavior of power promotors but may also emerge from another central characteristic of the promotor model. In order to leverage individual promotor contributions and to address interlinked sources of opposition it makes good sense to require promotor interaction. However, very cohesive promotor groups, exhibiting structural deficits due to their *ad hoc* informal character and operating under a "provocative context" determined by strong uncertainties and unclear objectives inherent in innovation tasks, may exhibit group-think effects. Such effects might lock the group to a specific but not necessarily optimal solution to the innovation problem.

Finally, arguing for the positive performance effects of promotor role behavior needs to take a contingency perspective. Promotor effects may in particular vary with degree of newness of the innovative task (Calantone, Chan, and Cui, 2006; Danneels and Kleinschmidt, 2001; Garcia and Calantone 2002; Gemünden, Salomo, and Hölzle, 2007). Empirical evidence is rather scarce, but data from highly innovative ventures suggest that in case of very strong technological uncertainty the "classical" troika constellation of power, expert, and process promotor does not improve performance (Gemünden, Salomo, and Hölzle, 2007). As externally available technological competence gains importance for project success, contributions from relationship promotors become a more valuable resource. Interestingly, in these situations the relationship promotor is best supported by a formally assigned experienced project manager. This indicates that informal innovator roles play a critical part in tackling innovation uncertainty and barriers to successful project execution. But this informal role behavior needs to be supplemented with formal instruments of project execution.

References and further reading

Aldrich, H., and Herker, D. (1977). "Boundary spanning roles and organization structure." *Academy of Management Review*, **2**(2), 217–230.

Allen, T. J. (1970). "Communication networks in R&D laboratories." *R&D Management*, **1**(1), 14–21.

Burgelman, R. A. (1983). "A process model of internal corporate venturing in the diversified major firm." *Administrative Science Quarterly*, **28**, 223–244.

Calantone, R., Chan, K., and Cui, A. S. (2006). "Decomposing product innovativeness and its effects on new product success." *Journal of Product Innovation Management*, **23**(5), 408–421.

Chakrabarti, A. K. (1974). "The role of champion in product innovation." *California Management Review*, **17**(2), 58–62.

Danneels, E., and Kleinschmidt, E. J. (2001). "Product innovativeness from the firm's perspective: Its dimensions and their relation with project selection and performance." *Journal of Product Innovation Management*, **18**(6), 357–373.

Dean, J. W. (1987). "Building the future: The justification process for new technology." In: J. M. Pennings and A. Buitendam (Eds.), *New Technology as Organizational Innovation* (pp. 35–58). Cambridge, MA: Ballinger.

Domsch, M., Gerpott, H., and Gerpott, T. J. (1989). *Technologische Gatekeeper in der industriellen F&E*. Stuttgart, Germany: Poeschel.

Ernst, H. (2001). *Erfolgsfaktoren neuer Produkte : Grundlagen für eine valide empirische Forschung*. Wiesbaden, Germany: DUV.

Folkerts, L., and Hauschildt, J. (2002). "Personelle Dynamik in Innovationsprozessen: neue Fragen und Befunde zum Promotorenmodell." *Die Betriebswirtschaft*, **62**, 1–23.

Garcia, R. and Calantone, R. (2002). "A critical look at technological innovation typology and innovativeness terminology: A literature review." *Journal of Product Innovation Management*, **19**(2), 110–132.

Gemünden, H. G. (1985). "'Promotors': Key persons for the development and marketing of innovative industrial products." In: K. Backhaus and D. Wilson (Eds.), *Industrial Marketing. A German–American Perspective* (pp. 134–166). Berlin: Springer.

Gemünden, H. G., and Lechler, Th. (1996). "Der bewußte Projektabbruch—ein verborgener Erfolgsfaktor." In: A. Schulz and C. Pfister (Eds.), *Strukturwandel mit Projektmanagement: Tagungsband Projektmanagement-Forum '96, Essen, October 25–29* (pp. 351–359). Munich, Germany: GPM Deutsche Gesellschaft für Projektmanagement.

Gemünden, H. G., Salomo, S., and Hölzle, K. (2007). "Role models for radical innovations in times of open innovation. *Creativity and Innovation Management*, **16**, 408–421.

Gemünden, H. G., and Walter, A. (1998). "The relationship promoter: Motivator and co-ordinator for inter-organizational innovation co-operation." In: H. G. Gemünden, Th. Ritter, and A. Walter (Eds.), *Relationships and Networks in International Markets* (pp. 180-197). Devon, U.K.: Pergamon Press/Elsevier Science.

Gilbert, C. G. (2005). "Unbundling the structure of inertia: Resources versus routine rigidity." *Academy of Management Journal*, **48**, 741–763.

Griffin, A., Price, R. L., Maloney, M. M., Vojak, B. A., and Sim, E. W. (2009). "Voices from the field: How exceptional electronic industrial innovators innovate." *Journal of Product Innovation Management*, **26**, 222–240.

Hauschildt, J., and Chakrabarti, A. (1989). "Division of labour in innovation management." *R&D Management*, **19**(2), 161–171.

Hauschildt, J., and Kirchmann, E. (2001). "Teamwork for innovation: The 'Troika' of Promotors." *R&D Management*, **31**(1), 41–49.

Hauschildt, J., and Salomo, S. (2007). *Innovationsmanagement* (Fourth Edition). Munich, Germany: Vahlen.

Howell, J. M., and Higgins, C. A. (1990). "Champions of technological innovation." *Administrative Science Quarterly*, **35**, 317–341.

Howell, J. M. and Shea, C. M. (2001). "Individual differences, environmental scanning, innovation framing, and champion behaviour: Key predictors of project performance. *Journal of Product Innovation Management*, **18**(1), 15–27.

Howell, J. M., Shea, C. M., and Higgins, C. A. (2005). "Champions of product innovations: Defining, developing, and validating a measure of champion behavior." *Journal of Business Venturing*, **20**(5), 641–661.

Katz, R., and Allen, T. (1985). "Project performance and the locus of influence in the R&D matrix." *Academy of Management Journal*, **28**(1), 67–87.

Kimberly, J. R., and Evanisko, M. J. (1981). "Organizational innovation: The influence of individual, organizational, and contextual factors on hospital adoption of technological and administrative innovations." *Academy of Management Journal*, **24**, 689–713.

Kleinschmidt, E. J., de Brentani, U., and Salomo, S. (2007). "Performance of global new product development programs: A resource-based view." *Journal of Product Innovation Management*, **24**, 419–441.

Leifer, R., McDermott, C. M., O'Connor, G. C., Peters, L. S., Rice, M. P., and Veryzer, R. W. (2000). *Radical Innovation: How Mature Companies Can Outsmart Upstarts*. Boston, MA: Harvard Business School Press.

Markham, S. K. (1998). "A longitudinal examination of how champions influence others to support their projects." *Journal of Product Innovation Management*, **15**(6), 490–504.

Markham, S. K. (2000). "Corporate championing and antagonism as forms of political behavior: An R&D perspective." *Organization Science*, **11**(4), 429–447.

Markham, S. K., and A. Griffin (1998). "The breakfast of champions: Associations between champions and product development environments, practices and performance." *Journal of Product Innovation Management*, **15**(5), 436–454.

Mirow, Ch., Hölzle, K., and Gemünden, H. G. (2007). "Systematisierung, Erklärungsbeiträge und Effekte von Innovationsbarrieren." *Journal für Betriebswirtschaft*, **57**, 101–134.

Moss Kanter, R. (1983). *The Change Masters: Innovation and Entrepreneurship in the American Corporation.*

Pinto, J. K., and Slevin, D. P. (1989). "The project champion: Key to implementation success." *Project Management Journal*, **20**(4), 15–20.

Rost, K., Hölzle, K., and Gemünden, H. G. (2007). "Promotors or champions? Pros and cons of role specialization for economic progress." *Schmalenbach Business Review*, **59**, 340–363.

Rothwell, R., Freeman, C., Horlesey, A., Jervis, V. T. P., Robertson, A. B., and Townsend, J. (1974). "SAPPHO updated: Project SAPPHO phase II." *Research Policy*, **3**(3), 258–291.

Rubenstein, A. H., and Schroder, H. (1977). "Managerial differences in assessing probabilities of technical success for R&D projects." *Management Science*, **24**, 137–145.

Scholl, W., Hoffmann, L., and Gierschner, H. C. (2004). *Innovation und Information—Wie in Unternehmen neues Wissen produziert wird*. Göttingen, Germany: Hogrefe-Verlag.

Schon, D. A. (1963). "Champions for radical new inventions." *Harvard Business Review*, **41**(2), 77–86.

Shane, S. A. (1994). "Are champions different from non-champions?" *Journal of Business Venturing*, **9**(5), 397–421.

Staw, B. M., Sunderlands, L. E., and Dutton, J. E. (1981). "Threat–rigidity–effects in organizational behavior: A multilevel analysis." *Administrative Science Quarterly*, **26**, 501–524.

Tushman, M. L., and Katz, R. (1980). "External communication and project performance: An investigation into the role of gatekeepers." *Management Science*, **26**(11), 1071–1085.

Tushman, M. L. (1977). "Special boundary roles in the innovation process." *Administrative Science Quarterly*, **22**(3), 587–605.

Walter, A., and Gemünden, H. G. (2000). "Bridging the gap between suppliers and customers through relationship promoters: A theoretical and empirical analysis." *Journal of Business & Industrial Marketing*, **15**(2/3), 86–105.

Witte, E. (1973). *Organisation für Innovationsentscheidungen: Das Promotoren-Modell*. Göttingen, Germany: Vahlen.

Witte, E. (1977). "Power and innovation: A two-center theory." *International Studies of Management and Organization*, **7**(1), 47–70.

Part Six

National Innovation Systems

39

Innovation at the National Level

Lois Peters, V. K. Narayanan,† Gina C. O'Connor,* and Mark Tribbitt†*

*Rensselaer Polytechnic Institute and †Drexel University

Innovation at the industry and firm levels of analysis is influenced—both facilitated and hindered—by developments taking place at the level of nation states. In this part of the book we will illustrate national-level influences through a series of case studies of a select set of nations. In the academic literature this level is typically captured by the concept of a national innovation system. Indeed, this literature emphasizes that national innovation systems are influenced by factors such as political philosophy, government support, and cultural factors. Accordingly, we have requested the contributors to this part of the book to follow a common structure as they describe the national innovation systems for each of their respective countries. The structure, provided in Table 39.1, is based on the broad themes that writers have found to influence innovation at the national level. We hope that readers will find the common structure easy to follow and that it aids readers in drawing comparisons across the cases.

To place the discussion of innovation at the national level, this introduction will unfold in two sections. First, we will discuss the notion of national innovation systems; and, second, we will provide a comparative summary of the featured case studies to enable the reader to place a specific country in its proper place.

Table 39.1. Structure for sample case entries: innovation at the national level.

I. A brief history and description of the specific NIS from the national point of view.
   a. when and how it was initiated;
   b. technology trajectories of current and recent importance;
   c. current institutional structure and evolution thereof;
   d. public and private institutions and their linkages.

II. Summary of output trends (R&D expenditures, patents, etc.).

III. Technology commercialization initiatives at the national level.

IV. National technology policy including current technology commercialization initiatives at the national level.

V. Relative strengths and weaknesses in technology.

VI. Funds flow for innovation.

VII. Cultural and political drivers influencing the national innovation system.

VIII. Educational system: level of general literacy as well as specific technical expertise.

IX. Openness and mobility in terms of trade, IP policy with other countries and within the country.

National innovation systems

The concept of national innovation systems (NISs) arose in the context of debates over international competitiveness. Its origins lie with a group of individuals who highlighted the shortcomings of mainstream economics in explaining the relationship between technical change and economic growth. They pointed out that social, political, and historical factors must be taken into account in macro-explanations of innovation. As such it is closely associated with the perspective of evolutionary economics (Nelson and Winter, 1982).

Since the beginning of the 1980s innovation systems and their importance for economic growth have increasingly captured the interest of researchers and policy-makers interested in understanding the dynamics of innovation on a national level. The phrase NIS was reportedly introduced in 1982 by Freeman in a report for the OECD that was later published in 2004. Nelson's two books (1982, 1984) laid the foundation for his cross-country comparison of national innovation systems (1993). Lundvall (1992), however, is generally recognized as the first to use the expression.

NIS focuses researchers on institutions, organizations, and linkages among them. There have been multiple and shifting interpretations of the NIS concept. It is variously defined as a system for generating and diffusing new technologies (Sharif, 2006), or as a tool to identify key linkages in the system of technical institutions and other organizations to create and manage innovation (Groenewegen and van der Steen, 2006; Guinet, 1995).

Evolution of NIS scholarship

Over time the NIS or systems approach to innovation has encompassed description about innovation capacity, innovation performance differences among nations, and diversity in institutional innovation structure. Early studies aimed at revealing country-specific innovation patterns (e.g., Nelson, 1993). Later on, policy-driven studies, as well as attempts to formalize the NIS concept, led to the identification of key indicators that allowed analytic comparisons across countries.

A recent trend among scholars is to go beyond comparison of highly industrialized nations' NISs and look at the NISs of newly industrialized or developing economies. This has led to a debate as to whether the NIS concept is applicable across all nations. The conflict raises the issue of how the concept of NIS is defined. In the case of those who say the concept applies to all nations, the definition includes the generation and diffusion of innovation. The more narrow definition of generating technical change leads to the claim that it is likely that many countries do not have an NIS. Park and Park (2003), for example, using a substantial amount of data on the structure of production and structure of R&D conclude that only when national R&D constitutes more than 2% of GDP can one identify a system of innovation. In general the debate highlights issues related to stage of economic development, the evolution of national innovation systems, and the importance of understanding dynamics. Examples of countries studied employing the perspective of NIS include Germany (Kaiser and Prange, 2004), the U.K., Italy (Visintin *et al.*, 2005), the U.S., France (Walsh and Le Roux, 2004), Latin America and the Carribbean (Alcorta and Peres, 1998), Ireland (Calliano and Carpano, 2000), South Korea (Chung, 2002), Russia (Deo Bardhan and Kroll, 2006), China (Liu and White, 2001), Argentina, Brazil, Mexico (Katz, 1987), India (Mehra, 2001), and Uganda (Wafula and Clark, 2005).

There is also disagreement among scholars and policymakers as to the importance of identifying the "national unit" as the delimiting criterion of an innovation system, as opposed to choosing the regional, sectoral, technological, or industrial system as the objective of study. Early on, the word nation was used in the development of the innovation system perspective because an objective of NISs was to bring into question what neoclassical economics had to say about macroeconomic (national level) theory and economic growth. Freeman states that the "national" domain better accommodates the policy dimension of the concept. Scholars in this camp argue that, even though regional, sectoral, or technological systems often transcend a country's borders, national characteristics and frameworks always have a role to play in shaping the system in question (Castellacci, 2009). Advocates of the opposing view argue that many interesting interactions occur across national borders in the context of modern innovation, particularly in an era of multinational companies (Sharif, 2006). Studies following a regional innovation systems perspective include those by Pinto (2009), Archibugi, Howells, and Michie, (1999), Cooke (2005), Morgan (2004), and Gerstlberger (2004). Those adopting a sectoral perspective include Chang and Shih (2005), Malerba (2005), Kobos, Erickson, and Drennen (2006), and Iammarino (2005). Scholars taking a technological perspective include Bartholomew (1997), Marsh (2003), Carlsson (1991), and Carlsson and Stankiewicz (1997). Finally, Dekkers (2009), Bossink (2009), and Guillou *et al.* (2009) represent those taking an industry perspective. The debate continues about whether or not a general systems approach may become more important than the national focus in innovation studies as questions arise about economic divergence within nations or potential economic integration across nations.

While many authors have considered the role of institutions in the innovation system, only a few authors have considered the organizational effect of these innovation systems on firms (Coriat and Weinstein, 2002; Kobos, Erickson, and Drennen, 2006; Senker, 1996; Teitel, 2006). Most studies focus on the importance of technological development and innovation, competitiveness, institutions, economic aspects, and policy but very few look at the NIS from the point of view of the firm (de la Garza Martinez, 2008 is a notable exception). There is opportunity for research in the field of innovation systems to fill this gap, especially from the point of view of multination firms in less developed counties.

Theoretical foundations of the NIS concept

An ongoing debate in the literature concerns whether the NIS concept should remain flexible and loose or should become more theoretically rigorous with formally operationalized variables. Descriptive NIS models focus on activities such as research, production, end use, linkages and education (see, e.g., Liu and White, 2001), or on structural specifics related to R&D expenditures, R&D performance, technology policy, human capital development, technology transfer, and climate for entrepreneurial behavior (e.g., Chang and Shih, 2005). An example of a more formalized approach is that by Furman, Porter, and Stern (2002), who investigate innovation capacity, which they define as a country's ability to produce and commercialize a flow of innovative technology over the long term. Based on three components—infrastructure, clusters, and linkages—they compared the innovative capacity of 17 countries. These varied methodological approaches can be construed as a weakness or absence of formalized methodology which in turn stems from lack of agreement about the fundamental principles or theoretical foundations of NIS.

NIS theoretical foundations have been associated with systems theory and evolutionary theory (Balzat and Hanusch, 2004). While it has been repeatedly proposed that evolutionary economic theory constitutes the theoretical foundation of NISs (Balzat and Hanusch, 2004; Saviotti, 1997; McKelvey, 1997; Lundvall, 2007), the system perspective is also at the heart of the concept. With the exception of Andersen, Metcalfe, and Tether (2000), however, the relationship between systems theory and the NIS approach has not been investigated.

Works such as that of Furman, Porter, and Stern (2002) tie the approach to industrial organization frameworks. But this has been criticized because it does not capture the dynamics fundamental to the NIS concept, and capacity comparisons can take on the character of a benchmark exercise.

Several studies examine the relation of the NIS concept with neoclassical, institutional economic theories and new-growth theories. Nelson and Nelson (2002) relate institutional analysis and evolutionary economic theory in the study of NIS. Groenewegen and van der Steen (2006) propose going beyond just identifying important institutions and investigating the interactions among the hierarchies of institutions. They argue that the NIS concept should be enhanced with elements of new and old institutional economics perspectives such as shared mental maps and interest group power. Nevertheless if ties to traditional economic theory are fully taken to heart, researchers are turning their backs on the original premise of the NIS concept (i.e., that mainstream economics is inadequate in explaining innovation and economic growth because it focuses on the wrong abstractions). Eparvier (2005) further suggests that the NIS concept is not very useful for neoclassical economic theories because non-optimality and radical uncertainty are not compatible with mainstream economics but are central to the NIS concept.

Edquist (2005, p. 186) takes a more contrary perspective in suggesting that the NIS is not a theoretical concept. Lundvall (2007), in contrast, proposes that the theory behind NISs is grounded theory informed by critical theory. He bases his argument on the fact that a long list of empirical studies show that innovation is an interactive process and that NISs arose out of criticism of national economic policies. Critical theory is grounded theory applied in contexts marked by social conflict.

Most scholars do agree that the NIS is a focusing device. Whether or not it is viewed as a theory has to do with one's perspective on theory being formal as opposed to being more loosely construed.

Impact on policy making

The NIS leads to innovation-enhancing policies as opposed to a focus on income and currency devaluation as a means to foster competitiveness. Freeman's (1982) book on Japan inspired the value of the NIS perspective in helping governments develop economic "catching up" policies. Indeed, NIS thinking has led to a structurally different view of how government can stimulate the innovation performance of a country. It shifts focus away from labor and wages and draws policy-makers' attention to "getting the institutional environment right." Its system-thinking point of view has played a significant role in pointing policy-makers away from the linear or pipeline model of innovation.

NIS policy studies aim to describe and compare the most important institutions, organizations, activities, and interactions of public and private actors that take part in or influence the innovation process of a country (Groenewegen and van der Steen, 2006). The growing number of policy-oriented studies of innovation systems signals that the creation of an innovation-enhancing framework has become a central target of policy-makers around the globe in both highly industrialized and industrializing nations (Balzat and Hanusch, 2004; Kleinknecht, 2000; de la Garza Martinez, 2008). The NIS perspective has had a positive impact in terms of what policy-makers and economists ascribe to the nature of "international competitiveness" by drawing attention to strategies that include positive sum gains. More recently it has been used to foster a movement from science and technology policy towards innovation policy (Nuur, Gustavsson, and Laestadius, 2009).

According to Lundvall (2007) the wider implication of the innovation and learning perspective of NIS for general economic policy needs to be more seriously studied. There is a need for further study of the NIS approach in linking innovation systems directly to growth and determining the link between other economic sub-systems (e.g., regional, sectoral, and technological) and national systems. In particular, the dynamic aspects of NISs need to be more closely studied. For example, without an understanding of innovation dynamics and path plasticity it is possible to miss how Japan became a leader in the global software game industry (Storz, 2008). In the future, as we focus more on developing nations, there will have to be attention paid to building effective innovation systems as opposed to analyzing them to determine appropriate policy. Finally, there is still a need to study innovation systems from the perspective of the firm and its strategic management.

A comparison of the featured case studies of nations

The case studies of innovation at the national level featured in the ensuing entries highlight both the uniqueness of the socioeconomic context and the trajectory of technology development, in spite of the attempts to discus the topic within a common framework (as displayed in Table 39.1).

Although innovation is difficult to quantify, the 14 countries represented in the case studies evidence significant differences in innovation output, measured by patents either on a *per capita* basis or as a percentage of GDP (see Tables 39.2 and 39.3). Not surprisingly, these data show that the United States, Japan, and Australia have the largest number of patents compared with their respective population levels (per million residents); Finland, Germany, Singapore, and The Netherlands follow these countries while emerging countries such as Russia, India, and China represent the smallest number of patents versus population. Notably these are three of the four countries thought to represent the highest level of growth over the next 50 years (Brazil, Russia, India, China). We see a similar breakdown when analyzing patent generation based on the level of gross national product in each country.

Table 39.2. Patents per million residents.

	Patent data by population (millions)			
	2000	*2005*	*2006*	*2007*
Germany	131.663	116.103	132.302	121.697
Ireland	35.958	40.625	46.479	36.842
Australia	225.722	248.077	361.033	353.547
Japan	259.494	249.151	308.477	281.295
India	0.129	0.368	0.456	0.514
Italy	34.539	27.146	32.219	30.925
Netherlands	88.512	73.529	100.734	97.436
Singapore	60.050	88.290	106.591	98.257
U.S.	343.779	279.101	342.305	310.623
Russia	1.265	1.076	1.235	1.358
China	0.128	0.433	0.740	0.937
Finland	125.290	143.048	190.702	178.261
Taiwan	NA	NA	NA	NA
U.K.	69.485	59.107	71.436	66.082

Sources: World Development Indicators database, April 2009; U.S. Patent and Trademark Office, 2009.

Table 39.3. Patents as a proportion of GDP

	Patent data by GDP (billions)			
	2000	*2005*	*2006*	*2007*
Germany	5.696	3.430	3.741	3.018
Ireland	1.421	0.841	0.903	0.622
Australia	8.922	5.138	7.017	5.965
Japan	7.054	6.993	9.006	8.198
India	0.285	0.497	0.553	0.491
Italy	1.793	0.896	1.023	0.874
Netherlands	3.662	1.896	2.457	2.084
Singapore	2.610	3.147	3.434	2.795
U.S.	9.935	6.673	7.787	6.813
Russia	0.712	0.201	0.178	0.150
China	0.134	0.253	0.365	0.385
Finland	5.326	3.838	4.796	3.854
Taiwan	NA	NA	NA	NA
U.K.	2.820	1.585	1.802	1.454

Sources: World Development Indicators database, April 2009; U.S. Patent and Trademark Office, 2009.

In addition to the differences in some visible indicators of output, we observe differences in innovation approaches among these countries with respect to three phenomena: (1) government influences, (2) technology diversification, and (3) socio-cultural influences.

Government influences

All of the countries represented in this work receive large amounts of support from their governments in fostering innovation. Yet, on a relative basis, the countries within the set are positioned along a continuum ranging from strong to moderate to minimal government support. China and India are representative of countries that consistently receive a great deal of support from their governments in fostering innovation. On the other hand, Italy and The Netherlands receive smaller amounts.

Government influences on innovation may be summarized along three factors: (1) structure and oversight imposed by the government, (2) tax incentives for channeling innovation efforts, and (3), direct investment in technologies within specific industries.

Structure and oversight. We have seen different forms of oversight and structure given to innovation in different countries.

- In the case of China, a major injection of Chinese government support began in the early 1950s and continues today through a series of 5-year plans established by the Chinese government to focus on various aspects of technological innovation. Additionally, a number of government organizations have been established to foster innovation. For example, the Ministry of Science and Technology has initiated a number of large-scale national research programs that are aimed at building the innovative capacity within China (Hu and Mathews, 2009).
- India has developed a number of state-run organizations that foster innovation, such as the Technopreneur Promotion Program (TePP), the Home Grown Technologies Program (HGT), and the Technology Development Board (TDB).
- Italy has made significant strides in improving its innovation at the national level through the development of government agencies like the CNR and ENEA, its primary support has come from local and regional governments (Virgillito, 2009).
- In The Netherlands there seems to be less government-structured innovation support; with the establishment of its second innovation platform, the Dutch Prime Minister has stepped in to chair the program in an effort to bolster innovation within the nation (van der Duin, 2008).

Tax incentives. In each of the outlined cases government has supported innovation by providing tax incentives. For example, the Indian government has instituted a number of financial incentives to those who pursue innovation. Some of these incentives include a 100% tax deduction on both revenue and capital expenditure, a weighted tax deduction of 125% for sponsored research in approved national laboratories, and a weighted tax deduction of 150% on government approved in-house R&D centers (Krishnan, 2010).

Direct investment in industries. In each of the outlined cases government has also lent their support by investing in technologies within specific industries. In China, the first plan was designed to improve "heavy industries" such as steel, coal, iron, and cement production with subsequent plans making incremental changes as existing technologies are improved and new technologies emerge. Most countries have made significant investments in their respective defense programs. In addition to defense, the United States has played a major role in developing its national space program (Simons and Wall, 2008). Other countries such as Germany, Japan, and China have focused investment in medium to high technologies like automotive and mechanical engineering, while Finland has invested funds in information and communications technologies and has lent financial support to the Internet security and virtual gaming industries (Allen, 2008; Hu and Mathews, 2009; Lovio and Välikangas, 2009; Watanabe and Fukuda, 2007). Additionally, almost all of the countries within this analysis have invested in biotechnology and pharmaceutical businesses although to varying degrees.

Diversity of technologies

The level of government funding has led to a wide array of technologies across this mix of countries. We can discuss these differences in terms of technology focus and technology maturity.

Technology focus. Some nations have focused more effort on fostering a narrow range of technologies while others have spent time developing more broad ranges. Ireland, Australia, and Italy are representative of countries that have kept their focus narrow (Cunningham and Golden, 2009; Kemmis and Matthews, 2009; Virgillito, 2009), while the United States, Singapore, Japan, and Taiwan demonstrate a more broad-based technological focus (Koh and Phan, 2009; Lu and Wann, 2009; Simons and Wall, 2008; Watanabe and Fukuda, 2007).

Technology maturity. In addition to the breadth of technological focus within the innovation system, countries determine whether to invest in developing mature or emergent technologies. Countries such as Australia, China, and Taiwan toil in more mature technologies such as manufacturing, agriculture, and consumer goods (Hu and Mathews, 2009; Lu and Wann, 2009; Kemmis and Matthews, 2009), whereas Singapore, the United States, Germany, Japan, and Ireland spend significant time developing emerging technologies such as pharmaceuticals, biotechnology, engineering, and information/communications technologies (Allen, 2008; Cunningham and Golden, 2009; Koh and Phan, 2009; Simons and Wall, 2008; Watanabe and Fukuda, 2007).

A mapping of the countries with entries in this section is given in Figure 39.1 and is based on the breadth of their technological focus and the nature of the technologies developed.

Socio-cultural factors

Government support and area of technological focus are just two of the components that contribute to the national

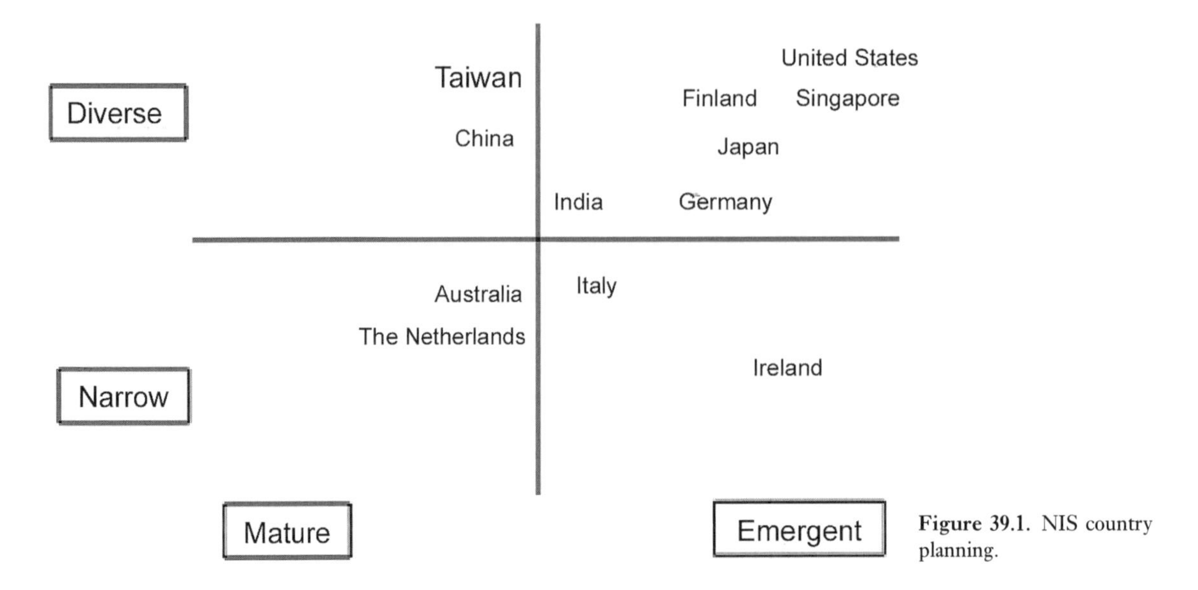

Figure 39.1. NIS country planning.

innovation of a particular country. Although not all-inclusive, other factors contributing to a nation's innovation are its level of education, its ability to commercialize innovation, and its overall social and cultural climate.

Education. In general, most of the countries represented acknowledge shortages in properly trained research personnel. In response to this issue most have worked at developing their respective university systems, not only to develop more technological capabilities, but to also better integrate both the academic and business communities. It is hoped that this integration will better enable the commercialization of innovation, something that has been successfully done within the United States (Simons and Wall, 2009).

Social and cultural climate. The longstanding cultures and social interactions within each of these countries have played a major role in influencing national innovation. For example, in The Netherlands the culture has been one which has fostered individualism. While many would anticipate this stance lending itself to an environment of entrepreneurialism it, in fact, has had the opposite effect leading to an environment where no one identifies very much with their country and where people who try to "stand up" are not encouraged (van der Duin, 2009). In Japan, a strong sense of tradition and culture has led to limited acceptance of foreign ingenuity. This approach has fostered a "not invented here syndrome" which might impact the development and enhancement of new technologies that arise (Watanabe and Fukuda, 2009). Additionally, within its borders Italy has seen significant regional focus, north versus south, which has caused the development of regional technology that may not be as efficient as necessary to improve innovation at the national level (Virgillito, 2009). These aspects of each country's culture can contribute in the development of both capabilities and rigidities and as such should be specifically addressed as they relate to fostering country innovation.

Conclusion

In this brief review we have introduced the reader to the concept of innovation systems at the national level, and provided a very brief comparison of those countries represented in this part of the encyclopedia. We leave it to the reader to draw further comparisons by examining each chapter. The contributing authors have drawn upon their expertise and have all written rich, interesting chapters that all scholars should find worthwhile.

References

Alcorta, L., and Peres, W. (1998). "Innovation systems and technological specialization in Latin America and the Caribbean." *Research Policy*, **26**(7/8): 857.

Allen, M. (2010). "The National Innovation System in Germany." In V. K. Narayanan and G. C. O'Connor (Eds.), *Encyclopedia of Technology and Innovation Management* (pp. 375–389). Chichester, U.K.: John Wiley & Sons [this book].

Andersen, B., Metcalfe, J. S., and Tether. B. (2000). "Distributed innovation systems and instituted economic processes." In: J. S. Metcalfe and I. Miles (Eds.), *Innovation Systems in the Service Economy: Measurement and Case Study Analysis.* Boston, MA: Kluwer.

Archibugi, D., Howells, J., and Michie, J. (Eds.). (1999). *Innovation Policy in a Global Economy*. Cambridge, U.K.: Cambridge University Press.

Balzat, M., and Hanusch, H. (2004). "Recent trends in the research on national innovation systems." *Journal of Evolutionary Economics*, **14**(2), 197.

Bartholomew, S. (1997). "National systems of biotechnology innovation: Complex interdependence in the global system." *Journal of International Business Studies*, **28**(2), 241.

Bossink, B. A. G. (2009). "Assessment of a national system of sustainable innovation in residential construction: Acase study from The Netherlands." *International Journal of Environmental Technology and Management*, **10**(3/4), 371.

Calliano, R., and Carpano, C. (2000). "National systems of technological innovation, FDI and economic growth: The case of Ireland." *Multinational Business Review*, **8**(2), 16.

Carlsson B. (Ed.). (1997). *Technological Systems and Industrial Dynamics*. Dordrecht, The Netherlands: Kluwer.

Carlsson, B., and Stankiewicz, R. (1991). "On the nature, function and composition of technological systems." *Journal of Evolutionary Economics*, **1**, 93–118.

Castellacci, F. (2009). "The interactions between national systems and sectoral patterns of innovation: A cross-country analysis of Pavitt's taxonomy." *Journal of Evolutionary Economics*, **19**(3), 321–347.

Chang, P.-L., and Shih, H.-Y. (2005). "Comparing patterns of intersectoral innovation diffusion in Taiwan and China: A network analysis." *Technovation*, **25**(2), 155.

Chung, S. (2002). "Building a national innovation system through regional innovation systems." *Technovation*, **22**(8), 485.

Cooke, P. (2005). "Regionally asymmetric knowledge capabilities and open innovation exploring 'Globalisation 2': A new model of industry organisation." *Research Policy*, **34**(8), 1128.

Coriat, B., and Weinstein, O. (2002). "Organizations, firms and institutions in the generation of innovation." *Research Policy*, **31**(2), 273.

Cunningham, J. A., and Golden, W. (2010). "The National Innovation System in Ireland." In V. K. Narayanan and G. C. O'Connor (Eds.), *Encyclopedia of Technology and Innovation Management* (pp. 431–443). Chichester, U.K.: John Wiley & Sons [this book].

Dekkers, R. (2009). "Endogenous innovation in China: The case of the printer industry." *Asia Pacific Business Review*, **15**(2), 243.

de la Garza Martinez, P. R. (2008). "Innovation projects by multinational companies in developing countries: The case of Mexico." PhD thesis, Rensselaer Polytechnic Institute, U.S.A.

Deo Bardhan, A., and Kroll, C. A. (2006). "Competitiveness and an emerging sector: The Russian software industry and its global linkages." *Industry and Innovation*, **13**(1), 69.

Edquist, C. (Ed). (1997). *Systems of Innovation: Technologies, Institutions and Organizations*. London: Pinter.

Edquist, C. (2005). "Systems of innovation: Perspectives and challenges." In: J. Fagerberg, D. Mowery, and R. R. Nelson (Eds), *The Oxford Handbook of Innovation*. Oxford, U.K.: Oxford University Press.

Eparvier, P. (2005). "Methods of evolutionism and rivalry with neoclassical analysis: The example of the National System of Innovation concept." *Journal of Economic Methodology*, **12**(4), 563–579.

Freeman, C. (1982). *The Economics of Industrial Innovation* (Second Edition, 250 pp.). London: Francis Pinter.

Freeman, C. (1995). "The 'National System of Innovation' in historical perspective." *Cambridge Journal of Economics*, **19**(1), 5–24.

Freeman, C. (2004). "Technological infrastructure and international competitiveness." *Industrial and Corporate Change*, **13**(3), 541.

Furman, J. L., Porter, M. E., and Stern, S. (2002). "The determinants of national innovative capacity." *Research Policy*, **31**(6), 899.

Gerstlberger, W. (2004). 'Regional innovation systems and sustainability: Selected examples of international discussion." *Technovation*, **24**(9), 749.

Groenewegen, J., and van der Steen, M. (2006). "The evolution of national innovation systems." *Journal of Economic Issues*, **40**(2), 277.

Guillou, S., Lazaric, N., Longhi, C., and Rochhia, S. (2009). "The French defense industry in the knowledge management era: A historical overview and evidence from empirical data." *Research Policy*, **38**(1), 170

Guinet, J. (1995). *National Systems for Financing Innovation*. Paris: Organization for Economic Cooperation & Development.

Hu, Mei-Chih and Mathews, J. A. (2010). "Evolution of China's National Innovation System and its challenges in technological development." In V. K. Narayanan and G. C. O'Connor (Eds.), *Encyclopedia of Technology and Innovation Management* (pp. 293–307). Chichester, U.K.: John Wiley & Sons [this book].

Iammarino, S. (2005). "An evolutionary integrated view of regional systems of innovation: Concepts, measures and historical perspectives." *European Planning Studies*, **13**(4), 497.

Kaiser, R.,, and Prange, H. (2004). "The reconfiguration of national innovation systems: The example of German biotechnology." *Research Policy*, **33**(3), 395.

Katz, J. M. (Ed.). (1987). *Technology Generation in Latin American Manufacturing Industries*. London: Macmillan Press.

Kleinknecht, A. (2000). "Indicators of manufacturing and service innovation: Their strengths and weaknesses." In: J. S. Metcalfe and I. Miles (Eds.), *Innovation Systems in the Service Economy: Measurement and Case Study Analysis*. Boston, MA: Kluwer.

Kobos, P. H., Erickson, J. D., and Drennen, T. E. (2006). "Technological learning and renewable energy costs: Implications for US renewable energy policy." *Energy Policy*, **34**(13), 1645.

Koh, W., and Phan, P. (2010). "The National Innovation System in Singapore." In V. K. Narayanan and G. C. O'Connor (Eds.), *Encyclopedia of Technology and Innovation Management* (pp. 327–339). Chichester, U.K.: John Wiley & Sons [this book].

Krishnan, R. (2010). "The Indian Innovation System." In V. K. Narayanan and G. C. O'Connor (Eds.), *Encyclopedia of Technology and Innovation Management* (pp. 341–349). Chichester, U.K.: John Wiley & Sons [this book].

Liu, X., and White, S. (2001). "Comparing innovation systems: A framework and application to China's transitional context." *Research Policy*, **30**(7), 1091.

Lu, Ta-Junk and Wann, Jong-Wen. (2010). "The National Innovation System in Taiwan." In V. K. Narayanan and G. C. O'Connor (Eds.), *Encyclopedia of Technology and Innovation Management* (pp. 309–326). Chichester, U.K.: John Wiley & Sons [this book].

Lundvall, B.-A. (Ed.). (1992). *National Systems of Innovation: Towards a Theory of Innovation and Interactive Learning*. London: Pinter Publishers.

Lundvall, B. (2007). "National innovation systems analytical concept and development tool." *Industry and Innovation*, **14**(1), 95–119.

Malerba, F. (2005). "Sectoral systems of innovation: Aframework for linking innovation to the knowledge base, structure and dynamics of sectors." *Economics of Innovation & New Technology*, **14**(1/2), 63.

Marsh, D. (2003). "Does New Zealand have an innovation system for biotechnology?" *Technovation*, **23**(2), 103.

McKelvey, M. (1997). "Using evolutionary theory to define systems of innovation." In: C. Edquist (Ed)., *Systems of Innovation: Technologies, Institutions and Organizations*. London: Pinter.

Mehra, K. (2001). "Indian system of innovation in biotechnology: A case study of cardamom." *Technovation*, **21**(1), 15.

Morgan, K. (2004). "The exaggerated death of geography: Learning, proximity and territorial innovation systems." *Journal of Economic Geography*, **4**(1), 3–21.

Nelson, R. R. (Ed.). (1982). *Government and Technical Progress: A Cross-industry Analysis*. New York: Pergamon.

Nelson R. R. (1984). *High-technology Policies: A Five-nation Comparison*. Washington, D.C.: American Enterprise Institute for Public Policy.

Nelson, R. R. (Ed.). (1993). *National Innovation Systems: A Comparative Analysis*. New York: Oxford University Press.

Nelson, R. R. (1995). "Co-evolution of industry structure, technology and supporting institutions, and the making of comparative advantage." *International Journal of the Economics of Business*, **2**(2), 171.

Nelson, R., and Winter, S. (1982). *An Evolutionary Theory of Economic Change*. Cambridge, MA: Harvard University Press.

Nelson, R. R., and Nelson, K. (2002). "Technology, institutions, and innovation systems." *Research Policy*, **31**(2), 265.

Nuur, C., Gustavsson, L., and Laestadius, S. (2009). "Promoting regional innovation systems in a global context. *Industry and Innovation*, **16**(1), 123.

Park, Yongtae and Park, Gwangman. (2003)."When does a national innovation system start to exhibit systemic behavior?" *Industry and Innovation*, **10**(4), 403–414.

Pinto, H. (2009). "The diversity of innovation in the European Union: Mapping latent dimensions and regional profiles." *European Planning Studies*, **17**(2), 303.

Saviotti, P. P. (1997). "Innovation systems and evolutionary theories." In: C. Edquist (Ed.), *Systems of Innovation: Technologies, Institutions and Organizations*. London: Pinter.

Scott-Kemmis, D., and Matthews, J. A. (2010). "Australia's National Innovation System." In V. K. Narayanan and G. C. O'Connor (Eds.), *Encyclopedia of Technology and Innovation Management* (pp. 279–291). Chichester, U.K.: John Wiley & Sons [this book].

Senker, J. (1996). "National systems of innovation, organizational learning and industrial biotechnology." *Technovation*, **16**(5), 219.

Sharif, N. (2006). "Emergence and development of the national innovation systems concept." *Research Policy*, **35**, 745–766.

Simons, K. L., and Wall, J. (2010). "The U.S. National Innovation System." In V. K. Narayanan and G. C. O'Connor (Eds.), *Encyclopedia of Technology and Innovation Management* (pp. 445–467). Chichester, U.K.: John Wiley & Sons [this book].

Storz, C. (2008). "Dynamics in innovation systems: Evidence from Japan's game software industry." *Research Policy*, **37**(9), 1480.

Teitel, S. (2006). "On semi-industrialized countries and the acquisition of advanced technological capabilities." *Economics of Innovation and New Technology*, **15**(2), 171.

Lovio, R., and Välikangas, L. (2010). "The National Innovation System in Finland." In V. K. Narayanan and G. C. O'Connor (Eds.), *Encyclopedia of Technology and Innovation Management* (pp. 391–402). Chichester, U.K.: John Wiley & Sons [this book].

van der Duin, P. (2010). "The Dutch Innovation System: Raising the lowland." In V. K. Narayanan and G. C. O'Connor (Eds.), *Encyclopedia of Technology and Innovation Management* (pp. 403–418). Chichester, U.K.: John Wiley & Sons [this book].

Virgillito, D. (2010). "The National Innovation System in Italy." In V. K. Narayanan and G. C. O'Connor (Eds.), *Encyclopedia of Technology and Innovation Management* (pp. 419–429). Chichester, U.K.: John Wiley & Sons [this book].

Visintin, F., Ozgen, B., Tylecote, A., and Handscombe, R. (2005). "Italian success and British survival: case studies of corporate governance and innovation in a mature industry." *Technovation*, **25**(6), 621.

Wafula, D., and Clark, N. (2005). "Science and governance of modern biotechnology in Sub-Saharan Africa: The case of Uganda." *Journal of International Development*, **17**(5), 679–694.

Walsh, V., and Le Roux, M. (2004). "Contingency in innovation and the role of national systems: Taxol and taxotere in the USA and France." *Research Policy*, **33**(9), 1307.

Watanabe, C., and Fukuda, K. (2010)."Inducing power of Japan's National Innovation System." In V. K. Narayanan and G. C. O'Connor (Eds.), *Encyclopedia of Technology and Innovation Management* (pp. 351–366). Chichester, U.K.: John Wiley & Sons [this book].

40

Australia's National Innovation System

Don Scott-Kemmis and Judy Matthews†*

*University of Sydney, NSW and †Queensland University of Technology, Brisbane

40.1 Introduction and historical overview

The Commonwealth of Australia was formed by the federation of the states in 1901, and has remained a federal system, with significant areas of public administration controlled at the state level. In 1901 the Australian population was less than 4 million, the (gross domestic product) GDP/capita was higher than the United States, and exports, which accounted for 25% of GDP, were almost entirely a (gradually widening) range of agricultural and mineral commodities, of which wool and gold constituted 60% (IRS, 2001). Manufacturing industry largely served the domestic market.

Much has changed over the past 107 years, although the level of dependence on commodity exports, while varied, has remained high. The population is now over 21 million and is far more ethnically diverse. The development of communications has greatly increased the extent and diversity of interaction with the world, but the majority of exports are still commodities—in 2005–2006 five primary products accounted for 40% of Australian exports: coal, iron ore, oil, gas, and gold.

By 1901 a broad foundation of industry, education, and research had been established. The Australian Museum, Botanic Gardens, Meteorological Observatory, Astronomical Observatory, and Geological Survey had been established in or near Sydney (New South Wales), and similar organizations were established in the other major states. Universities had been established in Sydney, Melbourne, Adelaide, and Hobart, and two major agricultural colleges had also been founded. There was even at that time a strong record of practical invention in agricultural and transport equipment, and mineral processing. The firms, universities, government organizations, and professional associations that had been formed also ensured extensive contact with Europe and North America and the rapid uptake of new technologies and ideas. While Australian industrial activity, research, and innovation has become far more diverse and substantial over the past century, with many areas of excellence, *a key role of the national innovation system has been and continues to be the acquisition, absorption, application, and adaptation of knowledge, ideas, and technologies from other countries.*

These organizational and institutional foundations were further developed in the period up to World War II. That period saw consolidation of the lines of research, invention, and innovation that have remained the major areas of Australia's strength: medicine and medical equipment; agriculture; and minerals exploration, mining, and processing. Importantly, the challenges of problem solving in these fields had become increasingly linked to organized research—largely in the public sector. The central features of the Australian innovation system were taking shape: knowledge-based primary industries and the major role of public-sector research. Australian science remained "knowledge-oriented and government-based" and as such was (and continues to be) linked to international science rather than to Australian technology (AATSE, 2000, pp. xx–xxi).

Two more major universities were established in the early years after federation (Queensland and Western Australia) but an important development was the formation in 1926 of what was to become the Commonwealth Scientific and Industrial Research Organization (CSIRO) with research in all fields of science and technology, except (to the extent that such demarcations can be made) defense and medical research. This was a key national commitment to research-based economic development. CSIRO has a strong reputation for the quality of its research and has played a major role in many industries and in addressing problems that limited development.[1] In the health and

[1] Based on Institute for Scientific Information (ISI) data on total citations of publications, as of June 2005 the CSIRO ranked in the top 1% of world scientific institutions in 12 of 22 research fields.

medical field, research was well established in many hospitals but an important development was the establishment in 1915 of the Walter and Eliza Hall Institute for Medical Research in Melbourne, which would develop into a leading national medical research organization. Today there are 29 independent medical research centers. In the defense field what was eventually to become the Defence Science and Technology Organization (DSTO) began to take shape with the establishment of the Laboratories of the Munitions Supply Board (initially Australian Arsenal Branch) in 1910, adding aeronautical research in 1939.

The 1939–1945 war had required a greater reliance on national capabilities in firms and research organizations, and the early post-war years were characterized by a strong drive to further develop national capabilities. One expression of this was the establishment of the Australian National University (ANU) in 1945 in the new national capital, Canberra. The ANU was directly funded by the Commonwealth government and focused on research and research training.

Over the period from the late 1930s to the early 1970s, government policy had sought to develop an Australian manufacturing sector behind a substantial tariff wall—a wall that was gradually built higher before it was dismantled in the 1980s and 1990s. This policy had several consequences. It encouraged international firms to establish local plants and stimulated local enterprise, but it did not encourage the development of exports and hence a drive to international competitiveness: "The post-war policy of diversified industrialisation produced a thin spread of technological evolution over many areas, with little concentration apart, perhaps, from the resource-based industries. Australian innovations were, therefore, thinly spread branches grafted onto the trees of international technology" (AATSE, 2000, p. xx).

The pace of economic growth after 1900 increased up to the late 1960s, with setbacks due to the wars and the depression. Over this period the contribution of manufacturing to GDP rose, particularly over the 20 years from 1935, but the export intensity of GDP declined from about 25% to about 12%.[2] By the 1970s and particularly the early 1980s, when international trade in manufactures grew rapidly, investment in plant and R&D by firms in the major OECD economies captured economies of scale and generated innovation in products and processes. The lack of competitiveness of the Australian manufacturing sector became evident not only in declining exports and increasing import penetration but also in a progressive decline in investment in manufacturing R&D. Consequently, in the 1970s the post-war boom came to an end in a period of slow growth, rising unemployment, and deteriorating terms of trade. Economic policy then changed direction, moving away from protection—tariffs were progressively lowered and the currency was floated. By 2000 Australia was one of the most open economies in the OECD (OECD, 2000). In parallel with this re-orientation of economic policy the renovation of the Australian innovation system also began. Following a range of policy initiatives in the mid-1980s manufacturing exports grew rapidly through the 1990s, increasing Australia's share of OECD exports.

Hence, the Australian economy has largely developed through the export of primary products and the growth of manufacturing for the domestic market. In 2006–2007 over 50% of Australia's exports were essentially primary products—coal, iron ore, gold, aluminum, and a wide range of other mineral and agricultural products. Elaborately transformed manufactures accounted for between 70% and more than 75% of imports but less than 20% of exports. In recent years the rapidly growing Chinese demand for raw materials has led to increasing export volumes and a sharp improvement in Australia's terms of trade, after a long period of decline (DFAT, 2007; Emmery, 1999). Nevertheless, by 2005–2006 the value of Australia's exports of goods and services as a share of gross domestic product (GDP) was still only 20.0%.

Over the period from 1930 to 1970 eight new universities were formed, the majority second or third universities in the growing cities of Sydney or Melbourne. Over the next 30 years (1970 to 2000) the number of universities in Australia doubled with the formation of 14 new universities and 14 institutes of higher education being converted into universities.

In renovating the innovation system the overall focus has been on broad measures for stimulating and supporting innovation rather than sectoral policies. However there have been some significant sectoral initiatives. The significance for regional employment of the highly protected textile, clothing, and footwear, and the automotive industries led to a program of gradual tariff reduction along with measures to increase innovation and exports. Various initiatives to boost exports to the fast-growing global high-tech markets were introduced; the most successful has been in the pharmaceutical industries area where Australia has research strengths and the government plays a major regulatory role in the domestic market.

By the 1970s the warp and weft of the Australian innovation system had been woven:

- A foundation of knowledge organizations and institutions had been built largely along European lines. The universities and research organizations, almost entirely funded by government, played an active role in the absorption and dispersion of new knowledge from overseas and developed close links with the primary industries and with knowledge-intensive activities in the

[2] A major exception to this trend was during the Korean war, when exports expanded and then declined rapidly.

government sector (e.g., health and natural resource management). Knowledge transfer from these organizations was largely through non-commercial channels: training and extension. Universities and research organizations have not generally been closely linked to or shaped by the knowledge demands of manufacturing or service industry.

- A strong knowledge-based primary sector, including a widening range of agricultural and mining activities, and a succession of leading sectors: these sectors were active innovators and developed close and generally effective links with public-sector advanced training and research organizations (i.e., they co-evolved).
- A fragmented manufacturing sector dispersed across states, sectors, firms and ownership types—with limited development of specialization and internationally competitive sectors and firms: there has been limited development of major firms or branches based on innovation in Australia—although there are many important exceptions at the firm level. The sector has a relatively high level of foreign ownership, particularly in those industries that are R&D-intensive in the home country, and a relatively low level of R&D and export intensity. For a range of reasons manufacturing industry based on the processing of primary products for export markets did not develop, and so the level of value adding to commodities has changed little over time. Hence, the manufacturing sectors evolved largely independently from the domestic science base, building on international technology.
- A development by Australian universities, research organizations, and firms of extensive links with (and organizational models based on) their U.K. counterparts. Links with the U.S. grew steadily from the mid-1800s, but expanded rapidly from the 1950s. Some 40% of Australia's S&T publications have international authorship, a level similar to Canada and the U.K.
- An innovation system, while lacking outstanding geographical or sectoral foci of research and innovation, that nevertheless ensured a high level of capability for absorbing and adapting new knowledge and technology for a diverse range of production and problem-solving activities.
- A well-educated and relatively prosperous population, and a strong research tradition, provided a background level of dispersed innovation generated by inventors, problem solving by firms, and research discoveries. But there has been a limited development of innovation-based entrepreneurship due to the limited scope of innovation in manufacturing, the lack of geographically focused high-tech research, institutional disincentives (e.g., cultural values and tax rates), and the isolation from world centers of innovation.
- A particularly strong foundation of health and medical research established in universities and independent non-profit research centers.

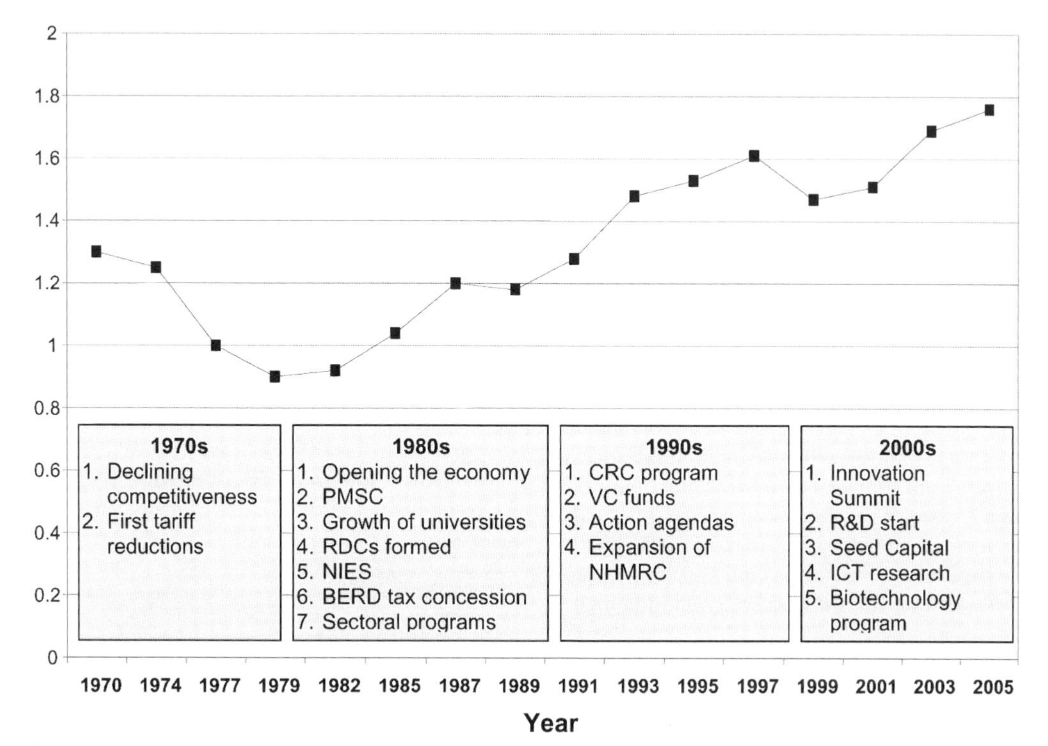

Figure 40.1. Progress of R&D (GERD/GDP) and major policy initiatives.

40.2 Summary of output trends

Innovation in a relatively small and remote resource-based economy

It is essential to make a basic point at the outset of an indicator-based overview of the Australian NIS. Because of the structure of the Australian economy the standard indicators of innovation inputs and outputs tend to provide a misleading perspective. Australia has a well-educated labor force, a high level of entrepreneurship, a foundation of world-class research and education organizations, and strong internationally oriented primary industries. But it has a relatively small high-tech sector, much of which is formed by the subsidiaries of international firms. This combination of characteristics leads to the rapid application and adaptation of new knowledge and technology, and a flexible economy with a constellation of small innovative firms, some of which develop into international suppliers of products and services.

Inputs to innovation[3]

Australia has a relatively high proportion of the working age population with higher education qualifications (25%) placing it among the top tier of OECD countries in this aspect of performance. However, a significant proportion (at least a quarter) of those with such qualifications have migrated to Australia, either with such qualifications or after attaining those qualifications in Australia.

In the early 1980s Australia's R&D intensity—gross expenditure on R&D (GERD)/GDP—was less than half the OECD average at that time. The introduction of an R&D tax concession in the 1980s stimulated a sustained growth in R&D (as shown in Figure 40.1, previous page). This rate of growth was faster than the OECD for most of the 20 years to 2004, by which time Australia's R&D intensity had reached close to 80% of the OECD average. In 2004 Australia was 16th of the 30 OECD countries in terms of R&D intensity (GERD/GDP). Australian GERD *per capita* in 2004 was at a level of US$567 which, adjusted to purchasing power parity (PPP), was lower than the OECD average ($627 PPP) and Korea ($588 PPP) and was about half the level of Sweden ($1,166 PPP).

In the early 1980s the majority of R&D in Australia was government-financed and carried out in major national R&D organizations. There has been a significant shift in the focus of government support for research and innovation over the past 25 years. The relative importance of the major national (and state) research agencies has declined (funding levels have been more or less static) and government support for innovation in the business sector and the higher education sector has increased.[4] Government funding of business-sector R&D is relatively low in comparison with other OECD countries (4.3% of BERD in 2004–2005 compared with an OECD average of 7.7%), but has grown over the past 20 years due to incentive policies. Expenditure on R&D in the higher education (HE) sector has grown steadily over the past 20 years, rising from 0.3% of GDP in the mid-1980s to 0.48% of GDP in 2004.[5] As a consequence of these trends, the proportion of GERD due to R&D in the business sector (BERD) and to a lesser extent the higher education R&D (HERD) sector has increased. Industry-funded contributions to HERD also grew substantially in the early 1990s, rising from 2.2% of HERD to 5.2%, stimulated in part by the Cooperative Research Centers program, but growth has since slowed, rising to 5.7% by 2004. Surprisingly, in view of the structure of industry, industry funding of HERD in Australia is close to the OECD average (5.7% compared with 6.1%) and significantly higher than, for example, Sweden 5.5%, Norway 5.0%, the U.S. 5.0%, Denmark 3.0%.

Characteristics of R&D activity in Australia

Overall, the structure of R&D in Australia is characterized by a relatively low level of BERD (0.91% of GDP and 54% of GERD, compared with the OECD average of 1.4% and 68%, respectively) and a relatively high level of government R&D funding and R&D performance. Compared with OECD averages relatively high proportions of BERD are carried out by:

- foreign affiliates (42% in 1999, the last year for which data are available)—higher than the level in Canada but lower than the level in the U.K. (45% in 2003);
- service sectors—which now account for almost half of BERD, the highest proportion in the OECD and almost twice the OECD average;
- Small and medium-size enterprises (SMEs)—Australia ranks 5th out of 13 countries in the OECD in terms of the share of BERD in GDP undertaken by small companies (fewer than 500 employees) and is the 3rd highest in the proportion carried out by firms with fewer than 50 employees;[6] and
- low and medium-technology industries—34.8% compared with the OECD average of 11.5%; and conversely a relatively low proportion of BERD carried out in high-technology sectors (27.4% compared with an OECD average of 52.8%).[7]

[3] This section is based on the compilation of indicators in DEST (2003, 2007a, b), ABS (2005), Garrett-Jones (2007), and Gregory (1993).

[4] In the mid-1970s expenditure by state governments accounted for about 15% of GERD, but by 2005 this expenditure accounted for only about 6% of GERD. However, over the past decade many states have increased their promotion of and investment in research and innovation initiatives.

[5] While the overall intensity of investment in R&D in the higher education sector (HERD) is higher than the OECD average (0.48% in 2004 compared with 0.39%), the level is substantially lower than in many smaller countries pursuing active innovation policies (e.g., Denmark 0.61%, Finland 0.69%, Canada 0.70%, and Sweden 0.87%).

[6] As not all OECD countries include firms with fewer than 10 employees in their R&D surveys there must be some doubt over the strict comparability of these data.

[7] This comparison is based on the OECD classification of high-tech sectors.

These characteristics lead to a relatively high focus on lower risk, short-term, process-related innovation. As in most OECD countries BERD in service sectors is growing significantly more rapidly than is BERD in manufacturing: over the period 1993–2002 BERD in manufacturing grew at an average annual rate of 5.7%, but in services at 10.7%. As a result of this sustained growth and the structure of the economy, Australia has a higher proportion of BERD carried out in the service sector than any other OECD economy (47% in 2002 compared with the OECD average of 25.4%). R&D in knowledge-intensive business services has grown particularly rapidly.

Innovation-related outputs

While there is much discussion within the policy and research community regarding Australia's innovation-related performance and a good deal of concern about some dimensions of that performance, the indicators that have dominated general political debate are those related to headline economic performance. Australia's GDP growth over 1995–2004 (3.6% annual average) has been among the highest in the OECD. Australia's GDP *per capita* (US$32,500 in 2004) is more than 85% of the U.S. level and well above the OECD average.

Australia has long been a strong contributor to the scientific literature, relative to its size. In 2003 Australian researchers accounted for 2.96% of total world scientific publications, a proportion that has risen from about 2.3% 20 years ago. The citation rate of Australian scientific papers is very close to the world average rate, and has been steadily rising above that average for the last 10 years. The fields of geoscience, clinical medicine, and plant and animal science are the areas of greatest strength in terms of the numbers of papers, the proportion of world papers in those fields, and citation rates. The output of scientific papers per million population (1,141) is significantly above the OECD average (697) and, while above most of the larger economies, is below the research-intensive Nordic countries (e.g., the level in Sweden is 1,741). However, the overall citation rate for Australian scientific publications is below that of most of the research-intensive OECD countries.

With a small high-tech sector and a relatively low level of BERD, Australia has not been a strong performer in patenting. However, Australia's contribution to world patents has grown steadily over the past 20 years. In the late 1980s Australia accounted for only 0.37% of total world triadic patent families, but by 2003 this proportion had more than doubled to 0.82%. Nevertheless, the number of Australian origin triadic patent families per million population (21.6) remains substantially lower than the OECD average (44.8) and far below the levels of top-performing countries.

Although Australian applications to the European Patent Office (EPO) have grown steadily over the past 20 years, the Australian share of EPO patents declined through much of the 1980s before largely recovering in the 1990s to levels around 0.9%. A very similar pattern is evident in Australian patenting through the USPTO, where in 2003 Australian patentees accounted for 0.73% of patents registered. These levels of patent intensity are low relative to comparable countries. For example, Australian applications to the EPO in 2003 (48.5 per million population) were half the OECD average level (95.7), and for USPTO patents granted in 2003 Australia had 80 per million population, again less than half the OECD average and substantially lower than R&D-intensive economies. In 1998–2002, the U.S. patents of countries other than Australia cited Australian patents at a ratio of 0.72, well behind technological leaders such as Singapore (1.55) and the United States (1.15).

Australia was not a strong patentee in either information and communication technologies (ICT) or biotechnology. Australia accounted for 0.42% of ICT patents registered by the USPTO in 2000, a level similar to Finland but substantially below Canada (1.84%). In the case of biotech, Australia accounted for an even smaller proportion of USPTO patents registered in 2000 (0.23%), again substantially below Canada (3.06%), but above Finland (0.13%).

As might be expected, Australia is a *net importer of technology*. Payments to overseas organizations for the use of their technology is three times higher (0.24% of GDP in 2004) than receipts from licence fees and royalties (0.08% of GDP).

As noted previously, Australia has a relatively small manufacturing sector and a particularly small high-tech manufacturing sector. Australia's exports are dominated by commodities and by low and low–medium technology goods—only 13.5% of goods exports in 2002 were high-tech. A range of innovative policies in the mid-1980s had stimulated exports from R&D-intensive industries, led by the pharmaceuticals, IT, and aerospace sectors, and their share of total OECD highly R&D-intensive exports grew from 0.19% to 0.58% over the 10 years leading up to the mid-1990s—although that has now dropped back to 0.42%.

The key indicators of Australia's innovation performance, in relation to OECD average levels, is shown in Figure 40.2.

40.3 Technology commercialization initiatives

The characteristics of the Australian innovation system, with a major role played by public-sector research and a relatively diffuse innovation activity in industry, suggest that effective commercialization will be particularly important. Over the past 20 years there have been four types of initiative to promote commercialization:

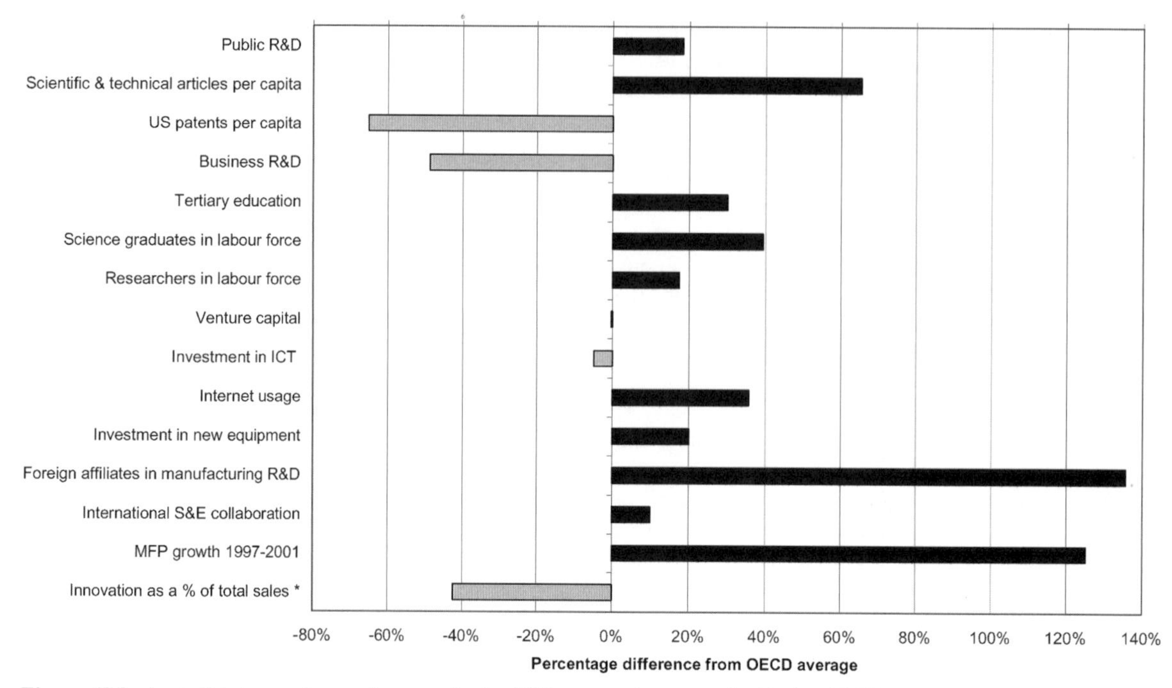

Figure 40.2. Australia's innovation performance in the 2004 scorecard compared with the OECD average.
Sources: ABS, OECD, *World Competitiveness Yearbook*, National Science Foundation, Thompson ISI, U.S. Patent and Trademark Office.
* Innovation as a % of total sales has not been updated from the 2002 scorecard due to a lack of new data (DEST, 2005).

- First, measures have been introduced to develop closer links between research organizations and industry. Among a range of initiatives the most significant has been the formation of the Cooperative Research Centers (CRCs) program in 1990 which builds collaborative research between universities, government R&D organizations, industry, and other research users. Government research organizations and universities have been expected to raise their commercialization performance, and this performance has been regularly assessed.
- Second, government has sought to stimulate the formation of venture capital (VC) funds. Again various initiatives have been introduced over time in response to identified problems. One of the most significant of these initiatives was the Innovation Investment Fund (IIF) program, which in effect provided subsidized capital to selected VC providers.
- Third, recognizing the many challenges in the early stages of new ventures, the Building IT Strengths (BITS) program supported the formation throughout the major cities in Australia of incubators for IT ventures.
- Fourth, there has been a succession of competitive funding programs over the past 25 years for the support of specific innovation projects in industry or in collaborations between industry and research organizations. One of the major programs was R&D Start. More recently, the challenges of early-stage ventures have been the focus for initiatives which have sought to provide pre-seed funds and to ensure the use of competent business support services. One of the major initiatives was the Commercializing Emerging Technologies' (COMET) program. This was subsequently replaced by the more comprehensive Commercial Ready program, but that in turn has been partially replaced by Commercialization Australia, a measure designed to provide funding and guidance to new ventures.

As in many countries, venture capital (VC) investment in Australia grew strongly through the 1990s, in part in response to reduced capital gains tax levels and to specific measures to stimulate the formation of venture capital funds. By the early 2000s the level of venture capital investment in Australia (0.1% of GDP), while substantially lower than the levels in the U.S. (0.35% of GDP), was close to the OECD median. VC investment grew slowly in the early 2000s but from 2004 has risen sharply, more than doubling the level of VC funds over three years. While almost 90% of VC in the U.S. is directed into the high-tech sectors of health/biotech and ICT, only about 20% of Australian VC goes into these sectors. There are no regions in Australia with the concentration of wealth creation and innovation as can be found in Boston or Silicon Valley.[8]

[8] Regan and Tunny (2008). As definitions of VC vary across countries and VC funding can be very volatile, caution is required in any international comparisons.

40.4 National innovation policy

The national innovation policy involves five components, each of which evolves over time in response to new priorities and identified shortcomings:

- *Research and innovation infrastructure*. The Commonwealth government supports four major national research organizations (CSIRO, DSTO, the Australian Nuclear Science and Technology Organization, and the national ICT research organization NICTA) and a wide range of smaller organizations (e.g., Australian Institute of Marine Science and GeoScience Australia). It also provides core funding for research and research infrastructure in the universities. The formation of new-enterprise incubators under the BITS program is a form of innovation infrastructure.

- *Framework measures*. Changes in tax (e.g., concessions for undertaking business R&D and reduced capital gains tax) and regulatory policy (essentially seeking to reduce business, product, and labor market regulation) have sought to increase flexibility and the incentives to undertake R&D. Initiatives that stimulate the development of VC funds, such as Venture Capital Limited Partnerships, also aim to increase involvement in higher risk innovative activity. Intellectual property support programs have also been introduced.

- *Competitive support programs*. Competitive funding programs have long been a component of the policy mix but their role has increased markedly over the past 20 years. The four major competitive funding mechanisms are the Australian Research Council (ARC) which funds research in universities, the NHMRC which funds research in universities and medical research centers, the Rural Research Corporations which fund research in rural and related downstream industries, and the mechanisms for funding innovation in industry (such as R&D Start). The CRC program is also a competitive funding scheme, as also is the International Science Linkages program. While Australian researchers have extensive international links, about 40% of government support for Australia's international S&T collaboration is allocated to collaboration with the U.S. and the U.K.

- *Thematic and sectoral programs and guidelines*. In the 1980s various sectoral programs were introduced to support large established industries (e.g., motor vehicles), sectors with potential for growth (e.g., pharmaceuticals), or emerging sectors (e.g., biotech). These sectoral programs have continued to be an important component of the policy mix. In the 1990s an Action Agenda program was introduced to facilitate joint government–industry–research reviews of sectoral issues and the development of joint development plans. In the 1990s a set of broad and high-level National Research Priorities were introduced, essentially as a set of indicative guidelines for research organizations and the competitive funding programs. The priorities are "an environmentally sustainable Australia", "promoting and maintaining good health", "frontier technologies for building and transforming Australian industries", and "safeguarding Australia".

- *Coordination mechanisms*. Coordination mechanisms have been established at several levels. The Commonwealth States & Territories Advisory Committee on Innovation seeks to improve coordination across levels of government. The Prime Minister's Science, Engineering and Innovation Committee (PMSEIC) provides a high-level coordination mechanism chaired by the Prime Minister and with the participation of key ministers and representation by the heads of major research agencies and professional associations. The Coordination Committee for Science and Technology (CCST) coordinates policy development and implementation among Commonwealth departments and agencies.

The evolution of policy has generally been conservative, adding or extending organizations and programs, with cautious reorganizations and redirections. The overriding theoretical framework has remained that of "market failure" and its conceptual twin in innovation policy, the "linear model". Economic orthodoxy has provided the legitimation, and organized public-sector science the primary focus of innovation policy. Within these confines, initiatives have been stimulated by the emergence of new priorities (e.g., renewable energy, water management, biotechnology, services), problems recognized through program reviews and broader policy reviews (e.g., low BERD, a lack of risk capital), powerful exemplars from other countries (e.g., technology incubators), and the political pressure from time to time to "do something", when perceived problems and "on the shelf" proposals may gain rapid policy traction. The rhetoric of "innovation systems" has been used in policy documents for some years but this perspective has had little real influence on policy. However, following a comprehensive independent review of the National Innovation System in 2008–2009 a new innovation policy was set out in a White Paper (see Parliament of Australia, n.d.: *http://www.innovation.gov.au/innovationreview/Pages/home.aspx*). While maintaining most of the existing programs and organizational arrangements, this policy framework sets out a broader and more systemic perspective on policy. It increases the emphasis on: raising capabilities at the firm level; stimulating R&D in industry; supporting innovation in services and in the public sector; raising the level of collaboration, and strengthening the mechanisms for coordination, in the innovation system.

40.5 Relative strengths and weaknesses in technology

The strengths and weaknesses of the overall national innovation system, including sectoral strengths, are summarized in Table 40.1. These are pervasive characteristics that shape both the response to new challenges and opportunities, and the development of innovation policy.

Australia performs best in research areas related to its resource and geographical endowments, where knowledge has been built over time through sustained investment and problem solving. Over the past 20 years, Australia has maintained a level of technological specialization similar to that of its scientific research publications. The revealed technological advantage has changed little, with the highest level of specialization in the field of agriculture—five times that of the world in the period 1994–2001 and increasing over the past 20 years. Australia also has increased its specialization in primary metals, mining, oil and gas, medical technology, and biotechnology, and has developed strengths in a few high-tech areas (Figure 40.3). Australia has broadened its research and commercial strengths in the biomedical field. Pharmaceutical exports have grown, and many firms have developed in the biotechnology and related fields.

40.6 Funds flow for innovation

The overall national expenditure on R&D in 2004–2005 was almost A$16 billion, 54% of which was by the business sector. Figure 40.4 shows the flow of funds for R&D in Australia in 2004-5.

Commonwealth government expenditure on research and innovation is distributed through a diverse range of mechanisms, with an increasing use of competitive funding schemes (as summarized in Figure 40.5).[9]

40.7 Cultural and political drivers influencing the NIS[10]

Australia is now an ethnically diverse and highly urbanized nation. Immigration continues to be a major source of population growth and an essential contributor of knowledge and skill (Withers, 1989). Despite its location in the Asia-Pacific region and the increasing significance of trade links with Asia, traditional cultural ties continue to shape international research collaboration and business investment. The growth of China and India as industrial economies and increasingly as centers of research, is leading to new initiatives to build research links. But it is also leading to some relocation to these countries of multinational R&D centers previously located in Australia.

As a result of the federal structure of government, substantial areas of public administration fall within the jurisdiction of the several state governments. These governments have become more active in the area of innovation and all have innovation "strategies" and some (e.g., Victoria and Queensland) have introduced substantial initiatives. This rivalry and diversity has been productive but sometimes leads to counterproductive competition (e.g., to attract investment).

Australia has a strong education system with high levels of participation and high levels of performance in international comparisons of standards.[11] The higher education sector expanded strongly through the 1980s, and a relatively high proportion of young Australians are tertiary-educated. Over 36% of the 25 to 34-year-old population in Australia are tertiary-level graduates, slightly higher than the OECD average. While the proportion of graduates who have engineering degrees is relatively low—perhaps not surprising in view of the structure of the economy—the proportion of graduates with science degrees is relatively high (14.3%)—higher than the U.S. (8.3%), Canada (11.1), and Finland (9.3). The sustained growth of the past 15 years and the surge of investment in infrastructure and facilities have led to shortages in almost all skill areas, but particularly skilled workers and engineers. Hence, these stocks are supplemented by skilled and professional immigrants.

Australia's history has shaped the characteristics of its human resources. In particular: "There [has been] a lack of a deeply rooted entrepreneurial culture within Australia which would see entrepreneurship becoming a more socially legitimate activity".[12] While cultural values are changing, Australians are often more likely to celebrate prowess in sport before success in entrepreneurship. However, the policy re-orientation that began in the 1980s, which progressively opened the economy to international trade, was supported by micro-economic reforms which reduced labor market and business regulation. These reforms have contributed to a more entrepreneurial and flexible economy with strong productivity growth. While Australian culture is individualistic it is also egalitarian. As a result policies seek to ensure that the costs of adjustment do not fall unfairly on some sections of society and that the opportunities from new policies and developments are not unfairly captured by other sections.

[9] Detailed information is available at DEST (2007c).
[10] See DEST (2003).

[11] For example, the OECD Program for International Student Assessment (PISA), a triennial survey of the skills of 15-year-olds (OECD, 2007)
[12] Prime Minister's Science, Engineering, & Innovation Council (PMSEIC) Working Group on Management Skills for High-growth, Start-up Companies).

Table 40.1. The overall strengths and weaknesses of the Australian innovation system.

NIS characteristic	*Strengths*	*Weaknesses*
Strong public-sector research system, but limited links with manufacturing and service sectors	• Some strong sectoral innovation systems • Effective diffusion of knowledge throughout industry	• Limited collaborative links and complementarity outside resource sectors
Knowledge-based primary sector	• Competitive production of commodities and effective response to new challenges • Many specialist equipment and service supply companies	• Capabilities not adequately used offshore to create and capture value
A fragmented manufacturing sector with high foreign ownership and relatively low R&D and export intensity	• Rapid knowledge diffusion • International participation in the Australian innovation system	• Low BERD • Limited demand for local R&D services • Little experience in major innovation • Public procurement oriented to proven imported systems
Distant from major world centers, but with extensive interaction with U.K. and U.S. research organizations and firms	• Rapid acquisition of knowledge • Collaboration in research and innovation	• Limited links with major trading partners • Scale, visibility, and distance affect Australia's capacity to attract FDI and international R&D investment
Strong capability for absorbing and adapting new knowledge	• Educated and flexible human resources • Rapid uptake of new technology (e.g., ICT) • Strong foundation for service sector growth • Strong productivity growth • Strong capabilities in designing and managing large projects	• Limited managerial and entrepreneurial skills in innovation-based new-business development
Dispersed innovation generated by inventors, problem solving by firms, and research discoveries	• Building new products and ventures	• Entry to export markets challenging • Lack of innovation-oriented clusters and early-stage finance • Weak research base for ICT in industry and public R&D centers
Strength in health and medical research	• IP and ventures based on new knowledge • Increased role in the international health and medical value chain	• Limited effective support for new-enterprise development

This assessment is based on a report produced by DEST (2003) and reports on Australian innovation systems study including Scott-Kemmis *et al.* (2006).

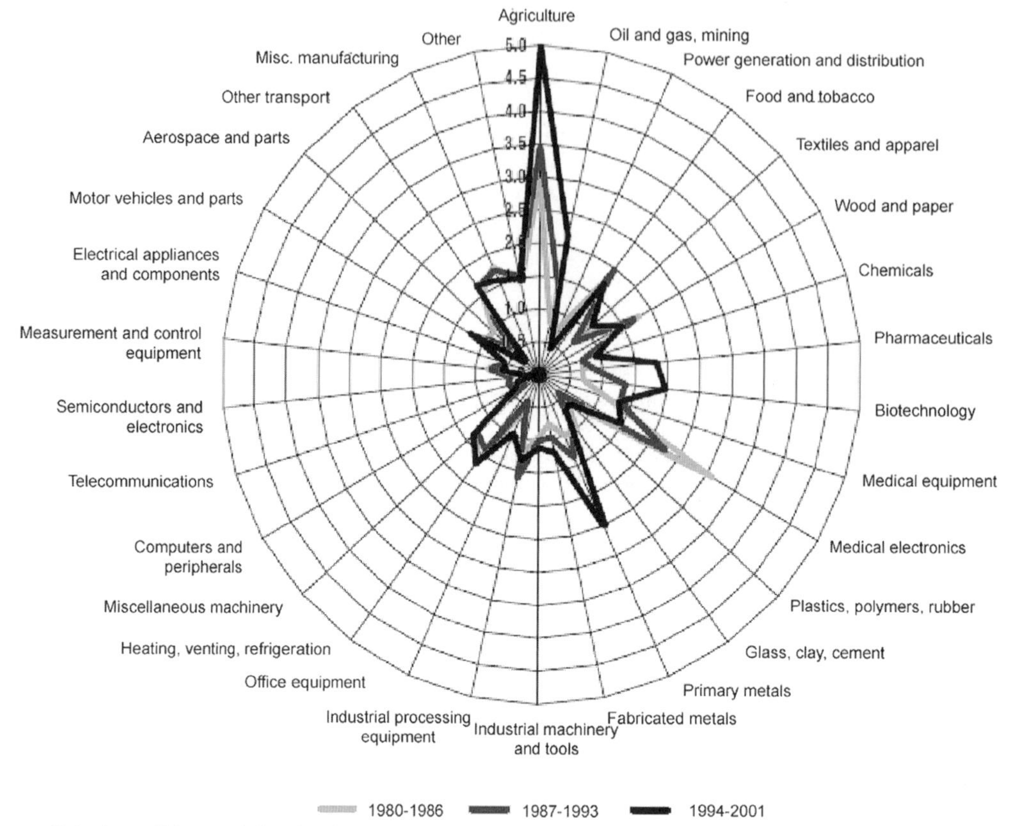

Figure 40.3. Australia's revealed technological advantage.
Source: Scott-Kemmis (2003).

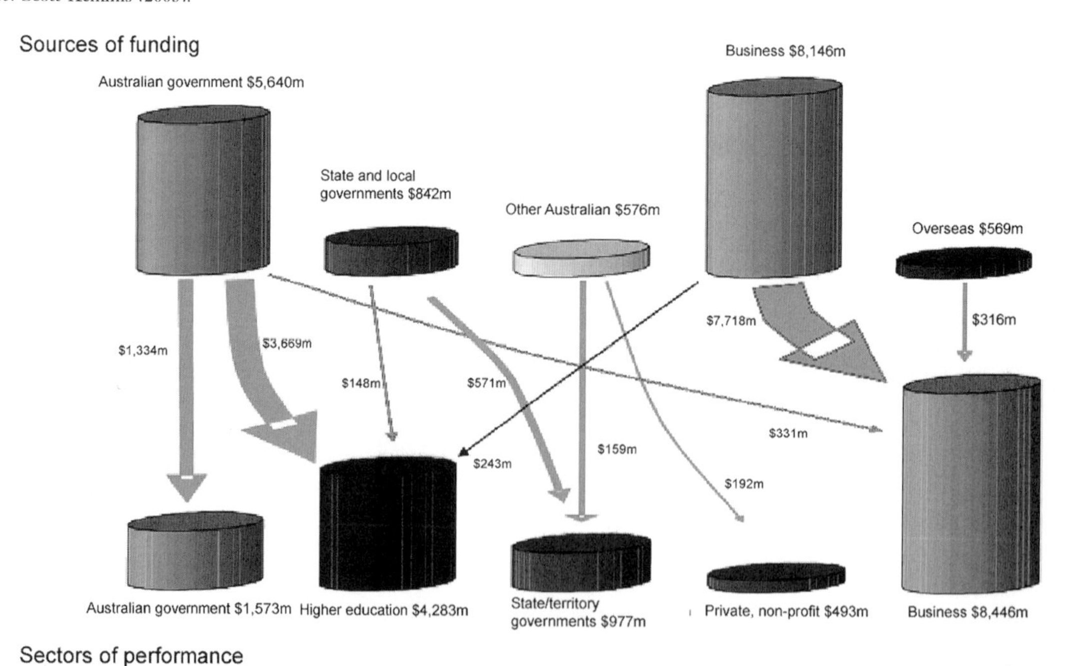

Figure 40.4. Major flows of funding for R&D in Australia, 2004–2005.
Source: DEST, based on ABS R&D data provided in October 2006.

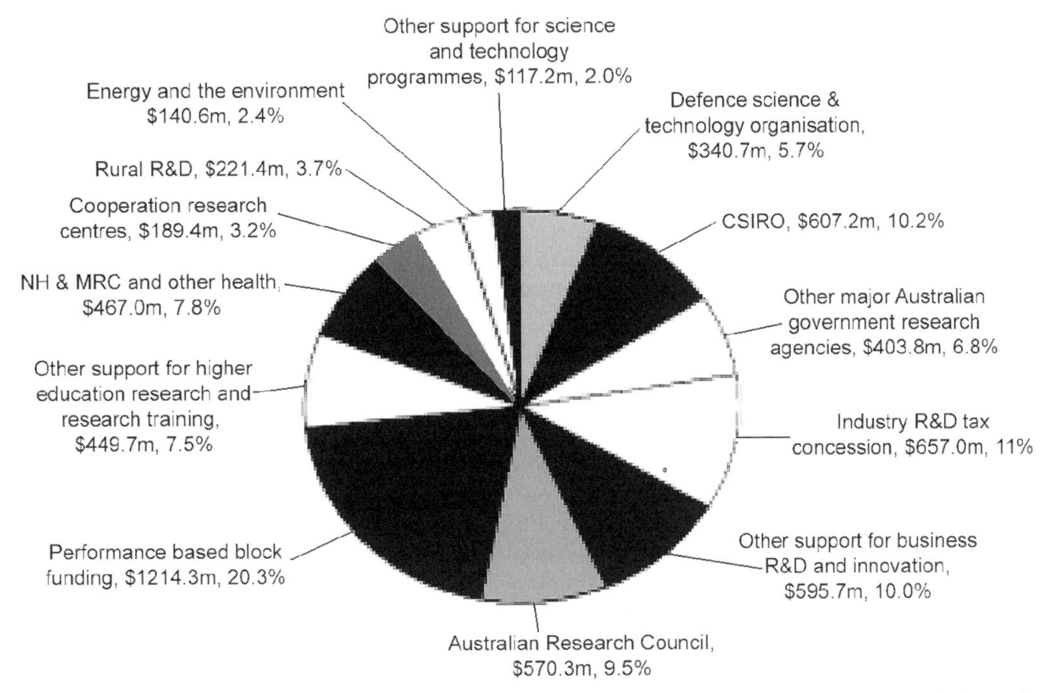

Figure 40.5. Distribution of Australian government support for science and innovation—by detailed component, budget estimates for 2006–2007.
Source: DEST (2007c).

40.8 Conclusion

Global industrial, technological, and economic change has been the main driver of the evolution of the Australian innovation system. But the influence of this driver has been shaped by three particular characteristics of Australia:

- institutional inheritance—the Australian NIS developed from the foundation of dynamic and robust first-world political, legal, science, and education institutions;
- natural resource base—the very large agricultural and mineral resources provide a strong production foundation to the economy and shape Australia's comparative advantage; and
- remoteness—a relatively small population dispersed across a large continent distant from world markets.

This particular combination did constrain the development of manufacturing to small-scale operations based on the domestic market and largely low or medium technology. It is also a large part of the explanation for the limited development of value-added processing of Australia's commodity exports.[13] The development of the NIS has tended to be conservative, cautiously adapting and extending organizations and programs. With little specialization and few large firms in the manufacturing sector, policy development has been dominated by orthodox economic perspectives and the strong influence and organized interests of the public research sector. Only in times of crisis has there been a preparedness to move beyond cautiously copying approaches implemented in other countries and to consider more radical initiatives. The 1983–1990 period was an era of major reform when there was a more vigorous process of reorienting the NIS.

The Australian NIS is now facing what may be a period of sustained change, although many of the specific challenges are much the same as those confronted by other countries. The rise of China and India is reshaping the global division of labor, the service sector is growing strongly, and the challenge of sustainability is becoming a major focusing device for research and innovation. Australia is now debating whether open trade markets, flexible labor markets, and strong competition policy, in the context of a foundation of public investment in knowledge creation and human resource development, are the necessary and largely sufficient conditions for effective innovation policy, or whether government has a role in leading the more rapid evolution of the innovation system. While there is at present no sense of crisis or urgency, there is a sense that opportunities may be lost without a more active and coordinated process of evolution.

[13] Various forms of protection in export markets is another.

Table 40.2. Policy initiatives and Australia's innovation system.

NIS characteristic	*Policy responses to leveraging strengths and reducing weaknesses*	*Specific examples*
Strong public-sector research system, but limited links with manufacturing sector firms.	• Improve linkages with the public sector • Improve levels of research commercialization • Attract MNC investment into R&D • Strengthen absorptive capacity in SMEs	• CRC program • National Industry Extension Service • Productivity centers
Knowledge-based primary sector	• Strengthen the research system • Promote exports of knowledge-intensive services	• RDCs • CRCs
A fragmented manufacturing sector with high foreign ownership and relatively low R&D and export intensity	• Incentives for industrial R&D • Sectoral programs • Support greater industry–research links • Use government procurement to influence investment and innovation	• Tax concession for R&D • Sectoral programs for pharmaceuticals, IT, and automotives • Action agendas
Extensive interaction with U.K. and U.S. research organizations and firms	• Widening bilateral and multilateral links	• S&T programs with China and India
Strong capability for absorbing and adapting new knowledge	• Promote upgrading capabilities in firms • Promote uptake of ICT	• Reducing regulation of business, labor markets, and trade
Dispersed innovation generated by inventors, problem solving by firms, and research discoveries	• Encouraging entrepreneurship and commercialization	• Formation of a national ICT research organization (NICTA) • VC support • Incubators—BITS
Strength in health and medical research	• Strengthen research • Improve commercialization • Link procurement and innovation	• Expansion of the NHMRC • Establish new research centers • Pharmaceutical industry program

40.9 References

AATSE. (2000). *Jan Kolm's Introduction: Technology in Australia 1788-1988.* Melbourne, Australia: Australian Academy of Technological Science and Engineering; *http://www.austehc.unimelb.edu.au/tia/titlepage.html*

ABS. (2005). *Patterns of Innovation in Australian Businesses.* Canberra: Australian Bureau of Statistics.

DEST. (2003). *Mapping Australia's Science and Innovation System.* Canberra: Department of Education, Science and Training, Parliament of Australia.

DEST. (2007a). *Australian Science and Innovation System: A Statistical Snapshot 2006.* Canberra: Department of Education, Science and Training, Parliament of Australia.

DEST. (2007b). *The Australian Government's 2007–08 Science and Innovation Budget Tables.* Canberra: Department of Education, Science and Training, Parliament of Australia.

DEST. (2007c). *Science and Innovation Budget Tables 2006–07.* Canberra: Department of Education, Science and Training, Parliament of Australia; *http://www.dest.gov.au/ministers/bishop/budget06/scitables.pdf*

DFAT. (2007). *Composition of Trade Australia 2006–07.* Canberra: Market Information and Analysis Section, Department of Foreign Affairs & Trade, Parliament of Australia.

Emmery, M. (1999). *Australian Manufacturing: A Brief History of Industry Policy and Trade Liberalisation* (Research Paper 7). Canberra: Department of the Parliamentary Library, Parliament of Australia.

Garrett-Jones, S. (2007). *Marking Time?: The Evolution of the Australian National Innovation System, 1996–2005* (Faculty of Commerce Papers). Wollongong: University of Wollongong.

Parliament of Australia. (n.d.). *Venturous Australia: Building Strength in Innovation* (report on the review of the National Innovation System), and *Powering Ideas: An Innovation Agenda for the 21st Century*. Canberra: Parliament of Australia; *http://www.innovation.gov.au/innovationreview/Pages/home.aspx*

Gregory, R. (1993). "The Australian innovation system." In: R. R. Nelson (Ed.), *National Innovation Systems: A Comparative Analysis* (pp. 324–352). New York: Oxford University Press.

IRS. (2001). *A Socio-Economic Profile of Australia at Federation* (Number 23, 2000-011901). Canberra: Information and Research Services, Department of the Parliamentary Library, Parliament of Australia.

OECD. (2000). *Economic Surveys 1999-2000 Australia*. Paris: Organization for Economic Cooperation & Development.

OECD. (2007). *PISA 2006: Science Competencies for Tomorrow's World* (the OECD Program for International Student Assessment). Paris: Organization for Economic Cooperation & Development.

PMSEIC. (2002). *Management Skills for High-growth, Start-up Companies* (Working Group report). Canberra: Prime Minister's Science, Engineering, & Innovation Council.

Regan, D., and Tunny, G. (2008). *Venture Capital in Australia Economic Roundup*. Canberra: Australian Treasury; *http://www.dest.gov.au/ministers/bishop/budget06/scitables.pdf*

Scott-Kemmis, D. (2003). *Australian Innovation Systems: An Evolutionary Context* (unpublished work commissioned by the Department of Education, Science and Training Innovation Management and Policy Program). Canberra: Australian National University.

Scott-Kemmis, D., Holmen, M., Balaguer, A., Dalitz, R. P., Bryant, K., Jones, A. J., and Matthews, J. H. (2006). *No Simple Solutions*. Canberra: Australian National University.

Withers, G. (1989). "The immigration contribution to human capital formation." In: D. Pope and L. Alston (Eds.), *Australia's Greatest Assets: Human Resources in the Nineteenth and Twentieth Centuries* (pp. 53–71). Sydney: Federation Press.

41

Evolution of China's National Innovation System and Its Challenges in Technological Development

Mei-Chih Hu and John A. Mathews*[†]

*Feng Chia University and [†]Macquarie University Sydney

This chapter utilized the data available in the various national and international S&T databases that combine firm-level, industry-level, and country-level observations to examine the progress and processes involved in China's national innovation system over the last 20 years. Four weaknesses and challenges in the process of building the national innovation system in China were identified as originating in China's historical separation of science and industry. By examining China's technology transaction, it is found that the dramatic rise in the demand for technology in the limited companies sector comprising SMEs is further evidence that, since 2000, SMEs have become an essential part of China's economy. In the process of building China's innovation infrastructure, the concurrence of internal and external factors plays an important role in affecting innovation capabilities. After the implementation of the 10th Five-year S&T Plan (2001–2005), new innovations and instruments are available in China's managed economy, in which the building of absorption capacity in the private sector is emerging and is evidence that China's innovation system is evolving and continuously moving forward.

41.1 Introduction

The performance of a national innovation system largely depends on how its constituents, including government, enterprises, bridging institutes, and other contributing institutions, function and interact with each other in the development and application of innovative knowledge. In the relevant literature, however, the functioning of technological systems of innovation has been addressed in different ways in the cases of developing countries (Alcorta and Peres, 1998; Arocena and Sutz, 1999; Cassiolato and Lastres, 1999, 2000; Dahlman and Sananikone, 1990; Katz 1995) and developed countries (Carlsson and Jacobsson, 1994; Carlsson and Stankiewicz, 1995; Wörner and Reiss, 1999). More specifically, differences between the so-called latecomer countries of the East and the more advanced countries of the West have been identified in the various means and processes of technological development and the diffusion of its results (Coe, Helpman, and Haffmeister, 1997; Eaton and Kortum, 1999; Hu and Mathews, 2005). Significantly, even in emerging, late-industrializing countries, such as China, India, and Russia, the distinctive patterns of industrial development and the accompanying unprecedented changes in social, legal, and economic institutions challenges the theoretical and practical models that have applied in the West (e.g., Buck *et al.*, 2000; Hoskisson *et al.*, 2000; Kash, Auger, and Li, 2004; Li and Atuahene-Gima, 2001; Zhou, Yim, and Tse, 2005).

In China a number of policies directed at reforming its national innovation system have progressively been implemented during the past 20 years. In 1985 the Resolution of the China Communist Party Central Committee (CCPCC) on the structural reform of the system of science and technology (S&T) was enacted. This represents a cornerstone of the nation's deviation from the Soviet model of an innovation system in which S&T activities at public research institutes (PRIs) and production at state-owned enterprises (SOEs) were kept completely separate (Xue, 1997). It is clear that such a system that separates science and industry does not work well and, consequently, the decision was made in 1985 to change the direction of China's innovation policy towards a science and industry linkage model. In this model the fundamental objectives of the reform encompassed three areas of S&T: S&T manpower utilization, S&T institutional reengineering designed to build up an effective technology transfer mechanism, and enabling S&T to become the core of economic development as a whole (Xue, 1997).

PRIs, universities, and enterprises have remained the three major players in China's innovation system since 1985. But, while most specific reform measures were targeted at PRIs in the first instance, since the late 1990s there has been a redirection of reform to focus on enterprises, particularly state-owned enterprises (Jefferson *et al.*, 2000; Motohashi, 2005). In the process a substantial number of PRIs became private technology service firms or were merged with large to medium-size enterprises. In these cases a series of market-based economic reforms have mobilized two-thirds of the R&D funding and personnel inputs directly, to serve the needs of national economic development and commercial activities (NRCSTD, 2003).

The present chapter examines the progress of and processes involved in China's changing national innovation system over the last 20 years. It utilizes the data available in the various national and international S&T databases that combine firm-level, industry-level, and country-level observations. The questions that structure this chapter are the following. What are the distinctive paths to achieve economic reform chosen by China? What are the consequences of these different reform paths for the global integration of China's production network? In what ways does the interaction of national regulations, involved institutions, and the evolution of economic development impact the performance of the emergent national innovation system?

This chapter is thus organized as follows. Section 41.2 reviews the evolution of China's innovation system since 1985, while the policies and effects of technology commercialization and science–industry linkage are discussed in Section 41.3. Section 41.4 examines the presented challenges in China's innovation system. The process of building innovation infrastructure under a strong policy-oriented regimen is addressed in Section 41.5, while China's skipping stage of industrial development is presented in Section 41.6. Concluding remarks and policy implications are made in Section 41.7.

41.2 Evolution of China's innovation system

Under the planned economic system based on the Soviet model that was built up in the 1950s, the development and transfer of technologies were the responsibilities of PRIs controlled by the government. State-owned enterprises were supposed to concentrate on production activities that applied the technologies made available by PRIs. As a result, PRIs had no incentive to understand the technological needs of enterprises, and state-owned enterprises had no incentive to undertake in-house R&D. As shown in Figure 41.1, China's innovation system began with a substantially increased national R&D expenditure aimed at facilitating technology diffusion from PRIs in a manner that would reinforce the technological capabilities of industries as a whole. Its R&D expenditure increased more than 16 times in 15 years, from a very low level of 15 billion RMB in 1991 to 245 billion in 2005.

China's PRIs, at both local and national levels, were the backbone of the nation's innovation system before the mid-1990s. During this time, more than half its R&D expenditure, funded mostly by the government, was allocated to the PRIs, with the universities taking the smallest proportion of 4% (SSTC, 1993). The role of universities was mainly confined to training engineers and scientists and to conducting certain basic and applied research. Productive enterprises were similarly constrained, and being mostly state-owned meant that the function of in-house R&D was substituted by the activities of the PRIs.

The data in Figure 41.2 provide evidence that the efforts involved in transforming China's innovation system embraced three specific characteristics: (1) the goal of the system in earlier years was to make PRIs more responsive to market needs by reducing their dependence on government funding; (2) since 1993 the declining R&D

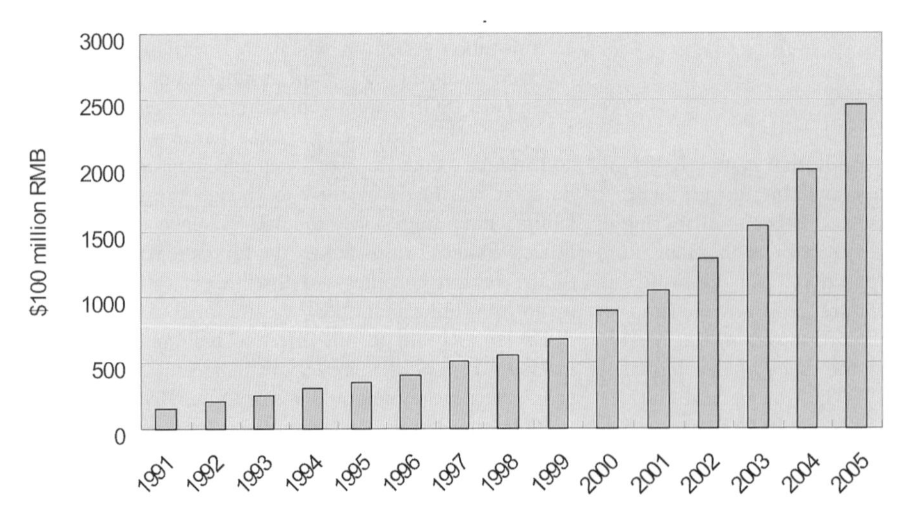

Figure 41.1. R&D expenditure.
Source: STS (2004–2006).

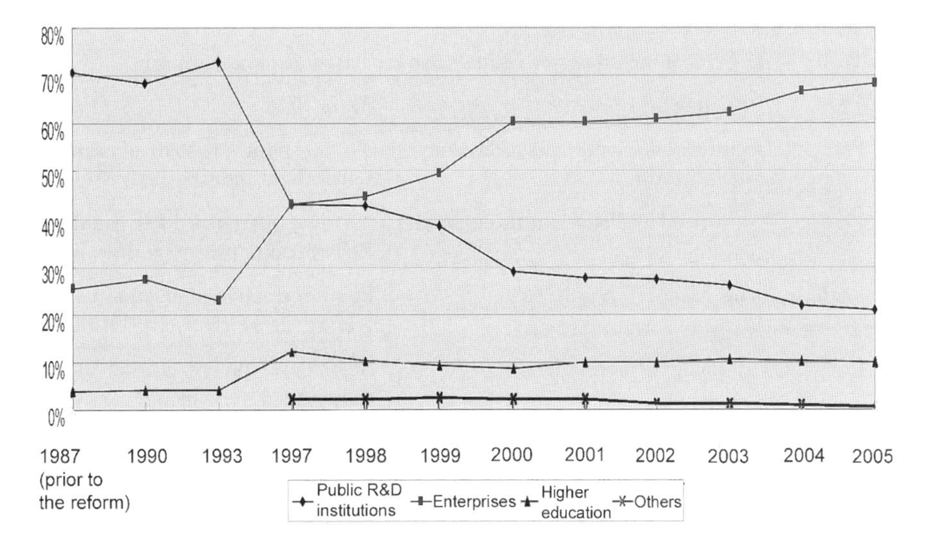

Figure 41.2. R&D expenditure, by sector. *Source*: STS (2004–2006).

expenditure in PRIs has been accompanied by rising expenditure in the private sector, a trend that reflects the effect of numerous policy instruments (see Table 41.1) aimed at structural reform designed to maximize integration within the science and industry linkage; and (3) while the reform was effected from the early 1990s, China's R&D expenditure in the private sector quickly reached the average level of approximately 60% found in industrialized countries in 2000. Undoubtedly, the substantial rise of private enterprise R&D expenditure was associated with the significant decline of such expenditure by PRIs since the beginning of the 1990s. This change has been government-oriented and has been achieved through the series of policy implementations and legislative enactments shown in Table 41.1. Consequent upon these policies, PRIs were obliged to focus on technology-licensing activities as the means of their survival. Alternatively, they were to transform themselves into private technology service enterprises (NRCSTD, 2003). These processes of structural transformation in national industrial policies recognize the previous deficiencies of China's innovation system relating to the efficacy of active technology transfer channels and of related knowledge flows.

China's entry into the WTO in 2001 has led to further market reforms and openness, with manufacturing enterprises realizing that they have to strive to survive in increasingly competitive markets. One means of their doing so is through establishing improved technological capabilities, particularly with respect to the acquisition of tacit knowledge rather than to concerns about the complexity of technology itself (Tranfield, Young, and Partingtond, 1991; Zhou and Leydesdorff, 2004). One of the best ways to acquire such tacit knowledge is to reinforce the input–output relationship within higher education. This is a strategy that has been used as the focus of similar developments in other East Asian countries such as South Korea and Taiwan. In 1999 it became central to one of the important policies in China known as the "211 Project" (Kebin, 2006). Recent ambitious projects comprise a series of reforms initiated through the policy of deregulation of outward foreign direct investment in 2002. These were intended to provide the means of utilizing China's ever-growing foreign reserves (US$870 billion by 2005) throughout the world, as well as opening up the channels for acquiring first-tier foreign technology and tacit knowledge.

In its earlier stages, China's economic reform has been mostly aimed at structural transformation with an essential focus on the appropriation system involved in funding S&T research, particularly as it concerns the funds provided to the PRIs. In order to transfer S&T research outcomes from laboratories to their application in industry, the government has gradually reduced the funding available to PRIs beyond that required to sustain basic R&D. The intention has been to push them to acquire funding from other sources. Such change in the appropriation system has meant that PRIs are no longer government-controlled. They now mainly act as technology providers concerned with applied R&D and experimental development. This strategy has turned PRIs towards competing for participation in specific national or local development projects while retaining their government orientation.

The significantly increased R&D expenditure in the industry sector has improved the innovation capability (in terms of patenting rate) of enterprises. But, as indicated in Figure 41.3, this did not show its significance until the ratio reached 60% in 2000. The evidence in Figure 41.3 is also indicative that foreign patentees dominated about 60% of invention patents, while domestic patentees comprised approximately 90% of new-utility models and new-design models listed in the China Intellectual Property Office (STS, 2006). The characteristics of China's patenting

Table 41.1. Critical milestones of China's market-based economic reforms.

Year	*Legislation or policies*	*Main effect*
1985	Decisions on science and technology system reform	To solve the problem of separation of the linkage between science and industry
1986	Categorized all PRIs into three types accordingly	Nearly a quarter of PRIs had disappeared and many surviving PRIs became private entities later
1992	Deng Xiaopin's South Talk	Reinforce active interaction between the science and industry sectors, in which open FDI policies were induced
1993	Technology Progress Law	S&T development acts as one of the priorities in China's economic development
1996	Technology Transfer Law	To encourage the science sector to transfer R&D outcomes and set rules for technology market transactions
1998	State development through promoting science, technology, and education	Establish the rule of property rights of technology and technology transfer
1999	Decisions on technology innovation, development of high-tech and industrialization	Public institutions were set up to promote technology diffusion and facilitate market-based technology transactions
2000	"211 Project" for strengthening 100 top universities for the 21st century	Aiming to strengthen the knowledge base nationwide through research and advanced teaching capabilities
2001	World Trade Organization membership	Reinforce international openness, foreign interaction, and FDI. Market-based legal system is further emphasized under international regulations
2002	Policy deregulation for outward direct foreign investment—"Go Global"	China's increased involvement in global economic activity with substantially international M&A and investments thereafter

Source: China Year Reports, various years.

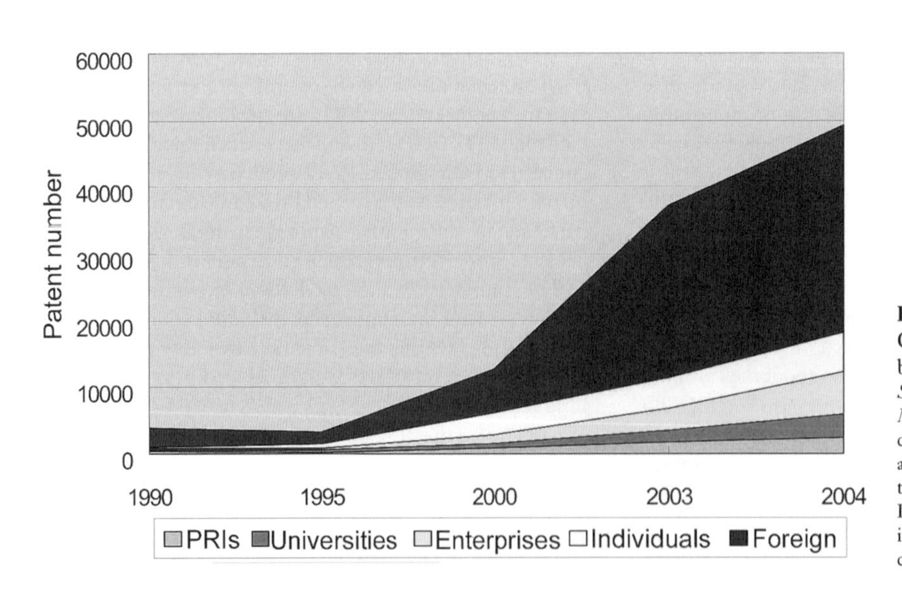

Figure 41.3. Patents granted in China's IPO, invention type—by sector.
Source: STS (2004–2006).
Note: China's patent system categorizes patents into three types, as most countries do (i.e., invention type, new utility, and new design). In this chapter we evaluate China's innovation capability by the most critical "invention" type.

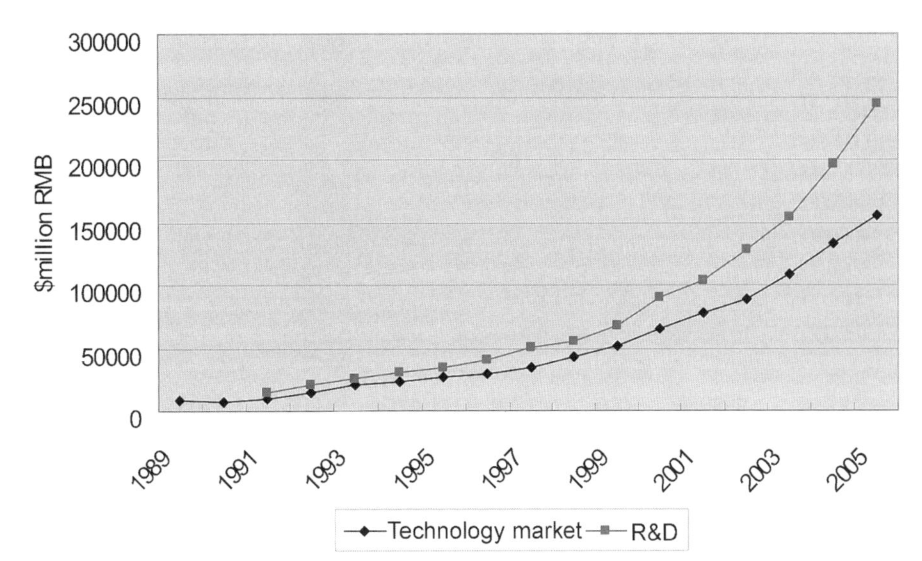

Figure 41.4. Technology market versus R&D expenditure, 1991–2005.
Source: NSB (Various years).

activity follow the linear evolutionary path found in other East Asian latecomer countries. That is, the process of building the national innovation system has been heavily reliant upon foreign technology in its earlier stages (Hobday, 1995; Mathews, 2002). In addition, the critical ratio of individual patentees since the mid-1990s reflects the encouragement of private enterprises to achieve levels of innovation that represent the equivalent of the innovation capabilities established in state-owned enterprises.[1]

41.3 Technology commercialization and science–industry linkage

China's first technology transaction center was regionally established as early as 1984 and involved 60 technology transfer offices in the Wuhan area. Since then, a series of technology transfer laws and policies have been activated nationally. These have included the establishment of the China Technology Market Association as a national organization in 1992, along with the provision of tax incentives by government for technology trade in the technology market. As shown in Figure 41.4, the former lack of any industry R&D enabled China's technology market to grow rapidly along with R&D expenditure, which increased 20 times from 1989 to 2005 and reached a peak of 155 billion RMB in 2005.

Regardless of the demands of the technology market, there has been a dramatically higher growth rate in this market than that in R&D expenditure in recent years (as shown in the Figure 41.4). This implies that the demand for market-based R&D knowledge has transcended the role of PRIs or science research institutions in the supply of technology at the current stage of China's transitional innovation system. Further evidence of this transition is shown in Figures 41.5 and 41.6 by the shift of participants in China's technology market. Enterprises with in-house R&D have become the primary technology providers as well as purchasers since the late 1990s. At the same time, the share of these activities undertaken by PRIs and state-owned enterprises has been significantly reduced as many of them have changed their role or mode of ownership (*China Statistical Yearbook*, 2006; Motohashi, 2005).

Figure 41.6 provides details of the types of buyers in industrial enterprises. The dramatic rise in the demand for technology in the limited companies sector comprising SMEs is further evidence that, since 2000, these SMEs have become an essential part of China's economy. They are now the key drivers of employment expansion and of economic growth and development. Indeed, up to the present date, China's SMEs account for almost half of GDP and provide 75% of urban and rural employment (ADB, 2007). The industrial output of SMEs has been increasing at an average annual rate of 30% compared with 20% for industry as a whole. This has occurred despite of the desire of the Chinese government to create world-class "national champions" from strategically redeveloped state-owned enterprises (Woo and Zhang, 2006).

Recent S&T data reflect the encouraging effect of technology policies on the establishment of science and industry linkages. Figure 41.7 indicates that the three main players in China's national innovation system represent different focuses on R&D activities. First, enterprises essentially concentrate on experimental development

[1] The high percentage of individual patentees is the unique characteristic of the Great Chinese Community (as also found in Taiwan). In general, the inventors are normally the founders and owners of high-risk and small-scale private enterprises so that they are likely to retain ownership of the patent rights personally for reasons of security and re-starting their companies (Mathews and Hu, 2007).

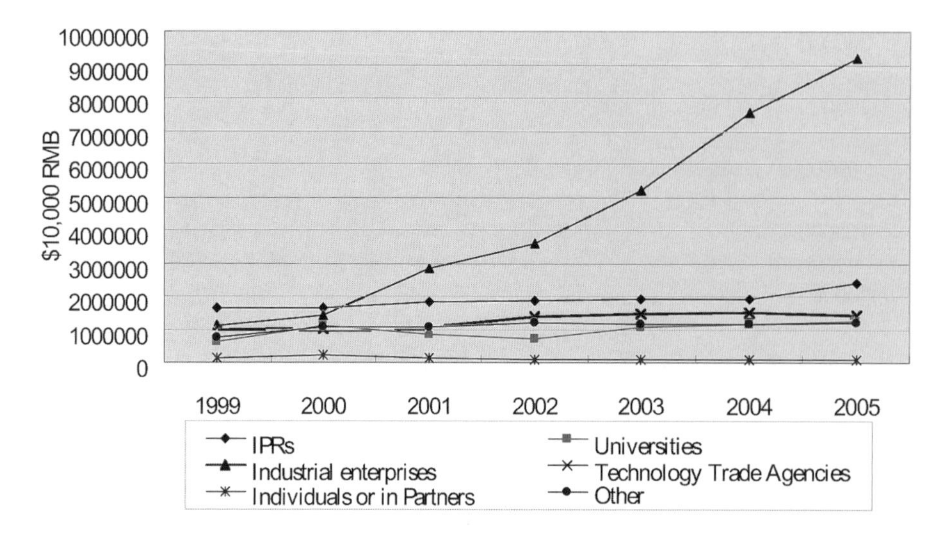

Figure 41.5. Technology transactions, by type of seller.
Source: compiled from NSB (Annual).

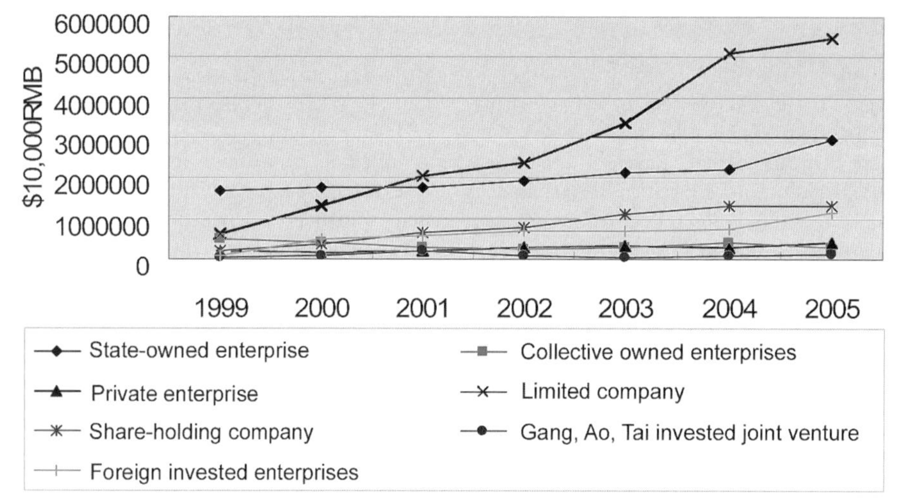

Figure 41.6. Technology transactions, by type of buyer.
Source: compiled from NSB (Annual).

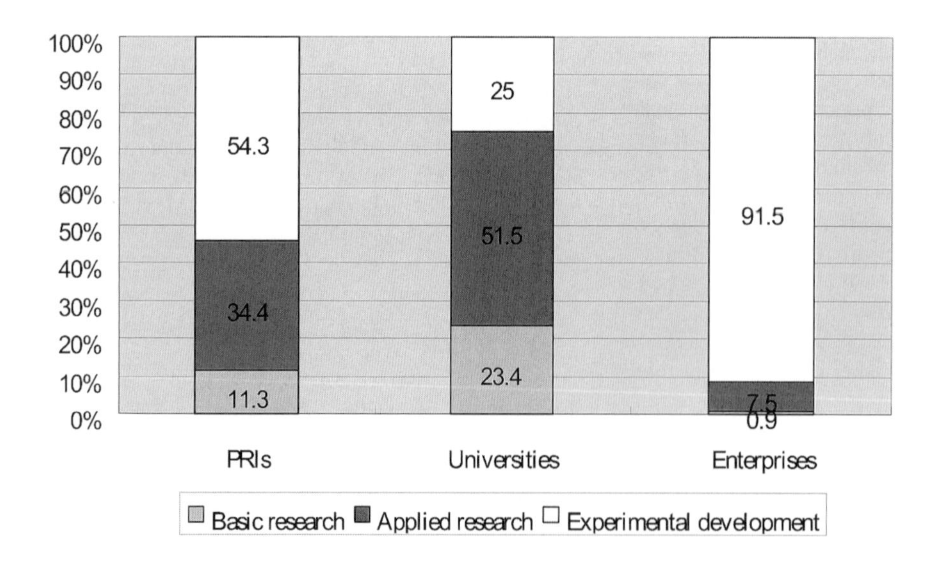

Figure 41.7. R&D activity in PRIs, universities, and enterprises, 2005.
Source: NSB (Annual).

(91.5%), which demonstrates a relatively lower level of technological capability. Second, universities primarily focus on applied R&D (accounts for 51.5%) and on basic research (23.4%). Finally, the main purpose of PRIs is to become closer to industry through focusing more on experimental development (accounts for 54.3%) and on applied research (34.4%). Consequently, it is not surprising to find that the surviving science institutions, PRIs, and universities have become popular partners for industry outsourcing, particularly for those enterprises in the science-intensive or knowledge-intensive sectors such as petrochemicals and drugs. This is especially so given the importance of government funding directed at PRIs and universities (*China Statistical Yearbook*, 2006).

Three important points should be noticed here. First of all, the increasing outsourcing activities that derive from China's weak industrial R&D create unique linkages between its science and industry sectors within which a substantial number of spinoffs are created by the surviving PRIs and universities (Eun, Lee, and Wu, 2006). In 1997, totals of 6,634 and 4,334 enterprises emerged as spinoffs from universities and research institutes, respectively. The accumulated number of university spinoffs reached 42,945 within seven years between 1997 and 2004 (MOE, 2002, 2005; THTIDC, 1999). It has been proposed by Eun, Lee, and Wu (2006) that the tendency of Chinese universities and PRIs to be linked to spinoffs rather than to other popular linking channels—such as collaborative research, and technology transfer and licensing prominent in other countries—needs to be understood in two ways. On the one hand, an effective science–industry linkage significantly depends on the internal absorptive capacity in each institution (Meyer-Krahmer and Schmoch, 1998). The historical separation of science and industry weakened this industrial absorptive capacity, whereas the forward integration initiated by universities with stronger research capabilities and abundant resources became the sole choice available, especially in the early stages of economic reform. On the other hand, the underdeveloped external environment—including intellectual property rights protection, the financial system, and the related laws and regulations—were inclined to stifle the transfer of scientific knowledge to industry. The transfer was even more difficult and inefficient without government support.

The second point that warrants attention here is that the lower level of technology capability in the private sector (mainly focused on experimental development), as well as the bank-based economy with imperfect financial markets (where heavy government regulation directs and affects prices of different financial assets), weakened the effective incentives for China's enterprises proactively to increase their investment in in-house R&D (Chan, Fung, and Thapa, 2006).[2] The third and final point to be emphasized is that where basic R&D attracted less attention and was primarily undertaken in universities, the ratio of S&T researchers allocated to universities has been decreasing over the years and reached only 16.6% in 2005. This indicates an overall lower level of basic R&D activity when international comparisons are made (OECD, 2006). These data certainly represent a challenge in the process of seeking an effective and long-term competitive advantage for China's innovation system.

41.4 Challenges of technology development

Corresponding to the outstanding economic growth rate of approximately 8% annually since 2000, the level of China's R&D expenditure as a proportion of GDP has increased nearly twofold from 0.7% in 1991 to 1.3% in 2005. However, as Figure 41.8 indicates, over the past 20 years China's innovation system has significantly relied on the increase in experimental development through financial and manpower inputs rather than through the higher value-adding creation of basic and applied R&D. In the period from 1992 to 2005, the proportion of R&D expenditure allocated to experimental development increased by 10 percentage points from 67% to 77%, while basic research remained at the level of approximately 5% and applied research fell by 10 percentage points from 27% in 1992 to 17% in 2005. The resource allocation devoted to both R&D expenditure and personnel reflect the fact that China's economic reform in the last 20 years has focused on internal organizational restructuring and the building of absorption capability in low-technology and medium-technology enterprises. This has been carried out by means of hundreds of legislative enactments and policy implementations.[3] It is also indicative that China's innovation system in this same period has concentrated on the diffusion of imported technology as befits its description in an OECD study as a large and rapidly catching-up country (OECD, 1999).

Four weaknesses in the process of building China's national innovation system can be identified as deriving from China's history of separating science from industry. First, the prior lack of industrial R&D and the function of direct support for S&T activities led to excessive demands

[2] In China, four big banks—Bank of China, China's Commercial and Industrial Bank, China's Agricultural Bank, and the People's Bank of China (The Central Bank)—control 70% of total banking assets. Moreover, 80% of total financing in China is bank-related (Fung, Tsiang, and Liu, 2007).

[3] In the last two decades, numerous legislative means and policy implementations were developed, along with many national programs. These mainly aimed at (1) internal organizational restructuring to improve the utilization of China's scientific and technological capabilities and resources, and (2) the utilization of foreign technology and knowledge through the venture capital policy of "exchange market for technology" (Steinfeld, 2004).

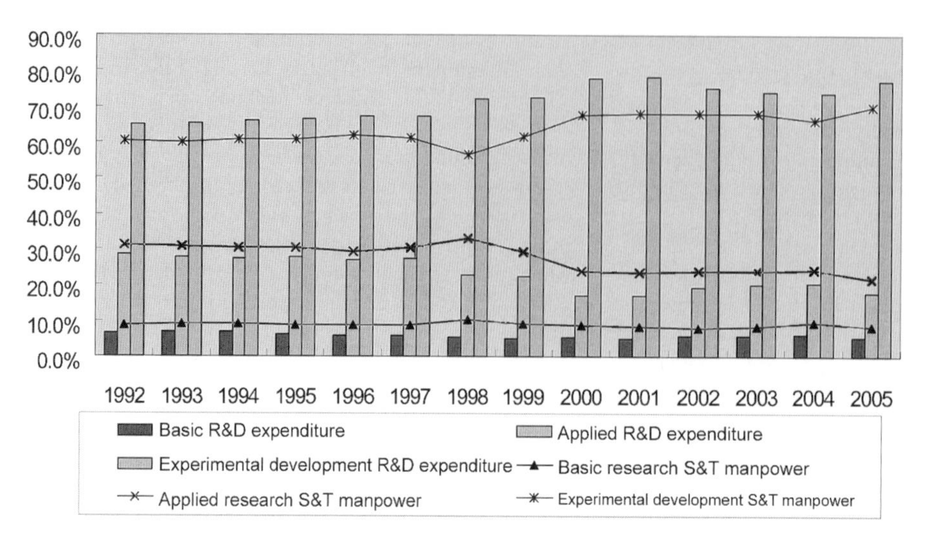

Figure 41.8. China's R&D expenditure and manpower, 1992–2005.
Source: STS (2004–2006).

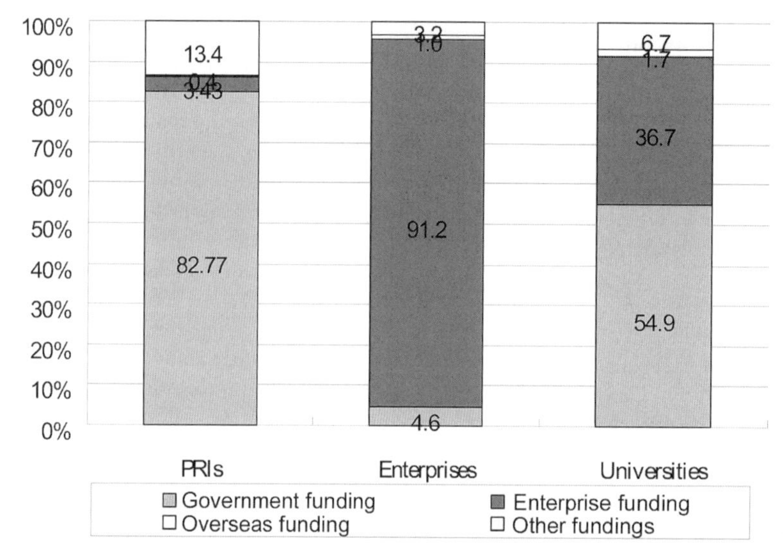

Figure 41.9. R&D expenditure, by source of funding, 2005.
Source: STS (2006).

on experimental development in order to respond to the requirements of mass production that would ensure enterprise profitability in an era of globalized and networked industrial integration. However, a great deal of emphasis was put on the reform of PRIs in the first half of the 1990s as a means of developing closer links between scientific research and industry. And, in the second half of the 1990s, such emphasis was particularly on the large and medium-size state-owned enterprises. However, it is still reported that between one-third and two-thirds of state-owned enterprises are running at a loss even though efficiency has been somewhat improved (Sun and Tong, 2003; Xue, 1997).[4] As shown in Figures 41.2 and 41.9, even though the proportion of total R&D expenditure allocated to PRIs has significantly decreased from 70% in 1986 to 21% in 2005, their funding by the government still accounted for 82.77% of their total revenue in 2005. A similar situation is found among China's enterprises. While R&D expenditure in industry has been increased to 68% in 2005, state-owned enterprises are still dominant in this sector and make up to 75% of the total expenditures (NSB, 2005). As suggested above, such strong government-led advances and the lack of industrial R&D due to the low absorptive capacity and biased financial markets, as well as the slow

[4] Corporate governance or managerial ownership in China may be one of the primary reasons for the unsuccessful reform of state-owned enterprises. While managers in state-owned enterprises are selected by the government, conflicting goals and fixed serving terms greatly reduce the incentives for pursuing profit and innovation (Li *et al.*, 2007 and Liu, 2001).

progress of enterprise reform, result in the burgeoning technology market being regarded as one of the major mechanisms involved in the rapidly growing science to industry linkages in China.[5]

The second weakness is the limited range of financing options available to China's large and medium-size state-owned and private enterprises. According to the World Bank's survey in 2001, Chinese enterprises relied heavily on retained earnings as their main source of financing because of the restrictive regulatory environment imposed by the government.[6] Relevant restrictions included, for example, quota restrictions on stockmarket listings and a regulated foreign exchange market. Consequently, the venture capital market is depressed and, overall, the financial infrastructure is unable to supply sufficient startup or entrepreneurial funding. Such funding is known to be one of the most essential engines driving the build-up of a healthy national innovation system.

The third weakness in the process of building China's national innovation system that should be cited here is the shortage of S&T research personnel. This is so despite the fact that technological advancement has become one of the priorities in China's economic development and that there exists a relatively large pool of R&D personnel in absolute terms. Measured in terms of the number of R&D personnel per 10,000 employees shows that China still lags well behind other countries; the ratio even increased from only 7.3 in 1999 to 18 in 2005. This outcome is to be compared with, for example, 135 in Japan, 129 in Russia, and 83 in South Korea (OECD, 2006), and it may well be due to China's education system giving greater emphasis to application-oriented studies rather than to basic education (Hu, 2000).

The fourth of the weaknesses to which reference should be made is the low level of technology capability in the private sector represented by the lack of absorptive and internalized capacities to collaborate with the science sector. This remains the case in spite of the rapid expansion of outsourcing activity in the technology market. Apparently, China's brilliant economic performance is mostly dominated by a dependence on low-end manufacturing of essentially standardized, non-differentiated products (Rosen, 2003; Steinfeld, 2004). In addition, due to the lack of adequate capabilities and the limited financing options in most Chinese enterprises, the linkages between science and industry are confined within the domestic economy and are geographically limited. The geographically constrained development within the global supply chain has caused the "shallow" integration of China's global production networking (Steinfeld, 2004). Despite the growing importance of foreign trade interactions, the R&D activities and supply chain networks among China's innovation institutions remain localized and domestic-oriented (WB, 2001). This is apparent in the overall scarcity and low level of involvements in international technology transfer, research collaborations, upstream supply networking, and downstream customer-based participations revealed in China's S&T database.[7]

41.5 Building the innovation infrastructure: policy-oriented

Different technological innovation capabilities and the impact of other relevant capabilities allow greatly varying competitive performance at the enterprise level as well as at the industry and the national levels. Successful technological innovation depends not only on technological capabilities, but also on other critical capacities that stem from the infrastructure in such areas as financial and legal systems, networking structures and capacities, organizational sophistication, strategic planning, learning capacities, and resource allocation (Yam, Guan, and Pun, 2004). In the process of building China's innovation infrastructure, the concurrence of internal and external factors plays an important role in the innovation capabilities that have emerged. The internal factors include, first of all, how China's financial market integrates within globalized markets, given its weak institutional environment in which credit ratings, high-quality appraisal and auditing services, and standardized reporting systems all remain underdeveloped. A second such internal factor is the formation of the legal system as it relates to intellectual property rights, especially the legislation of patents for China's flourishing manufacturing industry. The external factors include international openness and the utilization of external technology, both of which have become significant as incentives for China's policy-oriented national innovation system. These latter two factors are discussed in more detail below.

41.5.1 The development of intellectual property rights: patents

It is widely recognized in China today that the rule of law is essential for sustained growth. It is, though, far less clear how the rule of law can be achieved or even exactly what such a rule entails. Nevertheless, it is readily apparent that with technology diffusion regarded as one of the

[5] Three mechanisms of technology diffusion exist in China: technology markets, technology transfer contracts, and spinoffs (Chang and Shih, 2004).

[6] Financial restrictions thus result in an informal means of raising funds. Otherwise, business operations must run according to the orientation dictated by strong government policy (Chang and Shih, 2004).

[7] Although some exceptions are seen in several large-scale enterprises, such as Huawei in telecom equipment or Legend in electronics, active external networking within foreign interactions did not emerge in the later stages of their development until recent years (Fan, 2006).

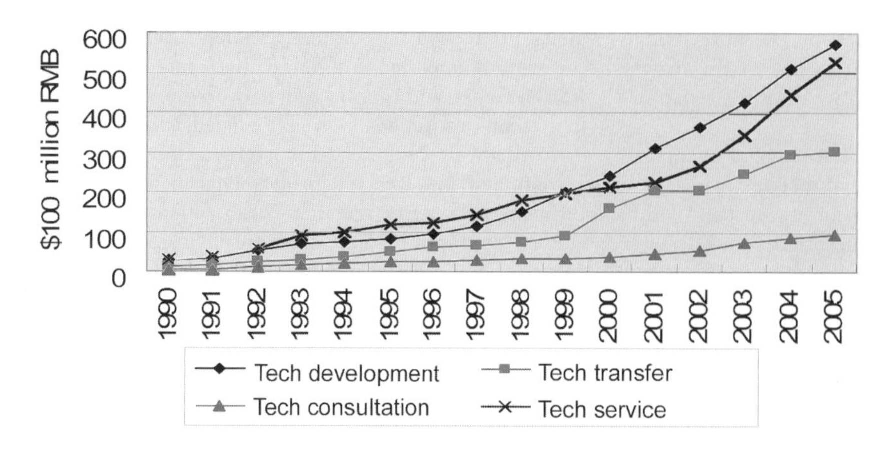

Figure 41.10. The types of technology transactions in industrial enterprises.
Source: compiled from NSB (Various years).

most important knowledge flows that can facilitate the integration of late-industrializing countries into the global economy, it is the law relating to intellectual property rights that must be given dominant emphasis.

China's patent law came into force in 1985 and has undergone two major rounds of modifications in 1992 and 2000 (Sun, 2003). In general, the Chinese patent system is, through its need to accord with national policy, more oriented towards promoting technology transfer and diffusion than towards protecting inventors' rights. Significant examples here are, first, the favoring of the single-claim system (even the more protected multiple-claim system is also acceptable). Second, the requirements for compulsory licensing according to which such licensing is possible where enterprises who have the capabilities to implement the patented invention and make reasonable offers to do so even though they fail to obtain licenses from the patentees (*The Patent Law of China*, 2000). Overall, the effect on the development of China's intellectual property rights, particularly as they concern patents, is represented in the significantly increased size of the technology market as shown in Figure 41.10. In this market, the share of technology development and services, together with the level of technology transfer, has grown over time (STS, 2006).

41.5.2 International openness and the utilization of external technology

China follows a policy of gradual advance in opening itself up to the outside world. However, the modes of open and modularized production that dominated in the earlier stages of development failed to allow Chinese firms effectively to connect with upstream and downstream supply chains. The result has been a "shallow" integration into the global economy. Indeed, most Chinese firms reported that the main inhibitors of export growth are costs of R&D and marketing that stem from the limited financing options imposed by government restrictions, as well as from the demands of meeting the requirements of legal product standards due to the protection of IPRs (Steinfeld, 2004). The impact of limited financial options for entrepreneurial firms restricts the diversification of investments and leads to an emphasis on maintaining liquidity. Enterprises thus remain focused on lower entry barrier activities, with the result being that the desirable skills and innovative capabilities have little opportunity to accumulate. These inhibitors are compensated for, slowly, by available foreign technology and other capabilities acquired consequent upon the state policy rubric of "exchange domestic market for foreign technology" that has been in effect since Deng Xiaopin's South Talk in 1992.

However, instead of waiting passively for "natural" transfers of external knowledge and technology, policymakers in Beijing have chosen proactively to attract foreign direct investment through industrial joint ventures and special economic zones. Such special economic zones as Shanghai's Pudong New Area have been leaders in China opening itself up to the outside world. They have made an enormous contribution to boosting the country's international interactions by establishing connections that strengthen its economic cooperation with foreign countries around the world. This is particularly so with respect to connections within the regional economic development area comprising Asian countries (with more than 60% of total FDI coming from these countries; see Table 41.2). For example, the main contributor to export growth is computers and electronics-related production; two-thirds of such export production rely on links to Taiwan-based IT firms (Chen, 2004; see also Figure 41.11). This corresponds with the fact that Taiwan has been the largest source of foreign patents in China since 1990, and it only became second to Japan in 2003. This trend is shown in Figure 41.12.[8] It is to be noted that the real scale of investment

[8] The counting of foreign patents granted in China includes all three categories: inventions, new-utility models, and new industrial designs. Most of Taiwan's patents granted in China are new-utility models, while Japanese patents are mostly of the inventions type (NBS, various years).

Table 41.2. Foreign direct investment in China, by area and country, 2000–2004 (unit: US$10,000).

	2000	*2001*	*2002*	*2003*	*2004*	*Ranking in 2004*
Asia	2,548,209	2,961,326	3,256,997	3,410,169	3,761,986	
Hong Kong	1,549,998	1,671,730	1,786,093	1,770,010	1,899,830	1
Japan	291,585	434,842	419,009	505,419	545,157	3
Singapore	217,220	214,355	233,720	205,840	200,814	6
South Korea	148,961	215,178	272,073	448,854	624,786	2
Taiwan	229,658	297,994	397,064	337,724	311,749	5
Africa	28,771	32,977	56,462	61,776	77,568	
Europe	476,539	448,398	404,891	427,197	479,830	
Germany	104,149	121,292	92,796	85,697	105,848	
Latin America	461,658	630,891	754,979	690,657	2,757,891	
North America	478,579	509,685	649,032	516,135	6,207,063	
United States	438,389	443,322	542,392	419,851	394,095	4
Oceanic and Pacific Islands	69,403	101,478	141,722	173,119	197,437	
Others	8,322	3,004	10,203	71,414	144,065	
Total	*4,071,481*	*4,687,759*	*5,274,286*	*5,350,467*	*6,062,998*	

Source: compiled from NSB (various years).

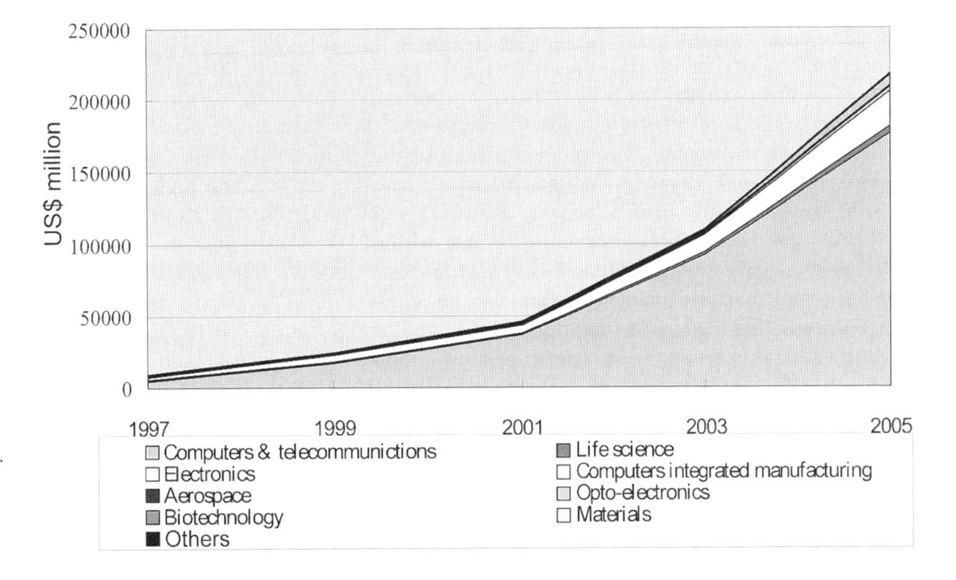

Figure 41.11. Exports of high-tech products, by industry.
Source: NSB (Various years).

from Taiwan is substantially larger than the official figures in Table 41.2 indicate because political restrictions have forced numerous Taiwanese sources of capital to appear to flow into China through different areas and channels, mainly through Hong Kong (Chang and Shih, 2004). This is due to the linguistic, cultural, racial, historical, and geographical similarities that provide links within the Great Chinese Community.

41.6 Skipping stage of industrial development

Despite the historical weaknesses and the relatively low intensity of technological development compared with that in advanced countries, China's continuously government-led reforms and industrial transformations have seen some notable progress in recent years. The nation's

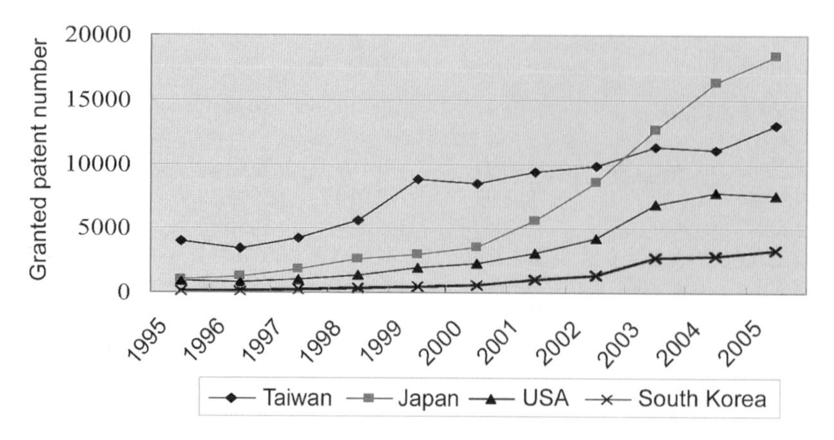

Figure 41.12. Top four foreign patentees in China, 1995–2005.
Source: compiled from NSB (Various years).
Note: (1) The patent number includes all three types in China's Intellectual Property Oce. (2) Taiwan's patenting activity in China is counted as domestic patents in the China Bureau of Statistics but we convert the data in this chapter.

technical innovation capacity has been reinforced by the accelerated increase of R&D inputs in terms of expenditure and manpower, and the resulting expansion of outputs when measured as patenting rates and technology transfer revenues.

Beginning with the implementation of the 10th Five-year S&T Plan (2001–2005), the Chinese government introduced a series of policies directed at (and enacted relevant legislative support for) linking the existing misalignments in the national innovation system. The measures taken include providing preferential policies and tax reductions (up to 50% of actually accrued expenditures) to encourage industrial R&D input, particularly for high value-added industries; supporting the development of venture capital and its related operations; enhancing the management of intellectual property protection; promoting cooperation among enterprises and the combination of S&T resources between industries and scientific research; and increasing the availability of S&T and higher education personnel, especially within the areas of essential sciences. Hundreds of such policies have been implemented regionally or nationally since the 10th Five-year S&T Plan was enacted, all targeted at supporting and promoting the linkages among innovation institutions (PRIs, universities, and enterprises) by means of building up their various innovation capacities. Along with the innovations and new instruments that are available in China's managed foreign exchange market, these efforts have been reflected in its recent export performance involving high-tech products. As indicated in Figure 41.13, China has relied heavily on high-tech imports to fuel its low value-added manufacturing activities (Steinfeld, 2004). For the first time in 2004, China's exports of high-tech products exceeded its imports in terms of product value, an outcome that may be argued implies the building up of absorption capacity in the private sector and provides evidence that China's innovation system is evolving and continuously moving forward.

It is also to be noticed that China's rapid catching-up in its rate of industrial development is more likely by way of a "skipping stage" or leapfrog development. That is, from the OEM mode it jumps directly to the OBM mode through strong government support for outward direct investment and international mergers and acquisitions (M&A) as found in the model deployed in South Korea (Lee, Jee, and Eun, 2007). In a manufacturing-based economy, such leapfrog development may lead to a fragmented global industrial supply chain which requires strong integration and complementary capabilities within the internal and external value chain to ensure its effective performance. China's "shallow" integration into the global production network has witnessed the challenges of such a fragmented supply chain, as well as of segmented financial markets (Steinfeld, 2004).[9] In South Korea, apart from chaebols[10] with their strong integration capacities, the "skipping stage" of industrial development weakened overall design capability, intellectual property rights activity, and multinational management capability, all of which were apparent in the diminishing significance of SMEs within the global industrial supply chain (Field, 2007; IBK, 2006). By contrast, the other catch-up industrial development model of Taiwan evolved stage by stage from OEM to ODM to OBM modes. In this case, the relevant capabilities were more well built and international M&A activities have only become prosperous in recent years with stronger SMEs having the capacity to integrate into the global industrial supply chain.

41.7 Concluding remarks

This study utilized the data available in the various national and international S&T databases that combine

[9] There are two classes of ownership-restricted shares in China's financial market: (1) A-shares, which can be owned and traded only by Chinese citizens; and (2) B-shares, which since February 2001 can be owned and traded by foreigners and by Chinese residents who hold foreign currencies.

[10] Large business conglomerates, typically family-owned, in South Korea.

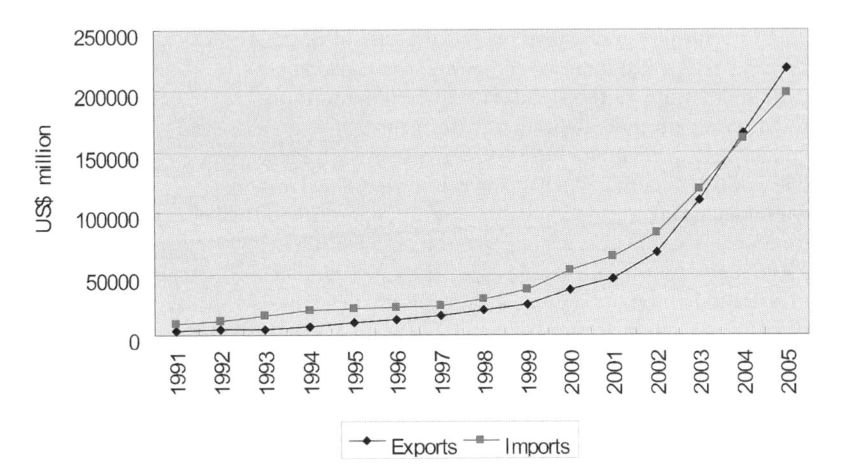

Figure 41.13. Imports and exports of high-tech products.
Source: STS (2006).

firm-level, industry-level, and country-level observations to examine the progress and processes involved in China's national innovation system over the last 20 years. China's PRIs at both the local and the national level were the backbone of its innovation system before the mid-1990s. However, increasing outsourcing activities derived from the weak state of industrial R&D led to the particular emergence of linkages between China's science and industry sectors comprising spinoffs rather than other forms such as technology transfer or licensing.

Four weaknesses in the process of building the national innovation system in China were identified as originating in China's historical separation of science and industry. First, the lack of industrial R&D and the function of direct support for S&T activities led to excessive demands on experimental development in order to respond to the requirements of profitable mass production in an era of globalized and networked integration. The second weakness has been the limited financing options for large and medium-size state-owned and private enterprises. Venture capital markets remained depressed, and there has been an overall inability to supply the startup or entrepreneurial capital that is most essential in building up an effective national innovation system. The third weakness consists in the shortage of S&T researchers that has emerged in spite of the priority given of technological development in China. There is a relatively large pool of R&D personnel consistent with the nation's large population, but it is relatively inadequate in terms of the proportion of the total labor force. Fourth, the low level of technology capability in the private sector represents a lack of the absorptive and internalized capabilities required to collaborate effectively with the science sector, even though outsourcing activity in the technology market has grown rapidly. Such insufficient capabilities and limited financing opportunities, which affect most Chinese enterprises, have meant that the linkages between science and industry have remained largely domestic and geographically limited.

By examining China's technology transactions, it is found that the dramatic rise in the demand for technology in the limited companies sector comprising SMEs is further evidence that, since 2000, SMEs have become an essential part of China's economy. They are now the key drivers of employment expansion and of economic growth and development. Indeed, up to the present date, China's SMEs account for almost half of GDP and provide 75% of urban and rural employment (ADB, 2007). The industrial output of SMEs has been increasing at an average annual rate of 30% compared with 20% for industry as a whole. This has occurred despite the desire of the Chinese government to create world-class national champions from strategically redeveloped state-owned enterprises (Woo and Zhang, 2006).

Successful technological innovation depends not only on technological capability, but also on other critical capabilities within the relevant infrastructure, including such areas as financial and legal systems, networking structures and capacities, organizational sophistication, strategic planning, learning opportunities, and resource allocation (Yam, Guan, and Pun, 2004). In the process of building China's innovation infrastructure, the concurrence of internal and external factors plays an important role in affecting innovation capabilities. The internal factors include, first, how China's financial market integrates with the globalized market given its weak institutional environment in which credit ratings, high-quality appraisal and auditing services, and standardized reporting systems remain underdeveloped. A second internal factor is the formation of the legal system relating to intellectual property rights, especially with respect to patent legislation in China's flourishing manufacturing industry. As for external factors, such as international openness and the utilization of external technology, these have turned into significant incentives for China's policy-oriented national innovation system. While China's rapidly catching-up industrial development is more likely to involve the skipping stage or leapfrog development (as was the case in South Korea), strong integration and complementary capabilities in the building up of internal and external factors are critical in delivering effective innovation performance.

After implementation of the 10th Five-year S&T Plan (2001–2005), hundreds of policies were introduced regionally or nationally with the intention of supporting and promoting the linkages among innovation institutions (PRIs, universities, and enterprises) through building up various innovation capacities. Along with the innovations and new instruments now available in China's managed market, the results of these efforts have become apparent in China's export value of high-tech products growing to exceed its imports of these products for the first time in 2004. This may well be indicative of the building up of absorption capacity in the private sector and provide evidence that China's innovation system is evolving and continuously moving forward.

41.8 References

ADB. (2007). *Asian Development Outlook 2006*. Asia Development Bank; *http://www.adb.org/Documents/books/ADO/2005/prc.asp*

Alcorta, L., and Peres, W. (1998). "Innovation systems and technological specialization in Latin America and the Caribbean." *Research Policy*, **26**, 857–881.

Arocena, R., and Sutz, J. (1999). *Looking at National Systems of Innovation from the South* (mimeo). Montevideo, Uruguay: Universidad de la República.

Buck, T., Filatotchev, I., Nolan, P., and Wright, M. (2000). "Different paths to economic reform in Russia and China: Causes and consequences." *Journal of World Business*, **35**(4), 379–400.

Carlsson, B., and Jacobsson, S. (1994). "Technological systems and economic policy: The diffusion of factory automation in Sweden." *Research Policy*, **23**, 235–248.

Carlsson, B., and Stankiewicz, R. (1995). "On the nature, function and composition of technological systems." In: *Technological Systems and Economic Performance: The Case of Factory Automation* (pp. 21–56). Dordrecht, The Netherlands: Kluwer Academic.

Cassiolato, J., and Lastres, H. (1999). "Local, national and regional systems of innovation in the Mercosur." *DRUID's Summer Conference on National Innovation Systems, Industrial Dynamics and Innovation Policies*; *http://www.business.auc.dk/druid*

Cassiolato, J., and Lastres, H.(2000). "Local systems of innovation in Mercosur countries." *Industry and Innovation*, **7**(1), 33–53.

China Statistical Yearbook. (2006). Beijing: China Statistics Press.

Coe, D. T., Helpman, E., and Haffmaister, A. W. (1997). "North–South R&D spillovers." *Economic Journal*, **107**, 134–149.

Chan, K. C., Fung, H. G., and Thapa, S. (2006). "China financial research: A review and synthesis." *International Review of Economics and Finance*.

Chang, P. L., and Shih, H. Y. (2004). "The innovation systems of Taiwan and China: A comparative analysis." *Technovation*, **24**(7), 529–539.

Chen, S. H. (2004). "Taiwanese IT firms' offshore R&D in China and the connection with the global innovation network." *Research Policy*, **33**, 337–349.

Dahlman, C. J., and Sananikone, O. (1990). *Technology Strategy in the Economy of Taiwan Province of China: Exploiting Foreign Linkages and Investing in Local Capability*. Washington, D.C.: World Bank.

Eaton, J., and Kortum, S. (1999). "International technology diffusion: Theory and measurement." *International Economic Review*, **40**(3), 537–570.

Eun, J. H., Lee, K., and Wu, G. (2006). "Explaining the 'university-run enterprises' in China: A theoretical framework for university–industry relationship in developing countries and its application to China." *Research Policy*, **35**, 1329–1346.

Fan, P. (2006). "Catching up through developing innovation capability: Evidence from China's telecom-equipment industry." *Technovation*, **26**, 359–368.

Field, A. (2007). "Seoul sleepwalk: Why an Asian export champion is at risk of losing its way?" *Seoul Financial Times*, March 19.

Fung, H. G., Tsiang, Y. S., and Liu, Q. (2007). "Financing alternatives for Chinese small and medium enterprises: The case for a small and medium enterprise stock market." *China and World Economy*, **15**(1), 26–42.

Hobday, M. (1995). *Innovation in East Asia*. Cheltenham, U.K.: Edward Elgar.

Hoskisson, R. E., Eden, L., Lau, C. M., and Wright, M. (2000). "Strategy in emerging economies." *Academy of Management Journal*, **43**(3), 249–267.

Hu, M. C., and Mathews, J. (2005). "National innovative capacity in East Asia." *Resaerch Policy*, **134**(9), 1322–1349.

Hu, Z. J. (2000). *National Innovation System: A Theoretical Analysis and an International Comparison* (in Chinese). Beijing: Social Science Literature Press.

IBK. (2006). *SME Leadership Should Drive Growth* (Russel Kopp Asian Banks Analyst). Seoul: Industrial Bank of Korea.

Jefferson, G. H., Tholmas, G. R., Wang, L., and Yuxin, Z. (2000). "Ownership, productivity change, and financial performance in Chinese industry." *Journal of Comparative Economics*, **28**(4), 784–813.

Kash, D., Auger, R. N. and Li, N. (2004). "An exceptional development pattern." *Technological Forecasting and Social Change*, **71**, 777–797.

Katz, J. (1995). "Industrial organization, international competitiveness and public policy in Lartin America in the nineties." *Revue d'Economies Industrielles*, **71**, 91–106.

Kebin, H. (2006). "Opportunities and challenges of China higher education in the era of globalization"; *http://www.apru.org/activities/ssm/2006/PowerPoint%20Presentations/Panel%20-Discussion%201%20(3)%20-%20Dr%20He%20Ke-bin%20(Tsinghua).pdf*

Lee, K., Jee, M., and Eun, J. H. (2007). "China's economic catch-up in a comparative perspective: Washington consensus, East Asia consensus or Beijing Model?" *World Forum on China Studies, Shanghai, China*.

Li, D., Moshirian, F., Nguyen, P., and Tan, L. W. (2007). "Managerial ownership and firm performance: Evidence from

China's privatization." *Research in International Business and Finance*.

Li, H., and Atuahene-Gima, K. (2001). "The adoption of agency business activity, product innovation, and performance in Chinese technology ventures." *Strategic Management Journal*, **23**, 469–490.

Liu, X. (2001). *Chinese Technology Innovation System in the 21st Century* (in Chinese). Beijing: Peking University Press.

Mathews, J. A. (2002). "Competitive advantages of the latecomer firm: A resource-based account of industrial catch-up strategies." *Asian Pacific Journal of Management*, **19**(4), 467–488.

Mathews, J. A., and Hu, M. C. (2007). "Enhancing the role of universities in building national innovative capacity in Asia: The case of Taiwan." *World Development*, **25**, 245–264.

Meyer-Krahmer, F., and Schmoch, U. (1998). "Science-based technologies: University–industry interactions in four fields." *Research Policy*, **27**, 835–851.

MOE. (2005). *Year 2004 Statistical Report of University-run Industry in China*. Chengdu, China: Ministry of Education.

MOE. (2002). *Year 2001 Statistical Report of University-run Industry in China*. Chengdu, China: Ministry of Education.

Motohashi, K. (2005). *China's Innovation System Reform and Growing Science Industry Linkage*. Tokyo: Research Center for Advanced Science and Technology, University of Tokyo.

NSB. (2005). *Science and Technology Statistics Yearly Report 2004*. Beijing: National Statistical Bureau; *http://www.stats.gov.cn/english/*

NRCSTD. (2003). *Report on China's S&T Development Research* (in Chinese). Beijing: National Research Center for Science and Technology Development.

OECD. (2006). *OECD Investment Policy Reviews: China*. Paris: Organization for Economic Cooperation and Development.

OECD. (1999). *Globalization of R&D: Policy Issue*. Paris: Organization for Economic Cooperation and Development.

Rosen, S. (2003). *Hollywood, Globalization and Film Markets in Asia: Lessons for China?* (pp. 115–143, in Chinese). Shanghai: Fudan University Press.

SSTC. (1993). *China's Law and Policy* (in Chinese). Beijing: State Science and Technology Commission.

Steinfeld, E. S. (2004). "China's shallow integration: Networked production and the new challenges for late industrialization." *World Development*, **32**(11), 1971–1987.

STS. (2004–2006). *China Science and Technology Statistics Data Book, 2004-2006*. Science and Technology Statistics Office; *http://www.sts.org.cn*

Sun, Q., and Tong, W. (2003). "China share issue privatization: The extent of its success." *Journal of Finance and Economics*, **70**, 183–222.

Sun, Y. (2003). "Determinants of foreign patents in China." *World Patent Information*, **25**, 27–37.

The Patent Law of China. (2000). "Article 48, Chapter 6." Beijing: China Law Publishing House.

THTIDC. (1999). *China New and High-tech Industrialization Development Report*. Beijing: Touch High-tech Industrialization Development Center/Science Press, China.

Tranfield, D., Young, M., and Partingtond, D. (1991). "Knowledge management routines for innovation projects: Developing a hierarchical process model." *International Journal of Innovation Management*, **7**(1), 27–49.

Woo, Y. P., and Zhang, K. (2006). "China goes global: The implications of Chinese outward direct investment for Canada"; *http://www.economics.ca/2006/papers/0892.pdf*

WB. (2001). *World Development Indicators Database: Data and Statistics*. Washington, D.C.: World Bank; *http://www.worldbank.org/data/countrydata/countrydata.html*

Wörner, S., and Reiss, T. (1999). *Technological Systems*. Brighton, U.K.: Brighton Science and Technology Policy Research; *http://www.sussex.ac.uk/spru*

Xue, L. (1997). "A historical perspective of Chin's innovation system reform: A case study." *Journal of Engineering and Technology Management*, **14**, 67–81.

Yam, R., Guan, J., and Pun, K. (2004). "An audit of technological innovation capabilities in Chinese firms: Some empirical findings in Beijing, China." *Research Policy*, **33**, 1123–1140.

Zhou, K. Z., Yim, C. K., and Tse, D. K. (2005). "The effects of strategic orientations on technology- and market-based breakthrough innovations." *Journal of Marketing*, **69**(2), 42–60.

Zhou, P., and Leydesdorff, L. (2004). "China's research output in nanotech takes the lead in the world in 2004: The world position of China's publications." *Chinese S&T Daily*, October 23; *http://www.stdao;y.com/gb/stdaily/2004-10/23/content_314274.html*

42

The National Innovation System in Taiwan

Ta-Jung Lu and Jong-Wen Wann

National Chung-Hsing University

Introduction

As they move farther into the 21st century, nations around the world all face global competition as the result of the rapid development of globalization and the knowledge economy. Taiwan, formally known as the Republic of China, is no exception, and it faces challenges not only because of intensifying globalization but also from emerging Asian economies, a declining birthrate, an aging population, and decreasing competitiveness. The weakening of competitiveness has resulted in lowered exports, a diminished GDP growth rate, and a high unemployment rate. In order to increase national economic growth and improve national competitiveness, Taiwan has been devoting great efforts to improving its innovation capabilities by investing heavily in basic research, accelerating research and development (R&D), attracting and cultivating science and technology (S&T) personnel, and developing key technology-intensive industries. Taiwan has successfully created its information communication technology (ICT)-based economic development and has begun to play instrumental roles in the world over the past two decades, thanks to well-planned industrial policies and the creation of a strong S&T infrastructure. However, Taiwan has yet to exploit its existing S&T strengths, upgrade industrial technology, and build new innovative capabilities to continue providing quality lifestyles for its citizens, to merge into the international community, and to build a model of sustainable economic and social development.

Because S&T development has been an important part of Taiwan's government policy, its national innovation system (NIS) is called the S&T system. For example, the government enacted the Fundamental S&T Act in 1999 to set guidelines and principles for the promotion of S&T development.[1] After the enactment of the Fundamental S&T Act, the Executive Yuan (a government branch similar to the Cabinet in the U.K.) has successively published National S&T Development Plans (2001–2004 and 2005–2008)[2] and White Papers on S&T (2003–2006 and 2007–2010),[3] each laying out the current status, visions, and strategies of S&T development to guide the promotion of S&T development in Taiwan.

The history of Taiwan's NIS development[4]

Realizing that S&T are the basis of national competitiveness and having observed the poor scientific environment and shortage of human resources in science, prominent scientist Dr. Ta-You Wu[5] proposed to the government in 1957 the formulation of a long-term academic development plan. In 1959, with the help of Dr. Shi Hu,[6] then President of the Academia Sinica, the government established the Long-range National Science Development Council, the earliest government agency in charge of promoting scientific development in Taiwan. During

[1] Fundamental Science and Technology Act: *http://web1.nsc.gov.tw/public/data/47149592271.htm*

[2] National Science and Technology Development Plan (2001–2004): *http://web1.nsc.gov.tw/public/data/47149583471.pdf* (2005–2008); *http://web1.nsc.gov.tw/public/data/571410244471.pdf*; for English version see (2001–2004): *http://web1.nsc.gov.tw/public/data/5831533771.htm* (Abstract); (2005–2008): *http://web1.nsc.gov.tw/public/data/581115222671.pdf*

[3] White Paper on Science and Technology (2003–2006): *http://web1.nsc.gov.tw/public/data/4714957471.pdf* (2007–2010): *http://web1.nsc.gov.tw/public/data/72261171171.pdf*; for English version see (2003–2006): *http://web1.nsc.gov.tw/public/data/47141051171.pdf* (Executive Summary); (2007–2010): *http://web1.nsc.gov.tw/public/data/74141117271. pdf*

[4] "The Story of Taiwan: Science and Technology": *http://www.gio.gov.tw/info/taiwan-story/science/tw_s03.html*

[5] Ta-You Wu was a Chinese-born atomic and nuclear theoretical physicist (1907–2000) who worked in the United States, Canada, mainland China, and Taiwan. He has been called the Father of Chinese Physics. He held various positions in Taiwan, including the President of the Academia Sinica (1983–1994). See: *http://en.wikipedia.org/wiki/Ta-You_Wu*

[6] Shi Hu was a Chinese philosopher and essayist (1891–1962). Hu is widely recognized today as a key contributor to Chinese liberalism and language reform in his advocacy of the use of vernacular Chinese. He was the President of the Academia Sinica (1958–1962). See: *http://en.wikipedia.org/wiki/Hu-Shi*

this period, the focus of research was primarily on meeting the everyday needs of the people, through such means as increasing the production of agricultural products.

Throughout the 1960s, Taiwan had established export-processing zones to increase employment opportunities, attract investment, and acquire foreign technology. The objectives of policy had shifted to strong emphasis on improving the investment environment, promoting labor-intensive industries, and establishing an export-oriented manufacturing sector.

In 1967, the Science Development Steering Committee was established as a policy research and advisory agency under the National Security Council. In the same year, the Long-range National Science Development Council was expanded and renamed the National Council on Science Development. Reorganized as the National Science Council, Executive Yuan (NSC) in 1969, the NSC became the main governmental promotion and funding body for S&T research in Taiwan, in charge of the planning, promotion, coordination, and execution of national S&T development policies. In 1968 the Science Development Steering Committee and the NSC jointly formulated a series of 4-year National Scientific Development Plans, to be implemented in three phases over a period of 12 years. In addition, the National S&T Development Fund was established in 1969 to improve S&T capabilities, encourage outstanding personnel, and enhance facilities in S&T R&D to facilitate timeliness and maximize effectiveness.

The 1970s were marked by strong economic growth and continued government encouragement of S&T. The government carried out, in three phases, a 12-year National Science Development Plan to strengthen the environment for both basic and industrial research. A variety of research and development centers were established during that period, such as the Asian Vegetable R&D Center, the Taiwan Plant Protection Center, the Precision Instruments Development Center, and the S&T Information Center, to improve the research infrastructures. In 1973 the United Industrial Research Institute, the Mining Research Institute, and the Metal Industries Institute were combined to become the Industrial Technology Research Institute (ITRI), which has acted as the engine for industrial innovation in Taiwan.

The Executive Yuan held the First National S&T Conference in 1978 to formulate a national science policy that would raise the nation's technological standards and further upgrade industry. Scholars, experts, business leaders, and heads of related government agencies attended the conference. Their decisions and recommendations were incorporated into the S&T Development Plan, which was promulgated in May of 1979 to provide guidelines for ministries, councils, and agencies.

In 1979, the Executive Yuan established the S&T Advisory Group (STAG) to implement the S&T Development Plan. Since its founding, the group has held annual meetings on S&T to promote the nation's technological development.

In 1982 the Second National S&T Conference not only reviewed the performance of earlier development projects but also revised and formulated the S&T Development Plan. In addition, the conference strengthened the cultivation and recruitment of advanced S&T personnel needed to meet the goals of social transformation and national development. During the 1980s, in order to build a stronger foundation for basic scientific research in a variety of fields, the Development Center for Biotechnology (DCB) and the Application Center for Precision Instruments were established. In 1986 the Third National S&T Conference created the 10-year Long-term S&T Development Plan to define objectives, policies, and areas of emphasis for S&T development. The Fourth National S&T Conference, convened in 1991, drafted a 12-year, Long-term National S&T Development Plan, to be implemented between 1991 and 2002, as well as a more detailed 6-year, Mid-term National S&T Development Plan to be in effect from 1991 to 1996. These plans explicitly set the nation's mid- and long-term goals for S&T development. In 1996 the Fifth National S&T Conference laid down the objectives of technological development and made solid proposals for related budgets, legislation of scientific development, advanced technological research, high-tech industrial development, and the balancing of science and the humanities. At the Sixth and Seventh National S&T Conferences, convened in 2001 and 2005, National S&T Development Plans for the periods of 2001–2004 and 2005–2008 were approved, to set the strategies and objectives to be accomplished by 2004 and 2008, respectively, as well as action plans to reach the goals.

In 1997, Taiwan's first White Paper on S&T was issued, and measures for promoting a sci-tech nation were approved one year later, to transform Taiwan into an S&T-based country. Coming into effect in 1999, the Fundamental S&T Act, which laid down the basic principles and directions for the promotion of S&T development in Taiwan, has had a profound impact on the environment and culture of research institutions. In accordance with this act, NSC has funded many of the 160+ universities and research institutions in Taiwan to set up the office of technology licensing, which has resulted in a greatly increased number of patent applications and technology transfer deals. Other government agencies, including the Council of Agriculture (COA), have followed suit and granted the ownership of intellectual property rights to the executing institutions of publicly funded research projects.

In summary, the timeline of Taiwan's S&T system and policy development is outlined in Figure 42.1. The organizational framework for S&T development in Taiwan is shown in Figure 42.2. For a detailed organizational chart of S&T development framework see Appendix 42.A.

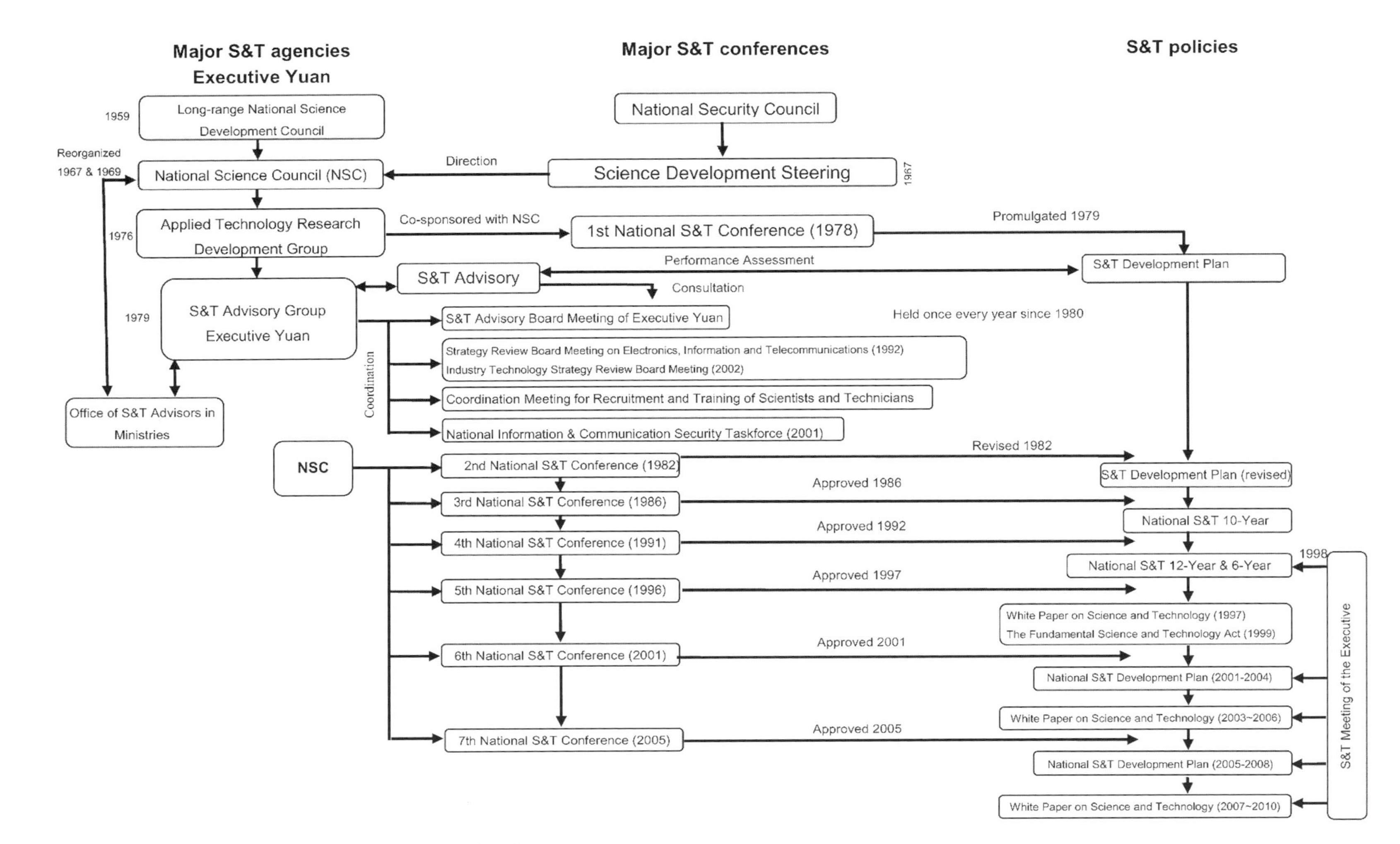

Figure 42.1. Timeline of Taiwan's S&T system and policy development.
Source: NSC (2007–2010).

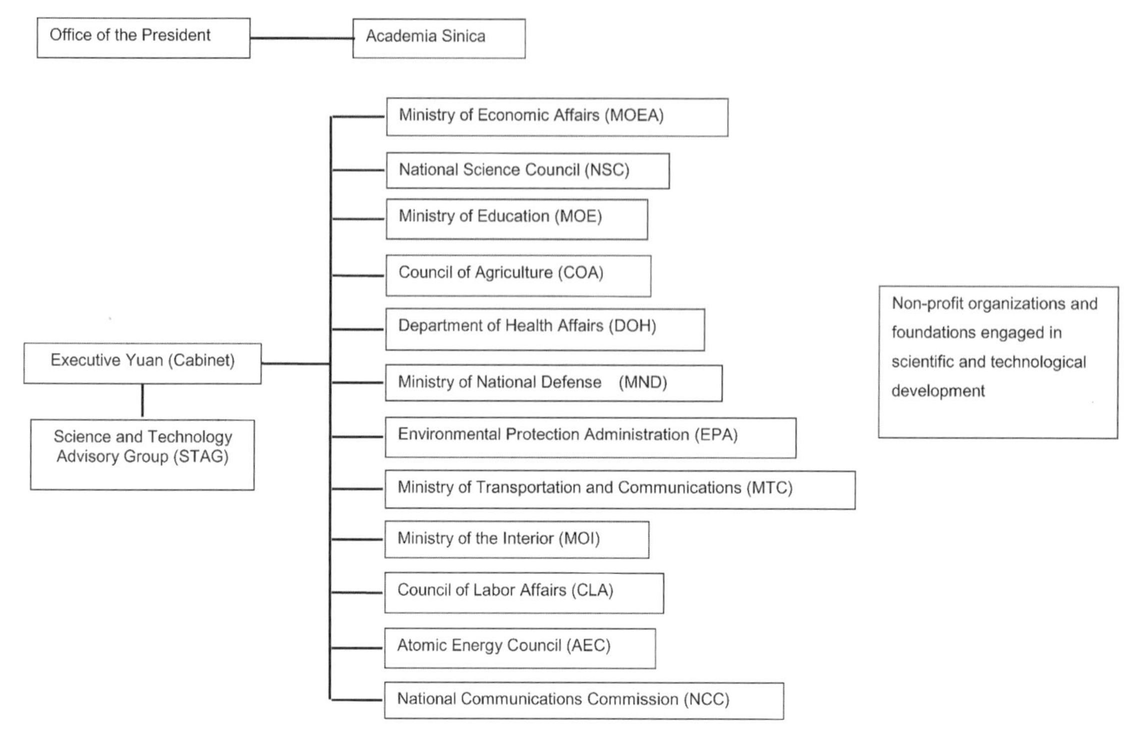

Figure 42.2. Organizational framework for S&T development in Taiwan.
Source: government agencies.

Taiwan's NIS infrastructure

The major organizational framework for S&T development in Taiwan can be divided into three groups: executive organizations, implementing organizations, and the planning and assessment system. Organizations responsible for executing and implementing S&T development, as well as the planning and assessment system, are described below.

Executive organizations

The responsibilities for promotion of S&T development in Taiwan are delegated among various government entities, of which 16 governmental ministries and agencies engage in R&D activities in S&T development. Among them are Academia Sinica under the Office of the President, the NSC, the Ministry of Economic Affairs (MOEA), the Ministry of Education (MOE), the Council of Agriculture (COA), the Department of Health (DOH), the Environmental Protection Administration (EPA), the Ministry of Transportation and Communications (MOTC), and the Atomic Energy Council (AEC). The government guides and implements policies on S&T development though budget planning and execution at each agency. Additionally, the mnister without portfolio in charge of S&T affairs is responsible for coordinating efforts across different agencies.

After STAG was established in 1979, agencies began designating their own S&T promotion units, responsible for coordinating and integrating programs within their respective organizations. The major S&T executive organizations are described below:

National Science Council of the Executive Yuan (NSC)[7]

The NSC, established in 1969, is the highest government agency responsible for promoting the development of S&T. The council, which consists of 12 departments and offices, along with 3 affiliated organizations, is the most important funding agency for basic S&T research carried out at universities and research institutions. In addition to funding general basic research, NSC co-sponsors many national S&T programs with other government agencies, in important fields such as nanoscience and nanotechnology, e-learning, biotechnology and pharmaceuticals, genomic medicine, agricultural biotechnology, telecommunications, disaster reduction, the Si-Soft National System-on-Chip Program (*http://140.113.34.204/english/news.asp*), and digital archives. The NSC also supports academia–industry joint research.

[7] Chinese website: *http://web1.nsc.gov.tw*; English website: *http://web1.nsc.gov.tw/mp.aspx?mp=7*; NSC Organization: *http://web1.nsc.gov.tw/ct.aspx?xItem=7317&CtNode=941&mp=7*

The NSC's annual S&T budgets for 2003–2007 were NT$21.38 billion, NT$23.05 billion, NT$25.77 billion, NT$28.23 billion, and NT$30.82 billion, respectively, accounting for more than one-third of the total Central Government S&T Budget of Taiwan.

The mission of the NSC includes promotion of national sci-tech development, support for academic research, and development of science parks. The NSC is responsible for drafting S&T development policies, strategies, programs, and mid/long-term plans; planning and implementing basic research and pioneering application research; improving the research environment; and training and recruiting S&T manpower.

Based on the urgency of certain research fields and the need for sound distribution of limited resources, seven priority areas for research have been identified: (1) interdisciplinary strategic research, advanced sciences, and international cooperation; (2) material chemistry, nanotechnology, high-energy physics, superconductivity and magnetic physics, mechanisms behind optoelectronic materials, biophysics, and soft matter; (3) environmental statistics, and seismic statistics/analysis; (4) the northern South Sea and Luzon Strait wave mechanism, the impact of Taiwan's land-originated substances on the peripheral seas of Asia; (5) exploration of nanocrystals; (6) active faults of southwestern Taiwan, evolution of the terrain structure of East Asia, earthquakes, and active faults; observation of plate peripherals in Taiwan; seismic belt structure and seismic activities of Taiwan and its adjacent areas, tectonic drilling in Taiwan; and (7) impact of Taiwan's topographic features on typhoon paths and torrential rains.

The Science and Technology Advisory Group of the Executive Yuan (STAG)[8]

STAG is responsible for evaluation of the performance of the National S&T Development Plan and the coordination of development efforts of government agencies, academic institutions, and private sectors. A number of world-renowned scholars, experts, and business leaders in various fields of S&T have served as members on the group's advisory board. Each year one or two STAG board meetings are held with government agencies, academic institutions, and the private sector to discuss strategic issues and assess the nation's S&T policies and development strategies. The conclusions of the meetings are implemented by STAG and all of the government agencies related to S&T through programs and projects.

The missions are (1) to provide recommendations concerning the nation's S&T policies and major S&T development plans, (2) to arrange and prepare for STAG board meetings and Strategy Review Board (SRB) meetings, (3) to get relevant S&T development programs and projects specified by the Executive Yuan underway, (4) to collect important S&T development information and provide advisory services, and (5) to perform other assigned matters.

The premier chairs the STAG board meetings, with the title of convener, and a minister without portfolio is designated to serve as deputy convener. The STAG board consists of 17 to 21 S&T advisors, one of whom serves as chief advisor. The selection of S&T advisors reflects the nation's S&T policy needs. Specialists in particular areas such as energy, information, and biotechnology have been added as the need for them arose. The first Executive Yuan S&T Meeting was held in April 1998 to strengthen implementation of S&T development matters in conjunction with the Action Plan for Building a Technologically Advanced Nation. The Executive Yuan S&T Meeting occurs once every three months, and the S&T Advisor Roundtable once every month.

The Executive Yuan Technology Strategic Review Board (SRB)

The Electronics Technology Review Board and the Telecommunication Technology Review Board meetings were first held in the 1980s. The Strategic Review Board Meeting on Electronics, Information, and Telecommunications was held in 1992 as a result of the rapid progress in electronic, information, and communication and the need for integration of efforts among government agencies, academia, and industry. Subsequently, the Biotechnology Review Board was set up. In response to the demand for development of a variety of emerging industries and the service industry, the Executive Yuan joined all the different review boards together to form the Executive Yuan Technology Strategic Review Board (SRB) in 2002. To achieve coherent actions in the development of important industrial fields, the Executive Yuan will invite specialists to an annual SRB meeting, focusing on one specific industry each year, to generate recommendations on national S&T development policy.

The Ministry of Economic Affairs (MOEA)[9]

The MOEA currently consists of 16 staff units, 14 administrative agencies, 6 state-run corporations, and 64 overseas commercial offices. As the agency in charge of administering the national economy, the MOEA has implemented various economic policies and measures to help Taiwan's industries reinforce their competitive advantage. The agencies responsible for industrial innovation activities are described below. The MOEA's annual S&T budgets for 2003–2007 are NT$22.69 billion, NT$24.74 billion, NT$23.32 billion, NT$25.88 billion, and NT$25.51 billion, respectively, accounting for about one-third of the total Central Government S&T Budget of Taiwan.

[8] Chinese website: *http://www.stag.gov.tw/index.php*

[9] MOEA website: *http://www.moea.gov.tw/*

Department of Industrial Technology (DoIT)[10]

In order to stay competitive, the MOEA initiated the Industrial Technology Development Program (TDP) in 1979 to upgrade traditional industries and develop high-tech, value-added industries. The Science & Technology Advisory Office, established in 1979 to manage the government's TDP, was restructured to become the DoIT in 1993. The missions of DoIT are to encourage frontier technology and innovation, to emphasize key technologies, to integrate digital technologies for smart living, to sustain and strengthen dominant industries, to increase the value of traditional industries, to promote innovative service industries, to develop science parks featuring industrial clusters, and to develop green industries.

The TDP program has gone through several stages of expansion and restructuring over the years, resulting in such programs as the Organization TDP, the State Enterprise TDP, the Industrial TDP, the Small-business Innovation Research Promoting Program (SBIR), the Program for Innovative Technology Applications (ITAP), the Industrial Technology Development Alliance Program, the Industrial Technology Innovation Center Program (MNCD), the Academic TDP, the Strategic Service-oriented R&D Program,[11] and the Multinational Innovative R&D Centers in Taiwan Program (MNCF). Total TDP expenditures were NT$17,216,579 million, NT$18,220,699 million, and NT$15,747,129 million for 2003, 2004, and 2005, respectively.

The Industrial Development Bureau (IDB)[12]

IDB was established in 1970 with the following missions and functions: to formulate policies, strategies, and measures for industrial development; to promote industrial upgrading; to develop and manage industrial parks: to formulate financial and tax measures for industrial development; to prevent industrial pollution, provide guidance on industrial safety, and regulate factories; and to administer general industrial affairs.

To promote industrial upgrading and improve competitiveness, IDB encourages and supports corporations engaging in innovative R&D through the Conventional Industry Technology Development Program, the New Leading Product Development Program, the Industry R&D Loan Scheme, the New-business Model Development Program, the R&D Service Development Program, the Industry Upgrading and Integration Service Program, the Taiwan Technology Exchange Mechanism Development Program, and the Industrial Knowledge Management Promotion Program.

The Small and Medium Enterprise Administration (SMEA)[13]

SMEA is responsible for helping and guiding SMEs, of which there are over a million in Taiwan, to thrive in the face of global competition. In addition, SMEA encourages, supports, and has established a coaching system for startup firms. The most successful measure that SMEA has taken in assisting SMEs was the promotion and development of innovation incubators, mainly in universities.

The Innovation Incubator Program began in 1996. As of the end of 2007, Taiwan had more than 100 incubators, which have incubated almost 3,000 companies, including 1,000 startup firms. The accumulated increase of investment has reached more than NT$44.9 billion and created more than 65,000 jobs.

The Council of Agriculture (COA)[14]

Originally, the Department of Agriculture and Forestry under MOEA was restructured into the Agricultural Bureau in 1981; it was then combined with the Council for Agricultural Planning and Development in 1984 to become the COA.

The COA is the authority on agricultural, forestry, fishery, animal husbandry, and food affairs in Taiwan. Its responsibilities include guiding and supervising provincial and municipal offices in these areas. The COA is the major sponsor of the agricultural R&D projects carried out by universities and its subordinate research institutes, such as the Taiwan Forestry Research Institute, the Taiwan Agricultural Research Institute, the Taiwan Livestock Research Institute, the Taiwan Agricultural Chemicals and Toxic Substances Research Institute, and the Taiwan Endemic Species Research Institute. These institutes have been the engine of Taiwan's agricultural innovations. The COA initiated its TDP in accordance with the conclusions of the 2002 S&T SRB Meeting, held primarily for agricultural research institutions. The COA's annual S&T budgets for 2003–2007 are NT$3.20 billion, NT$3.56 billion, NT$3.71 billion, NT$4.00 billion, and NT$4.26 billion, respectively.

The Ministry of Education (MOE)[15]

The MOE is in charge of drafting and implementing educational policies as well as promoting and implementing school education at all levels, including adult education, physical education, and other educational mat-

[10] DoIT website: *http://doit.moea.gov.tw/newenglish/00_whatsnew/whatsnew.asp*

[11] The Strategic Service-oriented R&D Program and the service sector of SBIR were combined in 2005 to become the Innovative Service R&D Program (ISP) to accelerate the development of tech-enabled and R&D service industry in Taiwan. The ITAP and ISP were merged in 2008 to become the Program for Innovative Technology Applications & Services (ITAS).

[12] IDB website: *http://www.moeaidb.gov.tw*

[13] SMEA website: *http://www.moeasmea.gov.tw/mp.asp?mp=2*

[14] COA website: *http://eng.coa.gov.tw/index.php*

[15] MOE website: *http://www.edu.tw/*

Relevant organizations / Research Levels	Promotion	Implementation		
	Government agencies	Universities and research organizations	Non-profit organizations	State corporations/ private enterprises
Basic research / **Applied research** / **Technology development**	▪ Academia Sinica ▪ Science & Technology Advisory Group ▪ National Science Council ▪ Ministry of Education ▪ Department of Health ▪ Environmental Protection Administration ▪ Ministry of Economic Affairs ▪ Council of Agriculture ▪ Ministry of Transportation and Communications ▪ Atomic Energy Council ▪ Ministry of the Interior ▪ Council of Labor Affairs ▪ Public Construction Commission ▪ Council for Cultural Affairs ▪ Ministry of National Defense	▪ Academia Sinica ▪ Universities and Colleges		
Applied research / **Technology development**			▪ National Health Research Institute (NHRI) ▪ National Applied Research Laboratories ▪ Industrial Technology Research Institute (ITRI) ▪ Institute for Information Industry (III) ▪ Development Center for Biotechnology (DCB) ▪ Pharmaceutical Industry Technology and Development Center ▪ National Synchrotron Radiation Research Center ▪ Animal Technology Institute Taiwan	
Technology development		▪ Architecture and Building Research Institute ▪ Chungshan Institute of Science and Technology ▪ Institute of Transportation ▪ Chunghwa Telecommunication Laboratories ▪ Institute of Nuclear Energy Research ▪ Agriculture Research Institutes ▪ Institute of Occupational Safety and Health		▪ State-run Corporations ◆ Aerospace Industrial Development Corporation (AIDC) ◆ CSBC Corporation, Taiwan ◆ Chinese Petroleum Corporation (CPC), Taiwan ▪ Private enterprises
Applied research				

Figure 42.3. Division of responsibilities among executing and implementing organizations.

ters. At present, the higher education system has more than 1.2 million students in more than 160 institutions.

In order to improve national competitiveness, the MOE encourages universities to pursue academic excellence. The MOE has brought together the necessary resources for developing research universities and for promoting strategic alliances and mergers between universities. The Program for Promoting Academic Excellence of Universities has been instituted to encourage universities to upgrade their research quality. In order to promote and upgrade the quality of Taiwanese academics and research, the MOE has set up a 5-year budget of NT$50 billion to support the Top International Universities and Research Centers Project. By doing so, it helps Taiwanese universities to improve their basic infrastructure, as well as the quality of their teaching and academic research, with the purpose of assisting these universities to become internationalized and world-rated. The MOE also assists promising students to pursue advanced studies abroad though government scholarships.

Implementing organizations

The development of technology in Taiwan follows the principles of overall planning and decentralized implementation. The NSC is responsible for overall executive tasks, the MND for defense technology, and the MOEA for industrial technology. The MOTC is responsible for transportation, telecommunications, and meteorological S&T as well as for manpower training and education.

S&T development can be divided according to the level of research into basic research, applied research, development, and commercialization. Apart from directing S&T policy, the government conducts upstream and midstream research in S&T development. Upstream research consists primarily of basic research and pilot applied research performed by Academia Sinica and colleges/universities. Midstream research mainly consists of applied R&D carried out by the research units of government agencies under the Executive Yuan, by the R&D departments of state-run businesses, and by specially commissioned nonprofit research institutes. Downstream research consists of development and commercialization and is conducted by private enterprises (see Figure 42.3 for a schematic view of the division of responsibilities among promoting and implementing organizations.) Several of the more important nonprofit research institutes are described below.

Industrial Technology Research Institute (ITRI)[16]

Founded in 1973 by the MOEA, ITRI is a nonprofit R&D organization engaged in applied research and technical service to address the technological needs of Taiwan's industrial development. ITRI is composed of six core research laboratories, four focus technology centers, five linkage centers, a business development unit, and other supporting units. The six core research laboratories are Biomedical Engineering, Electronics and

[16] ITRI website: *http://www.itri.org.tw/*

Optoelectronics, Energy and Environment, Information and Communications, Material and Chemical, and Mechanical and Systems. The four focus technology centers are Display; Medical Electronics and Devices; Photovoltaics, Identification, and Security; and System-on-Chip. The five linkage centers are the Center for Measurement Standards, the Creativity Lab, Industrial Economics & Knowledge Center, Nano-technology Research Center, and Technology Center for Service Industries.

ITRI has played a vital role in the transformation of Taiwan's economy from an agriculture-based to an industrial-based model. ITRI tackles both current industrial needs and the needs of future growth. Government-sponsored funding tends to be directed towards long-range development, while contract and project work from industry keep ITRI focused on what's happening right now. By the year 2007, it had grown to a 5,500-person operation, serving as the technical center for industry and an unofficial arm of the government's industrial policies in Taiwan. Backed by its broad research scope and close industrial ties, ITRI is becoming an increasingly active member in the global industrial R&D community.

ITRI's major industrial impact includes (1) establishing new high-tech industries, (2) upgrading traditional industries, (3) leading the drive for sustainable growth, (4) developing highly skilled human resources.

In addition to its direct contribution to the development of Taiwan's high-tech industry and the upgrading of traditional industry, ITRI has had many spin-offs that have become world leaders in their respective fields, such as United Microelectronices Corp. (UMC) in 1979, the Taiwan Semiconductor Manufacturing Company Ltd. (TSMC) in 1987, the Taiwan Mask Corporation (TMC) in 1988, and Vanguard International Semiconductor Corporation (VIS) in 1994.

Development Center for Biotechnology (DCB)[17]

The DCB is a nonprofit R&D organization established in 1984 with the support of the DoIT and MOEA. Its major missions are to establish internationally competitive R&D capabilities; to promote and upgrade the domestic biotechnology industry through the coordination of related government agencies, research institutions, academia, public and private enterprises; to select key projects and development plans; to establish a well-developed R&D infrastructure for, and to improve, the investment development environment; to amend and legislate development laws and regulations; and to promote international alliances and investments for Taiwan's biotechnology and pharmaceutical industries.

The DCB currently has about 400 employees operating three contract service facilities: the GPCR Drug Screening Facility, the Center of Toxicology and Preclinical Sciences (CTPS), and the "current" GMP (good manufacturing practice). The DCB's current R&D efforts focus on (1) environmental biotechnology; (2) herbal medicines; (3) pharmaceuticals such as small-molecule drugs and new formulations of existing drugs; and (4) biologics such as monoclonal antibodies, recombinant proteins, and vaccines.

The DCB also runs the facilities and programs of the Nankang Biotech Incubation Center, supported by the SMEA, designed to help early-stage biotech companies. The DCB assists in business planning, financing, networking opportunities, intellectual property and regulatory affairs, advertising, marketing and sales, human resources, and R&D resources.

National Health Research Institute (NHRI)[18]

The NHRI is a nonprofit, quasi-official autonomous research organization established by the government in 1995. Under the supervision of the DOH, the NHRI is dedicated to the enhancement of medical research and the improvement of health care in Taiwan. Scientists at the NHRI conduct mission-oriented medical research and investigate many aspects of the basic biomedical sciences, as well as specific areas of concern ranging from common problems—such as aging, cancer, infectious diseases, mental disorders, and occupational diseases—to health policy.

The NHRI has focused on and established research units to conduct state-of-the-art medical research in several major, mission-oriented target areas, including the National Institute of Cancer Research's eight divisions and six centers. The eight divisions are Molecular & Genomic Medicine, Clinical Research, Biotechnology & Pharmaceutical Research, Biostatistics and Bioinformatics, Environmental Health & Occupational Medicine, Medical Engineering Research, Mental Health & Substance Abuse Research, and Gerontology Research. The six centers are Cardiovascular and Blood Research, Health Policy Research & Development, Immunology Research, Nanomedicine Research, Stem Cell Research, and Vaccine Research & Development. The NHRI receives its operating funds from both the public and the private sector.

Institute for Information Industry (III)[19]

The III was incorporated in 1979 through joint efforts of the public and private sectors as a non-governmental organization to support the development and application of research in the information industry and information society in Taiwan. Since its inception, the III has been a source of vision, innovation, and technological excellence as well as a major contributor to Taiwan's development as

[17] DCB website: *http://www.dcb.org.tw/adminz/index.asp*

[18] NHRI website: *http://english.nhri.org.tw/*

[19] III website: *http://www.iii.org.tw*; English website: *http://140.92.88.47/english/index.asp*

a significant player in the global ICT area. The III has also helped to promote full utilization of ICT technologies, therby advancing the establishment of modern information society development in Taiwan. As a result, Taiwan ranked 13th in the WEF/NRI (Network Readiness Index) in 2006–2007.

The III comprises the Information Engineering Institute, the Networks & Multimedia Institute, the Innovative Digitech-enabled Applications & Services Institute, the Digital Education Institute, the Industry Support Division, the Project Resource Division, the Market Intelligence Center, the Science & Technology Law Center, and the Southern Information Service Division.

National Applied Research Laboratories (NARL)[20]
Established in June 2003, NARL has consolidated nine of the NSC's national laboratories into an independent nonprofit institute. Its mission is to construct, operate, and maintain the large-scale R&D facility or platform supporting academic research and to foster the necessary manpower in various advanced fields of national focus.

NARL comprises the National Nano Device Laboratories, the National Laboratory Animal Center, the National Center for Research on Earthquake Engineering, the National Space Organization, the National Center for High-performance Computing, the National Chip Implementation Center, the Instrument Technology Research Center, the Science & Technology Policy Research and Information Center, and the National Science & Technology Center for Disaster Reduction.

Planning and assessment system

Planning and approval
Besides formulating S&T policies and strategies for effective utilization of limited S&T development resources, the government establishes mid to long-range plans, evaluates the performance of continuing projects, and conducts impact assessments. Proposal evaluation, implementation control, and impact assessment are key to Taiwan's technology development.

The S&T unit in each government agency coordinates the S&T proposals within its agency and submits them to the NSC. The NSC and STAG are the supervisory agencies and form review boards consisting of scholars and experts from academia, government, and industry in different fields for evaluation and budget allotment. The S&T units are also responsible for the integration and coordination of proposals among different agencies, so as to minimize overlap of efforts. After approval of its recommendations by the Executive Yuan, each agency can start to implement its respective S&T projects. For several important fields, based on the recommendation of the 5th National S&T Conference, the NSC and STAG originated the National S&T Programs to meet the nation's major needs for economic and social development, as well as to integrate the R&D resources of upstream, midstream, and downstream industries in 1997.

Progress monitoring and performance assessment
The NSC and STAG are in charge of project appraisal. Experts are dispatched to inspect and evaluate the performance of S&T programs at the middle and end of the program. Compliance with the recommendations and improvements are checked at the end of the program as well as during the review process of any future programs the principal investigator (PI) may propose.

The policy-forming process of Taiwan's NIS

S&T policies are mainly formulated through important conferences such as the S&T Meetings of the Executive Yuan, NSC Council Meetings, National S&T Conferences, S&T Advisory Board Meetings, and the Industry Technology SRB Meetings. The consensus reached at these meetings is used to guide the formation of S&T policies. After the National S&T Development Plan submitted by the NSC Council Meeting has been approved by the Executive Yuan, relevant government departments and agencies implement the plan's recommendations. STAG regularly reviews and assesses policy, while providing its recommendations. Figure 42.1 depicts the evolution of S&T policy in Taiwan. The role played by major conferences in the formulation of S&T policy is described below.

National S&T Conference

Held once every four years since 1978, the National S&T Conference provides specific recommendations concerning Taiwan's S&T development, challenges, and vision. A National S&T Development Plan based on the conclusions reached in the First National S&T Conference was promulgated in 1979. Energy, material, information, and automation were chosen as the key technologies to be developed preferentially.

In 1982, as suggested by the Second National S&T Conference, biotechnology, optoelectronics, food, and hepatitis prevention were selected, together with the previous four, to become the eight key technologies. The National S&T Development Plan was revised according to the conference suggestions. The Advanced S&T Personnel Cultivating and Recruiting Program was also promulgated to help in building up the necessary qualified research manpower.

The Third National S&T Conference, held in 1986, resulted in the ratification of the National S&T 10-year

[20] NARL website: *http://www.narl.org.tw/en/*

Plan (1986–1995) and the addition of disaster prevention, synchrotron radiation, ocean, and environmental technologies to the key industry list. Together with the preceding eight technologies, these 12 key industries receive high priority when R&D grants are awarded. This conference also set key performance indicators.

In 1991 the National S&T 6-year Plan and National S&T 12-year Plan were approved following the Fourth National S&T Conference. The Fifth National S&T Conference, held in 1996, modified the S&T development indicators and recommended increasing the government S&T budget and building the legal infrastructure for S&T development.

The Sixth and Seventh National S&T Conferences,[21] held in January of 2001 and of 2005, respectively, achieved consensus on major general strategies and S&T development focal points and drafted the National S&T Development Plan for 2001–2004 and for 2005–2008.

STAG Advisory Board Meeting of the Executive Yuan

To ensure effective operation of the Advisory Board, STAG was established in December of 1979 as a mission-oriented organization to perform staff work. STAG does not directly implement projects; rather, it coordinates government agencies, industries, research institutes, and academia. Its duties include holding meetings and coordinating the implementation of conclusions and recommendations, forming and supporting inter-ministerial S&T committees, and coordinating inter-ministerial activities.

The S&T Advisory Board Meeting discusses policy recommendations submitted by invited advisors concerning Taiwan's S&T development plans and major interdepartmental R&D issues. It provides suggestions to national S&T development policy and periodically examines and evaluates policy implementation.

To map out the development of advanced high-tech industries, STAG holds SRB meetings on key industrial areas to solicit suggestions and attain consensus, which has contributed greatly to the formulation of many important industrial S&T policies. For example, the conclusions of the December 2000 STAG Board Meeting and the January 2001 Sixth National S&T Conference noted that nanotechnology will be one of the major directions of future industrial development in Taiwan. A conceptual plan for the National S&T Program for Nanoscience and Nanotechnology was approved in June of 2001. Formal implementation of the program began in January 2003. Funding of NT$23.2 billion has been committed to nanotechnology development under the program.

[21] For the respective websites of the Sixth and Seventh National S&T Conferences, see: *http://www.nsc.gov.tw/pla/tc/6th/6th_index.htm* and *http://www.nsc.gov.tw/pla/tc/index.htm*

The resources and outputs of Taiwan's NIS[22]

Expenditure

Central government's S&T budget

The Taiwan central government's S&T budget grew from NT$61.711 billion in 2003 to NT$81,853 billion in 2007, representing an average growth rate of 5.8% per year (see NSC, 2007–2010 and Table 42.1).

Gross domestic expenditure on R&D (GERD)

Taiwan's GERD as a percentage of GDP continues to rise, having reached 2.52% in 2006 and demonstrating stable growth (see Table 42.2). However, it did not arrive at the 3.0% target set by the White Paper on S&T (2003–2006). In addition, Taiwan still lags behind Japan, South Korea, the U.S. and others in this regard.

Regarding Taiwan's R&D spending by sector of performance, the highest level of R&D expenditure was seen in the business enterprise sector, followed by the government, higher education, and the private nonprofit sector (see Table 42.3).

R&D expenditure by type of R&D

Development has accounted for the greatest percentage of national R&D spending, followed by applied research, with basic research accounting for the smallest percentage (see Table 42.4).

Workforce

Human resources

Personnel trained through higher education represent a large cumulative pool of talent for Taiwan's scientific and technological development. The number of persons graduating from higher educational institutions has risen steadily because of the continued increase in the number of universities/colleges, to more than 160 currently. Although the number of students graduating in science and technology fields have been increasing, the percentage increase is the lowest of all fields. Master's graduates in particular grew the fastest over this period (see Table 42.5).

Total R&D personnel

R&D personnel at all levels increased over the past five years. Specifically, researchers holding master's degrees have outnumbered bachelor-level researchers since 2001 and are now the foundation of Taiwan's research force. Those with doctoral degrees have also increased each

[22] *Yearbook of Science and Technology*, Taiwan; *http://yearbook.stpi.org.tw/englishpdf.html*

Table 42.1. Taiwan Central Government Civil S&T Budget, by department (unit: million NT$).

Fiscal year ⇒ *Agency* ⇓	*2003*	*2004*	*2005*	*2006*	*2007*
National Science Council	21,384	23,051	25,773	28,226	30,817
Ministry of Economic Affairs	22,685	24,735	23,318	25,883	25,509
Academia Sinica	5,843	6,592	7,402	8,531	8,938
National S&T Development Fund (for inter-agency S&T programs)	3,126	3,232	3,447	3,483	4,190
Department of Health	2,832	3,146	3,609	4,215	4,396
Council of Agriculture	3,197	3,556	3,707	3,995	4,264
Atomic Energy Council	646	718	936	827	992
Ministry of Education	774	725	852	839	889
Ministry of Transportation and Communications	672	726	693	711	818
Others	552	682	684	894	40
Total	*61,711*	*67,163*	*70,421*	*77,604*	*81,853*

Source: Unit of Reviewing Government Sci-Tech Projects (NSC)—this is the team responsible for reviewing government sci-tech projects.

Table 42.2. R&D expenditure and national income (unit: million NT$).

Year ⇒ *Item* ⇓	*2002*	*2003*	*2004*	*2005*	*2006*
R&D expenditure (*A*)	224,428	242,942	263,271	280,980	307,037
GDP (*B*)	10,293,346	10,519,574	11,065,548	11,454,727	11,889,823
GNP (*C*)	10,535,848	10,848,447	11,437,647	11,745,593	12,201,522
NI (*D*)	9,227,042	9,478,427	10,001,162	10,260,406	10,659,523
R&D expenditure as a percent of GDP ($A/B \times 100$)	2.18%	2.31%	2.38%	2.45%	2.58%
R&D expenditure as a percent of GNP ($A/C \times 1$)	2.13%	2.24%	2.30%	2.39%	2.52%
R&D expenditure as a percent of NI ($A/D \times 100$)	2.43%	2.56%	2.63%	2.74%	2.88%

Source: Indicators of Science and Technology, Taiwan (2007, *http://www.nsc.gov.tw/tech/pub_data_main.asp*).
Notes: Wholesale and retail trade, financial and insurance, and real estate have been added to the surveyed industries since 2003. The figures in (*B*) and (*C*) are nominal GDP. GDP, GNP, and NI figures from 2002 to 2005 have been amended according to DGBAS, and R&D expenditure, as percentages of GDP, GNP, and NI, have also been revised.

year, but their proportion relative to all researchers has been gradually declining. Government agencies have much higher ratios of Ph.D.-level R&D personnel than the private sector, in which the figure is around 14.5% over this period. Master's degree personnel constitute the largest population of government R&D personnel, but bachelor degree personnel constitute the major work force in the private sector (see Table 42.6).

Output

National competitiveness

World Economic Forum

In the World Economic Forum's (WEF) *Global Competitiveness Report, 2007-2008*, Taiwan ranked 14th in the overall Growth Competitiveness Index, down one place from the previous year. Taiwan ranked 17th in efficiency enhancers in which higher education and training took

Table 42.3. Taiwan R&D expenditure by sector of performance (unit: million NT$).

Item	*2002*	*2003*	*2004*	*2005*	*2006*
Business enterprise sector	139,569	152,614	170,293	188,390	207,238
As a percent of total	62.19	62.82	64.68	67.05	67.50
Government sector	55,693	59,928	61,144	59,143	60,965
As a percent of total	24.82	24.67	23.22	21.05	19.86
Higher education sector	27,637	28,890	30,350	32,092	37,565
As a percent of total	12.31	11.89	11.53	11.42	12.23
Private nonprofit sector	1,530	1,510	1,484	1,355	1,270
As a percent of total	0.68	0.62	0.56	0.48	0.41

Source: Indicators of Science and Technology, Taiwan (2007, *http://www.nsc.gov.tw/tech/pub_data_main.asp*).
Note: defense R&D expenditures were first added in 2002, and other new industries were added in 2003.

Table 42.4. Government R&D expenditure by type of R&D (unit: million NT$).

Type of R&D⇒	*Total*	*Basic research*		*Applied research*		*Development*	
Year ⇓			*%*		*%*		*%*
2002	55,693	8,901	16.0	22,527	40.4	24,265	43.6
2003	59,928	11,967	20.0	22,663	37.8	25,298	42.2
2004	61,144	12,859	21.0	22,778	37.3	25,507	41.7
2005	59,143	12,110	20.5	22,887	38.7	24,146	40.8
2006	60,965	12,421	20.4	23,605	38.7	24,938	40.9

Source: Indicators of Science and Technology, Taiwan (2007, *http://www.nsc.gov.tw/tech/pub_data_main.asp*).
Note: including defense since 2002.

4th and 15th place in technological readiness, respectively. Taiwan was 10th in innovation and sophistication factors and 8th in innovation. Taiwan's outstanding performance in the field of innovation was mainly due to widespread university/industry research collaboration. With a global third-place ranking in average *per capita* utility patents and government procurement of advanced technology products, Taiwan's S&T strengths are being recognized around the world.

S&T development output indicators

Performance in academic papers and patents

The numbers of journal papers listed in the Science Citation Index (SCI) and Engineering Index (EI) have both been increasing steadily over the past five years. The ranking of SCI papers is between 18th and 17th, while that of EI articles is 12th or 11th.

The number of patent applications from Taiwan being approved by the U.S. Patent and Trademark Office (USPTO) has put Taiwan into 4th place in the world, beaten only by the U.S., Japan, and Germany (see Table 42.7).

The steady improvement in the quantity and quality of scientific and engineering papers from Taiwan academia, together with the strong performance in the number of U.S. utility patents obtained by Taiwan, demonstrates Taiwan's vigorous innovation capability.

Technology balance of payments

Despite the strong showing of large numbers of utility patents obtained from the U.S. and other countries, the technology trade performance of Taiwan has been poor. Tables 42.8 through 42.10 reveal the huge technology trade imbalance of Taiwan for the period 2000–2005, and the situation is worsening. It can be seen from analysis of the import value of technology that Japan and the U.S. are the two major sources of technology for Taiwanese industries, but the influence of the U.S. is rising. The export value of technology shows that China is the most important importer of Taiwanese technology,

Table 42.5. Taiwan students graduating from institutions of higher education (unit: persons).

Program	Field	Academic year (AY)								
		2001	2002	%	2003	%	2004	%	2005	%
Ph.D.	Humanities	235	216	−8.1	249	15.3	252	1.2	305	21.0
	Social sciences	218	192	−11.9	246	28.1	301	22.4	335	11.3
	Science and technology	1,010	1,093	8.2	1,265	15.7	1,411	11.5	1,525	8.1
	Subtotal	1,463	1,501	2.6	1,760	17.3	1,964	11.6	2,165	10.2
Master	Humanities	2,788	3,986	43.0	4,949	24.2	6,034	21.9	7,002	16.0
	Social sciences	4,912	6,472	31.8	8,225	27.1	9,850	19.8	12,149	23.3
	Science and technology	13,052	15,442	18.3	17,682	14.5	20,097	13.7	23,183	15.4
	Subtotal	20,752	25,900	24.8	30,856	19.1	35,981	16.6	42,334	17.7
Bachelor	Humanities	21,285	23,600	10.9	26,595	12.7	29,440	10.7	31,729	7.8
	Social sciences	38,750	67,363	73.8	62,614	−7.0	70,378	12.4	77,932	10.7
	Science and technology	57,395	72,308	26.0	86,835	20.1	93,036	7.1	101,102	8.7
	Subtotal	117,430	163,271	39.0	176,044	7.8	192,854	9.5	210,763	9.3
Total persons		*139,645*	*190,672*	*36.5*	*208,660*	*9.4*	*230,799*	*10.6*	*255,262*	*10.6*

Source: MOE (2007).
Notes: Humanities include education, arts, humanities, and other fields (including physical education). Social sciences include economic and social psychology, business and management, law, tourism and services, mass communication, and home economics (excluding food/nutrition). Science and technology include natural sciences, mathematics and computation, medicine and public health, industrial crafts, engineering, architecture and urban planning, agriculture/forestry/fishery/animal husbandry, transportation and communications, and food/nutrition.

because of almost one million Taiwanese businessmen currently reside in China.

The challenges of Taiwan's NIS

Structural problems

The disparity between impressive patent achievement and huge technology trade imbalance

For almost a decade, Taiwan has ranked 4th among foreign countries in the number of U.S. utility patents approved, but Taiwan does not seem to be able to capitalize on this innovation capability. One possible reason for this situation is that the strong original equipment manufacturer (OEM)-oriented industry tends to have many incremental process improvement patents but few fundamental platform patents. Consequently, the more products the industry produces, the higher royalty it has to pay to foreign patent holders. A solution to this dilemma is improving the science linkage of Taiwanese patents (i.e., raising the quality of patent by enhanced R&D efforts) and further exploring the immense innovation capability of universities through industry–academia cooperation (Chen, 2005, 2007; Chen and Pam Wen, 2006, pers. commun.; WI, 2005).

The inadequacy of Taiwan's innovation value chain

In order to exploit the potential of innovative capability, Taiwan not only has to focus on the originality of its inventions but also has to develop new-business models to extend the innovation value chain. In addition, Taiwan should nurture emerging breakthrough technologies instead of continuing to expand applications of current knowledge.

Insufficient R&D investment

Although the target of gross domestic expenditure on R&D was to account for 3.0% of GDP in 2006, as set out by the White Paper on Science and Technology (2003–2006), only 2.58% was realized. State-run enterprises, private research organizations, and nonprofit organizations currently invest relatively little in R&D. A staggering government deficit of NT$3.4 trillion, a failing economic environment, and a shift in focus by the current government have hampered government support of R&D activities.

Table 42.6. R&D personnel (FTE) by qualification (unit: person-years).

Year	*2002*	*%*	*2003*	*%*	*2004*	*%*	*2005*	*%*	*2006*	*%*
Government										
Ph.D.	3,526	14.5	3,572	14.6	3,470	14.1	3,671	14.3	3,901	14.6
Masters	8,425	34.7	9,312	38.1	9,525	38.6	10,099	39.3	10,437	39.1
Bachelors	5,534	22.8	5,508	22.5	5,579	22.6	5,760	22.4	6,106	22.9
Junior College	3,755	15.5	3,510	14.4	3,251	13.2	3,261	12.7	3,322	12.4
Others	3,059	12.6	2,546	10.4	2,849	11.5	2,882	11.2	2,918	10.9
Subtotal	24,298		24,449		24,674		25,673		26,684	
Enterprise										
Ph.D.	2,209	3.0	2,301	2.9	2,527	2.8	2,579	2.7	2,698	2.5
Masters	19,940	26.8	23,098	28.7	27,005	30.0	30,564	31.6	34,052	32.0
Bachelors	24,570	33.0	28,537	35.4	33,890	37.7	37,038	38.3	42,385	39.9
Junior College	22,143	29.7	20,528	25.5	20,797	23.1	20,595	21.3	20,866	19.6
Others	5,652	7.6	6,061	7.5	5,663	6.3	5,938	6.1	6,261	5.9
Subtotal	74,514		80,525		89,882		96,714		106,262	
Total										
Ph.D.	5,735	5.8	5,873	5.6	5,997	5.2	6,250	5.1	6,599	5.0
Masters	28,365	28.7	32,410	30.9	36,530	31.9	40,663	33.2	44,489	33.5
Bachelors	30,104	30.5	34,045	32.4	39,469	34.5	42,798	35.0	48,491	36.5
Junior College	25,898	26.2	24,038	22.9	24,048	21.0	23,856	19.5	24,188	18.2
Others	8,711	8.8	8,607	8.2	8,512	7.4	8,820	7.2	9,179	6.9
Grand total	*98,812*		*104,974*		*114,556*		*122,387*		*132,946*	

Source: Indicators of Science and Technology, Taiwan (2007, *http://www.nsc.gov.tw/tech/pub_data_main.asp*).

Table 42.7. Taiwan S&T development output.

	2002	*Rank*	*2003*	*Rank*	*2004*	*Rank*	*2005*	*Rank*	*2006*	*Rank*
Annual papers in SCI[a]	10,983	18	12,456	18	12,947	18	15,671	18	16,545	17
Annual papers in EI[a]	5,786	12	8,011	12	10,980	11	11,661	11	13,076	11
Number of U.S.-granted utility patents[b]	5,431	4	5,298	4	5,938	4	5,118	4	6,360	4

Source: [a]Compendex, U.S.A.
[b]U.S. Patent and Trademark Office.

Inadequate funding for higher education

The number of universities/colleges have grown from fewer than 50 ten years ago to more than 160 now. The budget for education has remained around 4% of GDP over the period in spite of the rapid increase in the number of universities. Most public universities have experienced developmental bottlenecks in recent years, as the continuing expansion of higher education has meant that educational resources are spread thin, and the quality of university education has continued to deteriorate. A severe consequence of the rapid expansion of university students is the sharp deterioration in student quality.

Table 42.8. Technology trade balance of payments (unit: million NT$).

Item ⇒ *Year* ⇓	*Export* (*A*)	*Import* (*B*)	*Total* (*A* + *B*)	*Net* (*A* − *B*)	*Ratio* (*A*)/(*B*)
2000	3,949	40,727	44,677	−36,778	0.10
2002	11,261	45,246	56,507	−33,985	0.25
2003	8,941	51,954	60,894	−43,013	0.17
2004	8,942	52,156	61,097	−43,214	0.17
2005	13,257	57,133	70,390	−43,877	0.23

Source: Industrial Statistical Survey Report, MOEA.
Note: MOEA's Factory Adjustment and Operation Survey was not conducted in 2001 and 2006.

Table 42.9. Import value of technology by nationality (unit: million NT$).

Country ⇒ *Year* ⇓	*Total*	*Germany*	*Japan*	*U.S.*	*Others*
2000	40,727	5,446	14,304	13,717	7,260
2002	45,246	2,243	15,938	16,404	10,661
2003	51,954	1,796	21,839	20,406	7,913
2004	52,156	2,806	21,403	20,428	7,519
2005	57,133	2,418	18,618	28,560	7,537

Source: Industrial Statistical Survey Report, MOEA.
Note: MOEA's Factory Adjustment and Operation Survey was not conducted in 2001 and 2006.

Table 42.10. Export value of technology by nationality (unit: million NT$).

Country ⇒ *Year* ⇓	*Total*	*P.R.C.*	*Thailand*	*U.S.*	*Others*
2000	3,949	1,142	28	620	2,160
2002	11,261	4,292	198	1,318	5,452
2003	8,941	4,329	56	773	3,783
2004	8,942	4,103	39	913	3,887
2005	13,257	6,510	36	1,071	5,640

Source: Industrial Statistical Survey Report, MOEA.
Notes: MOEA's Factory Adjustment and Operation Survey was not conducted in 2001 and 2006. Hong Kong is included in PRC.

Lack of qualified, high-level R&D personnel

The majority of people with Ph.D. degrees working in universities is close to 70% or in government research institutes around 20%. Therefore, only 10% are available to the private sector. In order to improve the quality and standard of industrial R&D, the relationships and interactions between industry and academia have to be tightened. Academic research results can lend impetus to industrial innovation, and sound relationships between universities and industry can also boost industrial innovation. The NSC has tackled this problem by initiating an industry–academia R&D program, which has seen some results, but there is still a long way to go (see Table 42.11, which follows Appendix 42.A).

Appendix 42.A Organizational framework for S&T development in the ROC.

Executive Yuan (Cabinet)

Science and Technology Advisory Group

Office of the President

Academia Sinica

- Institute of Mathematics
- Institute of Physics
- Institute of Chemistry
- Institute of Earth Sciences
- Institute of Information Science
- Institute of Statistical Science
- Institute of Atomic and Molecular Sciences
- Institute of Astronomy and Astrophysics
- Research Center for Applied Sciences
- Research Center for Environmental Changes
- Institute of Plant and Microbial Biology
- Institute of Cellular and Organismic Biology
- Institute of Biological Chemistry
- Institute of Molecular Biology
- Institute of Biomedical Sciences
- Agricultural Biotechnology Research Center
- Genomics Research Center
- Research Center for Biodiversity
- Institute of History and Philology
- Institute of Ethnology
- Institute of Modern History
- Institute of Economics
- Institute of European and American Studies
- Institute of Chinese Literature and Philosophy
- Institute of Taiwan History
- Institute of Sociology
- Institute of Linguistics
- Institute of Political Science
- Institutum Iurisprudentiae
- Research Center for Humanities and Social Sciences

Ministry of the Interior

- S&T Development Promotion Group
- Department of Civil Affairs
- Department of Population
- Department of Social Affairs
- Department of Land Administration
- Information Center

National Police Agency

Construction and Planning Agency

National Fire Agency

Conscription Agency

Central Police University

Children's Bureau

National Airborne Service

Architecture Research and Building Institute

Ministry of National Defense

Office of S&T Advisors

Chungshan Institute of Science and Technology

Ministry of Education

Advisory Office

Universities and colleges

National Research Institute of Chinese Medicine

Ministry of Economic

- Department of Industrial Technology
- Industrial Development Bureau
- Department of Commerce
- Bureau of Energy
- Water Resources Agency
- Central Geological Survey
- Intellectual Property Office
- Bureau of Standards, Metrology and Inspection
- Central Region Office
- Small and Medium Enterprise Administration

State-owned Enterprise Commission

- Chinese Petroleum Corporation Refining and Manufacturing Institute Exploration and Production Research Institute
- Taiwan Sugar Corporation Taiwan Sugar Research Institute
- China Shipbuilding Corporation Department of Research and Development
- Taiwan Power Company Taiwan Power Research Institute
- Aerospace Industrial Development Corporation

Ministry of Transportation and Communications

Office of S&T Advisors

Central Weather Bureau

- Meteorological R&D Center

Institute of Transportation

- Harbor and Marine Technology Center

Bureau of Taiwan High Speed Rail

Civil Aeronautics Administration

National Expressway Engineering Bureau

Directorate General of Telecommunications

Directorate General of Highways

Department of Health Affairs

Science and Technology Unit

Bureau of Medical Affairs

Bureau of Pharmaceutical

Bureau of Food Safety

Bureau of Nursing and Health Care

Bureau of Food and Drug Analysis

Information Management Center

Statistics Office

Bureau of National Health Insurance

Bureau of Controlled Drugs

Centers for Disease Control

Committee of Chinese Medicine and Pharmacy

Hospital Administration Commission

Nonprofit organizations and foundations engaged in scientific and technological

Ministry of the Interior (MOI)
Chinese Architecture & Building Center
Ministry of Economic Affairs (MOEA)
Industrial Technology Research Institute
Electronics and Optoelectronics Research Labs
Information and Communications Research Labs
Mechanical and Systems Research Labs
Material and Chemical Research Labs
Energy and Environment Research Labs
Biomedical Engineering Research Labs
Display Technology Center
SoC Technology Center
Photovoltaics Technology Center
Medical Electronics and Device Technology Center
RFIC Technology Center
Center for Measurement Standards
Industrial Economics & Knowledge Center
Nano Technology Research Center
Creativity Lab
Technology Center for Services Industries
Institute for Information Industry
China Productivity Center
CTCI Foundation
Taiwan Textile Research Institute
Taiwan Textile Federation
Chung-Hua Institution for Economic Research
China Data Processing Center
Sinotech Engineering Consultants
Taiwan Electric Research & Testing Center
Electronics Testing Center, Taiwan
Pharmaceutical Industry Tech & Development Center
Taiwan Electrical and Mechanical Engineering Services
Development Center for Biotechnology
Stone & Resource Industry R&D Center
Printing Technology Research Institute
Cycling & Health Tech Industry R&D Center
Tze-Chiang Foundation of Science and Technology
Automotive Research & Testing Center
Metal Industries R&D Center
Food Industry Research and Development Institute
Plastics Industry Development Center
Precision Machinery Research & Development Center
Footware and Recreation Technology Research Center
United Ship Design & Development Center
Taiwan Accreditation Foundation
Ministry of Transportation and Communications (MOTC)
China Engineering Consultants
Department of Health (DOH)
National Health Research Institutes
Center for Drug Evaluation
National Science Council (NSC)
National Synchrotron Radiation Research Center

Source: government agencies.
Note: data current as of December 2006.

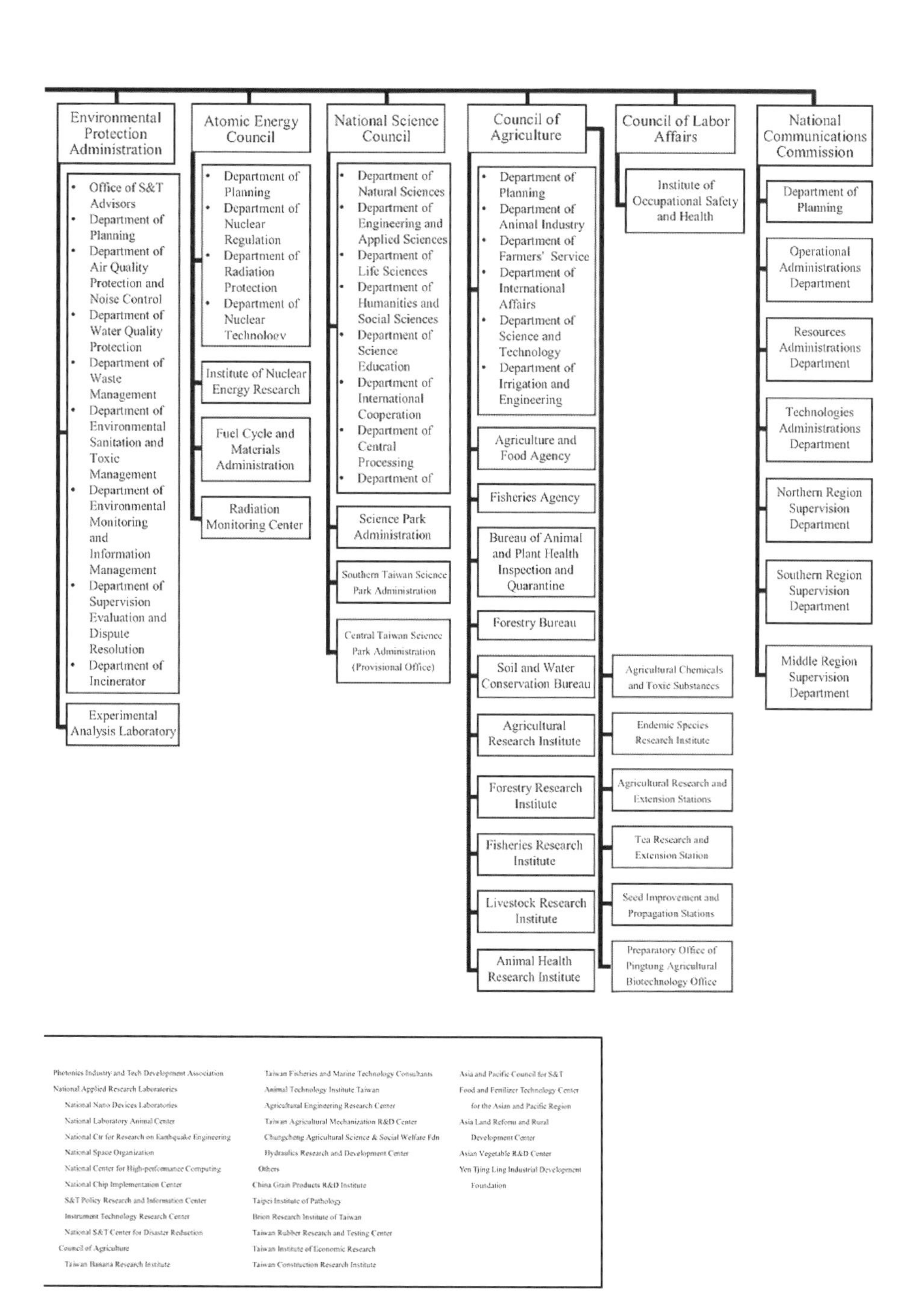
Environmental Protection Administration
Office of S&T Advisors
Department of Planning
Department of Air Quality Protection and Noise Control
Department of Water Quality Protection
Department of Waste Management
Department of Environmental Sanitation and Toxic Management
Department of Environmental Monitoring and Information Management
Department of Supervision Evaluation and Dispute Resolution
Department of Incinerator
Experimental Analysis Laboratory
Atomic Energy Council
Department of Planning
Department of Nuclear Regulation
Department of Radiation Protection
Department of Nuclear Technology
Institute of Nuclear Energy Research
Fuel Cycle and Materials Administration
Radiation Monitoring Center
National Science Council
Department of Natural Sciences
Department of Engineering and Applied Sciences
Department of Life Sciences
Department of Humanities and Social Sciences
Department of Science Education
Department of International Cooperation
Department of Central Processing
Department of
Science Park Administration
Southern Taiwan Science Park Administration
Central Taiwan Science Park Administration (Provisional Office)
Council of Agriculture
Department of Planning
Department of Animal Industry
Department of Farmers' Service
Department of International Affairs
Department of Science and Technology
Department of Irrigation and Engineering
Agriculture and Food Agency
Fisheries Agency
Bureau of Animal and Plant Health Inspection and Quarantine
Forestry Bureau
Soil and Water Conservation Bureau
Agricultural Research Institute
Forestry Research Institute
Fisheries Research Institute
Livestock Research Institute
Animal Health Research Institute
Agricultural Chemicals and Toxic Substances
Endemic Species Research Institute
Agricultural Research and Extension Stations
Tea Research and Extension Station
Seed Improvement and Propagation Stations
Preparatory Office of Pingtung Agricultural Biotechnology Office
Council of Labor Affairs
Institute of Occupational Safety and Health
National Communications Commission
Department of Planning
Operational Administrations Department
Resources Administrations Department
Technologies Administrations Department
Northern Region Supervision Department
Southern Region Supervision Department
Middle Region Supervision Department
Photonics Industry and Tech Development Association
National Applied Research Laboratories
National Nano Devices Laboratories
National Laboratory Animal Center
National Ctr for Research on Earthquake Engineering
National Space Organization
National Center for High-perfomance Computing
National Chip Implementation Center
S&T Policy Research and Information Center
Instrument Technology Research Center
National S&T Center for Disaster Reduction
Council of Agriculture
Taiwan Banana Research Institute
Taiwan Fisheries and Marine Technology Consultants
Animal Technology Institute Taiwan
Agricultural Engineering Research Center
Taiwan Agricultural Mechanization R&D Center
Chungcheng Agricultural Science & Social Welfare Fdn
Hydraulics Research and Development Center
Others
China Grain Products R&D Institute
Taipei Institute of Pathology
Brion Research Institute of Taiwan
Taiwan Rubber Research and Testing Center
Taiwan Institute of Economic Research
Taiwan Construction Research Institute
Asia and Pacific Council for S&T
Food and Fertilizer Technology Center for the Asian and Pacific Region
Asia Land Reform and Rural Development Center
Asian Vegetable R&D Center
Yen Tjing Ling Industrial Development Foundation

Table 42.11. NSC-sponsored Industry–Academia R&D Program.

Year ⇒ *Item* ⇓	*2003*	*2004*	*2005*	*2006*	*2007*
R&D Projects	1,018	1,144	1,040	1,046	1,084
Government grant (million NT)	544.02	596.36	512.95	532.14	495.46
Approved patents	202	321	396	482	397
Technology transfer (case)	975	1,301	1,336	1,072	1,242
Loyalty and licensing fee (million NT)	124	136	149	133	144

Source: NSC (Annual).
Note: 45 universities and/or research institutes have established technology transfer offices by February 29 of 2008.

International factors

The China factor

China has emerged as a global force that is driving the consumption and production of almost everything to record highs (WI, 2005); this has created an unfamiliar environment for Taiwan, which has to adjust quickly to cope with this situation and to maintain friendly relationships with China. The Taiwan government's emphasis in recent years on ideological issues has damaged the relationship between Taiwan and China. The resulting lack of confidence within the private sector has caused R&D investment to lose its momentum.

Global R&D outsourcing

The removal of trade barriers due to international trade agreements has allowed many emerging countries to enter the supply chain of developed countries. In particular, the trend of moving upstream to provide R&D services by these countries has become a threat to Taiwan's OEM-based industries. Taiwan has to upgrade its industry to compete effectively with these rising economies.

Conclusion

The S&T system in Taiwan has been built over the past several decades through government-coordinated planning and implementation. The system has helped to create the strength and capability of both basic scientific and applied industrial innovation. Many high-tech products ranked number one worldwide in volume are manufactured in Taiwan. Taiwan has become a crucial partner in the world supply chain of high-tech industries.

Although the national innovation system in Taiwan has worked well, problems associated with it have to be overcome to sustain Taiwan's long-term competitiveness. In particular, in this knowledge economy the NIS must be able to encourage knowledge-based innovation, boost academic research standards, develop distinctive academic fields, strengthen manpower, and overcome industrial development barriers by utilizing its limited resources effectively.

References

Chen, S.-H. (2005). "The problems of and the solutions to Taiwan's national innovation system." Discussion material for Topic Two of the *7th National S&T Conference*.

Chen, S.-H. (2007). "The national innovation system and foreign R&D: The case of Taiwan." *R&D Management*, **37**, 441–453.

Chen, S.-H., and Pam Wen, P.-C. (pers. commun.). "The trend of R&D internationalization and the challenges to S&T policy."

Chen, S.-H., and Pam Wen, P.-C. (2006). Globalization, knowledge-based innovation and the changing economic geography of innovation across the Taiwan Strait." *Macalester International Journal*, in press.

MOE. (2007). Ministry of Education; *http://english.moe.gov.tw/public/Attachment/610211595571.xls* (last accessed November 30, 2009).

NSC. (2007–2010). *Innovative Capabilities and Citizens' Quality of Life Will Reach the Level of a Developed Nation by 2015* (white paper on science and technology). Taipei, Taiwan: National Science Council.

WI. (2005). *Vital Signs*. Washington, D.C.: Worldwatch Institute.

43

The National Innovation System in Singapore

Winston Koh and Phillip Phan*[†]

*Singapore Management University and [†]Johns Hopkins University

History of technological development in Singapore

For two decades since independence in 1965, Singapore enjoyed a rate of economic growth that averaged 8.5% per annum. Today, according to the *Global Competitiveness Report* (WEF, 2006) Singapore's economy is consistently ranked in the top 10 globally, often ahead of Japan, the United Kingdom, Germany, and Australia. Singapore has no external debt, and her foreign exchange reserves, as of the third quarter 2007, is more than US$200 billion (S$1 = U.S.$0.69 as of November 2007), and ranks as one of the highest in the world (IMF, 2007). Singapore's economic development model combines an open-economy framework focused on trade and foreign investments, with strong government involvement in labor, land, and industrial development policies. A simple way to understand the structure of the economy is to think of the economic pie as being divided into thirds, in which one-third belongs to economic production carried out by government-linked corporations (primarily in transportation, telecommunications, and utilities), one-third by the more than 5,000 multinational corporations spread across more than two dozen service (primarily in banking and logistics) and manufacturing industries, and another third by local small and medium-sized enterprises (primarily in manufacturing, personal services, banking, and logistics). While this structure has enabled Singapore to move from Third World to First World status in the span of fewer than 40 years, there have been concerns lately that this approach is outdated now that the economy has to compete "close to the technological frontier" of the global economy—where capital, ideas, and talent are mobile—as opposed to the earlier, easier task of technological catch-up.

Government-linked corporations

A major feature of the Singapore economic landscape is the dominance of government-linked corporations (GLCs). The government invests in corporations through three vehicles: MND Holdings, Singapore Technology Holdings, and Temasek Holdings. From here, up to 70% of some GLCs are directly and indirectly controlled by the government while a smaller percentage of major non-GLCs in the banking, shipping, and technology sectors are controlled indirectly through inter-corporate equity shares between the GLCs and non-GLCs. At the end of the 1980s, GLCs comprised 69% of total assets and 75% of profits of all domestically controlled companies in Singapore. In the 1990s, through a program of privatization that dispersed the equity of these companies, those numbers have been reduced. However, the government continues to hold majority ownership, through her holding companies (Temasek Holdings, MND Holdings, and Singapore Technology Holdings), in these GLCs. Singh and Siah (1998) suggest that inter-firm competition in Singapore is tempered by cooperation and coordinated action in ventures that represent unrelated diversification strategies. This is particularly true for the GLCs, which have the social as well as economic objective of promoting the development of Singapore. For example, the regionalization of such GLCs as Keppel Corporation (shipbuilding and oil platforms) has often been achieved in concert with other companies, and in many cases, competitors.

Multinational corporations

Broadly speaking, Singapore's growth strategy in the 1960s and 1970s of attracting multinational corporations (MNCs) to locate in the city-state and produce for global export markets was initially born out of necessity. As an island economy with few natural resources, employment creation was an urgent task in the early years of Singapore's independence. Later, this strategy of attracting foreign direct investment by MNCs became part of the economic development strategy that seeks to position

Singapore as a major business hub in the global network of trade and capital flows. While the strategy was successful in accelerating growth, the side-effect is that Singaporean companies came to play a largely supporting role to the MNCs with little incentive to invest in indigenous capabilities in high-tech innovation.

From the mid-1960s to the late-1970s, post-independence rapid export-led economic growth was characterized by high dependence on technology transfer and diffusion from foreign MNCs. This was followed by a period, from the mid-1970s to the late-1980s, of local technological intensification, when the government initiated the development of science parks and investment in local technological infrastructure. These two periods constituted the investment-driven growth phase for Singapore, which also successfully developed herself as an air and shipping logistics hub. Together with Taiwan, Hong Kong, and South Korea, Singapore is one of the four successful tiger economies. Singapore's economic development model had combined free markets with strong state involvement in labor, land, and industrial development policies. This approach has enabled the country to now compete close to the technological frontier in the global knowledge economy. However, Singapore's relatively high levels of wages and domestic costs render it vulnerable to competition from lower wage economies, such as China and India, resulting in the need to reduce her reliance on MNCs and look for alternative sources of growth, which the government believes will be found in the small and medium enterprise sector.

Small and medium-size enterprises

The role of SMEs in Singapore's economy has generally been as recipients of technology transfer in manufacturing in the early 1960s and later in services in the 1990s. SMEs populate the economic niches considered too specialized or too small for MNCs to occupy and not strategic enough to influence the trajectory of economic development for GLCs to bother. More recently, however, there has been a focus on the development of the SME sector through the introduction of entrepreneurship in K-12 curricular, government support for SMEs through incubator agencies such as SPRING, formed by the merger of the National Productivity Board (NPB) and the Singapore Institute of Standards and Industrial Research (SISIR). By some accounts, the impetus for the shift in industrial policy toward a high-tech SME economy was the Taiwanese experience during the 1997 Asian Financial Crisis. Taiwan weathered the crisis with little consequence, due to a diversified and flexible economy based on high-tech SMEs.

In the late 1980s there was an intensification of efforts by the government to develop capabilities in basic research and technology commercialization. To do so, the Singapore government stepped up its efforts to remake the economy's institutions and technological infrastructure in order to foster an environment conducive to innovation and technology creation. This led to the formulation of the first 5-year National Technology Plan in 1991 (MTI, 1986, 1991). Since then, there have been several other National Science and Technology Plans that have provided road-maps for Singapore's transition to an innovation-based economy. These plans increasingly recognize the local high-tech SME sector as the engine of sustainable growth, with attendant policies aimed at bolstering their creation and survival.

Technology trajectory

Singapore's development strategy has shifted from emphasizing *using* technology to *creating* it. Broadly speaking, there were four main phases of Singapore's technological transition with respect to the global technological frontier (Figure 43.1):

(a) an industrial take-off phase from the 1960s to the mid-1970s, when there was high dependence on technology transfer from foreign MNCs;
(b) from the mid-1970s to the late-1980s there was rapid growth of local process technological development within MNCs and the development of local supporting industries;
(c) from the late-1980s to the late-1990s there was rapid expansion of applied R&D by MNCs and by publicly funded R&D institutions;
(d) from the late-1990s onwards there was emerging emphasis on high-tech startups and a shift toward technology creation capabilities.

Each successive phase of technological transition was built upon the resources and technological capabilities accumulated in the earlier phases. In particular, there was a phased building up of MNC local manufacturing enterprises (particularly in the electronics-supporting

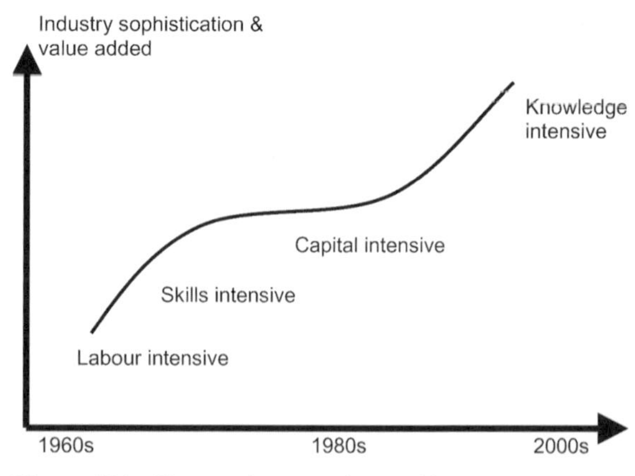

Figure 43.1. Singapore's economic transition.

industries), Public Research Institutes and Centers (PRICs), university-based research, and, finally, local high-tech startups pioneering new products. In terms of the development of technology capability, there was a sustained shift from learning to use (with high reliance on internal transfer by MNCs) to learning to adapt and improve (via "learning by doing" within MNCs as well as "learning by transacting" in local firms acquiring external technology), learning to innovate (mainly applied R&D in product or process), and, finally, learning to pioneer (creating indigenous technology and commercializing it in the marketplace through new ventures).

Over the past few decades, Singapore's technological capabilities have strengthened steadily. It is now consistently ranked by the World Economic Forum (WEF) and the Institute for Management Development (IMD) as among the top 10 countries in the world in areas such as quality of school science and technology education, adoption of information and communications technology (ICT), licensing of foreign technologies, use of advanced technologies in production and process management capabilities.

A major constraint faced by Singapore is her size. To augment the pool of local technical talent, the government has adopted a liberal immigration policy to attract foreign scientists and engineers to staff the research institutions. It has also introduced other policies aimed at strengthening her technological capabilities and expanding the pool of expertise. To build the pipeline of future scientists and research professionals, a National Science Scholarships (NSS) scheme was launched by A*STAR (Agency for Science, Technology, and Research) in 2001. The NSS scheme provides scholarships for Singaporeans to study and train in the biomedical sciences in top universities in the United States. In the 2003 IMD *World Competitiveness Report*, Singapore ranked number one in attracting top-flight foreign talent among 29 economies with a population of fewer than 20 million.

National innovation policy and policy-making agencies

National Science and Technology Plans

In the 1990s Singapore saw her economic future increasingly dependent on her ability to engage in technology creation and developing internal engines of growth, primarily through the SME sector. Critical to this transformation are the 5-year National Science and Technology Plans that began in 1991. These plans included recommendations for changes in the legal institutions governing property rights, intellectual property, high-risk financing, debt markets, and personal and corporate bankruptcy as well as specific technology sector targets for public and cluster (private/public syndicated) funding, and targeted research-based foreign universities and corporations for R&D joint ventures with local SMEs and Public Research Institutes and Centers (government research institutes).

The policy initiatives in the First 5-year National Technology plan, budgeted at S$2 billion, included accelerating the development of technology infrastructure, encouragement of private-sector R&D, and the development of technical manpower to support R&D. Ten key areas were identified for development: information technology and telecommunications; microelectronics and semiconductors; electronic systems; manufacturing technology; materials and chemicals technology; environmental technology; energy, water, and resources; biotechnology; food and agro-technology; and medical sciences.

Identification of these sectors for targeted development was based on a variation of the foresight program used in United Kingdom and other European countries. For each National Technology Plan, a large number of individuals were approached to provide their views and committees were formed to brainstorm and assess the potential of each sector in terms of Singapore's chances of success to emerge as a regional leader in the targeted area. To accelerate the development and strengthening of capabilities in basic research, the National Science and Technology Board funded the establishment of 13 research institutes in industry-specific areas.

In the Second National Science and Technology Plan (1996-2000), Singapore's technology strategy was "to build a world-class science and technology base in fields that match Singapore's competitive strengths and that will spur the growth of new high value-added industries." New milestones were set under the second Technology Plan, including (1) the ratio of R&D expenditure to GDP was to reach 2.6% in 2000, against 1.1% in 1994; (2) the number of scientists and engineers was to reach 65 per 10,000 workers. In 1998 the R&D expenditure to GDP ratio had reached 1.8% and there were 66 scientists and engineers for every 10,000 workers. Other targets were met as well.

Several new policy initiatives were announced in 1998, including a master plan to promote skills-upgrading and accelerate the attraction of foreign talents, and a Technopreneurship 21 (or T21 for short) initiative to foster high-tech startups. The T21 program represented the most visible policy shift by the government from its focus to promote technology adoption to one that supported both technology diffusion and technological innovation. The T21 initiative led to the liberalization in business regulations thought to stifle startups. Besides the amendment of bankruptcy laws, regulations and taxation governing company stock options were revised and new tax offset provision for losses incurred by investors in high-tech startups were introduced.

Under the Third Technology Plan for 2001-2005, the government set aside S$7 billion to develop additional infrastructure and to attract international talent to Singapore. Listing rules on the Singapore Stock Exchange were amended to allow technology to raise funds more easily. Entrepreneurs were allowed to incubate their ventures in their residences, heretofore a violation of residential housing statutes. A US$1 billion venture capital fund was set up to invest in top-tier international venture capital firms, to encourage them to set up regional headquarters in Singapore.

Reflecting the shift in focus to basic research under the Third National Science and Technology Plan, the NSTB was re-organized in 2000 to focus on promoting research and developing R&D manpower, taking on a role similar to that of the National Science Foundation (NSF) in the United States. Two research councils were set up: the Bio-Medical Research Council (BMRC) to award research grants and develop R&D manpower in the life sciences, and the Science and Engineering Research Council (SERC), to oversee research in selected scientific and technological fields. In 2000 the Singapore Government announced a strategic push to promote life sciences research and industry development, following the completion of the Global Genome Mapping Project and the ban on new stem cell lines in the U.S.

In the 2002–2003 WEF *Global Competitiveness Report*, Singapore ranked number one in terms of innovation policy, winning top scores for her effectiveness in protecting intellectual property, as well as for her support of R&D through various tax incentives and grants.[1] In 2004 the Singapore Government set aside more than US$1 billion in a fund to invest in government-backed venture capital funds focusing on technological entrepreneurship in the areas of soft and hardware computing, nanotechnology, alternative energy system, water purification, and biotechnology. In addition, funding for basic research at universities and research institutions increased substantially.

Singapore has consistently been ranked among the top 5 to 10 countries in the world in terms of technology use indicators in the *Global Technology Competitiveness Report* (WEF, 2006). These include the quality of school science and technology education, adoption of information and communications technology (ICT), licensing of foreign technologies, as well as the use of advanced technologies in production and process management capabilities.

[1] The efforts at building sustainable progress in technology creation capabilities have boosted Singapore's ranking in national innovative capacity from 13th to 10th from 2001 to 2002. The top 10 rankings in the 2002–2003 study are: United States, United Kingdom, Finland, Germany, Japan, Switzerland, Sweden, Taiwan, Canada, and Singapore.

Research, Innovation, & Enterprise Council

In August 2004 a ministerial committee was formed to review the R&D strategies and directions for Singapore. It was chaired by the then Deputy Prime Minister, Dr. Tony Tan, himself a medical doctor. The committee studied several small European countries such as Sweden, Finland, Denmark, and Switzerland, which have been successful in leveraging R&D for economic competitiveness, and recommended transforming Singapore into an R&D-driven, innovative, and knowledge-based economy. The committee also identified five strategic thrusts for the national R&D agenda, and recommended setting up a high-level Research Innovation & Enterprise Council (RIEC) to be chaired by the Prime Minister and supported by the National Research Foundation domiciled in the Prime Minister's Office. As a result, a Research, Innovation, & Enterprise Council, chaired by the current Prime Minister, Lee Hsien Loong, was created to provide a strategic overview of national R&D initiatives (see Figure 43.2 on p. 333). The council overseas the National Research Foundation, modeled after the National Science Foundation in the United States.

National Research Foundation

The National Research Foundation was formed in January 2006 as a department in the Prime Minister's Office. It was chaired by Dr. Tony Tan, with a board comprising ministers, senior civil servants, academics, and technology and business leaders. A Scientific Advisory Board (SAB) was formed with 15 eminent technologists, co-chaired by Dr. Curtis Carlson, President and CEO of SRI International, and Prof. Ulrich Suter, Professor of Polymer Materials and former Vice President (Research) of ETH Zurich. An initial endowment of S$5 billion was committed to the National Research Fund, which is managed by the National Research Foundation.

The functions of the National Research Foundation are:

- to coordinate the research of different agencies within the larger national framework in order to provide a coherent strategic overview and direction;
- to develop policies and plans to implement the five strategic thrusts for the national R&D agenda;
- to implement national research, innovation, and enterprise strategies approved by the RIEC, and to allocate funding to programs that meet the NRF's strategic objectives;
- to provide secretariat support to the RIEC.

The five strategic thrusts for the national R&D agenda are:

- intensify national R&D spending to achieve 3% of GDP by 2010;

- identify and invest in strategic areas of R&D;
- fund a balance of basic and applied research within strategic areas;
- provide resources and support to encourage private-sector R&D;
- strengthen linkages between public and private-sector R&D.

The NRF's Competitive Research Program (CRP) funding scheme aims to encourage high-impact research and enhance intellectual and human capital development for R&D. The CRP scheme aims to fund a broad base of research ideas at the program level through a competitive bottom-up approach. Each program may be comprised of several related projects sharing a common unifying (usually technology or application) theme. This is to allow a coordinated, integrated, and sustained way of supporting high-impact inter-disciplinary research because a larger budget can be allocated to fund a number of smaller related projects that address a given problem. The tight linkage among the projects in a program is also designed to facilitate the commercialization of research results. In addition, the CRP scheme provides a way to identify future strategic research areas.

The CRP is open to all areas of science and technology. Multidisciplinary research is strongly encouraged, as is partnership between industry and academia. The CRP scheme offers funding support of up to S$10 million per program over 3 to 5 years. There are two annual grant calls with supported programs from public-sector organizations being fully funded, while private-sector organizations are partially funded. The NRF has allocated S$250 million to the CRP scheme, and expects to grant up to S$50 million worth of program funding for the inaugural call in 2007. The granting process includes an initial submission of a White Paper describing the research proposal and objectives, with short-listed ones developed into full proposals, which are then put through an international peer review process for the consideration of the Evaluation Panel, comprised of local and foreign academic and practitioner experts, and chaired by Dr. Rita Colwell, a member of NRF's Scientific Advisory Board and a former Director of the U.S. National Science Foundation. Proposals are evaluated on research excellence, manpower development potential, economic impact, and industry involvement.

The NRF's Research Fellowship Program allows an NRF fellow complete freedom and independence to pursue his or her research program in Singapore. The fellowship comprises a 3-year research grant of up to US$1.5 million, with possible extension for 3 more years. It affords the grantee a competitive salary and the opportunity to be jointly appointed at a host university or research institution in Singapore. The NRF Research Fellowship is open to all scientists and researchers at or under the age of 35 years regardless of nationality.

The Scientific Advisory Board (SAB)

Together with the establishment of the NRF, the government established a Scientific Advisory Board to:

- highlight critical issues and emerging global trends in basic and investigator-led research where Singapore could fill a gap or meet a need;
- identify, with the NRF, new areas of research where Singapore can reap the benefits of cutting-edge science and build the foundation for enterprise and industry growth;
- review and give advice on the proposals and plans prepared by the NRF;
- assist and advise the NRF on the management of R&D, including the allocation of funding and the assessment of research outcomes; and
- recommend to the NRF R&D areas that Singapore can focus on to develop new growth areas.

SAB members are appointed by the Chairman of the NRF Board for 3-year terms. The SAB is co-chaired by two eminent scientists from Europe and the U.S. respectively and comprises 13 other members. It is a multidisciplinary international board with specialists in broad areas of science and technology.

Elements of the National Innovation System

Science parks

The core of Singapore's technological infrastructure and innovation capabilities can be traced to the development of the Singapore Science Park (SSP) in 1980.[2] The development of the SSP was part of a set of coordinated government policies on science and technology policy (including research and human capital formation), IT infrastructure, and promotion of entrepreneurship (see MTI, 1986, 1991). Each of these policies has been supported by a generous allocation of resources dedicated to specific goals. As an example, in the case of promotion of entrepreneurship, startup grants, venture capital, and a variety of government assistance have been provided.

Located adjacent to the National University of Singapore's medical and engineering schools, the Singapore Science Park now covers a total area of 65 hectares, comprising two adjoining facilities: Science Parks I and

[2] The SSP covers a total area of 65 hectares and employs more than 7,000 engineers, scientists, and support staff. It has signed formal alliances with the Sophia Antipolis Park of France and the Heidelberg Technology Park of Germany.

II. As of 2000 there were 307 tenant companies.[3] Physically, it was conceptualized as a business park, modeled after those in Silicon Valley, to attract science-based and new technology companies. About 7,000 engineers and scientists and support staff are employed in the Singapore Science Park. Formal alliances with the Sophia Antipolis Park of France and the Heidelberg Technology Park of Germany (both government-established science parks) have been established to provide information exchange on the management and development of such facilities.

An initial motivation of the SSP was to provide and upgrade local science and engineering *infrastructure* to house MNCs as well as new industries which require proximity to institutions of higher learning. Additionally, the SSP was to provide a focal point for research, development, and innovation in Singapore, with an emphasis on industrial R&D. A secondary objective of the Singapore Science Park is to *signal* to foreign firms and investors Singapore's readiness to promote and attract high-tech and knowledge-intensive industries. To position Singapore as a regional R&D hub, high-tech companies from Australia, New Zealand, and the U.S. are being courted to locate their R&D activities in the SSP, and to use Singapore as a gateway to penetrate the markets in China, India, Southeast Asia, and Indo-China. Given Singapore's long involvement with MNCs and her institutional orientation toward supporting and encouraging MNC engagement in the region, this was thought to be the easiest route to increasing technology transfer into the country's domestic SMEs.

Beginning in 2005, Singapore's science park strategy underwent another phase of evolution. Apart from the development of the planned phase 3 of the Singapore Science Park, the government had previously announced (in 2000) the development of a S$15 billion new science park—the One-North project—to strengthen the technological infrastructure as Singapore targets life sciences as a new-technology growth pillar for Singapore.[4] To be developed over 15 years, the objective of One-North, modeled after Silicon Valley, was to create the ambience of a multifaceted research community, with international schools, integrated public transport, and other supporting amenities. The objective was to integrate One-North, the Singapore Science Park, and other research centers into a "Science Habitat".[5] In the conceptualization and planning of One-North, studies were made on a number of science parks and incubators, including Hsinchu, Cambridge, Silicon Valley, Sophia Antipolis, the MIT Medialab, etc.

The development of the One-North Science Habitat comes at a time when the Singapore government is restructuring the Singapore economy, focusing on the creation of internal engines of growth, and placing renewed emphasis on the role that venture capital and entrepreneurship plays in creating local world-class companies. When completed in 2015, at a cost of S$15 billion in public funds with an equal amount of investment from private funds, One-North will house a state-of-the-art R&D infrastructure to provide a wider focal point for R&D and entrepreneurial activities in the biosciences and information technology. In addition, high-end residential development, libraries, primary and secondary schools, zero-emission transportation links, and other life style amenities such as sports clubs, theaters and cultural facilities, art galleries, and world-class shopping are included in the One-North Master Development Plan. In essence, the Science Habitat is designed to be a city within a city with its own economic and social dynamics revolving around the activities of knowledge creation and commercialization.

A new research facility, the Biopolis,[6] located in the One-North Science Park, has already commenced operations and now offers cutting-edge facilities for laboratory-based R&D activities tailored to biomedical sciences, particularly genetics and genomics companies. Generous financial incentives, in the form of tax relief, subsidized rents, R&D grants, and training subsidies, are given to companies to start up their operations. Together with efforts to develop high-tech entrepreneurship, these initiatives form part of the government's macro-strategy to transition the economy from technology use to innovation-based.

There are a number of areas that distinguishes the One-North masterplan from the original Singapore Science Park. First, it aims to provide an infrastructure that offers seamless connectivity, at both personal and business levels. In contrast, the sprawling complexes of Singapore Science Parks I and II did not facilitate close interaction. Second, "dynamic planning" is emphasized in One-North, to encourage vertical and horizontal integration of different uses. The aim is to encourage different companies to work closely together and to encourage the cross-fertilization of research ideas. Greater private-sector participation is encouraged with tax and technology transfer incentives. For instance, although biomedical sciences is a key sector in the first phase of the development of One-North—as shown in the investment in Biopolis—the current thinking is to create self-evolving structures that

[3] The largest group of tenant companies is in the field of IT followed by those in electronics. Tenants at the Singapore Science Pack include Sony, ExxonMobil Chemicals, Silicon Graphics, Lucent Technologies, Johns Hopkins, FujiXerox and Motorola Electronics. There are two venture capital firms located in the science park. *Source*: *http//:www.sciencepark.com.sg*

[4] The name One-North Science Park is chosen because Singapore is situated 1°N of the Equator. Further information on One-North can be obtained at *http//:www.onenorth.com*

[5] Based on interviews with officials involved in the One-North development project, we understand that the planning and conceptualization of the One-North project took more than a year. Details of the One-North project can be found at *http://www.one-north.com*

[6] A new medical research facility developed a short distance from the Singapore Science Park and the National University Hospital. The establishment of Biopolis is part of the government's current plan to develop Singapore into a biomedical hub.

will allow One-North to continue to catch the next wave of emerging technologies and be relevant to emerging technologies. More generally, the governing philosophy behind the development of this and future science parks in Singapore is to view them as part of an ecosystem of technology parks around the world, linked by high-speed communication infrastructures and resource-sharing frameworks, in order that their development is guided in partnership with those parks that have had extensive experience or complementary technologies. The intended result is the global sourcing and sharing of technology, talent, and markets with the science parks as conduits.

While there were substantial efforts over the years to define a global role for the Singapore Science Parks, it has only managed to achieve modest success in plugging Singapore's R&D community into the global network of technology clusters. Participation by the private sector in the development of Science Parks I and II was limited. There was also little interaction between the actors in the Singapore Science Park and with other regions, unlike the experiences of Silicon Valley and the Hsinchu Science Park. The One-North strategy represents a new attempt to address these deficiencies. While the development of state-of-the-art infrastructure remains a key draw for overseas companies to locate in One-North, it also aims to define a stronger regional role for itself by establishing Singapore as a regional center of R&D. The key constraint remains the limited pool of technical talent in Singapore. In the short term it is likely that foreign talent would need to be "imported" to provide the required technical expertise.

Government-sponsored research agencies

In 2005 the government earmarked a separate S$1 billion to be spent over 5 years to set up new research centers and populate them with globally renowned researchers. This invitation-only program is designed to kick-start the launch of new research initiatives by seeding local research efforts with the ongoing efforts of foreign researchers. For example, in July 2006 the government announced that the Massachusetts Institute of Technology, an early beneficiary of this program, will set up its first and largest international research endeavor by 2007. The SMART Center (Singapore–MIT Alliance for Research & Technology) will be headed by a senior professor from MIT and will eventually house up to 400 researchers. The plan annually brings in 10 tenured MIT professors from a broad cross-section of MIT faculty to Singapore to lead 10 inter-disciplinary research groups comprising up to 400 researchers from around the world.

In 2007 the Research Council identified three research areas—biomedical sciences, environmental and water technologies, and interactive and digital media—by earmarking them for funding of S$1.4 billion over the next 5 years. The three areas are expected to create 86,000 jobs and $30 billion in gross national value-added by 2015. In addition, S$1 billion will be spent on attracting the best research talent to Singapore, through a joint program that marries local research institutes with the best foreign research universities.

The Singapore government intends to use the MIT alliance and formation of the SMART center as a template to attract other top research institutions to Singapore. The objective is to establish Singapore-based campuses as sites of research excellence and technological expertise (as illustrated in Figure 43.3) in order to:

- develop a thriving ecosystem for high-tech innovation and entrepreneurship;

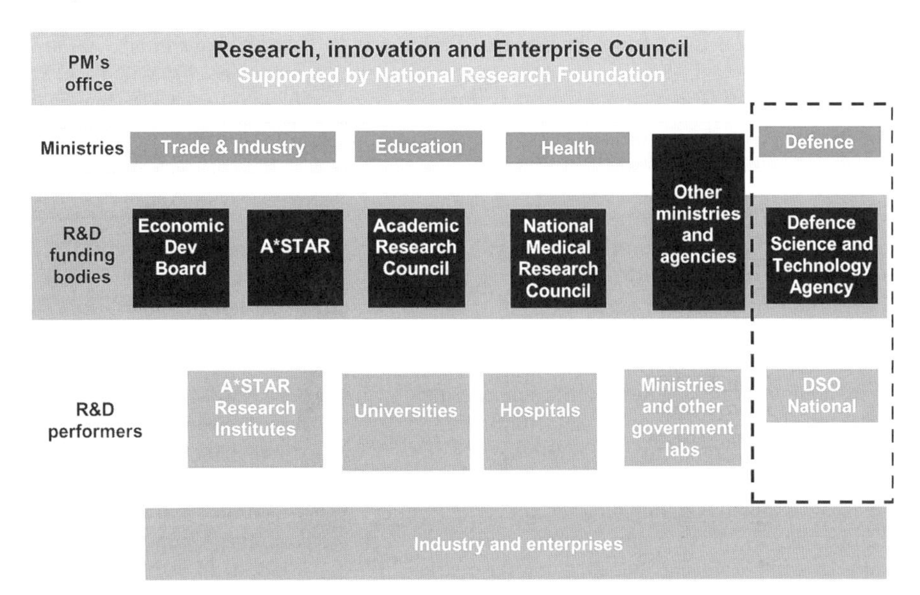

Figure 43.2. Agencies in the R&D landscape in Singapore.

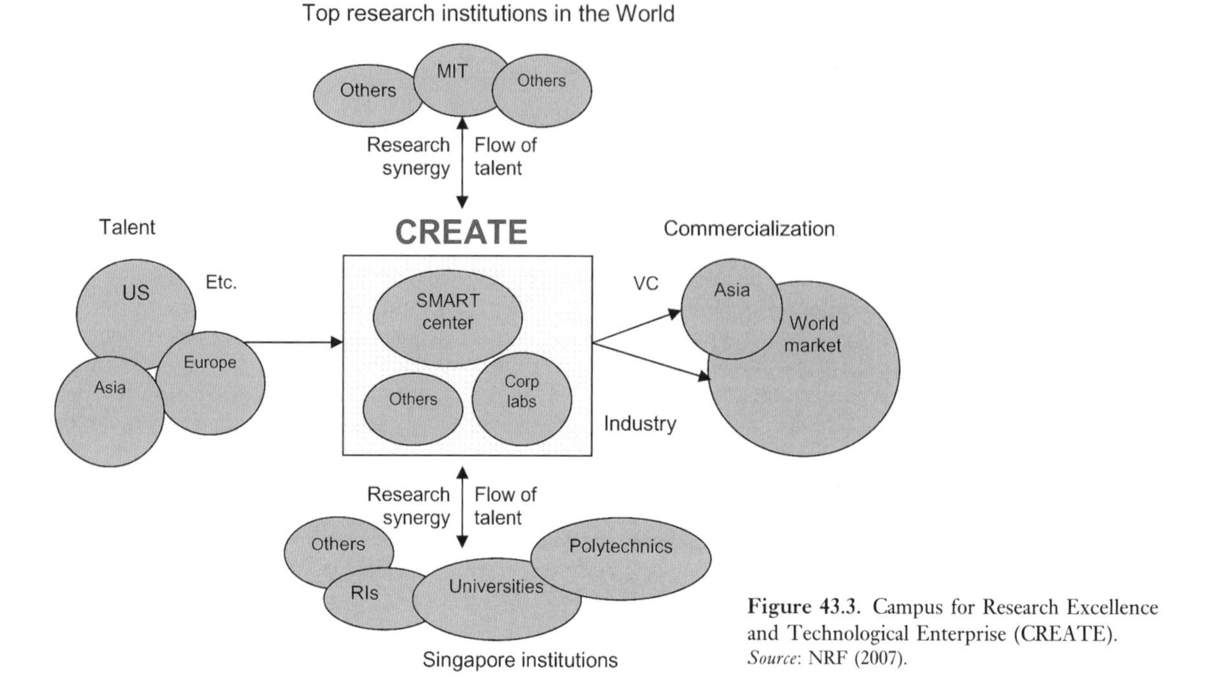

Figure 43.3. Campus for Research Excellence and Technological Enterprise (CREATE). *Source*: NRF (2007).

- infuse vigor and buzz into the local business environment;
- make high-tech, high-growth companies a sizable part of the economic landscape, so as to create jobs and new market opportunities;
- turn local universities into sources of commercializable intellectual property.

The Agency for Science, Technology, and Research

There are now a dozen public research institutes (PRIs) under the aegis of the Agency for Science, Technology, and Research (A*STAR):

- Bioinformatics Institute
- Bioprocessing Technology Institute
- Genome Institute of Singapore
- Institute of Bioengineering and Nanotechnology
- Institute of Molecular and Cell Biology
- Data Storage Institute
- Institute for Infocomm Research
- Institute of Chemical & Engineering Sciences
- Institute of High-performance Computing
- Institute of Materials Research & Engineering
- Institute of Microelectronics and Singapore
- Institute of Manufacturing Technology.

Each PRI is tasked with pursuing science at the frontier in its own scientific domain, and creating cutting-edge technologies with promising potential for the market. The initial mission of the PRIs was to develop the applied technologies deemed critical for the industrial clusters in existence in Singapore. In addition, some institutes had to develop core competencies in the new generic technologies (e.g., in molecular and cell biology and in wireless communications) needed to attract new high-tech industries.

From the late 1980s to the early 1990s, public-sector R&D conducted by universities and PRIs accounted for 40–50% of total R&D expenditure in Singapore. Since the 1990s the share of public-sector R&D expenditure declined to around 37% and has remained relatively stable since 1993. Only 13.6% of R&D expenditure in 1999 was in basic research, compared with 36.3% for applied R&D and 50% for experimental development. Over the 1990–1999 period, R&D expenditure by PRIs had expanded nearly ninefold, which is significantly faster than the growth of private R&D. By 1999 the 13 public research institutes had accounted for over S$370 million (or 14%) of aggregate R&D expenditure (NSTB, Annual).

Although public-sector R&D conducted by universities and government-funded research institutes accounted for close to 40% of total R&D expenditure, it accounted for a very small share of the U.S. patents granted to Singapore-based inventors: 4.5% during 1991–1995, 6.6% during 1996–2000, and 7.5% during 2001–2002. While public research institutes and universities in Singapore managed to spin off a number of companies in the late-1990s, the combined impact of these efforts has been relatively modest. In response to these problems, the government introduced an institutional restructuring in early 2001. The responsibility for implementing T21 was taken out of the hands of the National Science and Technology Board and transferred to the Economic Development Board. The NSTB was then re-organized to focus on promoting R&D manpower (and was renamed the Agency for Science, Technology and Research or A*STAR). The policies for technology commercialization were left to the EDB to implement.

The critical role of education

For Singapore, education has played a critical role in promoting social and economic change by expanding the pool of skilled workers and professionals (Kong, 2001). Early in her independence, Singapore embarked on a labor-intensive industrialization strategy with the relentless pursuit of MNCs. In the early days of independence, Singapore's educational priorities focused on enabling the country's economic development strategies. A key thrust of the curriculum was an emphasis on technical education throughout the system. Although there is now an even greater emphasis on technological entrepreneurship, a single motivation remains paramount in curriculum development—finding an educational emphasis that will provide the country with the requisite human resources to fuel economic development. Educational strategies have in large part been used to meet economic needs through the production of a suitably qualified workforce.

By the 1970s, economic development priorities gave new emphasis to faculties of engineering and commerce and a parallel growth in the polytechnics that specialized in engineering and technical skill training at the tertiary level. The Faculty of Engineering at the University of Singapore was established in 1969 and the Singapore Polytechnic and Ngee Ann Technical College were expanded and upgraded to 4-year diploma-granting institutions (Tham, 1989, p. 498). A Technical Education Department was established at the Ministry of Education in 1968 to administer a technical education and industrial training program. The emphasis on technical and technological skills-training meant that the vocational institutes had doubled their enrollments from 6,500 in 1973 to 11,500 in 1980. Growth in polytechnic enrollment was equally high, with a fourfold increase from 2,600 in 1966 to 8,300 by 1980. In the academic institutions, the area with most growth at the upper secondary levels (grades 11 and 12) was in the science classes, which experienced a fourfold increase from 6,250 in 1977 to 23,697 in 1980. This reflected the population's response to the government's message that science and technical education offered a ticket to personal economic advancement.

By the late 1990s, the convergence of breakthrough discoveries in six key technologies—microelectronics, computers, telecommunications, material science, robotics, and biotechnology—was thought to be the foundation for a new type of knowledge-based economy. Hence, Singapore intensified her investments in education and R&D with a curriculum that emphasized personal innovation, career flexibility, entrepreneurship, creativity, and a commitment to life-long learning. The role of science and technology remained important but increasing attention was given to broad-based multidisciplinary curricula. This resulted in the Thinking Schools, Learning Nation (TSLN) initiative in 1997. The initiative sought to foster a creative and critical thinking culture within the primary and secondary schools and a curriculum focused on developing students into active learners with critical thinking skills. The key features included

(i) explicit teaching of critical thinking skills;
(ii) reduction of subject syllabus content;
(iii) revision of assessment modes; and
(iv) emphasis on processes instead of outcomes when appraising schools.

National technology commercialization initiatives

Although Singapore's government-funded research institutions and local universities managed to spin off a number of companies in the late-1990s, the commercial impact of these efforts was relatively modest. University–industry collaboration in R&D is still weak. Fragmented regional markets, the lack of venture capital financing, as well as the shortage of good managerial expertise hampered the expansion of many of these spinoffs. Most of them had to rely on continued government support, through procurement contracting or outsourcing work, in order to survive; eventually, many of these startups were economically unviable.

In 2001, private-sector funding accounted for only 2.6% of total university R&D expenditure. In a survey of manufacturing firms at the end of 1999, the importance of collaboration with local universities and public research institutes was ranked behind collaboration with customers and suppliers, although some improvements were evident in a comparative survey 2 years earlier. The situation appears to be brighter in terms of the extent of university technology commercialization through licensing.

While the PRIs met their quantitative targets for R&D expenditures and the training of technical manpower, their effectiveness in developing national innovative capacity is unclear. Moreover, their changing roles over time have led to a number of problems. First, the PRIs were initially required by the NSTB to spin off high-tech startups as part of the push to develop technology entrepreneurship in Singapore. This policy appeared to have been hastily implemented without acknowledging the conflict between the PRI's original mission of licensing to existing companies and the requirement to spin off startups—a key role that the PRIs were supposed to play. As it turned out, the PRIs began to keep technologies they had developed from being licensed and started to encourage their staff members to start up companies to commercialize the technologies, in order that the personnel involved could receive sizable equity ownership in these startups. Second, the PRIs were not given incentives to cooperate with each other in research, technology, market intelligence gathering, or intellectual property management.

Funds flow for innovation

The lion's share of funds for innovation comes from public sources. In part, this is due to the relatively underdeveloped IPO equity market and along with it the nascent high-risk financing (venture capital and angel financing) market. Attempts to ramp up participation in high-risk financing by local and foreign funds, with the creation of a short-lived, high-tech equities stock market in 1995, were stymied by the 1997 Asian Financial Crisis, the effects of which have only recently been abated by the worldwide growth in commodity prices and China's double-digit economic growth.

Funds were made available for R&D investments with initiatives announced in October 2005 by Deputy Prime Minister, Dr Tony Tan, to more than double Singapore's R&D expenditure from S$4.1 billion in 2004 to S$6.9 billion by 2010. The government's funding plan is to increase R&D expenditure to 3% of GDP by 2010, about U.S.$4 billion, from the 2004 R&D expenditure of U.S.$2.5 billion (about 2.25% of GDP). It is hoped that a sustained high rate of public investment will create the psychological bedrock for private high-risk equity and debt financing of technology-intensive companies to follow.

In 2007, public-sector R&D budgets more than doubled to S$13.55 billion (i.e., from the 2005 level). The total budget comprised

- S$5 billion for the NRF;
- S$5.4 billion for the PRIs housed in A*STAR;
- S$1.05 billion for academic (university-based) research;
- S$2.1 billion for the EDB to promote private-sector R&D.

Exploit technologies

A*STAR's 12 research institutes have 2,000 researchers and scientists developing technologies and intellectual properties across multiple science and engineering disciplines and biomedical research areas. Within A*STAR there is a commercialization division, Exploit Technologies, which is tasked with supporting the institutes in transforming the Singapore economy through commercializing R&D. Exploit Technologies manages and consolidates all research institutes' intellectual properties under one roof and so functions as a one-stop technology transfer office for ,research institutes, industries, and enterprises.

Summary of innovation output trends

Table 43.1 illustrates that Singapore's expenditure in R&D, in absolute amounts and as a percentage of GDP, has increased steadily. Singapore had spent relatively little on R&D until the 1990s. Expenditure in R&D has increased steadily since then in both absolute amounts and as a percentage of GDP. In 2005, expenditure on R&D totaled S$4,582 million, which was 2.36% of GDP. Expenditure on R&D manpower was 42% (S$1,937 million) of total R&D expenditure, while other operating expenditure accounted for 40% (S$1,847 million) and capital expenditure for 17% (S$798 million) (A*STAR, 2005).

During the same period, Singapore made significant progress in her pool of scientific talent. In 2005 there were a total of 21,338 research scientists and engineers (RSEs) and 3,718 full-time postgraduate research students (FPGRSs) at the Master's and Ph.D. levels; 53% (11,264) of the RSEs had a bachelor degree, 26% (5,499) a Master's degree, and 21% (4,575) a Ph.D. as the highest formal qualifications; while 28% (1,043) of the FPGRSs were pursuing studies at the Master's level and 72% (2,675) at the Ph.D. level. The number of RSEs per 10,000 in the labor force was 90.1.

Table 43.2 shows that the total number of patents filed by Singapore-based organizations (including the local affiliates of MNCs). As shown, the total number of patents increased steadily over the years. As reported in the *Annual Survey of R&D* (A*STAR, 2005), the top 150 private-sector enterprises by R&D expenditure, in 2004, accounted for 85% (S$2,194 million) of private-sector R&D expenditure. Before the mid-1990s the rate of patenting in Singapore had grown slowly, but the situation changed markedly since the mid-1990s.

For the period 1996–2001, the growth rate of Singapore's utility patents in the United States, as measured by the number of patents granted by the United States Patents and Trademarks Office (USPTO) to Singapore-based inventors, was the highest among all OECD countries as well as the newly industrialized economies in Asia. However, as Mahmood and Singh (2003) found in a study of U.S. patent data, the patenting activity in Singapore, as well as Hong Kong, over the past 30 years has consistently been much lower when compared with South Korea and Taiwan. In addition, a large proportion of Singapore's patenting activity came from MNCs rather than domestic firms. On average, in the 1990s MNCs

Table 43.1. Singapore technological capabilities.

Year	*Private sector* (S$m)	*Higher education sector* (S$m)	*Government sector*	*Public research institutes* ($Sm)	*GERD* ($Sm)	*GERD/GDP* (%)	*RSEs*	*RSE/10,000 labor force*
1978	26	8	4	—	38	0.21	818	8.4
1981	44	24	13	—	81	0.26	1,193	10.6
1984	107	70	38	—	214	0.54	2,401	18.4
1987	226	95	54	—	375	0.86	3,361	25.3
1990	310	120	99	43	572	0.84	4,329	27.7
1991	442	147	97	71	757	1.00	5,218	33.6
1992	578	156	105	111	950	1.19	6,454	39.8
1993	619	157	107	116	998	1.07	6,629	40.5
1994	736	180	142	117	1,175	1.10	7,086	41.9
1995	881	193	110	181	1,367	1.16	8,340	47.7
1996	1,133	239	167	253	1,792	1.40	10,153	56.3
1997	1,315	278	216	296	2,105	1.50	11,302	60.2
1998	1,536	306	300	351	2,492	1.81	12,655	65.5
1999	1,671	310	305	371	2,656	1.90	13,817	69.9
2000	1,866	338	424	381	3,010	1.88	18,302	83.5
2001	2,045	367	425	396	3,233	2.11	18,577	72.5
2002	2,091	430	449	434	3,405	2.19	15,654	73.5
2003	2,081	457	435	449	3,424	2.15	17,074	79.4
2004	2,589	424	442	605	4,062	2.25	18,935	86.3
2005	3,031	478	443	630	4,581	2.36	21,338	90.1

Note: GERD = gross expenditure on R&D; RSE = research scientists and engineers.
Source: NSTB (Various years) and A*STAR (Various years).

accounted for more than half the R&D investment in Singapore.

Table 43.3 provides data on R&D expenditure from different types of research programs in 2005. Although local universities' primary focus is on basic scientific research, they are also under pressure to conduct more applied R&D and to foster industry linkages. In some cases the key performance criteria for university academic staff continued to be consistent publications in international journals, but the extent of technology licensing to the private sector and the number of R&D collaborations with industry were also used to evaluate R&D productivity in tertiary institutions by the Ministry of Education.

In 2005, private-sector expenditure on R&D accounted for 66% ($3,031 million) of total expenditure on R&D, and amounted to 1.56% of GDP in 2005. The government sector was 10%, higher education sector 10%, and the public research institutes 14% of total expenditure on R&D. Forty-seven percent of total R&D expenditure was on experimental development, 33% on applied research, and 21% on basic research. Fifty-seven percent of total R&D expenditure was in the fields of engineering and technology, 19% in the biomedical and related sciences, 10% in the natural sciences (excluding the biological sciences), 1% in agricultural and food sciences, and 14% in other areas. In the biomedical and related sciences, 39% of R&D expenditure was on basic research, 45% on applied research, and 15% on experimental development. In the natural sciences (excluding biological sciences), 27% of R&D expenditure was on basic research, 38% on applied research, and 35% on experimental development. In engineering and technology, 52% of R&D expenditure was on experimental development, 33% on applied research, and 15% was on basic research.

Conclusion

The National Innovation System in Singapore cannot be understood piecemeal in terms of just legal, economic, or technological institutions. As we have discussed the NIS comprises a set of strategic, far-ranging policy initiatives intended to capitalize on Singapore's advantage as a small nation-state with few natural resources to focus on building a sustainable and flexible economy domiciled in an evolving region dominated by the manufacturing prowess

Table 43.2. R&D output indicators for Singapore, 1993–2005.

	1993	*1994*	*1995*	*1996*	*1997*	*1998*	*1999*	*2000*	*2001*	*2002*	*2003*	*2004*	*2005*
No. of patents applied for in the year	142	263	242	316	490	579	673	1,268	1,096	936	1001	1257	1594
No. of patents awarded for the year[a]	52	58	51	91	132	136	161	285	461	451	460	599	877
Total no. of patents owned at December 31[b]	200	204	256	614	831	847	1,077	902	1,456	1,739	2,314	2570	3475
Revenue from royalties and licensing of patents/new technologies developed in Singapore (S$m)[c]	24.34	23.95	27.23	27.34	26.61	50.97	671.89	75	55.17	87.50	132.37	208.20	93.66

Note: [a]1993–1995 Singapore organizations only.
[b]1993–1995 Singapore organizations only.
[c]1993–1995 revenue from royalties and licensing of new-technology/products developed in-house ($m).
Source: NSTB (Various years) and A*STAR (Various years).

Table 43.3. R&D expenditure by sector and strategic focus (2005).

Year	*Private sector* ($Sm)	*Higher education sector* ($Sm)	*Government sector* ($Sm)	*Public research institutes* ($Sm)	*Total* ($Sm)
Pure basic research	364.01	90.29	0.40	180.66	635.36
Strategic basic research	—	164.22	19.36	126.37	309.95
Applied research	956.05	145.44	142.78	246.15	1,490.42
Experimental research	1,711.28	78.05	280.28	76.88	2,146.49
Total R&D expenditure	*3,031.34*	*478*	*442.82*	*630.06*	*4,582.22*

Source: NSTB (Various years) and A*STAR (2005).

of China, the technological monopoly of Japan and Korea, and the sophisticated capital markets of the West. Given the role that technology plays in Singapore's economic evolution, in contrast to the relative importance of capital and labor in agrarian economies transitioning to a manufacturing economy, it is not surprising that her industrial policy has consistently emphasized technology transfer and innovation. In short, the structural (science parks, research institutes, universities, and venture funds) and institutional (intellectual property regime, bankruptcy laws, capital markets, and education system) elements of the NIS can only be properly understood in the context of the country's self-perceived role as a major node in a networked global economy.

References

A*STAR. (Various years). *Annual Survey of R&D in Singapore*. Singapore: Agency for Science, Technology, and Research; *http://www.a-star.gov.sg*

IMD. (2003). *The IMD World Competitiveness Yearbook*. Geneva, Switzerland: Institute for Management Development.

IMF. (2007). *World Economic Outlook*. Washington, D.C.: International Monetary Fund.

Koh, F. C. C., Winston, T. H., and Koh, F. T. T. (2005). "An analytical framework of science parks and technology districts with an application to Singapore." *Journal of Business Venturing*, **20**(2), 217–239.

Koh, W. T. H. (2006). "Singapore's transition to innovation-based economic growth: Infrastructure, institutions and government's role." *R&D Management*, **36**(2), 143–160.

Koh, W. T. H., and Wong, P. K. (2005). "Competing at the frontier: The changing role of technology policy in Singapore's economic strategy." *Technological Forecasting and Social Change*, **72**(3), 255–287.

Kong, L. (2001). *Science and Education in an Asian Tiger* (White Paper). Singapore: National University of Singapore.

Mahmood, I. P., and Singh, J. (2003). "Technological dynamism in Asia." *Researxh Policy*, **32**(6), 1031–1054.

MTI. (1986). *Five-year National Technology Plan*. Singapore: Ministy of Trade & Industry.

MTI. (1991). *Five-year National Technology Plan*. Singapore: Ministy of Trade & Industry.

NRF. (2007). National Research Foundation; *http://www.nrf.gov.sg*

NSTB. (Annual). *National Survey of R&D in Singapore*. Singapore: National Science and Technology Board.

Reynolds, P., Camp, S. M., Bygrave, W. D., Autio, E., Hay, M. (2001). *Global Entrepreneurship Monitor: Executive Report*. Babson Park, MA: Kauffman Center for Entrepreneurial Leadership.

Singh, K., and Siah, H. A. (1998). *The Strategies and Success of Government-linked Corporations in Singapore* (Research Paper Series #98-06). Singapore: Faculty of Business Administration, National University of Singapore.

Tham, S. C. (1989). "The perception and practice of education." In: K. Sandhu and P. Wheatley (Eds.), *Management of Success*. Singapore: Institute of Southeast Asian Studies.

WB (various). *World Development Report*. Washington, D.C.: World Bank.

WB. (2002). *World Bank Development Indicators*. Washington, D.C.: World Bank.

WEF. (2002/2003). *Global Information Technology Report*. Geneva, Switzerland: World Economic Forum.

WEF. (2006). *Global Technology Competitiveness Report*. Geneva, Switzerland: World Economic Forum.

WEF. (2007). *Global Information Technology Report*. Geneva, Switzerland: World Economic Forum.

Wong, P. K. (2003). "From using to creating technology: The evolution of Singapore's National Innovation System and the changing role of public policy." In: S. Lall and S. Urata (Eds.), *Foreign Direct Investment, Technology Development and Competitiveness in East Asia*. Cheltenham, U.K.: Edward Elgar.

44

The Indian Innovation System

Rishikesha T. Krishnan

Indian Institute of Management Bangalore

Introduction

The evolution of the Indian Innovation System is best understood in two phases—the first from Indian independence in 1947 till about 1990 (the "self-reliance" phase); and the second from 1991 to the present (the deregulation phase).

The self-reliance phase

Before economic deregulation began in the early 1990s, India's dominant economic philosophy was one of self-reliance. The objective was to produce the country's requirements, to the extent possible, within the borders of the country. An elaborate process of economic planning created 5-year blueprints for the allocation of resources and the creation of capacities in the country.

The public sector was seen to be the fountainhead of industrial development and accounted for as much as two-thirds of the fixed capital investment in the factory sector. Public ownership was particularly stressed in those sectors where technology acquisition was expected to involve the evaluation of a range of non-commercial considerations (Tyabji, 2000).

Though India had a mixed economy (i.e., private Indian firms as well as multinational companies) there were tight regulations on inward capital flows, expansion, diversification and the import of capital goods, intermediates, and technology. Technology imports were regulated on a case-by-case basis, and companies permitted to import technology were often required to commit to progressive indigenization through a "phased manufacturing program". Physical constraints on imports and high import duties coupled with barriers to entry arising from industrial licensing meant that local firms were protected from competition to a large degree.

Successive governments encouraged small firms as they were seen as creating employment and lacked the monopoly power of large firms. The small-scale sector was provided reservation in hundreds of sectors and implicitly encouraged to make imitative products through reverse engineering and improvisation (Tyabji, 2000). Public-sector industrial enterprises were expected to support small-scale "ancillary" units that could produce small components and sub-assemblies for them. Small-scale industries enjoyed fiscal benefits like lower rates of excise duties and were largely outside the purview of industrial regulation.

Under Prime Minister Nehru, in the 1950s, newly independent India sought to create a network of national laboratories and institutions of higher learning in a planned effort to allow India to be an active participant in modern science and technology. This was at a time when few developing countries had a similar vision. Over 80% of the R&D done in India was financed by the government of India and conducted within government research laboratories (Forbes, 1999). Much of this was in the strategic sectors of atomic energy, defense, and space research, but the government also created a network of 40 laboratories under the aegis of the Council of Scientific & Industrial Research (CSIR) to do work of relevance to industry.

In parallel with the creation of the R&D institutional base, India created a network of engineering education institutions starting in the late 1950s. The best known of these, the Indian Institutes of Technology (IITs), provide world-class engineering education.[1] Five IITs were set up in the 1950s and 1960s, and a sixth IIT in the late 1980s. Seventeen Regional Engineering Colleges (now re-named as National Institutes of Technology) set up jointly by the Central Government with the state governments constituted the second tier of engineering education.

[1] A report on an Indian portal *rediff.com* quotes a *Times Higher Education Supplement* report that the IITs are the world's third best tech universities. See "IITs, World's 3rd best tech universities"; *http://in.rediff.com/money/2005/oct/10iit.htm* (last accessed March 23, 2006).

The institutional structure described above had decidedly mixed outcomes:

- —Economic growth remained stuck in the historical "Hindu rate of growth" of about 3.5% and was significantly lower than countries such as South Korea that at one time had comparable *per capita* incomes.
- —Self-reliance became an end in itself, leading to a very broad production base, but insufficient attention to efficiency and productivity (Forbes, 1999).
- —With a few exceptions, the public sector failed to drive the Indian industrial sector on to a higher growth trajectory and got bogged down by time overruns, high costs, and a lack of technological dynamism.
- —Private-sector industry felt little need to innovate as they enjoyed high levels of protection (Forbes, 1999; Krishnan and Prabhu, 1999). Constraints on growth also acted as a disincentive to innovative behavior (Forbes, 1999). With a protected market, and a high cost structure, very few firms pursued exports or targeted external markets aggressively. Such R&D as was done by industry was concentrated on import substitution and the creation of local sources for inputs.
- —Small firms had an incentive to remain small and there was a tendency to fragment capacities. There was no incentive to grow to exploit economies of scale or scope.
- —By the end of the 1980s, India had perhaps the strongest scientific and technological infrastructure and human resource base among developing countries, but little benefit of this was accruing to the industrial production system. The government R&D system tended to work in isolation and on its own agenda. Demand pull was weak—in a highly regulated economy, firms were under little pressure to improve their products and services or launch new ones; in fact, in many cases, government regulations forbade changes in products without government permission. Most firms preferred to source their technology as tried-and-tested packages from foreign sources. Foreign companies were willing to license their technology to Indian firms as they did not have direct access to the Indian market and technology licensing gave them at least some additional revenue from that market. An effort was made in the early 1970s to formulate a national science and technology plan that would dovetail with the economic planning process and help integration of the government's technology development efforts with industrial development, but this was short-lived.
- —As in the case of the national research laboratories, IITs had limited interaction with Indian industry. IIT graduates found few opportunities to use their technical knowledge in the industrial sector and tended to emigrate in large numbers, principally to the United States. Those that stayed behind went into the government research establishments or to management positions in the private sector.

The deregulation phase

While the trigger was an economic crisis caused by a serious decline in foreign exchange reserves due to the flaring up of oil prices, the Indian Government that took office in June 1991 attempted to address the structural problems underlying the crisis. While the broader objective was to stimulate economic growth by attracting foreign investment, removing licensing and monopoly controls, allowing imports and encouraging exports, an explicit focus of the new policies was the development of an innovative capability in the economy. The Industrial Policy Statement of the Government of India of July 24, 1991 had among its objectives "injecting the desired level of technological dynamism in Indian industry" and "the development of indigenous competence for the efficient absorption of foreign technology" and expressed the hope "that greater competitive pressure will also induce our industry to invest much more in research and development than they have been doing in the past . . ."

Successive governments have carried forward the reform process. Today, most industries do not require industrial licencing. Automatic approval is given for foreign investment, up to 100%, in many industries (Rathinasamy *et al.*, 2003). Physical constraints on imports like actual user conditions have been removed and duties have been reduced considerably, though they are still higher than in many other countries. Similarly, restrictions on technology imports have been removed.

The overall success of this deregulation process is reflected in the compounded growth rate of the Indian economy shifting to more than 6% since 1991.

Amendments to the Patents Act that came into effect from 2005 have brought Indian laws in line with the intellectual property rights (IPR) regimes prevalent in much of the industrialized world. With this framework in place, both Indian companies and multinational corporations are expected to step up R&D investments.

More than 15 years after the process of economic liberalization was initiated, government-owned and operated laboratories continue to be dominant players in the Indian Innovation System in terms of funding. But, in the last decade, enhanced innovation efforts by Indian companies (in select sectors), and by R&D laboratories set up by multinational corporations, are changing the contours of the Indian Innovation System.

Output trends

The total expenditure on R&D in India in 2002–2003, the latest year for which official figures are available, was

Table 44.1. Breakdown of total R&D expenditure as a percentage, 2002–2003.

Defence Research & Development Organization	17.10
Department of Space	12.02
Indian Council of Agricultural Research	7.62
Department of Atomic Energy	6.88
Council of Scientific & Industrial Research	5.30
Other central government agencies	13.67
Public-sector industry	4.50
State governments	8.50
Private-sector industry	20.30
Higher education	4.10
Total	*100.00*

Source: DST (2006).

Rs. 180 billion (approximately U.S.$4 billion). A breakdown of this expenditure is shown in Table 44.1.

As of April 1, 2000 about 296,000 people were employed in R&D in India, 63% of whom were employed in the institutional sector and 37% in the industrial sector (DST, 2006).

Even today, the Central Government continues to play a major direct role in R&D and accounted for a little over 62% of the national expenditure on R&D in 2002–2003; much of this funding is directed to work done in government laboratories. About 84% of the central government's spending on R&D was through 12 major scientific agencies. Amongst the major agencies, the Defence Research & Development Organization (DRDO) accounted for a little over 30% of the expenditure on R&D (DST, 2005). The other major central government R&D networks are, in decreasing order of funding, the Department of Space (DOS), the Indian Council of Agricultural Research, the Department of Atomic Energy (DAE), and the Council of Scientific & Industrial Research (CSIR). One estimate suggests that the government runs around 400 research establishments (FICCI, 2005).

While the higher education system is broad in scale and scope, its research output is poor. The higher education system accounted for just 4.1% of the national R&D expenditure in 2002–2003 (DST, 2005, p. 3). According to a prominent member of a task force set up by the government to look at basic scientific research in India's universities, "there has been a complete neglect of research culture in universities."[2] In spite of the large size of India's R&D and higher education sectors, India accounted for only 2.19% of the world's scientific publications in 1993–1997 and this declined to 2.13% for the period 1997–2001 (King, 2004). India's share of publications among the top 1% cited between 1996 and 2006 was 0.33% (Basu, 2007). Among the 31 countries accounting for 98% of the world's most cited scientific publications, India ranks No. 30 on citation rate per paper (King, 2004).

The Indian private sector survived the decades of a relatively closed economy and a cumbersome regulatory structure to blossom and flower after economic deregulation started. Propelled by a boom in services, and with a healthy growth rate in manufacturing, the Indian economy has maintained a growth rate in excess of 6% since economic liberalization began, riding on the back of growth of services and manufacturing.

Several companies have demonstrated major increases in productivity, the ability to develop and successfully manufacture new products, and have been the winners of top-quality awards like the Deming Prize (Krishnan, 2003). The share of the Indian private sector in industrial R&D spending has been rising slowly but steadily.[3]

Perhaps the most creditable achievement of India in recent years has been the creation and growth of an export-oriented software industry. Indian information technology and software exports grew from U.S.$3.4 billion in 1999–2000 to U.S.$40.4 billion in 2007–2008 (NASSCOM, 2008) and in the process became the largest constituent of India's export basket.

Another important sector is pharmaceuticals, which accounts for the largest R&D spend amongst Indian industrial sectors. While the total sales of the pharmaceutical industry reached $17 billion in 2007–2008, as much as $8.6 billion of this was exported (Anon., 2008). Pharmaceutical companies have consistently been among the top U.S. patent-holding Indian companies. These pharmaceutical companies are seeking to move from imitative research and reverse-engineering to the discovery of new molecules and drug delivery systems and their R&D intensity is more than 5%. Joint R&D initiatives with multinational drug companies, licensing of new discoveries to MNCs, sponsored research projects at national laboratories with government support, and the creation of international marketing networks in the hope of future exploitation of such networks to sell newly developed novel drugs are some of the developments in this area.

The success of the software industry and the growth in private-sector industrial R&D spending in sectors such as

[2] Professor Goverdhan Mehta, former Director of the Indian Institute of Science, quoted in "Research Booster Plan", *Education World*, July 2005; *http://www.educationworldonline.net* (last accessed March 23, 2006).

[3] According to DST statistics, the share of private industry in national R&D expenditure has gone up from 13.8% in 1990–1991 to 20.3% in 2002–2003. Sunil Mani (2006) has argued that if expenditure on R&D in industry is computed as the sum of the expenditure on R&D by public-sector enterprises, private enterprises, and government research institutes in areas relevant to industry, then the private sector's share of industrial R&D increased from 41% in 1985–1986 to 67% in 1998–1999.

pharmaceuticals, however, masks highly uneven technological capability-building. In many industries, while the most efficient companies operate at world-class efficiency levels, in aggregate terms India is considered to be at the low end of dynamic technology adopters (Dahlman and Utz, 2005, p. 78). A study by the McKinsey Global Institute shows that the labor productivity of the modern sectors of the Indian economy is only 15% of the globally highest levels (for a summary of the report see Krishnan, 2002). Efforts to upgrade indigenous technologies through imports show a downward trend—the number of technology transfer agreements signed by Indian companies for import of technology declined from 661 in 1991 to 307 in 2002 (Mani, 2006).

In the year 2001–2002 almost 89% of Indian companies did not report any spending on R&D (Bowonder, Kelkar, and Satish, 2003, p. 4). Much of the visible innovation is in the pharmaceutical and automobile industries, though even these industries have R&D intensities that are low by international standards.[4] High-growth sectors such as software are not really high-tech as they spend only about 1% of their sales on R&D, though there is a growing R&D services sector that provides R&D support services to companies in the international market.

With a strong human resource base, India seems well positioned to take advantage of the accelerated pace of the internationalization of R&D. While one estimate suggests that as of 2006 around 230 multinational corporations had set up R&D centers in India (NASSCOM, 2006), a study by the Technology Information Forecasting and Assessment Council (TIFAC) estimates that 135 foreign companies invested $1.3 billion to set up R&D centers in India between 1998 and 2003 with a total employment of 22,980 employees. A number of Indian firms and national laboratories undertake contract research for international clients. An Indian company, Wipro Ltd., claims to be the largest independent R&D service provider in the world—visit *http://www.wipro.com/pes/index.htm* (last accessed on December 1, 2007). An important area of captive and contracted R&D is the product development and engineering services segment—this accounted for exports of U.S.$6.4 billion in 2007–2008, and grew at 29% over the previous year (NASSCOM, 2008).

Commercialization of technology

Since the early 1990s government research networks have had significant strategic outcomes such as missiles, nuclear weapons, and aircraft (such as the Light Combat Aircraft and the Advanced Light Helicopter). However, the civilian spinoffs of strategic R&D programs have been limited.

To make the CSIR more business-oriented and to incentivize CSIR laboratories to transfer their know-how to industry, the government mandated the CSIR to raise one-thrid of its budget from outside the government grant and allowed individual laboratories to retain their surpluses from externally funded projects in a "Laboratory Reserve Fund" that could be used at the discretion of the laboratory. Though these policies predated deregulation by a few years, they began to have an impact after liberalization started. The CSIR leadership formulated a vision in the mid-1990s that included international competitiveness in technology development and a greater emphasis on obtaining international patents. The results were impressive—while the CSIR laboratories were granted just 25 U.S. patents between 1991 and 1995, they were granted 276 U.S. patents between 1996 and 2002 (Bowonder, Kelkar, and Satish, 2003). In 2003–2004 the CSIR was granted 191 U.S. patents, constituting 69% of the U.S. patents granted to Indians in India (CSIR, 2006). By 2006–2007 the CSIR was filing as many as 655 patent applications outside India (CSIR, 2008). In 2004–2005, CSIR's external earnings amounted to Rs. 2.58 billion (approximately U.S.$60 million).

Prior to 1991, there were no government schemes to support technology development and commercialization by private companies. Subsequently, there have been two major efforts to support technology commercialization by industry. The first is the setting up of a Technology Development Board (TDB) by the Ministry of Science & Technology. The TDB primarily provides low-cost loans to companies to support commercialization, though it occasionally takes an equity stake or makes outright grants. The second is the New Millennium Indian Technology Leadership Initiative (NMITLI) of the CSIR that supports joint work between Indian companies and the government laboratory network to create technology leadership positions in industries/technologies where India has a potential competitive advantage in global markets.

National Innovation Policy

In principle, the government recognized the importance of industrial research and development more than three decades ago. Even in the era of tight industrial regulation, companies could take a 100% write-off on R&D expenditure in the year the expenditure was incurred; a scheme of recognition of "in-house R&D units" allowed easy import procedures for equipment and consumables used in research and development; and the domestic Patent Law encouraged Indian firms to find new-process routes for existing drugs. However, prior to 1991, there were no

[4] The R&D intensity of the Indian automobile industry was 0.89% in 2002 and that of the pharmaceutical industry 1.89%. However, it is noteworthy that pharmaceutical industry R&D intensity was just 0.74% in 1998 (Bowonder, Kelkar, and Satish, 2003, Tables 3 & 4).

government schemes that provided financial support for R&D by private industry.

Since 1991 the government has created schemes for the financial support of local industrial research and development. Prominent among these are the Technopreneur Promotion Program (TePP), the Home Grown Technologies (HGT) Program, the Program Aimed at Technological Self Reliance (PATSER), and the Technology Development Board (TDB). While the TePP provides small grants to individual innovators, the other schemes essentially offer low-cost loans to industry for the development, scaling up, and commercialization of industrial technologies. The HGT and PATSER schemes explicitly encourage industry to work with national laboratories or science/engineering education institutions to adapt, improve, and implement technologies developed by the latter. Fiscal incentives for R&D include tax breaks for R&D expenditure and exemption from excise duty for products developed indigenously for which international patents have been obtained. Specifically (DSIR, 2000):

- Both revenue and capital expenditure on R&D are 100% deductible from taxable income under the Income Tax Act.
- A weighted tax deduction of 125% is allowed for sponsored research in approved national laboratories and institutions of higher technical education.
- A weighted tax deduction of 150% is allowed on R&D expenditure by companies in government-approved in-house R&D centers in selected industries.
- A company whose principal objective is research and development is exempt from income tax for 10 years from its inception.
- Accelerated depreciation is allowed for investment in plant and machinery made on the basis of indigenous technology.
- Customs and excise duty exemptions for capital equipment and consumables required for R&D.
- Excise duty exemption for 3 years on goods designed and developed by a wholly-owned Indian company and patented in any two of the following countries: India, the United States, Japan, and any European Union country.

Strengths in technology and innovation

India's strengths in technology development are in specific domains. While India has proven its capabilities in space technology and atomic energy, in the industrial domain India has strengths in pharmaceutical process development (reflected in Indian pharmaceutical companies' strong position in the production of generic drugs), computer software (much of the software in mobile phones, operating systems, and even Intel's latest chips is written in India), and automobiles and automobile components (the Tata Group announced the launch of the world's cheapest car, the Nano in early 2008). Indian companies have done well in "bottom of the pyramid" innovations such as sachets, mopeds, and very reasonably priced mobile phone services. The ability of Indian firms to engineer products and services at low cost has enabled Indian companies to see themselves at the forefront of the revolution of "frugal engineering".

Challenges faced by the innovation system and future directions

Challenges

There are challenges faced by the different organizational entities in the innovation system as well as in the links between them:

- The CSIR faces many challenges, as do other government laboratories. The most significant of these is retention of qualified scientific personnel. While the salaries of scientists and engineers in the government system are capped by overall civil service salary levels, the growth of the software industry and the emergence of India as an R&D hub (for export of R&D services) have opened up a number of job opportunities and these attract people from the government R&D sector. As a result, attrition (particularly among young scientists and engineers) from the government laboratory system is high.
- Another major challenge for the government is improving the effectiveness and efficiency of its investments in R&D. The government does not give enough flexibility to organizations under its control, and emphasizes procedures over results (*The Times of India*, 1999). Long decision cycles, inappropriate choices of priority areas (Chandrasekhar and Basavarajappa, 2001; Parthasarathi, 2002), and sub-optimal funding hamper the effectiveness of government investments in R&D (Krishnan, 2006).
- While there has been a rapid expansion of engineering education capacity in the private sector since the early 1990s, this has taken place without much attention to quality. The result is that engineering graduates from a large majority of these private engineering institutions have to be retrained by employers. Reflecting this gap, there are now efforts to offer "finishing schools" for engineering graduates. A recent report of a government committee noted that "barring some exceptions, there is scant regard for maintenance of standards" (Rao, 2003, p. 25). The accreditation process of the All-India Council of Technical Education (AICTE) is weak, and many of the colleges lack basic infrastructure or adequate

qualified faculty.[5] The existing policy environment offers few incentives to engineering colleges to upgrade their quality.[6] If teacher qualification and the teacher–student norms of the AICTE are applied uniformly, there is a shortfall of at least 18,000 engineering teachers with Ph.D. qualifications and 20,000 with Master's qualifications (Rao, 2003, p. 77). The number of Ph.D. qualifications in engineering created per year is small—two-thirds of the 6,714 science and technology doctorates produced in the country are in the sciences (DST, 2005).

—The high-quality end of the engineering education sector has problems of its own as well. The IITs currently constitute only a little over 1% of the undergraduate engineering education capacity.[7] Though the IITs have increased their earnings through sponsored research and consultancy,[8] they are still highly dependent on the government for their funding. The government has been reluctant to give the existing IITs the autonomy to chart their own future.

—So far, much of the investment in R&D by foreign companies has been directly or indirectly related to information technology. It is not clear to what extent this work involves significant research, development, or innovation content. However, what is evident is that the software development work undertaken by multinationals in India is largely for their global operations or products, and is not specifically targeted at Indian applications. Sensing of the market, and definition of the product is done largely outside India and these skills are rarely learned by the Indian employees of the development centers located in India. As a result, Indian employees enhance their technical skills and sometimes their general management skills, but are rarely exposed to the full gamut of skills across the different stages of product development and introduction that would enable them to gain an integrated business perspective. The main benefits of foreign R&D investments to the country seem to be in the area of human resource development.

—While the policy framework to support industrial innovation is quite comprehensive, government policy initiatives in India are undermined by implementation problems, and the government's initiatives in the arena of science and technology are no exception. The scale and scope of the schemes and programs to support industrial R&D remain small. The total outlay of the government on the TePP, HGT, PATSER, TDB, and NMITLI schemes is of the order of just U.S.$40 million per year and this support is spread too thin. Even this amount is tightly controlled by a centralized bureaucracy in New Delhi and difficult to access by small firms spread all over the country. In fact, some of the most prominent recipients of support under these schemes have been large companies such as Tata Consultancy Services (India's largest software company), and Tata Motors, both part of the highly profitable Tata Group. The TDB's single largest project is a rare outright grant to a civil aircraft development project of the National Aerospace Laboratories, a constituent of the CSIR. Many of the support schemes are biased against small firms and startups as they require a proven track record or R&D recognition by the Ministry of Science and Technology.[9]

—Though there are several small technology firms (Dahlman and Utz, 2005, p. 93), there is no critical mass of dynamic small firms that could form the bedrock of a strong technological capability. This is due to many factors—attractive employment opportunities with salaries increasing 25% year on year; absence of seed capital and venture capital support; lack of sophisticated local market demand; and inadequate government support. Further, there are constraints on academic professionals starting their own firms, barriers to movement from academia/research institutions to industry and back, and inadequate infrastructure for incubation and product development and testing.

—Established companies are reluctant to make significant investments in genuinely innovative R&D. Organizational barriers to innovation include a lack of ambition and vision at the top, a perception of loss of control by owner-managers in issues related to technology development, inadequate investment in plant and machinery,

[5] India's audit watchdog, the Comptroller and Auditor General of India audited 171 new institutions set up after accreditation by the All India Council of Technical Education and found that each one of these institutions was deficient on at least one of the prerequisites—classrooms, basic facilities, library, laboratory equipment, and faculty (Goswami, 2003, p. 5).

[6] Admissions to engineering colleges and the levels of fees that can be charged are controlled by state governments. The fees are fixed without any reference to the quality of education being offered. The AICTE has three schemes to upgrade infrastructure in engineering colleges: (1) a Modernization and Removal of Obsolescence Program to equip technical institutions, laboratories, and workshops with better infrastructure; (2) a Research & Development Scheme to promote R&D by faculty members; and (3) a Thrust Area Program in Technical Education to promote research in identified thrust areas. In 2002–2003, the total grant released under these schemes was Rs. 351 million (approximately U.S.$8 million). But only 11% of this amount went to self-financing colleges (see Rao, 2003, Chapter 6). In addition, the Central Government (through the Technology Information Forecasting and Assessment Council) has a program called REACH that part-finances the creation of centers of excellence within engineering colleges.

[7] In 2006 the IITs will admit 3,890 students through their Joint Entrance Examination. Computed from *http://www.jee.iitm.ac.in/seats.php* (last accessed March 19, 2006).

[8] In 2002–2003, IIT Kanpur earned Rs. 302.5 million (about U.S.$6.7 million) from sponsored research and Rs. 57.4 million (U.S.$1.3 million) from consultancy; in the same year, IIT Madras earned Rs. 180 million (U.S.$4 million) from sponsored research and Rs. 66 million (U.S.$1.5 million) from consultancy. These data were obtained directly from the two institutes.

[9] According to sources in the Ministry of Science & Technology, there is pressure from the audit wing of the government to require collateral security for loans under these schemes. If introduced, this will further bias the schemes against startups.

lack of the right people and skills, and hierarchical structures (Krishnan, 2006).

Since 2005, there has been increased recognition of the importance of strengthening the innovation system. In mid-2005 the Central Government formed a Knowledge Commission to advise the government on how India can promote excellence in the education system to meet the knowledge challenges of the 21st century, promote knowledge creation in S&T laboratories, improve the management of institutions generating intellectual property, improve protection of IPRs, and promote knowledge applications in agriculture and industry.

The government has taken initiatives towards enhancing the quality and size of higher education institutions (particularly in engineering and the sciences), providing additional funding for research by institutions of higher learning, and enhancing support for entrepreneurship and firm-level innovation. Some examples of these initiatives are given below:

—The Central Government budget for 2005 announced a grant of Rs. 1 billion (about U.S.$22 million) to the Indian Institute of Science (IISc) at Bangalore to help upgrading it into a world-class university. The IISc is one of only three Indian universities/institutions ranked among the top 500 in the world.[10] The Central Government budget for 2006 announced special grants to the historically important Bombay (Mumbai), Calcutta (Kolkata), and Madras (Chennai) universities and to the Punjab Agricultural University, a major site of India's green revolution in the 1960s and 1970s.
—In March 2005 the Ministry of Human Resource Development of the Central Government set up a task force on basic scientific research in universities. The report of the task force (accepted by the government) has recommended spending Rs. 6 billion (approximately U.S.$130 million) per year, two-thirds of which is to be spent on infrastructure development, and a quarter on setting up 10 networking centers in the basic sciences.
—The Central Government has announced its acceptance of the proposal to form a National Science Foundation–like body with funding of the order of U.S.$200 million per year. New Indian Institutes of Science Education & Research (IISERs) have been set up at Kolkata and Pune.
—In October 2005 the Department of Biotechnology of the Government of India announced a Small Business Innovation Research Initiative (SBIRI) scheme to provide early-stage funding to help create new technology-based businesses by science entrepreneurs.
—In 2007 the Indian Government committed itself to a six or sevenfold increase in allocations for higher education during the 11th 5-year Plan (2007–2012). This will be used to increase both access (reflected in a significant increase in gross enrolment ratio) and quality (it is proposed to set up several new IITs during this plan).

Future directions

The developments listed above suggest a growing realization that while India may have been able to benefit from the opportunities thrown up by the knowledge economy largely due to its large stock of qualified people, further growth will depend on the creation of a supportive policy framework. However, it is an open question as to what extent the government's initiatives will deliver results.

This skepticism is prompted by the absence of a systems approach to government thinking, the government's poor implementation record, and a lack of clarity in the government's philosophy. For example, consider the recent special grants announced for universities. The choice of universities has been *ad hoc* and not specifically linked to their performance; further, there is no process in place to recognize or incentivize universities that improve their quality or to penalize those that do not. Further, though most research in the field suggests that the locus of innovation and technology development should be the firm, the Indian Government continues to put its money on the network of government research institutions. Even if laboratories are to form the backbone of the government's strategy, the experience of other countries like Taiwan and China suggests that a more dynamic and purposive governance structure is required to achieve higher levels of performance with clear benefits to the wider economy.

The initiatives mentioned above will require sustained financial support from the government. But, economic policy makers in the government continue to have the dominant view that government's principal role is to create a climate conducive for business, and that it should not get involved in activities like backing technological development. Social-sector programs for health, primary education, and employment will compete for scarce resources. In these circumstances there is no guarantee that the funds required to support higher education, scientific research, and the creation of innovative capabilities will be forthcoming on a consistent basis.

In the absence of a clear philosophy, and a set of coordinated policies across ministries dealing with different subjects, the likelihood of a major transformation of the innovation system in the short run is small. The proportion of GNP spent on R&D will continue to hover around the 0.8% mark. The private sector's share of industrial R&D expenditure will rise slowly. The R&D services export activity is likely to continue to grow at about 30% per year, thus creating a bigger R&D services sector with a larger and deeper workforce. But domestic high-tech entrepreneur-

[10] The Indian Institute of Science, Indian Institute of Technology Kharagpur, and Calcutta University are the three Indian institutions listed in Shanghai Jiao Tong University's ranking of the top 500 universities in the world (for details see *http://ed.sjtu.edu.cn*)

ship will continue to be constrained by the non-availability of seed capital.

However, if economic growth continues at the present rate, the long-term prognosis of the innovation system is bright. The large number of MNC R&D centers may throw up new entrepreneurs as the years go by. Some among the new set of successful first-generation entrepreneurs may turn angel investors and venture capitalists.[11] Already, there is renewed interest in India from U.S.-based venture capitalists.

However, the full potential will be realized only if some key constraints are tackled. The government should move towards making the inception of new high-tech firms easier, reducing their mortality, and facilitating their growth. Some specific steps to facilitate the creation of high-tech firms include seed funding through incubators; allowing scientists and engineers from laboratories and universities to start firms with technologies they have developed; allowing national laboratories (or parts) to morph into industrial units; creating publicly-owned, but privately-managed high-tech infrastructure in selected industrial parks; making national laboratories and institutional facilities more easily available to small firms; and relaxing collateral requirements for firms. To reduce firm mortality, the government could expand/decentralize government support programs for technology development and commercialization; allow small firms to bid for research contracts from scientific agencies like the DST, ISRO, DRDO; modify public procurement rules to allow short-term monopolies on proprietary products; relax pre-qualification clauses in public procurement; encourage non-resident Indians to start firms in industrial parks; and increase the enforceability of IPRs.

Academic and research careers need to become more attractive, mobility between academia and industry easier, and research institutions more accountable for their output. There needs to be a fundamental change in the role of government in the support of research and higher education. While it should continue to support research and higher education financially, its primary role should be in terms of setting standards and accreditation. Instead, so far, the government has been inclined to "protect inefficiency in Indian higher education and restrict supply, rather than promote positive change and growth in a sector that is even more important than energy" (Singh, 2006). India needs to quickly enhance the availability of skilled resources that are increasingly in short supply. This can be done by removing the barriers that stand in the way of Indians and foreigners starting new institutions, improving accreditation procedures, creating new disclosure standards, and allowing government institutions to charge higher fees (Rajan and Subramanian, 2006).

[11] One of the founders of Infosys Technologies, Mr. N. S. Raghavan, has become an angel investor and supported more than 15 new enterprises.

The government will have to choose between a commitment to existing interests and structures resulting in a slow pace of evolution of the Indian Innovation System and a more ambitious process driven by drastic restructuring of it.

References

Anon. (2008). *Indian Pharmaceutical Industry Vision 2015*. Mumbai, India: Yes Bank and Organization of Pharmaceuticals Producers of India.

Basu, A. (2007). "Scientific productivity and citation of papers: Where do we stand?" *Current Science*, **93**(6), September 25, 750–751.

Bowonder, B., Kelkar V., and Satish, N. G. (2003). *R&D in India* (ASCI Issue Paper No. 9, March, p. 4). Hyderabad: Administrative Staff College of India.

Chandrashekar, S., and Basavarajappa, K. P. (2001). "Technological innovation and economic development: Choices and challenges for India." *Economic & Political Weekly*, **36**(34), August 25, 3239–3245.

CSIR. (2006). *Annual Report*. New Delhi: Council of Scientific & Industrial Research; *http://www.csir.res.in/csir/external/heads/aboutcsir/annual_report/2003-04/Overview.htm* (last accessed on March 23).

CSIR. (2008). *Annual Report*. New Delhi: Council of Scientific & Industrial Research; *http://www.csir.res.in/csir/external/heads/aboutcsir/annual_report/2003-04/Overview.htm* (last accessed September 19).

Dahlman, C., and Utz, A. (2005). *India and the Knowledge Economy: Leveraging Strengths and Opportunities*. Washington, D.C.: World Bank.

DSIR. (2000). *Research and Development in Industry: An Overview*. New Delhi: Department of Scientific & Industrial Research, Government of India.

DST. (2005). *Research and Development Statistics at a Glance 2004-05* (p. 3). New Delhi: Department of Science and Technology, Government of India.

DST. (2006). *R&D Statistics 2004-05*. New Delhi: Department of Science and Technology, Government of India.

FICCI. (2005). *India R&D 2005: The World's Knowledge Hub of the Future* (Background Paper, prepared by Evalueserve, p. 6). New Delhi: Federation of Indian Chambers of Commerce & Industry

Forbes, N. (1999). "Technology and Indian industry: What is liberalization changing?" *Technovation*, **19**, 403–412.

Goswami, U. (2003). "New engg colleges fail to meet AICTE norms." *The Economic Times*, Bangalore, May 27, p. 5.

King, D. A. (2004). "The scientific impact of nations." *Nature*, **430**, July 15, Table 1, p. 312.

Krishnan, R. T. (2002). "Prescription for 10 per cent growth: Food for thought." *Economic & Political Weekly*, March 23, pp. 1107–1108.

Krishnan, R. T. (2003). "The evolution of a developing country innovation system during economic liberalisation: The case of India." Paper presented at the *First Globelics Conference on Innovation Systems & Development Strategies for the Third Millennium, Rio de Janeiro, Brazil, November 3–6.*

Krishnan, R. T. (2006). *Barriers to Innovation and the Creation of a Knowledge Society in India* (Working Paper No. 243). Bangalore: IIM.

Krishnan, R. T., and Prabhu, G. N. (1999). "Creating successful new products: Challenges for Indian industry." *Economic & Political Weekly*, July 31, pp. M-114–M-120.

Mani, S. (2006). "Performance of India's innovation system since 1991." Paper presented at *India–Israel Workshop on Technology Innovation and Finance, Bangalore, February 23.*

NASSCOM. (2006). National Association of Software and Service Companies; *http://www.nasscom.org* (last accessed on March 25).

NASSCOM. (2008). National Association of Software and Service Companies; *http://www.nasscom.org* (last accessed on September 22).

Parthasarathi, A. (2002). "Priorities in science and technology for development: Need for major restructuring." *Current Science*, **82**(10), May 25, 1211–1219.

Rajan, R., and Subramanian, A. (2006). "India needs skill to solve the Bangalore Bug." *Financial Times*, March 17; *http://www.ft.com*

Rao, U. R. (Chairman). (2003). *Revitalizing Technical Education: Report of the Review Committee on AICTE* (p. 25). New Delhi: Ministry of Human Resource Development, Government of India.

Rathinasamy, R. S., Mantripragada, K. G., Krishnan, R. T., and Shivaswamy, M. K. (2003). "An insider's guide to doing business in India." *Journal of Corporate Accounting & Finance*, September/October, 17–33.

Singh, N. (2006). "The bottomline on learning from China." *The Financial Express*, March 23; *http://www.financialexpress.com/print.php?content_id=121290*

The Times of India. (1999). "Catalyst man cometh: Interview with Dr. Paul Ratnasamy, Director, National Chemical Laboratory, Pune." Bangalore edition, November 11, see p. 10 for a critique of the constraints placed by the government on the functioning of national laboratories.

Tyabji, N. (2000). *Industrialisation and Innovation: The Indian Experience*. New Delhi: Sage.

45

Japan's National Innovation System

*Chihiro Watanabe** and *Kayano Fukuda*[†]

*Tokyo Institute of Technology and [†]Japan Science and Technology Agency

Abstract

Japan's high-technology miracle as an industrial society up until the end of the 1980s prompts the postulate that co-evolutionary dynamism between innovation and institutional systems is decisive for an innovation-driven economy and that Japan indigenously incorporates a sophisticated capacity in such dynamism. Japanese industry's R&D intensity (R&D expenditure per GDP) ranks second after Sweden in 30 OECD countries, while its government ratio of R&D to industry R&D is the lowest. This demonstrates Japan's sophisticated capacity to induce vigorous R&D from its industry.

However, Japan's contrasting economic stagnation in an information society, resulting in a "lost decade" in the 1990s, prompts another postulate—namely, that such stagnation can be attributed to a system conflict between a new paradigm in an information society and the traditional business model molded by organizational inertia, and further that an innovation-driven economy may stagnate if institutional systems cannot adapt to innovations.

Although Japan's dynamism fell by the wayside in the 1990s, resulting in a lost decade, it re-emerged in the early 2000s. This is attributable to the co-evolution between indigenous strength developed in an industrial society and the effects of cumulative learning from competitors in an information society. This is based on the hybrid management of technology fusing the East (indigenous strength) and the West (learning from what can only be described as a digital economy).

45.1 Introduction

Japan's high-tech miracle as an industrial society up until the end of the 1980s demonstrates that the co-evolutionary dynamism between innovation and institutional systems is decisive for an innovation-driven economy. Although numerous studies have analyzed the structure and the role institutions play in bringing about such systems (e.g., North, 1990, 1994; Nelson and Sampat, 2001), no direct link was established with such co-evolutional dynamism. Watanabe *et al.* (2008) postulated that institutional systems sustain emerging innovation realized by means of a three-dimensional system consisting of (i) a national strategy and socio-economic system, (ii) an entrepreneurial organization and culture, and (iii) historical perspectives (Watanabe and Zhao, 2006, Figure 1).

This concept provides clear insight into Japan's explicit capacity to bring about sophisticated co-evolutionary dynamism in an industrial society. Japanese industry's R&D intensity ranks second after Sweden in 30 OECD countries (Figure 45.1), while the ratio of government R&D to industry R&D is the lowest (Figure 45.2). These observations demonstrate Japan's explicit capacity to induce vigorous R&D from its industry.

However, notwithstanding this sophisticated capacity, Japan's long-lasting economic stagnation in the information society that emerged in the early 1990s, resulting in a "lost decade", brings up another postulate—namely, that an innovation-driven economy may stagnate if institutional systems cannot adapt to innovations.

These observations prompt the hypothetical view that, while Japan indigenously incorporated a sophisticated capacity to induce vigorous R&D in its industry, leading to the co-evolutionary dynamism that enabled it to achieve conspicuous performance in a virtuous cycle between innovation and institutional systems in an industrial society up to the end of the 1980s, it went the opposite way in the 1990s because of a systems conflict between indigenous institutional systems and a new paradigm in an information society (Watanabe, Hur, and Lei, 2006). In addition, the reactivation of its capacity, which emerged in the early 2000s, was based on hybrid management that fused indigenous strength (East) and learning from what corresponds to a digital economy (West).

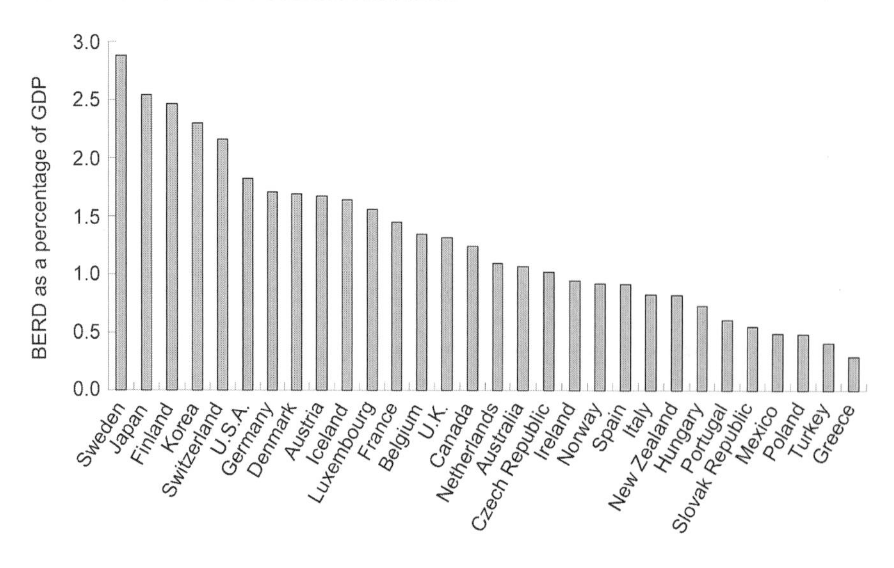

Figure 45.1. Industry's R&D intensity[a] in 30 OECD countries in 2005.[b]
[a]Expenditure on R&D in the business enterprise sector (BERD).
[b]Switzerland, Australia, and Turkey: 2004; New Zealand: 2003.
Source: OECD (2007).

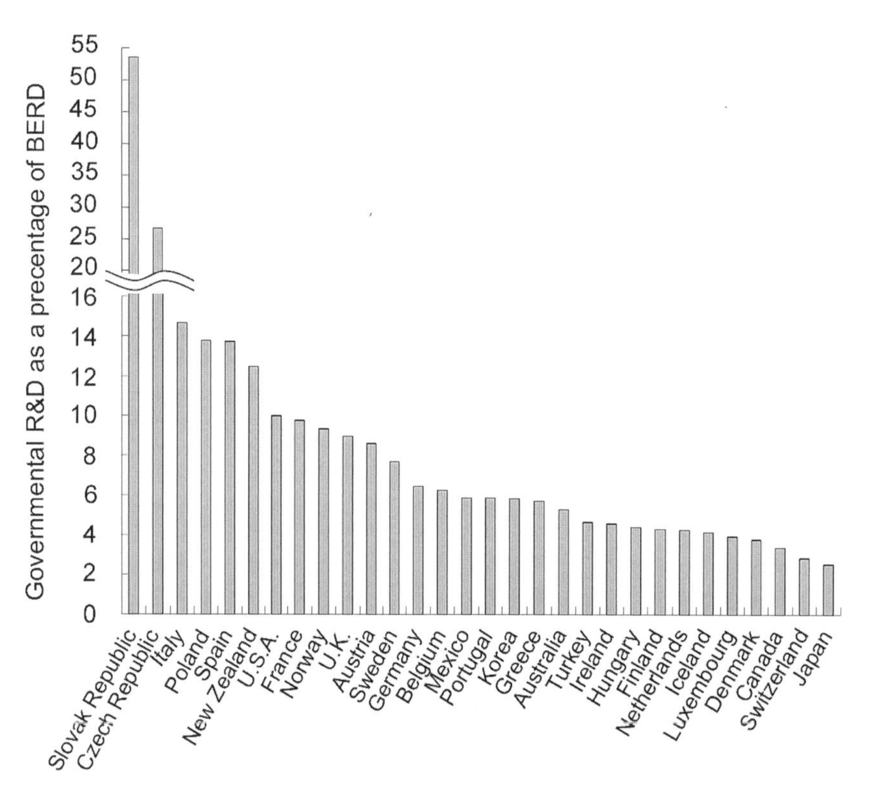

Figure 45.2. Percentage of BERD financed by government in 30 OECD countries in 2005.[a]
[a]Italy, Spain, France, Austria, Germany, Belgium, Australia, Turkey, and Switzerland: 2004; New Zealand, Sweden, Portugal, Greece, the Netherlands, Luxembourg, and Denmark: 2003.
Source: OECD (2007).

To demonstrate the foregoing hypothetical views, this chapter first reviews Japan's indigenous explicit capacity to co-evolve between innovation and institutional systems. Then, it demonstrates the effect of fusing East and West that led Japan to reactivate its indigenous explicit capacity.

Section 45.2 reviews Japan's indigenous explicit capacity. Section 45.3 identifies the mechanism underlying Japan's inducing policy. Section 45.4 analyzes Japan's mis-option in its growth trajectory and surge of the reactivation of co-evolutionary dynamism. Section 45.5 demonstrates hybrid management fusing East and West. Section 45.6 briefly summarizes new findings and policy implications.

45.2 Japan's indigenous explicit capacity

45.2.1 Japan's development trajectory

Japan naurally incorporates a sophisticated capacity—substituting technology for scarce resources—to enable firms to achieve conspicuous performance in a virtuous cycle between innovation and rapid economic growth that was seeb in the 1960s (Watanabe, Santoso, and Widyati, 1991). This was followed by substitution of technology for energy in the 1970s, leading to the world's highest energy efficiency and advances in manufacturing technology that were termed high-tech miracles in the 1980s (Watanabe, 1996, 1999). However, due to a system conflict between a new paradigm in an information society and the traditional business model in an industrial society this resulted in long-lasting economic stagnation in the 1990s.

The recovery that emerged in the early 2000s can be attributed to the fusion of indigenous strength in manufacturing technology (East) developed in an industrial society with the effects of cumulative learning from a digital economy (West) in an information society. This can also be attributed to Japan's explicit capacity to learn and fuse. These trajectories are illustrated in Figure 45.3.

Figure 45.4 illustrates the rise and fall of Japan's development trajectory over the past four decades, the continuous rise in the growth and wealth of its industrial society up until the end of the 1980s (except during 1974 just after the first energy crisis of 1973), the fall due to prolonged stagnation in the information society in the 1990s, and another rise from the beginning of the 2000s.

45.2.2 Government support for industry R&D

Rises and falls in the Japanese development trajectory can be seen as the consequence of the co-evolution and disengagement between innovation and institutional systems. Success in constructing a co-evolutionary dynamism can largely be attributed to an explicit capacity to induce vigorous R&D in its industry, as shown in Figures 45.2 and 45.3.

Figure 45.4 demonstrates the trend in Japan's government support for industry R&D investment (ratio of government R&D funds to industry R&D expenditures) over the period 1955–2005.

From Figure 45.4 we can see that, while Japan's government support ratio exceeded 10% of industrial R&D investment in the 1950s, this figure dramatically declined during the latter half of the 1950s and continued to decrease in the 1960s as its economic level increased. Consequently, this ratio decreased to 3–4% in the 1970s and the 1980s and is currently less than 2%. Table 45.1 and Figure 45.5 show a comparison of this ratio in five advanced countries in 2005. This comparison demonstrates Japan's remarkably low level of government support for industrial R&D investment (between one-fifth and one-eighth that of other countries).

Considering the vigorous R&D investment in Japan's industry (as demonstrated in Figure 45.1) and despite such a low level of government support, it is anticipated that there must be certain sophisticated system capacities within Japan's institutional systems that are responsible for inducing such vigorous R&D by industry.

45.2.3 Japan's indigenous institutional systems for innovation

On the basis of such a hypothetical view, Figure 45.6 illustrates a scheme showing institutional systems for innovation.

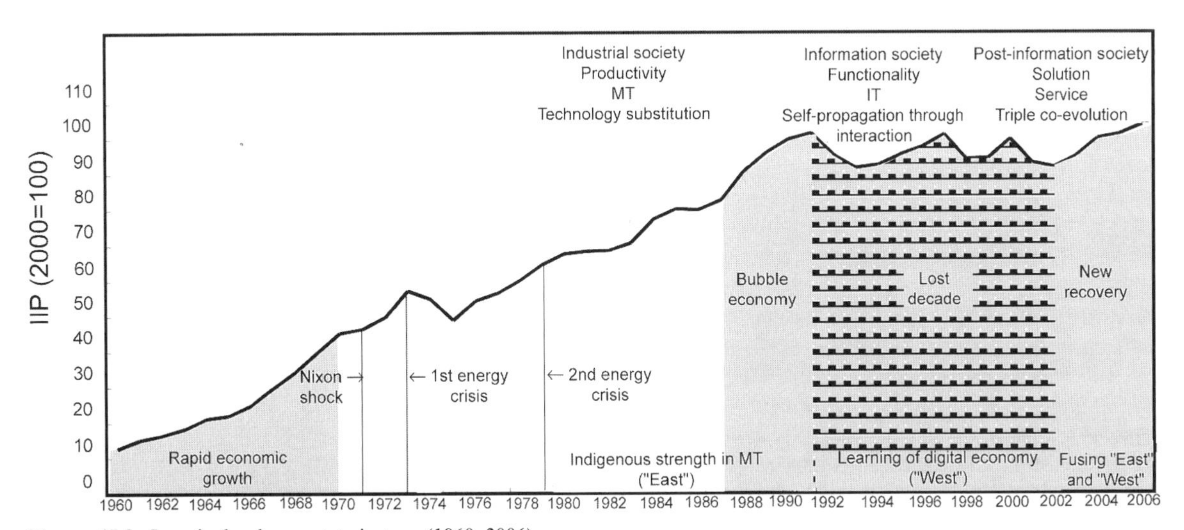

Figure 45.3. Japan's development trajectory (1960–2006).

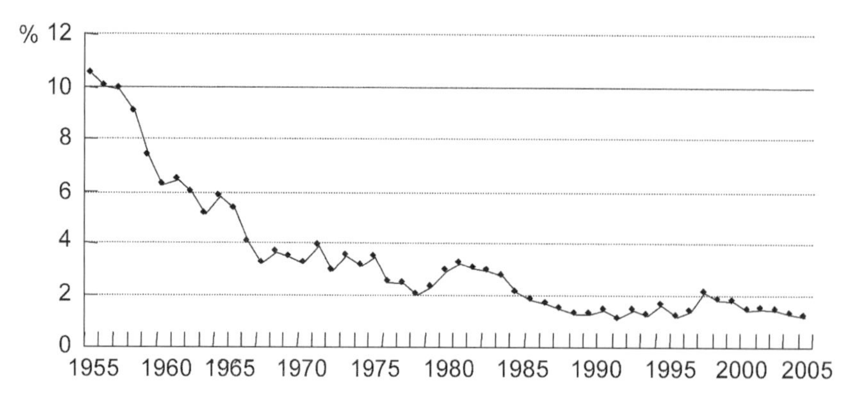

Figure 45.4. Trends in Japan's governmental support for industry R&D investment (1955–2005).[a]

[a]Ratio of government R&D funds to industry R&D expenditures.

Sources: Wakasugi (1986), MITI (Annual issues), STA (Annual issues), and OECD (2000–2006).

Table 45.1. Comparison of government support for industrial R&D investment in advanced countries (2005) expressed as a percentage.[a]

Japan	*U.S.*	*Germany*	*France*	*U.K.*
1.2	9.7	5.9[b]	9.3[b]	8.6

[a]Ratio of government R&D funds to industrial R&D expenditure.
[b]Figures for Germany and France are for 2004.
Source: OECD (2006).

Innovation itself improves institutional systems such as the economic environment, social/cultural environment, and natural environment, and these improvements in turn induce further innovation, leading to co-evolution. The innovation generation cycle leads to getting emerging innovations to market by means of the effective utilization of resources in innovation. This inducement may stagnate if institutional systems cannot adapt to evolving innovation, resulting in disengagement (Watanabe, Santoso, and Widyati, 1991).

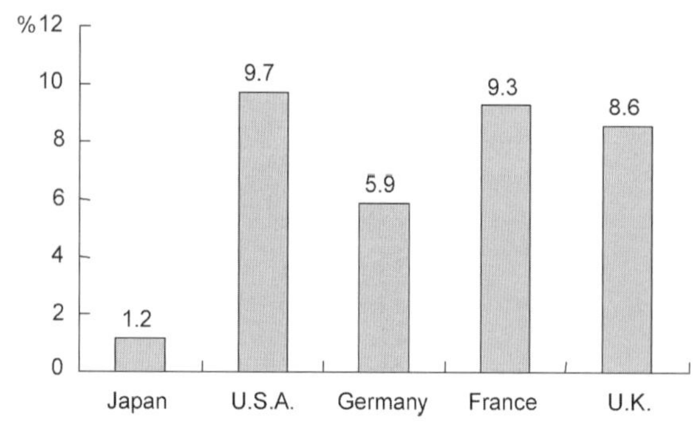

Figure 45.5. Comparison of government support for industry R&D investment in advanced countries (2005).[a]

[a]Ratio of government R&D funds to industrial R&D expenditures. The figures from Germany and France are for 2004.
Source: OECD (2006).

Japan's sophisticated capacity to bring about a co-evolutionary dynamism between innovation and institutional systems and the reasons for its success are found in its socio-cultural systems based on the nation's historical efforts (as outlined in Figure 45.7).

Historically, Japan has been flexible and selective in its way of introducing, adopting, assimilating, and developing Western technology into Japanese economic, social, and cultural systems without spoiling its indigenous culture and the homogeneity of the nation. During the *Edo* period (1603–1867) particularly, the "Sankin Kotai" system—which was the "Shogunate" system of periodic obligatory attendance at the capital city Edo (now Tokyo) for *Daimyo* (federal lords)—brought about information exchange among regions.

While Japan experienced three centuries of isolationist policy, the unexpected call by an American steamship in 1853 triggered an influx of Western civilization and culture, to which the Japanese reacted flexibly, both adopting and welcoming them. Following the Edo period, which ended with the Meiji Restoration in 1868, the Meiji Government (1868–1912) focused its policy on the wealth and military strength of the nation, strengthened industrial production, and intensively promoted Western learning while cultivating the Japanese spirit by establishing educational systems and emphasizing moral ethics.

These historical efforts greatly contributed to Japan's technological development before, during, and after World War II. The high-tech miracle of the 1980s can also be attributed to such historical efforts (Kondo and Watanabe, 2003). On the basis of such efforts, Japan constructed the following institutional foundations that brought about its economic development after World War II (Watanabe *et al.*, 1991).

All of these corresponded to the following social and cultural foundations:

(i) Social and cultural foundations

a. All parts of the nation are provided certain levels of education, and most people are very eager to get higher

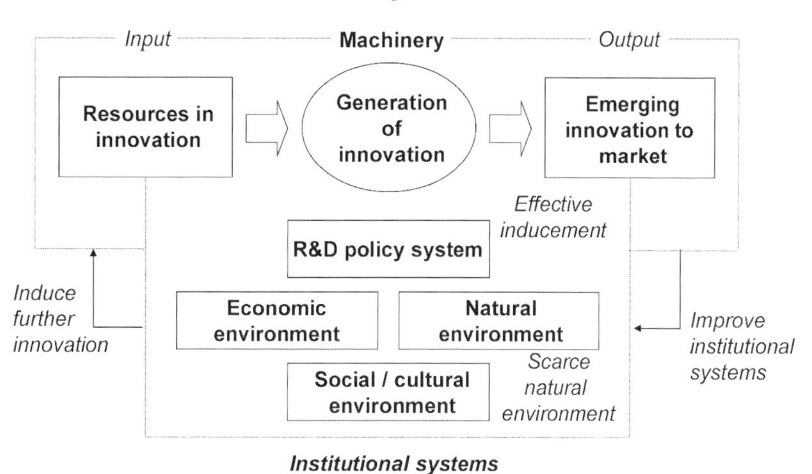

Figure 45.6. Scheme of institutional systems for innovation.

education. This is attributable to the intensive initiative by the Meiji Government which was established during the Edo period (1600–1867).

b. High levels of education have led both workers and managers to be diligent in their work and responsible for their commitments. They cooperate well in detecting quality issues and actively improve imported technology.
c. Japan's highly organized systems and customs are characterized by (i) a seniority system, (ii) life time employment, and (iii) enterprise unions that function well in gaining consensus, trust, and assimilation.
d. All the above factors contribute to bringing about an enlightened management strategy that is long term, adopts an active and flexible approach, and is dependent on government policy.

(ii) Economic environment

a. The highest degree of competition at multi-hierarchical levels (e.g., between government and industry, industry and consumers, rival firms, as well as rival countries) has been drummed into Japan's institutions, which results in increased productivity.
b. Users and consumers are very demanding about the quality, function, and design of their goods and never satisfied with anything less than the best quality.
c. Japanese industry actively interacts with relevant sectors, a typical example can be seen in the interaction between Japan's automobile industry and the iron and steel industry of the developing world leading to mutual benefits in the world market.
d. Change in the industrial structure from the heavy, chemical industrial structure of the 1960s to the knowledge-intensve industrial structure after the 1970s can be attributed to R&D advancement, and this change in the industrial structure brought about intensive R&D investment in industry, thus constructing a mutually inspiring virtuous cycle.

Clearly, a high level of education is a fundamental requirement for a society that has such a competitive nature and demands high-quality innovative goods. The commitment of workers and managers is also a key element, without which very little could be accomplished. Well-organized systems and customs bring about active inter-industry stimulation while responding to dynamic changes in industrial structure. Through the management strategies of firms, long-term considerations and long-term R&D investment decisions always keep structural change in industrial sectors in mind.

Japan's economic environment and social and cultural foundations brought together the factors that contributed to its economic development after World War II (Figure 45.8). Domestic factors fostered the economic environment while international factors acted as a catalyst to prepare for grave situations—such as energy crises and appreciation of the yen—which were capable of disrupting the favorable factors.

In the above review, a systematic view of the mechanisms underlying the vitality in Japan's industry has been developed (Figure 45.9). In this comprehensive organic system, we note that, on the basis of a strong economic environment and the corresponding social and cultural foundation, a strong desire exists for active R&D. Given such conditions, the role of government policy is to further motivate such strong potential desires so as to produce a chain reaction in industrial vitality similar to the role of a catalyst (Watanabe, 1997).

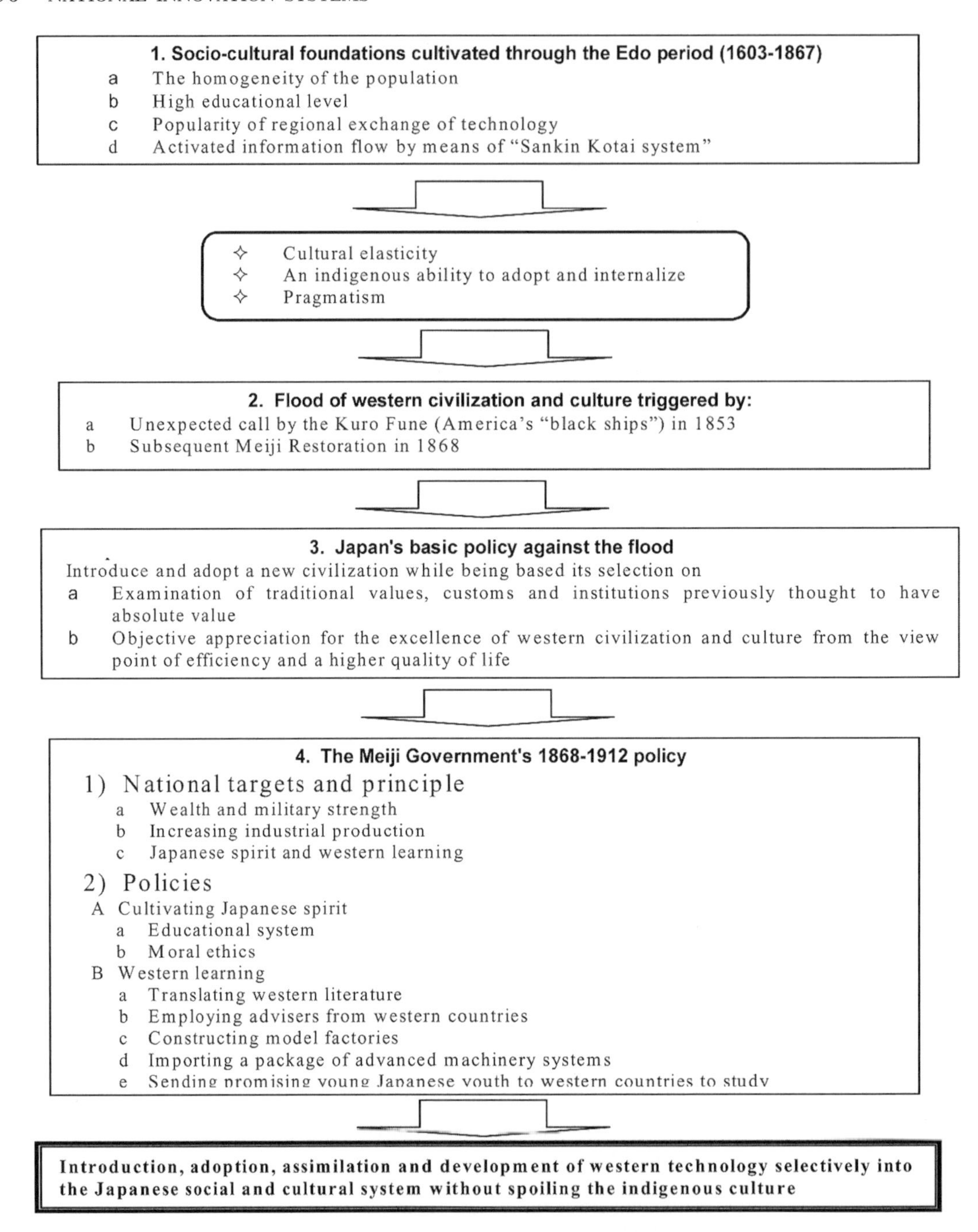

Figure 45.7. Sociocultural systems that enabled Japan's smooth and eective assimilation of external technology.

45.3 Industrial technology policy as a system

45.3.1 Science and technology policy

Figure 45.10 illustrates the science and technology organization in Japan. In this comprehensive organization, industrial technology policy is initiated by the Ministry of Economy, Trade, and Industry (METI).[1] While its share of the government science and technology policy

[1] The Ministry of International Trade and Industry (MITI) was renamed to METI in 2001 as a result of a structured reform by the Japanese Government.

A. External Factors

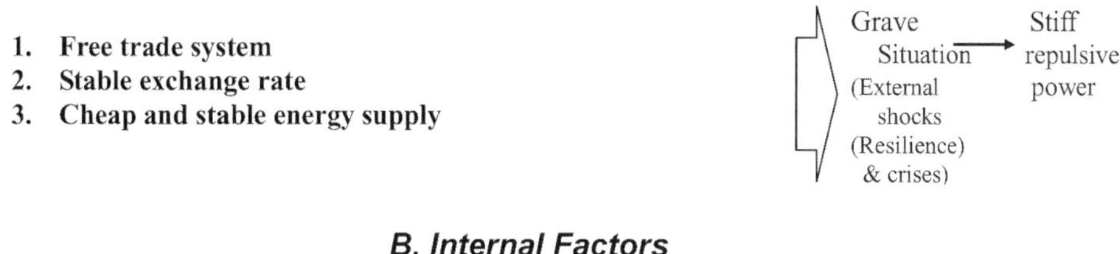

B. Internal Factors

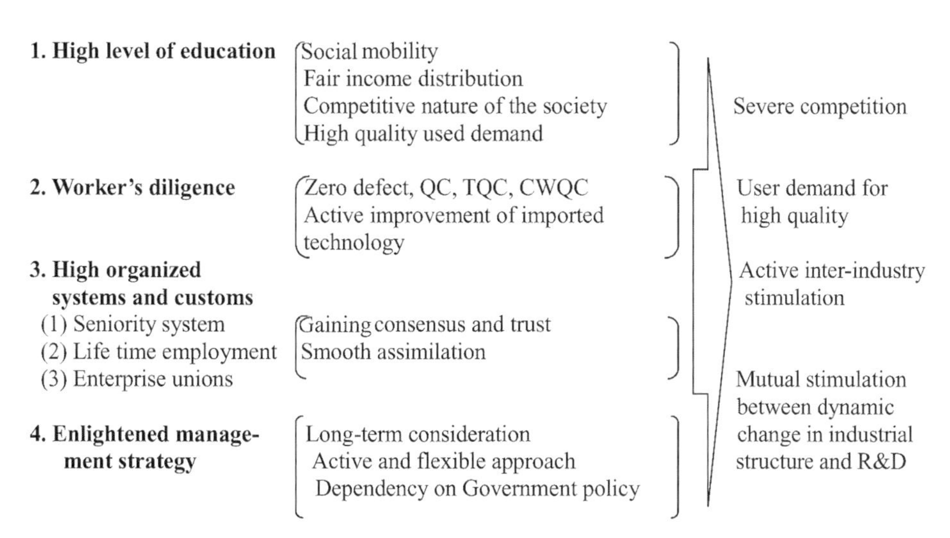

Political stability (1955-1993)

Successive trends in catch-up and growth (1945-1990)

Figure 45.8. Foundation of Japan's economic development after World War II.

budget as a whole is only one-eighth, it induces industrial R&D investment by acting as a catalyst (Watanabe *et al.*, 1991) and entrusting implementation to its affiliates NEDO (New Energy and Industrial Technology Development Organization) and AIST (National Institute of Advanced Industrial Science and Technology).

45.3.2 METI's inducing policy

Given its catalytic role in bringing about vigorous industrial R&D under conditions of limited government support (i.e., one-fifth to one-eighth that of other advanced countries), METI has established a sophisticated inducing policy with the following structure:

(1) Basic principle

(i) Activate free competition in the marketplace.
(ii) Stimulate the competitive nature of industry.
(iii) Induce the vitality of industry.

(2) Approach

(i) Use leading-edge technological foresight based on the vision approach.
(ii) Maintain close cooperation with related industrial policies such as the industrial technology policy or the policy for energy R&D.
(iii) Use an active and flexible approach by periodically reviewing visions that correspond to structural changes in external circumstances.
(iv) Best utilize innovative human resources in national research laboratories and universities to complement knowledge between fundamental science and engineering expertise.
(v) Organize ties between industries, universities, and government into a form of engineering research association on R&D consortium (see Table 45.2).

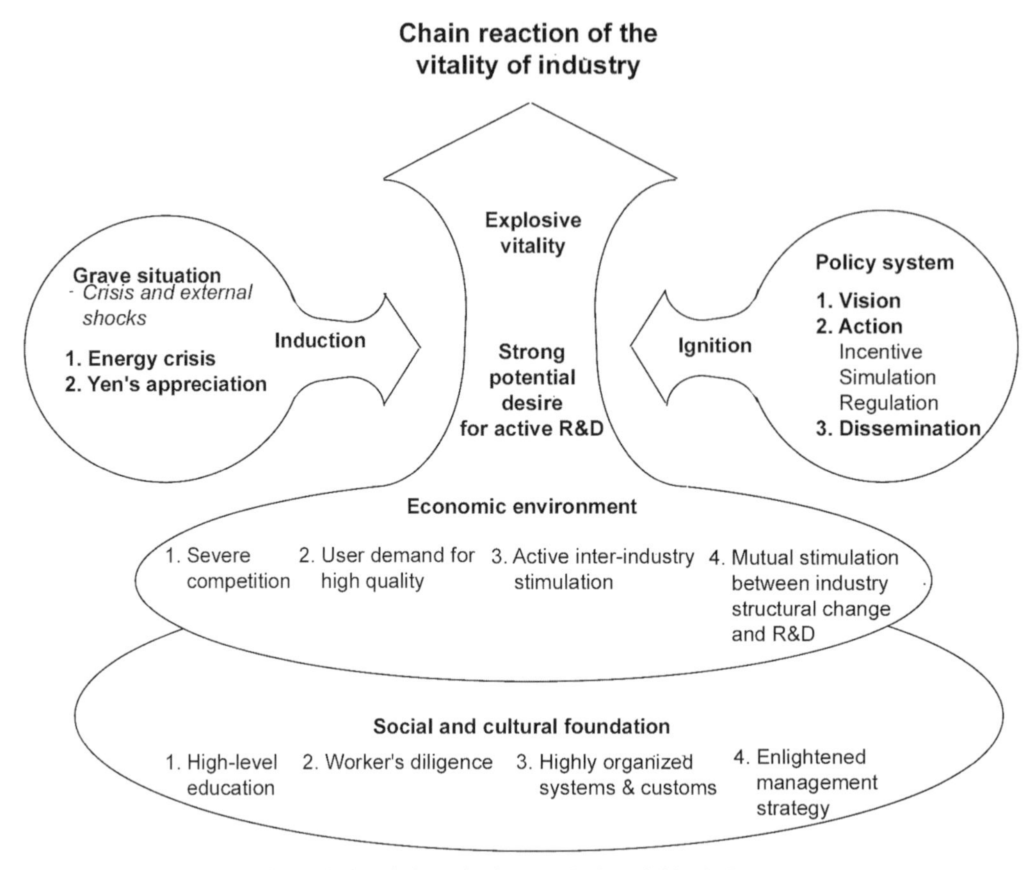

Figure 45.9. Scheme of the mechanism to induce industry's vigorous R&D activities in Japan.

(3) Policy formation/implementation

(i) Vision

a. penetration
b. identification
c. providing direction
d. instilling confidence
e. developing general consensus.

(ii) Action

a. Incentive: national research laboratory, R&D program, investment, conditional loans, financing, tax exemptions.
b. Stimulation: R&D consortium, publication, open tender.
c. Regulation: IPR, monopoly, accounting.

45.3.3 Vision as a soft-policy instrument

(1) The role of vision

In the foregoing policy system, leading-edge technology foresight aimed at identifying future social and economic needs and a corresponding flexible policy approach play key roles in bringing about a policy to prevent external shocks and crises leading to a chain reaction.

The industrial technology policy initiated by MITI and succeeded by METI focused on getting industry to respond to historic demands. Japan's success in constructing a virtuous cycle between technological development and economic growth in the face of numerous constraints can be attributed to a dynamic and flexible policy approach that promptly reacts to a dynamic change in domestic and international environments.

Visions, which can be defined as the soft technology of public administration, play a fundamental role in the identification of future social and economic needs and of a corresponding flexible policy approach. The role of visions in the identification of future prospects is neither a plan nor a prediction. They help to shape the direction of the future, which is not limited to expected futures, possible futures, or preferred futures alone, but rather is a synchronization of these three futures (as illustrated in Figure 45.11).

(2) Actors responsible for the formulation of visions

Visions are formulated by the government, through joint work and open discussion with representatives and experts from a broad spectrum including industrial circles, academia, financial institutions, small businesses, con-

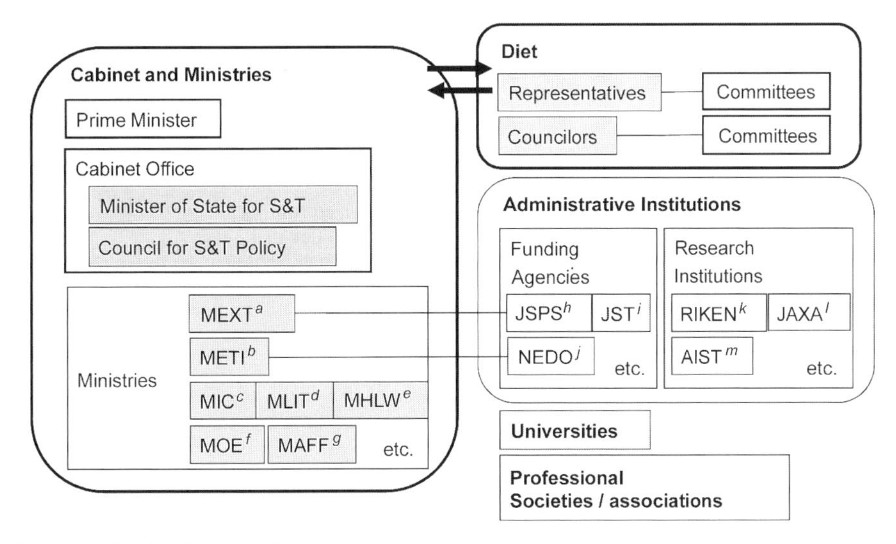

Figure 45.10. Science and technology organization in Japan.
[a]MEXT: Ministry of Education, Culture, Sports, Science, and Technology, whose role includes (i) planning basic science and technology policies, (ii) promoting basic research in such areas as IT and life sciences, and large-scale projects in such areas as space and the oceans, and (iii) improving the research environment and R&D infrastructure.
[b]METI: Ministry of Economy, Trade, and Industry, whose role includes (i) promoting industrial technology through improving business environments and innovation systems, (ii) setting up the Strategic Technology Roadmap for R&D, (iii) standardizing technology.
[c]MIC: Ministry of Internal Affairs and Communication, promoting policies and R&D for ICT.
[d]MLIT: Ministry of Land, Infrastructure, and Transport Japan, promoting policies and R&D for land, infrastructure, and transport.
[e]MHLW: Ministry of Health, Labor, and Welfare, promoting policies and R&D for the clinical, medical, and health sciences.
[f]MOE: Ministry of Environment, promoting policies and R&D for the environmental sciences.
[g]MAFF: Ministry of Agriculture, Forestry and Fishery, promoting policies and R&D for the agricultural, forestry, and fishery sciences.
[h]JSPS: Japan Society for the Promotion of Science, which promotes scientific and academic programs such as supporting young researchers, promoting international scientific cooperation, awarding Grants-in-Aid for Scientific Research.
[i]JST: Japan Science and Technology Agency, which conducts programs that promote R&D from basic research to commercialization, while upgrading the infrastructure of scientific and technological information.
[j]NEDO: New Energy and Industrial Technology Development Organization, which promotes the development of advanced industrial, environmental, new-energy, and energy conservation technologies.
[k]RIKEN: National Research Institute, which carries out high-level experimental and research work in a wide range of fields, including physics, chemistry, medical science, biology, and engineering, covering the entire range from basic research to practical application.
[l]JAXA: Japan Aerospace Exploring Agency, which promotes space development and utilization, and aviation research and development.
[m]AIST: National Institute of Advanced Industrial Science and Technology, which conducts research that transcends the barriers between disciplines and plays an active role in developing advanced expertise for tomorrow's industries.

sumers, labor, local public entities, and the media. Thus, visions can be seen as a vehicle for synchronizing possible, expected, and preferred futures by perceiving future directions, identifying long-term goals, creating consensus, instilling confidence, and establishing respective sharing of responsibilities by the broad sectors concerned.

(3) The significance of visions
The significance of visions can be categorized in the following five dimensions:

(i) Horizontal perspective. Visions are formulated across the board as a totally comprehensive system (general industrial policy), not as a simple subsystem (industrial technology policy).
(ii) Vertical perspective. Vision issues relevant to engineering are further considered by a special advisory committee with expertise on engineering which maintains a consistent and close interaction with general industrial policy.
(iii) Joint product. Visions are joint products resulting from joint work and open discussion between government and representatives from a broad spectrum, including industrial circles, academia, financial institutions, small businesses, consumers, labor, local public entities, and the media.
(iv) Prompt policy reaction. Prompt reaction to policy decisions by establishing national R&D programs in response to recommendations raised in visions.
(v) Fair return to contributors. Contributors to visions, particularly industry and academia, are given the opportunity to participate in R&D consortia and to conduct the R&D that they propose as essential to their future.

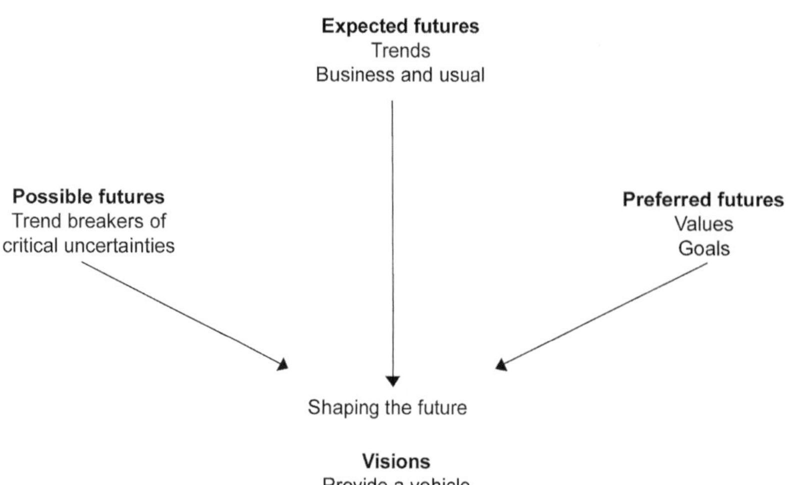

Figure 45.11. The role of visions—the soft technology of public administration.

(4) Characteristics of visions

The characteristics of visions can be identified by the following six dimensions:

(i) Concrete blueprint. Neither philosophical nor a general picture but concrete, specific plans.
(ii) Close interaction with total system. Not a sub-system consideration but one that maintains consistency and close interaction with general policy.
(iii) Soft technology for shaping the future. Neither a plan with a means of execution nor simple prediction but a soft technology of public administration for shaping the future.
(iv) Synchronization of three futures. The future to be shaped is not limited to expected futures, possible futures, or preferred futures alone but is a synchronization of all three.
(v) Shaping and realizing the future. Outcomes are promptly responded to through policy implementation in which contributors to the formulation are broadly involved.
(vi) Encourage joint research. Promote a joint effort regarding actions for shaping the future and realization of the vision.

(5) Important aspects of the foresight process

Important aspects of visions can be demonstrated by the following nine Cs:

(i) Communication. Bringing together disparate groups of people and providing a structure within which they can communicate.
(ii) Concentration on the longer term. Ensuring individuals concentrate seriously and systematically on the longer term.
(iii) Coordination. Enabling different groups to coordinate their future R&D activities.
(iv) Consensus. Creating a measure of consensus on future directions and research priorities.
(v) Commitment. Generating a sense of commitment to already attained results among those who will be responsible for translating them into research advances, technological developments, and innovations for the benefit of society.
(vi) Comprehensive analysis and consideration. Not a sub-system consideration but maintenance of consistency and close interaction with the total system.
(vii) Concrete perspective Not only a general macro-analysis and shaping but also a micro, in-depth analysis and concrete shaping in a vertical manner.
(viii) Consortia directing. New major long-term R&D efforts are generally preceded by establishing R&D consortia in which contributors to visions participate and realize the proposals they raised during the process of vision formulation.
(ix) Credibility. Cumulative evidence on the foregoing eight Cs have developed credibility regarding the

benefits of involvement in vision formulation and implementation.

45.3.4 National R&D program

The national R&D program has played a core role in accomplishing the inducing policy. The history of the development of this program can be summarized briefly as follows:

(1) Background to its establishment

With regard to the development of its economy, Japan confronted the following needs in the 1960s:

(i) Improvement of techno-economic levels.
(ii) Tightening of technology import conditions.
(iii) Urgency of improving its international competitiveness in face of the liberalization of foreign capital investment (starting in 1967).

(2) Policy steps taken for its inauguration

In response to the foregoing circumstances, MITI took the following strategic actions in order to enhance indigenous technological development capacity:

(i) Enactment of the Law of Engineering Research Association (ERA) facilitating incentives for participants in 1961.
(ii) Publishing MITI's vision of the 1960s drafted by the Industrial Structure Council of MITI in 1963 with the following key messages:
   a. shift from Japan's import technology–dependent structure to indigenous technology development;
   b. government initiatives; and
   c. tie-ups between industry, university, and national research institutes for priority R&D projects.
(iii) Introduction of the Expenses Commission System for R&D of Industrial Technology (the entrustment system) in 1964.
(iv) Establishing the National R&D Program in 1966 and giving it the following structure:
   a. a priority R&D program;
   b. the program to be initiated by the Engineering Research Association with ties between industry, university, and national research institutes; and
   c. subject to the entrustment system.

(3) Chronology

The National R&D Program was developed according to the national development focus in each respective era (as demonstrated in Table 45.2).

Corresponding to the increasing need for exploration of frontier innovative technologies, increasing environmental concerns with respect to global warming, and to advance effective National R&D Program projects, MITI integrated these projects into (i) the Industrial Science and Technology Frontier Program and (ii) the New Sunshine Program (R&D program on energy and environmental technologies) in 1993, at the same time entrusting the reorganization of its national research laboratories to these programs.

45.4 Miscalculation of the growth trajectory and then a surge of reactivation

45.4.1 Japan's contrast between co-evolution and its disengagement

In contrast to the high-tech miracle of the 1980s, Japan experienced a prolonged economic stagnation in the 1990s. This reversal can be attributed to co-evolution between innovation and institutional systems in an industrial society and co-evolution disengagement in an information society. Disengagement in an information society is due to a system conflict with a de-materializing society (Kondo and Watanabe, 2003). Japan's renowned capacity for substituting technology for scarce resources functioned well for materialized production factors. However, as the paradigm shifted to an information society (as shown in Table 45.3), the move from manufacturing technology to information technology led to a de-materializing society.

45.4.2 Surge in the reactivation of co-evolutionary dynamism

Although Japan's dynamism changed direction in the 1990s, resulting in a lost decade, this dynamism resurged massively in the early 2000s. This can largely be attributed to hybrid management fusing East (indigenous strength) and West (learning from its digital economy) typically observed in cellphone-driven innovation.

Cellphone-driven innovation triggered a surge in dynamic co-evolution. The result of a dramatic increase meant Japan's cellphone numbers exceeded the number of fixed phones in 1998. Furthermore, the i-mode service (NTT Docomo's cellphone with Internet access service) from February 1999 accelerated IP mobile diffusion, which stimulated the interaction with institutions. Intensive interaction with institutions increased the learning coefficient.

Cellphone-driven innovation can be attributed both to indigenous strength in manufacturing technology and the effects of cumulative learning from the digital economies of the West, which stimulated the reactivation of Japan's economy by playing a role as a catalyst in disseminating the spillover of core technologies (as demonstrated in Table 45.4).

Table 45.2. Trends in Japan's industrial structure policy and the chronology of MITI-initiated R&D programs.

1960s	*Heavy and chemical industrial structure*	
1951–	Financing for industry's new technology	Japan Development Bank
1963	MITI's vision for the 1960s	
1967–	Tax incentives for technological development	
1966–	Large-scale R&D projects	
1970s	*Knowledge-intensive industrial structure*	
1971	MITI's vision for the 1970s	
1974–	The Sunshine Project (R&D on new energy technology)	Oil-substituting energy technology
1976–1979	The VLSI Project (very large-scale integrated circuits)	Innovative computer technology
1976–	The R&D Program on Medical & Welfare Equipment Technology	Medical and welfare technology
1978–	The Moonlight Project (R&D on energy conservation technology)	Technology for improving energy productivity
1980s	*Creative knowledge-intensive industrial structure*	
1980	MITI's vision for the 1980s	
1981–	The R&D Program on Basic Technologies for Future Industries	Basic and fundamental technology
1982–1991	Fifth-Generation Computer Project	Innovative computer technology
1985–	Key Technology Center Project (industrial R&D on fundamental technology)	Fundamental technology initiated by industry
1990s		
1995	Science and Technology Basic Law	
1996	Basic Plan for Science & Technology (1996–2000)	
1997	Guideline for Technology Evaluation	
1998	TLO Act	
	Program for the Science & Technology Development for Industries that Create New Industries consists of (i) R&D projects at new industrial science and technology frontiers; (ii) R&D projects on application of industrial technologies; (iii) R&D projects on cooperation with academic institutions; and (iv) R&D cooperative projects with Industry (from 2000)	
1999	Industrial Competitiveness Council	
2000s		
2000	National Industrial Technology Strategy → flexibility, adoptability, and cooperation with industry, government, and universities	
	Industrial Technology Strengthening Act	
2001	Structural reform of the Central Government → MITI → METI, STA, and Ministry of Education → MEXT → MITI's 14 research institutes → AIST → Comprehensive Science & Technology Council	
	Second Basic Plan for Science & Technology (2001–2005)	
2002	21st Century COE Program	
2004	National University Corporation	
2005	Japan's National Innovation Ecosystem (Industrial Structure Council of METI)	
2006	Third Basic Plan for Science & Technology (2006–2010)	
2007	Innovation 25	

Table 45.3. Comparison of features between manufacturing technology and IT.

	–1980s	*1990s–*
Paradigm	Industrial society (materializing economy)	Information society (de-materializing—digital—economy)
Core technology	Manufacturing technology	IT
1. Optimization	Within firms/organizations	In the market
2. Key feature formation process	Provided by suppliers	Formed through interating with institutions
3. Fundamental nature	As given	Self-propagating
4. Actors responsible for features	Individual firms/organizations	Institutions as a whole
5. Objectives	Productivity	Functionality
6. Development trajectory	Growth-oriented trajectory	New functionality development–initiated trajectory

Table 45.4. Core technologies in cellphone innovation spillover.

Semiconductor	*Electronics*	*Sensor*	*Materials*
Battery	Wireless communication	IC card	Liquid crystal
Optics	Acoustic	Micro-devices	High density
Application	Platform	Security	Compression

45.5 Hybrid management fusing the East and the West

45.5.1 Japan's indigenous potential in fusing

"Japan is emerging from years of sluggish growth. Its firms appear to have produced something. Management method that incorporates lessons from US firms while preserving the practices that once made Japanese firms famous" (*International Herald Tribune*, August 31, 2006). These comments can be attributed to hybrid management achieved by fusing the East (indigenous strength) and the West (learning from digital economies).

Japan naturally incorporates an explicit fusing potential through its intensive, cumulative learning efforts and its unique capacity as a result of (i) xenophobia; (ii) using cumulative learning to stimulate assimilation of spillover knowledge; and (iii) being rich in curiosity, smart in assimilation, and thorough in learning and absorption. Based on this unique capacity, Japan's system of management of technology (MOT) has brought about co-evolutionary development through learning and assimilating innovation and advancement of its own institutional systems.

However, this natural strength (i.e., Japan's learning ability) experienced a dramatic decrease in the 1990s due to the NIH (not invented here) syndrome, a consequence of the "arrogance" of its success in the 1980s and also miscalculation of its development trajectory.

45.5.2 Surge in reactivation

Notwithstanding the unfavorable circumstances, certain high-tech firms have maintained efforts in fusing their indigenous strength with cumulative learning efforts for the advancement of the digital economy; these efforts include (i) increased digitalization of the manufacturing process, (ii) advanced digital infrastructures or alliances, and (iii) timely correspondence to the customer wishes in the digital economy.

The recent surge in the reactivation of indigenous explicit co-evolution between innovation and institutional systems can largely be attributed to such hybrid management efforts. Such noteworthy development in recent years corresponds to essential requirements in a ubiquitous economy characterized by the "on demand", "all actor participation and cooperation", "open-sourcing", and "seamless" community.

45.5.3 Global co-evolution for sustainable development

The co-evolutionary dynamism between innovation and institutional systems is decisive for an innovation-driven economy. The rise and fall of the Japanese economy over the past four decades can be attributed to the consequences of co-evolution and its disengagement between innovation and institutional systems.

Successful co-evolution in an industrial society by substituting manufacturing technology for labor and energy, which led to the high-tech miracle, changed to co-evolution disengagement in the information society of the 1990s, resulting in a lost decade for Japan.

The surge in new innovation in cutting-edge activities in certain high-tech firms can be attributed to the hybrid management of technology by fusing the indigenous strength developed in an industrial society (East) and the effects of learning in an information society (West). This surge suggests the possibility of reactivation of Japan's system of MOT and consequent revitalizing of its economy. This can be brought about by constructing a virtuous cycle in a revitalized world economy. In addition, such a surge suggests the significance of a hybrid system aimed at fusing indigenous strength with the learning of partners who have comparative advantages in certain fields.

Given the indigenous strengths and weaknesses in different countries, constructing a co-evolutionary dynamism between innovation and institutional systems in a global complementary scheme would be essential in accelerating the fusion between East and West from a global perspective.

The significance of global co-evolution for sustainable development is apparent. In this context, identification of similarities and disparities of institutional systems in leading countries would form the basis of effective hybrid systems in a global context to achieve effective fusion. International collaboration in this identification process and maximizing the effects of learning would be beneficial to global sustainable development.

45.6 Conclusion

The rise and fall of the Japanese economy over the past four decades prompts us to postulate that co-evolutionary dynamism between innovation and institutional systems is decisive for an innovation-driven economy, which may stagnate if institutional systems cannot adapt to innovations—Japan's economy in the 1990s is one such example.

In light of the significance of such co-evolution for the management of technology (MOT) toward a ubiquitous economy, this chapter analyzed the necessary conditions under which a technopreneurial trajectory with sustainable functionality can be developed by reviewing

(i) the systematic workings of Japan's indigenous explicit capacity;
(ii) the way in which dynamism in industrial technology policy can be made systematic;
(iii) miscalculation of growth trajectories followed by a surge of co-evolution reactivation; and
(iv) hybrid management fusing East and West.

In summary,

(i) Japan indigenously incorporates an explicit capacity to bring about co-evolutionary dynamism, enabling it to achieve a remarkable performance in a virtuous cycle between innovation and rapid economic growth in the 1960s followed by technology substitution for energy in the 1970s, leading to the world's greatest improvement in energy efficiency and broad advances in the level of manufacturing technology in the 1980s.
(ii) Such an explicit capacity can be attributed to the sophisticated policy that acts much in the same way as a catalyst for "chemical reactions" under "oxygen-rich" circumstances.
(iii) Although Japan's dynamism changed direction in the 1990s, resulting in a lost decade due to a systems conflict between indigenous institutional systems and a new paradigm in an information society, a surge of reactivation emerged in the early 2000s.
(iv) This surge can be attributed largely to hybrid management fusing East (indigenous strength) and West (lessons from IT-driven new economies) typically observed in cellphone-driven innovation.
(v) As a consequence of its fusing efforts, Japan's system of MOT is expected to become more resilient.

These findings provide the following important policy implications which can be used to support construction of a resilient policy aimed at bringing about a service-oriented economy:

(i) any shift of the point where innovation takes place must be identified;
(ii) consequently, switching from a growth-oriented trajectory to a new-functionality development trajectory must be attempted;
(iii) thus, innofusion function by integrating production function and diffusion function should be based;
(iv) sustainable functionality development should be key to a firm's competitiveness, and this can be attained by effectively utilizing potential resources in innovation; and
(v) inducing policy corresponding to an open innovation stream toward a ubiquitous economy is expected to how effectively induces industry and customer potential in maximizing their self-propagating dynamism.

45.7 References and further reading

Bass, F. M. (1969). "A new product growth model for consumer durables." *Management Science*, **15**(5), 215–227.

Brandenburger, A. M., and Nalebuff, B. J. (1997). *Co-opetition*. New York: Helen Rees Literary Agency.

Chen, C., and Watanabe, C. (2006). "Diffusion, substitution and competition dynamism inside the ICT market: A case of Japan." *Technology Forecasting and Social Change*, **73**(6), 731–759.

Chen, C., Watanabe, C., and Griffy-Brown, C. (2007). "The co-evolution process of technological innovation: An empirical study of mobile phone vendors and telecommunication service operators in Japan." *Technology in Society*, **29**(1), 1–22.

Cohen, W. M., and Levinthal, D. A. (1990). "Absorptive capacity: A new perspective on learning and innovation." *Administrative Science Quarterly*, **35**(1), 128–152.

Englander, A. S., and Mittelstadt, A. (1988). "Total factor productivity: Macro economic and structural aspect of the slowdown." *OECD Economic Studies*, **10**, 7–56.

Griliches, Z. (1979). "Issues in assessing the contribution of R&D to productivity growth." *Bell Journal of Economics*, **10**(1), 92–116.

Hobo, M.. and Watanabe, C. (2006). "Double spiral trajectory between retail, manufacturing and customers leads a way to service-oriented manufacturing." *Technovation*, **26**(7), 673–890.

IEA. (2006). *World Energy Statistics*. Paris: International Energy Agency.

Irvine, J. (1988). *Evaluating Applied Research: Lessons from Japan*. London: Pinter.

Jaffe, A. B. (1986). "Technological opportunity and spillovers of R&D: Evidence from firms' patents, profits, and market value." *American Economic Review*, **76**(5), 984–1001.

JRSC. (2006). *Trend in Domains of co.jp*. Tokyo: Japan Registry Services Co.

Komiya, R., Okuno, M., and Suzumura, K. (1984). *Industrial Policy of Japan*. Tokyo: Tokyo University Press.

Kondo, R., and Watanabe, C. (2003). "The virtuous cycle between institutional elasticity, IT advancement and sustainable growth: Can Japan survive in an information society?" *Technology in Society*, **25**(3), 319–335.

Kondo, R., Watanabe, C., and Moriyama, K. (2007). "A resonant development trajectory for IT development: Lessons from Japan's i-mode." *Journal of Advances in Management Research*, **4**(2), 7–27.

Kuchinskas, S. (2007). *How to Use Web 2.0 inside Your Company* (BNET Crash Course); *http://www.bnet.com/2403-13241_23-66093.html*

Mahajan, V., Muller, E., and Srivastara, R. K. (1990). "Determination of adopter categories by using innovation diffusion models." *Journal of Marking Research*, **27**, 37–50.

Mansfield, E. (1988a). "Industrial innovation in Japan and the United States." *Science*, **30**, 1769–1774.

Mansfield, E. (1988b). "Industrial R&D in Japan and the United States: A comparative study." *American Review*, **78**(2), 223–228.

Marten, G. (2001). *Human Ecology: Basic Concepts for Sustainable Development*. London: Earthscan.

McAfee, A. P. (2006). "Enterprise 2.0: The dawn of emergent collaboration." *Sloan Management Review*, **47**(3), 21–28.

Meyer, P. S. (1994). "Bi-logistic growth." *Technological Forecasting and Social Change*, **47**(1), 89–102.

Meyer, P. S., and Ausbel, J. H. (1999). "Carrying capacity: A model with logistically varying limits." *Technological Forecasting and Social Change*, **61**(3), 209–214.

MIAC. (2006). *White Paper on Japan's ICT 2006*. Tokyo: Ministry of Internal Affairs and Communications.

MITI. (Annual issues from 1955 to 2005). Tokyo: Ministry of International Trade and Industry.

MITI. (1988). *White Paper on Japanese Industrial Technology*. Tokyo: Ministry of International Trade and Industry.

Mitsuda, M., and Watanabe, C. (2008). "Accelerated interaction between firms and markets at ICT-based venture business: The case of mobile phone business in Japan." *Journal of Services Research*, in print.

Nelson, R. R., and Sampat, B. N. (2001). "Making sense of institutions as a factor shaping economic performance." *Journal of Economic Behavior & Organization*, **44**, 31–54.

North, D. C. (1990). *Institutional Change and Economic Performance*. Cambridge, U.K.: Cambridge University Press.

North, D. C. (1994). "Economic performance through time." *American Economic Review*, **84**, 359–368.

OECD. (Annual issues from 2000 to 2006). Paris: Organization for Economic Cooperation & Development; *www.oecd.org*

OECD. (1997). *Technology and Industrial Performance*. Paris: Organization for Economic Cooperation and Development.

OECD. (2007). *Main Science and Technology Indicators*. Paris: Organization for Economic Cooperation and Development.

Okimoto, D. I. (1989). *Between MITI and the Market: Japanese Industrial Policy for High Technology*. Stanford, CA: Stanford University Press.

O'Reilly, T. (2005). "What is Web 2.0." *Design Patterns and Business Models for the Next Generation of Software*; *http://hbswk.hbs.edu/archive/5102.html*

Ouchi, N., and Watanabe, C. (2009). "Co-evolutionary domestication for self-propagating functionality development." *Journal of Services Research*, **9**(1), 69–86.

Rogers, E.M. (1983). "Diffusion of Innovation," 3rd ed., New York, The Free Press.

Rossi, S. and Paolo, V. (2004). "Cross-country determinants of mergers and acquisitions." *Journal of Financial Economics*, **74**, 277–304.

STA. (Annual issues). *White Paper on Japanese Science and Technology*. Tokyo: Science and Technology Agency.

Utterback, J. M. (1994). *Mastering the Dynamics of Innovation: How Companies Can Seize Opportunities in the Face of Technological Change*. Boston, MA: Harvard Business School Press.

Wakasugi, R. (1986). *Economic Analysis of Technological Innovation and R&D*. Tokyo: Toyo Keizai Shimpo Co.

Watanabe, C. (1990). "Japanese industrial development." *Australian Journal of Public Administration*, **49**(3), 288–294.

Watanabe, C. (1992). "Trends in the substitution of production factors to technology." *Research Policy*, **21**(6), 481–505.

Watanabe, C. (1996). "Choosing energy technologies: The Japanese approach." In: IEA (Ed.), *Comparing Energy Technologies* (pp. 105–138). Paris: Organization for Economic Cooperation and Development/International Energy Agency.

Watanabe, C. (1997). "Predicting the future: Shaping the future of engineering." In: E. Shannon (Ed.), *Engineering Innovation and Society* (pp. 23–54). London: Royal Academy of Engineering.

Watanabe, C. (1999). "Systems option for sustainable development." *Research Policy*, **28**(7), 719–749.

Watanabe, C. (2000). "Visions in co-evolution: A Japanese perspective on the role and limitations of visions as a vehicle for effective governance under a new paradigm." Paper presented at *International Workshop on Science and Governance, Brussels, Belgium.*

Watanabe, C. (2005). "Japan's co-evolutionary dynamism between innovation and institutional systems." Paper presented at *SIMOT International Symposium, Tokyo.*

Watanabe, C., Akaike, S., and Shin, J. (2009). "Hybrid management of technology toward a service-oriented economy: Co-evolutionary domestication by fusing East and West." *Journal of Services Research*, **9**(2), 7–50.

Watanabe, C., and Asgari, B. (2004). "Impacts of functionality development on the dynamism between learning and diffusion of technology." *Technovation*, **24**(8), 651–664.

Watanabe, C., Asgari, B., and Nagamatsu, A. (2003). "Virtuous cycle between R&D, functionality development and assimilation capacity for competitive strategy in Japan's high-technology industry." *Technovation*, **23**(11), 879–900.

Watanabe, C., and Fukuda, K. (2006). "National innovation ecosystem: The similarity and disparity of Japan–US technology policy system toward a service-oriented economy." *Journal of Services Research*, **6**(1), 159–186.

Watanabe, C., Hur, J. Y. and Lei, S. Y. (2006). "Converging trend of innovation efforts in high technology firms under paradigm shift: A cse of Japan's electrical machinery." *Omega*, **34**(2), 178–188.

Watanabe, C., Kishioka, M., and Nagamatsu, A. (2004). "Effect and limit of the government role in spurring technology spillover: A case of R&D consortia initiated by the Japanese Government." *Technovation*, **24**(5), 403–420.

Watanabe, C., and Lei, S. (2009). "The role of techno-countervailing power in inducing the development and dissemination of new functionality: An analysis of Canon printers and Japan's personal computers." *International Journal of Technology Management.*

Watanabe, C., Santoso, I., and Widyati, T. (1991). *The Inducing Power of Japanese Technological Innovation.* London: Pinter.

Watanabe, C., and Honda, Y. (1991). "Inducing power of Japanese technological innovation." *Japan and the World Economy*, **3**(4), 361–390.

Watanabe, C., and Zhao, W. (2006). "Co-evolutionary dynamism of innovation and institution." In: N. Yoda, R. Pariser, and M. C. Chon (Eds.), *Chemical Business and Economics* (pp. 106–121). Tokyo: Chemical Society of Japan.

Watanabe, C., Takayama, M., Nagamatsu, A., Tagami, T., and Griffy-Brown, C. (2002a). "Technology spillover as a complement for high-level R&D intensity in the pharmaceutical industry." *Technovation*, **22**(4), 245–258.

Watanabe, C., Griffy-Brown, C., Zhu, B., and Nagamatsu, A. (2002b). "Inter-firm technology spillover and the creation of a 'virtuous cycle' between R&D, market growth, and price reduction: The case of photovoltaic power generation development in Japan." In: A. Gruebler, N. Nakicenovic, and W. D. Nordhaus (Eds.), *Technological Change and the Environment* (pp. 127–159). Washington, D.C.: Resources for the Future (RFF) Press.

Watanabe, C., Kondo, R., Ouchi, N., Wei, H., and Griffy-Brown, C. (2004). "Institutional elasticity as a significant driver of IT functionality development." *Technological Forecasting and Social Change*, **71**(7), 723–750.

Watanabe, C., Takahashi, H., Tou, Y., and Shum, K. L. (2006). "Inter-fields technology spillovers leveraging co-evolution between core technologies and their application to new fields: Service-oriented manufacturing toward a ubiquitous society." *Journal of Services Research*, **6**(2), 7–24.

Watanabe, C., Yamauchi, S., Shin, J., and Tou, Y. (2009). "Fusing 'East and West' for high-profitable resilient structure in mega-competition." *Journal of Services Research.*

Wikipedia. (2007). *The Free Encyclopedia*; *http://en.wikipedia.org/wiki/Web_3.0*

WEF. (2006). *The Global Competitiveness Report.* Geneva, Switzerland: World Economic Forum.

Yamauchi, S., Morisaki, S., Watanabe, C., and Tou, Y. (2007). "A resilience structure as a survival strategy for Japan's chemical industry amidst megacompetition: Suggestion to management research." *Journal of Advances in Management Research*, **4**(1), 29–48.

Zhao, W. (2006). "Co-evolution between software innovation and institutions: Elucidation of co-evolutionary dynamism between Japan and China through outsourcing." Master's thesis, Tokyo Institute of Technology.

46

The National Innovation System of the Russian Federation

Julian Cooper

University of Birmingham, U.K.

Introduction

As of 2008, it is debatable whether Russia possesses a national innovation system (NIS) in the normally understood sense of a coherent set of inter-related institutions promoting innovation as a natural outcome of their day-to-day functioning. Institutions and practices in the sphere of research and development still retain many features of the former Soviet system and it is not possible to understand the present-day situation without first exploring the Soviet legacy and the impact on the R&D system of the turbulent transition to a market economy during the 1990s.[1]

Features of the Soviet R&D system included the organizational separation of research from production, the dominant role not only in basic research but also in much applied work of the U.S.S.R. Academy of Sciences, which played a central role in the overall science policy of the country and occupied a position of prestige in society, and the relatively modest role in R&D of the higher educational sector (Zaleski *et al.*, 1969). In the business sector, all enterprises were state-owned and most R&D was undertaken by specialized applied research institutes, generally organizationally separate from enterprises which undertook little research. The Soviet R&D system was heavily militarized and successive attempts to transfer technology from the military sector to the civilian economy met with little success. The U.S.S.R. had a very substantial R&D system in terms of the number of people employed and reported spending on research as a share of economic output at levels high by international standards, 5% of national income in 1990.[2]

In the U.S.S.R. the innovation process was always understood, implicitly by government officials and often explicitly by economists, as a linear process (i.e., new products and processes are developed on the basis of ideas and inventions originating in basic and applied research, after which they are "introduced" into the sphere of production and then diffused more widely). Only in the very final years did some analysts become aware of the work of Chris Freeman and other Western science policy specialists who challenged the linear model and argued for a richer understanding involving feedback relationships.

Despite the scale of the R&D system, somewhat overstated by Soviet official statistics, overall innovative performance was poor. Soviet industry was characterized by an all-pervasive lack of responsibility, risk aversion, and weak incentives for successful innovation, coupled with mild penalties for failure in the absence of competition (Berliner, 1976; Amann and Cooper, 1982). The outcome of inadequate innovation was the inability of the Soviet Union to narrow the technological gap with leading Western countries (Amann, Cooper, and Davies, 1977). While the Soviet economic system was by its nature averse to innovation, some sectors showed better performance than others, above all the defence industries, where high-level political intervention and priority resource allocation to some extent overcame the inertia of dysfunctional economic institutions.

The technological capability of the U.S.S.R. Soviet was extremely uneven. There were strengths in military, space, and nuclear technology, but technologies associated with civilian activities, above all the sphere of consumption, were relatively weak by international standards. A weakness that over time had increasingly negative implications for the country's development was an inability to match Western innovations in electronics and computing.

Research undertaken by Russian and Western economists and science policy specialists reveals that, despite

[1] This chapter draws on the author's entry on Soviet and Russian research, development, and innovation in Durlauf and Blume (2008).

[2] For a comparative assessment of the Soviet R&D system, see Hanson and Pavitt (1987). Note that the Center for Science Research and Statistics, Moscow, has reassessed Russia's 1990 R&D spending using OECD methods and arrived at a figure of just over 2% of GDP.

reform measures, the Russian R&D system still retains many Soviet characteristics (Dezhina and Saltykov, 2005; Radosevic, 2003). There is still organizational fragmentation, with the majority of R&D organizations being remote from the business sector; a large proportion of research organizations remain in state ownership; budget spending predominates, with only a modest contribution from the private sector; higher education plays a limited role in R&D; the academy system, largely unreformed, still absorbs a significant share of total expenditure; and the military share of R&D remains substantial, almost as large in relative terms as in Soviet times. The inertia is such that it cannot be said that Russia yet possesses a coherent national innovation system. This situation has arisen in part because strong vested interests within the Academy of Sciences and industry have pursued their own survival strategies, resisting the government's reform initiatives.

While the Russian Government has overall responsibility for science, technologym and innovation policy, the institutional framework for policy making remains highly departmental, as in Soviet times. Formally, the Ministry of Education and Science has a leading role, but in practice its powers are limited. Overall economic policy, including policies to promote innovation and competition, is in the hands of the Ministry of Economic Development. Policy for industrial R&D and innovation is led by the Ministry of Industry and Trade, which also has responsibility for a large proportion of the state-owned research institutes linked to industry. Another important actor is the state corporation Rosatom, the successor to the Federal Agency for Atomic Energy, which oversees many R&D establishments of both civil and military orientation. The same applies, but on a smaller scale, to the Federal Space Agency. Another center of policy making for science and technology is the Ministry of Defense, which to only a limited extent coordinates its activities in the field with other government agencies. Last but not least is the Russian Academy of Sciences (hereafter RAS), a self-governing, non-state body, responsible for much research activity, in particular fundamental research, and influential in shaping overall research policy. In recognition of the inadequacy of overall coordination, from about 2003 a number of inter-agency bodies have been created, to the fore being the Interdepartmental Commission on Science and Innovation Policy, created in 2004 and chaired by the Minister for Education and Science. However, its powers are limited in so far as two key actors are not represented, the ministries for the economy and defense. In reality, policy for science, technology, and innovation is to a large extent shaped by the Ministry of Economic Development, charged with elaborating short and long-term scenarios for the economy, with policy inputs supplied by other government agencies.

There are three principal sectors of Russian R&D, their present-day features being shaped to a considerable extent by the Soviet legacy. The RAS remains a very strong actor having retained most of the network of research institutes it possessed in the past. It undertakes more than half of all fundamental research and accounts for a third of all budget spending on civil research. As a self-governing organization it has considerable autonomy in setting its own research priorities. As in Soviet times, the higher educational sector is a relatively weak actor. While some elite universities and technical institutes possess strong research capacities, less than 40% of higher educational establishments undertake research and the sector's share of total domestic spending in R&D is barely 5%. Much of the research is undertaken to contracts with government organizations and industry. However, while the RAS plays a modest role in training postgraduates, universities and other higher educational facilities, notwithstanding their limited research capacity, have basic responsibility for training researchers. The third sector is business, but here the economic difficulties of the 1990s have left their mark. In the U.S.S.R., industrial ministries possessed a large number of applied research institutes and little R&D was undertaken at the enterprise level. With the end of the ministerial system, severely limited funding, and privatization many of these institutes have either closed down or converted to other activities. Only the defense industry has retained a sizable proportion of these Soviet era institutes and design organizations. A few applied institutes remain in sectors of the civil economy but industrial companies have on the whole been slow in developing their own R&D capabilities. Russian industry is dominated by a relatively small number of very large companies associated with energy, metals, and the exploitation of other natural resources. Some of these companies, often highly profitable, such as Gazprom, Rosneft (oil), and Severstal (steel), have become active in R&D, but they are exceptions. Another feature of the Russian economy is the relative weakness of the small and medium enterprise (SME) sector, in many advanced countries so significant for their activity in innovation. In Russia the number of SMEs specifically devoted to research and innovation actually fell from some 31,000 in 2000 to 20,500 in 2005. They encounter many problems in operating in a rather unfavorable business environment, including difficulties in raising finance. There is a fourth sector of Russian science, but it is very weak, namely independent, non-commercial organizations engaged in research.

Not only is the basic institutional framework marked by the Soviet past, so also are the relationships between the basic sectors. The RAS is still to a considerable extent separate from the business sector and also, to a lesser degree, from the higher educational system. The latter, or at least its leading establishments, has links with the business sector but the scale of activity is modest. To a striking degree the Russian R&D system remains state-dominated in terms of ownership, funding, and employment, with a corresponding weakness of non-state actors.

Table 46.1. R&D personnel—number of researchers.

	1990	*1992*	*1995*	*1997*	*2000*	*2003*	*2006*
Researchers	1,227,388	804,043	518,690	455,108	425,954	409,775	380,100
Index, 1900 = 100	100	66	42	37	35	33	31

Source: Federal Agency of State Statistics RF, central database of statistics, *http:/www.gks.ru/dbscripts/Cbsd/DBInet.cgi* (last accessed April 15, 2008).

According to the leading Russian science policy specialist, Irina Dezhina, at present almost three-quarters of all organizations in the sphere of science and technology refer to the state sector being fully, or majority, owned by the state, and in them work 77% of personnel engaged in R&D. These state organizations are almost wholly funded from budget sources (Dezhina, 2007).

In Soviet times the R&D system was oriented to a considerable extent to the needs of the military and this remains the case in present-day Russia, although not to the same degree. In the U.S.S.R. in the late 1980s between 70 and 75% of all funded R&D was military in character. According to deputy prime minister (and former defence minister) Sergei Ivanov, over half of all scientists in Russia are employed in the defense industry, which means that within industry as a whole approximately three-quarters of all scientists work in the defense sector, although some will be engaged in predominantly civilian activities (Ivanov, 2006).

One of the most unsatisfactory features of the present-day Russian NIS is the age structure and quantitative trends in relation to research personnel. Ever since the late Soviet period, total number employed in the R&D sector, including the number of researchers, has steadily diminished and at the same time their average age has risen. The overall trends are shown in Table 46.1.

It can be seen that by 2006 the number of researchers had declined to less than one-third of the 1990 total. As of 2000 it was reported that 46% of all scientists were over the age of 50, including 57% of candidates of science and 80% of doctors of science; only one-fifth of scientists were in the age range 27 to 40 years. In the same year the average age of personnel within the RAS system was between 50 and 55 years.[3] By 2006 the average age of researchers had risen to 48 years and almost one-third of active researchers were of pension age (MEDRF, 2008a, p. 98). There is an inflow of younger people into science but many leave after a number of years, largely because salaries are relatively low and promotion prospects often poor because of the dominance of older scientists, frequently of a conservative disposition. Dezhina has estimated that the average period of time spent by a researcher in the sphere of science and technology is only 8 years, and that brief research careers are typical of those under 30 (Dezhina, 2007, p. 21).

R&D funding

During the 1990s domestic expenditure on R&D contracted to a very considerable extent both in absolute terms and as a share of GDP. Domestic funding was to a limited extent boosted by an increase in funding from foreign sources, which rose from virtually zero in 1991 to over 10% of total science funding by 1998.[4] The principal external sources were the European Union, in particular through the INTAS (International Association for the Promotion of Co-operation with Scientists from the New Independent States of the Former Soviet Union) program, the U.S. National Science Foundation, some European national research funding agencies, and non-governmental organizations, with the Soros Open Society Foundation to the fore. State budgetary funding has played the dominant role, mainly through the federal budget but also from budgets at the republican and regional levels. The share of non-state funding has grown steadily over time but as of 2007 was less than 40% of the total. As a share of GDP, total domestic spending on R&D fell from approximately 2% in 1990 to a low of 0.74% in 1992, but since 2000 has reached a plateau of 1.0–1.2%. In real terms, spending in 2006 reached only 45% of the level of 1990, even though the equivalent figure for the volume of the country's GDP was between 85 and 90%. Overall funding trends are shown in Table 46.2.

In 2006 budgetary sources accounted for 60.2% of total R&D expenditure, the business sector 19.7%, foreign sources 9.4%, and other sources (non-commercial organizations, means of higher education establishments, and extra-budgetary funds) 2.0%. The remaining 8.7% was funded from the internal resources of research organizations. By type of activity the pattern of spending has remained almost constant over time: the share of spending on fundamental research was 15.4% in 2004 (13.4% in 2000), on applied research 15.3% (16.4) and on develop-

[3] *Rossiiskaya gazeta*, November 28, 2000; *Delovaya tribuna*, May 11, 2000, p. 3.

[4] *http://www.kommersant.ru/Docs/Strat1.html* (last accessed June 7, 2000).

Table 46.2. Expenditure on R&D in Russia, volume in real terms, and share of GDP, 1990–2006.

	1990	*1992*	*1994*	*1996*	*1998*	*2000*	*2002*	*2004*	*2006*
Index, 1990 = 100	100	30	27	26	26	30	40	42	45
Percentage of GDP	2.03	0.74	0.84	0.97	0.95	1.05	1.25	1.15	1.07

Source: compiled from data of Federal Service of State Statistics RF and Center for Science Research and Statistics, Russian Academy of Sciences, Moscow.

ment 69.3% (70.2).[5] Russian official statistics on R&D expenditure by sector tend to overstate the role of the business sector in the performance of R&D. It is claimed that in 2006 the state sector accounted for 27.0% of spending, the business sector 66.6%, higher education 6.1%, and non-commercial organizations a mere 0.3%. However, state-owned businesses were probably responsible for a sizable proportion of R&D performed in the business sector.

Most funding of science in Russia takes the form of financial allocation to institutions regardless of their productivity in terms of the generation of new knowledge and contribution to economic and social development. Peer group review is still rather weakly developed. An exception is the work of funds created specifically to support research on a project basis, with allocations being made on the basis of peer group review. To the fore have been the Russian Fund of Fundamental Research and the Russian Humanitarian Research Fund. However, the volumes of funding have been modest and over time the management of the funds has come under increasing influence from the RAS.

According to the Ministry of Economic Development, citing expert assessments, Russia occupies 9th place in the world by the number of scientific publications, 15th place by level of citations, and 120th place by the number of citations per article (MEDRF, 2008a, p. 97). In 2005 Russia published 101 scientific and technical journal articles per million people, three times the level of China but about one-seventh of the figure for the U.S.A. (WB, 2007). This suggests that with a relatively large number of researchers Russia's contribution to scientific publications is strong in absolute terms, but the impact of these publications on the wider scientific community, as measured by the extent of citation, is rather modest.

Institutions of development

Over recent years the Russian Government has undertaken a number of initiatives to improve the performance of the R&D and innovation systems, including the creation of special economic zones focused on innovation and the formation of a number of venture capital funds. In the terminology of the Russian Government, these are "institutions of development". In addition, a number of regions and higher educational establishments have created technoparks—as of 2007 there were approximately 80 in the country (Dezhina and Kiseleva, 2008). Some regions in Russia have been especially active in creating technoparks and other institutional arrangements to promote innovation, to the fore being some "science cities" in the central part of the country, in particular Obninsk and Dubna, and also regions in Siberia such as Tomsk and Novosibirsk.

Programs and priorities

A central instrument of the Russian Government in seeking to advance the development of particular technologies and specific sectors of industry is the federal goal-oriented program. Programs of this type take several forms but have in common a set of activities over a period of time, usually 5 to 10 years, covering R&D and the commercialization of its results. They are inter-agency in character, usually led by the government ministry most directly concerned with the program. The bulk of funding is usually from the federal budget, but over time there has been increasing interest in developing public–private partnerships, with efforts to obtain funding commitments from the business sector (Dezhina and Kiseleva, 2008, pp. 124–129). Examples include federal programs for the development of microelectronics, the modernization of the production base of the defense industry, civil aviation, shipbuilding, and the pharmaceuticals industry. One federal program, on the "national technology base", is devoted to what are considered to be critical technologies from the point of view of the country's national security. A list of critical technologies was approved in the 1990s and is periodically updated. In August 2008 Prime Minister Vladimir Putin approved a new list of 35 categories of technology considered critical for the country's socioeconomic development and security.

In 2007 a new organizational form emerged for the development of priority activities, the state corporation, which in law has the status of a non-commercial organ-

[5] Data of Federal Service of State Statistics.

ization. One of the first to be created was a state corporation for the development of nanotechnology. Rosnanotekh was launched with substantial federal budget funding and will coordinate the activities of a large number of R&D organizations, many in the RAS, for the development of nanotechnologies and their application in the economy and society. Another state corporation, Russian Technologies, is based on the state arms export agency, Rosoboronexport, and incorporates over 400 enterprises, mainly in the defense industry, with the aim of modernizing their technologies. The interests of Russian Technologies include special metallurgies, electronics, and the development and manufacture of aeroengines and helicopters. As noted above, the nuclear industry, a significant performer of R&D, is now organized as the state corporation Rosatom.

Strengths and weaknesses

The overall patterns of strength and weakness in the development of technology have not changed fundamentally from those of the U.S.S.R. Relative strength is still found in some military-related fields, space, and the nuclear industry. According to the Ministry of Economic Development, in nuclear power 95% of applied technologies are approximately of world level, in the missile–space industry 85%, special metallurgy 70%, aviation industry 60%, chemical industry 55%, but in the machine tool industry only 35%, and in electronics, one of Russia's weakest sectors, only 20% (MEDRF, 2008a, pp. 97–98). In general, Russia is relatively strong in energy and materials-related technologies, but otherwise relatively weak in manufacturing technologies, especially those applied in consumer goods production. There are particular strengths in some fields of advanced software, and also in biotechnology. It may be indicative of the overall level of technology that, according to the Ministry of Economic Development, only 16% of technologies developed in Russia are put to commercial use and, of these, only half correspond to the world level (MEDRF, 2008a, pp. 97–98).

Innovation in Russia

Innovation activity in Russia's market economy remains relatively weak (for an overview of Russia's innovation system see EC, 2007). A problem in assessing performance is the inadequacy of Russian statistics on innovation. Official Rosstat data offer an extremely negative view: in 2007 little more than 10% of companies in industry were reported as realizing technological innovation. The predominant form of innovation was the renewal of production equipment. In industry outlays on innovation in 2005 amounted to 1.2% of sales. Some surveys by economists suggest that the situation may not be quite as negative as indicated by official data, but when compared with OECD economies the rate of innovation is still very modest. Explanatory factors include weak competition, inadequate managerial skills, underdeveloped technological capabilities at the company level, weak demand for new products and processes, inadequate finance, with little long-term bank lending and a lack of venture capital, and modest foreign direct investment in manufacturing (OECD, 2006; WB, 2006). In the view of many analysts, Russian and foreign, the root problem is not the nature of the NIS as such, but the weakness of competition: most companies simply do not experience competitive pressure to introduce new products and processes or undertake R&D. According to the World Economic Forum's annual Global Competitiveness Index for 2007–2008, Russia ranks 58 of 131 countries listed, compared with 48 for India, 34 for China and 27 for Estonia, the highest ranking ex-U.S.S.R. economy.[6] This is a problem that can only be tackled by determined government action to improve the business climate in the country. Until this issue is tackled increased R&D spending and attempts from above to create "institutions for development" are unlikely to have a significant impact.

Over the years the domestic intellectual property rights regime has been improved and brought more into line with international practice, but enforcement remains weak. The level of patenting remains relatively low and, given the number of researchers, Russia is not as prominent in international patenting as might be expected. According to data of the World International Property Organization, in filing international patents Russia's world share in 2007 was smaller than for China and India and if present trends are maintained China will very soon overtake Russia in the number of international patents filed per 1 million population (in 2007 this was 4.57 in Russia, 4.10 in China, but 173.2 in the U.S.A.).[7]

Looking to the future

By international standards Russia has a relatively well-educated population with a strong level of general literacy. However, demographic trends are unfavorable, with a declining and aging population. Looking to the future, the age structure of the population is not promising for a development strategy based on innovation: in 2006 fewer than 15% of the population were under the age of 14, compared with 32% in India, and 21% in both China and the U.S.A. The Soviet educational system

[6] *http://www.weforum.org/pdf/Global_Competitiveness_Reports/Reports/gcr_2007/gcr2007_rankings.pdf* (last accessed June 10, 2008).
[7] *http://www.wipo.int/ipstats/en/statistics/pct* (accessed June 2, 2008).

secured a relatively high level of general education but during the 1990s spending was severely limited leading to a failure to maintain modern standards in primary and secondary education. At 3.5% of GDP, in recent years total budget spending on education has fallen behind OECD member countries and emerging economies such as India and Brazil. While Russia has some elite universities and higher educational institutes of international standing, it is now recognized that the general level of higher education is inadequate. The Soviet higher educational system had a strong bias towards natural sciences, engineering, and technical disciplines in general, to a large extent to meet the needs of the defense industry, the principal employer of technically qualified personnel. In post-Communist Russia there has been a reorientation to management training (not always of the highest quality), law, economics, and the social sciences in general, with reduced emphasis on engineering.

Russia has a relatively open economy with exports amounting to one-third of GDP and with relatively low levels of import tariffs—somewhat higher than those of China but much lower than for India. However, the dominance of energy and other resource-based exports is such that manufactured exports account for less than one-fifth of total exports. The share of high-tech exports is extremely low, a matter of concern to the Russian Government. In 2006 high-tech exports accounted for 9.4% of manufactured exports and fewer than 2% of total exports. Combat aircraft, air defense systems, and some other types of arms are the principal high-tech export goods. According to the Ministry of Economic Development, in the civil sphere Russia accounts for 0.1% of the world high-tech market (MEDRF, 2008a, p. 97). In recent years Russia has had very few goods that are competitive on export markets apart from energy and other resource-based products (Cooper, 2007).

One aspect of international openness has been a matter of concern to the Russian authorities since the collapse of the Communist system—the mobility of scientists and educated personnel in general. There is a general perception within Russia that there has been a "brain drain" on a substantial scale to the U.S.A. and Western Europe and that this loss of expertise has had a very negative impact on Russia's research performance. The available data suggest that the number of scientists emigrating on a permanent basis is not as large as generally perceived—it has been estimated that in 2000 some 30,000 Russian scientists were working abroad.[8] But those leaving are predominantly young and talented by international standards, meaning that the loss to Russian science is probably more substantial than indicated by the numbers alone. Particularly acute losses have been experienced in mathematics, some branches of physics, and in high-level software skills. However, the evidence suggests that many Russian scientists working abroad maintain contact with their colleagues at home and as such provide links to the international scientific community.

For some time the Russian Government, in particular the Ministry of Economic Development, has been working on a draft conception of socioeconomic development of the country to the year 2020. Three scenarios have been outlined: one of inertia, with little future reform or diversification of the economy; a second based on the full exploitation of the country's resource wealth, with limited diversification; and a preferred third scenario setting out an "innovative" path of development, with gradual diversification away from resource-dependent economic growth and a focus on human capital. The third innovative scenario, projected to secure an annual average rate of growth of GDP of 6.5% between 2011 and 2020, likely to form the basis of a program of socioeconomic development to 2020, was adopted as official policy by the government in November 2008. The scenario envisages a serious improvement in Russia's innovative capability. Spending on R&D is to rise from 1.12% of GDP in 2007 to 1.2% in 2010 and between 2.5 and 3.0% by 2020, when half will be funded from non-state sources as opposed to 42% in 2010. It is envisaged that by 2020 the core of the science sector will be 50–60 large national research centers. Sectors for priority development include aviation, space services, nuclear power, shipbuilding, arms, software, and educational services. The proportion of companies undertaking innovation is forecast to rise from 13% in 2007 to 15% in 2010 and between 40 and 50% in 2020, when new, innovative products will account for between 25 and 35% of industrial output, compared with 5.5% in 2007 (MEDRF, 2008b, p. 101). Notwithstanding these developments, the forecast increase in Russia's share of world exports of high-tech goods is cautious, from 0.3% in 2009 to 2.0% in 2020. Total spending on education, state plus private, is forecast to increase from 4.8% of GDP in 2007 to between 6 and 7% in 2020. It is acknowledged that demographic factors will complicate the realization of what many Russian economists consider to be a highly, perhaps excessively, ambitious program. In particular, the decline in the number of personnel in R&D, and their aging, is likely to continue until some point between 2015 and 2020. Whereas only one-third of Russian researchers in 2007 were under the age of 40, in 2020 the share will rise to only 37%. The scenario envisages new technologies being derived primarily from domestic R&D and innovation, an essentially linear model being implicitly incorporated, but as some critics, notably the Finnish economist Pekka Sutela, have pointed out, given Russia's level of economic development there must be scope for further growth based predominantly on technological borrowing from more advanced countries, via import, foreign direct investment, and inter-

[8] *Inzhenernaya gazeta*, No. 30/31, 2000, p. 1.

national technological cooperation (i.e., "imitation" may be as important for Russian economic development as innovation—Sutela, 2008).

For Russia, promoting a more effective R&D system and improved innovative performance is becoming a policy priority in the face of competition from other dynamic emerging economies, in particular China and India (Cooper, 2006). Notwithstanding relative strength in human capital and the possession of a large research system in terms of personnel, the Soviet past still hampers Russia in the field of research, development, and innovation, and for science policy specialists and economists this represents an interesting case of the costs of institutional inertia. Almost two decades after the collapse of Communism, Russia still does not possess a well-structured and effectively functioning NIS and the available evidence suggests that it will take a few more years before this situation is rectified. The ambition to switch Russia to an innovative path of development is not yet matched by the possession of an appropriate institutional framework.

References

Amann, R., and Cooper, J. M. (Eds.). (1982). *Industrial Innovation in the Soviet Union*. New Haven, CT: Yale University Press.

Amann, R., Cooper, J. M., and Davies, R. W. (Eds.). (1977). *The Technological Level of Soviet Industry*. New Haven, CT: Yale University Press.

Berliner, J. S. (1976). *The Innovation Decision in Soviet Industry*. Cambridge, MA: MIT Press.

Cooper, J. (2006)."Of BRICs and brains: Comparing Russia with China, India, and other populous emerging economies." *Eurasian Geography and Economics*, **XLVII**(3), May/June, 255–284.

Cooper, J. (2007). "Can Russia compete in the global economy?" *Eurasian Geography and Economics*, **XLVII**(4), September/October, 407–425.

Dezhina, I. (2007). "Gosudarstvennoe regulirovanie nauki v Rossii." Aftoreferat dissertatsii na soiskanie uchenoi stepenii doktora ekonomicheskikh nauk, Moscow, p. 17.

Dezhina, I. G. and Saltykov, B. G. (2005), "The National Innovation System in the making and the development of small business in Russia." *Studies on Russian Economic Development*, **16**(2), 184–190.

Dezhina, I. G., and Kiseleva, V. V. (2008). *Gosudarstvo, nauka i biznes v innovatsionnoi sisteme Rossii*. Moscow: Institute of Economics of the Transition Period.

Durlauf, S. N., and Blume, L. E. (Eds.). (2008). *The New Palgrave Dictionary of Economics*. New York: Palgrave-Macmillan.

EC. (2007). *INNO-Policy Trend Chart: Policy Trends and Appraisal Report: Russia*. Luxembourg: European Commission, Enterprise Directorate-General.

Freeman, C. (1974). *The Economics of Industrial Innovation*. Harmondsworth, U.K.: Penguin Books.

Hanson, P., and Pavitt, K. (1987). *The Comparative Economics of Research, Development and Innovation in East and West: A Survey*. Chur, Switzerland: Harwood Academic.

Ivanov, S. (2006). *Voenni-promyshlennyi kur'er*, No.50, December 27, 2006–January 9, 2007; *www.vpk-news.ru* (las accessed December 27, 2006).

MEDRF. (2008a). *Osnovnye parametry prognoza sotsial'no-ekonomicheskogo razvitiya Rossiiskoi Federatsii na period do 2020–2030 godov—prilozhenie* (proekt, in Russian). Moscow: Ministry of Economic Development of Russian Federation.

MEDRF. (2008b). *Kontseptsiya dolgosrochnogo sotsial'no-ekonomicheskogo razvitiya Rossiiskoi Federatsii* (proekt, in Russian). Moscow: Ministry of Economic Development of Russian Federation.

OECD. (2006). *OECD Economic Surveys: Russian Federation*. Paris: Organization for Economic Cooperation and Development.

Radosevic, S. (2003). "Patterns of preservation, restructuring and survival: Science and technology policy in Russia in post-Soviet Era." *Research Policy*, **32**, 1105–1124.

Sutela, P. (2008). "The four i-words—and a fifth one." *Bank of Finland Institute for Economies in Transition (BOFIT)*, **1**, Focus/Opinion, February 27, 2008; *http://www.bof.fi/bofit*

WB. (2006). *Russian Economic Report* (No. 13, December). Washington, D.C.: World Bank.

WB. (2007). *World Development Indicators*. Washington, D.C.: World Bank.

Zaleski, E., Kozlowski, J. P., Wienert, H., Davies, R. W., Berry, M. J., and Amann, R. (1969). *Science Policy in the USSR*. Paris: Organization for Economic Cooperation and Development.

47

The National Innovation System in Germany

Matthew Allen

University of Manchester, U.K.

Brief history and description of the National Innovation System in Germany

Many aspects of Germany's innovation system have their roots in the 19th and 20th centuries. For instance, characteristics of apprenticeship schemes and universities as well as the origins of important research institutes, such as the Max Planck Society, and large and innovative industrial companies—for example, BASF, Daimler, Hoechst (part of sanofi-aventis since 2004), and Siemens—can be traced back to the first half of the 20th century and, indeed, in many cases to the latter half of the 19th century and beyond. It will not be possible here to describe the various changes that have taken place in the innovation system since then (for more detailed historical accounts see Grupp, Dominguez-Lacasa, and Friedrich-Nishio, 2005; Keck, 1993). This chapter will, instead, focus on more recent changes. This chapter will, of course, draw attention to the historical foundations of those institutions that have been part of the National Innovation System (NIS) in Germany for decades. It will, in addition, cover those aspects of the NIS that have been created or that have come to prominence more recently.

The importance of earlier periods should not, however, be underestimated as Germany's innovative strengths often still lie in those industries that came to prominence in the 19th century. For instance, and as will be shown, Germany continues to have strengths in vehicles, mechanical engineering, and certain electrical and chemical-related industries. However, in other areas, most notably pharmaceuticals, in which early innovations provided the impetus to the establishment of successful companies, Germany has fallen behind similarly advanced economies. The organizations that helped to create these successful companies, such as research institutes, and a strong vocational training and education system, have had to adapt to changing economic and political pressures. This is especially true today as politicians seek to adjust Germany's innovation system to meet heightened competitive pressures in order to ensure the continued strength of that system.

It should be noted at the outset that, because of Germany's Federal political structure, many policies at the national level are influenced by the concerns of the governments of the Federal states, or *Länder*. In addition, the *Länder* can supplement national-level policies with their own at the regional level. This is particularly true in relation to the education system. Therefore, it should be borne in mind that, although what follows depicts the characteristics of the national system, there may be substantial variation between the *Länder* in key areas. Moreover, the still relatively recent unification of Germany in 1990 has meant that many research institutes and industries are, on the whole, less well embedded in eastern Germany than they are in western Germany. There are, however, exceptions as, for example, the *Länder* of Saxony has managed to focus on promoting the establishment of innovation-oriented organizations within its borders. In some instances, it has done this more successfully than many of its peers, regardless of their geographical location.

Current institutional structure and its evolution

There are many important research institutes in Germany. Changes over the last couple of decades, which have arguably accelerated in pace, have sought to streamline the institutional structure in order, first, to promote research excellence; second, to gain the most from those resources that have already been invested; and, finally, to target funds to researchers and institutes that are most likely to produce the desired results. In terms of policy coordination and the channeling of research resources into certain institutes or areas, the Science Council and the German Research Foundation occupy key positions. The most important research institutes are the Hermann von Helmholtz Association of Research Centers, the Max

Planck Society, the Fraunhofer Society, the Leibniz Science Association, and the Centre for Advanced European Studies and Research (CAESAR) Foundation.

The Science Council

The Science Council plays an important role in Germany's innovation system. Established in 1957 following an agreement between the Federal Government and the *Länder* or Federal States, it has a coordination and advisory function with regard to the development of institutions of higher education, science, and research. Some of the most important recommendations that have influenced policy have included the introduction of new degree structures in German universities (see below), the system by which the activities of the Helmholtz Association should be evaluated, as well as proposals regarding the future role of universities in Germany's innovation system. In addition, the Science Council has the task of evaluating research institutes and accrediting, where warranted, newly established private institutes of higher education. It therefore provides guidance within the overall system. Furthermore, it monitors and helps to ensure high research standards within universities, an important element of the NIS.

The German Research Foundation

The German Research Foundation, or *Deutsche Forschungsgemeinschaft* (DFG), is the central, self-governing grant-awarding body in Germany. Its task is to provide financial support for research projects that are carried out, first and foremost, by researchers within higher education. It promotes research into all branches of the sciences and humanities at, primarily, universities and, secondarily, other publicly financed research institutes. The Foundation, furthermore, seeks to facilitate cooperation amongst researchers, to support the development of early-career researchers, and to promote links between German research centers and those abroad.

The German Research Foundation can trace its roots back to 1920 when its predecessor organization, the Emergency Association for German Science, or *Notgemeinschaft der Deutschen Wissenschaft*, was established. Refounded in 1949, this organization was, following a merger with the Research Council in 1951, renamed the German Research Foundation. Its current members include 69 institutions of higher education, 15 non-university research establishments, 7 academies, and 3 industrial associations.

Following a 2002 agreement, 58% of the funds provided by the DFG come from the Federal Government and 42% from the *Länder*. In 2006 the Foundation awarded research grants that totaled €1,588 million. Just over half of this (€817 million) was invested in coordinated programs. The Foundation awarded €568 million under its individual research grants program, and €16 million in prizes. A further €105 million was invested to support early-career researchers. Approximately 3% (€56 million) of the Foundation's budget supported research infrastructure projects. In 2006, €577 million (39%) of the Foundation's research budget for coordinated programs, the individual grants program, and schemes to support early-career researchers was used to fund projects in the life sciences; €388 million (26%) in the natural sciences; €313 million (21%) in engineering; and €211 million (14%) in the humanities and social sciences.

The Hermann von Helmholtz Association of Research Centers

In 2001, 15 research centers that focused on various aspects of biomedicine, science, and technology came together to form the Helmholtz Association. It is the largest research organization in Germany. In 2006 its budget of €2,349 million was largely met by government funds (two-thirds). The Federal Government's share of this funding was 90%; the *Länder* provided the remainder. The approximately one-third of funding for individual Helmholtz Centers that does not come directly from government sources includes support from both the public and private sectors, and the European Union (EU). In 2007 the Association employed approximately 26,500 persons, of whom 8,000 were senior researchers.

Despite focusing on different technological fields, the research centers are united by a commitment to the pursuit of long-term objectives that are of benefit to society. The Association therefore seeks to link research and technology development with measures both to prevent medical ailments and to apply innovations in various areas. In doing so, it identifies and conducts research into highly advanced areas that are of major strategic and programmatic importance to society, science, and industry. Such research often involves major capital expenditure on both equipment and facilities.

The year 2001 marked an important change in the allocation of funds within the Association. Since 2001, finances flow to the Association rather than the individual Centers, as had been the case up until then. The Association then awards funds to research programs that are carried out by the Centers, which are legally independent entities, in cooperation with one another. The change has therefore facilitated a move towards greater collaboration between the Centers. This is intended to enhance the Association's strategic importance and its research performance. The reform is hence intended to promote not only the development of researchers' capabilities, but also innovation. The collaborative research undertaken at the Centers is carried out with other national and international partners.

In order to facilitate knowledge transfer both between the various Helmholtz Association Research Centers and between the Centers and industry, the long-term mission and work priorities of the Centers, drawing on their key strengths, have been streamlined to focus on six major areas. These are energy, the environment, health, key technologies, the structure of matter, and, finally, transport and space. The decisions to fund individual projects are taken by the Federal Government and the *Länder*; their opinions are, however, based on the assessments of project proposals by international groups of experts.

The Max Planck Society for the Advancement of Science

The Max Planck Society for the Advancement of Science was founded in 1948 as an independent, not-for-profit research organization. Although founded after World War II, its roots can be traced back to before World War I, as it is the successor organization to the Kaiser Wilhelm Society, which was established in 1911. The Max Planck Society has grown from 25 research institutes in 1948 to 78 institutes and research centers in Germany in 2007. In addition, it has three overseas institutes and several branches abroad. In total, the Max Planck Society employs approximately 23,400 people. Its 2007 budget of €1,433 million was funded to a large extent (82%) by the Federal German Government and the *Länder*. The remainder was met by donations, externally funded projects, and members' contributions.

The common goal of the various Max Planck Society research institutes in the natural sciences, life sciences, and the humanities is to perform basic research in the interests of the general public. By conducting such research, these institutes seek to pursue innovative research agendas that German universities may lack the resources in terms of both finances and personnel to carry out. Moreover, the Max Planck Society seeks to perform research that is of a more inter-disciplinary nature than that often performed at German universities. This is not to suggest, however, that the activities of the research institutes of the Max Planck Society are wholly divorced from those of German universities. Indeed, in many areas, the Max Planck Society institutes complement research performed elsewhere. Moreover, some institutes make their equipment and facilities available to a wide array of researchers.

The Fraunhofer Society

Founded in 1949, the Fraunhofer Society initially undertook a largely advisory and administrative role to channel public funds to researchers who were conducting research projects that could benefit industry. In the 1970s, its role changed as it began to receive funding from the Federal Government, which matched that from industry, to perform its own research. This emphasis has continued to the present day as the Society aims to undertake applied research that is a direct benefit to private and public enterprises and that also aids society as a whole. It conducts contract research for those in the private (both manufacturing and services) and public sectors. There are 56 Fraunhofer institutes in Germany; they employ approximately 13,000 people. In addition, it has research centers in other European countries, the U.S.A., Asia, and the Middle East. Funding for the Fraunhofer Society reflects its main aim. In 2006 its revenues amounted to €1,186 million. The lion's share of this funding (€787 million, or 66%) came from public sources, which included revenues from the Federal and *Länder* governments, and the Ministry of Defense. Industry provided approximately one-third of the Society's revenues (€399 million).

The Society's remit is to fill a gap in Germany's research structure. For instance, university research (see below for the contribution of the education system to Germany's NIS) often focuses on basic science. It is funded almost entirely from public sources. By contrast, industrial research and development seeks to generate commercial opportunities from research, most of which is financed by private enterprise. Therefore, the Fraunhofer Society, which relies on both public and private funds, aims to pursue not only more application-oriented research than that conducted at universities and other research institutes in Germany, but also studies that are of a more "basic research" nature than those undertaken by commercial organizations. Its links to industry are, as a result of its objectives, stronger than those of other research institutes in Germany.

The Leibniz Science Association

The 84 institutes of the Leibniz Science Association (formerly "Blue List Institutions", which were initiated in 1977) are funded by the Federal Government and the *Länder* as independent research centers. Their two main roles are to conduct their own inquiries and to provide supporting services, which can include advice on knowledge transfer and the use of equipment, to other researchers and research institutes. This latter function means that they play an important part in carrying out university-led research projects. They thus form a cardinal and uniquely close link between the wider research system and university-instigated research. This does not, however, mean that studies conducted within the Leibniz Science Association Centers are only carried out in collaboration with university-based research: support is also provided to researchers based elsewhere, such as those at the Max Planck Society, the Fraunhofer institutes, and, indeed, national and international companies. As the research activities of the Association's Centers lie between basic and applied research, the Association aims to form a link between the two. In order to facilitate innovation as well as cooperation between various research centers, the institutes of the Leibniz Science Association focus on:

- regional collaboration with universities in an attempt to form clusters;

- inter-disciplinary inquiries into areas that are likely to be of increasing prominence in the future (infectious diseases, learning research, environment and climate change research, marine research, and optical technologies); and
- inter-disciplinary working groups.

The focus on inter-disciplinary research reflects the broad focus of the Association's work. The five sections of the Association are:

- humanities and educational research;
- the social sciences and regional infrastructure research;
- the life sciences;
- mathematics, the natural sciences, and engineering; and
- environmental sciences.

Of the Association's total budget of €1,102 million in 2006, €756 million came, in equal measure, from the Federal and *Länder* governments. Other sources of funding include EU research grants, the private sector, and income from licences and services.

The Center for Advanced European Studies and Research (CAESAR) Foundation

The Caesar Foundation is a relatively new addition to Germany's innovation system. Established as part of the Bonn Berlin Compensation Law of 1994, which was designed to offset some of the expected job losses as a result of the decision to move the Federal capital and the majority of ministries and embassies to Berlin, the Foundation conducts basic and application-oriented research in nanotechnology, biotechnology, and neuroscience.

Uniquely amongst the major public research institutes in Germany, the Foundation does not receive an annual grant from either the Federal or *Länder* governments. Instead, it is financed from returns from its endowment fund (totaling €383 million, of which €350 million came from the Federal Government and €33 million from the state of North Rhine Westphalia) and from research conducted on behalf of industry. As a result of its funding structure, the Foundation focuses strongly on linking science and research to innovations that are likely to be commercially viable, to cooperating with the private sector, and to gaining research contracts.

Summary of output trends (R&D expenditure, patents, etc.)

Although research and development (R&D) expenditure in Germany still falls below the target of 3% of gross domestic product (GDP) as outlined by the EU's Lisbon Agenda, it still invests more than many other European countries, such as the U.K. Indeed, as Table 47.1 shows, gross domestic expenditure on R&D (GERD) as a percentage of GDP grew between 1994 and 2004 in Germany, whilst it fell in the U.K. Moreover, in 2004 just over two-thirds of GERD was financed by industry in Germany, whereas in the U.K. under half was funded by the domestic private sector. The German figure is comparable with the share of R&D supported by U.S. and Japanese industry. Between 1994 and 2004 the share of GERD that came from the government fell in Germany from 37.5% to 30.4%. As is discussed in greater detail below, the German Federal Government in association with the *Länder* has announced a number of measures that, in part at least, can be seen as attempts to redress this imbalance, particularly in research areas that may be neglected by the commercial sector (for other measures of technology output trends in Germany see BMBF, 2007).

In terms of patents, Germany has often been seen, as Japan has, as strong in medium to high-tech industries, such as automobiles, mechanical engineering, and certain electricity-related sub-sectors. Indeed, it can be argued that Germany is to a far greater extent reliant on innovation from these sectors than any other country, including Japan (Frietsch, 2007). Put another way, whilst innovation in Germany is undoubtedly strong in medium to high-tech industries, patents in cutting-edge technologies are on the whole weak.

Table 47.2 shows the relative patent advantage (RPA) for selected countries and high-tech sectors. The RPA scores are calculated by comparing the number of patents in a particular sector in a certain country with the total number of patents for all sectors of the economy for that country; this figure is then compared with the same ratio for the world as a whole. Once transformed to make the score symmetrical around zero, the RPA indicates the degree to which a country specializes in patents in the individual sectors. Positive RPA figures show that a country specializes in that sector to a greater extent than the "global average"; negative figures, that a country is less focused on that sector.

Table 47.2 reveals that Germany's innovation system remains strong in areas such as motors and engines; vehicles, vehicle engines, and parts; precision instruments; machine tools; agricultural equipment; and trains and trams. By contrast, Germany is comparatively weak in the following sectors: data-processing equipment; electro-medical equipment; biotech, pharmaceuticals, and medicines; medicaments; radio and television equipment; communications technology; office machines; and optical and photographic equipment. As many of the latter sectors are expected to be amongst the key drivers of economic growth and employment in the future, reforms within Germany's innovation system have been designed to rebalance innovation activities towards these sectors.

Table 47.1. Expenditure on R&D, total, and by funding source.

	1994	*1995*	*1996*	*1997*	*1998*	*1999*	*2000*	*2001*	*2002*	*2003*	*2004*
Gross domestic expenditure on R&D (GERD)—percentage of GDP											
Germany	2.2	2.2	2.2	2.2	2.3	2.4	2.5	2.5	2.5	2.5	2.5
U.K.	2.0	1.9	1.9	1.8	1.8	1.9	1.9	1.8	1.8	1.8	1.7
U.S.	2.4	2.5	2.5	2.6	2.6	2.6	:	:	:	:	:
Japan	2.8	2.9	2.8	2.9	3.0	3.0	:	:	:	:	:
Percentage of GERD financed by industry											
Germany	60.4	60.0	59.6	61.3	62.4	65.4	66.0	65.7	65.5	66.3	66.8
U.K.	50.3	48.2	47.6	49.9	47.6	48.5	48.3	45.6	43.6	42.3	44.2
U.S.	58.5	60.2	62.4	64.0	64.8	66.5	68.6	66.6	64.6	61.4	:
Japan	68.2	67.1	73.4	74.0	72.6	72.2	72.4	73.0	73.9	74.5	:
Percentage of GERD financed by government											
Germany	37.5	37.9	38.1	35.9	34.8	32.1	31.4	31.4	31.6	31.2	30.4
U.K.	32.7	32.8	31.5	30.7	30.6	29.2	30.2	28.8	28.8	31.6	32.8
U.S.	37.0	35.4	33.2	31.5	30.1	28.4	25.8	27.5	30.3	30.4	:
Japan	18.1	19.4	18.7	18.2	19.3	19.6	19.6	18.6	18.2	17.7	:
Percentage of GERD financed from abroad											
Germany	1.7	1.8	2.0	2.4	2.5	2.1	2.1	2.5	2.4	2.3	2.5
U.K.	12.3	14.5	16.3	14.6	16.9	17.3	16.0	19.8	21.6	20.4	17.2
U.S.	:	:	:	:	:	:	:	:	:	:	:
Japan	0.1	0.1	0.1	0.3	0.3	0.4	0.4	0.4	0.4	0.3	:

Source: Eurostat.
Notes: Rounding errors may prevent the relevant column totals summing to 100%.
":" signifies that the data are not available.
Germany's RPA score is calculated thus:

$$\text{RPA}_{kj} = 100 * \tanh \ln\left[\left(P_{kj} \Big/ \sum_j P_{kj}\right) \Big/ \left(\sum_k P_{kj} \Big/ \sum_{kj} P_{kj}\right)\right]$$

where P_{kj} represents the number of patent registrations of country k in sector j. Positive values mean that a sector has a greater weight within the relevant country than it does within the world. Negative values indicate that the country has a below-average specialization in that technological field.

In summary, the RPA scores reveal that Germany's innovation strengths do indeed often lie in those medium to high-tech industries that emerged in the 19th century. This patent specialization frequently leads to comparative advantages for Germany in the same sectors (for information on the sectors of the German economy that have a comparative advantage see Allen, 2006). It should, however, be noted that the RPA scores are aggregated at the sectoral level; this may mean that they mask strengths in sub-sectors within those sectors. For instance, within the biotech, pharmaceuticals, and medicine sector, Germany has a negative RPA score. However, it has been shown that within the biotech sector, those innovation activities—such as platform-enabling biotechnologies that are related to greater levels of organizational complexity and appropriability risks—may be facilitated by Germany's innovation system (Casper and Whitley, 2004):

Technology commercialization initiatives (national level)

In 2001 the Federal Government launched an "action scheme" that was designed to improve technology commercialization initiatives (BMBF/BMWi, 2001). After identifying deficits, the Federal Government launched "offensives" in the following four areas:

- exploitation, which focuses on transferring research results more rapidly into commercial products and services;

Table 47.2. Patent specialization for selected countries and high-tech (cutting-edge and medium to high-tech) sectors, for the period 2002 to 2004.

	Germany	*U.S.*	*Japan*	*EU*
Aircraft and spacecraft	8	28	−82	15
Data-processing equipment	−51	27	7	−31
Electro-medical technology	−41	35	−10	−29
Inorganic chemicals	9	−15	29	−4
Biotech, pharmaceuticals, medicines	−31	35	−34	−15
Engines and motors	41	−46	28	15
Other speciality chemicals	15	10	-3	3
Medicaments	−38	36	−45	−15
Vehicle, vehicle engines, and parts	63	−71	20	36
Organic pest control	−4	28	−47	−9
Measuring equipment	−1	4	14	−6
Warships, weapons, etc.	47	−36	−89	36
Lamps, batteries, etc.	12	−38	54	−13
Radio and television equipment	−81	−32	48	−50
Medical equipment	−45	48	−57	−25
Machines, n.e.s.[a]	43	−49	−30	28
High-value instruments	34	−23	−23	20
Machine tools	45	−41	-5	25
Communications equipment	−39	5	−11	−13
Office machines	−48	−14	75	−60
Power generation and distribution	26	−54	40	3
Climate, filtration, and air conditioning	15	7	−42	12
Dyes and pigments	24	−16	40	−3
Agricultural machinery and tractors	52	−44	−83	44
Polymers	12	1	37	−4
Optical equipment	−50	1	53	−34
Electronics	−38	7	52	−40
Optical and photographic devices	−56	−20	68	−53
Rubber manufactures	−36	−28	58	−27
Organic chemicals	−9	26	−32	−5
Trains and trams	68	−94	−60	49
Pyrotechnics	31	−3	−71	23
Photochemicals	−88	52	48	−73
Radioactive materials, nuclear reactors	12	21	−60	15
Essential oils and surfactants	6	19	−22	19

Source: Frietsch (2007, p. 21).
Note: author's translation.
[a]Not elsewhere stated.

- spinoffs, which intends to increase the number of research-related startups;
- partnerships, which concentrates on improving incentives for collaboration between research institutes and the private sector; and
- competence, which aims to facilitate the use of research results in firms' innovation processes.

As part of the "exploitation offensive", the Federal Government initiated moves to establish patent and exploitation agencies (PVAs), which would be dedicated to patenting innovations that emerge from universities and other publicly-funded research institutes in Germany. As individual universities may lack the resources and expertise to establish their own PVA, each PVA is responsible for the patenting activities of several universities within a region. A further change under the "exploitation offensive" has been to the so-called university teachers' privilege. This privilege, which granted university teachers the sole authority to decide whether or not to patent their inventions, was abolished in February 2002. Now, in general, inventions are owned by the university. If an invention is patented and if that patent generates revenues, the university researcher receives 30% of the gross income. Other measures that are designed to facilitate cooperation between researchers have been undertaken.

In order to support researchers who wish to set up their own business, the Federal Government has increased publicly-available funds for this purpose (see below). Furthermore, the Federal Government is seeking to create a more favorable environment for spinoffs and startups; for instance, by establishing associations that enable experienced entrepreneurs to mentor new ones, by creating awards for entrepreneurs, and by investing in professorial chairs in entrepreneurship at 18 higher education institutes.

In its efforts to encourage greater collaboration between research institutes, the Federal Government has streamlined its funding to them and has acted upon recommendations to draw research centers together into broader associations (see below). It is hoped that organizational barriers that impede cooperation will be reduced as a result. In addition, the Federal Government has changed its funding regulations. Now, if a project receives Federal financial support, it must contain an exploitation plan and that plan must be implemented. As part of its "competence offensive", the Federal Government has reformed the *Meister* qualification in certain vocations and has upgraded vocational training centers so that greater use can be made of information and communication technologies.

National technology policy

Technology policy in Germany has three main strands. The first is the focus set by the government on establishing objectives for researchers in both the public and—through the use of incentives—the private sectors. The second element within technology policy concentrates on improving the research and development "infrastructure" (research institutes and equipment that requires major capital outlays). Finally, technology policy seeks to improve the skills and capabilities of scientists and researchers who either work in Germany or may be about to embark on a career in an innovation field. To be sure, in practice, the distinctions between these three elements are not wholly discrete, and, for instance, funds used to increase Germany's innovation infrastructure have implications for the development of individuals. Despite this, the categories of strategy, infrastructure, and people are useful ones to structure the following portrayal of Germany's technology policy. The latter part of this trio is covered in the education section below.

Innovation strategy

In an attempt to create the conditions in Germany that will enable researchers and organizations to gain leading positions in markets that are both technologically advanced and likely to grow in importance in the future, the Federal Government announced, in August 2006, a High-Tech Strategy for Germany. This is the first time that a national strategy has been developed that spans all ministries in Germany. Its contents have been shaped, during extensive consultation exercises, by representatives from industry and science. It is hoped by the Federal Government that the Strategy will give renewed impetus to its efforts to turn Germany into the country that provides the most conducive conditions in the world for research and innovation. By doing so, the Federal Government aims to be able to attain high rates of environmentally sustainable economic growth.

The High-Tech Strategy for Germany concentrates on altering technology policy in four main ways.

1. The High-Tech Strategy defines goals for 17 technology fields that are likely to be important in terms of both jobs and prosperity in the future. For each one of these 17 areas, a number of initiatives are planned. These initiatives focus on the promotion of research and the framework conditions within which it takes place. In addition, the aim of the Strategy is either to establish new markets for innovative products and services or to increase the economic importance of existing markets. Three of the technology fields that are deemed to be of cardinal importance in the future are health, security, and energy.
2. The Federal Government aims to harness the innovation capabilities of both science and the private sector in its High-Tech Strategy. In order to do so, cooperation and collaborative projects will be promoted to a much greater extent than has previously been the case, a research incentive will be introduced that should encourage publicly-funded institutes to gain more contracts from the private sector, and greater support will be provided to facilitate the formation of clusters in cutting-edge technologies.
3. The High-Tech Strategy aims to enhance efforts to turn research results more rapidly into innovative products, services, and processes. In order to achieve this goal, new measures have been introduced that are designed to simplify the assessment of the economic viability and value of research ideas and results. The High-Tech Strategy, furthermore, supports the efforts of the private sector to establish industry norms and standards more quickly. This, in turn, should increase the competitiveness of commercial organizations. Public

procurement will also be adapted so that the possibility of purchasing innovative products and technologies will be evaluated.

4. In pursuing its High-Tech Strategy, the Federal Government aims to improve the conditions for innovation-oriented startups and SMEs. The Strategy aims to ease the access to markets for company founders, improve the links between commercial entities and research institutes, facilitate the transfer of SMEs' own innovation-focused activities into new products, and simplify various schemes used to aid SMEs. One concrete measure that has been taken to improve the general framework conditions within which startups and SMEs operate is the reform of corporation tax. Other planned measures include the promotion of venture capital in Germany.

It is too soon to judge the effects of the High-Tech Strategy on Germany's innovation system.

Germany as a location for research and innovation

In response to heightened competitive pressures and the desire to improve Germany's innovation capabilities and hence economic performance, the Federal Government has initiated a number of programs and measures that are designed to create the framework and incentives that are needed to promote research and technological advances. The most important individual measures and programs, which form part of the Federal Government's Campaign for Innovation and Growth, that have recently been implemented include

- the Joint Initiative for Research and Innovation;
- the €6,000 million program for research and development (2006–2009);
- the Federal Government and the *Länder*'s Initiative for Excellence to promote science and research at German institutes of higher education (see section on education below).

The Joint Initiative for Research and Innovation

On June 23, 2005 the Federal Government and the *Länder* adopted the Joint Initiative for Research and Innovation. As a result of this Joint Initiative, most of the major science and research institutes mentioned above (the Hermann von Helmholtz Association, the Max Planck Society, the Fraunhofer Society, the Leibniz Science Association, and the German Research Foundation) have since 2006 received greater financial support. This increase is designed to enhance their performance, facilitate stronger and more extensive cooperation, and promote the development of early-career researchers. In addition, the Joint Initiative contains provisions that should enable new and unconventional projects to receive higher levels of funding. A yet further aim of the Joint Initiative is to enhance the formation of clusters that include researchers from both the public and private sectors. Support for women in research and science will also be increased. In order to achieve these objectives, the Federal Government and the *Länder* have decided to increase the budgets of the research institutes by at least 3% (or an additional €150 million) per year until 2010.

This additional funding, which will flow to research and innovation-focused activities in Germany, has been aided by the Lisbon Strategy of the EU. This Strategy was initiated by the European Council in 2000. One of its objectives is to increase the expenditure on research and development amongst the EU member states to 3% of GDP by 2010. In 2007 this figure was 2.5% in Germany.

The €6,000 million program for research and development

In order to stimulate innovation further, the Federal Government intends to invest an additional €6,000 million in research and development projects between 2006 and 2009. This increased funding has been earmarked to promote promising innovations that can be used to increase economic efficiency and hence economic and employment growth. In order to maximize the benefits, cooperation across Federal ministries will be emphasized. One of the reasons for enhancing the amount of public money available to research-related activities in Germany is, the Federal Government would contend, the move away from long-term research, the success and economic benefits of which may be difficult to predict, by private organizations. This has arguably led to an even greater onus on publicly-funded research for projects that have highly uncertain outcomes, but that if successful have clear benefits for society.

The €6,000 million program can be seen as one of the practical consequences of the High-Tech Strategy for Germany. That strategy seeks to prioritize research activities that are largely publicly-funded and attempts to maximize the gains from existing resources. The €6,000 million program focuses on providing additional funding in areas that promise the highest returns in terms of both economic and employment growth. Therefore, by allocating more resources to long-term, yet market-oriented research in the €6,000 million program, the Federal Government aims, partially at least, to anticipate and to be at the forefront of the markets of the future. Moreover, as a supplementary measure to the €6,000 million program, the Federal Government's entire research budget will be pooled and merged with the intention of producing greater benefits for society.

The technologies that have been prioritized under the €6,000 million program are information and communication technology, energy and security-related technology, and biotechnology and nanotechnology. The additional funds will also be used to strengthen research facilities and capabilities in Germany in the areas of medical technologies and pharmaceutical products. Such emphasis aims to build on existing strengths in these areas. For instance, Germany is the second largest exporter in the world of medical equipment; yet, it is relatively weak in terms of patents in this area. Some of the funding will be used to promote research into diagnostic and therapeutic compounds and procedures.

As the Federal Government sees the current quality and quantity of clinical trials as factors that restrict the competitiveness of Germany's pharmaceutical industry, research funding and the framework conditions for clinical research will be improved. Some of the additional funds that are being made available under the €6,000 million program will be used to conduct research into and the development of new sources of energy that are secure and economically and environmentally sound. In total, €2,000 million will be spent in this area between 2006 and 2009.

The Federal Government also plans to enhance the innovation capabilities of small and medium-sized enterprises (SMEs) as part of the €6,000 million program. In 2006, Federal funds to develop innovations and to intensify the exploitation of research findings increased by €62.5 million. Similarly, the Federal Government also launched a High-Tech Startup Promotion Fund in 2006. It is designed to encourage the establishment of new ventures that are based on cutting-edge technologies. The Fund aims to close the perceived gap in Germany for seed financing. Under the scheme, startups can receive a maximum of €1 million in equity financing. Between 2006 and 2009 the Federal Government plans to invest a total of €262 million under this scheme. Other programs that will benefit the innovation capabilities of SMEs are the Program to Promote the Innovation Capabilities of SMEs (PRO INNO II) Fund and the Cooperative Industrial Research Scheme. These programs saw increased funding, respectively, of €19 million and €6 million in 2006.

Relative strengths and weaknesses in technology

As noted above, Germany is relatively strong in medium to high-tech industries, but weak in cutting-edge technologies. Many of the policies adopted by the Federal Government are designed to address this shortcoming.

Funds flow for innovation

As noted elsewhere in this chapter, the majority of the funds for the activities of public research centers come from the Federal or *Länder* governments. However, as Table 47.1 also shows, most R&D expenditure comes from the private sector. It is worth noting here that one of the areas that has been seen as a weakness of Germany's innovation system has been the relative dearth of venture capital, which can cover seed, startup, expansion, replacement, turnaround, and bridge funding. As highlighted above, the Federal Government has announced a number of initiatives that are designed to address this shortcoming. According to the German Private Equity and Venture Capital Association, or *Bundesverband deutscher Kapitalbeteiligungsgesellshaften* (BVK), €50 million was invested as seed funding by venture capitalist investors in 2007. This represented approximately 6% of the €840 million invested by venture capitalists in that year. The majority of the funding went on expansion (c. 50%) and startups (approximately 36% or €300 million) (BVK, 2008). By comparison, according to the British Private Equity and Venture Capital Association (BVCA), £242 million was invested by venture capitalists in 2006 in the U.K. in startups; nearly £3,000 million was used to finance the expansion of companies by venture capitalists in the U.K. in 2006 (BVCA, 2007). Therefore, venture capital in Germany does not appear to be as available for the expansion of existing businesses as it does in the U.K.

In more general terms, it has often been noted that the financing of companies is shifting from a bank-based to a market-based system. This may have ramifications for the types of innovation that companies in Germany are able to carry out successfully. Put simply, it has been argued that banks are able to provide companies with "patient" capital that is focused on long-term returns. In contrast, the financial resources that are provided by markets or institutional investors mean that short-term returns are emphasized. Therefore, the provision of funding by banks may enable companies to carry out activities that they would not be able to do if they were funded largely by equities that are bought and sold by institutional investors who are focused on companies posting good financial returns in the short term. The ability of companies in Germany to adopt a more long-term approach, which can help to incease employees' firm-specific skills, has been said to be an advantage in those industries, such as vehicles and mechanical engineering, that, because they require employees with in-depth skills and knowledge about the firm's routines and products, rely on incremental rather than radical innovations (Hall and Soskice, 2001; see also Vitols, 2003).

Cultural and political drivers

The main cultural and political drivers in Germany's innovation are its education system and its vocational education training programs.

Education system

Education policies lie primarily within the responsibility domain of the *Länder*. This means that there can be substantial variation between the Federal states on, for instance, the length of time spent in different schools and the amount of emphasis on different school subjects. Therefore, the first part of this section provides a broad overview of the education system in Germany; it does not provide details on policies and practices in individual Federal states. It should also be noted that, although the Federal Government cannot direct the *Länder* to pursue certain policies, it can provide incentives to encourage them to implement certain measures.

Until the age of approximately 10, all pupils, regardless of ability, attend the same sort of school. After that age, pupils are streamed, based on intellectual ability, into one of three schools. The choice of schools at this age largely determines the type of education and training that is available to people later in life. For instance, secondary general schools, or *Hauptschulen*, prepare pupils for vocational education and training (VET) (for more on VET see below), whilst pupils at grammar schools, or *Gymnasien*, receive a more academic education and will, if they so wish and pass the relevant examinations, be able to go to university. This is not possible for pupils at a *Hauptschule*. Although pupils can move between the three types of schools, the majority of pupils do not do so (Germany's education system is portrayed in Figure 47.1).

The attainment levels of pupils in German schools were often assumed to be high. The results of the OECD's Program for International Student Assessment (PISA) study in 2000 came therefore as an unwelcome surprise as German pupils fared less well than those in other developed countries. As a result of this "PISA shock",

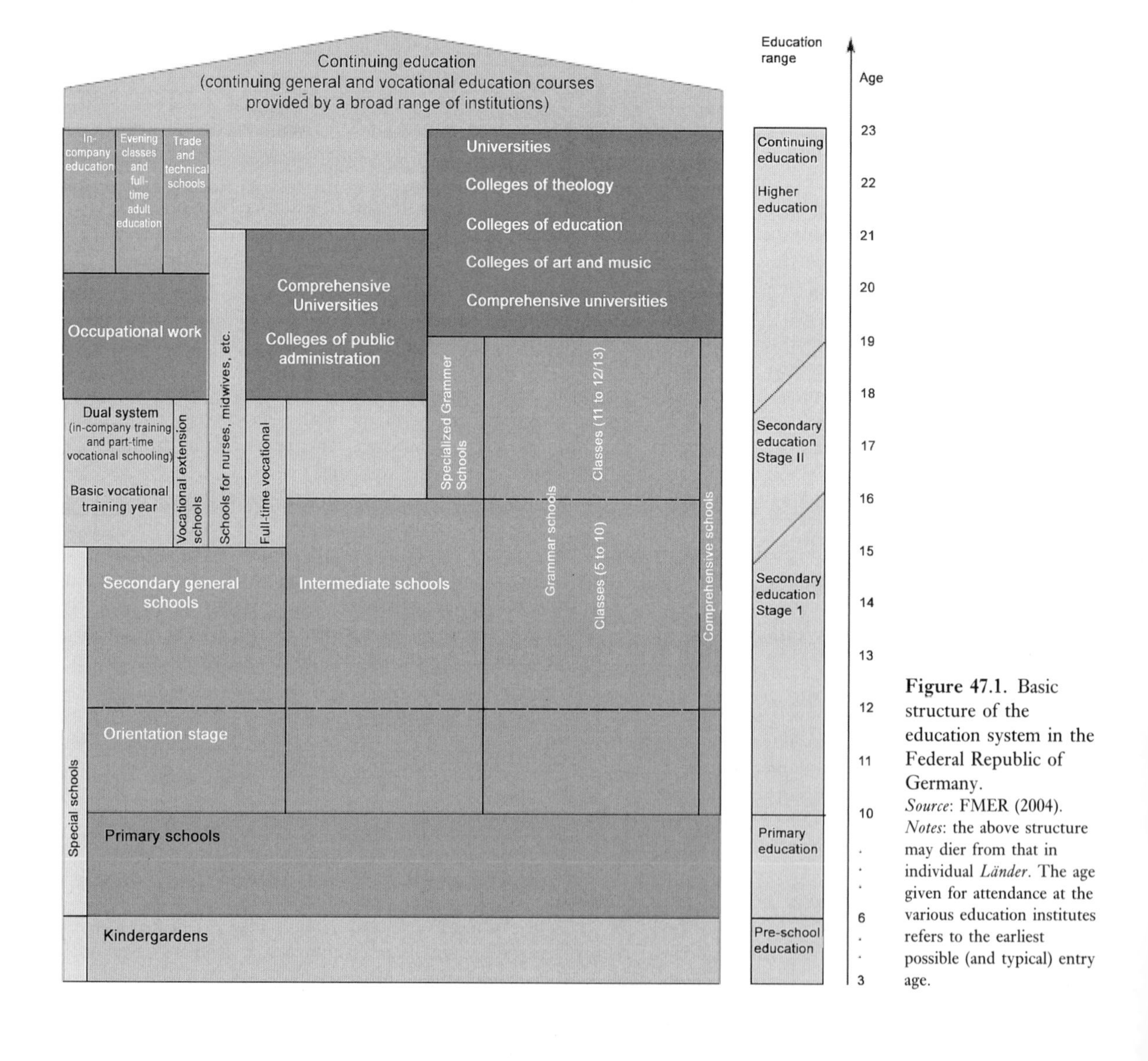

Figure 47.1. Basic structure of the education system in the Federal Republic of Germany.
Source: FMER (2004).
Notes: the above structure may dier from that in individual *Länder*. The age given for attendance at the various education institutes refers to the earliest possible (and typical) entry age.

Table 47.3. PISA mathematics and reading scores.

Mathematics score		*Reading score*			
2006		*2006*		*2000*	
Finland	548	Korea	556	Finland	546
Hong Kong-China	547	Finland	547	Ireland	527
Korea	547	Hong Kong-China	536	Hong Kong-China	525
The Netherlands	531	Canada	527	Korea	525
Japan	523	Sweden	507	Japan	522
Germany	*504*	The Netherlands	507	Sweden	516
Sweden	502	Japan	498	Norway	505
Ireland	501	United Kingdom	495	France	505
OECD average	*498*	*Germany*	*495*	*OECD average*	*498*
France	496	Denmark	494	Denmark	497
United Kingdom	495	*OECD average*	*492*	Switzerland	494
Hungary	491	France	488	Italy	487
Norway	490	Norway	484	*Germany*	*484*
Spain	480	Czech Republic	483	Hungary	480
United States	474	Hungary	482	Poland	479
Italy	462	Italy	469	Greece	474

Source: OECD (2000).

the *Länder* and Federal Government have undertaken steps to improve the performance of schools and hence the ability of pupils. For instance, in order to increase the number of all-day schools, or *Ganztagsschulen*, the Federal Government made funding available for this purpose. Before this measure was introduced, nearly all schools in Germany closed at midday. By the end of 2007, approximately 3,000 schools had benefited from this Federal program. It is hoped that the increased school day length will improve pupils' performance.

Some of the PISA scores in reading and mathematics are provided in Table 47.3 for selected countries. In terms of reading, German pupils performed less well than those in many other countries. Indeed, with an average score of 484, German pupils fell below the average for the OECD as a whole (498), and Germany was ranked in 20th position out of 38 participating OECD and non-OECD countries. It should be noted that many countries, including the U.K. and the U.S., did not receive scores on this measure as too few pupils participated to ensure statistically robust results. By 2006 the performance of pupils in Germany in terms of reading had improved. Within the PISA 2006 study, German pupils fared better than the average for the OECD. In 2006, Germany was ranked joint 15th out of 45 participating OECD and non-OECD countries. In mathematics, pupils in Germany fared better. In 2006, pupils in Germany attained an average score of 504 against an OECD average of 498. This placed Germany 17th out of 40 participating countries.

Table 47.4 shows the annual expenditure on public education institutes per pupil/student compared with *per capita* GDP. In terms of spending on all levels of education, this figure has remained relatively static in Germany, whilst it has grown in other EU member states. This may signify that Germany will need to spend more per pupil if it is to ensure that its workers have higher or comparable skill levels with those in other European countries. (The level of public expenditure on all education as a percentage of GDP fluctuated marginally around 4.6% between 1995 and 2004.) The annual public expenditure on education institutes at all levels masks significant variation within the three different sectors. Whilst Germany spends relatively less per pupil at the primary and secondary levels than the 27 current member states of the EU do, it spends more at the tertiary level. This latter figure declined markedly, however, between 1999 and 2004.

Higher education

In common with other aspects of Germany's innovation system, the origins of the university system can be traced back to the 19th century. There is insufficient space to cover the many changes that have taken place within this system since them. Therefore, this section will concentrate on the major characteristics of the system as well as the important changes that have occurred within the last 5 years. Whilst there have been significant changes of late to the German higher education system, one of the enduring characteristics of it has demonstrated comparatively high numbers of graduates in engineering, manufacturing, and construction. As Table 47.5 shows, as a share of all

Table 47.4. Annual expenditure on public education institutes per pupil/student compared with GDP *per capita*.[a]

	1999	*2000*	*2001*	*2002*	*2003*	*2004*
For all levels of education						
EU (27 countries)	23.2[b]	23.3[b]	24.3[b]	24.2[b]	24.6[b]	26.1[b]
Germany	22.3	21.7	21.6	22.2	22.6	22.3
Primary level of education (ISCED 1)						
EU (27 countries)	16.7	17.1	18.1	18.6	19.2	20.8[b]
Germany	16.3	15.7	16.2	16.6	16.7	16.4
Secondary level of education (ISCED 2–4)						
EU (27 countries)	24.7	24.8	25.7	25.1	25.1	26.4[b]
Germany	19.9	19.3	19.1	19.2	19.2	19.0
Tertiary level of education (ISCED 5–6)						
EU (27 countries)	37	36.6	38	37.2	38.4	39.5[b]
Germany	43.1	41.9	41	41	42.4	40.9

Source: Eurostat.
Notes: [a]Based on full-time equivalents. [b]Eurostat estimate.
The ISCED levels relate to the International Standard Classification of Education (ISCED). Level 1 covers primary education; levels 2 to 4 comprise lower secondary, upper secondary, and post-secondary non-tertiary education; and levels 5 and 6 encompass tertiary education.

Table 47.5. Graduates (ISCED 5–6) by discipline—as a percentage of all disciplines.

	1999		*2005*	
	EU	*Germany*	*EU*	*Germany*
Education and training	10.9[a]	8.7	9.9	7.5
Humanities and art	12.3[a]	10.0	11.4	10.5
Social science, business, and law	32.6[a]	20.7	36.2	24.3
Science, mathematics, and computing	10.2[a]	9.5	10.0	10.9
Engineering, manufacturing, and construction	14.5[a]	17.9	12.7	16.3
Agriculture and veterinary	2.0[a]	2.4	1.6	2.3
Health and welfare	14.1[a]	26.9	14.2	24.2
Services	3.4[a]	3.9	3.9	3.9
Unknown	2.6[a]	0.3	0.5	0.4
Total graduates (ISCED 5–6)—absolute numbers	12,511,189[a]	2,087,044	16,342,307[a]	2,268,741

Source: Eurostat.
Notes: "EU" relates to all current 27 members; rounding and estimation errors may prevent the relevant columns summing to 100%.
[a]Eurostat estimate.

graduates those in engineering, manufacturing, and construction and those in health and welfare comprised a far higher percentage in Germany than they did in the 27 current member states of the EU in 1999 and 2005.

The freedom of research and teaching, which is anchored in Germany's constitution (*Grundgesetz*), has meant that universities, in general, and university professors, in particular, are relatively more autonomous than their counterparts in many other developed countries. Whilst the freedom that university professors have enjoyed may in some respects have enhanced innovation, as it has meant that their choice of research areas has been less directed by a governing body, it can also be seen as hindering flexibility within the innovation system. This lower degree of flexibility may, in turn, hamstring efforts to pursue new research possibilities. In short, research within German universities has tended to be conducted as discrete projects that are controlled by one or a couple of professors who also have administrative responsibilities. This, in turn, may hamper competition within departments, which are led administratively and intellectually by a professor, and lead to lower levels of innovative outcomes. By contrast, where administrative authority is separated from intellectual authority, the power of departmental heads to direct research may be diminished. This may result in a diversity of research goals and approaches (Whitley, 2007). The Initiative for Excellence (see below) may have increased the competition between universities for increased funding, but it may not lead to an increase in the plurality of research projects that are conducted.

In addition to the selection of "elite universities" (see the section on the "Initiative for Excellence" below), one of the main changes in the higher education sector that has recently been implemented is the 2004 reform of the Framework Act for Higher Education. This reform has for the first time given universities in Germany the opportunity to award bachelor's and master's degrees. This reform means that the broad contours of the German higher education system will resemble those in the U.S. and U.K. The Federal Government hopes that this restructuring will boost the numbers of those on business studies and technology-related degree programs. A further change has been the introduction of a new salary structure for those who are appointed as professors for the first time. Such appointees receive a lower basic salary than those who were already professors before the reform. However, the former, unlike the latter, can supplement their basic salary by performance-related payments that are based on research and teaching. It is hoped that this reform will help professors intensify their research activities, which, in theory, will spur innovation. It is too early to judge the effects of these reforms.

Initiative for Excellence

Marking a significant shift in Germany's higher education policy, the Federal Government in cooperation with the *Länder* announced an Initiative for Excellence that, it is hoped, will enable selected universities to increase their expertise and international renown in research areas in which they are already strong. Before the Initiative, all universities were regarded as equal by the Federal Government. This, in turn, reflected a desire that emerged in the 1960s amongst the populace to eschew the idea of elitism within the higher education sector. The Initiative therefore represents an important caesura in policy, as it intends to facilitate the emergence of "elite universities" that are able to conduct more extensive and more advanced research than those not selected. The Initiative can be seen as a response, first, to increased competition in the areas of research and innovation from both developed and developing countries; second, to calls to address the problem of chronic underfunding within tertiary education, and, finally, to concerns about Germany's reputation abroad as a research location.

Within the framework, €1,900 million will be made available to the selected universities between 2007 and 2011. The Federal Government will contribute 75% of this sum; the remainder will come from the *Länder*. In order to select the universities that will receive the additional funding, an exercise to evaluate current research activities as well as the ability to develop the talents of early-career researchers was conducted. That exercise was led by the German Research Association and the Science Council. In October 2006 the first three "elite universities" were announced. They are Munich's Ludwig-Maximilian University, Munich's Technical University (TU), and the University of Karlsruhe. These three universities will receive approximately €120 million in additional funds between 2007 and 2011. In a second round of the Initiative, Berlin's Free University, RWTH Aachen University, and universities in Freiburg, Göttingen, Heidelberg, and Constance were, in October 2007, recognized as further "elite universities". Each of these six universities will receive approximately €100 million in additional funding between 2007 and 2011.

In a further measure to support important research projects and to increase the number of university students, the Federal Government and the *Länder* concluded, in 2007, the Higher Education Pact 2020. It will channel increased funds to selected projects that have gained funding from the German Research Foundation. In 2008, €242 million was available under this scheme. By 2010, €1,270 million will have been invested.

Vocational education and training

Within the area of vocational education and training (VET), the Federal Ministry of Education and Research (BMBF) together with the Federal Institute for Vocational Education and Training play in conjunction with employer and employee representatives key roles in establishing the broad parameters within which employers, training providers, and employees operate. The provision of VET is underpinned by the principle of dual training.

Within Germany, the Federal Ministry of Education and Research (BMBF) is responsible for general policy issues that relate to vocational education and training (VET). As part of its remit, it is legally accountable for the supervision and funding of the Federal Institute for Vocational Education and Training (BIBB). In addition, its tasks include the implementation of measures that are designed to improve the quality of VET. The BMBF does not, however, have the power to recognize individual occupations that require formal training; that responsibility lies with the individual ministries that oversee the relevant occupational area. In practice, this means that the majority of occupations that require formal training are recognized by the Federal Ministry of Economics and Technology. In the mid-2000s, there were 343 recognized training occupations that covered all sectors of the economy. The origins of the current occupation-focused and industry-focused VET system in Germany can be traced back to the 19th century when large companies in industrial sectors established their own training programs. The input of employers in designing training programs has remained a key feature of the system and has ensured the continuing relevance of the skills provided by it.

Founded in 1970, the Federal Institute for Vocational Education and Training (BIBB) is a national and international center of excellence for research not only into initial and continuing VET, but also into the development of VET. Its research, development, and advisory work focus on identifying the future demands that VET is likely to face and the ways in which training can meet those demands. Moreover, it seeks to develop practical solutions for initial and continuing VET. The activities of the BIBB have ramifications for those organizations involved in the development of VET. These include Federal ministries, *Länder* ministries, and peak-level employers' associations and unions. In addition, guidance and recommendations made by the BIBB are directed towards influencing the activities of universities, colleges of further education, and vocational training schools. The BIBB also seeks to shape the training activities within firms, as its activities have implications for those people within organizations who have an influence over training, such as personnel managers and works councillors. In many of its activities and recommendations, however, the BIBB consults with and is influenced by employers' associations and unions.

Dual training

The principle of dual training underpins many VET measures. This means that two partners share the responsibility for providing VET to trainees. In the first instance, a company concludes a training contract with an apprentice. This contract includes details of the training measures that the apprentice will undertake. Much of this learning is performed within the company. The apprentice usually spends 3 or 4 days a week at the firm. For the remainder of the working week, apprentices attend vocational training schools, the other partner in the dual-training program. The material studied there is of both a theoretical and practical nature; it is designed to support the primarily practice-oriented knowledge acquired within the company. The dual-training system therefore promotes the provision of firm-specific and industry-specific skills.

The continuing relevance of the skills acquired within the dual-training system to companies is maintained by the important contribution of employers' representatives and unions in the design of and influence over changes to VET schemes. The actions of the Federal Government, the *Länder*, employers' associations, and unions are governed by the provisions of the Vocational Training Act, which was amended most recently in 2005. The Chambers of Industry and Commerce, which cover companies within a particular sector, perform advisory and monitoring functions for individual training contracts. They also verify that companies and instructors involved in VET have the necessary skills to do so.

In terms of the NIS, VET schemes in Germany provide firms and industries with the workers with the skills that can help to maintain competitive advantages (Culpepper and Finegold, 1999; Hall and Soskice, 2001; Thelen, 2004; Whitley, 1999, 2007). The role of state agencies, employers' associations, and unions in organizing and controlling the vocational skills system leads to the development of highly valued, standardized skills for a large section of the workforce. This therefore enables workers to make a contribution to the innovation capabilities of companies. This is likely to be especially true in situations in which lengthy job tenures, which are in part encouraged by Germany's system of industry-wide collective wage agreements, promote firm-specific skills in addition to the industry-specific skills already acquired (Whitley, 2007). The structure of VET in Germany is thought to be an important source of competitive and hence comparative advantage in certain sectors, such as vehicles and mechanical engineering, as workers' skills are well suited to the innovation patterns in those industries.

Most school leavers (approximately 60%) embark upon a dual-training course. Others attend full-time vocational schools, for which the *Länder* are solely responsible. Such schools provide training in the health and laboratory sectors. Attendance at these schools does not preclude training placements within companies. The companies that provide the training contracts contribute the largest share to the financing of the dual-training system. In 2007, companies are thought to have spent nearly €15,000 million on dual training. The *Länder* spend close to €3,000 million on part-time vocational schools.

Conclusion

Germany's innovation system should be seen as a relatively coherent set of policies and practices that support the emergence of certain organizational capabilities. Those capabilities are particularly important in the medium to high-tech industries. It is, however, the case that the

Federal German Government is aware that economic growth and employment are overly reliant on those sectors. For that reason the government has undertaken several steps to promote innovation and growth in other economic sectors. Indeed, by creating "elite universities" the Federal Government has shown that it is willing to break long-held taboos. The interlocking nature of the innovation system means, however, that other aspects of the framework that remain unchanged are likely to continue to exert an influence over the types of capabilities that firms are able to create. This is not to suggest though that the reforms are likely to be futile. It does, instead, indicate that competences in German organization may be suited to certain forms of innovation activities than others. These competences are likely to be prerequisites in many sectors and sub-sectors of the economy, including those that are based on cutting-edge technologies.

References

Allen, M. M. C. (2006). *The Varieties of Capitalism Paradigm: Explaining Germany's Comparative Advantage?* London: Palgrave Macmillan.

BMBF (2007). *Report on the Technological Performance of Germany*; *http://www.bmbf.de/pub/tlf_2007_summary.pdf* (last accessed January 28, 2008).

BMBF/BMWi. (2001). *Knowledge Creates Markets Action Scheme of the German Government*; *http://bundesforschungsministerin.net/pub/wsm_englisch.pdf* (last accessed January 11, 2008).

BVCA (2007). *Report on Investment Activity 2006*; *http://www.bvca.co.uk* (last accessed February 3, 2008).

BVK (2008). *BVK Statistik: Das Jahr 2007 in Zahlen*; *http://www.bvk-ev.de/media/file/163.BVK_Jahresstatistik_2007_final_210208.pdf* (last accessed February 3, 2008).

Casper, S., and Whitley, R. (2004). "Managing competences in entrepreneurial technology firms: A comparative institutional analysis of Germany, Sweden and the UK." *Research Policy*, **33**, 89–106.

Culpepper, P. D., and Finegold, D. (Eds). (1999). *The German Skills Machine: Sustaining Comparative Advantage in a Global Economy*. Oxford, U.K.: Berghahn Books.

Frietsch, R. (2007). "Patente in Europa und der Triade: Strukturen und deren Veränderung." *Fraunhofer Institut für System- und Innovationsforschung Studien zum deutschen Innovationssystem*, **9**.

Grupp, H., Dominguez-Lacasa, I., and Friedrich-Nishio, M. (2005). "The National German Innovation System: Its development in different governmental and territorial structures." In: K. Dopfer (Ed.), *Economics, Evolution and the State: The Governance of Complexity* (pp. 239–273). Cheltenham, U.K.: Elgar.

Hall, P. A., and Soskice, D. (2001). "Introduction." In: P. A. Hall and D. Soskice (Eds.), *Varieties of Capitalism: The Institutional Foundations of Comparative Advantage* (pp. 1-68). Oxford: Oxford University Press.

Keck, O. (1993). "The national system for technical innovation in Germany." In: R. R. Nelson (Ed.), *National Innovation Systems: A Comparative Analysis* (pp. 115–157). New York: Oxford University Press.

OECD. (2000). *Program for International Student Assessment*. Paris: Organization for Economic Cooperation and Development.

Thelen, K. (2004). *How Institutions Evolve: The Political Economy of Skills in Germany, Britain, the United States and Japan*. Cambridge, U.K.: Cambridge University Press.

Vitols, S. (2003). "From banks to markets: The political economy of liberalization of the German and Japanese financial systems." In: K. Yamamura and W. Streeck (Eds.), *The End of Diversity? Prospects of German and Japanese Capitalism*. Ithaca, NY: Cornell University Press.

Whitley, R. (1999). *Divergent Capitalisms: The Social Structuring and Change of Business Systems*. Oxford, U.K.: Oxford University Press.

Whitley, R. (2007). *Business Systems and Organizational Capabilities: The Institutional Structuring of Competitive Competences*. Oxford, U.K.: Oxford University Press.

48

The National Innovation System of Finland

Raimo Lovio and Liisa Välikangas

Helsinki School of Economics

Introduction[1]

Finland, a Nordic republic with 5.2 million inhabitants bordering Sweden, Norway, and Russia, has been a member of the European Community since 1995. Finnish economic performance has been lauded as highly competitive (WEF, 2006a). This has been attributed to respect for the law (TI, 2006), competent government (Lopez-Claros, cited in WEF, 2006b), effective collaboration between public institutions and private companies (EU/CIS, 2007), and a high level of general education (Dahlman, Routti, and Ylä-Anttila, 2006). Equality is a closely held value in the Finnish nation-state. Women have played an important role in public life since emancipation in 1906; 66% of 15 to 64-year-old women work full-time outside the home (as compared with 56% in other EU countries). Social equality is sought through everyone's right to education and health care[2] together with redistributive wealth policies.[3]

Finns are considered pragmatic (Lilja, 2005, p. 79), hard-working (work is seen as intrinsically important although Finns also like their long vacations), determined, and ethical. Innovation and creativity are regarded as key qualities to future success. Design is valued: its manifestations can be seen in Alvar Aalto's functionalistic architecture or in Marimekko's colorful textiles. Finland has produced a number of world-leading musicians, including orchestra conductors such as Esa-Pekka Salonen and rock bands such as HIM. Also, many Finns are seen as "keen on technological modernism and front-line experimentation" (Lilja, 2005, p. 79). From an early age, Finns are some of the most devoted users of cellphones in the world. Broadband Internet connections are common. Linus Torvalds, the Finnish founder of the Linux Operating System, is well-known at least among software engineers and open-source enthusiasts.

Until the 1970s the Finnish economy was forest-based although forest-related metal and engineering have also been strong industries (see Figure 48.1). The dominant pulp and paper industry guided the economic policy for decades. Its cycles led to a series of devaluations between 1945 and 1991 to increase price competitiveness. In the 1970s and 1980s government policy intent emerged to diversify the industrial base of the Finnish economy, strengthening electronics and other high-tech products. This resulted in Tekes, the national technology development agency, being founded with a remit to increase public R&D funding in 1983. In the 1990s the economic landscape was dominated by the rise of Nokia, today the world's largest cellphone manufacturer, which gave birth to the thriving information and communications industry (ICT). The ICT industry cluster includes 6,000 firms, 300 of which are first-tier suppliers to Nokia (Dahlman, Routti, and Ylä-Anttila, 2006, p. 16).

However, the seeds of this spectacular transformation from a relatively inwardly-closed,[4] resource-based economy to an open, knowledge-intensive society were sown long before the emergence of cellphones as a "killer

[1] This chapter draws heavily on recent publications edited by Pekka Ylä-Anttila and some of the figures are reprinted or updated from them (Dahlman, Routti, and Ylä-Anttila, 2006; Ylä-Anttila and Palmberg, 2007).

[2] Finland has long had one of the lowest child mortality rates in the world: in 2006 there were 2.8 infant mortalities (undera year old) per 1,000 live births (according to Statistics Finland; *http://www.stat.fi/til/kuol/tau.html*).

[3] The Gini Index, a measure of (in)equality, for Finland is 26.9 according to UNDP (2006). Marginal tax rates for individuals can be up to 60%. However, there is no wealth tax in Finland (abolished in 2006). Incentives for early retirement also seem to even out income distribution (DeNardi, Ren, and Wei; *http://www.chicagofed.org/publications/economicperspectives/2000/2qep1.pdf*).

[4] Restrictions on foreign ownership of Finnish corporations were eliminated in 1993. In 2006, 51% of the Helsinki Stock Market was owned by foreign investors. Nokia is currently almost 90% foreign-owned (*http://www.porssisaatio.fi*). Outwardly, the export of goods, labor, and capital was not restricted even before deregulation in the early 1990s.

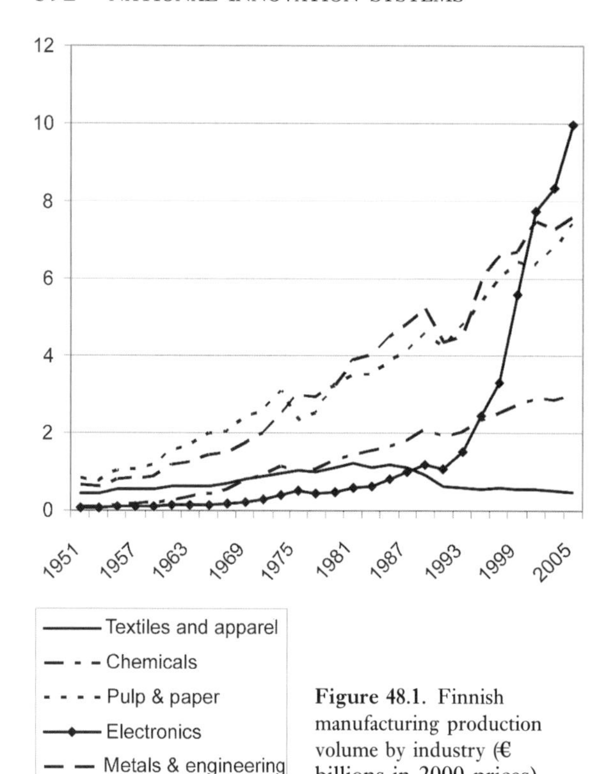

Figure 48.1. Finnish manufacturing production volume by industry (€ billions in 2000 prices).

application". The economic transformation was fueled by financial deregulation, by a deep recession (unemployment in 1993 was 17% whereas in 1991 it was 3%), and by the collapse of Finland's trade with the falling Soviet Union in the early 1990s.[5] This set the stage for a national innovation system as an explicit government policy approach.

In this chapter we discuss the emergence of the current National Innovation System in Finland from the perspective of its supporting institutions, initiatives, and its comparatively good performance. We conclude with a view on future innovation challenges.

National technology policy: emergence

The origins of a national innovation policy in Finland can be traced back to the Prime Minister's Science & Technology Policy Council which in its review in 1990 made R&D investment a national priority and explicitly called for a national innovation system to be developed (see Table 48.1). The call led to an absolute increase in public R&D funding that was sustained even in the deep recession of the early 1990s. In 1996 the government lifted public R&D funding by 25% through Tekes, the National Technology Agency, now renamed the National Technology & Innovation Agency. Further, MEE (1993) suggested the formation of Porterian industry clusters: information and communications as well as health care were proposed as growth clusters whereas forestry was seen as a stable or more mature industry together with basic metals, construction, energy and transport. These continued to be important yet received less focus in public R&D funding.

As opposed to the 1970s and 1980s when efforts were made to "pick the winners", technology and industrial policy in the 1990s moved to improve the overall operating conditions of business enterprises. Policies supporting chosen technologies turned out to be difficult as fast technological advances and foreign direct investment made the Finnish business environment less predictable. Recently, attention has focused on the globalization of Finnish firms and strategic renewal of forestry-based industries. There is increasing competition for public R&D funding, and the economic utilization of academic research has become a criterion for research funding.

Social innovation is called for, and business management is presenting a priority area for innovation in one of the publicly-funded R&D programs (Tekes' research program on innovative business competence and management). Thus, the focus on technology is increasingly yielding to management and organizational innovation. The role of social capital and networking, for example, feature prominently in public discussions and research interests: the VTT Technical Research Center of Finland founded by the state, for example, has a research program on social media. Recently, the application of open source–like innovation methods has received a lot of attention as many Finnish firms seek to invest in open innovation platforms (e.g., Nokia's research centers seek to foster open local collaboration; *http://research.nokia.com/*).

Institutions

The Finnish National Innovation System includes a number of relatively autonomous actors (see Figure 48.2).

The Science & Technology Policy Council develops guidelines for innovation policy under parliamentary supervision.

The Academy of Finland (*www.aka.fi*) supports scientific research through its research councils, which prioritize and coordinate research in biosciences, culture and society, natural sciences and technology, and health care. The Academy also appoints research professors and fellows dedicated to their area of research for a limited period and selects world-class research units for special funding (see Table 48.2). It funds major research programs, such as the 1998–2003 technology convergence research program that aimed to strengthen Finnish electronics and telecommunications research internationally with an investment of

[5] Exports to the Soviet Union constituted about 20% of Finland's outward trade at the time of the collapse.

Table 48.1. The emergence of a national innovation system as government policy.

1970s and 1980s	Technology development as a government priority in the 1970s and 1980s to diversify the industrial base.
	The foundation of Tekes, the national technology agency, in 1983 to grow public funding of R&D. Renamed the National Technology & Innovation Agency in 2007.
	Foundation of the Prime Minister's Science & Technology Council in 1987 to provide policy guidance.
1990s	Recommendation for a national innovation system approach in 1990 as a science and technology policy framework (by the Science & Technology Council)
	Adaptation of the Porterian industry clusters approach as an economic and industrial policy framework in 1993 (by the Ministry of Trade & Industry).
	Internationalization of the Finnish economy since the early 1990s: ● financial deregulation in 1993; ● EU membership in 1995; ● growing foreign direct investments by Finnish corporations; ● becoming a member of the European Monetary Union and the introduction of the euro as the currency in 1999.
	25% increase in public funding of R&D in 1996 (through Tekes).
	Focus on regional development in the late-1990s through access to EU structural funds, giving regions a strong incentive to formulate industrial policy strategies of their own.
	The emergence of the strong Nokia-driven Internet and telecommunications industry in the late-1990s.

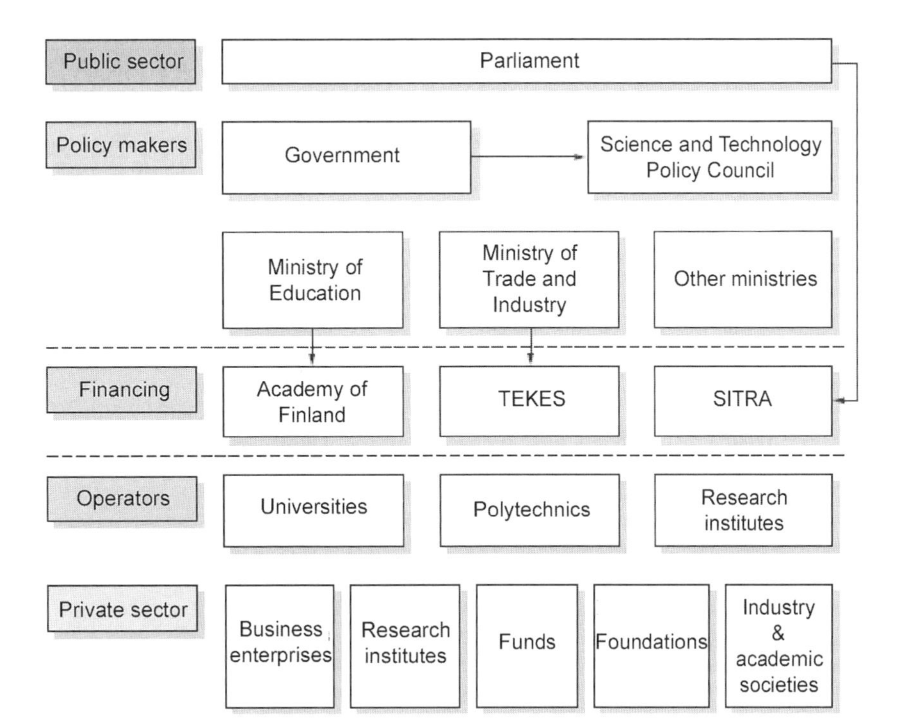

Figure 48.2. The Finnish Innovation System.

Table 48.2. World-class research units/programs as nominated by the Academy of Finland, 2002–2007.

- Applied Microbiology Research Unit
- Bio- and Nanopolymer Research Group
- Center for Environmental Health Risk Assessment
- Center of Excellence for Research in Cardiovascular Diseases and Type 2 Diabetes
- Center of Population Genetic Analyses
- Developmental Biology Research Program
- Finnish Research Unit for Mitochondrial Biogenesis and Disease (FinMIT)
- Formal Methods in Programming
- From Data to Knowledge Research Unit
- Helsinki Brain Research Center (HBRC)
- History of Mind Research Unit
- Research Program on Male Reproductive Health
- Research Unit of Geometric Analysis and Mathematical Physics
- Research Unit on Economic Structures and Growth
- Research Unit on Physics, Chemistry, and Biology of Atmospheric Composition and Climate Change
- Radios and Wireless Research (SMARAD)

Source: Academy of Finland (*www.aka.fi*).

€77.7 million, and participates in building the European Research Area of joint research projects across countries. Researcher mobility is one focus area: the Academy seeks to foster international collaboration by supporting Finnish researchers abroad and internationally acclaimed scholars in research sabbaticals in Finland. The Academy's total budget is €250 million annually.

Tekes, the Technology & Innovation Agency under the Ministry of Trade & Industry (renamed Ministry of Employment & the Economy in 2008), is the major funding arm for public R&D through its technology development programs (*http://www.tekes.fi/eng*). It has a budget of €500 million annually and a staff of 300 people. Tekes typically funds collaborative research projects between research institutions and private corporations (some 2,000 projects every year). Tekes is the largest public-funding source for applied research in Finland. Tekes maintains an international presence through its offices in Europe, the U.S., China, and Japan.[6]

Sitra (*http://www.sitra.fi/en/About+Sitra/sitra.htm*), a publicly-endowed fund established in 1967 to commemorate 50 years of Finnish independence with an annual budget of €40 million, supports the exploration of potentially important new areas to the Finnish economy and acts as an early-stage venture capitalist. Sitra also hosts forums for public discussion of innovation policy and facilitates the diffusion of innovation ideas throughout society.

The state's VTT Technical Research Center of Finland (*http://www.vtt.fi/?lang=en*) has 2,700 researchers and an annual budget of more than €200 million of which more than half is commercial income, the rest is funded by the Ministry of Trade & Industry under which it operates as a non-profit research organization. Its areas of expertise range from Internet and communication technologies, electronics, machinery, biotechnology, paper and pulp, energy through to building.

Technology and science parks

Worthy of mention are Finland's technology and science parks as platforms for public–private interchange. Perhaps the most well-known is Otaniemi's connection to the University of Technology in Helsinki. There are 26 parks connected to universities across Finland, hosting 2,400 enterprises and related organizations in the fields of ICT, health care and medical technology, biotechnology, environmental and food technology, materials research, and digital media (see *http://www.tekes.fi/english*). Support programs for business networking and development are funded by the Finnish Government, Tekes, and the European Union.

Educational system

Finland has a highly lauded public education system that has received accolades for student performance and teacher qualifications. For example, OECD's Program for International Student Assessment (PISA) ranked Finland as the highest performing country on science

[6] Tekes also funds the R&D of small and medium-sized (SME) companies. A significant number of these companies employ fewer than 10 people. In its SME enterprise projects Tekes seeks to share risk so that its funding provides opportunities for greater innovation. Also, Sitra in its venture capital capacity funds startups. It is estimated that there are 800–1,000 startup companies seeking venture capital every year (Maula *et al.*, 2006).

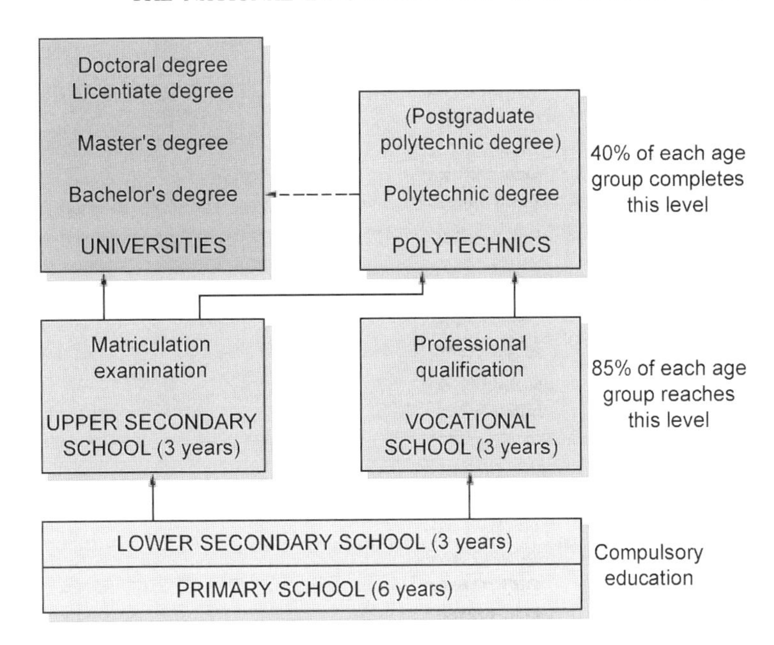

Figure 48.3. The Finnish education system.

scale in 2006. The education system is often viewed as part of the Finnish Innovation System due to the high skills required by advanced industries. The system is founded on the principle of equal and free access to education; nevertheless, students must pass entrance examinations to get into a university. Forty percent of each age group gain university degrees. Alternative routes to a profession exist through vocation schools (see Figure 48.3 for the structure of the education system). Note that Finland occupies third place in the number of graduate students in mathematical and scientific disciplines in the world (after Korea and Germany) (Ylä-Anttila and Palmberg, 2007, p. 170; Koski *et al.*, 2006, p. 61). Finns are also avid readers of books and newspapers.

Finland's 20 universities are publicly-funded and state-governed.[7] This situation is currently under review as the national innovation strategy calls for the concentration and autonomous governance of resources for world-class research. The Helsinki School of Economics, the School for Art & Design, and the University of Technology will be merged into a cross-disciplinary, autonomous research and teaching institution in the next 2 years.

Toward research corporations

To reach the next level of innovation investment, the government is currently forming research corporations with both private companies and public research organizations as investors in key industrial sectors including forestry, metal and engineering, information and telecommunications, energy and environment, and health and welfare. These new research corporations have the autonomy, together with their investors, to direct research funding to areas deemed to have future growth potential.

Funding

By means of the above actors of the innovation policy and in its funding of research in public universities and research institutions, the Finnish Government, to motivate basic and applied research, has opted for direct R&D subsidies instead of tax incentives (tax incentives were tried out but abandoned in the 1987 tax reform) amounting to, in total, €1.6 billion in 2005 (see Table 48.3). In 2004 total R&D expenditure in Finland, including private investment, was about €5 billion, amounting to one of the highest ratios to GDP (3.5%) in the world (Koski *et al.*, 2006, p. 46). The R&D expenditure of leading companies in Finland is presented in Table 48.4. Finland participated in 1,021 EU-funded research projects during 2002—2006, with VTT, Helsinki University, the University of Technology, and Nokia as leading partners. Altogether, R&D funding originating from the EU amounted to €345 million (Tekes, 2006, p. 23) and focused on ICT, material sciences, industry automation, and health care.

A characteristic of the Finnish Innovation System is its decentralized nature (coordinated through the institutions discussed above), frequent interactions among decision makers (such as study trips and discussion forums where political and business leaders exchange views) and a systemic view of innovation policy that acknowledges the existence of frequent innovation spill-overs from one

[7] There are also 31 high-level professional schools that focus on the application of science and technology.

Table 48.3. Government research funding in 1991–2005 in Finland (€ million, current prices).

	1991	*1995*	*1996*	*2000*	*2005*
Academy of Finland	75.6	77.1	84.4	153.8	223.5
Tekes	156.5	243.9	246.2	390.8	448.4
Subtotal	*232.1*	*321.0*	*330.6*	*544.6*	*671.9*
Universities	226.3	220.4	258.6	346.4	416.7
Research institutions	209.9	194.6	196.1	215.8	259.4
Other research funding	131.5	158.0	153.4	189.1	248.7
Total	*799.7*	*894.0*	*938.8*	*1,295.9*	*1,596.7*

Source: STPC (2006).

Table 48.4. Leading Finnish corporations as measured by international turnover in 2006 by industry.

Corporation	*Main business areas*	*Staff in 2006 (% abroad)*	*R&D investments in 2006* (€ million)
Internet and telecommunications (ICT)			
Nokia	Cellphones and networks (global leader)	68,483 (65)	3,897
Elcoteq	Electronics manufacturing services (EMS)	23,292 (97)	7
TietoEnator	Information systems	14,597 (58)	70
Pulp and paper			
Stora Enso	Paper and board (biggest in the world)	43,887 (71)	79
UPM	Paper and board (third biggest in the word)	28,704 (48)	44
Metsäliitto	Forest products (fourth biggest in Europe)	25,007 (65)	30
Huhtamäki	Packaging	14,792 (94)	19
Ahlstrom	Special paper	5,677 (87)	25
Myllykoski	Printing paper	3,539 (77)	5
Metal and engineering			
Kone	Elevators (fourth biggest in the world)	29,321 (94)	50
Metso	Paper and mineral machines (global leader)	25,678 (64)	109
Wärtsila	Ship and power plant machinery	14,346 (80)	85
Rautaruukki	Metal-based components and systems (leader in Nordic markets)	13,303 (46)	22
Cargotec	Cargo-handling solutions	8,516 (83)	31
Outokumpu	Stainless steel (global leader)	8,159 (66)	17
Others			
YIT	Construction (leader in Nordic markets)	22,311 (49)	21
SanomaWSOY	Media (leader in Nordic markets)	18,929 (63)	11
Kemira	Special chemicals	9,237 (67)	51
Fortum	Energy (leader in Nordic markets)	8,134 (63)	17
Amersport	Equipment (global leader)	6,553 (94)	59
Uponor	Plumbing and heating systems	4,325 (89)	17
Orion	Pharmaceuticals	3.061 (11)	84

Source: Lovio (2007).

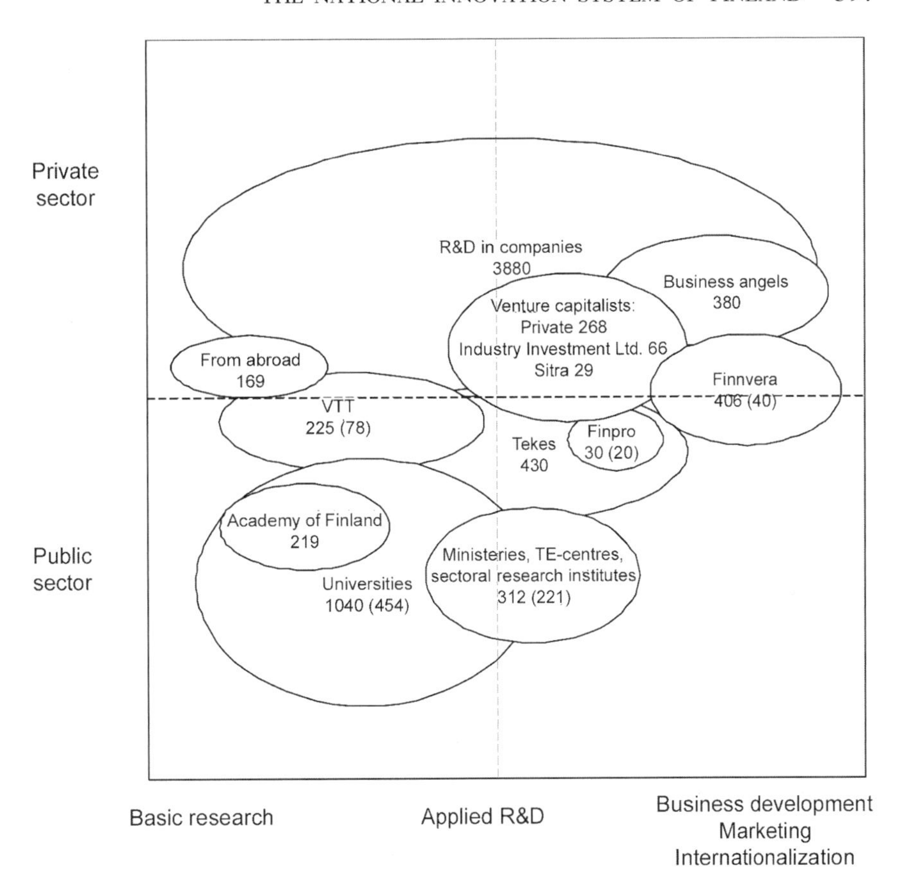

Figure 48.4. Innovation environment in Finland.

domain to another. The aim is to promote innovation "industry pull" (Ylä-Anttila and Palmberg, 2007) rather than "technology push" (i.e., fund research that has industry appeal) by finding an industry partner for state-funded research projects. An important criterion for Tekes funding is collaboration among firms and between firms and research organizations: Finland has the highest share of collaborative research projects in the European Union (EU/CIS, 2007; Koski *et al.*, 2006, p. 50). Further, another European survey found large firms in Finland are actively engaging in innovative activities and benefiting from public R&D funding (Koski *et al.*, 2006, p. 51). For overall innovation funding in Finland, see Figure 48.4.

The government plays an important role in the Finnish economy. The public sector creates more than 50% of GDP. Nevertheless, this does not seem to have crowded out private activity in the innovation markets. While the government has continued to invest in R&D, private investment has grown even more such that the government's relative contribution has decreased over time, even though it has increased in absolute terms. Nokia has played a major part: its R&D investment has amounted to one-third of national R&D expenditure in the last few years. Nokia is also an example of public–private crossover: in 1980, Tekes funded more than 25% of Nokia's R&D (Ali-Yrkkö and Hermans, 2002, p. 8). Nokia also gained access to the Finnish education system for the supply of qualified engineers[8] and benefited from participation in free telecommunications markets.

Output trends and performance compared internationally

Based on patent activity, the government's R&D policy appears successful. Finland is one of the leaders in terms of patenting activity *per capita* according to the European Patent Office and ranks fourth in the world in terms of patents granted in the U.S. Patent & Trademark Office. The propensity to patent in the Internet and telecommunications industry contributes to this—one-third of all patents submitted to the European Patent Office originate from the electronics industry where Finland is the leader among EU countries. However, patenting activity in-

[8] Remarkably, the public education system responded quickly to the need to educate more engineers for the growing ICT industry in the 1990s. There was an increase of 126% over 5 years of student admissions to ICT programs at universities between 1993 and 1998 (Koski *et al.*, 2006, p. 62).

Table 48.5. Forward-looking, ongoing technology development programs by Tekes.

Name	*Description*
ClimBus	Business opportunities in the mitigation of climate change 2004–2008
COMBIO	Commercialization of biomaterials 2003–2007
DENSY	Distributed energy systems 2003–2007
Drug 2000	Biomedicine, drug development and pharmaceutical technology 2001–2006
FinNano	Nanotechnology program 2005–2010
FinnWell	Future health care 2004–2009
	Fuel cell 2007–2013
GIGA	Converging networks 2005–2010
Liito	Innovative business competence and management 2006–2010
MASINA	Technology program for mechanical engineering 2002–2007
MASI	Modeling and simulation 2005-2009
NORDITE	Nordic collaboration project in the ICT industry 2005–2010
NewPro	Advanced metals technology—new products 2004–2009
Sara	Value networks in construction 2003–2007
NewPro-Serve	Innovative services 2006–2010
SISU 2010	Innovative manufacture 2005–2009
SymBio	Industrial biotechnology 2006–2011
	Tourism and leisure services 2006–2009
Ubicom	Embedded systems 2007–2013
VAMOS	Value-added mobile solutions 2005–2010
Verso	Vertical software solutions 2006–2010

Source: Tekes (2006; *www.tekes.fi*).

creased in the 1990s in other industries such as pulp and paper as awareness of the importance of intellectual property grew as a source of competitive positioning (Koski *et al.*, 2006, pp. 51–52). As part of the European Union, Finland is governed by the EU's intellectual property rules and regulations.

Perhaps another sign of R&D effectiveness, the export of electronics and high-tech products—with R&D investment exceeding 4% of company turnover—is significant, amounting to 18% in 2006 (Statistics Finland).

The European Innovation Scoreboard (2006) ranks Finland as one of the innovation leaders (published by the Directorate for General Enterprise and Industry of the European Union). The WEF's *Global Competitiveness Index* puts Finland in second position (after Switzerland) in its 2006–2007 ranking, continuing its good performance throughout the 2000s. Finland is praised for the ethical behavior of its firms, its technological readiness, the quality of its education system, and the effectiveness of its antitrust policy. Intellectual property protection is also considered good. The most problematic factors for business are judged to be high tax rates[9] and restrictive labor regulations. The IMD's *World Competitiveness Report* ranks Finland number 17 in 2007 (down from 10 the year before). While structural unemployment continues to be an issue, Finland has successfully created new jobs despite globalization and the related trend toward outsourcing. In 2007 the number of people employed (2.5 million) is a record only equaled in 1990, the historical lead year.

Finnish companies are world leaders in cellphones and networks (Nokia) and many other information and communications technology-related products, forestry (Stora-Enso, UPM-Kymmene, Metsäliitto), elevators (Kone), paper and mineral machines (Metso), Outokumpu (stainless steel) and Amer (sports equipment). Leaders in Nordic markets include Rautaruukki (metal-based products), YIT (construction), SanomaWSOY (media and publishing), and Fortum (energy). For the major companies and leading industries see Table 48.4.

Looking to the future

The private service sector employs fewer people in Finland (about 35%) than in other European countries where on average this figure is 41%.[10] This reflects the strong

[9] The corporate tax rate in Finland is currently 26%, which is close to the European Community average. However, individual income tax rates can be relatively high as noted earlier.

[10] See *http://www.stat.fi/tup/tietotrendit/tt_06_06_palvelut.html* The main service sectors in Finland are business services, transportation, and information technology-related services.

rise of the ICT industry in the last 10 years and perhaps the customary practice of self-service in Finland. As global outsourcing gains momentum in the ICT industry, in particular, the service sector may rise in relative importance. The service sector provides an opportunity for entrepreneurship, which is considered a challenge in Finland due to the difficulty in fostering startup growth. The public sector continues to be dominant as a service provider or payer, as reflected in the tax regime (taxes amount to more than 40% of GDP). Recently, an interesting trend toward state-paid but privately produced well-being services has emerged, improving the quality of life of those in need of special care or attention (Välikangas, 2006). Thus, the demographic challenge of an aging population is also a business opportunity.

Other promising development areas, according to Tekes (2006), include the development of management and business know-how, Internet and telecommunications, nanotechnology and biotechnology, and materials science (see Table 48.5 on ongoing programs in 2006). Growth is expected from ICT applications, innovative service business models, well-being and health care, pleasant working and living environments, environmentally friendly business models, and smart materials. Finland occupies a strong position in Internet security services and has a number of innovative companies in virtual game development (a nascent but fast-growing industry that was supported by Tekes R&D funds in early 2000). Europe's leading virtual community among the youth is Habbo Hotel by Sulake Corporation of Finland.

Perhaps the most pressing challenge for the Finnish economy is coping with the globalization of its economy as it integrates with international finance, labor markets,[11] and knowledge flows. So far, however, the new R&D centers established abroad (e.g., in China and India) by companies such as Nokia have not reduced the number of research staff in Finland (Lovio, 2006). The percentage of foreign-owned subsidiaries has increased steadily in the last 10 years, Sweden taking the lead, followed by the U.S. and Germany, with the acquisition of Finnish companies being the dominant form of entry (rather than greenfield investments). This has drawn attention to the potential issue of retaining in-company technical competence in Finland despite foreign ownership.

[11] The number of foreigners living in Finland is comparatively very low, though growing—some 120,000 people in 2006 (see Statistics Finland; *http://www.stat.fi/tup/suoluk/suoluk_vaesto_en.htm*).

Appendix 48.A Stages of industrial and economic development in Finland.

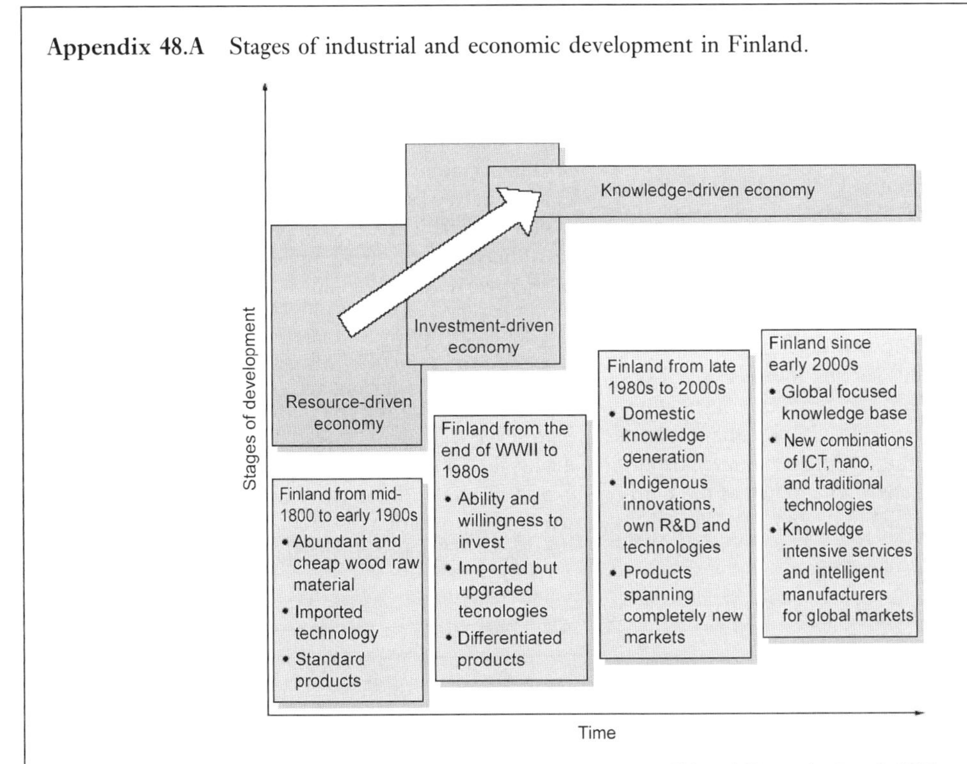

Source: Dahlman, Routti, and Ylä-Anttila (2006, p. 96). Adapted from Porter (1990) and Hernesniemi *et al.* (1996).
Note: the annual productivity growth in Finland between 1992 and 2004 has been better than the OECD average at 2.6% (McKinsey & Co., 2007, p. 17).

Appendix 48.B Key public organizations in new industrial policy formulation and implementation.

Major industrial policy actors in Finland

Finnvera is a state-owned specialist financing company administered by the Ministry of Trade & Industry (MTI). It is also Finland's official Export Credit Agency and acts as an intermediary between the EU's financing programs and Finnish SMEs. It focuses on the promotion and development of SME operations, corporate internationalization, and export operations, by offering financing services. Finnvera's business financing includes loans, guarantees, and export credit guarantees.

Tekes is the main financing organization for R&D in Finland, and was established in 1983. Tekes provides funding and expert services for R&D projects to companies registered in Finland and Finnish research institutes and universities, and promotes national and international networking. Tekes' services are also available via the network of the TE centers (see below). In addition to funding, Tekes provides various expert services and organizes technology programs in selected strategic areas.

Finnish Industry Investment (FII) is a state-owned equity investment company administered by the MTI. Its task is to improve the business environment for SMEs, in particular, by investing equity in venture capital funds. FII can also make equity investments directly in companies, particularly business ventures involving long-term risk. Regional funds target companies at various stages of development across the country. FII also engages in direct investment, together with other investors and financial institutions.

TE centers are public offices under ministerial supervision, consisting of a network of 15 regional offices with business departments serving the needs of SMEs by providing business support services, consultation, and advice, as well as finance. In addition, they serve as a regional network for other organizations and channel their services to the regions. Close to half of the funds come from EU Structural Funds and are directed to EU objective areas.

The Foundation of Finnish Inventions supports and promotes the development and exploitation of inventions. The FFI's services and funding provide a chain of support for individuals and micro-enterprises throughout the invention process up to commercialization. Free information on the development of inventions, patenting, and commercialization is offered through the FFI's invention agents, and through various invention fairs and events where the FFI provides general advice. The FFI can also help establish links with businesses locally and abroad, as well as provide legal and contractual assistance.

Finpro is an expert service organization, partly financed from public funds, providing business support services for internationalization, in the shape of market information and advice, business development, consulting and marketing services, and organizing innovation programs. Companies can purchase services for their international marketing needs from Finpro Marketing Ltd., which functions as a private company.

The Finnish National Fund for Research and Development, *Sitra*, is an independent public foundation under the supervision of the Finnish Parliament. Sitra's tasks include providing research information on Finnish society for the basis of decision making, organizing innovative operations to create new cooperative networks and training for decision-makers, media representatives, and professionals, as well as providing corporate funding for technology companies in their early stages

The Academy of Finland is an expert organization in research funding and science policy subordinated to the Ministry of Education. Its object is to promote high-level scientific research through long-term, quality-based research funding, science, science policy expertise, and efforts to strengthen the position of science and scientific research. The Academy's operations cover all scientific disciplines.

Source: Ylä-Anttila and Palmberg (2005, p. 11).

Appendix 48.C SWOT analysis of the Finnish Innovation Environment.

Strengths, opportunities, and means

—Finland having become an active partner in international co-operation quite recently but rapidly; a very high rate of participation in the activities of international organizations.
—Science and technology policy implemented on a long-term basis; investment in R&D regarded as important.
—Well-functioning education, research, and innovation systems.
—Openness, intensive co-operation, and competitiveness of the innovation system.
—A high proportion of competitive R&D funding.
—A high level of education among the population.
—Brain drain relatively small.
—A high proportion of women among researchers and Ph.D.s by international standards.
—A large number of researchers, who make up a large percentage of the employed.
—Research volume, quality, and impact at a good international level.
—Active international patenting.
—Finland's good reputation: reliable, safe.
—Knowledge-intensive businesses remaining in Finland.
—Good co-operation between business enterprises and public research.
—Finnish enterprises being internationally networked.
—Effective and efficient national innovation environment boosting competitiveness ↔ internationalization of the activities and organizations of the innovation system.
—An enhanced knowledge base and R&D environment, attracting new foreign investments and intellectual resources to the country and improving Finland's position as an attractive region for business operations.
—Looking for competence where it is best: global and diverse international cooperation, going beyond the EU.
—Compensating for the small size and geographical remoteness by active, strategically sound cooperation.
—Prioritized pooling of limited, fragmented resources.
—Open-minded and sufficient support for creativity and innovation.
—Enhancing foresight activities and their linkage to decision making and strategic steering.
—Implementation and productization of social innovations.
—Enhancing positions in international cooperative institutions and R&D organizations.
—Improving the organizational and functional structure of the innovation system and the division of tasks.
—Developing business and marketing competence.
—Creating a favorable business environment and promoting entrepreneurship.
—Supporting the creation and growth of businesses that focus on R&D and exploitation of leading-edge expertise.

Weaknesses, framework conditions, and threats

—Strong dependence on global trends.
—Remote location from global market centers, geographically distant from the center of Europe.
—Difficulties in relation to attractiveness and growth: a small domestic market area, a limited number of inhabitants, a small language area, and severe climate.
—A relatively low level of internationalization by European standards.
—Limited economic and intellectual resources: a low volume of knowledge and competence in many fields and the cutting edge of scientific research in the hands of a select few.
—Problems with venture capital (amount, availability, matching of demand and supply).
—Deficiencies in marketing and business competence and in knowledge and innovation management.
—A small number of spinoff businesses from universities and research institutions.
—Fragmented research activities: resources allocated to a large number of small units.
—A small number of highly educated foreign experts, students, and researchers.
—A small number of growth-oriented enterprises.
—Enterprises and parts of their operations moving abroad.
—Low inflow of foreign direct investments; negative balance of investment.
—There is an international economic recession and decline in Europe.
—Finland does not attract foreign direct investments, R&D investments, researchers, and students.
—Finland is less active in the EU and global R&D cooperation.
—The operational foundations of the EU become weaker: more internal conflicts and less commitment and cooperation.
—National interests are overemphasized in international cooperation.
—Focus is missing: participation in too many projects with scarce resources.
—Links between research and economic development, employment, well-being, and innovations grow weaker.
—Diminishing age groups and an aging population undermine the balance of the public economy, the room for economic maneuver, and the supply of highly skilled labor.
—The regulatory framework does not support the transfer of research results from R&D organizations to businesses and the commercialization of results.
—Availability of competence in the labor market is insufficient: education does not meet labor market needs.
—The number of new R&D-intensive businesses declines.
—The favorable development of public R&D funding stagnates.
—Business R&D expenditure starts to decline.
—Businesses increasingly move their operations abroad.
—Brain drain increases: high competence moves abroad.

Source: STPC (2006).

References

Academy of Finland; *http://www.aka.fi*

Ahlback, J. (2005). *The Finnish National Innovation System.* Helsinki: Helsinki University Press.

Ali-Yrkkö, J., and Hermans, R. (2002). *Nokia in the Finnish Innovation System* (ETLA Discussion Paper No. 811). Helsinki: Elinkeinoelämän Tutkimuslaitos.

Dahlman, C. J., Routti, J., and Ylä-Anttila, P. (Eds.). (2006). *Finland as a Knowledge Economy: Elements of Success and Lessons Learned.* Washington, D.C.: World Bank Institute.

DeNardi, M., Ren, L., and Wei, C. (2000). "Income inequality and redistribution in five countries." *Economic Perspective*, Second Quarter; *http://www.chicagofed.org/publications/economicperspectives/2000/2qep1.pdf*

EU/CIS. (2007). *Fourth Community Innovation Survey* (Eurostat News Release 27). Luxembourg: Eurostat.

Hernesniemi, H., Lammi, M., and Ylä-Anttila, P. (1996). *Advantage Finland: The Future of Finnish Industries* (ETLA Series B 113). Helsinki: Elinkeinoelämän Tutkimuslaitos.

IMD. (2007). *World Competitiveness Report.* Lausanne, Switzerland: IMD.

Koski, H., Leijola, J., Palmberg, C., and Ylä-Anttila, P. (2006). "Innovation and education strategies and policies in Finland." In: C. J. Dahlman, J. Routti, and P. Ylä-Anttila (Eds.), *Finland as a Knowledge Economy: Elements of Success and Lessons Learned.* Washington, D.C.: World Bank Institute.

Lilja, K. (Ed.). (2005). *The National Business System in Finland: Structure, Actors, and Change* (Publication B-60). Helsinki: Helsinki School of Economics.

Lovio, R. (2006). "Globalisaatioprosessin piirteitä suomalaisesta näkökulmasta: Nokia-klusteri maailmankiertueella 1990–2006." *Kansantaloudellinen aikakauskirja*, **102**, 339–358 [in Finnish].

Lovio, R. (2007). *Suomalaisten monikansallisten yritysten kotimaisen sidoksen heikkeneminen 2000-luvulla* (Working Paper W-420; in Finnish). Helsinki: Helsinki School of Economics.

Maula, M., Ahlström, J., Haahkola, K., Heikintalo, M., Lindström, T., Ojanperä, H., and Tiainen, A. (2006). *The Prospects for Successful Early-stage Venture Capital in Finland* (Report No. 70). Helsinki: Sitra.

McKinsey & Co. (2007). *Suomen Talous: Saavutukset, Haasteet ja Prioriteetit* (a report; in Finnish). Helsinki: McKinsey & Co.

MEE. (1993). *National Industry Strategy for Finland* (in Finnish). Helsinki: Ministry of Employment and the Economy.

OECD. (2006). *Program for International Student Assessment (PISA).* Paris: Organization for Economic Cooperation & Development.

Porter, M. (1990). *The Competitive Advantage of Nations.* New York: Simon & Schuster.

Statistics Finland; *http://www.stat.fi*

STPC. (1990). *Review 1990: Guidelines for Science and Technology Policy in the 1990s.* Helsinki: Science & Technology Policy Council of Finland.

STPC. (2006). *Science, Technology, Innovation.* Helsinki: Science & Technology Policy Council.

Tekes. (2006). *Annual Report 2005.* Helsinki: Tekes.

Tekes. (2007). *Annual Report 2006* (in Finnish). Helsinki: Tekes.

TI. (2006). *Global Corruption Report.* Berlin: Transparency International.

U.N. (2006). *Human Development Report* (U.N. Development Program). New York: United Nations.

Välikangas, K. (2006). *Kuntien toiminta ikääntyneiden kotona asumisen ja palveluiden kehittämisessa* (Suomen ympäristö 21; in Finnish). Helsinki: Ympäristöministeriö.

WEF. (2006a). *Global Competitiveness Report.* Geneva, Switzerland: World Economic Forum.

WEF. (2006b). *Finland Ranks 2nd in the World Economic Forum's 2006 Global Competitiveness Rankings* (press release); *www.weforum.org/pdf/pressreleases/finland.pdf*

Ylä-Anttila, P., and Palmberg, C. (2005). *The Specificities of Finnish Industrial Policy: Challenges and Initiatives at the Turn of the Century* (ETLA Discussion Paper No. 973). Helsinki: Elinkeinoelämän Tutkimuslaitos.

49

The Dutch Innovation System: Raising the Lowland?

Patrick van der Duin

Delft University of Technology

49.1 A brief history of the Dutch NIS from a national point of view

To discuss the entire development of the Dutch economy and society here would require a book of its own. We do, however, want to address a number of aspects that we feel are relevant to the Dutch innovation system. We emphasize that, in the first years after World War II, there was not really an *innovation* system as such, and that even today there is no fully fledged and national innovation system. In light of the fact that the Dutch economy and society is very open, the significance of the Dutch innovation system as a geographical phenomenon must not be overrated, even though it has a right to exist.

Dutch economic history after World War II

Like many other Western European countries, the economic structure of the Netherlands suffered tremendously during World War II. Because of its strategic location, the port of Rotterdam was hit particularly hard. The Dutch economy benefited hugely from the Marshall Plan, the aid provided by the U.S. at the end of the war. After the sober years following the war, the country experienced high economic growth between 1951 and 1973 (Van der Zanden and Griffiths, 1989, p. 210). After that, the global economic malaise also affected the Netherlands until the 1980s, with particularly high unemployment figures. An important turning point was the 1982 Wassenaar Agreement between government, employers, and employees, to moderate wages to provide the business community more room to invest and provide economic growth.

The fact that the war left its traces is demonstrated by a housing shortage that lasted until the 1980s. Even today, there is a shortage, as can be witnessed from high house prices. Another cause for this state of affairs is the fact that the Dutch Government has implemented fiscal measures to encourage home ownership and the inability on the part of the housing market to match supply and demand (in terms of quality).

Another legacy of World War II is the large number of baby-boomers who were born right after the war, and who will soon reach retirement age. The resulting aging population, on the one hand, means that there will be a large claim on the country's pensions and government-related retirement provisions, while, on the other hand, there will be a shortage in the labor market. Especially in light of the fact that the Dutch economy is predominantly a service economy, this will be a problem because services are relatively labor-intensive and because it is extremely difficult to realize labor productivity growth in this sector. This puts even more of a premium on innovation since innovation, especially process and organizational innovations, can help in increasing labor productivity and increase economic growth on a macro-level thereby raising sufficient financial funds to finance increasing pensions and health costs.

The modernization of the Netherlands after World War II can be seen, among other things, in the expansion of the physical infrastructure. In the late 1940s, there was a steady growth with regard to roads, railways, and waterways. Virtually every area of the country has been unlocked, even though the economic and social emphasis remains firmly on the "Randstad" region,[1] in the west of the country. Nevertheless, this growth has not matched the pace of the country's growth in economic terms and mobility. Highways are overflowing, the country has the largest number of inhabitants per square mile in the world, and the railway network is one of the busiest in the world. The statement "full is full" may have been considered racist in the 1980s, nowadays even politicians on the left of

[1] This is the region between Amsterdam, The Hague, Rotterdam, and Utrecht. These "nodes" and the connections along them form the economic heart of the Netherlands. In recent years, we have also seen a shift or widening of economic activity towards the southeast (Eindhoven, Arnhem, Nijmegen).

the political spectrum seem unwilling to disagree. All this warrants the conclusion that the Dutch economy is characterized by scarcity and that quantitative growth is difficult to realize. This is exactly why innovation is so important to the Dutch economy.

Innovation in the Netherlands after World War II

The trends in innovation have not passed the Netherlands by. The four generations of innovation (Niosi, 1999; Rothwell, 1994, among others) can also be seen in the Dutch business landscape. Many of the country's R&D institutes have a technological origin, and in the 1970s they discovered the value of market research in the innovation process, while in the 1980s they began developing innovation networks, and in 2004 fourth-generation innovation processes are very much in vogue, as witnessed by the high level of attention to "open innovation". The development of R&D and innovation at Dutch multinational Philips illustrates this generational development beautifully. In the beginning, there was Philips NatLab (founded in 1914), which excelled in scientific discoveries that were often laid down in patents. Later, the company started focusing more and more on the market, as demonstrated by the creation of Philips design and the hiring of social scientists at NatLab. After that, the ties with suppliers became closer, which led to co-production as well as co-development. Finally, based on the company's "open innovation" philosophy, NatLab has recently been transformed into Philips High-Tech Campus, where other companies are allowed to operate, in the hope and expectation that this will lead to more and better innovations. In the development of the Senseo,[2] Philips worked together with Douwe Egberts, the nation's leading coffee producer. Although the Senseo is by no means a unique technological achievement, this collaboration between two companies from different sectors is indeed an achievement, and this innovation has a business model that is unique for Philips: the company derives its turnover and profits not only from the sales of the coffee machine itself, but receives a percentage of the coffee pad sales as well.

A specific description of Dutch innovation policy (and of the country's innovation system) is provided by Smits and Kuhlmann (2004), who describe three trends that have made the "innovation landscape" what it is today: (1) the end of the linear innovation model; (2) the rise of the systems approach to innovation; and (3) the recognition of inherent uncertainty of innovation and (therefore) the need for a learning approach to innovation. Smits and Kuhlmann have investigated what these three trends mean for the (new) function of innovation policy with regard to innovation systems; namely, managing interfaces (between sub-systems), setting up and organizing innovation systems, providing "platforms" for learning and experimenting, providing an infrastructure for strategic intelligence, and the promotion of demand, strategy, and the development of vision.

Smits and Kuhlmann specifically address the development of Dutch innovation policy since the 1970s. There are four different aspects that can be distinguished: supply side; demand side; intermediary; infrastructure (institutions, mechanisms, and organizations that improve the interface and exchange between demand and supply); and supportive infrastructure (e.g., educational systems, material and immaterial infrastructure, availability of risk capital). In 1980 the Dutch innovation policy was mainly focused on supporting the supply side and promoting the increase of R&D expenditure. The next step (in the late 1980s) was to bring supply and demand closer together, which made it necessary to develop the intermediary infrastructure, including measures to bring universities and the business community closer together. In the mid-1990s it became clear that bringing supply and demand together was no longer enough, nor was the emphasis on the development and distribution of knowledge. It is precisely the market that can determine what is needed for innovation, which is why innovation policy addresses, for example, how to help companies adopt more innovative products and make the interaction between supply and demand more intensive from both ends. In the latter phase, the number and diversity of the actors involved in innovation policy and the innovation system has grown considerably, and there are many intermediary organizations that concern themselves with establishing contacts and streamlining the activities between all these actors. Smits and Kuhlmann (2004, p. 14) summarize the development of Dutch innovation policy between 1980 and the present as follows: "... the direction in which the innovation policies and related instruments in the Netherlands are heading is quite clear: from supply-oriented and one-to-one interactions toward demand-oriented and a more systemic approach."

Concluding remarks on the historical development of the Dutch NIS

The Dutch economy has made a transition from an emphasis on agriculture and industry towards one on services, while maintaining the strong focus on trade and distribution. The Dutch ambition to become more innovative is virtually synonymous with the ambition to become a knowledge country.[3] The traditionally open character of the Dutch economy and society fits well with the current globalization trend. There is a reason

[2] A coffee machine that makes it possible to make coffee extremely rapidly, using so-called "coffee pads".

[3] It would be better to have an ambition to become an "innovation country", but due to the fact that these two terms are often interchanged, that is most likely what is meant here.

the Netherlands was among the most ardent supporters of the European idea. Whether or not the Netherlands will manage to benefit from the opportunities offered by globalization and "Europe", of course, remains to be seen. It is clear that having an international outlook will help. As far as the international orientation is concerned, it has to be said that many people in the Netherlands are beginning to recognize the downside of globalization. Problems with the integration of foreigners into Dutch society and doubts about the actual added value of "Europe" has made many people skeptical, which became clear when the referendum regarding the European constitution was met with a resounding "no".

Dutch innovation policy, in as far as it existed immediately after World War II, has developed by and large in parallel with the developments in innovation itself. This does not automatically mean that the country's innovation policy has always been successful. It is often said, somewhat cynically, that the country owes its wealth primarily to the natural gas reserves near Slochteren, and that the economic weaknesses will become apparent once that runs out. Many of the natural gas proceeds are reinvested in the nation's infrastructure and new scientific knowledge. In summary, it can be said that, after World War II, the country entered a new phase of economic and social modernization, in which globalization, social integration, knowledge, and innovation are the main themes, and earlier eras like World War II and the Golden Age (17th century) can no longer be used as reference points.

49.2 Summary of output and trends (R&D expenditure, patents)

In this section we provide a description of the Dutch innovation system by presenting a number of core data and by presenting a graphical representation that was made by the Ministry for Economic Affairs itself. Although, nowadays, innovation reaches beyond R&D and technology alone, we want to mention a number of technology-oriented indicators here, because in the literature a national innovation system is still predominantly viewed as a macro-economic and technological phenomenon rather than at micro-level. If we look at the gross domestic expenditure on R&D (GERD) financed by industry—in other words the expenditure on private R&D as a percentage of overall R&D expenditure—we see that, between 1995 and 2003, it has risen from 45% to 51.1% (Eurostat, 2006) in the Netherlands. That is not a stunning growth rate, and it is below several European averages (e.g., the Eurozone). As far as the overall expenditure on R&D in relation to GDP (gross domestic product) between 1995 and 2006 is concerned, we see a fall from 1.97% to 1.72%. Again, this is below various European averages. The number of patents awarded by the United States Patents and Trademark Office to the Netherlands grew from 60,922 per million inhabitants to 96,369 in 2000 (Eurostat, 2006). This is better than a lot of other European countries, and it is a number that is well above several European averages. The number of technical patents per million inhabitants also grew from 16,455 in 1993 to 48,673 in 2003 (Eurostat, 2006). If we look at employment in knowledge-intensive sectors (as a percentage of overall employment), we see that it was 3.75% in 1995 against 2.51% in 2006 (Eurostat, 2006). This is also below the European average. This figure shows that the Netherlands performs relatively poorly compared with European averages. It is only with regard to technical patents that the country scores relatively well. If we take into account that the Netherlands performs meagerly with regard to employment in knowledge-intensive sectors, the only possible positive conclusion is that the R&D expenditure and employees in knowledge-intensive sectors have a high level of efficiency.

The figures presented above are confirmed by more recent Organization for Economic Cooperation and Development data (OECD, 2006). With regard to gross domestic expenditure on R&D as a percentage of GDP and total number of researchers (FTE; full-time equivalent) per thousand employees, the Netherlands falls below the OECD average. However, with regard to the number of patents per million inhabitants, the country falls above the OECD average. And although, in comparison with other European countries the private R&D expenditure is lower, this falls above the average of many OECD countries.

Private expenditure on R&D in the Netherlands in 2006 was higher then than Government expenditure (51.1% versus 36.2%) (OECD, 2006). With regard to the execution of R&D, the business community also takes the lead (57.4%), followed by universities (28.1%), and then the government (14.5%). According to the OECD, in 2006 the Netherlands had 40,442 FTE researchers, while €9,991.8 million was spent on R&D.[4]

According to the World Economic Forum (WEF), the Netherlands occupies 13th place in the Global Competitive Index 2007–2008. When we look specifically at innovation,[5] we see that the country comes in 10th, although it has to be added that the only indicator being used in this area is the quality of scientific research institutions. Other indicators that, although they do not fall under the heading "innovation" but are certainly relevant, are intellectual property rights, infrastructure, higher education and training, venture capital availability, and technological

[4] The figures mentioned in this section are largely identical to a study by the Dutch Observatory of Science & Technology (DOST, 2005).

[5] One of the indicators (pillars) being used to determine competitiveness. Other indicators are, for example, macro-economic stability, infrastructure, and labor market efficiency.

readiness. On all these indicators, the Netherlands performs better than its listing on the competitiveness index would indicate, sometimes even better than its score on innovation. It is only with regard to technological readiness that the scores are lower:

—FDI (foreign direct investment) and technology transfer: 31;
—firm-level technology absorption: 27;
—mobile telephony subscribers (hard data): 22;
—laws relating to ICT (information and communication technology): 15;
—availability of latest technology: 14.

The scores with regard to these indicators show that Dutch companies and governments are somewhat hesitant when it comes to new technologies. The lower score on "laws relating to ICT" is an indication that the Dutch Government often waits to see what "Europe" decides in this area. The lower score on "availability of latest technologies" probably has to do with the relatively small size of the Dutch market. The lower scores on "FDI and technology transfer" and "firm-level technology absorption" are more serious. The former indicator has to do with the "technological attractiveness" of the Netherlands and the latter with the degree to which Dutch companies adopt new technologies. It is hard to explain where that comes from exactly, especially since the Netherlands is a relatively attractive place for businesses to settle, at least if we are to believe the country's listing on the global competitiveness index. The low score on "firm-level technology absorption" is particularly hard to explain. And although the score on "mobile telephony subscribers is based on hard data, this low score is remarkable, because the Netherlands has traditionally been seen, after the Scandinavian countries, as the most modern country with regard to the adoption and use of ICT (see also Section 49.4). The findings listed above are broadly confirmed by a recent study by Porter *et al.* (2008), in which the technology-based competitiveness of 33 countries is compared on the basis of expert panels, in areas like national orientation, socioeconomic infrastructure, technological infrastructure, and productive capacity. In virtually all these areas, the Netherlands end up somewhere in midfield, even though experts do predict that the overall high-tech competitiveness of the Netherlands will be better than it is today in 15 years' time (which, incidentally, is what they expect for most countries).

49.3 Technology policy and commercialization initiatives at the national level

An important initiative in optimization of the Dutch innovation system is the Innovation Platform (*Innovatieplatform*). In September 2003 this Platform was founded with an eye on the objectives laid down in the Lisbon Agreement[6] and the realization that the Netherlands was not only lagging with regard to realizing those objectives, but also because of a growing awareness that, at a global level, the Netherlands should not pursue a low-cost strategy, and that it is important to maintain profit margins through innovation. The self-appointed aim of the Platform is to "take a critical look at the functioning of the knowledge and innovation system and to provide breakthroughs. It is the Innovation Platform's task to create the conditions, establish the connections and develop the vision that is needed to provide an impulse to innovation and entrepreneurship in the Netherlands" (*www.Innovatieplatform.nl*). With the fourth Balkenende Government, the second Innovation Platform has become operational. The fact that Prime Minister Balkenende himself is chairman of the Platform, and that it also includes a number of important ministers, is an indication that the Dutch Government takes the Platform very seriously.

Generally speaking, people are not overly enthusiastic about this initiative. Many ask themselves what in essence is the difference between a traditional industrial policy and this initiative. In that context some point out that a genuine innovation policy is given shape by (new) "innovative forces". In the current and previous Innovation Platform, there are many representatives of large Dutch companies that may be more interested in maintaining the *status quo*, even displaying "rent-seeking" behavior because taking part in the Platform increases the likelihood that they may get their hands on a part of the budget associated with it. The fact that the Platform also concerns itself with the content of innovation—by selecting key areas (like water technology, flowers and nutrition, creative industry, and chemical technology) as areas on which innovation should focus—is not appreciated by everyone. Some feel that this places the government on the entrepreneur's chair, which is something governments rarely do well. The government should focus exclusively on creating the right conditions for encouraging entrepreneurship and innovative behavior. Finally, there is criticism with regard to the picture the platform has painted of the future, which is said to smack of a utopia that is built on wishful thinking, which means that the ambition to inspire entrepreneurs will not be realized (van der Duin and Sabelis, 2007).

According to the Dutch Ministry for Economic Affairs, the Dutch innovation system can be portrayed as shown in Figure 49.1.

This picture shows the different actors in the innovation system and how they are connected. In addition, the figure

[6] The agreement was signed in 2000 by the European Council and contains the ambition to be the world's most competitive and dynamic knowledge economy by 2010. This agreement was specified during the Barcelona summit in March 2002 in an R&D and innovation expenditure of 3% of GDP, two-thirds of which are to funded by private businesses.

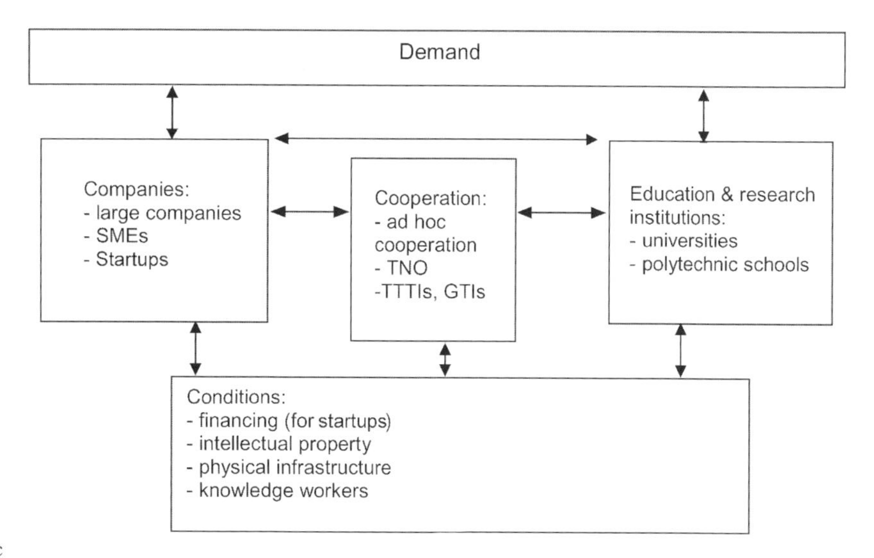

Figure 49.1. The Dutch dynamic innovation system.

contains the conditions of the innovation system, placing it in an international environment, of which the EU is the most important component. According to the Ministry, this figure illustrates the complex interplay between end-users, companies, knowledge suppliers, and intermediaries, and this interplay determines the innovation strength of the Netherlands. Although the figure is in itself accurate and no doubt complete, we feel that it offers precious little in the way of insight. It only shows which actors play a role in the innovation system and that they are (almost) all interconnected, which diminishes the figure's explanatory power considerably. Why this innovation system is called *dynamic* in not clear, and it does make one wonder what a *static* innovation system would look like.

49.4 Relative strengths and weaknesses in technology

In this section we assess the (relative) strengths and weaknesses of the Dutch innovation system using two models: (1) the Cyclical Innovation Model, in which the innovation system is presented as a system of cyclical interactions between actors in science, technological research, product development, and market transitions, and (2) the second model has been developed by Hekkert *et al.* (2007) and consists of the seven conditions a well-functioning (national) innovation system must meet. But first we do a more *ad hoc* analysis by referring to different studies that have researched good and bad aspects of the Dutch innovation system.

An *ad hoc* analysis of the Dutch innovation system

The operation of the Dutch innovation system is often investigated. Below, we present the findings of a few studies.

A report from 2003 from the Ministry for Economic Affairs about the Dutch innovation system mentions the following shortcomings:

—the R&D intensity of Dutch companies is low by international comparison;
—a growing shortage of knowledge workers, in particular scientists, technicians, and R&D people;
—not enough innovative entrepreneurship;
—insufficient exploitation of the results of scientific research;
—weak points in the interaction between the knowledge infrastructure and the business community;
—funding of innovation is problematic (venture capital is used insufficiently for innovation, among other things).

The good points are:

—high-quality scientific research;
—the Netherlands scores well on scientific research;
—relatively high co-funding of publicly-applied research by the business community;
—access to ICT and the use of ICT are good;
—high number of knowledge workers in general.

According to Kleijn (2007), the Dutch innovation system has the following good qualities:

—high-quality scientific research;
—diversified industrial networks;

—active in the area of patents;
—access to and use of ICT (although the effect of ICT on economic growth, according to Nobel laureate Robert Solow, is negligible or not noticeable for the time being).

In addition, Kleijn also identifies a number of shortcomings:

—the R&D intensity of Dutch companies is lagging;
—not enough innovative entrepreneurship;
—insufficient use of scientific research ("innovation paradox");
—shortage of people in science and engineering.

The lists presented above by and large are in agreement. Although the Netherlands potentially has enough knowledge and funding, the country does not succeed in translating those resources into a larger innovative ability.

Many also point at the low levels of innovativeness of Dutch SMEs. And because 99% of all Dutch companies fall into this category, this can be seen as a serious obstacle on the road to realizing the desired "knowledgeland" or "innovationland". Based on a survey of relevant studies in this area, we have concluded that most SMEs have an informal and unsystematic approach to innovation and are generally speaking less innovative than large(r) companies (Van der Duin, 2006, p. 83). The Ministry for Economic Affairs has a separate regionally organized department with the aim of aiding SMEs to innovate.

There are two bottlenecks that are identified in the "Innovation Letter" written in 2003 by former Cabinet Minister Brinkhorst:

—the Dutch culture of "act normally, that is crazy enough" does not breed courage and entrepreneurial spirit;
—due to the Dutch consultation culture, the Dutch culture is complex and opaque, which is reflected in the way people deal with innovation.

In the same letter, the former minister mentions a number of ambitions, like increasing the specific budget for R&D among SMEs by 30%, setting up the Innovation Platform (see below), and making it easier to attract foreign knowledge workers. Brinkhorst's analysis shows that, generally speaking, the Dutch culture is not very conducive to innovation, at least not if elements like excelling, distinguishing oneself, and taking risks are considered "soft" success factors.

The Dutch innovation system from a cyclical perspective

According to Berkhout (2000) and Berkhout *et al.* (2007), innovating is predominantly a cyclical interaction between two actors in the "innovation arena" exchanging knowledge and information. These actors can be clustered into four groups: science, technology, product, and market. It is important for these four groups to be closely connected and for each one of them to be able to initiate innovation. The value of the Cyclical Innovation Model (CIM) is that it does not consider an innovation system just as a collection of actors and stakeholders, but that it clearly positions them relative to each other. In addition, CIM values the various flows between the actors, which is an essential feature of innovation systems (OECD, 1997, pp. 9 and 11). In schematic terms, CIM can be portrayed as shown in Figure 49.2.

In CIM, scientific exploration, technological research, product creation, and market transition are cyclically connected, thus creating an "innovation arena" in which an innovation process can commence at each point. Knowledge and information are exchanged between the "nodes". According to CIM, the information system functions optimally when all the nodes are connected to each other, without being hindered or cut off. The entrepreneur (central node) serves as a kind of "innovation director" who has to make sure that all the nodes are occupied, that actors contribute to the system, and that all flows function properly.

With the help of CIM, we take a look at two problems regarding innovation in the Netherlands. First of all the inability to translate knowledge into innovation. According to CIM, the left-hand node (scientific exploration) and the right-hand node are not connected.

In this situation, scientific exploration receives too few signals from product creation, and as a result scientific insights do not become available to product creation. Figure 49.3 visualizes the well-known innovation paradox, whereby a high level of scientific knowledge does not lead to more and better products. There is a reason that the Dutch Government, among others, spends a great deal of time and money on intermediaries who find a better match between supply and demand in the innovation system. At business level, we can think of the Palo Alto Research Center (PARC), Xerox's research lab, where fantastic things are invented (like the graphical user interface and the high-speed laser printer), which were, however, never used to make a lot of money by the company itself, because it did not utilize an intelligent business model which is required to bring the product to market.

In addition, there can be a problem with the match between technology and market. Often, this has to do with too much emphasis either on technology push or on marker pull. According to CIM, this can be portrayed as shown in Figure 49.4.

In this situation, products are being developed on the basis of new technological insights that the market does not want. On the other hand, technological development is insufficiently inspired by market needs. This reveals a

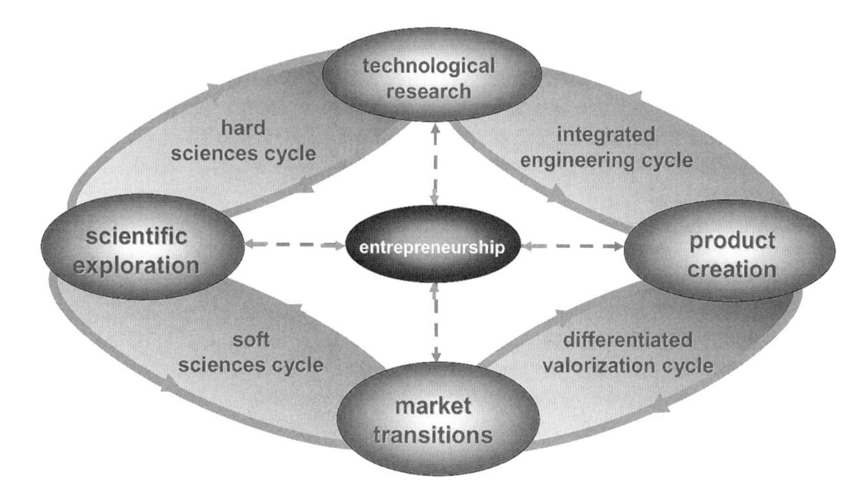

Figure 49.2. The Cyclical Innovation Model.
Source: Berkhout *et al.* (2007, p. 39).

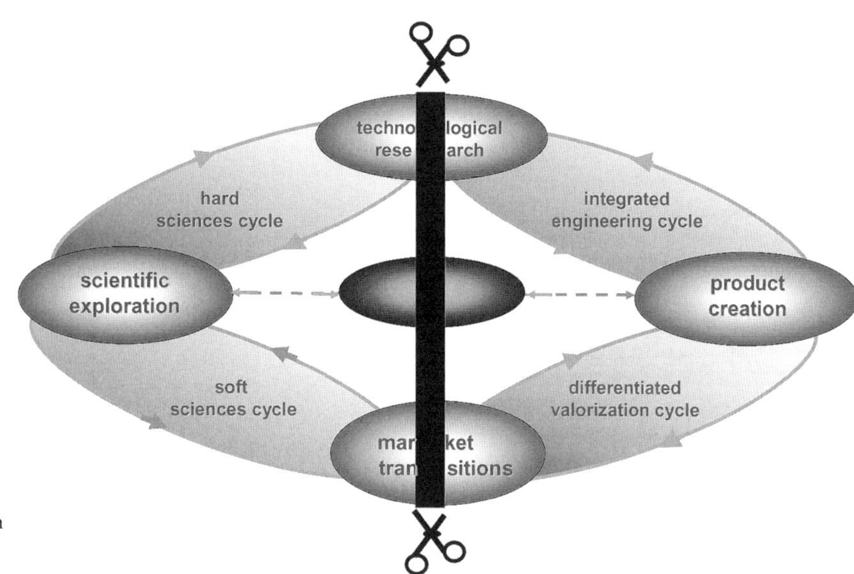

Figure 49.3. The lack of communication between scientific exploration and product creation in the Dutch innovation system.

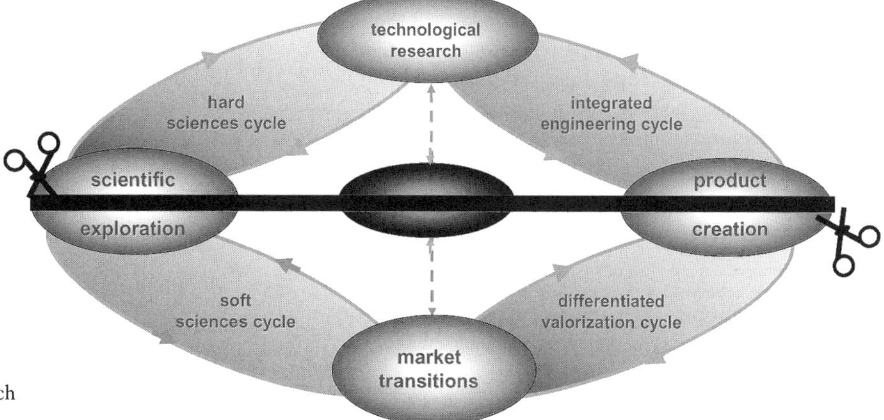

Figure 49.4. The lack of communication between technological research and market transitions in the Dutch innovation system.

classic market research problem; namely, the difficulty users have to formulate what they want in such a way that it is clear what this requires in technological terms. There is a risk that innovation policy in the Netherlands, with its emphasis on technology, focuses too much on the top element of CIM, forgetting that the bottom element is also an important part of the innovation system.[7] At business level, this situation applies to Philips, even though the company now has a more even balance between technology and market. However, it is a fact that, for a long time, Philips developed many and good new technologies that for some reason did not match market demand. From an organizational perspective, this was reflected in the distance between NatLab (technological research) and Philips Design (market transitions).

As we mentioned earlier, according to CIM, a (national) innovation system functions optimally when the flows between the nodes are not obstructed and entrepreneurs have enough freedom, time, and opportunities to manage the innovation system, with an emphasis on bringing together various actors from different nodes and making sure that they communicate well and often. In CIM, government is located at the center (almost literally), acting as a facilitator (or director) of the innovation system. That the Dutch government is not quite ready for this role is illustrated by the fact that innovation advisors at Syntens do not (yet) have enough time to help set up an innovation system (Van der Duin, Ortt, and Kok, 2007). They still get paid for advising individual companies, and their time is limited to avoid unfair competition towards commercial advisors.

An analysis of the Dutch innovation system on the basis of the functions of innovation systems

According to Hekkert *et al.* (2007, p. 413), innovation systems are "a very important determinant of technological change." They also argue that "the emergence of a new innovation system and changes in existing innovation systems co-evolve with the process of technological change." Although these authors focus primarily on the technical or technological aspects of innovation systems,[8] we nevertheless want to apply their insights to the Dutch innovation system, because technology is an important driver for this innovation system.

Hekkert *et al.* (2007, p. 421) have devised a framework designed to map processes that are very important to the way innovation systems function. They suggest the following seven "functions to be used when mapping the key activities in innovation systems, and to describe and explain shifts in technology specific innovation systems":

1. Entrepreneurial activities: both starters who start up their own business and existing companies that adapt their business to benefit from new opportunities.
2. Knowledge development: new knowledge lies at the basis of virtually every innovation or innovation system. This not only applies to explicit and/or codified knowledge, but also knowledge that is acquired through learning.
3. Knowledge distribution through networks: knowledge can only be used when it is communicated and exchanged.
4. Guidance of the search: not all technologies can be explored and developed, which means that choices will have to be made. It is important to establish long-term objectives that provide guidance to the technology development process.
5. Market formation: a new technology can only succeed when a market has been found for it. To give new technologies a chance, it may be necessary to realize a kind of niche market in which the technology can be applied.
6. Resource mobilization: to develop an innovation system, there have to be enough financial and human resources.
7. Creation of legitimacy and counteracting resistance to change: new technologies always fight against "old" technologies. New technologies have to become a part of an " 'incumbent regime', or it even has to overthrow it" (Hekkert *et al.*, 2007, p. 425).

Although the framework discussed above is predominantly aimed at innovation systems surrounding a technology or industry, we nevertheless want to apply them to the Dutch (national) innovation system. Below, we apply the seven functions outlined above to the Dutch innovation system:

(1) Entrepreneurial activities: on the basis of the frequently heard complaint that the Dutch are not good at translating (new) knowledge into (new) business, it is possible to establish that the Dutch innovation system does not score well in this area. Dutch people are not prepared to take too much entrepreneurial risk,)like having a long-term contract with a single employer, while going bankrupt is considered a sin. In addition, successful entrepreneurs are often seen as show-offs who have become rich off the hard work of others (their hard-working employees) or, even worse, with a lot of luck. In recent years, all this is slowly changing. The taboo on setting up your own company

[7] In the framework designed by Hekkert *et al.*, which we discuss below, "market transitions" can be seen as "market formation", the difference being that CIM is concerned with actual market developments, whereas the framework designed by Hekkert *et al.* focuses more on the deliberate creation of a market.

[8] Innovation systems are still primarily approached from a technological perspective. Carlsson *et al.* (2002, p. 325), for instance, in an overview on innovation systems, argue that "(T)he function of the innovation system is to generate, diffuse, and utilize technology."

is disappearing. More and more people set up their own business, are hired by the company for which they used to work, or develop an idea of their own that their former employer does not want to develop. Figures from the CBS (Dutch Statistical Office) show that the number of companies with zero employees (meaning independent one-man companies) has risen from 330,000 in 2000 to 413,000 in 2006. The numbers of companies with more employees has remained fairly constant over the years or even shrunk a little.

What is striking is that many non-natives also set up their own company, although in many cases that does not involve new business on the basis of a new technology, but rather the deployment of entrepreneurial activities in the retail sector. The native entrepreneurs mentioned above are for now working in an advisory capacity and, as we said earlier, do the same work, not as an employee, but as the director of their own company. There is a risk that entrepreneurship in the Netherlands is becoming too much a matter of outsourcing existing work, rather than of developing new activities. Nevertheless, it would appear that ideas about entrepreneurship are changing, to a considerable extent due to increased interests at Dutch universities. More and more universities are involved in knowledge exploitation—in other words, they are trying to commercialize their scientific knowledge. In particular, technological universities have embraced this subject, because they operate in an area where knowledge is often laid down in patents. Dutch students are also increasingly interested in entrepreneurship, which is presently considered a serious career, in addition to working for the government or at a (large) company. A good example of combining knowledge exploitation and the growing interest among students for entrepreneurship is a university course called "Turning Technology into Business", which is taught at the Delft University of Technology. The students taking this course are given a patent that is owned by the university, with the aim of trying to turn it into a potentially successful business plan. After that, the students can start their own business on the basis of their business plan, which is something a growing number of students are interested in.

(2) Knowledge development: there is a fierce discussion going on about the quality and effectiveness of the Dutch educational system. In 2008 a government committee evaluated the education policies of the last 35 years, and its conclusions were not extremely positive. Virtually all educational innovations have failed completely or partially and the schools themselves have not been involved very much in the various innovations or improvements. That is the reason many political parties argue that education requires a special emphasis and that a kind of "Deltaplan"[9] is needed to get the Dutch education up and running again. This is certainly necessary, in light of the ambition to turn the country into a genuine "knowledge country". The Dutch economy has not enough industry and the country's wages are too high to maintain a "physical" economy. For the country's international competitive position it is better to focus on providing high-quality services that compete on quality rather than costs. To do so, a good education system is indispensable.

According to Hekkert *et al.*, indicators like R&D projects, patents, and investments in R&D provide information about the level of knowledge development. On this basis, knowledge development in the Netherlands is not in a very good state. As became clear in Section 49.2, the Netherlands only scores relatively well when it comes to patents. We want to point out that not all innovations and innovation systems are based on R&D and/or technological knowledge, and that technological knowledge is by no means in all cases the deciding factor. It is possible, however, to draw the conclusion that, for a country and an innovation system that is so dependent on the development of new knowledge and that has the ambition to become a knowledge country and develop a knowledge economy, an average level of knowledge development is not good enough.

(3) Knowledge diffusion through networks: in determining to what extent knowledge is diffused through networks, what springs immediately to mind is the innovation paradox from which many European countries suffer: many people are surprised about the fact that, although Europe has a high level of (scientific) knowledge, it is somehow unable to translate that knowledge into a high level of innovation.[10] It is true that in the Netherlands knowledge is not used as well as it could be, but it remains to be seen whether this has to do with the lack of networks or of poorly functioning networks that share knowledge insufficiently. Science and the business community are much too separated in the Netherlands. As far as scientists are concerned, there are very few stimuli to establish contacts with the business community, while the business community see universities as a separate world

[9] This term refers to a large-scale plan that was designed in the 1950s to protect the Netherlands better from water, in the aftermath of the flood that hit the province of Zeeland in 1953, when 1,800 people drowned. Nowadays, many of the large-scale proposals to solve major social problems are referred to with this term.

[10] Supporters of the innovation paradox actually display a traditional opinion about innovation, because the paradox is based on the idea that high levels of R&D and/or scientific knowledge equal innovation. That is why the paradox (which in reality is not a paradox and not even a contradiction) has more to do with amazement about the lack of a direct connection between R&D and innovation.

through which they find it hard to navigate.[11] It has to be said that an overly close relationship between science and business may have a negative impact on innovation in the Netherlands. In practice, the existence of a closer relationship would not only mean that there is a better match between what scientists are investigating and what businesses need. It would also mean that the business community would have a major influence on the research agenda of universities. This commercialization of universities, as indicated by the principle of knowledge exploitation, could lead to more applied research and less fundamental research, which is good news for incremental innovations but bad news for radical innovations. So academic freedom is not only a cultural value, but also an economic one. Fundamental scientific research that does not have a clear potential for knowledge exploitation, which is fairly common in light of the unpredictability of the future, deserve a large degree of independence. Perhaps the problem is not so much that scientific knowledge is translated sufficiently or insufficiently into innovation, but rather that it simply takes too long. This is another reason why it may be bad for networks to have relationships that are too close. In the Dutch "polder" tradition, where consultation, consensus, and securing a broad support base are considered essential, cooperation takes a lot of time, as a result of which the innovation process and the creation of an innovation system are delayed. A case in point concerning the development of Thixomolding® in the Netherlands shows that, although the Netherlands was quick in developing this Canadian patent, much time was lost, allowing others to catch up (including France and Austria) who were able to make decisions more quickly (Van der Duin, Ortt, & Kok, 2007).

(4) Guidance of the search: above, we argued that innovations are difficult to predict, due to future uncertainty. In fact, the moment something can be predicted, it is no longer genuinely innovative. However, it is precisely that uncertain future that demands a process in which the future is explored (see Van der Duin, 2006). In the area of scientific futures research, the Netherlands is among the most active countries in the world, together with the U.S., the U.K., Australia, Finland, and Japan. In the Netherlands there are four futures research professors as well as many institutes, often closely connected to the national government, involved in futures research. The typically Dutch notion that society can be shaped is the foundation of this. However, exploring the future does not always mean that futures studies will be used in innovation processes or in setting up innovation systems. A recent study among all the Dutch ministries about the way futures explorations are used to devise strategies (including policy regarding innovation, knowledge, and technology) shows that, although setting up futures exploration is not very difficult, applying it is. An important explanatory factor in this respect is that the short term often prevails, at the expense of the long term (see also AWT, 2004, p. 17). In light of the fact that innovating is a long-term affair, this is not a good thing. A case in point where the quality of "guidance of the search" left something to be desired is the vision of the future that was developed by the Innovation Platform (see Section 49.3). An analysis of this vision by van der Duin and Sabelis (2007) reveals the following shortcomings: it is an almost utopian vision, there is too much wishful thinking, no distinction is being made between the vision and the way it should be realized through innovations, and the vision does not apply different time horizons to the various subjects. Another example of bad "guidance of the search" is given by Prent (2008) who criticizes the Dutch innovation system on photovoltaics (i.e., solar cell technology) of lacking an image of the future.

(5) Market formation: basically, this has to do with the degree to which certain promising technologies are given a protected place where they can be developed further on the basis of information that emerges from the use of the technologies. This often requires special government measures. It is difficult to determine the extent of these niche markets in the Netherlands. Existing policy on competition is at any rate strict, and cannot be used to shield certain markets. The presence of many cartels in the Dutch economy (which are increasingly the focus of competition policy) could be beneficial in this respect, because companies can agree to apply new technologies in certain (new) market segments. For the time being, this does not seem likely, and cartels are mainly concerned with dividing certain markets. It would take up too much space in this paper to investigate the effect of competition policies on the Dutch innovation system (in addition to being very difficult), but for now it would appear that stricter policies lead above all to an emphasis on price-related competition. Although process innovations can contribute to the realization of lower costs and lower prices, it is doubtful whether this would improve the Dutch innovation system. Price-related competition often leads to a focus on the

[11] A fairly recent policy measure that came from the Innovation Platform is the "SME innovation voucher" which businesses can use to purchase knowledge from knowledge institutions. The idea is to kill two birds with one stone: to improve the relationship between the business community and science and increase the innovative ability of Dutch SMEs. The innovation vouchers are generally seen as a success, often especially with regard to the number of vouchers handed out. Whether or not the vouchers are indeed a success depends, of course, on the extent to which they help improve innovation among SMEs.

shorter term and on defending market shares in existing markets, rather than to opening up new markets. In addition, Dutch politics when formulating new laws has a tendency to "look to Europe first". One of the consequences of this approach is that it is unlikely that specifically Dutch laws aimed at giving certain Dutch markets an advantageous position will ever see the light of day. As a result, the international positions with regard to windmills and solar cells have been lost—Thixomolding® is an example (Van der Duin, Ortt, and Kok, 2007). A good example where the Dutch government has made a positive contribution with regard to market formation is the PC market. Although it is not completely comparable (because it was above all a demand issue), measures to promote PC use by Dutch citizens have made a substantial contribution to the position that the country now occupies in the area of ICT.

(6) Resource mobilization: this is a question of whether the Dutch innovation system has sufficient financial and human capital at its disposal. With regard to human capital, we can once again point to the number of R&D workers and employees in knowledge-intensive sectors (as mentioned in Section 49.2), and in this area the Netherlands does not score well, as we said earlier. The quality of the Dutch education system is simply not high enough to distinguish itself from other countries in this respect. There is a distinct shortage of science students. It is often heard that most Dutch students prefer the "softer" sciences. Government has launched various campaigns to remedy this state of affairs, but thus far it has failed to make the subject attractive enough to make any major impression. In addition, not all science students move on to technology-oriented careers. In many cases, they end up in managerial jobs. Fortunately, the country has not yet been the subject of a "brain drain", otherwise the problems might be even worse. At the moment, attempts are being made to make it more attractive for foreign students to study in the Netherlands and go on to technical positions. Unfortunately, this movement is at odds with the country's immigration policy, which has become much stricter in nature. Help may come from the fact that the country's population is aging, which means that there may be a serious labor shortage in the future. A more flexible immigration policy may help remedy the problems associated with the aging population, as well as make sure that more computer experts find their way to the Netherlands. Again, we want to point out that the degree to which this is seen as a problem depends on the extent to which technology is the driving force behind (and in) the Dutch innovation system. If we are convinced that technology is the crucial nexus of the innovation system, then the shortage of technological knowledge is indeed worrying. If, on the other hand, we believe technology to be somewhat less important, the problems are likewise smaller.

The availability of financial capital in the Netherlands is in principle well-organized. There is an excellent banking system, in which both citizens and businesses have great confidence, a number of large international banks, and the average citizen has a high savings quota. In its innovation policy, the government also has included many financial instruments to support businesses in their innovation. This can be seen as a positive legacy from the 1970s and 1980s, when individual companies were supported financially, especially when times were bad. It is only in the area of venture capital that less financial capital is available. However, due to the international face of capital, this need not be a problem in principle for Dutch entrepreneurs.

(7) Creation of legitimacy and counteracting resistance to change: it is perhaps in this area of innovation systems that contradictions in Dutch culture are most evident. On the one hand, the Netherlands is a progressive country with many individual freedoms and very liberal laws. On the other hand, there are conservative forces that place an emphasis on maintaining the *status quo*. As mentioned earlier, the many cartels are an expression of economic conservatism and they often stand in the way of vital innovation. To realize a genuine innovation of the Dutch innovation system, there has to be room for new actors with new ideas. Of course, it is possible for existing actors to come up with new ideas, but they are often less interested in changing the innovation system. In the abovementioned Innovation Platform, which can be seen as a kind of innovation taskforce, most positions were awarded to people from the same well-known companies. Although we feel that open innovation (Chesbrough, 2003) is being hyped enormously, a more open innovation system, with plenty of room for newcomers and that is not all about "backing winners", would be a welcome change.

49.5 Funds flow for innovation[12]

In Section 49.2 a few statistics were given about Dutch gross expenditure on R&D relative to other European countries. In this section we provide a few statistics on funds for innovation which are in principle more difficult to find since innovation is a broader term than R&D making the boundary between money streams related to innovation, in particular, and industrial policy, in general,

[12] This section is based on the budget for the Ministry of Economic Affairs for 2008 (version September 2007).

difficult to draw. However, the Dutch Ministry of Economic Affairs does provide some "statistics" which can quite easily be classified as funds flow related to innovation. First of all, there are so-called "innovation programs" which are directly subsidized by the Ministry. Examples of these innovation programs are: Food Nutrition Delta (food industry); Water Technology; Point-One (nanoelectronics and embedded systems); High Tech Automotive Systems (HTAS, automotive); and Maritime Innovation (the maritime industry). Possible new innovation programs are: chemistry, life sciences and health, and materials. These innovation programs are part of key areas that were established in 2004 by the Innovation Platform. The amount of money that the government will devote to these innovation programs depends on how many Dutch companies are eager to come up with ideas for such programs, relevant ideas, and initiatives for innovation. The Dutch Ministry has also a special *arrangement* for SMEs, called the Small Business Innovation Research (SBIR) programs specifically for projects and ideas that are not specifically R&D but aim at *renewal* and *change* of processes, products, and services. Other policy instruments that are aimed at improving innovative capacity are:

—Innovation achievement contracts:[13] an arrangement for SMEs to set up long-term innovation processes between different companies.
—TechnoPartner: an organization funded by the Dutch Ministry of Economic Affairs whose goals are to (financially) support *technostarters* (pioneering small companies) that want to set up a new business founded in new technology and organizing cooperation between knowledge institutes and companies.
—Law to promote research and development:[14] a fiscal arrangement to lower the labor costs of employees who do research and development and can be applied for by any type of company.
—Subsidies for investments in knowledge infrastructure (Dutch acronym: BSIK[15]): a very broad investment program to stimulate scientific research and development that is being developed into new products, processes, and societal concept. BSIK consist of 37 projects that have a lead time of 4 to 6 years, and a total budget of €802 million. The 37 projects are on the following five topics:

1. Sustainable system innovation.
2. Excellent spatial planning and use.
3. Information and communication technology.
4. Microsystem and nanotechnology.
5. Health, food, genetic, and biotechnological breakthroughs (including genomics).

Increasingly, attention (and probably money) goes to projects about developing innovation to tackle or possibly solve societal issues. The Innovation Platform in 2008 initiated a project called *Nederland Ondernemend Innovatieland* (NOI) ("Making Holland an entrepreneurially innovative country") that is aimed at improving the use of knowledge and entrepreneurial drive to address societal issues. This project is organized between several governmental departments and the choice of societal issues will be based on a long-term strategy that is yet to be developed.

It is not easy to quantify the arrangements and decisions mentioned above, but from the 2008 budget of the Ministry of Economic Affairs some statistics can be provided. The amount of money spent on innovation vouchers (see Section 49.4 and footnote 11) in 2008 was €8 million and is expected to grow to €11 million in 2011. In 2008, €5 million were spent on the SBIR and IPC and this will double by 2011. A total of €2.5 million on innovation credits was spent in 2008 and this will grow to €10 million in 2011. The funds flowing to BSIK in 2008 reached €9.6 million. In the 2008 budget no growth figure was given for BSIK for 2011 but considering the rise in funds aimed at other projects it can be expected to rise as well.

The specific goal of the Ministry of Economic Affairs is not only to enlarge the amount of government spending on innovation but also by doing this to spur companies to devote more financial resources to innovation. The policy instruments and arrangements to do this mentioned above are just a small sample of a broader portfolio of policy initiatives to promote innovation in the Netherlands.

49.6 Cultural and political drivers influencing the NIS

Relatively speaking, the Netherlands contains a large number of multinational companies for a country its size, like Philips, Akzo, Shell, Océ, and Unilever. These companies spend by far the most of all Dutch companies on R&D. Traditionally, the country has always been focused strongly on trade and transport, with Rotterdam as one of the world's largest ports and Schiphol Airport as a European mainport as prime examples. The added value of these two distribution nodes has to be called into question, however. Although Schiphol Airport is more labor-intensive than the port of Rotterdam, it remains doubtful that simply moving people and goods, with additional activities or processing, will provide much in the way of added value and thus economic growth. In an

[13] In Dutch: *Innovatieprestatiecontracten* (IPC).
[14] In Dutch: *Wet Bevordering Speur- en Ontwikkelingswerk* (WSBO).
[15] BSIK stands for *Besluit Subsidies Investeringen Kennisinfrastructuur.* BSIK is a follow-up to another research and innovation program that was funded by reveneues generated from the exploitation of the Dutch gas inventory.

extension to this, the Dutch economy is characterized as a transaction economy (Den Butter, 2007). According to den Butter, this means that, as far as the Dutch economy is concerned, trade innovations are as important as product innovations, and combining and communicating are as important to innovation as creating. Den Butter mentions the growing "creative industry" in Amsterdam as an example and as an "essential element of the 'transaction industry'."

From the point of view of innovation, the most relevant features of Dutch society are the importance of personal freedom, a great faith in consultation and consensus, and a strong international orientation. Personal freedom expresses itself in a very liberal legislation compared with most other countries, with regard to abortion, euthanasia, and soft drugs, to mention a few examples. The faith in consultation and consensus is demonstrated by the large number of advisory bodies that surround and penetrate the political system. This is also known as the "Polder Model" which has its roots in the widely recognized need for a collective approach to tackling problems concerning water. The strong international orientation is reflected in the active role played by the nation's government in the development of the EU and the perceived importance of the role played by international organizations like the U.N.

These sociopolitical features also have a downside. The strong individualistic attitude means that Dutch citizens do not identify themselves very much with their country. Patriotism is considered an exaggeration, and we might even say that national pride is a rather "un-Dutch" phenomenon.

As far as the consultation structure is concerned, this often considerably prolongs the decision-making process. In addition, the desire to reach a consensus does not lead to clear choices, but to a compromise that may be beneficial to the relationships between the parties involved (like trade unions, employers' organizations and consumer organizations), but that does not lead to decisions that benefit everyone. The most perverse expression of the Polder Model are the numerous cartels that exist in the Netherlands, the reason the country is sometimes referred to as "Kartelland". Another negative example of the consultation structure is construction fraud based on so-called "preliminary consultations" between companies regarding price levels. Although the country's progressive position on values is in itself commendable, the Dutch are willing to sacrifice this stance when it damages their economic interests. "All values come at a price" could be a new Dutch saying.

The Dutch economist Geert Hofstede compares cultures and countries with each other on the basis of five indicators: distance to power, degree of individuality, degree of masculinity, risk avoidance, and time orientation. The Netherlands rate as follows: small distance to power, high degree of individuality, low level of masculinity, average degree of risk avoidance, and a long-term focus that is slightly below average (and leans towards to the short term).

Dutch culture is above all focused on reaching a consensus that takes all interests and opinions into account and tries to accommodate them. In addition, the Dutch see little reason to distinguish themselves in any area whatsoever. In this flat country, people who try to stand up are not encouraged. There is the reason that "act normally, that is crazy enough" is such a typically Dutch saying. Looking for a consensus and defining compromises is the outcome of the ancient battle against water in a time when it was hard for people to act on their own. That battle is something that brought (and continues to bring) the Dutch together and there is a strong emphasis on the collective nature of society. Although there are few countries in the world where people are more individualistic than the Netherlands, this individualism has not yet reached the institutional level, which means that the world-famous Polder Model has not yet been buried. This model implies that decision-making processes surrounding political issues involve consultation with various interest groups and advisory committees to gather information. This is not only a top-down process. The various advisory bodies also give advice on their own initiative. Often, government members are obliged to ask for advice and they need to realize that ignoring the advice that they are given requires a thorough explanation. In other words, this model is a formalized consultation at the institutional level.

A direct consequence of the Polder Model is that decision-making processes are extremely slow and rarely lead to a drastic change of course. A large degree of influence almost automatically means that various interests have to be served. This is not only a time-consuming process, it also often results in a kind of "average" solution, where everybody gets a little of what they want. Needless to say, it is doubtful that this approach to decision-making is suitable for the increasingly dynamic world of innovation.

49.7 Concluding remarks

Improving innovation, in general, and the Dutch innovation system, in particular, is not only a matter of adapting a policy, changing an economic incentive, or extending certain subsidies. The sub-structure, to use one of Karl Marx's terms, also needs to be adapted, and is arguably foremost in determining the innovative ability of the Netherlands. Economic structure and culture need to be such that innovation is worthwhile and appreciated. The problem with these factors is that it hard to influence them through policy. The Dutch have little faith left in their ability to mold society, which has affected political primacy. However, innovating in the "new economy" requires the government to take the lead at certain times and stimulate innovative activities. Dutch economic policy in general and with regard to innovation is still founded too much on neoclassical economic principles, which do not sit well with innovation, because innovation is too important to be left to invisible market forces. What is perhaps the most persuasive proof that this argument is correct is that Dutch entrepreneurs themselves often want to play a more active role in improving the nation's innovative ability.

An interesting example of the ratio between macro and micro with regard to innovation is the current discussion surrounding a more flexible approach to labor contracts. It is very doubtful that this development will lead to more innovation. According to neoclassical theories, a flexible labor market—for instance, with short-term contracts, a limited right to fire people, and non-binding collective labor agreements—is an important condition for economic growth and growth in employment. The Dutch (but German-based) economist Alfred Kleinknecht disagrees with that assessment, arguing that a rigid labor market leads to more innovation and with it more economic growth. He argues that a more flexible approach shortens the time horizons of employees, which means they identify less with their company (see, among others, Kleinknecht and Naastepad, 2005). Because innovation is a long-term affair, it clashes with the time horizons that are preferred by entrepreneurs.

For the Dutch government, it is important to make sure that the current innovation policy is significantly different from the industrial policy of the 1970s and 1980s. To do so, it is necessary to shift the emphasis of innovation policy from *content* toward *process*. In other words, rather than having a policy that is aimed at investing and providing support to industrial sectors, it is wiser to create specific conditions that allow entrepreneurs to innovate. This kind of innovation policy does not have to stimulate entrepreneurs and businesses, but it has to create the freedom entrepreneurs need to innovate. The Dutch Government needs to be aware that, given the diversity of the Dutch business community, its innovation needs to have a similar diversity. Innovation is almost by definition contextual in nature. Businesses will innovate differently in different contexts (Ortt and van der Duin, 2008). That is why the government's innovation policy must adapt to various contexts. For example, some rules may have an effect on the agricultural sector, but have an adverse effect on the construction sector. And stimulating radical innovation processes requires a different approach from encouraging incremental innovation processes. Based on this, the conclusion has to be that innovation policy has become more microscopic in nature, in which business aspects, in general, and innovation aspects, in particular, need to be given a larger space than classical macro-economic subjects. That is why it is a good thing that the Ministry for Economic Affairs has embraced the open innovation concept by Henry Chesbrough (2003) (see also AWT, 2007), even though at the same time it has fought the hype surrounding this concept. Another expression of the application of a management concept is the idea of a "launching customer", which shows the clear influence of ideas from Eric von Hippel (2005) surrounding lead users. By letting ministries act as launching customers, a diffusion of innovative products and services can be realized.

The problem that keeps returning is the poor exploitation of scientific and technological innovations (failing knowledge transfer). The gap between knowledge and entrepreneurship seems impossible to close adequately. It is quite possible that one of the reasons the Dutch innovation system fails in this respect is the strong influence of the wrong metaphor. Often people speak of a "gap" between science and the business community, or of "bridging" the gap between science and business, or of "shortening" the road from idea or patent to new product. Although metaphors in themselves may be innocent, they express a view in which the innovation process is erroneously seen as a linear process. In this process, ideas, patents, and scientific knowledge are seen as input to an innovation process (hroughput), whereby input is transformed into innovations (output). Unfortunately, this does not do justice to reality. Innovation processes are rarely linear in nature, but more often than not they are chaotic, and it is hard to determine the beginning and end of the innovation process. In addition, there are many feedback loops and relationships with parallel innovation processes. Innovation policy that is based on a linear process and in which the errors in the innovation system are largely considered from within a linear framework is an innovation policy that will perform poorly and tackle the wrong problems. For instance, although the idea to invest money in R&D and pay more attention to patents makes perfect sense from a linear approach, it will have little effect on innovation processes at companies where second, third, or fourth-generation innovation models apply. Putting more money into R&D in that kind of situation can be compared with solving the problem of drinking yogurt through a straw by putting more yogurt in the carton. Dutch innovation policy seems to be based on the principles of first-generation innovation management (i.e., linear and technological), while at the same time using a concept like *innovation systems*, which is part of the fourth-generation innovation management.

The widening of the approach to innovation to include non-tangible innovations like new business models and other forms of organization, as well as the increasing importance of non-economic factors as explanations of innovativeness, means that innovation should not only be viewed through an economic lens and the responsibility for innovation policy should not only lie with the Ministry for Economic Affairs. In recent decades the Ministry for

Education has already become more closely involved in innovation policy, because education as a result of problems with the number of science students is increasingly seen as a determinant of innovation policy. Also, the increasing importance of social factors (e.g., regarding the ongoing discussion surrounding labor rights) makes it necessary for the Ministry for Social Affairs and Employment to contribute to innovation policy. And in realizing its "launching customers" policy, the Ministry for Health, Welfare, and Sport can play a useful role (e.g., by making hospitals aware that they are an important instrument for the innovative ability of the Netherlands). The fragmentation of the Dutch government is not helpful to innovation policy (see also SER, 2003, p. 7). The Dutch innovation system needs an integral and, therefore, interdepartmental approach to take the diversity in innovation and diversity in the innovation system into account.

In this context, we can also point to "social innovation", which means the "innovation of the labor organization and a maximum utilization of competencies, aimed at improving the business performance and deployment of talent" (SER, 2006, p. 13). Thus, improving innovation and the innovation system are not merely seen as a matter of more money and/or technologies, but rather as a smarter way of working and organizing that allows businesses both to adopt and to produce innovations better. However, no matter how logical and desirable social innovation may be, it is hard to formulate an innovation policy that promotes social innovation. It is clear that the SER also struggles with this problem when we read their advice to the government, in which all that is mentioned is that the government needs to play a facilitating role in the promotion of social innovation, without specifying how it should do that.

In summary, innovating in the Netherlands not only has an economic significance, but also contains social and cultural aspects. Improving the Dutch innovation system, and in particular the accompanying innovation policy, needs to address these aspects. The importance that is contributed to innovation across a wide spectrum is a step in the right direction, but it can only be meaningful if the right policies are designed and implemented. For the Dutch economy and society to have a healthy future, innovating the Dutch innovation policy is crucial.

49.8 References

AWT. (2004). *Kennisbeleid bij de Nederlandse overheid: Een inventarisatie van het kennisbeleid bij de Nederlandse ministeries* [Knowledge Policy at the Dutch Government: An Inventory of the Knowledge Policy at Dutch Ministries]. The Hague: AWT.

AWT. (2007). *AWT-advies nummer 68, Openheid van zaken: Beleid voor Open innovatie* [AWT Advice 68, Openness of Affairs: Policy for Open Innovation]. The Hague: AWT.

Berkhout, A. J. (2000). *The Dynamic Role of Knowledge in Innovation: An Integrated Framework of Cyclic Networks for the Assessment of Technological Change and Sustainable Growth.* Delft, The Netherlands: Delft University Press.

Berkhout, A. J., van der Duin, P. A., Hartmann, L., and Ortt, J. R. (2007). *The Cyclic Nature of Innovation: Connecting Hard Sciences with Soft Values.* Oxford, U.K.: Elsevier.

Boschma, R. A., Frenken, K., and Lambooy, J. G. (2002). *Evolutionaire economie: Een inleiding* [Evolutionary Economy: An Introduction). Bussum, The Netherlands: Uitgeverij Coutinho.

Brinkhorst (2003). *Innovation Letter.* The Hague.

Butter, F. A. G., den. (2007). "Innoveren in de Nederlandse transactie-economie" ["Innovating in the Dutch transaction economy"]. Lecture presented at *SenterNovem, May 14, Bergen, The Netherlands.*

Carlsson, B., Jacobsson, S., Holmén M., and Rickne, A. (2002). "Innovation systems: Analytical and methodological issues." *Research Policy*, **31**, 233–245.

CBS. (2007). *Statistics.* Voorburg/Heerlen, the Netherlands: CBS (Dutch Statistical Office).

Chesbrough, H. W. (2003). *Open Innovation: The New Imperative for Creating and Profiting from Technology.* Boston, MA: Harvard Business School Press.

DOST. (2005). *Science and Technology Indicators.* Leiden, the Netherlands: Dutch Observatory of Science & Technology

Duin, P. A., van der (2006). *Qualitative Futures Research for Innovation.* Delft, The Netherlands: Eburon Academic Publishers [published PhD thesis].

Duin, P. A., van der, and Sabelis, I. (2007). The future revisited: An application of lessons learned from past futures. The socio-cultural domain and innovation policy in the Netherlands. *Foresight*, **9**(2), 3–14.

Duin, P. A., van der, Ortt, J. R., and Kok, M. (2007). "The Cyclic Innovation Model: A new challenge for a regional approach to innovation systems?" *European Planning Studies*, **15**(2), February, 195–215.

Duin, P. A., van der, van Oirschot, R., Kotey, H., and Vreeling, E. (2009). "To govern is to foresee: An exploratory study into the relationship between futures research and strategy and policy processes at Dutch ministries." *Futures*, forthcoming.

Eurostat. (2006). *Statistics.* Luxembourg: Eurostat.

Hekkert, M. P., Suurs, R. A. A., Negro, S. O., Kuhlmann, S., and Smits, R. E. H. M. (2007). "Functions of innovation systems: A new approach for analysing technological change." *Technological Forecasting & Social Change*, **74**, 413–432.

Hofstede, G. (1991). *Allemaal andersdenkenden: Omgaan met cultuurverschillen* [Cultures and Organizations: Software of the Mind]. Amsterdam, The Netherlands: Uitgeverij Contact.

Kleijn, M. (2007). "Innovation policy in the Netherlands." Guest lecture presented at *Delft University of Technology, March 22.*

Klein Woolthuis, R., Lankhuizen, M., and Gilsing, V. (2005). "A system failure for innovation policy design." *Technovation*, **25**, 609–619.

Kleinknecht, A., and Naastepad, C. W. M. (2005). "The Netherlands: Failure of a neo-classical policy agenda." *European Planning Studies*, **13**(8), December, 1193–1203.

MEA. (2003). *Analyse van de Nederlandse innovatiepositie* [Analysis of the Dutch Innovation Position]. The Hague: Ministry of Economic Affairs.

MEA. (2007). *De overheid als launching customer: Geef innovatie een kans om door te breken!* [The Government as Launching Customer: Give Innovation a Chance for a Breakthough!] (Publicatie No. 07DC06). The Hague, The Netherlands: Ministry of Economic Affairs.

Niosi, J. (1999). "Fourth-generation R&D: From linear models to flexible innovation." *Journal of Business Research*, **45**, 111–117.

OECD. (1997). *National Innovation Systems*. Paris: Organization for Economic Cooperation & Development.

OECD. (2007). *Statistics*. Paris: Organization for Economic Cooperation & Development.

Ortt, J. R., and van der Duin, P. A. (2008). "The evolution of innovation management towards a contextual approach." *European Journal of Innovation Management*, **11**(4), 522–538.

Porter, A. L., Newman, N. C., Jin, X. Y., Johnson, D. M., and Roessner, J. D. (2008). *High-tech Indicators: Technology-based Competitiveness of 33 Nations* (2007 Report to the Science Indicators Unit, Division of Science Resources Statistics). National Science Foundation.

Prent, M. (2008). "Innovative solar cell technologies in the land of the rising sun: The promoting and limiting factors for PV development and diffusion in Japan and what we can learn for the Dutch case." Master's thesis, Delft University of Technology, The Netherlands.

Rothwell, R. (1994). "Towards the fifth-generation innovation process." *International Marketing Review*, **11**(1), 7–31.

SER. (2003). *Interactie voor innovatie* (Publicatienummer 11). The Hague: SER (Socio-Economic Council).

SER. (2006). *Welvaartsgroei door en voor iedereen: Thema Social innovatie* [Welfare Growth by and for Everybody: On the Theme of Social Innovation]. The Hague: SER (Socio-Economic Council).

Smits, R., and Kuhlmann, S. (2004). "The rise of systemic instruments in innovation policy." *International Journal of Foresight and Innovation Policy*, **1**(1/2), 4–32.

Von Hippel, E. (2005). *Democratizing Innovation*. Cambridge, MA: MIT Press.

WEF. (2007). *Global Competitiveness Report*. Geneva, Switzerland: World Economic Forum

Zanden, J. L., and Griffiths, R. T. (1989). *Economische geschiedenis van Nederland in de 20e eeuw* [Economic History of the Netherlands in the 20th Century]. Utrecht, The Netherlands: Uitgeverij Het Spectrum.

50

The National Innovation System of Italy

Daniele Virgillito

University of Catania

Introduction

The aim of this chapter is to answer the following question: What kind of national system of innovation lies at the base of the Italian economy?

Italy represents one of the most successful examples of post–World War II economic growth. Over the past 45 years, productivity and *per capita* income have risen rapidly. Recently, Italian industry has shown a high degree of internationalization; manufacturing exports, for example, have increased approximately 9.7% between 2003 and 2007. In a relatively short period of time, Italy has been transformed from a semi-industrialized country to an advanced industrial economy, despite the fact that Italian international specialization remains closely tied to traditional products. To gain a wider understanding of the Italian system of innovation, we must consider the distinction, in terms of business facilities, between the developed northern and central areas and the weaker south. The structures of Italian companies (mostly small firms) must also be taken into account. The network of small firms, historically developed on local, regional, and vocational bases and characterized by capabilities accumulated through productive experience, has worked effectively and performed successfully over the past decades up to the present day. The success of the system is based on network relations between a large number of firms linked to each other by economic, environmental, cultural, and social factors. Networks made up of small firms across the country have generated virtuous cycles of wealth creation which have been the foundation of the successful performance of small and medium-size Italian firms over recent decades.

Throughout this chapter various examples of how the Italian networks evolved will be illustrated. In Italy the presence of numerous industrial districts composed of small and medium-sized firms continues to dominate the Italian economy.

A brief history of the Italian NIS: the dual-economy dilemma

Italy is a frequently cited example of a country with non-homogeneous economic and social development. The Italian "regional problem" is conceivably at least a thousand years old. The rapid post–World War II economic growth occurred within a dual economy in terms of firm size and geography.

Italy did not develop a modern industry until the 1950s. Advanced technological and productive capabilities, as well as managerial skills, began only in the past 40 years (Malerba, 1993). Between 1950 and 1970, Italy was a low R&D-intensive country and a technological follower. During the 1980s the distance between Italy and other major countries, in terms of total R&D expenditure, decreased (Archibugi, Cesaratto, and Sirilli, 1991). National R&D expenditure between 1980 and 1990 grew at an annual rate of approximately 10%, a value higher than that of most OECD countries (Malerba, 1993).

High-tech sectors such as electronics and aerospace significantly increased their importance, reaching 24% and 15%, respectively, in 1990. R&D funding from the public sector grew at an annual rate of 12.5% between 1980 and 1990 (Malerba, 1993). The public sector became a major source of funds for R&D. However, government support was not spread evenly across all industries and across all parts of the country. Major differences persist, in terms of R&D and technological innovation, between the north and the south (see Figure 50.1). For example, the south produces approximately 17% of the total Italian value-added, and the northwest 34% (ISTAT, 2006).

In the late 20th century the northeast regions went through one of the most impressive processes of socio-economic change to occur in Italy, the so-called "miracle of the Third Italy", with rapid transformation of its areas into fast-growing entrepreneurial regions (Bagnasco, 1977). Small firms are numerous in traditional and specialized

Figure 50.1. R&D expenditure across Italy.
Source: adapted from ISTAT (2006).

supplier sectors, which constitute a major part of Italian industry. With more than 80,000 units (ISTAT, 2006), Italy's small and medium-sized companies are the backbone of the Italian economy. On the other hand, the core of Italian industry is made up of large firms, active mainly in scale-intensive and high-tech sectors.

The central and northern regions show some similarities in the historical character of their local knowledge bases. Local governance, entrepreneurship, and local culture were firmly rooted in Emilia and Tuscany, comparable with the Lombardy area. Today, regions in the north of Italy, such as Lombardy and Piedmont, are a symbol of the technological heart of Italian industry. The economic prosperity of Tuscany started in the 11th century, based on the unbeatable cultural traditions of its most important cities, such as Lucca, Pisa, Pistoia, Siena, and Arezzo. Resources such as iron from Elba and marble from Carrara were the basis for related manufacturing activities. The strength of local communities and social ties, supported by local authorities, gave rise to specialized dynamic clusters, which became a model for the "industrial district".[1]

Currently, in all these regions knowledge flows through systematic interactions between small firms, which are particularly intense in those industrial areas. Such technological links, largely informal and unstructured, are enhanced by spatial proximity and by an economic and cultural homogeneity. These industrial clusters employ about 2.2 million workers and account for almost one-half of all manufacturing employment in Italy. These firms are major generators of export revenue, accounting for U.S.$35 billion out of a total turnover of U.S.$80 billion. In this area the role of institutional actors (e.g. specialized business services, technology transfer agencies, private business associations, chambers of commerce, and training agencies) creates a favorable, specific environment that supports the entrepreneurial process (Zamagni, 1990).

Historically, the whole central area was under the political influence of the Roman Catholic Church.[2] Whereas in Umbria and Marche agriculture and ceramics industries were the major strengths, Lazio's economic and social activities were basically linked to the bureaucratic structure (administrative, legal, and representative services). The economy of the region of Lazio is mostly related to the services sector. This state of affairs has persisted since the period of Italian unification[3] (1815–1861), and was further reinforced after by the choice of Rome as the national capital.

Approximately 73% of the working population are employed in the services sector; this is a considerable proportion, but is easily explained by Rome being the core of public administration, banking, tourism, insurance, and other sectors. Many national and multinational corporations, public and private, have their headquarters in Rome. As a result of its past and present role, a large proportion of the national public R&D infrastructure and expenditure is currently concentrated in Lazio. The most frequent relationships are those between a restricted number of science-based firms and public and private research institutes. In the whole central area, systemic interactions between firms and public institutions do not play a major role, and support from local governments is neither proactive nor particularly effective; regional social environments, more generally, are not particularly cohesive, nor are their economic structures oriented towards technological change and institutional reformism. The collaborative relationships between firms, as well as other forms of knowledge and technological connections, have become intense in the last few years. The south of Italy is characterized by little industrialization, low R&D intensity, and limited dissemination of advanced technologies (Camagni and Capello, 1999).

The so-called Italian *Mezzogiorno* (i.e., southern Italy) had its golden age under the rule of the Normans (Frederick II), who by 1266 had made the area economically and culturally advanced and created the magnificence

[1] For example, the case study of Prato's textile industry (see Becattini, 1987).

[2] After the short Roman Republic and the region's annexation to France by Napoleon I, Latium again became part of the Pontifical States. In 1870, when the French troops abandoned Rome, General Cadorna entered the pontifical territory, occupying Rome on September 20, and Latium was enclosed within the Kingdom of Italy.

[3] Italian unification (*risorgimento*) was the political and social process that unified different states of the Italian peninsula into a single nation. It is difficult to identify the exact dates of the beginning and end of the Italian reunification process. However, most scholars assert that it began with the end of Napoleonic rule and the Congress of Vienna in 1815, and ended with the Franco-Prussian war in 1871 (Zamagni, 1990).

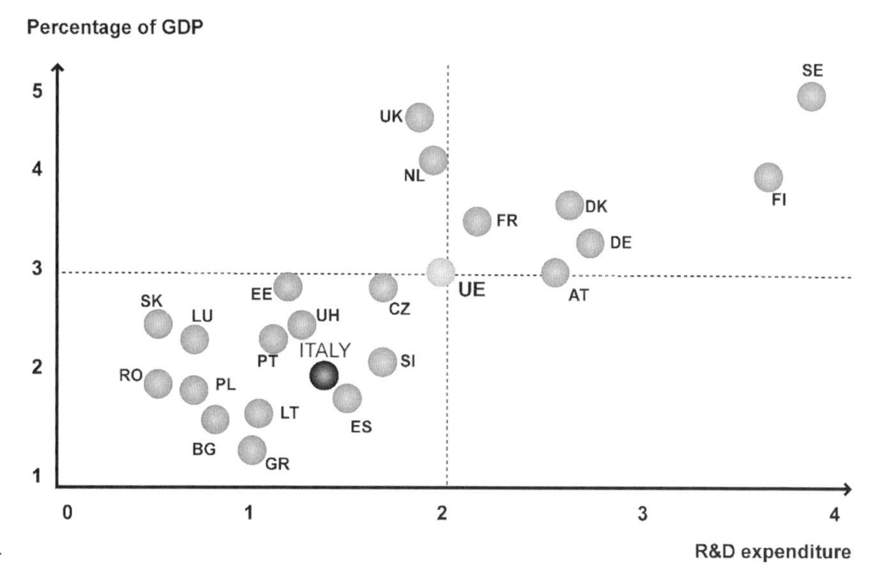

Figure 50.2. Business R&D expenditure as a percentage of GDP.
Source: adapted from Eurostat (2006).

of such cities as Naples and Palermo.[4] Sardinia's development was strongly affected by being subjected for long periods to colonization, first by the Spanish, then by the Savoyards from Piedmont.[5]

Various historical factors contributed to the weakness of territorial systems and the lack of social capabilities for institutional change in the whole of southern Italy. The southern part of Italy to this day is considered a backward area, particularly in terms of innovation and technology indicators, when compared with the rest of the nation and the European Union (EU). Its problem mainly relates not only to the weak technological performance of its firms, but also to the nearly complete absence of any systemic dimension of innovation processes, at least at the regional level.

However, various empirical studies suggest that firms are carrying out innovation activities separately, through other firms and R&D public institutions. Contemporary economic literature, particularly in the last decade, suggests that in the area of southern Italy, clusters of high-tech related enterprises are forming, thanks to the financial support that derives from specific European Community programs (Evangelista *et al.*, 2002; Guerrieri and Iammarino, 2002; Iammarino, Jona Lasinio, and Mantegazza, 2004; Schillaci and Virgillito, 2004).

Summary of output trends

With respect to business R&D expenditure as a percentage of GDP, the Italian figure is approximately 50% lower than the EU average (see Figure 50.2).

The low level of business R&D expenditure thus reflects both the de-specialization of Italy in these sectors and the predominant role of small firms in manufacturing in general (Balconi, Breschi, and Lissoni, 2004).

As was highlighted earlier, Italy is characterized by strongly marked technological and economic regional differences. The northwest of the country received over 38% of total national public R&D expenditure in 2006, followed by the central area (24%). The northeast and southern areas received a mere 39% of total R&D expenditure. The bulk of R&D activity in Italy is, in fact, concentrated in just three regions: Lombardy, Lazio, and Piedmont. The contribution of medium-sized regions, such as Emilia Romagna, Tuscany, Liguria, and Veneto, is much lower than that of the northeast. Each of these regions receives between 6% and 4% of total R&D expenditure. Lastly, all southern regions (with the exception of Campania and Sicily), as well as Umbria and Marche (in the center), represent only a small share of national R&D. More than one-quarter of total public R&D is concentrated in Lazio, while the strongest concentration of R&D by the private sector is in Piedmont and Lombardy. With the exception of

[4] Since then, the "Southern Kingdom" was ruled by Spain; the colony was administered on the basis of undisputed privileges given to the nobility, and exploitation of local resources and the peasantry. The Kingdom became "independent" under the Bourbons (1734), who traditionally have been blamed for the backwardness of the area, but who in fact were in power for only 126 years (roughly between 1734 and 1859, with the inclusion of a few years of French rule) (see Zamagni, 1990).

[5] Unfortunately, the Savoyards exploited the natural resources of the island without instituting any structural reforms or innovation processes (agricultural productivity in Sardinia in 1840 was approximately one-tenth that of Lombardy). Economic factors include the late advent of a feudal agricultural system (at a time when the society of communes had already been established in the north), a sharp separation between financial capitals and the management and organization of production, unstable and hierarchical economic relations, a rather concentrated urban geography, and more recently a process of state-led industrialization based on resource and scale-intensive sectors.

Table 50.1. Primary institutions and number of R&D personnel.

Number of R&D personnel	*2002*	*2003*	*2004*
Public institutions	39,343	42,610	44,061
Universities	122,358	120,736	123,266
Non-profit organizations	5,696	5,354	6,386
Firms	85,687	81,189	81,822
Total	*253,084*	*249,889*	*255,535*

Source: adapted from ISTAT (2006).

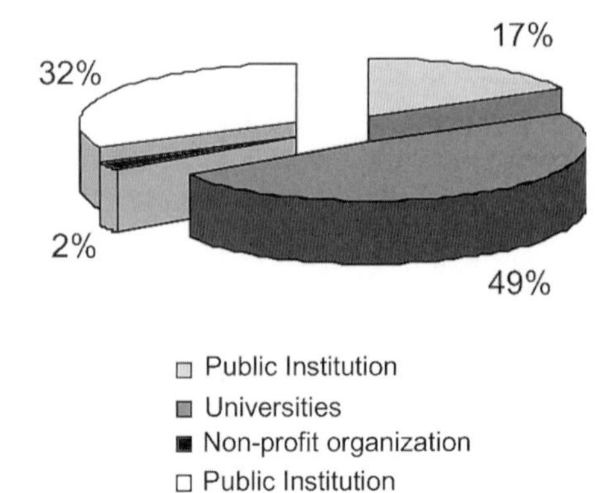

Figure 50.3. Percentage of R&D workforce per sector.
Source: adapted from ISTAT (2006).

Campania, southern regions play a very marginal role with respect to private R&D, although they show a relatively more significant contribution in terms of public resources devoted to R&D.

In absolute terms the number of employees involved in R&D in the main institutional sectors from 2002 up to the present has grown from 253,084 to 255,535 (see Table 50.1).

Sixty-six percent of all R&D workers are employed in public institutions and universities, while a mere 32% are work in private enterprises. These results confirm that R&D is mostly financed and carried out by public institutions (see Figure 50.3).

In terms of the number of patent applications in high-tech classes per million population, Italy is positioned well below the average, scoring 70% below the EU mean (Eurostat, 2003).[6] The leading technological role of the north of Italy is also confirmed by regional breakdown of patent applications submitted to the European Patent Office.

At the national level, Lombardy dominates the scene, with one-third of all Italian applications (Eurostat, 2002): Emilia Romagna (15%), Piedmont (14%), and Veneto (10%). Catania (Sicily) is an exception because of STMicroelectronics. In fact, this company has contributed towards the growth of a significant cluster of ICT-related companies and has developed a large number of industrial patents, approximately 200 in 2002 (Schillaci and Virgillito, 2004).

Italy appears to be specialized in industrial automation (which is strongly correlated with the automotive industry), with 5.6% of total EPO (European Patent Office) patent applications in this sector.[7] This positive result is also confirmed for the sub-sectors of motors and machine tools and particularly for industrial robots (see Table 50.2). In high-tech industries most patents are registered by companies with more than 1,000 employees.

[6] *http://www.istat.it/servizi/infodati/ESDS/dationline.html*
[7] *http://www.epo.org/patents.html*

Table 50.2. Number and percentage of patents per high-tech industry.

Sectors	*Italian*	*patents*
Computer hardware, semiconductor devices, and electronic components	671	23.9%
Consumption electronics and telecommunication (hardware)	399	14.2%
Pharmaceuticals	384	13.7%
Precision instruments, measurement, and control apparatus	372	13.2%
Fine chemicals	223	7.9%
Industrial automation	222	7.9%
Plastics	203	7.2%
Optical instruments	136	4.8%
Electromedical instruments	134	4.8%
New materials	48	1.7%
Aerospace	17	0.6%
Total	*2,809*	*100%*

Source: adapted from ISTAT (2006) data.

Technology commercialization initiatives

In order to better comprehend the evolution of the National Innovation System (NIS) in Italy, considering Italy's heterogeneous social and economic development, it is necessary to map the potential and actual initiatives linked to the innovation system, in various geographical areas (see Figure 50.4).

A Aerospace
F Food industry
C Cultural heritage
B Biotech
IT ICT and TLC
L Logistic
IA Industrial automation
N Nanotechnologies
NA New materials and advanced materials
E Environmental risks
M Microelectronics and semiconductors

Figure 50.4. The Italian districts.
Source: adapted from Bonaccorsi (2005).

The region in Italy with the highest innovation potential rate is Lombardy. Activity in this region is distributed among different sectors, mainly electronics, biotech, fine chemistry, and industrial automation. The central and northern regions of the country have significantly higher potential in the high-tech sector than the southern regions, and this is especially true with regard to industrial research. Academic research is also stronger and more efficient in these areas (see also Balconi, Breschi, and Lissoni, 2003; MURST, 1999). In particular, the regions of Lazio and Rome have a high concentration of academic research activity. A few significant local systems, in terms of high-tech activity, exist in Italy. These include electronics in the Milan and Turin areas; biotech (pharmaceuticals) in Milan and Rome; industrial automation in Milan, Turin, and Bologna; and an important high-tech and semiconductor cluster in Catania.

Even though Italy is a large market for electronics and telecommunication businesses and is considered to be the largest European market for cellphones, the country produces very little in this sector. Moreover, the computer sector, in terms of hardware and components, has been heavily affected by closure of the Olivetti Computer Production Company, which was one of the largest manufacturers in Europe in the 1980s. Currently, to our knowledge, no significant computer producer is operating in the country.

In the motor vehicle industry, the major automotive group in the country is Fiat-Auto and Ferrari.

High-tech activities and innovation involving this industry range from new production methods to machines for metal working, from robotics to electronics. A variety of innovative and internationally competitive small firms producing equipment, many serving the automotive industry, are present in the north of Italy. This is a highly dynamic industry; moreover, many new entrepreneurs are spinoffs from established equipment firms or from large users such as Fiat. Patenting activity in the mechanical industry is relatively high.

The following sections provide further insight into the Italian NIS from the viewpoint of the ICT (information, communication, and technology), biotech, and pharmaceutical industries.

National technology policy, including current technology commercialization initiatives at the national level

The information and telecommunication industry (ICT) absorbs a large share of total national expenses, accounting for 20% in 2005 (ISTAT, 2006). Some dominant firms in this field are Telecom Italia, Telecom Italia Mobile, and Tiscali. In the past few years, Tiscali has grown rapidly to become a pan-European Internet service provider through a wide number of acquisitions. In the field of semiconductors, the most important company is STMicroelectronics, whose main R&D centers are located in Milan and Catania (Sicily). STM currently owns around 80% of the patents in the sub-sector of electronic components. The activity in Milan and Catania in this field has been widely studied in order to test the spinoff potential in both areas. The Milan area (including Pavia, Bergamo, Brescia, and Varese) has the highest activity for the electronics sectors. The technical schools of Milan and Pavia account for around 500 researchers in electronics-related fields. The R&D centers of large multinationals, including Alcatel, Bull, Ericsson, Pirelli, Siemens, and STMicroelectronics, as well as the smaller centers of Agilent, Lucent, and Philips, account for around 5,300 R&D employees (Gattoni, Modena, and Vita-Finzi, 2001). The Catania area registers approximately 1,500 engineers and scientists. Fewer than 1,000 of these are currently working at STMicroelectronics, which dominates the industry in the region (Schillaci and Virgillito, 2004). In 2001 almost 200 firms related to the ICT sector at various levels of research intensity were located in the area; a few are STM spinoffs related to the semiconductor sector. Today, to our knowledge, no venture capital fund is established in the southern regions. The advantageous political and financial policy for this area during the past few years has attracted R&D departments from some large multinational firms, for reasons mainly related to access to EU structural funds. Given that Sicily is an "Objective 1 Area",[8] STM and other significant R&D departments such as Alcatel, Siemens, Bull, Ericsson, Telecom Italia, Magneti Marelli, Nokia, and IBM operate in the region.

Regional agencies, local public authorities, and universities (especially in the north) are effective in supporting the needs and requirements of small firms in the area. The full range of contacts and interactions that form the framework of an innovation system in this area is highly developed in terms of structured relationships among firms as well as between firms and other organizations (universities, research institutes, industry associations, etc.); good scientific and technological infrastructure; R&D intensity; diffuse networks of technological services; attractiveness for external sources of technology; effective regional innovation policies; relatively high institutional flexibility and adaptability to change; and structured social ties and networks.

The Italian pharmaceutical industry ranks fifth in the world for sales of finished products, sixth for number of employees, and seventh for exports. Unfortunately, the relatively good performance in production and distribution activity is not matched by consistent research activity. Italy spent only around $0.8 billion in 2000, compared with $15.4 billion by the U.S., $4.6 billion by Japan, $3.2 billion by the U.K., $2.7 billion by Germany, $2.5 billion by France and $1 billion by Switzerland (Farmindustria, 2001).

Research expenditure as a percentage of total revenues in the sector was about 6.05% in Italy in 2001. The average in the European Union was approximately 11.75% and in the U.S. 15.91%. These data testify to the scarcity of research activity in Italy, relative to its market.

The pharmaceutical sector is characterized by a high level of concentration. The top 25 companies, which account for 58.51% of total turnover, can be categorized as branches of international "big pharmas" such as Glaxo, Merck, Novartis Bayer, and Ciba, and Italian medium-sized firms such as Geigy and Menarini & Bracco. In Italy, research and patenting activity in this field are mostly carried out by large firms.

In the past 10 years the biotech sector has consistently provoked more interest among investors, largely because of the scientific revolution that has occurred primarily in the field of genetics. Despite being the fifth largest world market for pharmaceuticals, Italy has not been involved in the industrial boom that has characterized this sector. However, in the past few years Italy has demonstrated strength in academic research in this sector. It was found that as many as 10,000 researchers are working in related fields (genetics, medicine, biotech, or pharmaceuticals). These are, for the most part, evenly distributed throughout the country. It is also worth noting that the number of physicians *per capita* in Italy is twice as high as the average of other European countries. The most active centers in Italy are Milan and Rome.

Other significant concentrations of academic activity are developing throughout the country in cities such as Turin, Padova, Bologna, Pavia, and Naples. In spite of a large market, at the moment very few biotech spinoffs are present. Italy, in fact, has one of the lowest numbers of biotech enterprises in Europe, approximately 50, extremely low when compared with other European countries, such as the U.K. with 325, Sweden with 200 and Finland, with approximately 95 (see Ernst & Young, 2003).

Relative strengths and weaknesses in technology

A large part of Italian industry is composed of many small and medium-sized firms, mostly operating in traditional industries. The small firm has grown in importance to become the most typical configuration of the Italian industrial system. More than 80% of productive units and 80% of employees involved in industrial activities are concentrated in the central and northern regions. Note that the spread of small firms is not concentrated in larger towns; about 80% of industrial productive units and 80% of employees in industry are situated in towns with fewer than 100,000 inhabitants. This means that the choice of location for production has been in the majority of cases on the basis of local traditions or knowledge and on the need to maintain close contact within the social context.

These firms are highly profitable and successful internationally. These small and medium-sized firms create a highly dynamic disparate learning network and are characterized by advanced capabilities for absorbing, adapting, and improving new technologies. Innovation primarily originates not by formal R&D agreement, but by informal networking. Engineering skills, product know-how, and understanding of customer requirements are the major sources of incremental innovations and product customization by this network.

The origins of the current growth model are founded on the industrial district model. This type of production system foresees geographic proximity and the sharing of values and knowledge among the stakeholders of the value

[8] Regions with economic and industrial underdevelopment, which include Sardinia, Sicily, Campania, Basilicata, Puglia, and Calabria.

chain. Within industrial districts, characterized by both cultural and social homogeneity and developed historically on a vocational basis, innovation occurs among a large number of small and medium-sized firms (Becattini, 1987). These districts are active in several industries and are located in various Italian regions: textiles in Prato, Como (silk), Biella, and Carpi; footwear in Vigevano, Barletta, and Casarano; furniture in Brianza and Udine; ceramic tiles in Sassuolo and Caltagirone; gold jewelry in Valenza Po and Arezzo; household products in Lumezzane; and high-tech products in Catania. Some of these districts have been in existence for decades, while others grew up during the 1960s and 1970s, such as Sassuolo, Prato, and Valdarno Inferiore and, most recently, Catania for microelectronics (see Figure 50.4).

In these districts, productive flexibility and adaptability to changing market demands at the final product level are significant. Most firms are specialized in one stage of the production process, while a few firms operate in more than one phase of the value chain and eventually sell the final product.

Local institutions and associations play a major role in supporting the external network of the district. Regional and local governments, banks, and professional schools provide public support, financial resources, and a qualified labor force for firms. Export and distribution associations help overcome the problems faced by small firms in selling their final products on international markets (Guerrieri and Iammarino, 2002).

Dissemination of process technology within the district is rapid. Technical change is circulated within the district by widespread transmission of knowledge among a large number of producers who share a common culture. Their similarity is also the basis for transmitting and assimilating explicit (sometimes very valuable) non-codified knowledge. In this respect, personal contacts and workforce mobility between firms play major roles in developing the system. Note that a number of large Italian firms have maintained some of the attributes of the industrial district by using a very decentralized productive organization, in which a large number of small local firms are specialized in a specific stage of the production process or in the supplying specific inputs. Hence the reason the emergent network model is based on the existence of the district model.[9]

In Italy, in spite of relevant quantitative growth in terms of number of high-tech companies during recent years, a number of qualitative elements of the National Innovation System are not well developed. Proper regional systems of innovation are found in only in a few well-defined areas. In most regions, systemic interactions and knowledge flows between the relevant actors are simply too sparse and too weak to reveal a strong innovation system. The Italian context is being shaped by such crucial points as strong territorial differences in terms of both heterogeneity and historical contingency; lack of a system connecting its numerous administrative regions; highly problematic measurement issues due to persistent and significant interregional and intraregional differentiation; relatively low technological intensity of specialization patterns; and copious evidence of policy failure, particularly with respect to innovation and technology policies.

Funds flow for innovation

The government is the major sponsor of scientific and technical research in Italy. National R&D expenditure from 2002 to 2004 recorded a positive growth: from €14,599,933 in 2002 to over €15,200,000 in 2004 (ISTAT, 2006). Firms get 53% of the total, 66% of which is given to private enterprises. Public administration, including that of research institutions, universities, and other public institutions, covers 47% of total expenditure. Firms' R&D expenditure is highly concentrated in large firms, which are normally part of industrial groups. Around 9% of the firms regularly involved in R&D account for over 80% of the total R&D expenditure by firms (ISTAT, 2006), thus confirming that R&D is mainly carried out by large firms. Around 15% of national R&D funding is provided by public administration, 75% by firms, and 11% by international funds. Almost 90% of public administration R&D expenses have been used directly by public institutions and the rest by private firms (ISTAT, 2006).

The central institutional point of reference for Italy's science and technology system lies basically with the Ministry of Education & Scientific Research and the Ministry of Industry. The Ministry of Education & Scientific Research determines R&D and higher education policy, formulating development plans for universities and scientific institutions. The Ministry of Industry promotes strategic industrial research and oversees the research carried out by such specialist agencies as ENEA (National Agency for New Technology, Energy, & Environment), INFN (National Institute for Nuclear Physics), INFM (National Institute for Physics & Materials), and other related agencies.[10]

---

[9] The Benetton Group is often cited because it has successfully matched a decentralized production organization (typical of Italian textile firms) with an advanced electronic sales network. It created a hierarchical system composed of independent medium-size firms assembling and controlling the production of a large number of specialized independent subcontractors. Almost 80% of Benetton's production is handled by external firms and artisan shops specialized in labor-intensive and non-technologically progressive operations. Benetton's distribution, on the other hand, is characterized by an advanced network linking Benetton with a decentralized sales structure composed of independent agent firms and numerous selling points. This network allows Benetton to keep in close contact with customers, and to maintain control over information and market demand in different countries (Becattini, 1987).

[10] *www.miur.it*

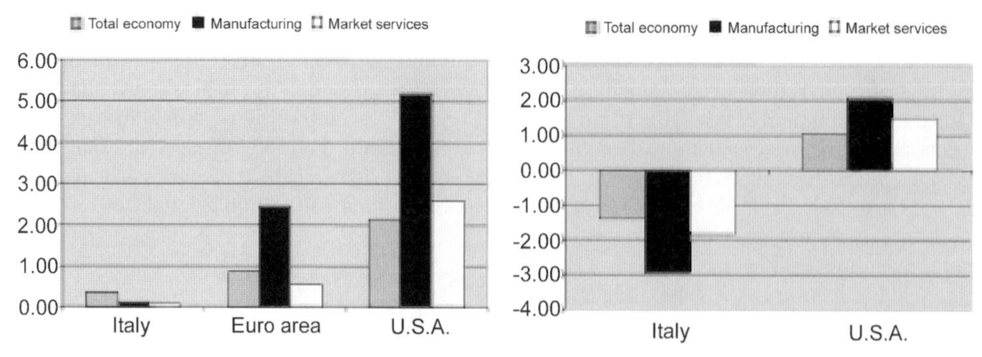

Figure 50.5. Comparative tables: (*left*) productive growth from 1995 to 2005; (*right*) productivity growth acceleration (1995–2005) over (1989–1995).
Source: OCSE (2006).

The bulk of financial government sponsorship is delivered through agencies of the Italian government, such as CNR (National Council for Research) and ENEA. ENEA directly engages in a wide range of research projects, most of which concern alternative energy and biotechnology today. ENEA's core business is to carry out applied research of benefit to Italian industry. It also conducts contract research in various areas, such as materials testing for Italian and foreign organizations. ENEA's budget is devoted almost exclusively towards the salaries of its technical and scientific workforce and administration costs. ENEA also promotes and participates in research consortia at both the national and international levels, and it owns shares in a number of high-tech companies. The major focus of research activity in these consortia and companies is agri-biotech, renewable energy, and environmental protection. ENEA also provides training and technical support to SMEs and startups. The CNR accounts for more than 350 research institutes and centers, most of which are closely connected to universities all over Italy. Its fields of research consist primarily of economics and social sciences.

Cultural and political drivers influencing the NIS

Data on Italy's National Innovation System are difficult to come by: studies are neither recent nor numerous. From a general perspective, Italian innovation indicators are well below the EU average strictly regarding high-tech activity, whereas they are roughly average or above-average regarding innovativeness in all manufacturing sectors (see Figure 50.5).

However, when studying R&D indicators, we must take into consideration the nature of the Italian political system. In fact, innovation in Italy is kept alive primarily thanks to financial stimuli coming primarily from the public sector, and only in part from private corporations (see Table 50.3). The share of R&D government-financed support relative to total business enterprise R&D is approximately 77%.

In the enlightening work of Malerba (1993) a clear distinction is made between the "core R&D system", which is mostly managed by the main actors dealing with R&D (i.e., universities, large publicly-owned research centers such as ENEA and the CNR), large public and private firms, and the "small firms system". In his view, the system is supported by three major actors:

- the university system;
- large publicly-owned research agencies, such as the CNR, ENEA, INFN, INFM, ASI (Italian Space Agency), and INS (National Institute of Health) which report to various ministries of the national government;
- large public and private enterprises, which have their own R&D infrastructures.

Over the past decade, Italy's university system has undergone many changes related to the structure of the teaching system and research activity. Teaching reforms were created in order to solve two problems: the marked difference between education systems in Italy and the rest of Europe (the Italian *laurea* is a year longer than a European B.A. or B.Sc.), and the duration of undergraduate studies which, on average, require 60% more time than for the rest of Europe (see Table 50.4). Because of this disparity, Italian graduates/undergraduates enter the workforce a few years later than their European counterparts. In 2000, for example, only 7.5% of graduates received their degree on time (at approximately age 23), whereas more than 50% were 28 years old or older. In order to solve these problems, a "first-level" degree was introduced in 2001. This degree will be followed by a "second-level specialization" after a further 2 years of study.

The university research system recently became completely autonomous. At the present time, universities can

Table 50.3. Public versus private R&D expenditure.

Country	*Private R&D expenditure*		*Public R&D expenditure*		*Total R&D expenditure*	
	Euros	*GDP (%)*	*Euros*	*GDP (%)*	*Euros*	*GDP (%)*
Italy	6,646	0.55	7,502	0.58	14,148	1.13
France	17,777	1.18	13,103	0.87	30,880	2.05
Germany	35,899	1.69	18,890	0.79	54,789	2.48
U.K.	13,207	0.83	9,416	0.76	22,623	1.59

Source: adapted from NVC (2006).

Table 50.4. League table of the first 30 countries.

Ranking of the first 30 countries	*Country*	*a*	*b*	*c*
1	Canada	1,503	45	84
2	U.S.	1,339	39	88
3	Japan	1,579	37	84
4	Sweden	1,765	35	83
12	U.K.	2,190	29	65
18	Germany	1,190	25	84
19	France	2,796	24	65
29	Italy	1,227	11	48
	Mean OCSE	*1,608*	*25*	*67*
	Mean EU19	*1,530*	*23*	*56*

Source: OCSE (2006).
Notes: [a]Number of citizens between 25 and 34 years of age with science degrees per 100,000 population.
[b]Percentage of the population between 25 and 64 years of age with degrees.
[c]Percentage of the population between 24 and 65 years of age at the undergraduate level.

autonomously decide on their regulations regarding their links with industry, their participation in university spin-offs, the intellectual proprietary rights of researchers, and the possibility of their taking part in industrial activities. In the past few years, many institutions, especially in the north of Italy, have set up technology incubators and liaison offices. However, the external university network with private firms remains the primary weak point of the Italian Innovation System (Iammarino, Jona Lasinio, and Mantegazza, 2004). On the other hand, this is not the only weak element in the system. In fact, its lack of competitiveness is predominately related to the scarcity of public financial resources associated with the development of scientific research (see Table 50.5).

The main consequence of this situation is the phenomenon of skill shortage. One need only compare the number of Ph.D.s in Italy with the average number at the European level to understand the dimensions of this problem. In 2003 the average number of Ph.D.s in the UK, for example, was approximately 140% more than in Italy during the same year.

The distribution of scientific publications on R&D indicates a scarcity of research activity in vast areas of the nation. In fact, there are few centers of excellence, except those located, for the most part, in the central and northern areas of the country. Despite distribution numbers of scientific publications being below the European average, there are a few centers of excellence in the southern part of Italy, an area famous for being particularly backward from an economic point of view (see Figure 50.6).

In any case, a significant connection between the presence of universities and the formation of clusters of firms exists and can be considered the foundation for the development of external networks between universities, research centers, and industries (Balconi, Breschi, and Lissoni, 2003).

Conclusion

The traditional north–south divide does not fully account for the wider spectrum of regional patterns in Italy. In particular, regional innovative patterns differ not only according to the specific strategies and technological performances of firms, but also to the relevance of systemic interactions and the presence of contextual factors favorable to innovation. A "portrait" of the contemporary Italian regional division consistently reproduces these broad historical pathways. Economic differentiation, even more pronounced than political fragmentation, has

Table 50.5. Financial resources and scientific research.

	Public and private expenditure for public universities		*Public and private expenditure on GDP*		
Country	*Public resources*	*Public resources*	*Public resources*	*Public resources*	*Total*
France	81.3	18.7	1.1	0.2	1.4
Germany	87.1	12.9	1	0.1	1.4
Italy	72.1	27.9	0.7	0.2	0.09
Japan	39.7	60.3	0.5	0.8	1.3
U.K.	70.2	29.8	0.8	0.3	1.1
U.S.	42.8	57.2	1.2	1.6	2.9
OECD mean	*76.4*	*23.6*	*1.1*	*0.4*	*1.4*
EU19 mean	*84.3*	*15.7*	*1.1*	*0.2*	*1.3*
OECD total			*1*	*0.9*	*1.9*

Source: adapted from OCSE (2006).

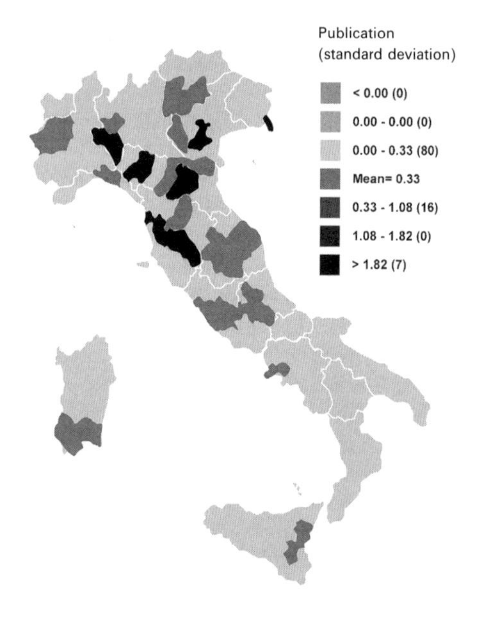

Figure 50.6. Distribution of scientific publications.
Source: adapted from Bonaccorsi, Piscitello, and Rossi (2005).

become consolidated over the centuries, preventing Italian economic development from achieving a single "natural dimension", until quite recently. What emerges from current attempts to apply the NIS concept to the Italian case (Camagni and Capello, 1999; Evangelista *et al.*, 2001, 2002; Iammarino, Prisco, and Silviani, 1996) is the variety of regional patterns, differing not only in terms of firms' strategies and performance, but also in terms of contextual and systemic characteristics, density, and quality of interactions between public and private institutions.

References

Antonelli, C. (1988). *New Information Technology and Industrial Change: The Italian Case*. Dordrecht, The Netherlands: Kluwer.

Archibugi, D., Cesaratto, S., and Sirilli G. (1991). "Sources of innovative activities and industrial organisation in Italy." *Research Policy*, **20**, 299–313.

Asheim, B. T. (1995). *Industrial Districts as Learning Regions: A Condition for Prosperity?* (STEP Report No. 3). Oslo: STEP Group.

Asheim, B. T., and Gertler, M. S. (2003). "Regional innovation systems and the geographical foundations of innovation." In: *Oxford Handbook of Innovation* (TEARI WP No. 11, October). Oxford, U.K.: Oxxford University Press.

Asheim, B. T., and Isaksen, A. (1997). "Location, agglomeration and innovation: Towards regional innovation systems in Norway?" *European Planning Studies*, **5**(3), 299–330.

Asheim, B. T., and Isaksen, A. (2002). "Regional innovation systems: The integration of local 'sticky' and global 'ubiquitous' knowledge." *Journal of Technology Transfer*, **27**, 77–86.

Audretsch, D. B., and Feldman, M. P. (1996). "Knowledge spillovers and the geography of innovation and production." *American Economic Review*, **86**(3), 630–640.

Aydalot, P. (Ed.) (1986). *Milieux Innovateurs in Europe*. Paris: Gremi.

Bagnasco, A. (1977) *Tre Italie: La problematica dello sviluppo italiano*. Bologna, Italia: Il Mulino.

Balconi, M., Breschi, S., and Lissoni, F. (2003). "Il trasferimento delle conoscenze tecnologiche dall'università all'industria in Italia: nuova evidenza sui brevetti di paternità dei docenti." In: A. Bonaccorsi (a cura di), *Il sistema della ricerca in Italia*. Milano, Italia: Franco Angeli.

Balconi, M., Breschi, S., and Lissoni, F. (2004). "Networks of inventors and the role of academia: An exploration of Italian patent data." *Research Policy*, **33**, 127–145.

Becattini, G. (Ed.). (1987). *Mercato e forze locali: Il distretto industriale*. Bologna, Italia: Il Mulino.

Bonaccorsi, A., Piscitello, L., and Rossi, C. (2005). "Local disparities in ICT adoption: Spatial heterogeneity and spatial dependence." *Italian Journal of Regional Science*, **5**(1), 11–29.

Bonaccorsi, F. (2005). *Les districts technologiques en Italie* (Osservatorio Nazionale sui Distretti Tecnologici, ONDT, research report 2005-068). Paris: Ministre de l'éducation nationale, de l'enseignement supérieure et de la recherche.

Boschma, R. A. (2003). *Social Capital and Regional Development: An Empirical Analysis of the Third Italy*. Utrecht, The Netherlands: University of Utrecht.

Breschi, S., and Lissoni, F. (2001). "Knowledge spillovers and local innovation systems: A critical survey." *Industrial and Corporate Change*, **10**(4), 975–1005.

Camagni, R., and Capello, R. (1999). "Innovation and performance of SMEs in Italy: The relevance of spatial aspects." In: M. M. Fischer, L. Suarez-Villa, and M. Steiner (Eds.), *Innovation, Networks and Localities*. Berlin: Springer-Verlag.

Cantwell, J. A., and Iammarino, S. (2003). *Multinational Corporations and European Regional Systems of Innovation*. London: Routledge.

Cooke, P. (2001). "Regional innovation systems, clusters, and the knowledge economy." *Industrial and Corporate Change*, **10**(4), 945–974.

Cooke, P., Boekholt, P., and Todtling, F. (2000). *The Governance of Innovation in Europe: Regional Perspective on Global Competitiveness*. London: Pinter.

Cooke, P., Gomez Uraga, M., and Etxebarria, G. (1997). "Regional innovation systems: Institutional and organisational dimensions." *Research Policy*, **26**, 475–491.

Edquist, C. (1997) *Systems of Innovation: Technologies, Institutions and Organisations*. London: Pinter.

Ernst & Young. (2003). *European Biotechnology Report*. San Francisco, CA: Ernst & Young.

Eurostat. (Annual). *Eurostat Regional Yearbook*. Luxembourg: Eurostat Statistical Books.

Evangelista, R., Iammarino, S., Mastrostefano, V., and Silvani, A. (2002). "Looking for regional systems of innovation: Evidence from the Italian innovation survey." *Regional Studies*, **36**(2), 173–186.

Evangelista, R., Iammarino, S., Mastrostefano, V., and Silvani, A. (2001). "Measuring the regional dimension of innovation: Lessons from the Italian innovation survey." *Technovation*, **21**(11), 733–745.

Farmindustria. (Annual). "Fatti e cifre"; *www.servizi.farmindustria. it/fatcifre/fatcifre.htm*

Fritsch, M. (2001). "Cooperation in regional innovation systems." *Regional Studies*, **35**, 297–307.

Gattoni, P., Modena, V., and Vita-Finzi, P. (2001). *The Italian Innovation System* (paper prepared for Project IFISE, June). Pavia, Italy: University of Pavia.

Gordon, I. R., and McCann, P. (2000). "Industrial clusters: Complexes, agglomeration and/or social networks?" *Urban Studies*, **3**, 513–532.

Guerrieri, P., and Iammarino, S. (2002) "Vulnerabilità e regioni nell''Unione Europea: un esercizio sul Mezzogiorno italiano." *Italian Journal of Regional Sciences*, **2**, 5–28.

Iammarino, S., Jona Lasinio, C., and Mantegazza, S. (2004). *Labour Productivity, ICT and Regions: The Revival of the Italian Dualism?* (SEWP No. 127, SPRU). Brighton, U.K.: University of Sussex.

Iammarino, S., Prisco, M., and Silviani, A. (1996). "La struttura regionale dell'innovazione." *Economia e Politica Industriale*, **23**(89), 187–229.

ISTAT. (2006). *Tavole annuali*. Rome: Collana Informazioni; *www.istat.it*

Lundvall, B. A. (1988). "Innovation as an interactive process: From user–producer interaction to the National System of Innovation." In: G. Dosi, C. Freeman, R. Nelson, G. Silverberg, and L. Soete (Eds.), *Technical Change and Economic Theory*. London: Pinter.

Lundvall, B. A. (1992). *National Systems of Innovation: Towards a Theory of Innovation and Interactive Learning*. London: Pinter.

Malerba, F. (1993). "The National System of Innovation: Italy." In: R. R. Nelson (Ed.), *National Innovation Systems: A Comparative Analysis*. London: Oxford University Press.

Mansell, R., and Steinmueller, W. E. (2000). "Competing interests and strategies in the information society." In: *Mobilizing the Information Society: Strategies for Growth and Opportunity*. London: Oxford University Press.

Markusen, A. (1985). *Profit Cycles, Oligopoly, and Regional Development*. Cambridge, MA: MIT Press.

Markusen, A. (1996). "Sticky places in slippery space: A typology of industrial districts." *Economic Geography*, **72**, 293–313.

Morgan, K. (2004). "The exaggerated death of geography: Learning, proximity and territorial innovation systems." *Journal of Economic Geography*, **4**, 3–21.

Moulaert, F., and Sekia, F. (2003). "Territorial innovation models: A critical survey." *Regional Studies*, **37**(3), 289–302.

MURST. (1999). *Il Sistema Universitario Italiano: La Popolazione Studentesca—Il Personale a.a. 1998/1999*. Rome: Sistema Statistico Nazionale.

Nelson, R. R. (Ed.). (1993). *National Innovation Systems: A Comparative Analysis*. London: Oxford University Press.

Nelson, R. R., and Rosenberg, N. (1993). "Technical innovation and national systems." In: R. R. Nelson (Ed.), *National Innovation Systems: A Comparative Analysis*. London: Oxford University Press.

NVC. (2006). *Università: I Numeri che Devono Cambiare*. Rome: Nucleo di Valutazione Confindustria.

Oakey, R. P., and Cooper, S. Y. (1989). "High technology industry, agglomeration and the potential for peripherally sited small firms." *Regional Studies*, **23**(4), 347–360.

OCSE. (2006). *Education at a Glance*. Rome: OCSE.

Patel, P., and Pavitt. (1994). "National Innovation Systems: Why they are important, and how they might be measured and compared." *Economic Innovation and New Technology*, **3**, 77–95.

Perroux, F. (1950). "Economic space: Theory and applications." *Quarterly Journal of Economics*, **1**, 89–104.

Prakash, A., and Hart, J. (1999). *Globalisation and Governance*. London: Routledge.

Saxenian, A. (1994). *Regional Advantage: Culture and Competition in Silicon Valley and Route 128*. Cambridge, MA: Harvard University Press.

Schillaci, C. E., and Virgillito, D. (2004). "Catania e ICT: Le ragioni di un successo." *L'impresa*, October.

Segal, Q. (1985). *The Cambridge Phenomenon*. Thetford, U.K.: Thetford Press.

Steinmueller, W. E. (2001). "Seven foundations of the information society: A social science perspective." *Journal of Science Policy and Research Management*, **16**(1/2), 4–19.

Storper, M. (1992). "The limits of globalisation: Technology districts and international trade." *Economic Geography*, **68**, 60–92.

Storper, M. (1998), *The Regional World: Territorial Development in a Global Economy*. New York: Guilford.

Todtling, F. (1994). "Regional networks of high-technology firms: The case of the greater Boston region." *Technovation*, **14**(5), 323–343.

Zamagni, V. (1990). *Dalla periferia al centro: La seconda rinascita economica dell'Italia (1861–1990)*. Bologna, Italia: Il Mulino.

51

The National Innovation System of Ireland

James A. Cunningham and William Golden

National University of Ireland, Galway

"Ireland's ambition is to become a leader in innovation. Our goal is to develop an innovation-driven economy that maintains competitive advantage and increases productivity"

Mary Coughlan, TD, Táiniste and Minister for Enterprise, Trade, & Employment (Coughlan, 2008)

A brief history

The Irish economy over the last two decades has been one of the top-performing economies in the world with sustainable growth rates, success in attracting leading global companies through foreign direct investment (FDI), flexible labor markets, and a fiscal policy based on Irish business attitudes towards competitiveness and globalization. The cornerstones of this success and the development of a national innovation system were laid down in the 1960s when Ireland adopted an economic development strategy which had as its central focus the attraction of foreign direct investment. The predominant focus of industrial policy right through to the late 1990s was increased employment which was achieved by attracting and retaining multinational corporations by means of tax incentives and leveraging the skilled English-speaking workforce. In addition, there was an emphasis on building an indigenous industry base that was export-focused. The two policy platforms of FDI and indigenous company growth have resulted in what some have called a "dual economy" within Ireland.

Investment in science, technology, and innovation (STI) prior to 2000 was extremely limited with researchers in Ireland largely dependent on the EU Framework Programs (see *cordis.europa.eu*), and other international competitive funding sources to sustain research activities. The investment in STI by the Irish Government between 1994 and 1999 was €0.5 billion. The shortage of national investment in STI was viewed as a major stumbling block that was in large part responsible for the lack of a sustained critical mass of research competence. A technology foresight group was created in the late 1990s under the auspices of the already established Irish Council for Science, Technology, and Innovation. The report they produced, along with an increasing recognition within public policy circles that there was a need to increase the innovative capacity of Ireland, resulted in the commitment of €2.5 billion to STI in the National Development Plan for 2000–2006.

An additional key policy document with respect to the role of enterprise in Ireland was the Enterprise Strategy Group (ESG) report published in July 2004 (O'Driscoll, 2004). That document issued a number of recommendations to sustain Ireland's competitiveness against a backdrop of unemployment of 4.7% and government debt at 34% of GNP at the time. The characteristics of competitive advantage that the ESG identified for Ireland were based on expertise in markets, technology products, and service development; world-class skills, education, and training; attractive tax regime; and effective, agile government. Moreover, the report identified the essential conditions for this to occur: cost competitiveness, a physical and communications infrastructure, innovation and entrepreneurship, and management capability (see Figure 51.1).

O'Driscoll (2004) argued "... that enterprise in Ireland, while having highly developed manufacturing ability, lacks capability in two essential areas: international sales and marketing and the application of technology to develop high value products and services. The report points to areas of activity in services and high value manufacturing which, if enabled by expertise in markets and technology, would significantly enhance the enterprise base." Specific recommendations of O'Driscoll (2004, pp. xv–xxi) included:

- a market intelligence and export promotion structure;
- 1,000 sales and marketing personnel;

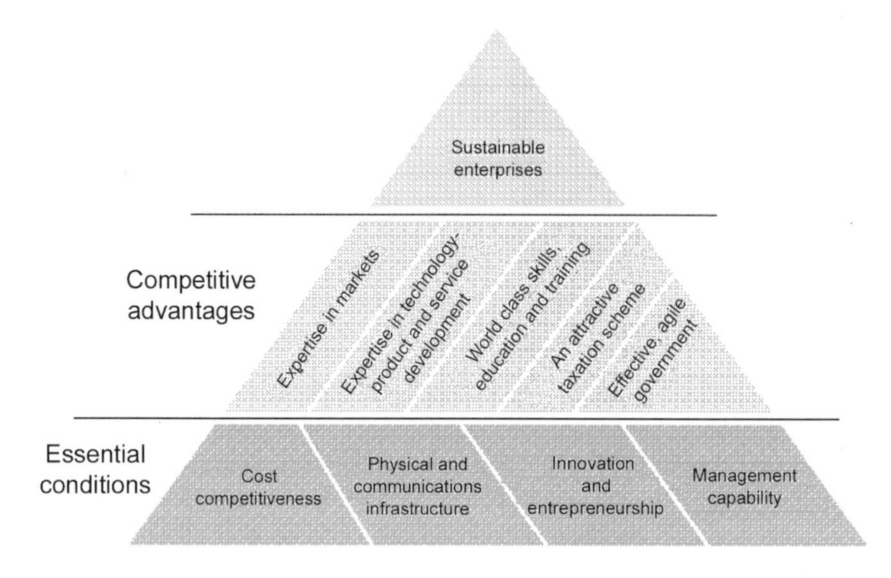

Figure 51.1. Enterprise Strategy Group: making it happen.

- target sales and marketing and European headquarters projects;
- an R&D and innovation coordination structure;
- increase applied R&D funding;
- investment in enterprise-led networks;
- the "One Step Up" initiative, facilitated by the National Framework of Qualifications to encourage greater participation and learning;
- quantity and quality of graduates and post-graduates;
- commitment to 12.5% corporate tax rate;
- cabinet enterprise review process.

The Strategy for Science, Technology, and Innovation (SSTI) published by the government in 2006 represents another policy milestone in getting the economy to evolve towards a knowledge-based one, and builds on investments made through the National Development Plan (2000–2006), the Program for Research in Third Level Institutions (PRTLI), and the Science Foundation Ireland (SFI). The strategy covered such themes as building world-class research; capturing, protecting, and commercializing ideas and know-how; research and development for enterprise; innovation and growth; science, education, and society; research in the public sector; all island and international R&D and implementation issues (see Table 51.1).

The current cycle of the National Development Plan covers the years 2007–2013 and has set very ambitious output targets and spending commitments for STI—which give concrete public funding commitments to aspirations as set out in SSTI (2006). During this period over €8.2 billion will be spent overall on STI. This spend will be accounted for on the basis of the following: €3.42 billion will be spent on world-class research in STI, €1.5 billion in R&D expenditure on higher education, €1.29 billion on enterprise STI and €1.35 billion on sector-specific research in areas such as agri-food, energy, marine, geoscience, health and environment (NDP, 2007–2013). This targeted spending will enable the achievement of the Strategy for Science, Technology, and Innovation which embraces the following key objective:

> "Ireland by 2013 will be internationally renowned for the excellence of its research, and will be to the forefront in generating and using new knowledge for economic and social progress, within an innovation-driven culture" (SSTI, 2006).

Technology trajectories

Through sector-targeted initiatives by the Industrial Development Authority (IDA) with respect to FDI, Ireland has built up substantial capabilities as a result of attracting leading multinational companies including some of the most successful in computer and software design, communications technology, pharmaceuticals, and medical devices. Cognizant of the existence of these multinational companies and in an attempt to further grow the science and technology base of the country a technology foresight exercise was carried out in 1999 which concluded that biotechnology and information and communications technology represent "the engines of future growth in the global economy" and that Ireland should, as a matter of priority, seek to create a world-class research capability in selected niches of these two enabling technologies as an essential foundation for future economic growth in Ireland. This document was the precursor to the establishment of the Science Foundation Ireland (SFI) in 2000 which was charged with providing financial research support for ICT and biotechnology. In 2008 the strategic area of scientific endeavor to be funded by the SFI was

Table 51.1. Main aims of the Strategy for Science, Technology, and Innovation 2006–2013.

Academic research	• significantly increase the number of research teams led by internationally competitive principal investigators; • upgrade existing research infrastructure and develop new facilities; • develop sustainable career paths for researchers; • enhance the mobility of researchers; and • double the number of Ph.D. graduates in science, engineering, and technology to 1,000 per annum by 2013
Graduate schools	• establish a number of graduate schools to provide high-quality training of researchers, and equip them with generic and transferable professional skills that are relevant to a modern knowledge-based enterprise economy; and • accommodate industrial placements to facilitate development of enterprise expertise
Commercialization	• increase the output of economically relevant knowledge, know-how, and patents from third-level institutions; and • strengthen intellectual property/commercialization functions within higher education institutes and provide them with expertise to translate research into applications
Industrial research	• transform the quality and quantity of research undertaken by enterprise in cooperation with third-level institutions; • grow business annual expenditure on R&D from €1 billion in 2003 to €2.5 billion; and • develop a number of industry-led research-driven competence centers with research facilities in third-level institutes
Sectoral research	• enhance the contribution of research to economic and social development across relevant areas of public policy; and • provide a competitive fund to encourage excellent research in areas of social, economic, or environmental need, such as sustainable agriculture, treatment of specific medical conditions, and energy security
Public awareness	• increase public awareness and appreciation of the role of science in society, with particular focus on schoolchildren and those that influence them; and • increase the number of schoolchildren taking science subjects
Cross-border and international cooperation	• increase international cooperation in science and technology and participation in transnational research activity; and • encourage Irish researchers to collaborate internationally and to avail themselves of EU Framework Program funding as well as leveraging complementary strengths in institutions and enterprises in Ireland and Northern Ireland through increased cross-border cooperation.

Source: DETE (2008).

extended to include sustainable energy and energy-efficient technologies.

Current institutional structures and evolution

There are a number of different public institutions and statutory bodies that play important roles in the Irish National System of Innovation. The first four—Forfás, IDA, the Science Foundation Ireland, and Enterprise Ireland—are the responsibility of DETE, the Government Department of Enterprise, Trade, & Employment. The other two—the research councils and the Higher Education Authority are overseen by the Department of Education & Science.

Forfás: established in 1994, Forfás is the body in which the legal powers of the state for the promotion and development of industry, science, and technology are largely vested. It is the national policy and advisory board for enterprise, trade, science, technology, and innovation. It provides DETE and other stakeholders with analysis, advice, and support on issues related to enterprise, trade, science, technology, and innovation. It also provides administrative and/or research support to a number of independent bodies including the Advisory Council for Science, Technology, & Innovation; the Expert Group on Future Skills Needs (EGFSN); the Management Development Council (MDC); the National Competitiveness Council (NCC); and the Small Business Forum. It also hosts the Office of the Chief Scientific Adviser to the Government (*www.forfas.ie*).

Industrial Development Authority (IDA) Ireland: IDA is a state-sponsored agency funded primarily through government grant aid. Broadly, the key objective of IDA is to attract and retain foreign direct investment (FDI) in Ireland, and in so doing contribute to Ireland's economic development. Specifically, it works to develop the strong base of over 1,000 overseas companies already located in Ireland and also to attract new investment. IDA's success to date in attracting inward investment is a major driving force behind the growth of the Irish economy. As of 2005 FDI accounts for 35% of GDP and over 85% of manufactured exports. IDA has and continues to win a disproportionate share of FDI from Europe. Independent reviews confirm that Ireland's market share of new U.S. greenfield manufacturing projects that want to locate in Europe is consistently strong relative to its share of the EU population and GDP. The U.S. is and continues to be the most significant source of inward investment—accounting for 70% of FDI (IDA, 2006, p. 9).

Science Foundation Ireland (SFI): The SFI is the national foundation for excellence in scientific research, was established in 2000, and is modeled on the National Science Foundation in America. It is responsible for the management, allocation, disbursement, and evaluation of expenditure for investment in academic researchers and research teams who are most likely to generate new knowledge, leading-edge technologies, and competitive enterprises in the fields underpinning two broad areas: biotechnology, and information and communications technology (*www.sfi.ie*).

Enterprise Ireland: Enterprise Ireland (EI) provides intensive supports for Irish-owned businesses (employing between 10 and 250) involved in manufacturing and internationally traded services. They also support start-ups and micro-businesses (fewer than 10 employees) in the same sector, provided they have the potential to achieve rapid growth and international expansion. These latter businesses are referred to as high-potential startups (HPSUs), about 70 of which receive Enterprise Ireland support each year.

Research councils: coinciding with the increased attention that was focused on research and development in the 2000–2006 National Development Plan two core research councils were established. The Irish Research Council for Humanities and Social Sciences (IRCHSS) was constituted in 2000 and the Irish Research Council for Science, Engineering, & Technology in 2001. Both research councils operate and manage numerous research schemes such as graduate and post-graduate fellowships, and research grants for new and established academic researchers.

HEA: the Higher Education Authority is a statutory body with responsibility for the planning and policy development of higher education and research in Ireland. The HEA has wide advisory powers throughout the whole of the third-level education sector. In addition, it serves as the funding authority for universities and technology institutes. One of the key research funding programs administered by the HEA is the Program for Research in Third-Level Institutions (PRTLI) which was launched in 1998. It has invested €866 million to date over four funding cycles (see Table 51.2) in strengthening national research capabilities via investment in human and physical infrastructure. The aim of the program is to advance Ireland in its bid to establish itself internationally as a

Table 51.2. PRTLI funding cycles.

	Year	*Funding period*	*Buildings and equipment* (€ million)	*Research programs and people* (€ million)	*Total* (€ million)
Cycle 1	1999	2000–2003	177.5	28.6	206.1
Cycle 2	2000	2001–2004	48.8	29.7	78.5
Cycle 3	2001	2002–2006	178.0	142.4	320.4
Cycle 4	2007	2007–2010	108.8	120.3	261.2
Total			*513.1*	*321.0*	*866.2*

Source: *www.hea.ie*

premier location for carrying out world-class research and development in all academic disciplines. PRTLI awards are evaluated by an international panel of distinguished researchers and scholars on the basis of excellence in strategic planning and focus, inter-institutional collaboration, research quality, and impact of research on teaching and learning (*www.hea.ie*).

Public and private institutions and their linkages
The main public institutions were discussed in a previous section. All of these public institutions have extensive linkage with private institutions. Both IDA and EI have, by virtue of their remits, extensive interactions with businesses. Forfás creates multiple linkages with private institutions by appointing a number of industrialists to the Forfás Board. In addition, the major reports commissioned by Forfás and its associated bodies are overseen by task forces that have substantial private institutional representation.

The Science Foundation Ireland (SFI) ensures collaboration with industry by the funding structures that it administers. For example, one of its major funds—the Centers for Science, Engineering, & Technology (CSETs)—has as a requirement that industry partners are willing to contribute and provide financial support to the research program. In return, key industrial partners have the first call on IP exploitation.

Summary of output trends

In terms of innovation activity the top-four active innovative sectors are chemicals, medical precision, rubber and plastics, and food and beverages. The top-four service sectors in terms of activity include computer-related services, communications, engineering and technical services, and transport. For firms the total estimate of spending of innovation activities was estimated to be €5.72 billion, with 60% being spent on machinery and equipment, 24% on in-house R&D, and 12.7% on external knowledge (Forfás, 2006).

A strong contributor to the growing R&D has been expenditure within third-level educational institutions. The rate of R&D expenditure in higher education has tripled since 1998 from €169 million to €568 million and between 2004 and 2006 the rate of growth of R&D expenditure increased by 23.1% (see Table 51.3).

Business expenditure on R&D (BERD) in real terms has shown strong growth in Ireland—nearly tripling in the 10 years between 1995 and 2005 (Table 51.4). The majority of

Table 51.3. Main higher education R&D (HERD) indicators 2000–2006.

	2000	*2002*	*2004*	*2006*
Higher education expenditure on R&D (€ million)	238	322	492	601.4
HERD as a percentage of GNP (Ireland)	0.27%	0.31%	50.40%	0.40%
HERD as a percentage of GDP (EU25 average)	0.37%	0.40%	0.39%	0.40%
Ireland's rank among 29 OECD countries	22nd	19th	16th	14th
Total researchers in the HE sector (FTE)	2,148	2,695	4,152	4,689
HE researchers per 1,000 labor force	1.2	1.5	2.2	2.2
Ireland's rank among 29 OECD countries	24th	23rd	14th	13th

Source: Forfás (2008, p. 6).

Table 51.4. Business expenditure on R&D (BERD).

	1995	*2001*	*2003*	*2005*
BERD Ireland (current prices €m)	470	900	1,105	1,329
BERD Ireland (constant prices 2006 €m)	658	1,059	1,201	1,380
BERD (%GNP) Ireland	1.00%	0.92%	0.94%	0.98%
BERD (%GDP) EU	1.05%	1.17%	1.15%	1.14%
BERD (%GDP) OECD	1.39%	1.57%	1.53%	1.54%
BERD (%GDP) Ireland	0.89%	0.77%	0.80%	0.82%

Source: Forfás (2007a).

Table 51.5. Patent filing.

Patent filing and grants by office

	Resident direct filing 2005	*Non-resident direct filing*	*PCT national phase entries*	*PCT international applications 2006*	*Grants to residents 2006*	*Grants to non-residents 2005*
Ireland	789	75	NA	144	300	210

Patent filing by country and territory of origin

	Non-resident direct filing	*PCT national phase entries 2005*	*PCT international applications 2006*	*Grants to non-residents 2005*	*Patents in force 2005*
Ireland	815	910	407	572	2,882

Patent filing by population, GDP, and R&D expenditure

	Resident filings per million population 2005	*Resident filings per $ billion GDP 2005*	*Patent filing per $ million R&D expenditure 2005*
Ireland	190.09	3.34	0.48

Source: WIPO (2007, pp. 46–51).

the expenditure goes to foreign-owned firms, who in 2005 spent €939 million or 71% of the total BERD expenditure. The source of funding for business R&D is overwhelmingly private funding (95% of the 2005 total) with the bulk of the remaining 5% accounted for by Irish Government funding. The extent of R&D expenditure is highly concentrated—with 50 firms accounting for 57.7% of the total R&D spend in 2005. The make-up of the R&D being carried out is also changing with 12% of all spending being classed as basic research in 2005, as against only 4% in 2001 (Forfás, 2007a).

In terms of patent activity, Ireland's activity level is growing but remains low (see Table 51.5). One explanation, as Forfás (2004, p. 21) outlines, is "Ireland's dual industry structure and a foreign-owned sector that, with few exception, does not carry out high value-added activities in Ireland." The growth rate in patent filing at the European Patent Office (EPO) between 1995 and 2000 was 26% per annum. Nevertheless, Ireland's share of European and U.S. grants has been increasing but still remains low. The top-three for technology classification patent activity by sector 1999–2001 were health (including medical devices), electronics, and medicines and toiletries (Forfás, 2004, pp. 23–24).

Technology commercialization initiatives (national level)

In the last 15 years Ireland has made some significant strides in responding in a coherent manner to developing technology commercialization initiatives targeted at the dual economy. Nevertheless, Ireland's track record of exploiting intellectual property has been poor given the low levels of patent registration in Ireland compared with EU countries. In addition further weaknesses lay in staffing levels in technology transfer offices within the third-level sector as Forfás (2004, p. 34) noted: "Not all Irish research institutions have clearly identified the role and the function of IP management and have created specific positions to manage it. In these institutions approximately 62 people, (22 full-time equivalents (FTE)), are regarded as being involved in commercialisation activities—providing an average of 0.96 staff per research institution." Such staffing levels are well below the international norms required to achieve the appropriate level of IP exploitation.

Clarity about IP ownership and management of strategy-funded research is seen as another barrier to technology transfer and further research collaboration between industry and academia. The publishing of a National Code of Practice in 2004 (ICSTI, 2004) with respect to IP ownership and management for purely publicly-funded research removed uncertainties around IP ownership. The National Code of Practice clarified issues to do with invention disclosures, share of income and IP assignments, and the state's right to retain an interest in IP management.

The publication of the *Strategy for Science, Technology and Innovation* is designed to scale and coordinate national efforts in commercialization and technology transfer and set out clear key actions to enhance technology commercialization within the Irish economy. Key actions with respect to IP management and exploitation include (SSTI, 2006, p. 17):

- Ensure that higher education institutions (HEIs) encompass IP management and commercialization as a central part of this mission, equal in importance to teaching and research.
- Strengthen institutional competence at the Technology Transfer Office (TTO) level and among researchers.
- Establish a competitive fund administered by EI to assist strengthening of the IP management function.
- Establish a new function in EI providing centralized support to HEIs thereby maximizing the commercialization of IP.

The ongoing investment, implementation, and enhancement of technology commercialization initiatives led by Enterprise Ireland (EI) with third-level institutions and firms are designed to address the dual-economy focus. To date, EI is responsible for the administration of six schemes that are designed to bring technology to the marketplace. These include commercialization expertise; a commercialization fund; a patent fund and advice; campus incubation centers; an applied research enhancement program and initiatives to strengthen technology transfer.

Commercialization expertise: Enterprise Ireland provides expertise to companies as well as researchers in helping them access new technology. The focus of this is to seek to improve company competitiveness on an international level by concentrating on higher value market arenas. This expertise is provided in three domain areas: biotechnology (*BiotechnologyIreland.com*), industrial technologies, and informatics. Technology commercialization activities supported by EI in 2006 in the biotechnology arena equated to an investment of €5.5 million in 14 new biotechnology research projects, 9 technology licences, and 3 startups. This support is valued at €10 million, bringing total investment in biotechnology by EI to €40 million since 2001. In 2006, the Industrial Technologies Group supported technology commercialization which accounted for 41 proof-of-concepts projects, 23 technology development projects, 8 technologies were licenced, and 3 startups were formed. Informatics commercialization supported by EI in 2006 led to 24 proof-of-concept projects, 14 technology development projects, 7 technologies were licensed, and 1 startup was formed, in addition to holding of an Informatics Technology Commercialization Showcase (Enterprise Ireland, 2006).

Commercialization fund: this fund is administered over three phases—proof-of-concept phase, technology development phase, and business development phase (CORD, which stands for Commercialization of Research and Development)—and is designed to bring new technologies to the marketplace by improving technical and business cases. The proof-of-concept phase is aimed at researchers within third-level institutions who seek to explore commercial applications for their technology. The grants cover 100% of eligible costs (personnel, equipment, material, and travel) and range from €50,000 to €100,000 over a 12-month period. Since the launch of the proof-of-concept fund in 2003 over 350 proposals have been funded to the tune of over €28 million. The next phase of support is through the Technology Development Fund which is targeted at scaling a technology aimed at a particular market arena, through licensing or new business startups. The support levels available under this scheme to a researcher's institution range from €100,000 to €400,000 over an 18 to 36-month period. In 2006 alone, EI supported and funded 155 research projects through the proof-of-concept and technology development phases which equated to €29.7 million (Enterprise Ireland, 2006, p. 27) of investment. Finally, the business development phase is designed to support bringing new technologies from third-level institutions to the marketplace. Typical types of support under this scheme include market research, product trials/market assessment, cost analysis, financial projections, and establishing potential joint venture partners. The level of support is an approved grant which can be as much as 50% of eligible expenditure with a ceiling of €38,000 per grant (see Table 51.6 for details of maximum expenditure limits).

Patent Fund and advice: Enterprise Ireland provides professional advice and some financial assistance towards the cost of patenting for client companies. The advice provided includes the use of IPR (intellectual property rights), confidentiality agreements, licensing, and technology acquisition. EI also administers the IP Fund for Higher Education Sectors, designed to provide support to protect IP that has marketplace potential. This support is open to all third-level colleges and associated teaching hospitals. Funding can come in three stages, as follows:

- Stage 1: up to €7,000 to assist with the costs of preliminary patent protection;
- Stage 2: up to €20,000 to support patenting costs arising in the continuing prosecution of an already filed initial patent application or extension of patent coverage to other countries.
- Stage 3: funding to provide support for later stages of the patenting process.

The amount is determined by Enterprise Ireland for each case but normally is no more than €50,000. Funding is restricted to costs directly associated with the protection

Table 51.6. Business development phase: maximum level of funding.

Expenditure item	*Level of funding*
Salaries/wages	€950 per week (none if employed by the college)
Consultancy	€650 per day
Travel	Economy-class fare
Mileage within Ireland	€0.40c per mile
Subsistence	€100 per night within Ireland, €200 per night outside Ireland
Prototype	50% of costs
Promotional materials	€6,350

Source: Enterprise Ireland (2006).

of the invention concerned. It will normally cover 100% of such costs (see Enterprise Ireland, 2007a).

Campus incubation centers: to date there are 14 campus incubation centers with 6 bio-incubation facilities offering resources and facilities to startup companies. Since 1998 Enterprise Ireland has invested €47 million in campus incubation centers and between 2001 to 2006 181 companies have used them. This equates to 67 university-linked startups and 114 companies located in incubation spaces within 16 institutes of technology. For instance, Changing Worlds, a company founded in 1999 by Professors Barry Symth and Paul Cotter of University College Dublin, commercialized their own research from UCD's Smart Media Institute. Based at NovaUCD the company is a provider of intelligent personalization technology for telecommunication. In 2007 over 200 companies were located in campus incubation centers employing more than 900 people (Enterprise Ireland, 2007a, p. 53).

Applied Research Program: This research program is aimed specifically at enhancing the level of applied research in institutes of technology. These institutes have achieved particular success in applied research areas against a backdrop of resource research deficits when compared with Irish universities. Funding of €1.25 million is available for up to a 3-year period. To date, 11 centers of excellence have received funding under the program.

Technology transfer strengthening initiatives: this intervention is the implementation of the SSTI recommendation to enhance the capabilities and operations of Technology Transfer Offices within the third-level system in response to the identified TTO staffing deficit. Enterprise Ireland was tasked with rolling out this initiative which has a budget allocation of €30 million from 2007 to 2011. To date 9 universities have received funding of €15.6 million which has led to an additional 29 trained technology transfer professionals being hired by TTOs.

One of the most significant developments on technology commercialization initiatives has been the establishment of Centers of Science, Engineering, & Technology (CSETs) funded by Science Foundation Ireland, which are located in Irish universities. These centers are designed to build a critical mass of excellence in biotechnology and ICT which can be exploited by their commercial partners, who have first right to intellectual property exploration (see Table 51.7). In addition to IP exploration and exploitation, these CSETs also have a strong outreach element to their activities, for purposes of raising awareness of science and highlighting career

Table 51.7. Centers of science, engineering, and technology (CSETs).

Alimentary Pharmabiotic Center
€16.5 million SFI funding (2003)
48 SFI-funded staff

Biomedical Diagnostics Institute
€16.5 million SFI funding (2005)
Industrial partners: Amic, Analog Devices, Becton Dickinson, Enfer Scientific, Hospira, Inverness Medical Innovations/Unipath

Center for Research on Adaptive Nanostructures and Nanodevices (CRANN)
€10 million SFI funding (2006)
Industrial partner: Intel

Center for Telecommunications Value Chain Research
€20 million SFI funding (2004)
97 academic staff

Digital Enterprise Research Center
€12 million SFI funding (2003)
80 full-time staff
Industrial partner: HP

LERO (The Irish Software Engineering Research Center)
€9.1 million SFI funding (2005)
40 researchers and Ph.D. students

Regenerative Medicine Institute (REMEDI)
€14.9 million SFI funding (2003)
Industrial partner: Medtronic

Source: SFI (2006, pp. 9–13).

and education opportunities.

Private firm–level state support of SSTI

At the firm level, Enterprise Ireland has again been the lead agency tasked with encouraging and supporting greater R&D and technology commercialization. To this end, EI provides R&D funding, new R&D collaboration, R&D management, and technology management. In 2006, EI reported that 601 client companies invested in excess of €100,000 in R&D and 40 client companies invested in excess of €2 million in R&D development. During 2006, EI approved 194 in-company R&D projects which represents financial support of €52.9 million in research funding (Enterprise Ireland, 2006, p. 25).

R&D support: there are four strands to the support that EI offers: (a) stimulating research and innovation grants, (b) R&D fund, (c) collaboration, and (d) innovation expertise. The stimulating research and innovation grants are specifically targeted at firms with an export orientation who have little or no R&D but who wish to build this capability within their firm. The aim of support at the firm level is achievement of at least one of the following outcomes:

- at least one full-time person dedicated to R&D;
- a continuing R&D budget of between €100,000 and €200,000 per annum;
- a written plan for R&D projects being or about to be carried out (Enterprise Ireland, 2007b).

The R&D fund is aimed at supporting research, development, and innovation at all stages of firm development. The fund is designed to increase the breadth and depth of research and development among Irish firms which will equip them to scale and compete in international markets. To avail themselves of this fund companies must be an Irish-based manufacturer or an internationally traded company, and the company must demonstrate that it has sufficient resources to implement the proposed project. The outcomes this funding attempts to achieve at the firm level include the establishment of R&D budgets, recruiting of dedicated R&D resources, investment in R&D facilities, formal R&D management procedures, and the development of a firm-level innovation culture. The maximum R&D grant that a company can receive is €650,000 (Table 51.8 provides a breakdown of grants per size).

New R&D collaboration: Enterprise Ireland provides a number of different types of support and enhancements to encourage the development of new R&D collaboration. These include innovation partnerships between firms, third-level institutions, and competence centers.

Table 51.8. R&D fund grants according to size (i.e., maximum project funding).

	Small company	*Medium company*	*Large company*
Innovative projects that are technically challenging and involve significant risk	45%	35%	25%
Innovative projects where there is collaboration between two companies	+15%	+15%	+15%

Source: Enterprise Ireland (2006).

Innovation partnerships are designed to provide support of between 50% to 70% of eligible costs of research projects for collaboration between a firm and third-level institutions. In 2006, Enterprise Ireland supported 63 innovation partnerships (Enterprise Ireland, 2006, p. 25). Innovation partnership projects are managed by Technology Transfer Offices within the third-level sector.

The competence center concept and ancillary support was formed in 2007. It is designed to achieve greater research and development collaboration between a number of firms and research institutions through undertaking industry-led research which has a clear market orientation. As a result of the establishment of competence centers it is envisaged that research organizations will deepen their expertise in particular domains oriented towards addressable market problems as a research provider. For Irish firms this means that they can direct the research, access the industry partner, and enhance their competitiveness in the international market through technology making which is normally high risk in nature.

R&D management: the focus of this initiative is to provide training at a firm level that is needed to support R&D development initiatives at a firm level by typically subvention of participation fees of up to 70% for SMEs and 50% for non SMEs eligible companies. Training programmes include Introduction to Innovation and R&D Management, Innovation to Profit, BA in Technology Management, M.Sc. in Technology Management, M.Sc. in Technology Management distance education, and the Champion of Innovation Program.

Table 51.9. Growth fund maximum support limits.

	Maximum funding	*Funding type*	*Minimum company spend*
Capital investment	€300,000	Grant/50% repayable	€75,000
Technology acquisition	€300,000	Grant	€50,000
Training and management development	No specific limit	Grant	€25,000
Consultancy	25% of total project	Grant	NA
Recruitment of key managers	€200,000	Grant	NA
Workplace innovation and management development	No specific limit	Grant	€25,000
Overall maximum support for submission		€650,000	

Source: Enterprise Ireland (2007).

Technology acquisition: technology acquisition is supported and managed through the TechSearch web portal (*http://www.enterprise-ireland.com/TechSearch*) which provides information and overviews of technology that is available to licence through the Innovation Relay Network (from 33 countries with 2,500 technology solutions) into a firm. In 2006, this intervention assisted 160 client companies, resulting in 35 licence agreements which equates to over €2 million of investment with a sales potential of €15 million based on these licences (Enterprise Ireland, 2006, p. 25). EI also runs several technology-mapping briefing sessions designed to support technology acquisition.

Enterprise Ireland also provides additional support to improve the productivity and competitiveness of client companies through their business innovation offer support which comprises three elements, namely stimulating business innovation, growth expertise, and a growth fund.

Within the stimulating business innovation supports EI provides supports such as supply chain management (SCM), ebusiness management initiative (eBMI), green technologies, feasibility studies, and innovation vouchers. Growth expertise support is focused on providing best practice support to client firms in the areas of automation, benchmarking, ebusiness, environmental management, and supply chain management. The growth fund is specifically designed to improve the innovation and productivity of client companies. This fund supports capital investment, technology acquisition, talent recruitment, management development, and training (see Table 51.9 for maximum support limits). In 2006, 200 EI client companies participated in management development programs such as Leadership 4 Growth delivered by Stanford Graduates School of Business.

Relative strengths and weaknesses in technology

The policy agenda being pursued by the Irish Government and its stakeholders is designed to build on the strengths of the NIS and to address some of the current deficits. Under the aegis of the Office of the Chief Science Adviser a SWOT analysis was conducted of Ireland's National R&D Action Plan in 2006 (see Table 51.10). From this analysis it is evident that strengths in the NIS lie in the ability to be adaptive and responsive to change given the size of the economy. Key weaknesses are in developing sustainable collaboration and linkages between research organizations and Irish business.

The Irish NIS has acknowledged key strengths other than those outlined in the SSTI report (SSTI, 2006). FDI in Ireland has meant that the country has many Tier-1 players in the ICT and biomedical industries. In essence, Ireland has a significant presence of world innovative firms located in Ireland, through FDI initiatives. In real terms the life science sector in Ireland accounts for €5 billion in exports and has over 35,000 employees. U.S. FDI into Ireland between 2000 and 2006 totaled $44.3 billion in comparison to $15.4 billion into China and $5.3 billion into India (Hamilton, 2008). Consequently, this provides Ireland with a key strength as many of these companies have expanded their initial remit in Ireland into value-adding activities. In addition, the talent and flexibility of Irish management teams and employees also supported and underpinned this expansion of activities (Cunningham, 2008).

Table 51.10. Strengths and weaknesses of Ireland's research in technology and innovation (RTI) performance.

Strengths	*Weaknesses*
Government's commitment to driving Ireland as a knowledge-based economy and strong government commitment to research	Historic absence of a fully developed national strategy for science, technology and innovation (STI), and integration of sectoral and socio-economic research within that framework
Success in attracting high-quality, high-tech foreign direct investment	Research Capacity in universities/institutes of technology (IoTs) and industry (numbers; quality supervision) in context closing output/quality gap with competitors
Highly adaptive manufacturing base	Lack of research and technology absorption capabilities by companies and weak commercialization structures
Importance of engineering and the quality of Irish engineers	Number studying science subject to leaving certificate level
Government support for enterprise	Structural weaknesses in universities/institutions
Positive fiscal environment	Lack of funding for research support disciplines
European Union Framework Program 7 (FP7): ability to organize ourselves and influence the make-up of FP7	Low availability of seed capital
Government responsiveness to changing competitive environment	
Emerging whole-of-government approach to science, technology, and innovation	

Source: SSTI (2006—2013, pp. 89–90).

Some weaknesses still pervade Ireland's NIS in addition to those outlined in Table 51.10. The NCC (2007, p. 3) notes that there are significant weaknesses in prices and costs, domestic competition, infrastructures, innovation and R&D, and sustainability of the environment which are undermining national competitiveness. In addition, several operational barriers and weaknesses to research commercialization still remain within the third-level sector. These include lack of space, research, and the funding treadmill among the contract research community for definable career paths for contract researchers (Cunningham and Harney, 2006). Other weaknesses in the Irish NIS include the absorptive capacity of SMEs, the breadth and depth of expertise in IP management and exploitation, IP registration and exploitation, firm and public investment in R&D, early-stage seed capital, and life-long learning.

The *European Innovation Scorecard* (EU, 2007) (see Figure 51.2) captures the key strengths and weaknesses for Ireland in relation to EU25. Clearly, investments in the public funding of innovation, entrepreneurship levels, and high-tech exports reflect NIS strengths but the medium to average levels of IP exploitation is currently a deficit within the NIS.

Funds flow for innovation

Ireland's total science budget reached €2.5 billion in 2007, with education and training accounting for 45.9% of spending in 2007, and the number of R&D personnel employed by the government sector reaching 1,344. Science budget allocations for 2007 included education and training €907 million; R&D €907 million; technical services €231 million; technology transfer €123 million, and other S&T activities €96 million. The top-three government departments or agencies funding R&D activities were the Department of Education & Science at €497 million, the Department of Enterprise, Trade, & Employment at €243 million, and the Department of Agriculture & Food at €83 million (Forfás, 2007c, p. 10) .

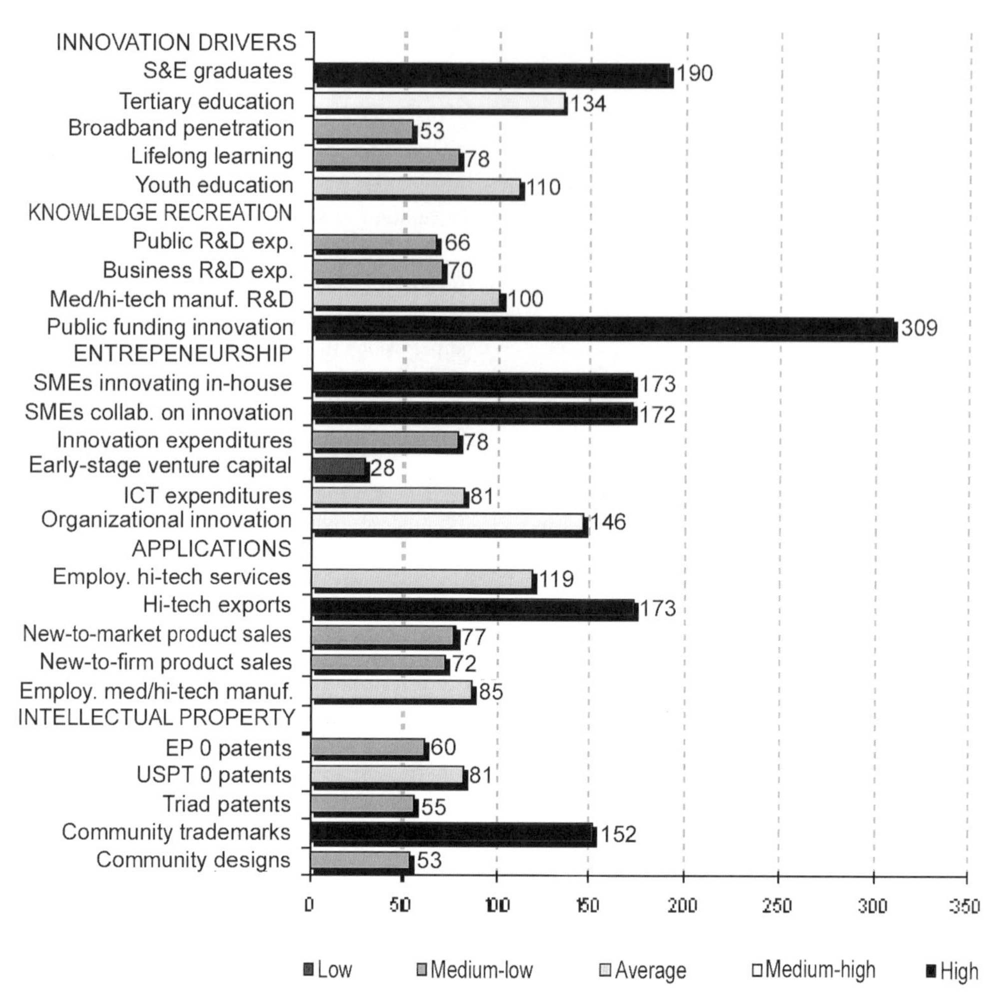

Figure 51.2. EU (2007) innovation performance.
Source: EU (2007).

Cultural drivers

We see four main cultural drivers within the Irish NIS that have supported its evolution. The first cultural driver has been the increasing levels of entrepreneurship within the Irish economy, where entrepreneurship is accepted within society as a legitimate career path. The success stories of Irish entrepreneurs have pervaded all parts of Irish society. As Forfás (2007c, p. 44) notes "Entrepreneurs are held in very high esteem in Irish society and this high regard is re-enforced by a very supportive media in which positive stories, about entrepreneurs and their successful companies abound."

The second driver is the interconnection through social networks of policy-makers, private industry, and third-level sectors which has led to an agility and capability in responding to change within the economy. Such interconnections provide a nimbleness of response and action which other NISs would find difficult to replicate given their greater scale. This is further enabled by the strong and established tradition of social partnership, and the successive agreements that have been the bedrock for negotiated job conditions, wage increases, as well as social issues. Moreover, the medium-term focus on innovation at the heart of policy making is summarized in *Innovation in Ireland* (DETE, 2008) published by the Department of Enterprise, Trade, & Employment: "In short, our ambition is to put innovation at the core of our policies and strategies for the future, so that Ireland becomes a leader in innovation."

The third cultural driver has been the responsiveness of policy initiatives through the auspices of Forfás and its ability to convene specialized task forces made up of economic and social pillars of the country.

The fourth and final cultural driver is the importance of educational attainment within Irish society, where education is still seen as an enabler for individuals. This is reflected in the abolition of fees for undergraduate programs at third-level institutions in the mid-1990s and various policy initiatives targeted at life-long learning, educational access, and disadvantage.

References

Coughlan, M. (2008). *Foreward Innovation in Ireland*. Dublin: Department of Trade, Enterprise, & Employment.

Cunningham, J., and Harney, B. (2006). *Strategic Management of Technology Transfer: The New Challenge on Campus*. Cork, Ireland: Oak Tree Press.

Cunningham, J., and Harrington, D. (2008). "Management 2.0." In: J Cunningham and D Harrington (Eds.), *Irish Management 2.0: New Managerial Priorities in a Changing Economy*. Dublin: Blackhall.

DETE. (2008). *Innovation in Ireland*. Dublin: Department of Enterprise, Trade, & Employment.

Enterprise Ireland. (2006). *Annual Report*. Dublin: Enterprise Ireland; *www. enterprise-ireland.com/ResearchInnovate/Research+ Commercialisation/Campus_Enterprise.htm*

Enterprise Ireland. (2007a). *Annual Report*. Dublin: Enterprise Ireland; *http://www. enterpriseireland. com / ResearchInnovate / Research + Commercialisation/Intellectual + Property + Fund + for + the + Higher + Education + Sector.htm*

Enterprise Ireland. (2007b). *http://www.enterpriseireland.com/ResearchInnovate/R+and+D+in+your+Enterprise/RandD+Stimulation+Grant htm*).

Enterprise Ireland. (2007c). *www.enterprise-ireland.com/Grow/Finance/Growth +Fund.htm*

EU. (2007). *European Innovation Scorecard*. Brussels: Pro Inno Europe, EU Commission; *http://www.proinno-europe.eu/*

Fitzsimons, F., and O'Gorman, C. (2007). *Irish Annual GEM Report for 2006*. Dublin: Business School, City University [cited in Forfás, 2007b].

Forfás. (2004). *From Research to the Marketplace: Patent Registration and Technology Transfer in Ireland*. Dublin: Forfás; *http://www.forfas.ie*

Forfás. (2006). *The Fourth Community Innovation Survey: First Findings* (Forfás Innovation Survey). Dublin: Forfás; *http://www.forfas.ie*

Forfás. (2007a). *Research and Development Performance in the Business Sector, Ireland 2005/6*. Dublin: Forfás; *http://www.forfas.ie*

Forfás. (2007b). *Towards Developing an Entrepreneurial Policy for Ireland* (p. 44). Dublin: Forfás; *http://www.forfas.ie*

Forfás. (2007c). *The Science Budget 2006/2007: The First Findings* (p. 10). Dublin: Forfás.

Forfás. (2008). *The Higher Education R&D Survey 2006 (HERD)*. Dublin: Forfás.

Hamilton, D. (2008). "The transatlantic economy." In: *American Business Directory* (p. 40). Dublin: American Chamber of Commerce Ireland.

ICSTI. (2004). *National Code of Practice for Managing Intellectual Property from Publicly Funded Research*. Dublin: Irish Council for Science, Technology & Innovation.

IDA. (2006). *Freedom of Information Manual* (p. 9). Dublin: Industrial Development Authority; *http://www.ida.ie*

NCC. (2007). *Annual Competitiveness Report: Volume 1, Benchmarking Ireland's Performance*. Dublin: Forfás/National Competitive Council.

NDP. (2007). *National Development Plan*. Dublin: Government Publications.

O'Driscoll, E. (Chair.). (2004). *Ahead of the Curve: Ireland's Place in the Global Economy*. Dublin: Enterprise Strategy Group/Forfás.

SFI. (2006). *Annual Report*. Dublin: Science Foundation Ireland; *http://www.sfi.ie*

SSTI. (2006). *Strategy for Science, Technology and Innovation 2006-2013*. Dublin: Government Publications.

WIPO. (2007). *The Patent Report: 2007 Edition Activities* (pp. 46–51). Geneva, Switzerland: WIPO.

52

The U.S. National Innovation System

Kenneth L. Simons and Judith Walls*[†]

*Rensselaer Polytechnic Institute and [†]University of Michigan

The United States of America (U.S.) has been heralded as the world's premier engine of technological advance. However, recent criticisms suggest that its technological advance may slow relative to other nations. This chapter describes the U.S. National Innovation System, how it is evolving, and factors that threaten or reinforce its ability to generate technological innovations.

52.1 Nature and development of the U.S. National Innovation System

The primary philosophy supporting innovation in the U.S. has been to establish conditions that let innovation flourish on its own. While particular institutions contribute to the flow of innovation, any analysis of the U.S. innovation system must begin with the underlying conditions that aid innovative progress.

52.1.1 Underlying conditions for innovation

The preconditions for U.S. innovative success are many, and identification of the most important factors is a subject of ongoing research. Nonetheless, it seems safe to identify, albeit crudely, four cornerstones of U.S. success at innovation.

52.1.1.1 Incentives

One cornerstone of U.S. success at innovation is incentives. Businesses and entrepreneurs in the U.S. have, to a reasonable extent, the right incentives to propel innovation. The incentives include intellectual property rights and, typically more importantly, other means to capture monetary returns from invention.

Intellectual property rights are enshrined in the U.S. constitution, written in 1787, which gives the U.S. Congress the power "To promote the progress of science and useful arts, by securing for limited times to authors and inventors the exclusive right to their respective writings and discoveries." Current U.S. law grants patents for a period of (generally) 20 years from date of filing. It grants copyrights for the author's life plus 70 years, for work by an identified non-hired author, or the shorter of 95 years from publication or 120 years from creation, for other works. Both patent and copyright laws have changed over time, with some changes described below.

Nevertheless, corporate technology managers report that they rely mainly on mechanisms other than patents in order to protect the potential profits of their technology. More often used mechanisms of protection include lead time in innovation, steadily decreasing production costs, secrecy, and sales or service efforts (Levin *et al.*, 1987, pp. 794–795). In contrast, patents often fail to protect ideas because competitors can readily "patent around" them by developing and patenting modified technologies that achieve the same purpose, or because existing legal systems and competitive business practices make it impractical to enforce patent rights. Additionally, "thickets" of large numbers of patents may provide their owners with the ability to participate in an industry by reaching cross-licensing agreements when single patents do not provide this ability.

Hence U.S. success at innovation is aided by a patent (and copyright) system that allows inventors to gain limited protection. In practice, the benefits of patents are greatest in particular industries, and the uses and effects of patents are sometimes more subtle than would be expected from the idealized myth of how patents should work. Indeed it is important that patents are not too easily obtained and too strong, for large numbers of trivial but easily defended patents would create immense transaction costs that might bring innovation to a standstill. More generally, other incentives that increase the financial returns to research and development (R&D) and innovation, or decrease the costs, are likely to spur greater R&D and innovation, as long as they do not create transaction costs or other disadvantages that outweigh the benefits.

One such incentive is the corporate R&D tax credit, which in most years since 1981 has allowed deduction from U.S. federal taxes of a modest portion of research and experimentation expenses above a previous norm for the firm. On average this has allowed companies using the credit to deduct close to 6% of the cost of R&D before federal taxes are calculated.[1] State R&D tax credits in most U.S. states may double the savings to firms. Recent careful studies of the effects of R&D tax credits suggest that they spur as much extra research as they cost governments (cf. Bloom, Griffith, and Van Reenen, 2002), and since the social benefits to R&D are typically at least several times the private (company) benefits, this suggests that R&D tax credits are highly worthwhile. R&D tax credits are larger and more sustained in many nations other than the U.S., helping to attract firms and their R&D away from the U.S.[2]

52.1.1.2 *Government support*

A second cornerstone of U.S. success at innovation has been its government support. Inventors almost never capture the full monetary value of their inventions, so without public funding, inventors' limited incentives would yield less invention than is socially optimal. The solution is to support basic and applied R&D activities through appropriately disbursed government funds. Federal R&D funds, including for government laboratories, constituted 27.7% of U.S. R&D in 2006 ($92.4 billion of $340.4 billion).

At least as important as the amount of federal funding is the means by which federal funding is disbursed. Publicly-transparent, merit-based proposal systems open to all areas of technology would seem to yield the greatest potential for advance. Merit-based systems are used by agencies such as the National Science Foundation and the National Institutes of Health, but in other agencies including the Department of Defense funding is driven by more complex procurement decision processes. The U.S., like many governments, has targeted billions of dollars to particular technological projects such as supersonic passenger aircraft, light-water civilian nuclear reactors, synthetic fuels, the superconducting super collider, and, recently, hydrogen fuel cells.[3] The high rate at which giant targeted government projects have failed has led to critique of such projects, with problems (beyond ordinary new-venture risks) said to include poor assessments of technological potential, politicking for allocation of funds, and mid-stream project cancellations given alternative interests of successive government administrations. Nonetheless, even for targeted projects and defense technology procurement, Congressional and other oversight tends to ensure substantial evaluation of costs and benefits of national R&D funding and to select, albeit imperfectly, more rather than less worthwhile R&D spending priorities.

52.1.1.3 *Mix of entrepreneurial and large-firm capitalism*

A third cornerstone of U.S. success at innovation has been its mix of entrepreneurial and large-firm capitalism. While evidence is still lacking on the extent to which major inventions are best generated by entrepreneurial startups and independent inventors, it is clear that small new ventures or independent inventors often develop new technologies that apparently would emerge much less readily from major firms. At the same time, large established firms have often developed and commercialized major new technologies, and moreover they have carried out the extensive product and process improvements that refined new technologies to the point at which they become widely used. One example is penicillin, discovered by an individual, Alexander Fleming, but made a practical product by several large U.S. firms (notably Merck, Squibb, and Pfizer) as a World War II priority. Depending on the circumstances and technologies, it would appear that entrepreneurs and large established firms are at different times best placed to deliver new technology. U.S. government funding is in part reserved for small businesses, under the Small Business Innovation Research program, and legislation generally has not impeded small-business formation (although there are recent concerns, for example, with the Sarbanes–Oxley Act, which deters new business ventures from raising funds through stock offers because of high accounting costs for public companies).

52.1.1.4 *Societal institutions*

A fourth cornerstone of U.S. success at innovation is well-functioning legal, social, and infrastructure systems that do not impede business formation, layoffs, and operation; that maintain reasonably stable regulations that are not too burdensome; that provide appropriate educational opportunity; and that maintain a culture in which R&D and innovation are appreciated and respected. Only a few of these broad requirements of a well-functioning economy are addressed in this chapter. Education, immigration, and social attitudes have been important in recent policy formulation and will be discussed in the context of policy proposals.

52.1.2 Breadth of technology

The U.S.'s technological base is broad, with U.S. firms

[1] The almost 6% figure comes from a potentially biased source, the R&D Credit Coalition; *http://www.investinamericasfuture.org/PDFs/Coalition_Interntl_RD_tax_5-18-07.pdf*

[2] Wilson (2009) analyzes inter-state competition for R&D within the U.S., and the relatively low R&D tax credits in the U.S. relative to other nations have been emphasized by trade associations that seek a more technology-friendly and consistent U.S. policy.

[3] The first three of these examples are discussed by Lambright, Crow, and Shangraw (1988).

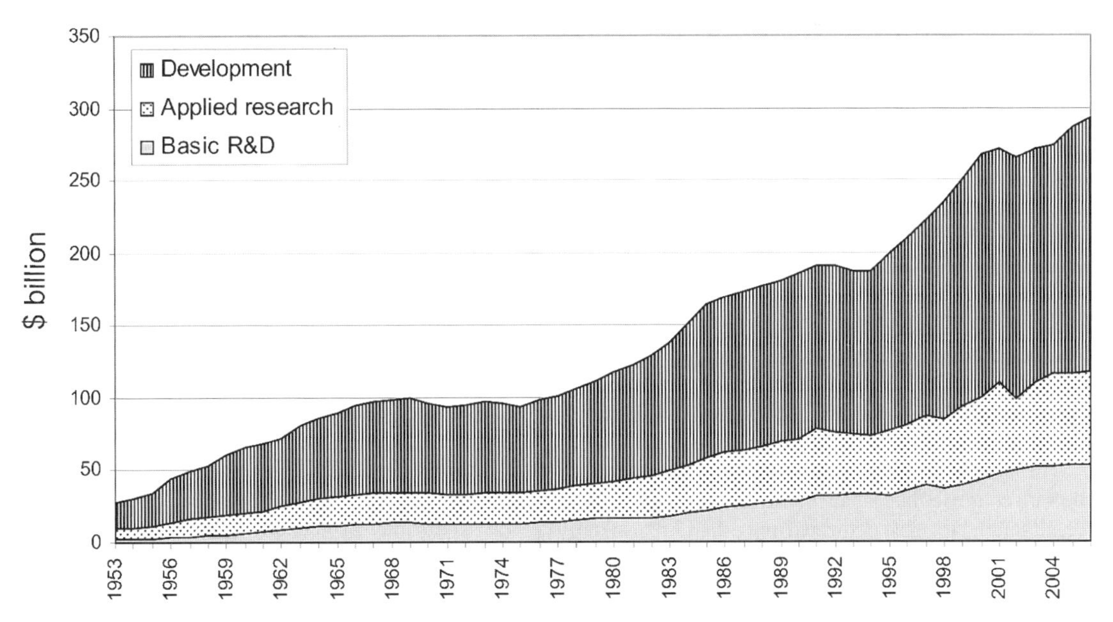

Figure 52.1. Overall trend in U.S. research and development expenditures (in 2000 $ billion).
Source: NSB (2008, vol. 2, based on data in tables 4-8, 4-12, and 4-16, pp. A4-13/14, A4-21/22, and A4-29/30).

and universities having leading-edge expertise in nearly every broad area of technology. This expertise does not always hold for narrowly defined areas of technology relating to details of specific products and their methods of production: in many industries, such as manufacture of televisions and other consumer electronics products, U.S. firms did not manage to sustain the pace of product and process innovation kept up by their foreign competitors and consequently lost control of the industry, in many cases with every U.S. company exiting a specific product market. Nonetheless, within broad technology classes the U.S. retains the ability to generate novel kinds of products and to keep up with the scientific state of the art. Despite this, U.S. producers do not retain market dominance of every industry.

52.1.3 Institutions

U.S. R&D occurs in, and is funded by, companies, universities, government agencies and laboratories, and non-profit organizations. Companies performed 71.8% of R&D (by dollar value) in 2006, compared with only 16.0% in universities, 7.2% in government settings, and 5.0% in non-profits. Of course, the Federal Government funds much of this activity, with the government being the ultimate source of 27.7% of R&D expenditures in 2006. Although industry's role has long been important, the Federal Government's share of funding has fallen dramatically, from 53.9% of R&D funding in 1953. Federal funding has increased by less than half since the late 1960s, while industry funding grew more than fivefold.

Universities have long been an important source of talent for companies and have also been a source of basic technologies. The flow of inventions from universities to industry has been encouraged by the Bayh–Dole Act of 1980, which allowed universities to retain patent rights from federally funded R&D. The Act has greatly increased commercial licensing from universities, and has encouraged more applied instead of basic research. At the same time, university technology transfer offices face new challenges trying to earn money from licensing, and most actually lose money. This is perhaps due to too great an emphasis on trying to extract the full value of inventions causing many firms to abandon attempts to work with them.

52.2 Trends in R&D and R&D output

Research and development takes place on a huge scale in the U.S. The level of national investment in R&D is larger than that of all other OECD countries combined and rests on the three pillars of industry, universities, and Federal Government. The roles of these three sectors have changed substantially over time.

52.2.1 R&D expenditures

In general, U.S. research and development expenditures, adjusted for inflation, show a tenfold increase from 1953 to 2006 (Figure 52.1). Basic research has grown the most over the period, from $2.5 billion in 1953 to $53.0 billion in 2006 (in 2000 dollars). However, the largest proportion of funding is attributed to development. In particular, R&D spending on "infratechnology"—technologies that facilitate technological infrastructure for R&D, production, and marketing in industries, such as for calibration equipment or production control equipment—has an important impact on innovation, and both federal laboratories and manufacturing industry have made significant investments in this area. In 1990, for example, 37.6% of federal and 16.3% of manufacturing R&D investments were dedicated to infratechnology (Leyden and Link, 1992).

The U.S. spends about 2.6% of its GDP on research and development. Since the 1960s, U.S. expenditure on R&D has been consistently between 2.0% and 3.0% of GDP, and it was a little lower in the 1950s. Over the last two decades, the U.S. proportion of GDP spent on R&D has been comparable with countries like Japan, Germany, the U.K., and France, although Japan has dedicated more than 3.0% of its GDP on R&D in the last decade or so, and the U.K. a little less than 2.0%. In contrast, countries like Russia, Italy, and Canada typically dedicate only about 1.0–2.0% of their GDP towards R&D (NSB, 2008). Hence U.S. R&D relative to GDP has been typical of the highly developed economies.

Table 52.1. Federal R&D by agency, 2007.

	$ million	*Percent*
Defense (DOD)	79,009	54.9
Health & Human Services (HHS)	29,621	20.6
NASA	11,582	8.0
Energy (DOE)	9,035	6.3
National Science Foundation (NSF)	4,440	3.1
Agriculture (USDA)	2,275	1.6
Other	1,113	0.8
Commerce	1,073	0.7
Homeland Security (DHS)	996	0.7
Veteran Affairs (VA)	819	0.6
Transportation (DOT)	767	0.5
Interior	647	0.4
Environmental Protection Agency (EPA)	557	0.4

Source: AAAS (2008, table I-1).

52.2.1.2 Defense and civilian programs

Federal research programs in the U.S. have historically been strongly tied to national defense. Navy-sponsored research programs, for example, date back to 1789. Since World War II, direct government support of R&D increased in response to military needs. For example, total R&D expenditure increased by a factor of 15 from 1940 to 1945 and military services have dominated the federal R&D budget for decades (Mowery and Rosenberg, 1993).

The majority of federal R&D is spent on defense, although this proportion has decreased dramatically since the post-World War II era. The 1960s saw an increase in spending on R&D for space programs. R&D for civilian programs increased somewhat during the 1970s but has been relatively stable as a proportion of spending since World War II. By 2007, the largest proportion of federal R&D expenditures remained allocated for defense, which accounted for $79.0 billion and 54.9% of the entire budget.[4] Health and human sciences accounted for 20.6% ($29.6 billion), and the National Aeronautics and Space Administration (NASA) as well as the Department of Energy and the National Science Foundation accounted for significant proportions of federal R&D expenditure (Table 52.1).

52.2.1.3 Industry, government, universities, and non-profits

Industry has performed (though not always paid for) 67.6% to 78.0% of U.S. R&D, by value, during each year between 1953 and 2006. Industry R&D rose tenfold from $19.9 billion in 1953 to $210.6 billion in 2006 (in year 2000 constant dollars, Table 52.2). The proportion of R&D performed by the Federal Government decreased substantially over the time period. In 1953 the federal

Table 52.2. U.S. R&D expenditure by performing sector (in constant 2000 $ million).

Year	*Federal*	*Industry*	*University*	*Non-profit*
1953	5,563	19,896	2,212	611
1958	7,350	40,925	3,934	1,039
1963	11,738	57,946	8,064	2,629
1968	14,037	69,959	11,677	3,335
1973	15,186	66,719	11,913	3,366
1978	15,216	72,785	14,790	3,685
1983	16,608	100,094	17,010	4,234
1988	18,948	128,167	24,838	4,917
1993	18,704	132,835	29,166	6,808
1998	17,997	175,367	32,889	8,484
2003	21,383	190,954	44,883	14,410
2006	21,025	210,666	46,828	14,735

Source: NSB (2008, vol. 2, table 4-4, pp. A4-5/6).
Note: industry, university, and non-profit figures include federally funded research and development centers (FFRDCs).

[4] War-related supplemental funds to be enacted in late 2008 were expected to further increase federal R&D for 2008 and 2009, according to a report of the American Association for the Advancement of Science (AAAS, 2008) published shortly before this article was completed. Total defense R&D, including supplemental funds, is anticipated to reach $84.5 billion in 2009.

Table 52.3. U.S. R&D expenditure by source of funds (in constant 2000 $ million).

Year	*Federal*	*Industry*	*University*	*Non-profit*
1953	15,251	12,313	200	301
1958	34,023	18,084	271	520
1963	53,429	25,030	440	904
1968	60,066	36,156	885	1,162
1973	52,081	41,756	1,077	1,323
1978	53,355	49,078	1,484	1,589
1983	63,568	69,416	2,080	1,872
1988	79,438	89,805	3,338	2,750
1993	68,482	109,243	4,196	3,833
1998	68,803	153,252	5,351	5,288
2003	78,565	174,969	7,188	8,334
2006	81,160	192,417	8,009	9,034

Source: NSB (2008, vol. 2, table 4-6, pp. A4-9/10).
Note: data exclude "other non-federal" sources of funding.

sector performed 20.1% of all (excluding non-profit) R&D, and by 2006 its share had fallen to 7.2%. Significant growth occurred in the university and non-profit sectors; both performing sectors increased R&D expenditures by a factor of 20 since 1953. In 1979 the university sector overtook the federal sector. By 2006 the federal sector performed 7.1% of R&D, universities 16.0%, and non-profit organizations 5.0%. Companies dominated R&D performance with 71.8%.

While industry is the largest spender of R&D, it was not always the largest source of funding. In 1953, federal sources accounted for 53.9% of R&D expenditure. The proportion of R&D funding from federal sources increased steadily up to 66.8% in 1964. This year appears to have been a turning point at which industry sources of R&D expenditure represented a low of 30.8% of the total. Industry sources of funding increased steadily and federal sources stagnated until, in 1980, industry became the largest source of R&D funding. Funding from industry continued to strengthen to 69.6% of R&D expenditure in 2000, while federal sources fell to a low of 24.8%. After 2000, the proportion of R&D expenditure from federal sources rose again to 29.6% in 2004, dropping to 27.7% in 2006. Industry remains the largest source of R&D expenditure at 65.6%, but the amounts contributed by universities and non-profit organizations, while small in comparison, have grown the most since 1953 (Table 52.3).

The nature of R&D differs across organizational types. Organizations' funding of 2006 R&D appears in Table 52.4, while organizations' performance of 2006 R&D appears in Table 52.5, both of which delineate basic R&D versus applied research or development. Federal

Table 52.4. R&D expenditure by source of funds (2006 $ million).

Type of R&D	*Federal and government labs*	*Industry*	*Universities and colleges*	*Non-profit organizations*	*Total*
Basic R&D	36,161	10,568	6,415	6,266	59,410
Applied research	24,892	43,960	2,363	2,660	73,875
Development	33,164	168,843	519	1,561	204,087
Total	*94,217*	*223,371*	*9,297*	*10,487*	*338,320*

Source: NSB (2008, vol. 2, based on tables 4-13, 4-15, and 4-17, pp. A4-23/24, A4-27/28, and A4-31/32).
Note: data exclude "other non-federal" sources of funding, therefore the total figures are less than totals in Table 52.5 which includes all R&D expenditure.

Table 52.5. R&D expenditure by performing sector (2006 $ million).

Type of R&D	*Federal and government labs*	*Industry*	*Universities and colleges*	*Non-profit organizations*	*Total*
Basic R&D	4,938	9,395	38,498	8,688	61,519
Applied research	7,750	49,923	11,855	5,125	74,653
Development	11,720	185,237	4,008	3,293	204,257
Total	*24,408*	*244,555*	*54,361*	*17,106*	*340,429*

Source: NSB (2008, vol. 2, based on tables 4-7, 4-9, and 4-11, pp. A4-11/12, A4-15/16, and A4-19/20).

agencies funded more basic R&D than applied or development work, but federal agencies' in-house R&D was largely development work. Universities and non-profit organizations largely focused on basic R&D. Industry focused overwhelmingly on development.

The sources of funding differ for each of the sectors in industry, universities, or non-profit organizations. The majority of R&D expenditure in industry is funded by industry itself. This has changed over time; federal funding of R&D was dominant for industry during the 1950s and 1960s. For universities, in contrast, federal funding represents the largest source of R&D funds. Increasingly, universities and colleges are contributing to their own R&D funding.

52.2.1.4 R&D by industry

R&D expenditure by industry (from all sources of funding) appears in Table 52.6. Some figures are unavailable because the National Science Foundation suppresses data on particular industries to avoid disclosing confidential information. Nevertheless, available data on R&D expenditure are revealing. Overall, manufacturing industries have higher expenditure on R&D than non-manufacturing industries. Among the industries for which information is available, the largest expenditure takes place in chemicals (19.0%), computer/electronics (19.4%), transportation equipment (15.8%), and professional, scientific, and technical services sectors (14.2%).[5] The fastest rate of increase in R&D was in the beverage and tobacco industry.[6]

Most of the funding for R&D in industry comes from non-federal sources, but federal sources are important for some sectors. The construction industry, for instance, has the highest funding share from federal sources (23.7%), although the total amount of R&D is small in this sector. Other industries with substantial federal R&D funding include computers/electronics (16.7%), transportation equipment (18.9%) and professional, scientific and technical services (18.2%).[7]

R&D expenditure per 1,000 employees in 2005 was highest in the chemicals industry, which spent twice as much on R&D per employee as any other industry. The chemicals industry also had the highest R&D per scientist and engineer. The food industry, miscellaneous manufacturing, and health care services also spent substantial amounts on R&D per scientist and engineer, with over $200 million per 1,000 scientists and engineers.[8]

[5] R&D expenditure for computers/electronics and transportation equipment is calculated from 2003 figures, since 2005 data were not available.
[6] Based on 2001–2005 figures because data for 1999 were not available.
[7] Figures for computers/electronics, transportation equipment, and construction were calculated based on 2003 figures, since 2005 data were not available.
[8] Data on employees and funding by industry for 2005 came from Wolfe (2007).

52.2.1.5 R&D by firm size

R&D performed by industry during 2005 totaled $226.2 billion, of which 37.6% was performed by companies with more than 25,000 employees. Small companies had the highest R&D per scientist and engineer, whereas the largest companies had the highest R&D per employee. The largest percentage growth of R&D from 1999 to 2005, 99.7%, was for companies with 25 to 49 employees.[9] Companies largely relied on non-federal sources of funding to perform R&D: only 9.7% of company-performed R&D came from federal sources. Companies with more than 25,000 employees enjoyed the largest share of federal R&D funding, 60.9%. Companies with 5 to 24 employees received only 3.3% of R&D funding, and relied more heavily on federal sources (14.6% of their R&D funding came from federal sources).

52.2.1.6 R&D by state

R&D funding has been affected not only by national policy, but also by individual U.S. states. National industrial policy has rarely been closely tied to innovation issues, in contrast to the strategic design more common in Europe and Japan. State policy in many ways has done more to set up industry-targeted innovation programs, perhaps because some state governments have been relatively willing to influence directly their educational, industrial, and financial sectors (see Lindsey, 1988).

R&D expenditures vary significantly across individual states. In 2005, Michigan had the largest R&D intensity (state R&D expenditure divided by gross state product) of all U.S. states, as Table 52.7 shows. Although California's R&D expenditure is more than double that of Michigan, it comes fifth in terms of intensity. Nearly all R&D funding (98.8%) in Michigan comes from non-federal sources. In contrast, Massachusetts, the second most R&D-intensive state, receives about 19.1% of its R&D expenditure from federal sources. Federal funding, as Table 52.8 shows, is particularly important for states with low R&D intensity, such as Alabama, Virginia, Maryland, and New Mexico, plus Washington, D.C. For these states, federal sources account for over 30% of total R&D expenditure.

These state R&D patterns have many causes, but a role of technological investments is apparent from comparison of California and Massachusetts with Michigan. Both California and Massachusetts benefited from several private-sector innovation initiatives, stimulated by federal R&D programs that date back to World War II. These efforts contributed to innovative industrial growth areas such as Silicon Valley and the Boston area's Route 128. Early developments were spurred by the creation of specialized

[9] Data from the U.S. Census Bureau and the U.S. Small Business Administration suggest that the number of companies with 20–99 employees grew by 3.8% from 1999 to 2005. Thus, the increase in R&D for companies with 25–49 employees does not appear to be a result of overall growth in this sector.

Table 52.6. U.S. R&D by industry.

Industry	*1999*	*2001*	*2003*	*2005*	*% of total R&D*	*% of R&D from federal funding*	*R&D as % of sales*	*R&D ($m)/ employees (thou)*	*R&D ($m)/ scientists and engineers (thou)*	*Change in R&D 1999–2005 (%)*
All industries	***184,129***	***202,017***	***200,724***	***226,159***		***9.7***	***3.7***	***14.1***	***206.03***	***22.8***
Manufacturing industries	*118,339*	*124,217*	*120,858*	*158,190*	*69.9*	*9.9*	*4.0*	*16.76*	*227.35*	*33.7*
Food	1,132	1,819		2,716	1.2	0.2	0.7	2.73	230.17	139.9
Beverage and tobacco products		152	173	539	0.2	0.0				254.6[b]
Textiles, apparel, and leather	334			816	0.4	0.6	1.6	4.00	104.62	144.3
Wood products	70	182			0.1[a]	0.0				160.0[b]
Paper, printing, and support activities										
Petroleum and coal products	615				0.3[a]					
Chemicals	20,246	17,892	23,001	42,995	19.0	0.4	6.9	40.03	364.36	112.4
Plastics and rubber products	1,785		1,764	1,760	0.8	0.7	2.0	4.67	160.00	−1.4
Non-metallic mineral products		990	474	894	0.4	0.7				−9.7[b]
Primary metals	470	485	530	631	0.3	3.5				34.3
Fabricated metal products	1,655	1,599	1,374	1,375	0.6	3.8	0.8	2.20	82.34	−16.9
Machinery	6,057	6,404	6,304	8,531	3.8	1.3	3.7	10.25	143.14	40.8
Computer and electronic products	37,350	50,591	39,001		19.4[a]	16.7[a]		0.00	0.00	4.4[b]
Electrical equipment, appliances, and components		4,980	2,073	2,424	1.1	4.2	2.4	7.55	135.42	−51.3[b]
Transportation equipment	33,965	25,965	31,747		15.8[a]	18.9[a]				−6.5[b]
Furniture and related products	248	301		400	0.2					61.3
Miscellaneous manufacturing	3,851	6,606	7,455	5,143	2.3	1.6	6.2	15.22	237.00	33.5
Nonmanufacturing industries	*65,790*	*77,799*	*79,866*	*67,969*	*30.1*	*9.2*	*3.2*	*10.30*	*169.12*	*3.3*
Utilities	421	133		210	0.1	14.3				−50.1
Construction	691	320	333		0.2[a]	23.7[a]				−51.8[b]
Wholesale trade			25,092		12.5[a]	0.5[a]				
Retail trade			1,488		0.7[a]	1.7[a]				
Transportation and warehousing	460	1,848	272		0.1[a]					−40.9[b]
Information (publishing, broadcast, telecom, internet)	15,389			23,836	10.5	0.9	5.4	15.97	177.62	54.9
Finance, insurance, and real estate			1,455	3,030	1.3	0.0	0.5	2.40	100.33	108.2[b]
Professional, scientific, and technical services	18,994	27,704	27,967	32,021	14.2	18.2	12.2	27.09	187.70	68.6
Management of companies and enterprises		381	67		0.0[a]	0.0[a]				−82.4[b]
Health care services	642	1,149	717	989	0.4	0.7	3.9	10.63	210.43	54.0
Other non-manufacturing		1,259	1,679	2,137	0.9	4.2				69.7[b]

Sources: NSB (2008, vol. 2, table 4-19, pp. A4-35/36), Wolfe (2007, tables 2 and 3).

Notes: blank cells are data suppressed by the NSF to avoid disclosure of confidential information, or represent unavailable data (wholesale and retail trade in 1999 and 2001, "management of companies and enterprises" in 2005). "Mining, extracting, and support" industry is excluded because all data were suppressed.

[a] Percentages calculated from latest available year if 2005 data are unavailable.

[b] Change in R&D calculated for available years if 1999 and 2005 data are unavailable.

Table 52.7. Top-10 states' R&D intensity in 2005.

Rank	*State*	*R&D intensity**
1	Michigan	4.50
2	Massachusetts	4.20
3	Connecticut	4.08
4	Washington	3.56
5	Rhode Island	3.22
6	New Jersey	3.11
7	California	3.10
8	Minnesota	2.73
9	New Hampshire	2.68
10	Delaware	2.64

Sources: NSB and Bureau of Economic Analysis.
*R&D/gross state product × 100%.

Table 52.8. States with largest federal R&D funding in 2005.

Rank	*State*	*R&D intensity**	*% federal*
1	Alabama	0.94	50.7
2	D.C.	0.20	44.0
3	Virginia	1.25	38.7
4	Maryland	1.52	33.8
5	New Mexico	0.59	31.6
6	Florida	0.62	28.6
7	Hawaii	0.31	27.4
8	Mississippi	0.24	24.2
9	Massachusetts	4.20	19.1
10	Connecticut	4.08	18.3

Sources: NSB and Bureau of Economic Analysis.
*R&D/gross state product × 100%.

organizations including the Massachusetts Technology Development Corporation in 1978 and the California Commission on Industrial Innovation in 1981. California and Massachusetts also received the largest share of venture capital funding in 2006 and 2007; both states also had the highest venture capital per dollar of gross state product (see Table 52.19 and NSB, 2008, table 8-45, p. 8-97). Michigan, in comparison, has traditionally had a strong industrial base in automobiles and not in industries with high federal R&D funding. The state has tried to reverse this trend (e.g., by establishing a task force on high technology in 1981 to guide Michigan toward leadership in robotics, molecular biology, and related fields). Michigan's Industrial Technology Institute, established in 1982, and Molecular Biology Institute, established in 1983, combined public and private funds to aid Michigan firms in computer-based production and biotechnology, respectively.

52.2.2 R&D workforce

The U.S. workforce in science and engineering (S&E) grew tremendously over the last half of the 20th century, at an average annual growth rate of 6.4% (NSB, 2008). The proportion of S&E jobs grew from 2.6% in 1983 to 4.2% in 2006. In conjunction, degrees in science and engineering at bachelor's, master's, and doctoral levels also increased steadily from 1950 to 2000. In spite of the growth in the number of researchers employed in the U.S., the level of growth is still a third less than that of other OECD countries.

There is high and growing demand for S&E skills in the U.S. labor force. The S&E demand extends to occupations not directly related to science and engineering; during 2003, 66% of S&E degree holders who held non-S&E jobs, such as positions in management or marketing, indicated that their degree was related to their job. The supply of potential S&E workers tightened from 1993 to 2003, judging from median salaries. Median salaries increased more rapidly in those years for S&E bachelor's degree holders than for persons with non-S&E bachelor's degrees. The largest increases were for those who held degrees in engineering (34.1%), computer and mathematical sciences (28.0%), and life sciences (24.5%). Differences in median salary increases were less at the master's level and negligible at the doctoral degree level.

Foreign-born scientists and engineers are becoming increasingly important to U.S. S&E enterprises, and their presence continues to grow. As of 2003, 25% of all college-educated workers and 40% of doctoral degree holders in S&E occupations were foreign-born. Areas in which foreign-born doctorate holders dominate in the U.S. include computer science, electrical engineering, civil engineering, and mechanical engineering. The U.S. has undoubtedly benefited from the inflow of international knowledge and personnel, but competition for skilled labor is intensifying as other countries are increasing their investments in research and making high-skilled migration part of their economic strategy. Nations such as Germany, Canada, and Japan have changed immigration laws accordingly. The U.S. provides temporary work visas, typically in the form of H-1B visas. During 2001, it was estimated that the majority of these visas went to S&E-related occupations, particularly in computing (57.8%).

A massive drop in the annual cap on the number of H-1B visas granted, from 195,000 in 2003 to 65,000 in 2004, occurred when the Federal Government allowed relevant legislation to expire. Consequently, on October 1, 2004, the U.S. Citizenship and Immigration Services Bureau announced that the ceiling of H-1B visa applica-

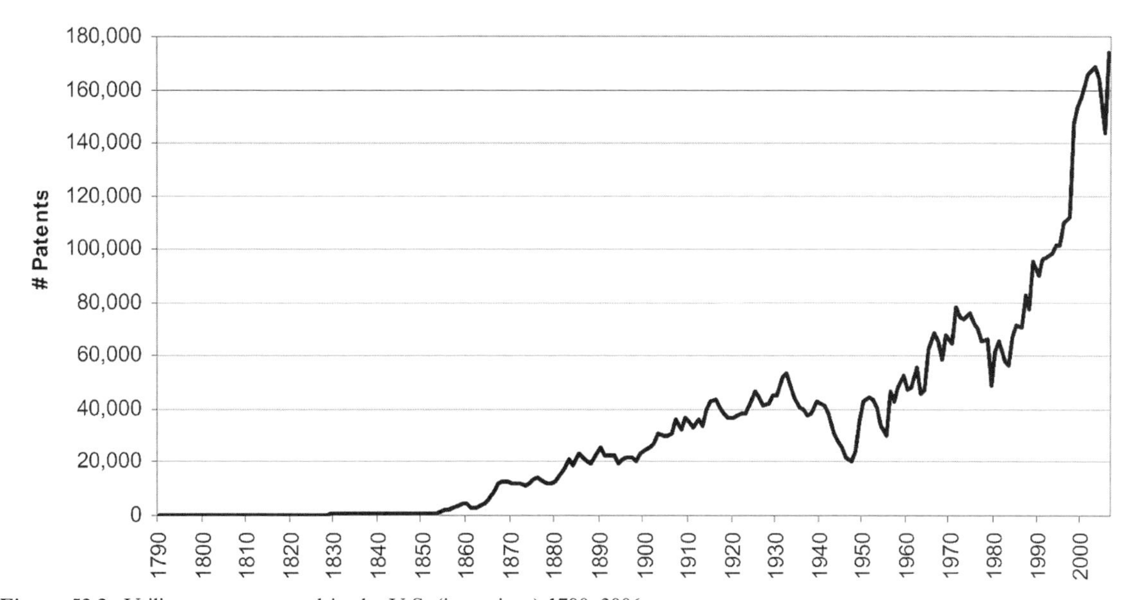

Figure 52.2. Utility patents granted in the U.S. (inventions) 1790–2006.
Source: U.S. Patent and Trademark Office (USPTO TAF database); older statistics from USPTO (1977).

tions had already been reached for 2005, on the first day of the fiscal year to which the quota applied.[10] Some U.S. Senators have proposed an increase of the cap, but as of mid-June 2008, such legislation has yet to be passed by Congress.

52.2.3 R&D output as indicated by patents

Patents are indicative measures of R&D and innovation output. USPTO, the U.S. Patent and Trademark Office, tracks various types of patents applied for and granted. A patent of invention, called a "utility" patent under U.S. law, is a patent issued for the "invention of a new and useful process, machine, manufacture, or composition of matter, or a new and useful improvement thereof."

52.2.3.1 Long-term U.S. patent trend

In 1790, three U.S. utility patents were granted. This number has grown tremendously to more than 170,000 utility patents granted in 2006 (Figure 52.2), with more than twice that number of applications. The enormous surge in patenting in the 1980s through 2006 corresponds to the establishment of the Court of Appeals for the Federal Circuit, which made patents easier to get and easier to defend in court, and to shifts in practices at the USPTO that have been claimed to make the office more beholden to applicants' interests (Jaffe and Lerner, 2004).

52.2.3.2 Internationalization of U.S. patent ownership

In 1963, the majority of utility patents granted were to U.S. inventors or organizations. However, with increasing internationalization of business in recent decades, the proportion of utility patents granted to U.S. inventors or organizations fell, as Figure 52.3 shows, to just over half in 2005. The overall number of utility patents to U.S. inventors or organizations increased 142% from 1963 to 2006.

52.2.3.3 Organization types of U.S. patent assignees

Historically, only a small percentage of patents granted have been assigned to universities and government organizations, although this has been changing (Table 52.9). Corporations in 2004 received 73.5% of patents. Individual inventors account for 17.0% of patents in 2004, and the frequency with which individuals are assignees has declined over much of the 1900s, although it rose from 2001 to 2004. The contribution of government has been decreasing steadily, while patents granted to universities are becoming more common.

There has been a tremendous growth in academic patenting and licensing since 1991, as Table 52.10 shows. The number of U.S. patents granted to major universities was 2,944 in 2005, compared with 1,307 in

[10] Academic institutions are exempt from the ceiling, and exemptions also apply to students with master's or doctoral degrees received at U.S. schools. Alternative visa options do exist for scientists and engineers (L-1), high-skilled workers under the North American Free Trade Agreement (TN-1), individuals with outstanding abilities (O-1), and other programs. Students may additionally apply for F-1 or J-1 visas.

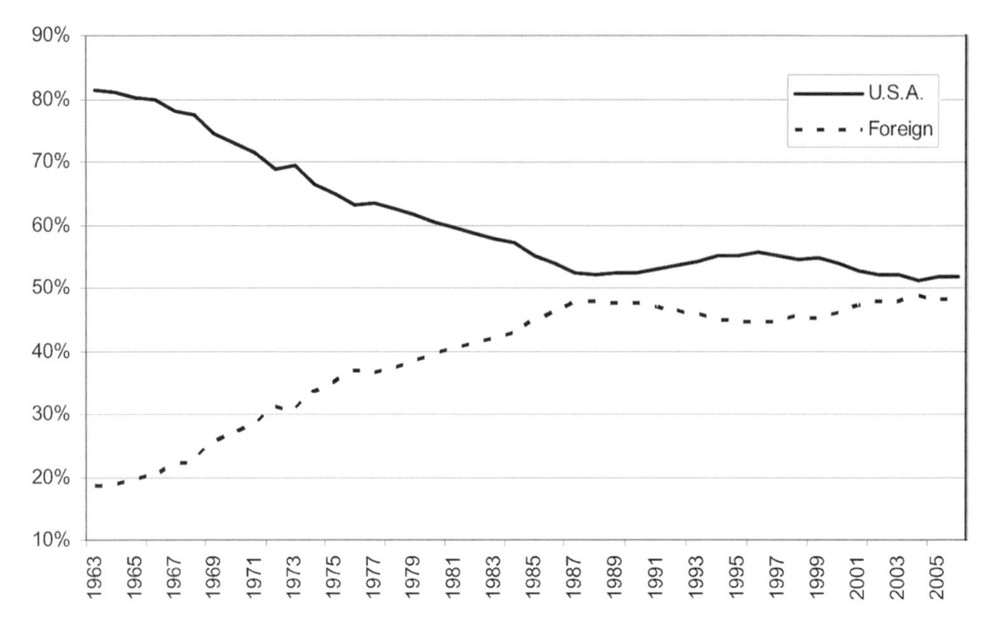

Figure 52.3. Proportion of utility patents granted, U.S. versus foreign, 1963–2006.
Source: USPTO TAF database.

1993, a growth of 125%.[11] Increases also occurred in receipt of royalties and in new research funding received as a result of licenses. The number of startup companies formed from universities increased from 169 in 1995 to 418 in 2005.

Table 52.9. Percent of U.S. utility patents granted to corporations, government, individuals and universities (1993–2004).

	Corporation (%)	*Government (%)*	*University (%)*	*Individual (%)*
1993	74.2	1.9	2.5	21.4
1994	75.1	1.5	2.5	21.0
1995	77.9	1.4	2.2	18.6
1996	77.9	1.1	2.6	18.4
1997	79.1	1.0	2.5	17.4
1998	79.3	1.0	3.3	16.5
1999	80.2	0.8	3.3	15.6
2000	81.2	0.9	3.0	14.9
2001	82.0	1.0	3.2	13.9
2002	80.7	0.9	3.6	14.8
2003	78.5	0.9	4.8	15.7
2004	73.5	0.9	8.6	17.0

Source: USPTO TAF database.

52.2.3.4 Top patenting corporations

As of 2006, IBM was the U.S. corporation with the most U.S. utility patents granted with 3,621 patents. Hewlett-Packard came in second place with 2,099 patents, and Intel occupied the third spot with 1,959. IBM has not always held the first place slot; in 1969, for example, GE held the Number-1 spot, followed by AT&T and DuPont. Nevertheless, IBM has long remained strong in patenting, occupying fourth place in 1969. While strong, the top-10 U.S. corporations do not dominate relative to international firms in the number of utility patents granted by the USPTO. For example, Hewlett-Packard ranks second in the U.S., but fifth overall. Foreign firms Samsung, Canon, and Matsushita ranked second, third, and fourth, respectively, in 2006 utility patents granted.

52.2.3.5 Technology classes

Although rankings of the most active technology classes of U.S. firms depend on class definitions, they indicate at least some of the most actively researched types of technology. Patenting by U.S. firms in the leading classes in 2006 is shown in Table 52.11. The most utility patents granted to U.S. owners were within the "drug, bio-affect-

[11] The figures reported here are close approximations because the data in Table 52.10 are for AUTM members, and new (generally less technologically active) universities joined AUTM over time. Perhaps one-tenth of the growth from 1993 to 2001 for statistics reported in Table 52.10 results from increases in AUTM members. This figure results by using USPTO data that indicate 1,610 university patents in 1993 and 3,423 in 2001, implying that AUTM data captured 81% of university patents in 1993 versus 91% in 2001. This implies a corrected estimate for the true growth g from 1993 to 2001 in any statistic n used in the table of about $g = ((100/91) * n_{2001})/((100/81) * n_{1993}) = 0.89 * (n_{2001}/n_{1993})$, yielding the one-tenth rule mentioned here.

Table 52.10. Academic patenting and licensing 1991–2005.

	1991	*1993*	*1995*	*1997*	*1999*	*2001*	*2003*	*2005*
No. institutions reporting	98	117	127	132	139	139	165	158
				Millions of dollars				
Net royalties	NA	195	239.1	391.1	583	753.9	866.8	1,588.1
Gross royalties	130	242.3	299.1	482.8	675.5	868.3	1,033.6	1,775.0
Royalties paid to others	n/a	19.5	25.6	36.2	34.5	41	65.5	67.8
Unreimbursed legal fees expended	19.3	27.8	34.4	55.5	58	73.4	101.3	119.1
				Number				
Invention disclosures received	4,880	6,598	7,427	9,051	10,052	11,259	13,718	15,371
New U.S. patent applications filed	1,335	1,993	2,373	3,644	4,871	5,784	7,203	9,306
U.S. patents granted	NA	1,307	1,550	2,239	3,079	3,179	3,450	2,944
Startup companies formed	NA	NA	169	258	275	402	348	418
Revenue-generating licenses/options	2,210	3,413	4,272	5,659	6,663	7,715	11,118	10,251
New licenses/options executed[a]	1,079	1,737	2,142	2,707	3,295	3,300	3,855	4,201
Equity licenses/options	NA	NA	99	203	181	328	316	278

Sources: NSB (2008, vol. 2, table 5-42, p. A5-81); USPTO TAF database.
[a]Data prior to 2004 may not be comparable with data for 2004 and beyond due to change in survey wording.

Table 52.11. Top technology classes among U.S.-owned utility patents granted in 2006.

Technology class	*U.S.-owned patents 2006*
Drug, bio-affecting, and body-treating compositions	3,262
Multiplex communications	2,297
Semiconductor device manufacturing: process	2,067
Telecommunications	1,958
Chemistry: molecular biology and microbiology	1,920
Multi-computer data transferring (electrical computers and digital processing systems)	1,780
Active solid-state devices (e.g., transistors, solid-state diodes)	1,699
DP: database and file management or data structures (data processing)	1,437
Communications: electrical	1,432

Source: USPTO TAF database; patent counts based on original and cross-reference classifications with duplicate patents eliminated within classes.

ing and body treating compositions" class (USPTO Class 514), which had 3,262 patents in 2006. Other areas of high utility patenting include multiplex communications, semiconductor device manufacturing, telecommunications, molecular and microbiology, multicomputer data transfer, solid-state devices, database management, and electrical communications.

52.2.3.6 State differences

California was the largest producer of utility patents in 2006, with 24.3% of all U.S. patents being granted in the state, followed by Texas (6.9%) and New York (6.1%). Historically, California has dominated utility patents in the U.S. (Table 52.12). However, from 1963 to 2006 the fastest growing state in terms of utility patents granted was Idaho which grew 31-fold and ranked 18th in 2006, followed by a 14-fold growth of Vermont (ranked 32nd in 2006).

52.2.4 R&D output as indicated by publications

Publications are an alternative measure of R&D output. In 2005 the U.S. held about 28.9% of the world's science and engineering article output by publishing country. By comparison, as Table 52.13 indicates, European nations together produced about 36.2% of publications in 2005. The largest portion of U.S.-originated articles were in the medical and biological sciences fields, as Table 52.14

Table 52.12. Top states, 2006 utility patents.

State	*1963*	*1968*	*1973*	*1978*	*1983*	*1988*	*1993*	*1998*	*2003*	*2006*
California	4,357	5,637	7,040	5,504	4,427	6,070	8,165	15,790	19,688	22,275
Texas	1,340	1,772	2,080	1,930	1,873	2,388	3,389	5,576	6,029	6,308
New York	4,437	4,956	5,413	3,921	2,731	3,315	4,692	6,319	6,234	5,627
Massachusetts	1,647	1,910	2,203	1,732	1,321	1,622	2,207	3,413	3,908	4,011
Michigan	2,347	2,816	2,923	2,348	1,677	2,196	2,873	3,506	3,857	3,758
Illinois	3,288	3,713	4,073	3,109	2,139	2,369	2,846	3,725	3,296	3,294
Washington	337	395	508	495	450	654	896	1,774	2,285	3,286
New Jersey	2,847	3,769	4,129	3,452	2,632	2,619	2,912	3,767	3,522	3,172
Minnesota	710	881	973	831	776	1,075	1,547	2,473	2,953	2,957
Pennsylvania	2,752	3,379	3,658	2,900	2,172	2,426	2,677	3,370	3,182	2,842
Ohio	2,917	3,207	3,328	2,568	1,905	2,142	2,531	3,272	3,183	2,630

Source: USPTO TAF database.

Table 52.13. Percent share of science and engineering article output.

Region	*1995*	*1997*	*1999*	*2001*	*2003*	*2005*
Europe	37.0	38.0	38.2	37.7	36.8	36.2
U.S. and Canada	38.4	36.1	34.4	33.8	33.2	32.6
Asia	13.6	15.0	16.4	17.7	19.1	20.6
Central/South America	1.7	2.0	2.3	2.6	2.8	2.9
Australia/New Zealand	2.8	2.8	2.8	2.8	2.7	2.7
Africa/Middle East	2.4	2.4	2.4	2.4	2.5	2.5

Source: NSB (2008, vol. 2, table 5-34, pp. A5-57-59), which draws on articles in Thompson ISI's Science Citation Index and Social Sciences Citation Index.

Table 52.14. U.S. science and engineering article output (percent by field).

	1995	*1997*	*1999*	*2001*	*2003*	*2005*
Medical sciences	26.7	27.5	27.6	27.7	27.7	28.0
Biological sciences	27.8	27.3	27.0	27.3	26.2	25.9
Physics	10.1	9.5	9.5	9.0	9.2	9.6
Chemistry	7.7	7.6	7.6	7.4	7.8	8.0
Engineering	6.5	6.2	6.5	6.5	6.7	6.7
Geosciences	5.1	4.9	5.2	5.2	5.7	5.2
Social sciences	5.0	5.2	5.1	5.2	5.2	5.0
Psychology	4.0	4.3	4.1	4.0	3.9	4.0
Mathematics	1.6	1.6	1.9	1.9	1.9	1.8
Other life sciences	1.2	1.6	1.5	1.6	1.5	1.7
Agricultural sciences	1.9	1.9	1.5	1.7	1.9	1.7
Astronomy	1.3	1.3	1.5	1.4	1.3	1.3
Computer sciences	1.0	1.0	1.1	1.1	1.1	1.1

Source: NSB (2008, vol. 2, table 5-36, pp. A5-63-65), which draws on articles in Thompson ISI's Science Citation Index and Social Sciences Citation Index.

shows, and the strongest growth from 1995 to 2005 was in other life sciences (57.0%), followed by computer sciences (24.9%) and mathematics (22.9%).

52.3 Technology commercialization in small businesses

Technology commercialization patterns and policy have been evident above, but relevant programs affecting small businesses remain to be discussed.

In 1982, the Small Business Innovation Research (SBIR) program was established under the Small Business Innovation Development Act, which was reauthorized until 2008. The program mandates that federal agencies with extramural R&D budgets over $100 million must administer SBIR programs and set aside 2.5% of their budget for innovative research to be conducted at small companies. As of January 2007, $12 billion was awarded to small businesses through SBIR. Eleven federal agencies participate in the SBIR program: the Departments of Health and Human Services (DHHS), Agriculture, Commerce, Defense (DOD), Education, Energy (DOE), Homeland Security, and Transportation; the Environmental Protection Agency; NASA; and the National Science Foundation (NSF).

A second program, Small Business Technology Transfer (STTR), supports joint R&D of small businesses with non-commercial research institutions. In this program, federal agencies with extramural R&D budgets over $1 billion are mandated to set aside 0.3% of their budget to administer STTR programs. Five agencies participate in this program: DOD, DOE, DHHS, NASA, and NSF. The program was established in 1992 by the Small Business Technology Transfer Act and was reauthorized until 2009.

The objective of these programs is to stimulate technological innovation in small businesses and their participation in meeting federal R&D needs. In addition, an important aspect of the program is to increase commercialization of innovations within the private sector. The programs are structured in three phases in which the first phase is to establish the technical merit, feasibility, and potential of commercialization of the proposed R&D efforts. If successful, the project proceeds to a second stage in which more money is awarded toward commercialization. In the final phase, the business pursues non-SBIR/STTR funds to achieve its commercialization objectives. Within the SBIR program, the research must be based at a small-business concern in order to receive the award. Within the STTR program, research partners at universities and non-profit institutions must have formal collaborative relationships with small businesses.

52.4 National innovation policy

Elements of a national innovation policy are rooted in the U.S. constitution. As early as 1790, the Patent Act was established, acknowledging that inventions had value and granting inventors temporary monopoly control over their ideas. Two major government policy instruments support innovation: antitrust statutes and military R&D and procurement. The U.S. Navy, for instance, sponsored research programs as far back as 1789, and the U.S. Department of Agriculture has been involved in the land grant college system since the mid-1800s. In fact, the Hatch Act of 1887 focused on the development and widespread use of new seeds, equipment, fertilizers, and farm practices. An important early development to establish the private property rights of plant varieties was the Plant Patent Act of 1930. Living organism patent rights were extended to engineered organisms by a 1980 U.S. Supreme Court decision in *Diamond v. Chakrabarty*.[12]

World War II brought an extra emphasis on federally funded R&D, and the National Science Foundation Act of 1947 focused on basic and applied R&D. Total R&D expenditure rose from $1.0 billion in 1940 to $15.8 billion in 1945 (in 2006 dollars). The Manhattan Project created a research and weapons production complex. The Office of Scientific Research and Development (OSRD), which entered into contracts with the private sector and relied on universities to do research, was another legacy of the war. After World War II, the innovation policy emphasis shifted and instruments were established specifically to protect small businesses and entrepreneurs with the Small Business Act of 1953. New firms and small startups were particularly important in microelectronics, computer hardware and software, robotics, and biotechnology. Several policy initiatives have changed the relationship between military and civilian technologies. They include the National Center for Manufacturing Sciences, the Defense Advanced Research Projects Agency, and Sematech.[13]

52.4.1 Programs and policies

While most of the innovation in the U.S. takes place in the private sector, it is clear that national programs and institutions play an important role, as do individual state programs. The Internal Revenue Code in 1954 allowed

[12] *Diamond vs. Chakrabarty* pertained to an oil-eating bacterium (used to clean oil spills) created by modifying a preexisting bacterium with new genetic mechanisms. This helped pave the way for the USPTO's later decision to grant gene patents.

[13] Sematech was an industry–government program to strengthen the U.S. semiconductor industry. The program aimed to leverage resources and share risks of both government and non-government semiconductor technologies because foreign control of essential computing resources threatened military and economic security. It has since expanded into an international body to include non-U.S. firms cooperating on standards and specifications.

businesses to treat R&D expenditures as current expenses for tax purposes. The Bayh–Dole Act of 1980 created a uniform patent policy enabling small businesses, non-profit organizations, and universities to retain title to inventions made under federally funded research programs. Other direct government involvement includes the Economic Tax Recovery Act (1981), the National Cooperative Research Act (1984), the Technology Transfer Act (1986), the Omnibus Trade and Competitiveness Act (1988), plus the programs discussed above including R&D tax credits and the SBIR.

Nevertheless, the U.S. national innovation policy is not based on any one economic strategy. Instead, the structure of R&D programs, financing, and administration to stimulate innovation is fragmented, diverse, and often issue-oriented. For example, the National Advisory Committee on Aeronautics, founded in 1915, invested in engineering and innovation in commercial aviation. The National Cancer Institute began in the 1930s with federal investment in biomedical research. A Civilian Industrial Technology Program (CITP), proposed in 1964 by the Kennedy Administration amid signs of lagging innovation in the U.S., would have supported industrial research at universities and provided incentives for industry to take on costly and risky R&D initiatives; although it was defeated, the CITP reflected ongoing consideration of government funding for non-military industrial R&D. The National Cooperative Research Act in 1984 strengthened enforcement of intellectual property protection and reduced antitrust restrictions on research collaborations.

A New Technologies Opportunity Program (NTOP), set up in 1971, considered tax incentives for private R&D and changes in antitrust laws. It also suggested large increases in federal spending for applied civilian research. Often too, the plug was pulled on such national program initiatives. Congress, for example, approved only a portion of the CITP program but later eliminated funding, and funds for NTOP were impounded. R&D and innovation issues were reverted to the Office of Science & Technology and later, under the Carter Administration, to the Domestic Policy Review (1978). Following that, in 1983, under President Reagan, the Commission on Industrial Competitiveness was charged with reviewing federal R&D and innovation policies.

52.4.2 State programs

Without a strong cohesive national innovation policy, much of the emphasis on innovation in the U.S. has been in state-level programs, with the exception of national defense and space programs. In the 1960s, state-level programs were implemented to influence the decisions of private firms to relocate plants, to attract venture capital, and to direct R&D results from government research labs to universities. Governors of all states are directly concerned with technological innovation (Lindsey, 1988). States have the potential to affect several components of society, including educational institutions, industrial firms, financial firms, and other private sectors. Strong state-level innovation initiatives exist particularly in California, Massachusetts, and Michigan (see Section 52.2). North Carolina and Connecticut also have strong programs. North Carolina, for example, created in the 1950s the Research Triangle Park that has attracted more than 50 major federal and private R&D facilities and over $150 million in funding between 1977 and 1988. The Connecticut Product Development Corporation, established in 1973, later transitioned into Connecticut Innovations, which operated much like a venture capital fund and as of 2008 had received more than $1 billion in private equity investments.

52.4.3 2004–2007 policy initiatives

More recently, new innovation policies have been considered. In 2004, the Bush Administration proposed a program with three major public policy initiatives, focused on hydrogen fuel technology, health care information technology, and broadband technology. In 2005 federal R&D investment reached a record $132 billion, doubling the funding for nanotechnology to $1 billion and increasing information technology funding to $2 billion. However, the Administration also proposed to cut science funding for 21 of 24 agencies.

A National Innovation Act was proposed in 2005 by senators John Ensign and Joseph Lieberman. The proposal would have established a President's Council on Innovation as well as an Innovation Acceleration Grants Program. It sought to nearly double the NSF research funding level by 2011 and expand graduate research, fellowship, and trainee programs, to require that the National Institute of Standards and Technology support research for advanced manufacturing systems to enhance productivity and efficiency, and to make permanent the research and experimentation credit for qualified expenses. Furthermore, a goal of the program was to mandate that the Department of Defense allocate at least 3% of its budget towards science and technology, with at least 20% going to basic research. This Act was introduced to the House of Representatives in January 2006, but was not enacted.

Congress and President Bush did, however, enact in 2007 the American Competitiveness Initiative, as the America COMPETES Act. The goals of the initiative were to double, over 10 years, funding for innovation-enabling research at federal agencies, modernize research and experimentation tax credits, strengthen K-12 math and science education, reform the workforce-training system, and reform immigration in order to retain highly skilled foreign workers in the U.S. The legislation retained the massive increase in NSF funding of the National Innovation Act. However, the American Competitiveness Initiative was enabling legislation that allowed but did

Table 52.15. U.S. advanced technology products: import to export ratio.

Industry	*2003*	*2004*	*2005*	*2006*	*2007*[a]
Opto-electronics	2.13	2.22	2.63	3.87	4.33
Info/communication	2.07	2.24	2.30	2.32	2.40
Nuclear technology	1.74	1.44	2.44	1.85	1.78
Life science	2.38	2.26	1.84	1.80	1.73
Advanced materials	1.46	1.58	1.56	1.55	1.42
Biotech	0.76	0.53	0.92	0.94	0.89
Flexible manufacturing	0.75	0.58	0.75	0.70	0.71
Electronics	0.54	0.57	0.56	0.52	0.54
Weapons	0.32	0.29	0.42	0.37	0.41
Aerospace	0.46	0.44	0.41	0.34	0.37

Source: U.S. Census Bureau, U.S. Trade in Advanced Technology Products.
[a]2007 data are January–October only.

Table 52.16. U.S. exports and imports in advanced technology products January–October 2007.

Industry	*Exports ($bn)*	*Imports ($bn)*
Biotech	6.48	5.78
Life science	17.56	30.32
Opto-electronics	4.56	19.77
Information/communication	61.28	147.27
Electronics	41.91	22.56
Flexible manufacturing	12.49	8.90
Advanced materials	1.38	1.96
Aerospace	75.69	28.27
Weapons	1.53	0.63
Nuclear technology	2.21	3.94

Source: U.S. Census Bureau, U.S. Trade in Advanced Technology Products.

not provide funding. The federal budget approved in December 2007 failed to allocate more than a minor fraction of the intended funds for the American Competitiveness Initiative.

52.5 Technological strengths and weaknesses

To assess the relative strengths or weaknesses of U.S. innovation, Table 52.15 shows import-to-export ratios of advanced technology (AT) products. The U.S. is a major exporter of aerospace technology, weapons, and electronics. It also exports more flexible manufacturing technology and biotechnology than it imports. In contrast, over the five years 2003–2007, the U.S. increasingly imported optoelectronics and to a lesser extent information/communication technology. The U.S. is also predominantly an importer of nuclear technology, life science, and advanced materials.

The bulk of advanced technology product exports during January to October 2007, totaling more than half of all AT product exports, was composed of aerospace (33.6%) and information/communication technology (27.2%) (Table 52.16, Figure 52.4). Electronics is a significant export industry within AT products at 18.6% of all AT product exports (Figure 52.4). The information/communication technology sector nevertheless also represents the largest proportion of total AT product imports, with 54.7% during 2007 (Figure 52.5).

52.6 Innovation

Innovation costs money, so one way to assess trends in innovation is to examine funding flows to new businesses. Venture capital has a minority role, but information on venture capital allocations is readily available, so figures for venture capital help provide clues to innovation trends. Table 52.17 shows the relative contribution of venture capital compared with other funding sources. Private partnerships form the largest proportion of all new capital commitments made in the U.S. in terms of number of firms, funds, and total capital provided. However, the average (mean) capital provided per fund is highest for investment banks ($306.9 million).

In the U.S., venture capital peaked in the year 2000 with $105.8 billion of new capital committed to the industry. Since then, the amount of new capital committed to venture capital has been greatly reduced, although estimates of the amount of available funding range enormously from $3.8 billion to $37.9 billion during 2001–2007.[14] Nevertheless, capital committed to new ventures appears to be rising and the number of new funds created each year has been steady during 2002–2007 (Figure 52.6).

At its peak in 2000, the largest amount of venture capital disbursed in the U.S. went to the Internet industry, with some $53.5 billion. The communications ($25.6 billion) and computer software ($21.9 billion) industries also received substantial capital during 2000. Since then, ven-

[14] Data on venture capital vary depending on the source. We used data reported for "venture capital raised" in Goldfisher (2007), which cites data from Thompson Financial and the National Venture Capital Association. Other figures indicate that "new capital commitments" dropped to $7.7 billion during 2002 (Thompson Financial), or that venture capital "disbursements" dropped to $21.3 billion in 2002 (reported by the NSF who cite Thompson Financial as a source). All sources concur that the peak during 2000 was around $105 billion.

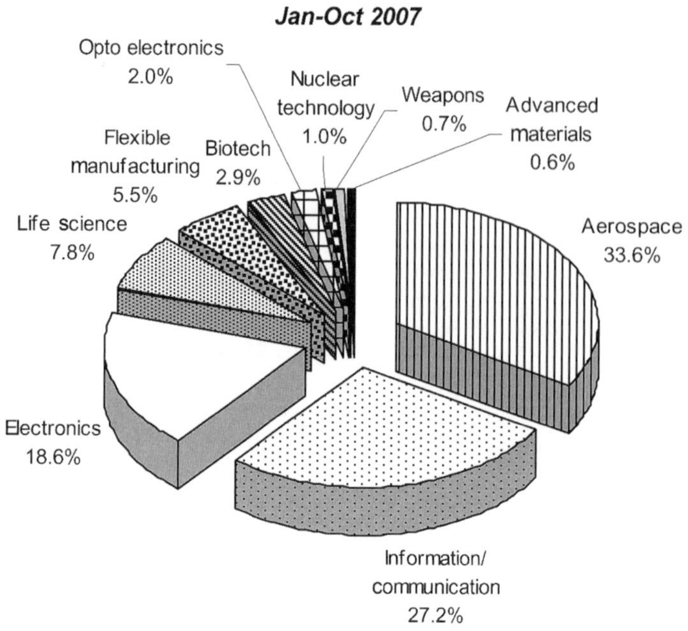

Figure 52.4. U.S. exports of advanced technology products January–October 2007.
Source: U.S. Census Bureau, U.S. Trade in Advanced Technology Products.

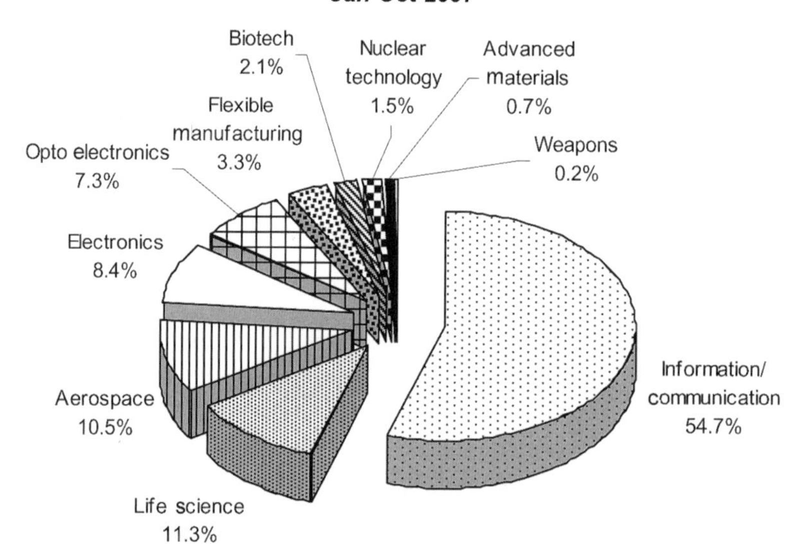

Figure 52.5. U.S. imports of advanced technology products January–October 2007.
Source: U.S. Census Bureau, U.S. Trade in Advanced Technology Products.

ture capital investment has declined overall, but appears to be picking up again. During 2007 the "other" industry category received the largest share at $33.3 billion; this included $13.0 billion for transportation, $7.7 billion for financial services, $3.9 billion for business services, and $2.9 billion for manufacturing. The second largest industry to benefit from venture capital in 2007 was the medical/health (excluding biotech) industry ($15.3 billion), followed by the consumer-related (i.e., food and beverages, entertainment, clothing, house and garden products, restaurants, travel, etc.) industry ($12.9 billion) (Table 52.18). Venture capital to the transportation, medical/health, and consumer-related industries did not peak in 2000, but instead grew steadily over the years.

During 2007, Californian companies attracted $22.3 billion of venture capital, taking up 24.8% of total U.S. venture capital investment (Table 52.19). While California attracted the most venture capital investment, VC funding

Table 52.17. Total new capital commitments in the U.S. by type of institution.

Type of funding	*No. of firms*	*No. of funds*	*Total capital ($bn)*	*Average capital ($m)*
Private partnerships	3,644	9,273	1,828.5	197.20
Investment banks	398	1,225	375.9	306.90
Fund of funds and sec.	317	900	247.7	275.20
Other financial institutions	574	1,314	178.1	135.60
Corporate (venture capital)	263	448	48.0	107.20
Advisors non-FOF	27	58	11.6	199.30
Corporate (direct)	84	113	7.8	68.90
Government programs	53	75	5.3	70.30
Public venture funds	33	42	4.6	108.60
Individuals	34	49	2.9	59.00
SBICs	29	36	2.1	58.90
Development programs	43	69	1.8	26.20
Universities	18	22	0.9	39.40
Pension funds	7	9	0.6	67.80
Endowments/foundations	2	2	0.1	49.80
Other	4	4	0.3	65.90

Source: Thompson Financial's VentureXpert database.

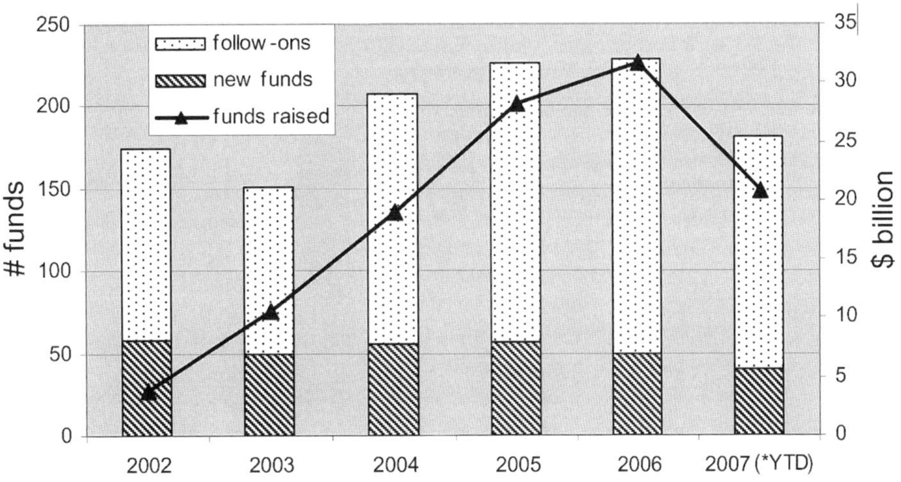

Figure 52.6. Historical trend in U.S. venture capital funds.
Source: Goldfisher (2007).
*2007 data are up to October, 2007.

was largely raised (i.e., funded) in New York, which is a center of the finance industry. New York raised $139.7 billion through VC funds in 2007, and also had the highest venture capital funds under management during 2007 at $140.1 billion.

One measure of the success of startup businesses is whether or not they undergo an initial public offering (IPO) on the financial market and become publicly-owned. Since 1980, the number of IPOs of private equity firms, the total offer amount, and the total post-offer value have varied considerably from year to year (Table 52.20). Nevertheless, the most IPOs occurred in 1999, one year before the peak in venture capital investment. When considering the average post-offer value, a small peak occurred in 1989 when IPOs averaged $194.6 million. Subsequently average post-offer values dropped a little and peaked again in the year 2000 at $539.7 million on average. While in the following years the number of IPOs dropped sharply, those firms that did undergo IPOs maintained fairly high post-offer values, and in 2006 the average post-offer value overtook 2000 values, reaching a new height in 2007 with an average value of $610.7 million.

Table 52.18. U.S. venture capital disbursements ($m) by industry 1980-2007.

Industry	*1980*	*1985*	*1990*	*1995*	*2000*	*2005*	*2007*
Communications	72.5	720.5	2,262.7	2,101.3	25,683.0	7,603.0	5,580.4
Computer hardware	146.2	720.8	425.7	959.6	6,626.2	1,629.0	4,448.8
Computer software	15.9	413.0	581.6	1,291.6	21,878.5	9,268.0	8,249.8
Internet-specific	0.1	5.1	47.2	879.4	53,530.9	4,299.4	7,360.7
Computer (other)		5.9	11.9	19.0	147.9	64.9	43.1
Semiconductor/Electronics	76.7	454.9	302.6	601.1	10,280.4	3,403.6	4,188.9
Biotech	52.3	145.2	299.2	719.9	3,787.0	3,398.1	5,446.6
Medical/Health	53.6	353.0	862.2	1,633.7	5,855.9	7,853.0	15,301.0
Consumer-related	105.1	523.2	1,041.7	2,610.0	6,259.5	7,321.5	12,925.8
Industrial/Energy	156.5	498.9	365.7	1,914.5	4,831.8	4,155.3	9,050.7
Other	69.1	471.2	2,418.3	2,030.0	22,465.5	15,944.1	33,345.1
Transportation	26.2	81.8	234.8	285.7	3,880.4	4,236.8	13,010.4
Financial services	15.4	182.3	940.3	798.9	8,359.2	6,270.5	7,698.8
Business services	14.3	53.4	279.5	319.2	4,882.1	1,784.6	3,902.1
Manufacturing	11.4	81.0	878.3	333.3	1,687.2	473.1	2,923.6
Agriculture	0.3	13.5	7.9	30.6	600.2	206.4	1,128.3
Construction	0.7	55.0	43.4	107.7	1,741.9	314.4	1,068.0
Utilities	0.8	0.6	25.6	25.6	450.9	610.7	79.2
Other	0.0	3.6	8.5	129.1	863.5	2,047.8	3,534.7
Total	*747.9*	*4,311.7*	*8,618.7*	*14,760.0*	*161,346.4*	*64,939.7*	*105,940.7*

Source: Thompson Financial's VentureXpert database.

Table 52.19. U.S. venture capital by state during 2007.

State	*Companies*	*VC investments ($m)*	*Percent of U.S. investments (%)*
California	1,730	22,278.6	24.8
Massachusetts	478	8,179.5	9.1
Ohio	117	7,265.8	8.1
New York	336	6,444.3	7.2
Michigan	72	5,825.8	6.5
Texas	310	4,978.0	5.5
Nevada	17	3,834.1	4.3
Maryland	145	2,569.0	2.9
Florida	149	2,498.7	2.8
Pennsylvania	244	2,422.6	2.7

Source: Thompson Financial's VentureXpert database.

During 2007, Californian companies raised the most capital through venture-backed IPOs at $4.4 billion. Texan companies raised $2.8 billion and companies in Massachusetts raised $1.6 billion.

52.7 Culture and politics

52.7.1 Drivers of aggregate federal R&D funding and funding areas

World War II, Nikita Khrushchev's (misleadingly translated) "We will bury you!", Sputnik, and the Cold War all stimulated U.S. public perception of the need for innovation. This public perception came at a convenient time, since U.S. industry was flush with production after World War II, and the other formerly strong economies of Europe and Japan had their businesses in ruins. It is perhaps no surprise, then, that the U.S. found both the political will and the funds to support massive national investments in R&D and innovation.

The result was, as Table 52.2 has reported, a rapid growth of federal R&D funding through the late 1960s. In the 1970s through the 1990s, coincident with slower economic growth, federal R&D funding grew much more slowly. Mikhail Gorbachev's efforts to end the Cold War, and the 1989 fall of the Berlin Wall, helped reduce the tensions that stimulated a public sense of need for technological investment. Defense R&D as a fraction of total federal R&D outlays grew to more than two-thirds by 1985, but that fraction quickly shrank thereafter to just above one-half in 2001. The Al-Qaeda attacks of September 11, 2001 again changed the perceived needs, and defense R&D grew rapidly.

Table 52.20. U.S. venture capital IPOs 1980–2007.

Year	*No. of IPOs*	*Total offer price*	*Total post-offer value ($m)*	*Average post-offer value ($m)*
1980	61	678	3,652	59.9
1981	98	1,070	4,892	49.9
1982	39	577	2,663	68.3
1983	206	4,071	19,698	95.6
1984	89	1,098	5,143	57.8
1985	84	1,670	8,053	95.9
1986	389	4,168	26,831	69.0
1987	147	2,917	12,846	87.4
1988	66	971	4,175	63.3
1989	83	2,469	16,151	194.6
1990	79	1,767	7,558	95.7
1991	176	6,782	25,486	144.8
1992	224	9,010	37,607	167.9
1993	250	8,469	28,873	115.5
1994	181	5,102	19,159	105.9
1995	222	9,124	37,687	169.8
1996	285	12,688	60,178	211.2
1997	172	7,933	35,360	205.6
1998	99	6,282	28,843	291.3
1999	296	26,972	150,422	508.2
2000	260	27,470	140,318	539.7
2001	51	5,590	26,031	510.4
2002	41	5,784	20,944	510.8
2003	45	5,834	22,832	507.4
2004	138	19,559	81,973	594.0
2005	108	18,700	59,194	548.1
2006	112	19,126	67,862	605.9
2007	112	17,130	69,049	616.5

Source: Thompson Financial's VentureXpert database.
Note: the post-offer value represents the value of all shares outstanding at the offer date.

Given the military impetus that drove U.S. federal R&D during this period, as well as industrial needs to maintain technological leadership in order to remain competitive, it is no surprise that historically the U.S. Government's R&D expenditures and programs to support innovation have focused on the physical sciences and engineering. What may surprise some readers, however, is that the focus has shifted dramatically. Medical sciences (life sciences plus psychology) by 2007 made up 54.0% of federal R&D expenditures, whereas the physical sciences, engineering, mathematics, and computer science together made up only 33.2% of federal R&D outlays. Environmental sciences made up 6.6% of federal R&D outlays in 2007, and the social sciences made up 2.2% (the small remainder was not classified or fitted to other categories).[15]

This suggests that changing national interests may drive the allocation of federal R&D funding in future. Global climate change plus the increased need for energy security might lead to a rise in environmental and energy-related research in future. Commonplace deviations between policies intended to be successful versus policies that would actually achieve intended goals also suggest that a dramatic rise in funding of serious social science has the potential to yield dramatic benefits for the economy, well-being of individuals, and functioning of businesses and other organizations.

52.7.2 Education

U.S. elementary and secondary education provides youth with broad-based education at a fairly high level, although with substantial disparities and with average achievement levels below those in some nations. Literacy is high according to U.S. Government statistics, with 99% of adult Americans judged to be literate based on census self-reports, although actual literacy may be somewhat lower.[16] Mathematics and science skills of U.S. students are below those in various other developed nations, with specific differences depending on the age of students and the nature of the test used for comparison. Using the Program for International Student Assessment's 2003 test, which surveys 15-year-olds and has an applied bent, U.S. students performed near the lowest among industrialized nations at mathematics questions, and well below the median at science questions. Some of the strongest performers, in contrast, were Finland, South Korea, Japan, the Netherlands, Canada, and the Czech Republic (NSB, 2006, Vol. 2, tables 1-13 and 1-14, pp. A1-15 and A1-16; NSB, 2008, Vol. 1, pp. 1-14 to 1-15).

More favorable findings for U.S. fourth and eighth-grade students result from tests of the Trends in International Mathematics and Science Study, but even these results do not rank U.S. students especially highly. Figures 52.7 and 52.8 report the eighth-grade test scores for participating countries in 1995, 1999, and 2003, and make clear the U.S.'s continuing unimpressive educational achievement in mathematics and science. This poor performance has continued despite considerable efforts to do better; for example, in 1990 then-President George Bush Senior and state governors established the goal that U.S. students should lead the world in mathematics and science achievement by 2000. Efforts to improve the educational system have rarely been successful, perhaps because they

[15] The breakdown of federal R&D outlays in 2007 is drawn from NSB (2008, Vol. 2, table 4-31, p. A4-55).

[16] More detailed studies show that 12–14% of Americans above age 15 can at best locate easily identifiable information in simple prose or documents and follow simple instructions in informational documents (see NCES, 2006).

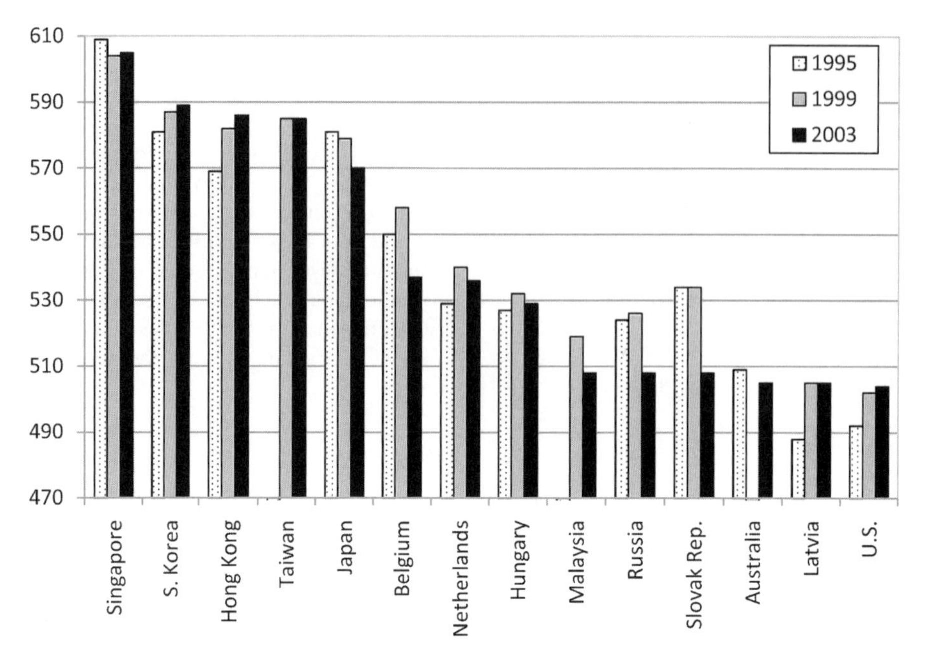

Figure 52.7. International mathematics scores of eighth-grade students of the top-14 countries (1995–2003).
Source: Gonzales *et al.* (2004, p. 78, table C4).
Note: for Belgium, only the Flemish-speaking educational system participated.

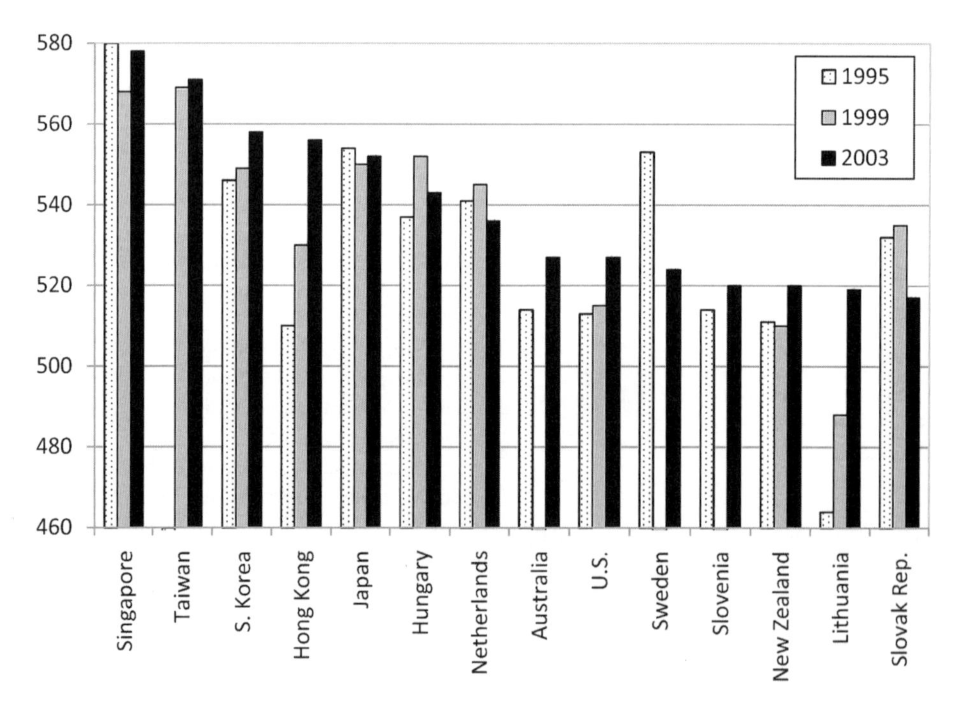

Figure 52.8. International science scores of eighth-grade students of the top-14 countries (1995–2003).
Source: Gonzales *et al.* (2004, p. 90, table C14).

have failed to raise the average quality of educators. Educators on aggregate exhibit below-average university performance, and rarely received university degrees in the subjects that they ended up teaching. One recent analysis suggests that teacher quality would be enhanced by creating a highly selective teacher certification system, in which only very strong university graduates (through any degree rather than mainly teaching degrees) could earn the honor of being selected for certification (*Economist*, 2007).

Cultural attitudes and priorities may be important in shaping the type of education that elementary and secondary students receive. In mathematics, Americans often exhibit a misinformed belief that they and their children innately lack the ability to gain mathematics skills, whereas South Koreans and Japanese more often have confidence in children's innate mathematics potential and push their children to study. Such self-image drivers of education are coupled with perceived need; with little sense that the study of mathematics and science is either innately interesting or economically important, it is perhaps only to be expected that Americans do not ensure an educational system that provides strong mathematics and science skills. Recent policy suggestions have also included, in reaction to a perception that schools train students to be effective but not creative, making education more entrepreneurship-friendly.

The U.S. has had, in the last half-century, the world's preeminent international university system. Educational achievement is high by international standards. Nearly two-thirds of U.S. high-school graduates go on to enroll in 2-year or 4-year colleges. Among those who receive bachelor's degrees, about one-third do so in science or engineering (15% in social and behavioral sciences, 12% in natural sciences, and 5% in engineering). At the graduate level, 110,000 students entered science and engineering master's and doctoral study in 2004 (up from 81,000 in 1983; NSB, 2008, Vol. 2, table 2-21, p. A2-28). Foreign students constituted 25.5% of all 583,000 science and engineering graduate students enrolled in 2005 (up from 19.3% in 1985; NSB, 2008, Vol. 2, table 2-22, pp. A2-29 to A2-37). Since about half of foreign graduate students have remained in the U.S. after graduation (NSB, 2008, Vol. 1, pp. 2-34 to 2-35), they have been an important contributor to human capital in the U.S. workforce.

The continuing excellence of U.S. university education and research depends on salaries to hire quality faculty as well as research facilities and funds. Labor force competition among universities, driven in part by students' exhibited desire to attend universities with high rankings reflecting real faculty quality, has contributed to a rise in earnings of university faculty that has helped to fuel faster-than-inflation increases in the cost of education. At the same time, U.S. state funding for universities has declined, again contributing to rising educational costs. These rising costs have not, however, reduced attendance.[17] A government system of loans and grants, coupled with reduced tuition granted by many universities to needy students, have helped to allow low-income students to continue to enroll in university study.

52.7.3 Immigration

The influx and retention of highly educated foreigners has contributed greatly to U.S. innovation and U.S. society more generally. This influx of foreign talent may slow as developing nations' economies improve and careers become more attractive in other nations. At the same time, restrictions on immigration reduce the influx of foreign citizens into the U.S. Awards of visas to foreign students and others, even highly educated persons with the means for financial security, can be lengthy and haphazard. This has been especially true after September 11, 2001, given the apparent need to guard against the entry of potential terrorists. There is much that the Federal Government could do (and in part is doing) to enhance immigration especially among highly educated science and engineering workers who are particularly needed to fill skilled jobs. Apart from adding resources to more quickly process backlogged visa applications, the U.S. Government could—as other nations do—make it more attractive for skilled foreign individuals to come to or remain in the U.S. Visas could be processed especially quickly in appropriate cases, and spouses and family members could be granted both the same rapid processing and permission for employment.

52.7.4 Openness, mobility, and international relations

The U.S. has been a champion of international free trade, and has generally had relatively low trade barriers in the form of tariffs. (To be fair, the growth of U.S. free trade policy coincided with U.S. industries' ability to compete successfully internationally, and some U.S. policies such as anti-dumping laws have been somewhat protectionist.) Labor and capital are highly mobile within the country and internationally. Moreover, the U.S. remains a hub for international R&D, as companies worldwide increasingly spread their R&D across the globe (NSB, 2008, vol. 1, pp. 4-50 to 4-54). International relations support continued cooperation in international trade, science, and industry.

52.7.5 Recent policy initiatives

The low growth of federal R&D funds in recent decades has generated some degree of alarm among analysts, academic researchers on innovation, and U.S. businesses and trade associations. These groups have generated a steady stream of policy reports showing the importance of science and technology funding and initiatives by the U.S. Government. However, these reports and the political pressure they have generated have been insufficient to yield serious gains in U.S. technology policy. The lack of success may change. With an increasing perception that U.S. business and jobs are being lost to overseas

[17] U.S. higher education attendance rose from 12.7 million in 1986 to 16.9 million in 2004, increasing in every year during this period except 1993–1995, despite a decrease in the U.S. college-age population during the 1990s (NSB, 2008, vol. 1, p. 2-18 and vol. 2, table 2-13, p. A2-14).

competitors, there is growing realization of the need to strengthen the U.S. technological base in order to retain the viability of the U.S. economy.

One of the more recent policy reports, the National Innovation Initiative, helped to build momentum for the America COMPETES Act, bipartisan legislation enacted in August 2007 that enabled a range of U.S. technology-enhancing policies. As mentioned above, the intended policy changes from the Act included doubling basic funding in the physical sciences, doubling funding through the National Science Foundation over a 4-year period, and mathematics and science education improvement programs. However, with budget pressures and concerns raised by the President and others, the budget passed by the U.S. Congress in December 2007 failed to provide the funds to support the changes enabled by the America COMPETES Act. Innovation policy remains important on the political agenda, with candidates for the 2008 political election developing thought-out positions on innovation policy, so these and other innovation policies may still be implemented in the next few years.

Intellectual property policy has also been evolving. For patents, excessively strong and broad rights have been blamed for slowing technological progress when large numbers of patent licenses must be negotiated with many different companies, and for allowing "hold-up" of large-scale manufacturers by charging them in the courts with violation of patents that are irrelevant or obvious. The U.S. Supreme Court in 2006 and 2007 made several decisions that mitigate some problems of breadth, obviousness, and strength of patent rights, reversing trends that followed the 1982 establishment of the patent-related Court of Appeals for the Federal Circuit. These decisions strengthened the criterion of non-obviousness (patents must be non-obvious to inventors skilled in the art), made it harder to obtain court injunctions to halt firms' activities involving an infringed patent, and allowed software that infringes a patent to be reproduced and sold without royalties in nations where the patent is not in force.[18]

For copyrights—perhaps less important to technology than patents but nonetheless determinants of knowledge creation and diffusion—politics and a recent ideal of ownership rights have pressured change toward longer duration of rights. The Copyright Term Extension Act of 1998 added 20 years to U.S. copyrights. This change pertained not only to newly granted copyrights, but also to works still under copyright, effectively taking away rights from the public and giving them to existing copyright holders. Oddly, the U.S. Supreme Court held that the change was constitutional even for preexisting works, despite that it did nothing to promote creation among previously copyrighted works. Given the minor benefits to authors of 20 extra years of copyright beyond an initial 75 or more years (even with a constant income stream the added value to authors is only 0.03% assuming a typical discount rate of 10%), this seems indicative of how political processes can drive policy in favor of particular interest groups.

[18] *KSR International v. Teleflex*, *eBay v. MercExchange*, *Microsoft v. AT&T*.

52.8 Conclusion

The U.S.'s preeminent economic status grew from strengths in science, technology, and innovation. Yet aggregate R&D as well as federal R&D and innovation policies have been weakening compared with other nations. Future U.S. economic conditions therefore are endangered. U.S. policy-makers, and the U.S. citizenry whose concerns sway policy, need to prioritize R&D and innovation in a manner consistent with the nation's diverse social groups and (current and potential) business composition. Getting the funding and priorities right is the easier part, and will be enormously beneficial. Determining and enacting the right guiding policies are more subtle challenges, since the consequences of R&D and innovation policy have often proved counterintuitive. However, greater social science research and clever policy initiatives by well-placed individuals can do much to construct a policy environment that enhances U.S. innovation and economic outcomes. With the right efforts, the U.S. can remain an important contributor to worldwide technological advances needed to support the world's economic prosperity, including health, security, environmental quality, education, and quality of life. U.S. citizens' economic prosperity hangs in the balance, along with enormous potential gains across the globe.

52.9 References and further reading

AAAS. (2008). *AAAS Report XXXIII: Research and Development FY 2009* (Intersociety Working Group, AAAS Publication 08-1A). Washington, D.C.: American Association for the Advancement of Science.

Audretsch, D. B., Keilbach, M. C., and Lehmann, E. E. (2006). *Entrepreneurship and Economic Growth*. Oxford, U.K.: Oxford University Press.

Baumol, W. J., Litan, R. E., and Schramm, C. J. (2007). *Good Capitalism, Bad Capitalism, and the Economics of Growth and Prosperity*. New Haven, CT: Yale University Press.

Bloom, N., Griffith, R., and Van Reenen, J. (2002). "Do R&D tax credits work? Evidence from a panel of countries 1979–1997." *Journal of Public Economics*, **85**(1), 1–31.

Economist. (2007). "How to be top." *Economist Magazine*, October 20–26, pp. 80–81.

Goldfisher, A. (2007). "VC fund-raising slows in Q3." *Private Equity Week*, **14**(41), October 22, p. 2.

Gonzales, P., Guzmán, J. C., Partelow, L., Pahlke, E., Jocelyn, L., Kastberg, D., and Williams, T. (2004). *Highlights from the Trends in International Mathematics and Science Study (TIMSS) 2003* (NCES 2005-005). Washington, D.C.: U.S. Department of Education, National Center for Education Statistics.

Jaffe, A. B., and Lerner, J. (2004). *Innovation and Its Discontents: How Our Broken Patent System Is Endangering Innovation and Progress, and What to Do about It*. Princeton, NJ: Princeton University Press.

Lambright, W. H., Crow, M., and Shangraw, R. (1988). "National projects in civilian technology." In: J. D. Roessner (Ed.), *Government Innovation Policy: Design, Implementation, Evaluation* (pp. 63–74). New York: St. Martin's Press.

Levin, R. C., Klevorick, A. K., Nelson, R. R., Winter, S. G., Gilbert, R., and Griliches, Z. (1987). *Appropriating the Returns from Industrial Research and Development* (Brookings Papers on Economic Activity No.3, special issue on microeconomics, pp. 783–831). Brookings Institution.

Leyden, D.P., and Link, A. N. (1992). *Government's Role in Innovation*. Boston, MA: Kluwer Academic Publishers.

Lindsey, Q. W. (1988). "Technological innovation policy: The role of states." In: J. D. Roessner (Ed.), *Government Innovation Policy: Design, Implementation, Evaluation* (pp. 91–104). New York: St. Martin's Press.

Mowery, D. C., and Rosenberg, N. (1993). "The U.S. national innovation system." In: R. R. Nelson (Ed.), *National Innovation Systems: A Comparative Analysis* (pp. 29–75). New York: Oxford University Press.

NCES. (2006). *A First Look at the Literacy of America's Adults in the 21st Century* (NCES 2006-470). Washington, D.C.: National Center for Education Statistics.

NSB. (2006). *Science and Engineering Indicators 2006* (two volumes: vol. 1, NSB 06-01; vol. 2, NSB 06-01A). Arlington, VA: National Science Board/National Science Foundation.

NSB. (2008). *Science and Engineering Indicators 2008* (two volumes: vol. 1, NSB 08-01; vol. 2, NSB 08-01A). Arlington, VA: National Science Board/National Science Foundation.

Roessner, J. D. (Ed.). (1988). *Government Innovation Policy: Design, Implementation, Evaluation*. New York: St. Martin's Press.

USPTO. (1977). *Technology Assessment and Forecast Seventh Report*. U.S. Patent and Trademark Office.

USPTO TAF database. U.S. Patent and Trademark Office's TAF database; *http://www.uspto.gov/web/offices/ac/ido/oeip/taf/reports/htm*

Wilson, D. J. (2009). "Beggar thy neighbor? The in-state, out-of-state, and aggregate effects of R&D tax credits." *Review of Economics and Statistics*, **91**(2), 431–436.

Wolfe, R. M. (2007). *Expenditures for U.S. Industrial R&D Continue to Increase in 2005: R&D Performance Geographically Concentrated* (Info Brief, October, NSF 07-335). Arlington, VA: National Science Foundation.

53

Cooperative R&D Agreements (CRADAs)

Paul Olk

University of Denver

CRADAs, or cooperative research and development agreements, emerged in the U.S. in the 1980s as part of an effort to enhance country competitiveness and the social benefits received from publicly-funded research. Similar to the efforts surrounding university licensing (see Chapter 54), CRADAs developed from legislation designed to encourage collaborative research. The Federal Technology Transfer Act of 1986, passed by the U.S. Congress, provided incentives for government-owned and government-operated laboratories (GOGOs) to commercialize inventions. The follow-up legislation, the National Competitiveness Act of 1989, extended these policies to government-owned and contractor-operated laboratories (GOCOs).

The purpose behind these legislations was to encourage the transfer of technology developed in a GOGO or GOCO federal laboratory to private-sector firms. Many of the U.S. Federal Government departments (e.g., Department of Defense, Department of Energy, U.S. Department of Agriculture) have dedicated research laboratories. Prior to these legislations, most research conducted in these laboratories helped the agency's agenda but there was limited effort devoted towards commercialization. According to Adams, Chiang, and Jensen (2003), the push for technology transfer of laboratory research stemmed from three related reasons. The first is that with the decline in the Soviet Union and the reduced need for a strong defense against U.S.S.R. weaponry, there was speculation that many of the laboratories' budgets—particularly those dedicated to military defense projects—would be cut. Demonstrating successful technology transfer and commercialization would show the additional benefits of research and help justify preserving a laboratory's budget. Second, legislatures hoped the technology transfer might translate into greater economic productivity. Finally, it was argued that the public was not receiving adequate benefits from the research.

These legislations permitted the formation of CRADAs. A CRADA provides a relatively flexible legal framework for collaboration between a federal laboratory and a non-federal partner, usually a private-sector firm, to engage in collaborative research. While providing some specific requirements, CRADAs offer participants the option to define the relationship as narrowly or broadly as is needed to conduct the research. These legislations stipulated that the agencies managing a laboratory, or the contractors operating a laboratory, should standardize and expedite the process for establishing such agreements, and that special consideration be given to small businesses and business units located in the U.S. As summarized by Rogers *et al.* (1998), one of the provisions in a CRADA is that the two parties can share research personnel, equipment. and property rights. While the non-federal partner can contribute funds, under the laws the federal agency is prohibited from providing direct funding to the partner. The *federal laboratory* can protect from public disclosure any information developed from a CRADA for up to 5 years, and federal employees can help commercialize the innovation and receive royalties. In exchange, the partner may receive the rights to any intellectual property developed from the collaboration.

Since the passage of the acts, the number and types of CRADAs have proliferated. According to a recent review (Munson and Spivey, 2006), the number of active CRADAs have stabilized at around 3,000–3,500, with the Department of Defense being the most active user of CRADAs, followed by the Department of Energy. These two agencies account for about two-thirds of all the CRADAs formed. In terms of impact on commercialization and budgets, however, the record on CRADAs is more mixed. Mowery (2003) noted that there is a lack of adequate data to evaluate CRADAs. Most evaluation efforts have had a narrow focus (e.g., a single agency) or provide incomplete information on CRADA outcomes. The findings that exist reveal some effectiveness—for example, Adams, Chiang, and Jensen (2003) report evidence that CRADAs have stimulated industrial patents and company-funded R&D—but to date there has not been a comprehensive evaluation of the impact of CRADAs. In particular, there is a need to evaluate CRADAs in terms of whether the effort expended has produced adequate benefits, as well as how effective they have been relative to the other options available in the broad array of channels for federal technology transfer.

References

Adams, J. D., Chiang, E. P., and Jensen, J. L. (2003). "The influence of federal laboratory R&D on industrial research." *Review of Economics and Statistics*, **84**, 1003–1020.

Mowery, D. C. (2003). "Using cooperative research and development agreements as S&T indicators: What do we have and what would we like." *Technology Analysis & Strategic Management*, **15**, 189–205.

Munson, J. M., and Spivey, W. A. (2006). "Take a portfolio view of CRADAs." *Research-Technology Management*, July/August, 39–45.

Rogers, E., Carayannis, E. G., Kurihara, K., and Allbritton, M. M. (1998). "Cooperative research and development agreements (CRADAs) as technology transfer mechanisms." *R&D Management*, **28**, 79–88.

54
University Licensing

Paul Olk

University of Denver

Commercializing patented research through licensing agreements has become more important to U.S. and European universities. While some universities have long supported practically oriented research and licensed inventions (Mowery and Sampat, 2001), historically most have not been interested in commercialization. This reflected a general bias in the U.S. towards more basic research and in Europe norms about university researchers, but not the universities, engaging in patenting activity (Verspagen, 2006).

This situation started to change in the 1980s. In the U.S., a defining event was the passage of the Patent and Trademark Law Amendment Act of 1980, better known as the Bayh–Dole Act. Concerns that commercial benefits from publicly-funded research were not being realized led to an effort to realign incentives. Previously, if research was funded by federal grants, the findings were considered part of the public domain. No one company or individual could patent them. Central to the Act was the provision that a university could maintain ownership—via patenting—over technology developed from federally funded research. This encouraged universities to grant an exclusive license to an interested company and to share a portion of the licensing fee (typically one-third) with the researcher. It also led to many universities setting up Technology Transfer Offices (TTOs) to encourage and coordinate the licensing of university-based research (Thursby and Thursby, 2007). In Europe, where the innovation system and the research funding agreements differ from the U.S., universities recognized the financial potential and began to pursue patents. In the last 10 years there has been an increase in the number of university-owned patents (Geuna and Nesta, 2006).

The impact of this increased patenting activity has been mixed. In the U.S., according to an Association of University Technology Managers (AUTM) survey reported in Thursby and Thursby (2007), most TTOs have not covered their basic costs from the received licensing fees. Explanations for why include that there may be other, better ways to transfer technology to companies than by using patents, and the goals of a TTO may be more than just licensing. They may also include helping universities fulfill a mission of engaging in research that will encourage economic growth and to obtain sponsored research grants (Thursby and Thursby, 2007). In Europe there is considerable variation across countries and academic disciplines in patenting levels. While there have been some successes, there is no evidence that licensing is profitable for universities (Geuna and Nesta, 2006). Evaluating the impact of university licensing, though, will require more research because of the current lack of reliable data that capture European universities' involvements in patents (Verspagen, 2006).

References

Geuna, A. L., and Nesta, J. J. (2006). "University patenting and its effects on academic research: The emerging European evidence." *Research Policy*, **35**, 790–807.

Mowery, D., and Sampat, B. (2001). "Patenting and licensing university inventions: Lessons from the history of the Research Corporation." *Industrial and Corporate Change*, **10**, 317–355.

Thursby, J. G., and Thursby, M. C. (2007). "University licensing." *Oxford Review of Economic Policy*, **23**, 620–639.

Verspagen, B. (2006). "University research, intellectual property rights and European innovation systems." *Journal of Economic Surveys*, **20**, 607–626.

Part Seven

Emerging Technologies

55

What Are Emerging Technologies?

V. K. Narayanan and Gina C. O'Connor†*

*Drexel University and †Rensselaer Polytechnic Institute

Introduction

"Emerging technologies" is a term used to cover the description of novel technologies, the processes by which they begin to infuse themselves into the commercial marketplace, and the nature of their impact over time. Today's key "emerging" technologies include nanotechnology, biotechnology, information technology, and a variety of energy technologies. These have the power to disrupt some industries and instigate the birth of new ones. According to the *Merriam-Webster Dictionary*, "emerging" means coming forth into view, and emerging technologies may thus be considered technologies coming into prominence. Although the exact definition of this term is rather vague, most will agree that emerging technologies have the potential to create a new industry or transform an existing one, to provide investment opportunities, and to change the world in terms of offering new benefits and transforming standards of living.

According to Perez (2002), technological revolutions, spurred by the emergence of a novel technology or combination of multiple technologies, have occured approximately every 50 years over the past 250 years in the history of the civilized world. Indeed, a sequence of (a) technological revolution, (b) financial bubble, (c) collapse, (d) golden age, and (e) political unrest can be traced to occur repeatedly, caused by emerging technologies and the follow-on availability of financial capital for investment. The five technological revolutions are identified as

(1) The Industrial Revolution (1771), with the launch of the first textile mill: a process technology that enabled automation for the first time.
(2) The age of steam and railways (1829) with the test of the *Rocket* steam locomotive.
(3) The age of steel, electricity, and heavy engineering (1875) was introduced with Carnegie's first steel plant in Pittsburgh, Pennsylvania;
(4) The age of oil, the automobile, and mass production (1908) with the frist Model-T automobile by Henry Ford.
(5) The age of information and telecommunications (1971) with the introduction of Intel's first microprocessor.

One might argue that the biotechnology revolution is underway, initiated by the sequencing of the human genome, and that the discovery of the buckyball launched the age of nanotechnology.

Each such surge of development is characterized by a four-phase cycle interspersed with a turning point period, Perez observes, as follows:

(1) irruption, which occurs just after the key invention, when the new products and technologies, backed by financial capital, are showing their future potential and making powerful inroads in the economy that is still basically shaped by the previous paradigm;
(2) the frenzy phase, when financial capital drives the intense build-up of the new infrastructure and the new technologies, so, in the end, the potential for the new paradigm is strongly installed in the economy and ready for full deployment.

The turning point, usually as a result of the recession that follows the collapse of the financial bubble that arises in the frenzy phase, when the required regulatory changes are made to facilitate and shape the next two phases (the period of deployment):

(3) a synergy phase, when all conditions are favorable to production and to the full flourishing of the new paradigm; and, finally

(4) the maturity phase, when the last industries, products, technologies, and improvements are introduced while signs of dwindling investment opportunities and stagnating markets appear in the main industries of the revolution.

One could say that the first two phases described above, irruption and frenzy, constitute the emergence of a technology. Irruption is characterized by a mass entrance of potential investment money into the market, generated by the firms of the old paradigm. As the technology improves and more applications emerge, so too do more and more investors, seeking returns higher than those yielded on the incremental innovations of the most recent technology that has been exploited to the end of its life. New entrepreneurs enter, and incumbent firms face threatened futures. As a result, this period is marked by increasing unemployment, and even political standoffs driven by the downturn of the dominant technological paradigm and, with it, a number of industries.

The frenzy phase is a time of new millionaires. Financial capital takes over; its immedeate interests are opportunistic profits. The paper economy decouples from the real economy (such as what happened during the dotcom boom), and there is a growing rift between the forces in the economy and the regulatory frameworks that must eventually drive what is and what is not commercializable. It is a time of speculation and intense exploration of all the possibilities opened up by the technological revolution. Through diversified trial-and-error investment the potential of the emerging technology for creating new markets and rejuvenating old industries is fully discovered and firmly installed in the economy. The productivity explosion reaches more and more activities, inducing a process of restructuring in the productive sphere where the new or renewed prosper and the old wilt or die. This is a phase of fierce "free" competition. Individualism flourishes both in business and in political thinking, sometimes confronted by anti-technology ideas or groups. Tensions exist. The wealth that has grown and concentrated in relatively few hands is greater than can be absorbed by real investment. Much of this excess money is poured into furthering the technological revolution, especially its infrastructure (such as the Internet's "last mile" to the home) . . . often leading to overinvestment that might not fulfill expectations. The late frenzy period is a financial bubble time.

Once the economy reaches this stage, the technology not only has emerged, but the economy must correct itself due to overinvestment, and the technology must be deployed synergistically into productive pursuits. As this occurs there is a settlement period (the synergy phase) and a focus on incremental improvements until the next investment opportunity, fueled by a new advanced technology, begins the cycle again.

One way of talking about emerging technologies is with the help of the trajectory of technology evolution. Technology evolution is often described by means of the S-curve, and technologies in the early stages of the S-curve are candidates for being labeled "emerging". However, in the very early stages, technologies are visible mostly to some scientists and technologists, not to the markets at large, and their industrial potential is unknown. Once they "irrupt", meaning their scientific benefits are becoming clearer and the technology has evolved to a stage where its potential is visible even dimly can we call it emerging. Thus the vantage point of when a new technology is labeled "emerging" is not merely technical—when scientific breakthroughs are achieved or prototypes are built—but when the technology shows promise of industrial and market potential and invites investment capital and entrepreneurial activity.

The concept of emerging technologies may also be distinguished from other terms we have encountered. For one, they include innovations derived from radical innovations (e.g., micro-robots) as well as more evolutionary technologies formed by the convergence of previously separate research streams (e.g., fax machine or the Internet). They embrace not merely "disruptive" technologies that may upset the competitive position of incumbents but also technologies that open up new markets and industries. Finally, they lie at the cusp between embryonic and growth stages of technology since they are beginning to be noticed by various players in the market.

Why are emerging technologies important?

Emerging technologies are important for several reasons. First, they offer the opportunity for changes in the quality of life or standard of living. Perez's documentation of the technology revolutions described above enabled transportation, communication, mass production, climate control, information processing, widespread availability of food and clothing, and improved health care. Second, they offer the potential for technology substitution such that outsiders can compete for an incumbent's position in an industry. For example, fiber optics presented Sprint the opportunity to upgrade its telephone network for greater clarity and speed, but also enabled it to provide data communications lines in addition to voice communication. Third, emerging technologies often enable firms to develop products for latent consumer needs, such as GPS in cars for people who have trouble with directions. Fourth, they allow for new processes for greater efficiency, and speed. For example, robotics has changed the face of manufacturing processes in many industries.

Finally, they offer new investment opportunities for financial markets due to the new-business creation opportunities they generate. platforms from which many a new product can be launched.

Because of the reasons cited above, emerging technologies are arenas in which major competitive battles are fought, won, or lost. When firms fail to detect and act on their knowledge of emerging technologies, their products are often rendered obsolete, as in the case of disruptive technologies, or they may miss major opportunities for new growth platforms.

Managing emerging technologies: roles and responsibilities

The startup. Many start-ups build their businesses around emerging technologies, which offer them either the possibility of leapfrogging established players, or the potential to create new industries. These startups often have the advantage of being close to the origins of the new technology, as in the case of university-based startups. Being in the midst of technology creators, and often being the creators themselves, these startups do not have to concern themselves with the challenges of detecting emerging technologies. Their challenge is to build a viable product and bring it to the market (as is true of all startups).

Mature firms. Existing firms face a different challenge. They are typically viewed as having a more defensive stance toward novel technology. Viewed in this manner, their role is to detect emerging technologies given the technologies' competitive implications. The technology base of established firms is both a help and a hindrance to their ability to detect emerging technologies. These firms will find it relatively easy to detect emergence in domains close to their technological competence. However, surprises lurk for them from areas they are not familiar with. Traditional technical and competitive intelligence functions, if existent within the firm, will suffice to detect emergence in familiar areas. However, the firms should be prepared to be surprised in the case of technologies far from their competence. Corporate venturing efforts, strategic alliances, and/or mergers are some mechanisms established companies use to give them early indicators of technology that may threaten their core businesses.

Alternatively, large firms can be proactive about technology development. Given their rich resource pools of knowledge, talent, networks, and capital, they are potential generators of novel technology.

Universities play a very major role in technology emergence. Through the basic research function that is a requisite part of the university's role, many scientific discoveries occur in this setting. Publication is one mechanism scholars use to communicate their discoveries, but technology licensing is another method universities use to begin the commercialization process.

Government is involved in emerging technology through funding mechanisms. The National Science Foundation, the Department of Energy, and the National Institutes of Health are but a few of the many agencies in the U.S. devoted to supporting early-stage technology development for specific purposes. Other countries have similar governmental organizations whose role is to support the development of novel technology for social benefit.

Venture capital firms operate in the private markets to fund technology emergence as well. However, VC funds are not interested in investing in the earliest stages of development, since their objectives are purely financial. Later-stage, proven technologies with identified market promise are those most likely to attract venture funding.

Detection of emerging technologies

Detection of emerging technologies, although necessary, is complicated by several factors. First, since technology emergence is characterized by failures in the beginning, firms are likely to make errors in understanding emergence, erring sometimes on the optimistic side if the firm is involved in development, and sometimes suffering from blindness to the technology. Second, in the case of emergence of new technology, the terms used to describe the technology may sound esoteric precluding easy intelligibility. Third, many of the players involved in emerging technology are likely to be obscure and unknown and, finally, publications and trade journals discussing the technology may not have come into existence at this stage of technology development.

Understanding the overall process enables detection, and Perez (2002) provides a convenient way of desribing the process by her description of technology surges.

Perez model. Each such surge of development is characterize by a four-phase cycle interspersed with a turning point period, Perez observes, as follows:

(1) irruption, which occurs just after the key invention, when the new products and technologies backed by financial capital are showing their future potential and

making powerful inroads in the economy that is still basically shaped by the previous paradigm;
(2) the frenzy phase, when financial capital drives the intense build-up of the new infrastructure and the new technologies, so, in the end, the potential of the new paradigm is strongly installed in the economy and ready for full deployment.

The turning point, usually the result of a recession that follows the collapse of the financial bubble that arises in the frenzy phase, when the required regulatory changes are made to facilitate and shape the next two phases (the period of deployment):

(3) a synergy phase, when all conditions are favorable to production and to the full flourishing of the new paradigm; and, finally,
(4) the maturity phase, when the last industries, products, technologies, and improvements are introduced while signs of dwindling investment opportunities and stagnating markets appear in the main industries of the revolution.

One could say that the first two phases described above, irruption and fenzy, constitute the *emergence of a technology*. Irruption is characterized by a mass entrance of potential investment money into the market, generated by the firms of the old paradigm. As the technology improves and more applications emerge, so too do more and more investors, seeking returns higher than those yielded on the incremental innovations of the most recent technology that has been exploited to the end of its life. New entrepreneurs enter, and incumbent firms face threatened futures. As a result, this period is marked by increasing unemployment, and even political standoffs driven by the downturn of the dominant technological paradigm and, with it, a number of industries.

The frenzy phase is a time of new millionaires. Financial capital takes over; its immediate interests are opportunistic profits. The paper economy decouples from the real economy (such as happened during the dotcom boom), and there is a growing rift between the forces in the economy and the regulatory frameworks that must eventually drive what is and what is not commecializable. It is a time of speculation and intense exploration of all the possibilities opened up by the technological revolution. Through diversified trial-and-error investment the potential of the emerging technology for creating new markets and rejuvenating old industries is fully discovered and firmly installed in the economy. The productivity explosion reaches more and more activities, inducing a process of restructuring in the productive sphere where the new or renewed prosper and the old wilt or die. This is a phase of fierce "free" competition. Individualism flourishes both in business and in political thinking, sometimes confronted by anti-technology ideas or groups. Tensions exist. The wealth that has grown and concentrated in relatively few hands is greater than can be absorbed by real investment. Much of this excess money is poured into furthering the technological revolution, especially its infrastructure (such as the Internet's "last mile" to the home) . . . often leading to overinvestment that might not fulfill expectations. The late frenzy period is a financial bubble time.

Once the economy reaches this stage, the technology not only as emerged, but the economy must correct itself due to overinvestment, and the technology must be deployed synergistically into productive pursuits. As this occurs there is a setlement period (the synergy phase) and a focus on incremental improvements until the next investment opportunity, fueled by a new advanced technology, begins the cycle again.

Pointers for detection. Perez's model suggests that by tracking the flow of funds a firm could detect many of the emerging technologies since venture capital funding is a clear signal that technology is ripe for emergence. In many large firms, which have technology-scanning units, it is useful to track emerging technologies by monitoring university alliances, venture funds, goverment mandates, and newly emerging firms. These sources may enable firms to get an insight into the scientific developments that are reaching the stage of commercialization.

Summary

The chapters that follow describe several emerging technologies of our era. Each chapter details the origin of the technology, the influence it has had so far, and the potential it offers for the future. Some are brand new on the horizon, and others might be characterized as in the latter part of the frenzy period. Each is surely exciting to contemplate.

Reference

Perez, C. (2002). *Technolgoical Revolutions and Financial Capital: The Dynamics of Bubbles and Golden Ages.* Cheltenham, U.K.: Edward Elgar.

56

Biotechnology: The Technology of the 21st Century

Shreefal Mehta

Rensselaer Polytechnic Institute

Introduction

The 21st century has been called "the Biotech Century": a reference to the potential of emerging biotechnologies and their applications to significantly impact our lives and economies in the coming decades. Biotechnology is an emerging technology that is marked by the creation, accumulation, and commercial exploitation of new knowledge.

Knowledge-based industries flourish in countries that have fostered and established specific supportive environmental factors including mechanisms that protect and respect intellectual property rights, long-term public and private support of robust, independent academic research institutions and innovation systems that support the transfer of technology. Such established knowledge-based economies have already seen the impact of the bio-economy, an example being the exponential rise in revenues and market value of publicly-traded biotechnology companies in the U.S. from the early 1990s to 2007.

What exactly is meant by biotechnology?

Biotechnologies are the set of tools and insights that allow us to manipulate and use cellular, genetic, and biomolecular processes to solve problems or make useful products. The applications of these biotechnologies mostly involves manipulation and selective utilization of various biological processes or molecules, either in isolation from, or engineered in concert with their living cells to derive new solutions and products.

Biotechnologies in use before the 1950s largely focused on harnessing the efficiency or specific production capabilities of entire living organisms (carefully bred animals or plants or whole cells). The new generation of biotechnologies that continue to be developed even today, starting from the 1950s, made it possible to manipulate and use the smallest parts of organisms—their biological molecules—in addition to entire organisms.

The promise of biotechnology and biomolecular processes and molecules lies in their specificity and efficiency which is often difficult to achieve with human-designed tools and processes. For example, the production machinery in a cell can be engineered to yield large quantities of specific biomolecules that cannot be made synthetically. In addition, isolated biological enzymes can be used to drive chemical or other industrial processes with far greater efficiency than possible with man-made reagents or processes. The recent attention to "biotechnology" is due to this ability of various biotechnologies to capitalize the attributes of cells and biological molecules in a very precise, specific, and reproducible way, leading to the industrialization of biotechnology and its application to many different areas.

Therefore, as summarized in the Biotechnology Industry Organization's *Guide to Biotechology* (*http://bio.org/speeches/pubs/er/*): "Biotechnology is a collection of technologies that capitalize on the attributes of cells, such as their manufacturing capabilities, and put biological molecules, such as DNA and proteins, to work for us."

Since the term "biotechnology" is used rather broadly, it is important to note that the term "biotechnology company" or "biotechnology industry" has the following adjacent sectors or technologies/applications which are also sometimes self-referenced as "biotech", but do not have any specific biological processes or products involved:

—synthetic chemical–manufacturing processes that do not use biomolecules;
—medical devices that do not incorporate biological molecules or cells;
—in particular, in the pharmaceutical industry, most drugs are produced using synthetic chemical processes and are separate from "biotechnology drugs", which refer to either extracted biomolecule therapeutics or molecules produced using biotechnologies in the manufacturing processes

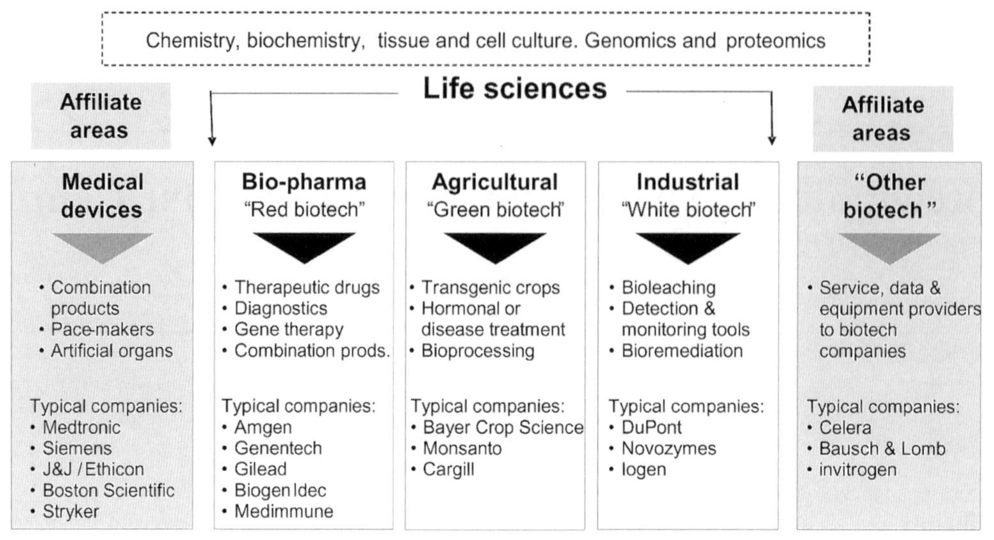

Figure 56.1. Biotechnology, based on basic tools and science of life sciences, has three key product application areas. Other application sectors that are often involved in various applications or supporting tool and product developments, are shown on either side in the graph. Typical applications and companies that are representative of the sector are shown at the bottom of the graph.
Source: adapted from original presented at a Departmental Biotechnology Seminar at the University of Texas Austin, 2004.

—life science laboratory equipment or reagent manufacturers who do not use biotechnology processes for their products;
—food-processing companies that do not use biotechnology-based processes in production;
—information technology (bioinformatics) based companies that develop software applications in life sciences but do not develop or use specific biotechnologies in production.

The impact of biotechnology on various industries and aspects of human existence will be explored in the next section, followed by a look at the historical drivers of innovation, and then a discussion of the emergence of trends in the biotech industry.

Applications for biotechnology—past, current, and future[1]

Applications of biotechnology can be classified into three major areas—health, industrial, and agricultural applications. These application spaces and associated specific technologies are also known as "red biotech", "white biotech", and "green biotech", respectively. Figure 56.1 shows this classification as well as two general areas that are affiliated but do not typically fall into the "biotechnology industry" (as described in the previous section).

Biotechnologies have changed our lives from the beginning of human civilization. We began growing crops and raising animals 10,000 years ago to provide a stable supply of food and clothing. We have used the biological processes of microorganisms for 6,000 years to make useful food products, such as bread and cheese, and to preserve dairy products. The harnessing of plant growth and livestock breeding allowed us to live in large communities and develop the interdependent societies that led the way to progress in other technologies.

The more precise and industrialized biotechnologies available today perform essentially the same conceptual task of previous selective breeding and hybridization techniques in animal husbandry and plant cultivation, respectively, as practiced by humans for centuries. However, today's biotechnologies can be integrated into many different processes and areas and have had an impact on diverse industrial and functional sectors (as shown in Figure 56.2).

"Red biotech"—applications of biotechnology in human and animal health

Knowing the molecular basis of health and disease leads to improved methods for treating and preventing diseases. In human health care, biotechnology products include quicker and more accurate diagnostic tests, therapies with fewer side-effects, and new and safer vaccines.

Diagnostics

The application of biotechnologies to diagnostic tests includes the following benefits (examples follow).

[1] Certain segments of this section have been reproduced from the Biotechnology Industry Organization, with permission.

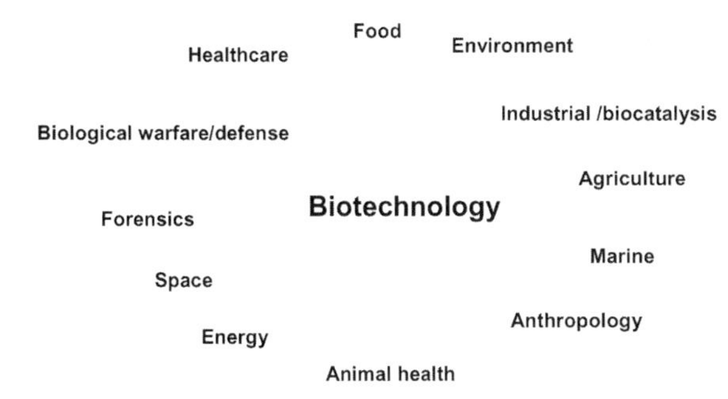

Figure 56.2. Biotechnologies are used in diverse applications today.

Improved speed and accuracy

A familiar example of these benefits is the new generation of home pregnancy tests that provide more accurate results much earlier than previous tests. Tests for strep throat and many other infectious diseases provide results in minutes, enabling treatment to begin immediately, in contrast to the 2-day or 3-day delay of previous tests.

Lower costs of diagnosis

We now use biotechnology-based tests to diagnose certain cancers, such as prostate and ovarian cancer, by taking a blood sample, eliminating the need for invasive and costly surgery.

Earlier diagnosis leading to better prognosis and prevention measures

Proteomics researchers are discovering molecular markers that indicate incipient diseases before visible cell changes or disease symptoms appear. The wealth of genomics information now available will greatly assist doctors in early diagnosis of hereditary diseases, such as type-I diabetes, cystic fibrosis, early-onset Alzheimer's disease, and Parkinson's disease—ailments that previously were detectable only after clinical symptoms appeared. Genetic tests will also identify patients with a predisposition to diseases, such as various cancers, osteoporosis, emphysema, type-II diabetes and asthma, giving patients an opportunity to prevent the disease by avoiding triggers such as diet, smoking, and other environmental factors. Physicians will someday be able to immediately profile an infection being treated and, based on the results, choose the most effective antibiotics.

Public health–related preventive measures are now possible

New detection tests are used to screen donated blood for the pathogens that cause AIDS and hepatitis.

Therapeutics

Already, biotechnology is applied to improve today's therapeutic regimes as well as to create innovative treatments that would not be possible without these new techniques. Biotechnology therapeutics approved by the U.S. Food and Drug Administration (FDA) to date are used to treat many diseases, including leukemia and other cancers, anemia, cystic fibrosis, growth deficiency, rheumatoid arthritis, hemophilia, hepatitis, genital warts, and transplant rejection. All the therapies given below as examples make use of biological substances and processes designed by nature. Some use the human body's own tools for fighting infections and correcting problems. Others are natural products of plants and animals. The large-scale manufacturing processes for producing therapeutic biological substances also rely on nature's molecular production mechanisms.

Many biotech products are *biologics*, meaning they are derived from living sources such as cells. Biologics are complex mixtures whose active ingredients—usually proteins—are hundreds of times larger than the compounds found in most pills. These products usually must be injected or infused directly into the bloodstream to be effective. Biologics include blood and blood-derived products and vaccines, as well as biotechnology-based recombinant proteins and monoclonal antibodies. Most biologics are regulated by the FDA under the Public Health Service Act and require approval of a biologic license application (BLA) prior to marketing. Through the late-1990s, biotechnology was closely associated with recombinant and antibody-based biologics, but increasingly biotech companies are using genetic and other biological discoveries to develop so-called small-molecule drugs. These are the chemically simple compounds that are so familiar on pharmacy shelves. They are often formulated as pills (although small-molecule products may also be injected or infused). The list below covers a diverse range of therapeutic products, including biologics, but do not include the small-molecule drugs which had biotechnology tools and processes applied only in the discovery stage of the drug development process.

Using natural products

Many living organisms produce compounds that have therapeutic value for us. For example, many antibiotics are produced by naturally occurring microbes, and a number of medicines on the market, such as digitalis, are made by plants. Plant cell culture, recombinant DNA technology, and cellular cloning now provide us with new ways to tap into natural diversity. Byetta™ (exenatide), an incretin mimetic, was chemically copied from the venom of the gila monster (a North American lizard) and approved in early 2005 for the treatment of diabetes. PRIALT® (ziconotide), a recently approved drug for pain relief, is a synthetic version of the toxin from a South Pacific marine snail.

Replacing missing proteins

Some diseases are caused when defective genes don't produce the proteins (or enough of the proteins) the body requires. Today we are using recombinant DNA and cell culture to produce the missing proteins as drugs consisting of the biomolecules, typically delivered as an infusion into the blood stream to the desired level. Replacement protein therapies include:

—factor VIII—a protein involved in the blood-clotting process, lacked by some hemophiliacs;
—insulin—a protein hormone that regulates blood glucose levels (diabetes results from an inadequate supply of insulin).

Gene therapy

Gene therapy presents an opportunity using DNA, or related molecules such as RNA, to treat diseases. For example, rather than giving daily injections of missing proteins, physicians could supply the patient's body with an accurate instruction manual—a non-defective gene—correcting the genetic defect so the body itself makes the proteins. Other genetic diseases could be treated by using small pieces of RNA to block mutated genes. Currently, only genetic diseases caused by the lack of protein, such as hemophilia, are amenable to correction via replacement gene therapy. In late 2003, China licensed for marketing the first commercial gene therapy product, Gendicine, which delivers the P53 tumor suppressor gene. The product treats squamous cell carcinoma of the head and neck, a particularly lethal form of cancer. Clinical trial results were impressive: 64% of patients who received the gene therapy drug, in weekly injections for 3 months, showed a complete regression and 32% attained partial regression.

Cell transplants

There are not enough organs for people who need organ transplants and people die each day waiting for organs to become available for transplantation in the U.S. To circumvent this problem, scientists are investigating ways to use cell culture to increase the number of patients who might benefit from one organ donor. For example, liver cells grown in culture and implanted into patients kept them alive until a liver became available.

Modulating the immune system

Stimulating: cancer vaccines that help the immune system find and kill tumors have also shown therapeutic potential. Other cellular signaling biomolecules can now be produced in large enough quantities to try and trigger specific responses in specific cell types in the immune system.

Suppressing: currently we are using monoclonal antibodies to suppress, very selectively, the type of cell in the immune system responsible for organ transplant rejection and autoimmune diseases, such as rheumatoid arthritis and multiple sclerosis.

Personalized medicine

In the future, our individual genetic information will be used to prevent disease, choose medicines, and make other critical decisions about health. This is personalized medicine, and it could revolutionize health care, making it safer, more cost-effective, and most importantly more clinically effective. A few examples of the progress of personalized medicine applications today are seen in the therapeutic treatment and diagnosis of cancer.

Cancer. The biotech breast cancer drug Herceptin® is an example of a pharmacogenomic drug. Initially approved in 1998, Herceptin® targets and blocks the HER2 protein receptor, which is overexpressed in some aggressive cases of breast cancer. A test can identify which patients are overexpressing the receptor and can benefit from the drug. Other new tests and therapeutics are now looking for combinations that improve the efficacy of specific drugs for specific patients.

Regenerative medicine

Tissue engineering combines advances in cell biology and materials science, allowing us to create semi-synthetic tissues and organs in the lab. These tissues consist of biocompatible scaffolding material, which eventually degrades and is absorbed, plus living cells grown using cell culture techniques. Ultimately the goal is to create whole organs consisting of different tissue types to replace diseased or injured organs.

Vaccines

Vaccines help the body recognize and fight infectious diseases. Most of the new vaccines consist only of the antigen, not the actual microbe. The vaccine is made by inserting the gene that produces the antigen into a manufacturing cell, such as yeast. During the manufacturing process, which is similar to brewing beer, each yeast cell makes a perfect copy of itself and the antigen gene. The antigen is later purified from the yeast cell culture. By isolating antigens and producing them in the laboratory, it is possible to make vaccines that cannot transmit the virus or bacterium itself. This method also increases the amount of vaccine that can be manufactured because biotechnology vaccines can be made without using live animals. Biotechnology is also broadening the vaccine concept beyond protection against infectious organisms. Various researchers are developing vaccines against diseases such as diabetes, chronic inflammatory disease, Alzheimer's disease, and cancer.

Plant-made pharmaceuticals

The flexibility provided by biotechnology presents many

opportunities for using plants in new ways. Advances in biotechnology have made it possible to genetically enhance plants to produce therapeutic proteins essential for the production of a wide range of pharmaceuticals—such as monoclonal antibodies, enzymes, and blood proteins. Therapeutic proteins, including antibodies, antigens, growth factors, hormones, etc. have been grown in field trials in a wide variety of plants, including alfalfa, corn, duckweed, potatoes, rice, safflower (an Old World herb), soybeans, and tobacco. In addition, scientists have made excellent progress in using plants as vaccine-manufacturing and delivery systems. Because protein-producing plants require relatively little capital investment, and the costs of production and maintenance are minimal, they may provide the only economically viable option for independent production of therapeutic proteins in underdeveloped countries.

Thus the application of biotechnology to human health has already started to deliver on its promise of a new age of medicine, with even greater potential and promise driving continuing interest in this sector.

"Green biotech"—agricultural production and processing, animal health applications

Biotechnology can help meet the ever-increasing need by *increasing yields, decreasing crop inputs* such as water and fertilizer, and providing *pest control methods that are more compatible with the environment*. Applications of biotechnology to animal health are used to improve animal health, enhance animal products, gain conservation and environmental benefits, and help advance human health.

Agricultural applications

Stone Age farmers selected plants with the best characteristics and saved their seeds for the next year's crops. By selectively sowing seeds from plants with preferred characteristics, the earliest agriculturists performed genetic modification to convert wild plants into domesticated crops long before the science of genetics was understood. The tools of biotechnology allow plant breeders to select single genes that produce desired traits and move them from one plant to another. The process is far more precise and selective than traditional breeding in which thousands of genes of unknown function are moved into our crops. Biotechnology also removes the technical obstacles to moving genetic traits between plants and other organisms. This opens up a world of genetic traits to benefit food production. We can, for example, take a bacterium gene that yields a protein toxic to a disease-causing fungus and transfer it to a plant. The plant then produces the protein and is protected from the disease without the help of externally applied fungicides.

Improving yield. The crop production and protection traits agricultural scientists are incorporating with biotechnology are the same traits they have incorporated through decades of crossbreeding and other genetic modification techniques: increased yields; resistance to diseases caused by bacteria, fungi and viruses; the ability to withstand harsh environmental conditions such as freezes and droughts; and resistance to pests such as insects, weeds, and nematodes. An example is the incorporation of specific protein-producing genes from the Bt bacterium (used since the 1930s as several of its proteins are lethal to the corn borer and other insects but not to humans) into the corn plant. The plant that once was a food source for the insect now kills it, lessening the need to spray crops with chemical pesticides to control infestations. In addition to increasing crop productivity by using built-in protection against diseases, pests, environmental stresses, and weeds to minimize losses, scientists use biotechnology to improve crop yields directly. Researchers at Japan's National Institute of Agrobiological Resources added maize photosynthesis genes to rice to increase its efficiency at converting sunlight to plant starch and increased yields by 30%.

Decreasing crop inputs. Genes found in any organism can be used to improve crop production by making them resistant to drought or extreme temperatures. Often, herbicides must be applied several times during the growing cycle, at great expense to the farmer and possible harm to the environment. Using biotechnology, it is possible to make crop plants tolerant of specific herbicides. When the herbicide is sprayed, it will kill the weeds but have no effect on the crop plants. This lets farmers reduce the number of times herbicides have to be applied and reduces the cost of producing crops and damage to the environment.

Environmental and direct human health benefits. Monsanto has donated virus resistance technologies to Kenya for sweet potatoes, Mexico for potatoes and Southeast Asia for papaya, and technology for pro-vitamin A production in oilseed crops to India. As described above, biotech crops allow us to increase crop yields by providing natural mechanisms of pest control in place of chemical pesticides. These increased yields can occur without clearing additional land, which is especially important in developing countries. In addition, because biotechnology provides pest-specific control, beneficial insects that assist in pest control will not be affected, facilitating the use of integrated pest management. Herbicide-tolerant crops decrease soil erosion by permitting farmers to use conservation tillage. According to the National Center for Food and Agricultural Policy's (NCFAP) 2004 report, in 2003 the 11 biotech crop varieties adopted by U.S. growers increased crop yields by 5.3 billion pounds, saved growers $1.5 billion by lowering production costs, and reduced pesticide use by 46.4 million pounds. Based on increased yields and reduced production costs, growers realized a net economic impact or savings of $1.9 billion.

Environmental benefits that biotechnology is providing to the forestry industry include enzymes for

—pretreating and softening wood chips prior to pulping;
—removing pine pitch from pulp to improve the efficiency of paper-making;
—enzymatically bleaching pulp rather than using chlorine;
—de-inking of recycled paper;
—using wood-processing wastes for energy production and as raw materials for manufacturing high-value organic compounds;
—remediating soils contaminated with wood preservatives and coal tar.

Animal health

Improved characteristics for animals as food for human consumption. Livestock cloning is the most recent evolution of selective assisted breeding in the ancient practice of animal husbandry. Cloning does not change an animal's genetic makeup: it is simply another form of sophisticated assisted reproduction. Cloning allows livestock breeders to create an exact genetic copy of an existing animal—essentially an identical twin. In December 2006, the U.S. Food and Drug Administration published a draft risk assessment that concluded that meat and milk products from animal clones and their offspring are safe for human consumption. As a new assisted reproductive technology, cloning can consistently produce healthier animals and a healthy meat and milk supply.

A transgenic animal is one that has had genetic material from another species added to its DNA. This breakthrough technology allows scientists to precisely transfer beneficial genes from one species to another. Transgenic technology can improve the nutritional value of animal products through enhanced genes. In addition, the technology promises improved animal welfare and productivity—a critical capability in meeting the food demands of a growing global population.

Biotech animals for improved human health. For decades, animals have been used to produce human pharmaceuticals. Horses, pigs, rabbits, and other species have been enlisted to produce such products as anti-venoms, biologics to prevent organ transplant rejection, and the blood thinner heparin. Biotechnology now allows us to modify genes in these animals so that the drug proteins are more compatible with human biochemistry. In 2006, the European Commission approved the first pharmaceutical product manufactured with ingredients derived from biotech goats. The drug's ingredients include proteins from the milk of biotech goats. The pharmaceutical treats the rare blood-clotting disorder antithrombin deficiency.

Endangered species conservation. Biotechnology is providing new approaches for saving endangered species. Reproductive and cloning technologies, as well as medicines and vaccines developed for use in livestock and poultry, can also help save endangered mammals and birds.

Borrowing biotechnology techniques used by livestock breeders, veterinarians at the Omaha Zoo recently used hormonal injections, artificial insemination, embryo culture, and embryo transfer to produce three Bengal tiger cubs. A Siberian tigress served as the surrogate mother for these embryos.

Summary of benefits from agricultural applications of biotechnology

- More food for the growing world population
  —increased yield through disease resistance, hardiness.
- Better food
  —increased nutrition;
  —improved taste;
  —less exposure to chemicals.
- Better for environment
  —less land tilled;
  —reduced used of pesticides or fertilizers.

"White biotech"—industrial bioprocessing and environmental applications

Biotechnology in industry employs the techniques of modern molecular biology to reduce the environmental impact of manufacturing. Industrial biotechnology also works to make manufacturing processes more efficient for industries such as textiles, paper and pulp, and specialty chemicals. Some observers predict biotechnology will transform the industrial manufacturing sector in much the same way that it has changed the pharmaceutical, agricultural, and food sectors. Industrial biotechnology will be key to achieving industrial and environmental sustainability.

An industrially sustainable process should, in principle, be characterized by

—reduction or elimination of toxic waste;
—lower greenhouse gases;
—low consumption of energy and non-renewable raw materials (and high use of carbohydrate feedstocks, such as sugars and starch);
—lower manufacturing cost.

Living systems manage their chemistry more efficiently than man-made chemical plants, and the wastes that are generated are recyclable or biodegradable. Biocatalysts, and particularly enzyme-based processes, operate at lower temperatures and produce less toxic waste and fewer byproducts and emissions than conventional chemical processes. They may also use less purified raw materials (selectivity). Use of biotechnology can also reduce energy required for industrial processes.

Material and energy inputs

Biotechnology provides ways to produce cleaner products and processes by reducing the use of petroleum inputs.

Industrial biotechnology instead uses natural sugars as feedstocks.

Through biotechnology, the use of renewable, biomass-based feedstocks will increase. Bio-feedstocks offer two environmental advantages over petroleum-based production: production will be cleaner, in most cases, and less waste will be generated. By 2006, at least 5 billion kilograms of commodity chemicals were produced annually in the U.S. using plant biomass as the primary feedstock.

Efficient industrial manufacturing processes

Unlike many chemical reactions that require very high temperatures and pressures, reactions using biological molecules work best under conditions that are compatible with life (i.e., temperatures under 100°F, atmospheric pressure, and water-based solutions). Therefore, manufacturing processes that use biological molecules can lower the amount of energy needed to drive reactions. Manufacturing processes that use biodegradable molecules as biocatalysts, solvents, or surfactants are also less polluting. The biotechnology techniques of protein engineering and directed protein evolution maximize the effectiveness and efficiency of enzymes. They have been used to modify the specificity of enzymes, improve catalytic properties, or broaden the conditions under which enzymes can function so that they are more compatible with existing industrial processes.

Biofuel

Biotechnology tools promise increasing efficiencies in the conversion of biomass to ethanol by improved enzymes and bioprocessing methods. There are other biotechnology contributions that are being explored to aid energy production or capture in different ways, but the production of biofuels is the most advanced to date.

In his January 2006 State-of-the-Union address, President Bush declared: "America is addicted to oil, which is often imported from unstable parts of the world. The best way to break this addiction is through technology." One of his key technological solutions was "research in cutting-edge methods of producing ethanol, not just from corn, but from wood chips and stalks, or switch grass." He announced a national goal to make this new kind of ethanol practical and competitive within 6 years, pledging $150 million in FY 2007 for biomass research, development, and demonstration. Advances in industrial biotechnology and development of new integrated "biorefineries" are at the heart of ethanol production from all sources. April 2004 saw the first commercial production of ethanol from cellulose, made from wheat straw using biotech enzymes.

The 2005 Energy Policy Act authorized over $3 billion in funding for biofuels and bio-based products and established a national renewable fuels standard. The bill established a goal of displacing 30% of today's gasoline consumption with ethanol or other biofuels by 2030. A recent Natural Resources Defense Council report suggests that that potential could be even higher.

Green plastics

Biotechnology also offers us the prospect of replacing petroleum-derived polymers with biological polymers derived from grain or agricultural biomass.

In 2001 the world's first biorefinery opened in Blair, Nebraska, to convert sugars from field corn into polylactic acid (PLA)—a compostable biopolymer that can be used to produce packaging materials, clothing, and bedding products. Price and performance are competitive with petroleum-based plastics and polyesters. Several national retailers, including Whole Foods and Wal-Mart, are now using PLA packaging.

Environmental biotechnology

Environmental biotechnology is the use of living organisms for a wide variety of applications in hazardous waste treatment and pollution control. For example, a fungus is being used to clean up a noxious substance discharged by the paper-making industry.

The vast majority of bioremediation applications use naturally occurring microorganisms to identify and filter manufacturing waste before it is introduced into the environment or to clean up existing pollution problems. Some microorganisms, for example, feed on toxic materials such as methylene chloride, detergents, and creosote. The bacteria then "eat" the hazardous waste at the site and turn it into harmless byproducts. After the bacteria consume the waste materials, they die off or return to their normal population levels in the environment.

Environmental monitoring

Companies have developed methods for detecting harmful organic pollutants in the soil using monoclonal antibodies and the polymerase chain reaction, while scientists in government labs have produced antibody-based biosensors that detect explosives at old munitions sites.

Summary of industrial and environmental applications of biotechnology

The various industries in which the benefits of biotechnology applications are established include:

- The chemical industry: using biocatalysts to produce novel compounds, reduce waste byproducts, and improve chemical purity.
- The plastics industry: decreasing the use of petroleum for plastic production by making "green plastics" from renewable crops such as corn or soybeans.
- The paper industry: improving manufacturing processes, including the use of enzymes to lower toxic byproducts from pulp processes.

- The textiles industry: lessening the toxic byproducts of fabric dying and finishing processes. Fabric detergents are becoming more effective with the addition of enzymes to their active ingredients.
- The food industry: improving baking processes, fermentation-derived preservatives, and analysis techniques for food safety.
- The livestock industry: adding enzymes to increase nutrient uptake and decrease phosphate byproducts.
- The energy industry: using enzymes to manufacture cleaner biofuels from agricultural wastes.

Technology trends in biotechnology—innovations, impact, and convergence

Currently the largest revenue-generating applications are in human health (red biotech), and biotechnology has most visibly impacted this sector of applications. Therefore innovations and new technologies and future trends will be largely discussed from the viewpoint of the health care sector. However, it should be noted that the economic impact of green and white biotech is projected to grow to an even greater extent than health care; the choice to focus on health care is largely driven by popular attention and historical trends.

Drivers for technological innovation in knowledge-based sectors

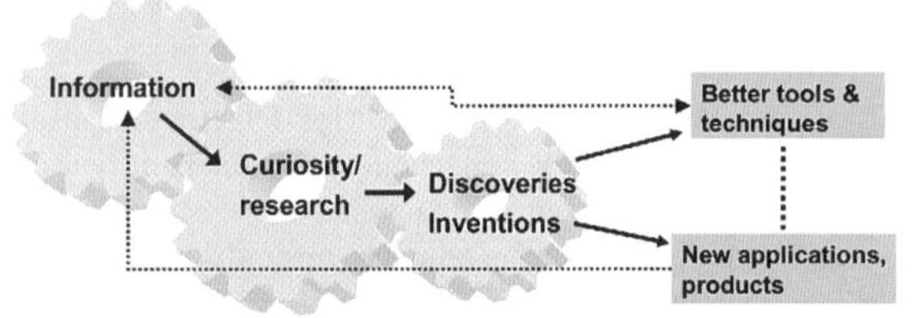

Figure 56.3. Technologies link curiosity, discoveries, and new applications in a cycle of innovation.
S. Mehta, *Commercializing Successful Biomedical Technologies*. Cambridge, U.K.: Cambridge University Press, 2008, with permission.

In-depth information in an area builds momentum as multiple iterations are made to better understand a phenomenon or a technology, ultimately leading to better tools and new applications and products. These new applications, tools, or products eventually lead to new information that enters the cycle shown in Figure 56.3. The spark of curiosity of humans and the intensified, globally-competitive research activities of this century are the drivers for innovations, new technologies, and applications entering the market.

Technology trends and innovation milestones in red biotech

Technology has played an important part in drug development and discovery over the years, either by opening new pathways for better treatments or by speeding up the process of developing drugs. Most early drugs were derived as extracts from natural sources. The components of these extracts, when purified, were identified and synthesized using chemical synthesis methods to yield a reproducible compound. However, as discussed earlier, drug technologies have seen a big change in the methods of production with the advent of biotech drugs (biologics). These biotech drugs, typically proteins that are enzymes or antibodies (monoclonal antibodies), are produced using genetically engineered living cells.

The biotech industry really got off the ground with two basic technologies in 1975, recombinant DNA (rDNA) and monoclonal antibody (mAB) production from hybridomas. These technologies allowed the industrialization of biotechnologies by providing precise and reproducible tools for large-scale applications in industrial production or in large-scale biotechnology research. These initial technological innovations have now cascaded into many more breakthroughs in new biotechnology platforms, leading to an ever-increasing range of applications going beyond basic manufacturing techniques to enhance the entire supply chain in drug and diagnostic development. Figure 56.4 overlays the actual revenue figures for biotechnology drugs with a few selected technology milestones. These technology and commercial milestones are meant to be representative and not comprehensive.

The next set of emerging technologies include stem cells, tissue engineering, gene therapy, siRNA and *in silico* biology. This new generation of human therapeutics will likely require the development of new production technologies.

Pharmaceutical companies are still working on integrating large number of new technologies and methods that have emerged over the last decade and have been shown to individually have great promise in discovering targets or drugs that can lead to a cure for diseases. It is possible that the expected outcome of these new insights will not emerge in the form of real curative medications for some years hence. However, it is clear from looking at the number of targeted therapies (monoclonal antibodies, drugs that target specific cell-signaling mechanisms) and therapies that require preselection of patients using genomic diagnostics that an era of new medicine is emerging. It is, in fact, in the diagnostics area of health care that some remarkable breakthroughs have resulted in more effective targeted therapeutics and earlier discovery of the presence of disease.

In the mature *in vitro* diagnostics (IVD) markets of blood chemistry clinical diagnostics, technology development has slowed, with a focus on cost and operational efficiency

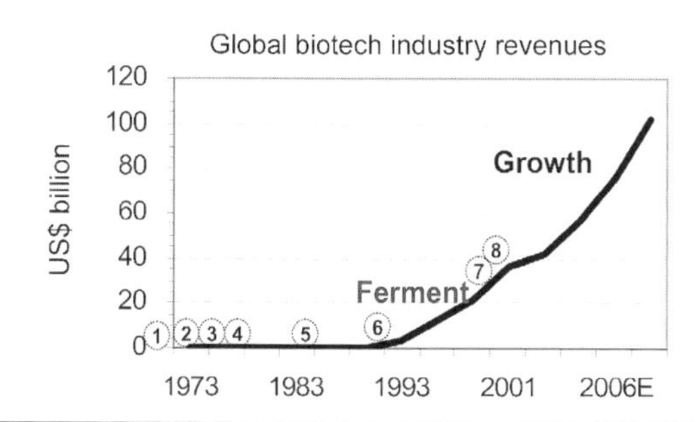

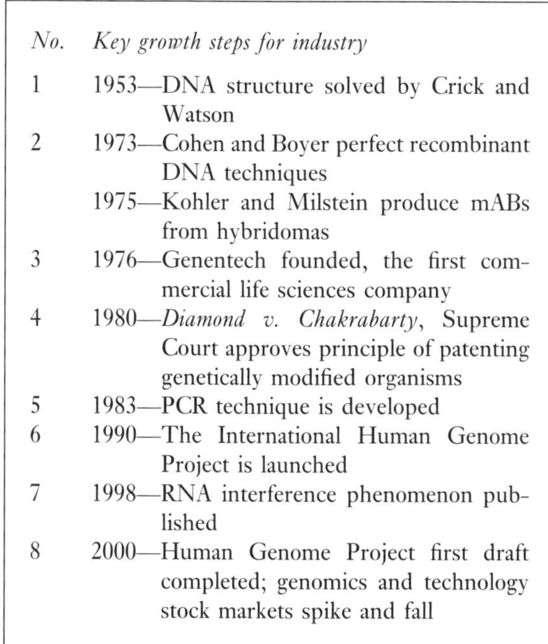

No.	*Key growth steps for industry*
1	1953—DNA structure solved by Crick and Watson
2	1973—Cohen and Boyer perfect recombinant DNA techniques
	1975—Kohler and Milstein produce mABs from hybridomas
3	1976—Genentech founded, the first commercial life sciences company
4	1980—*Diamond v. Chakrabarty*, Supreme Court approves principle of patenting genetically modified organisms
5	1983—PCR technique is developed
6	1990—The International Human Genome Project is launched
7	1998—RNA interference phenomenon published
8	2000—Human Genome Project first draft completed; genomics and technology stock markets spike and fall

Figure 56.4. Biotech industry development technical milestones.
S. Mehta, *Commercializing Successful Biomedical Technologies*. Cambridge, U.K.: Cambridge University Press, 2008, with permission.

that has driven incremental innovation in automation and simplification of existing tests. However, the nucleic acid testing (NAT) or DNA-based genomic diagnostics segment has seen a rapid pace of development of innovative technology platforms and revenue growth driven by key biotechnology innovations. These enabling biotechnologies include genetic sequencing, polymerase chain reaction (PCR) technologies, and DNA microarray devices, and are leading to new discoveries linking the human genetic code and disease.

Selected emerging technologies and materials in the nucleic acid diagnostics field

- Alternative amplification technologies—new techniques for copying genetic material for increased sensitivity or speed of testing (compared with PCR) are being developed.
- Bio-chips and lab-on-a chip—DNA, protein, glycosaccharides, and lipid array chips with multiple probes arrayed on a chip can provide large amounts of information from a single sample. Microfluidic technology and micro-electromechanical system (MEMs) technologies combined with standardized semiconductor industry silicon chip fabrication and opto-electronic technologies have made possible the creation of various versions of a "lab-on-a chip". Materials that have been used so far include silicon substrate, rubberized silicone, gallium nitride, and other electronic industry–based materials. Additionally, nanoscale materials (quantum dots, etc.) are being developed as markers and readouts for various assays. It is anticipated that these technologies will shift the industry from the current techniques of detection of single analytes to the large-scale, parallel testing of tens, hundreds, or perhaps ultimately thousands of genes and/or proteins in the same multi-analyte test. Together with increased automation, these new lab-on-a-chip devices could also shift the focus from central labs to testing at the patient bedside.
- Other multi-probe technologies—novel techniques such as mass spectrometry with simultaneous probes to identify a "molecular signature" that is indicative of disease, rather than a single molecule at a time.

Convergence of technologies with biotechnology

In the biotechnology industry, the level of information that is now available at the molecular level is increasing rapidly, and that knowledge is spreading rapidly at all scales of study of biotechnology processes.

As shown schematically in Figure 56.5, early observations many centuries ago were made at the phenotype level usually for individual traits. Emerging from the processes of European scientific inquiry, the reductionist approach to biology took shape, where individual elements were studied in isolation from their organism or system to determine the parameters of functional activity and interaction. For example, blood was first studied as a system, then at the level of isolated vessels, cells, and individual serum components to study how they each changed and behaved in healthy or diseased states. Further reductionist approaches led to isolation of individual receptors on cell membranes, elaboration of intracellular signal transduction pathways, genes, and DNA codons. The goal was always to take this information and knowledge and put the individual pieces back together to be able to understand the complex organism and system—like a child with a box of gears trying to build the mechanism of a complex and delicate clock. However, this approach has seen limited success until the recent technological developments of computational speed and data storage and the development of algorithms that can combine disparate types of data.

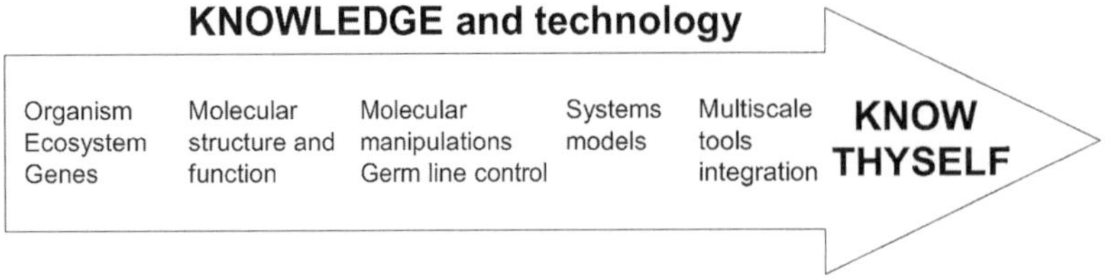

KNOWLEDGE and technology

Organism Ecosystem Genes | Molecular structure and function | Molecular manipulations Germ line control | Systems models | Multiscale tools integration | **KNOW THYSELF**

Figure 56.5. Convergence of various technologies towards personalized medicine.
S. Mehta, *Commercializing Successful Biomedical Technologies*. Cambridge, U.K.: Cambridge University Press, 2008, with permission.

By putting together computational simulations derived from this increased knowledge, the ultimate goal would be to develop a multiscale model of behavior of any given biological system. Knowing and predicting at once the activity of the organism and the molecular or cellular drivers of that particular state of activity (healthy or disease), will allow the true advent of personalized medicine, where a disease state is recognized, understood at the level of specific protein/gene dysfunction, and a therapy that specifically treats the cause of the disease in that individual. This still remains many years away.

Thus, ultimately, the drive to biotechnology developments and innovations is contained in the dictum—"know thyself"—towards a future where personalized medicine and individualized therapy are the norm, made possible by integrating technologies at various scales to achieve detailed knowledge of the inner workings of each individual. This is what we as society and as individuals would eventually like to have, a complete knowledge of conscious and unconscious, macroscopic and microscopic body processes that allow us to make the best decisions about our health.

In order to truly gain that broad and deep understanding, current research is examining the organism through a variety of different perspectives and techniques that include technologies more closely related to traditional engineering disciplines. For example, we are using robotics and informatics to analyze genetic coding differences among individuals. The increase in information and knowledge in biological sciences is now critically integrated with technological improvements in basic engineering disciplines and in physical sciences.

Additional convergence of technologies is seen across traditional industrial boundaries (depicted in Figure 56.6), when new tools developed for human medicine find applications in other areas and may in fact be enhanced in interactions with traditional engineering processes in other fields.

Applications of biotechnology that include considerable interaction with other emerging or advanced technologies such as advanced materials and nanotechnology, include some of the following areas:

- biomass renewable energies (e.g., biodiesel);
- biosensors;
- diagnostic medical devices;
- bioremediation;
- digitization of health care information;
- bioinformatics;
- biotherapeutics;
- biometrics;
- biosecurity.

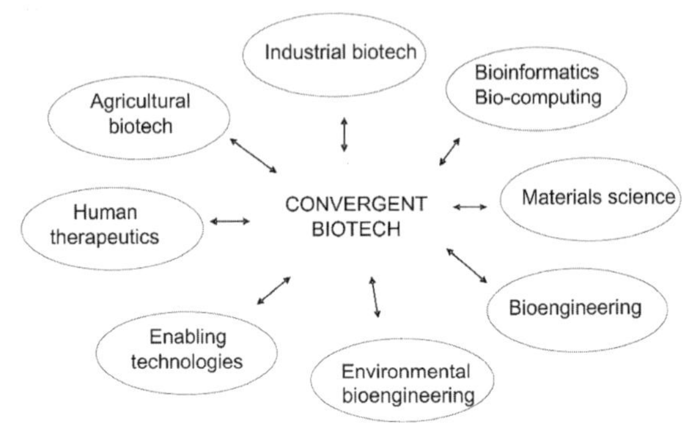

Figure 56.6. Technologies from various disciplines interact in biotechnology today.
Source: adapted from a slide presentation at the BIO 2004 Industrial Biotechnology conference.

Continuing along this trend is an increase in multi-disciplinary research and development, with advances in other disciplines impacting developments in energy and biotechnology. As represented in Figure 56.7, the integration of various emerging technologies into the energy

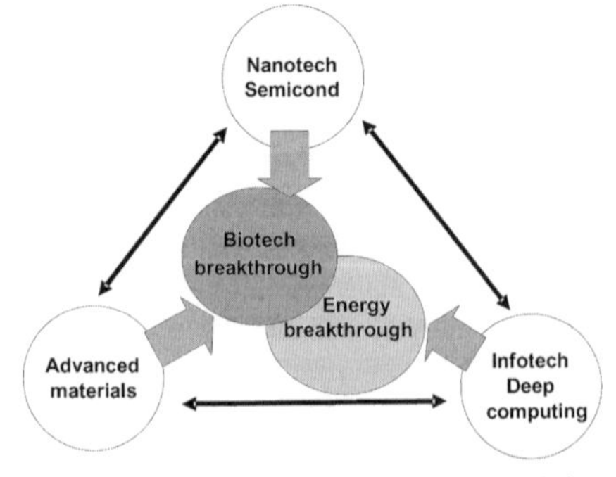

Figure 56.7. Convergence across industries and inter-disciplinary studies will result in breakthrough applications at the interfaces, in the focus areas of energy and health care.
S. Mehta, *Commercializing Successful Biomedical Technologies*. Cambridge, U.K.: Cambridge University Press, 2008, with permission.

and biotechnology industry makes it likely that breakthroughs in performance or reduction in cost or both will result.

References and further reading

Bronzino, J. (Ed.). (2006). *Biomedical Engineering Handbook* (Third Edition). Boca Raton, FL: CRC Press.

Burns, L. (Ed.). (2002). *The Health Care Value Chain*. San Francisco, CA: Jossey-Bass.

Burns, L. (Ed.). (2005). *The Business of Healthcare Innovation*. Cambridge, U.K.: Cambridge University Press, 2005.

Burrill, S. J. (Ed.). (2008). *Biotech 2008: Life Sciences—A 20/20 Vision to 2020* (Annual Report). San Francisco, CA: Burrill & Company.

Chiesa, V., and Chiaroni, D. (2004). *Industrial Clusters in Biotechnology*. London: Imperial College Press.

Collins, S. W. (2004). *The Race to Commercialize Biotechnology*. London: Routledge.

DiMasi, J. A., Hansen, R. W., and Grabowski, H. G. (2003). "The price of innovation: New estimates of drug development costs." *Journal of Health Economics*, **22**, 151–185, 2003.

Estrin, N. F. (1990). *Medical Device Industry*. New York: Marcel Dekker.

FDA. (2006). U.S. Food & Drug Administration; *www.fda.gov*

Kayser, O., and Müller, R. H. (Eds.). (2004). *Pharmaceutical Biotechnology: Drug Discovery and Clinical Applications*. New York: John Wiley & Sons.

KI. (2002).*The Worldwide Market for In Vitro Diagnostic Tests*. Hartford, CT: Kalorama Information.

McKelvey, M. D., Rickne, A., and Laage-Hellman, J. (Eds.). (2004). *The Economic Dynamics of Modern Biotechnology*. Cheltenham, U.K.: Edward Elgar Publishing.

Mehta, S. (2008). *Commercializing Successful Biomedical Technologies: Basic Principles for the Development of Drugs, Diagnostics and Devices*. Cambridge, U.K.: Cambridge University Press; *www.commercializingbiotech.com*

Moore, J. E., and Zouridakis, G. (Eds.). (2004). *Biomedical Technology and Devices Handbook*. Boca Raton, FL: CRC Press.

NCFAP. (2004). *Biotechnology-derived Crops Planted in 2005: Impacts on US Agriculture*. Washington, D.C.: National Center for Food & Agricultural Policy.

NRDC. (2006). Natural Resources Defense Council and Climate Solutions; *http://www.nrdc.org/air/transportation/ethanol/ethanol.pdf*

Porter, M. (1985). *Competitive Advantage: Creating and Sustaining Superior Performance*. New York: The Free Press.

Ratledge, C., and Kristiansen, B. (Eds.). (2001). *Basic Biotechnology* (Second Edition). Cambridge, U.K.: Cambridge University Press.

Roco, M. C., and Bainbridge, W. S. (Eds.). (2004). *Converging Technologies for Improving Human Performance*. New York: Springer-Verlag.

Sasson, A. (2006). *Medical Biotechnology*. New York: United Nations University Press.

Websites

www.bio.org—Biotechnology Industry Organization.

www.phrma.org—Pharmaceutical Research and Manufacturers Association.

www.medicaldevices.org—Medical Device Manufacturers Association.

57

The Continuing Economic Potential of Nanotechnology

Lois Peters

Rensselaer Polytechnic Institute

Introduction[1]

Nanoscience and nanotechnology involve studying things smaller than 100 nanometers which is at the level of atoms and molecules. One nanometer is one-billionth of a meter and a single human hair is around 80,000 nanometers in width while a strand of DNA is 2.5 nanometers wide. A gold atom is one-third of a nanometer. Working at the nano level is of interest because the physical and chemical properties of materials at the nanoscale can be novel in ways that have economic potential. For example, the strength-to-weight ratio of carbon nanotubes is superior to diamonds, a crystalline carbon material. Nanotubes are 100 times stronger than steel. In the realm of nanotechnology the position of a single atom can make all the difference—whether a material functions as a semiconductor or an insulator, whether it triggers a vital chemical process or stops it cold.

More specifically, nanoscience is the study of phenomena and manipulation of materials at atomic, molecular, and acromolecular scales, where properties differ significantly from those at a larger scale. Nanotechnology refers to a field of applied science and technology whose theme is the *control of matter* on the atomic and molecular scale, generally 100 nanometers or smaller, and the fabrication of devices or materials that lie within that size range. It is nanotechnology that is of most immediate interest to business.

According to policy-makers, the development of nanotechnology is the latest mega-trend in science and engineering. It will bring a wave of radical innovation and, by enabling fundamentally new means of production, spark a new industrial revolution (ICON, 2008; Siegel, Hu, and Roco, 1999; WWCIS, 2008). The potential for broad-based societal impact stems from predictions that nanotechnology will become the basis for remarkably powerful and inexpensive computers, enabling cost-effective alternate energy technologies, fundamentally new medical technologies that could save millions of lives, water treatment and remediation technologies that will address the global problem of water scarcity, making sensors that are important in military applications as well as environmental protection, and development of new zero-pollution manufacturing methods that could create greater material abundance for all (BIAC, 2009; Roco, Williams, and Alivisatos, 1999). As of August 21, 2008 the Project on Emerging Nanotechnologies (PEN) estimated there were 803 nano-enabled products on the market as compared with 212 in March 2006 (PEN, 2009). This reality and the pace of scientific and technological development makes observers increasingly convinced that nanotechnology will have a significant impact on society, business, and economic growth.

Scope of nanotechnology

Nanotechnology is highly inter-disciplinary and draws on chemistry, biology, molecular biology, quantum physics, biochemistry, materials science, as well as electrical engineering and chemical engineering among other fields. The inter-disciplinary nature of the field has implications for the importance and complexity of scientific and business collaborations, education and training, as well as market entry. The concept of nanotechnology covers a group of technologies or tools involved in the design, characterization, production, and application of structures, devices, and systems by controlling shape and size at the nanometer scale. The scanning tunneling microscope (STM) and atomic force microscope (AFM) are examples of such tools. They can be used to manipulate atoms and produce three-dimensional surface images at the nanoscale.

[1] This work was supported primarily by the Nanoscale Science and Engineering Initiative of the National Science Foundation under NSF Award Numbers DMR-0117792 and DMR-0642573.

Materials at the nanoscale are often a function of their surface properties, and this can result in changes in conductivity, absorption, bioavailability, stability, strength, and reactivity as compared with materials structured at larger sizes. Quantum-mechanical effects come into play at nanoscale dimensions and lead to new physics and chemistry (e.g., in optical and magnetic properties). Basically, there are two approaches of interest for producing nanotechnology-enabled products. The first is the top-down approach, meaning the refinement of practices and techniques such as cutting, carving, and molding to the point that they operate at the nano level. The second is the bottom-up approach that focuses on building materials atom by atom. Modifying semiconductor techniques, such as lithography, to go from constructing transistor circuits or features at the micrometer level to the nano level is an example of the top-down approach, and constructing a sensor by placing a bioactive molecule on a nanotube is an examples of the bottom-up approach.

Another distinction within the nanotechnology domain is an evolutionary line of attack to its development versus a revolutionary one. Much of nanotechnology development can be viewed as recasting of research in surface chemistry, colloids, materials, and semiconductors that leads to improved product materials and tools. Research within this tradition can be characterized as nanoscale technology defined as *anything* with a size between 1 and 10 nanometers with novel properties. The more controversial domain is molecular manufacturing or engineering of functional systems at the molecular scale. Molecular manufacturing is defined as the use of programmable chemistry to build exponential manufacturing systems and high-performance products. It is controversial because in the past scientists have claimed it is theoretically impossible and because early in its inception much was made of its potential for creating out-of-control self-replicating robots that could consume all matter on Earth (see the section on issues and challenges). Over the past decade researchers have attempted to shore up theoretical arguments for molecular manufacturing. This has led some analysts to state that major progress in this area might come in the next 5 years. Others believe major breakthroughs are at least three decades away and still others hold firm on its impossibility (Pandze and Holt, 2007).

History and scientific developments

Key milestones

Not all nanotechnology is new. Carbon black, a mix of nanoscopic carbon particles, has been used in rubber for decades to make it tougher. Recognition that much of nanotechnology is about the study of colloids (e.g., coffee is a colloid) and solid gel phase changes, leads to the claim that the pathway to nanotechnology started with research on aspects of the production of fertilizers such as potash or on ceramic science. Others point to research in the early 1970s on the workings of the biological cell as early nanotechnology research. This research characterizes biomotors within the cell, such as mitochondria (organelles in the cell that produce energy, often referred to as the powerhouse of the cell) that create new molecules and immediately place them in a structural position that allows unique functioning. Still others cite investigations by Joseph Proust in 1799 demonstrating that chemicals tended to combine in particular atomic ratios as the starting point of nanotechnology . Some even date the early Roman ages as the beginning of nanotechnology applications. The Lycurgus Cup is one example from this period whereby gold nanoparticles present in the glass cause this ancient object to change color when subjected to different angles of light.

The origin of recent scientific and commercial interest in nanotechnology, however, is most often traced back to a lecture entitled "There is plenty of room at the bottom" given by Richard Feynman (1959). This validated the idea that it was important to explore novel material properties at the nanoscale level. Norio Taniguchi first used the word nanotechnology in 1974 to refer to "production technology to get the extra high accuracy and ultra fine dimensions, i.e. the preciseness and fineness on the order of 1 nano meters." Independently, Eric Drexler introduced the term nanotechnology in 1986 in his book *Engines of Creation* which drew on his concepts of molecular nanotechnology developed in the late 1970s at MIT. Drexler's book popularized the term nanotechnology and drew attention to one area of nanotechnology research—molecular manufacturing—which we described as the revolutionary approach to nanotechnology in the previous section.

The challenge of not being able to reach down into the nano-world was met in 1981 with the invention of the scanning tunneling microscope (STM) by IBM Corp.'s Heinrich Röhrer and Gerd Karl Binnig. They received the Nobel Prize 5 years later. Then, in 1986 the development of the atomic force microscope (AFM) allowed production of nanoscale topographical maps. In 1989, IBM scientists in Zurich, using the tip of a scanning tunneling microscope, demonstrated that it is possible to precisely position 35 xenon atoms to spell "IBM". A demonstration of nanocompounds with unique properties came about when the fullerene was discovered in 1985 by Robert Curl, Harold Kroto, and Richard Smalley at the University of Sussex and Rice University. Smalley named this 60-atom cage of carbon a Buckminster fullerene or a buckyball after Richard Buckminster Fuller, whose geodesic domes it resembled. In 1991, Sumio Lijino a researcher at NEC, Tsukuba, Japan, discovered carbon nanotubes while attempting to synthesize variants of buckyballs by passing an electrical discharge between graphite rods in a chamber filled with helium. Since Lijino's discovery, carbon nanotube tech-

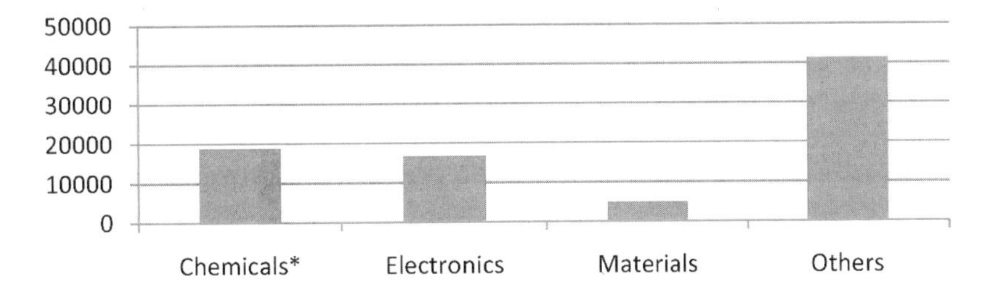

Figure 57.1. Number of nano-patents 1976–2002 by industry sector.
Source: Huang *et al.* (2003).
*Includes chemical catalysts and pharmaceuticals.

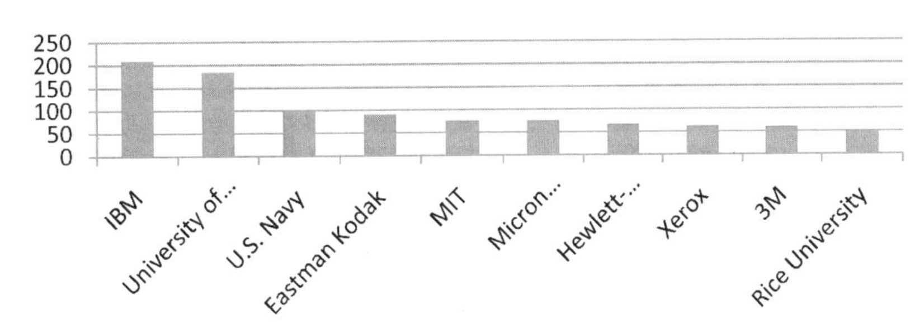

Figure 57.2. Number of nano-patents by institution 1976–2006.
Source: Chen *et al.* (2008).

nology has overtaken buckyballs in terms of commercial potential. Stiffer than any other known material, carbon nanotubes also come with electrical properties that demonstrate the potential for future development of nanoelectronic devices, sensors, and computers.

Colloidal semiconductor nanocrystals, or quantum dots, are important as molecule-sized LEDs and can be used as probes to track antibodies, viruses, proteins, or DNA within the human body. A key discovery with regard to quantum dots came in 1983 when Louis E. Brus noticed how conductivity changed with the particle size of materials. In 2008, researchers from IBM and Purdue University discovered that tiny structures called silicon nanowires might be ideal for manufacturing in future computers and consumer electronics because they form the same way every time.

Recent achievements in the fields of micro-systems and nanotechnology continue to demonstrate their value as technology platforms for a broad array of applications. One example is tiny sensors suitable for detecting a multitude of environmental parameters. Another development is that of radio-sensors that can transmit pressure or temperature changes across several meters without any source of power. Much current attention has been on biomimetic nanotechnology fueled in part by a book called *Soft Machines: Nanotechnology and Life* (Jones, 2004). It proposes the development of nanomachines by using the lessons learned from biology on how things work. Biomimetic nanotechnology is poised to make important contributions to the new field of synthetic biotechnology.

Parallel to technological developments there have been important supportive and infrastructure initiatives. The first university course in nanotechnology was given at Stanford University in 1988, and the *Journal of Nanotechnology*, the first nanotechnology journal, was published by the Institute of Physics in 1990.

Patents

Evidence of the increasing importance of nanotechnology research results, as a stimulus for economic growth, is reflected in the increase of "nano" patents. Early nanotechnology patent analysis involved searching the entire issued patent description for the term nano, because at the time the U.S. Patent and Trademark Office (USPTO) had no separate nanotechnology classification. One such analysis (Huang *et al.*, 2003) using nano and nanotechnology terms, such as quantum dot, adopted in previous NSF database searches identified 89,153 nano-patents in the USPTO database issued over the period 1976–2002. Tullis (2005) conducting a similar search but for the years 1976–2005 found 96,312 nano-patents. Using the term nano in a patent search from 1976 to March 17, 2009 returns 141,679 patents.

Huang's study indicates that the majority of patent classifications fall into chemistry, pharmaceuticals, and electronics. The high number in the "other" classification, however, represents the great variety in the areas in which patents are filed (Figure 57.1). For the years 1976–2002 U.S. assignees held the majority of patents. Both these results continue to hold.

As of 2006 IBM had the greatest number of nano-patents. Universities have been quite active in obtaining nano-patents (as shown in Figure 57.2) where out of the top-10 nano-patenting organizations 3 are universities.

While key nanotechnology patents can frequently originate in universities—individuals, high-tech startups, and multinational firms such as IBM and NEC are also key players (Tables 57.1 and 57.2).

Table 57.1. Examples of key nanotechnology patents.

Organization/Assignee	*Technology*	*Patent No.*
IBM	Scanning tunneling microscope	4343993
Northwestern University	Dip pen nanolithographic printing	6827979
NEC	Nanotubes	5747161
Berkeley and MIT	Quantum dot	5505928 and 6322901
Cambridge	Thin films of conjugated polymers	5247190
Berkeley	Biomimicry adhesive based on gecko lizard observations	6737160
Zyvex	Non-biological self-replicators	6510359
James Dempsey, Oshkosh, Wi	Space elevator	6981674

Source: Foley & Lardner LLP (2007).

Table 57.2. Projected government funding for nanotechnology by country/region.

Region/country	*Projected funding period*	*Billions of dollars*
U.S.	2007–2009	4.44
EC	2007–2013	5.1
Japan	2006–2010	5.36
China	2006–2010	3.3
South Korea	2001–2010	1.5

Source: compiled from public sources such as the NSF, the EC, the OECD, and national newspapers.

Recently the USPTO came out with a nanotechnology classification which returns 5,396 issued patents from 1976 to March 17, 2009 (Figure 57.3).

This analysis also shows growth in nanotechnology patents, but the absolute numbers are much smaller than earlier studies. Because there is no standard definition of nanotechnology, estimation of the number and quality of nanotechnology patents varies considerably.

Nanotechnology funding

Worldwide

Over 35 countries have national initiatives in nanotechnology (Peters and Sundarajaran, 2008). Global funding for nanotechnology research reached $25 billion in 2008 according to figures from Científica (*The Nanotechnology Opportunity Report*, 2008) as compared with the $4 billion worldwide R&D spending in 2001 reported by the same group. Lux Research (2009a) provides a somewhat lower estimate of global R&D in 2008, $18.2 billion. Broad investment by all sectors associated with nanoscience or nanotechnology has been increasing globally. According to Lux Research (2009a), government spending was $8.4 billion, corporate funding was $8.6 billion, and venture capital provided $1.2 billion in 2008. This compares with their figures for 2004 indicating that worldwide spending was $8.6 billion with government, corporate, and venture capital spending at $4.46 billion, $3.69 billion, and $412.5 million, respectively. Of note is the fact that corporations now spend more on nanotechnology R&D than governments. Table 57.2 shows projected government funding for nations/regions with the most significant investment in nanotechnology research.

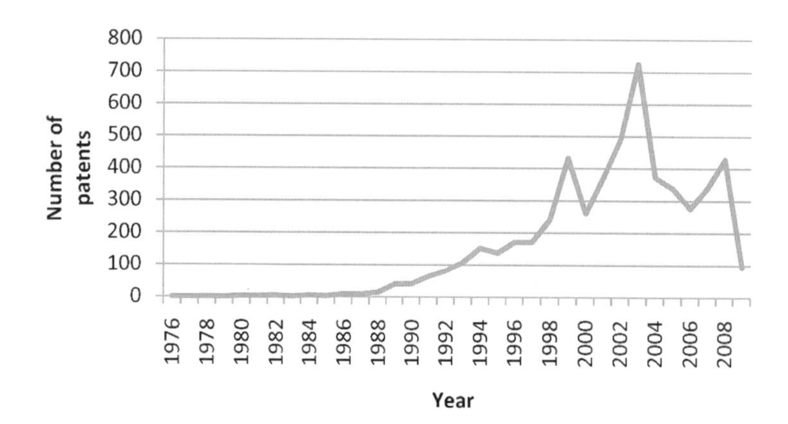

Figure 57.3. Nanotechnology patents by year using USPTO classification: 977. *Source*: USPTO.

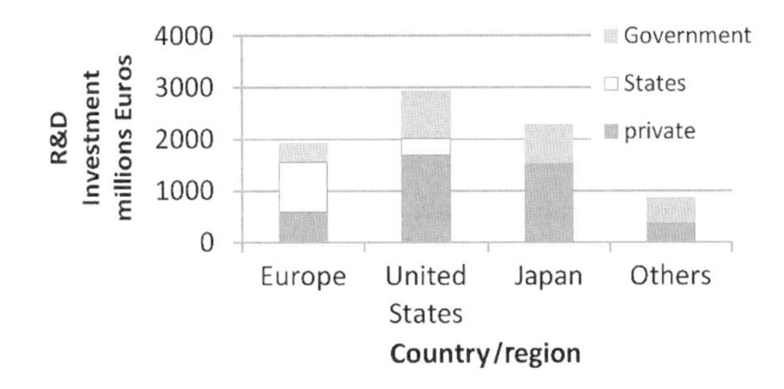

Figure 57.4. Worldwide public and private expenditures in 2004.
Source: Lux Research.

India kicked off a nanotechnology initiative in 2001, and Taiwan launched a 6-year plan in 2003. The Indian Government initiated a 5-year national mission called the NanoScience and Technology Mission (NSTM) with an allocation of $245 million over 5 years starting in 2007 (*www.nanowerk.com/news/newsid=1887.php*, last accessed March 23, 2009). In 2007, Russia announced that it will spend over U.S.$1 billion over the next 3 years on equipment for nanotechnology research. While Russia has been accused of nanotechnology posturing there is evidence that this money will actually be spent (Trader, 2007).

In 2004 the U.S. was clearly the leader in nanotech spending in terms of both public and private spending (Figure 57.4). Recent analysis, however (Frost and Sullivan, 2005; Jayarajah, 2008; Lux Research, 2009a) suggests this may be changing.

According to these analysts, in 2006 Europe (which includes more than just EC spending) outspent the U.S. in government nanotechnology R&D, and Asian companies spent the most on corporate nanotech R&D. Lux Research suggests that the growth rates for corporate and government spending in both Europe and Asia exceeds that of the U.S.

International comparisons, however, should proceed with caution for a number of reasons. Among them is lack of consensus about what to include in the domain of nanotechnology research. For example, some would include all research at the level of 100 nanometers or less, others would only include research on new materials such as carbon nanotubes, quantum dots, and nanowires. Others confine nanotechnology research to molecular manufacturing. Another complication is lack of R&D-specific exchange rates. This requires using consistently compiled and available data for a large number of countries. The choice is therefore between using market exchange rates (MERs) and purchasing power parities (PPPs). The quality of PPPs for developing nations is questionable because R&D performance is often concentrated geographically and the costs of goods or services in these regions are substantially greater than for the country as a whole. Finally differences in national systems of innovation may make one country more effective than another in translating investments in science and technology into economic growth or other social benefits (NSB, 2008). For example, there can be significant investment in basic R&D but the appropriate links to commercial entities are weak and therefore place a roadblock to innovation and commercial development.

U.S. funding

In 2001 the National Nanotechnology Initiative (NNI) was established to coordinate the U.S. Government's work in nanotechnology. At the time this large-scale integrated program made the U.S. effort unique. Since its inception the NNI has channeled about $10 billion federal into nanotechnology research performed by universities, research institutes, government laboratories, and companies. The NNI coordinates the nanotechnology-related activities of 25 federal agencies, 13 of which have budgets for nanotechnology research and development. The NNI was up for reauthorization in 2008 but got sidelined in the Senate because of the 2008–2009 financial crisis. It was reintroduced in the House and passed in January 2009. The NNI bill is currently pending approval in the Senate and is expected to pass. The reauthorization mandates a greater focus on nanotechnology environmental, health, and safety topics and provides for increased research funding in this area. Figure 57.5 shows that U.S. federal funding for nanotechnology is set to rise to an estimated $1.527 billion in 2009; about three times that of the estimated $464 million spent in 2001. Currently there are eight program areas, and investment by program area is shown in Figure 57.6. Figure 57.7 shows the growth in U.S. small-business innovation awards.

Institutions and infrastructure

Overview

The growth of organizations engaged in nanotechnology has been considerable (as seen in Figure 57.8). Based on our own data collection, through NSF support (Peters,

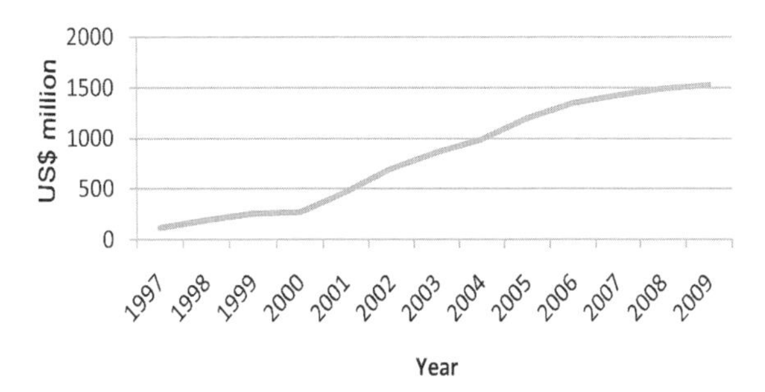

Figure 57.5. U.S. Government nanotechnology R&D funding by year.
Source: Huang *et al.* (2003); NNI (2009).

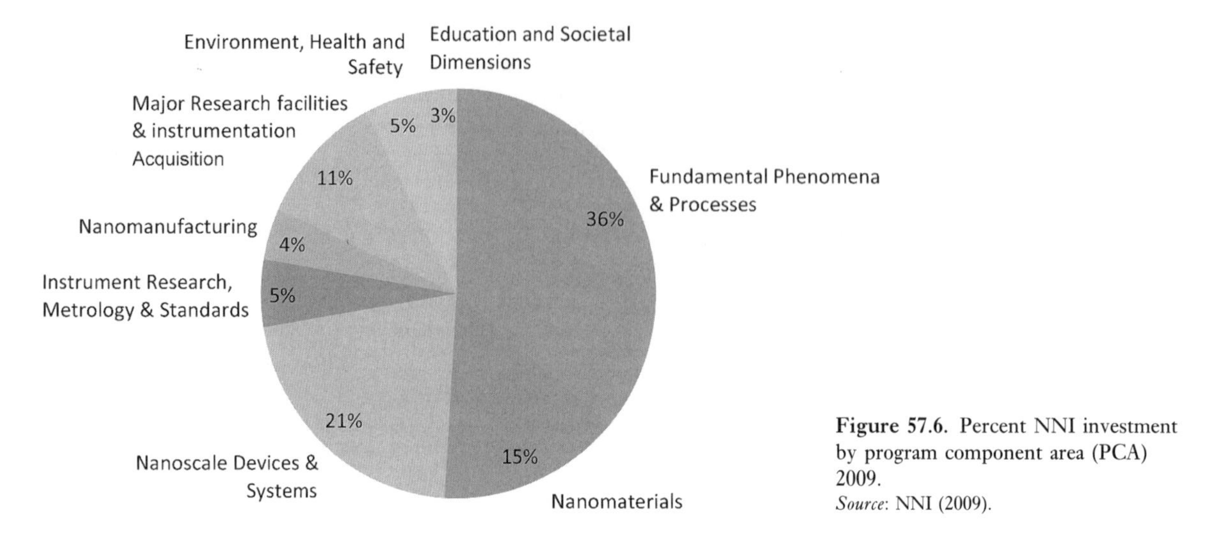

Figure 57.6. Percent NNI investment by program component area (PCA) 2009.
Source: NNI (2009).

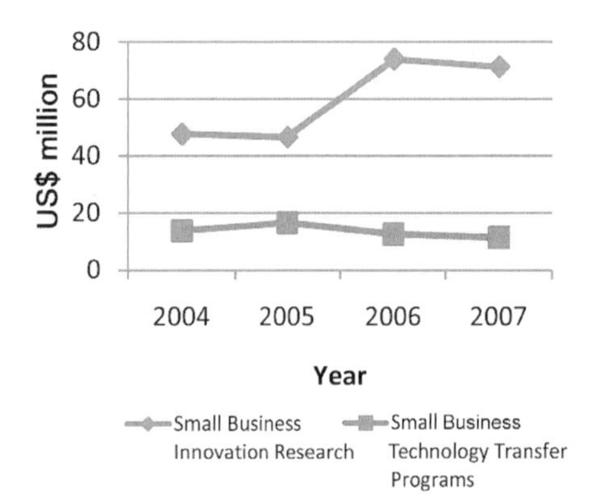

Figure 57.7. 2004–2007 agency SBIR and STTR awards.
Source: NNI (2009).

2009), the Nanotechnology Company Directory (*http://www.nsti.org/companies*, last accessed March 16, 2009) and the NanoVip Nanotechnology Directory (*http://www.nanovip.com*, last accessed March 16, 2009), we estimate there are over 2,366 organizations worldwide with significant involvement in nanotechnology endeavors.

Like other nanotechnology estimations, numbers can vary widely because of variations in the definition. There are a great variety of players including consultants, investment firms, information organizations, government laboratories, research institutes, universities, legal organizations, watchdog institutions (such as the Foresight Institute and the Center for Responsible Nanotechnology) as well as startups and multinational firms. It should also be noted that many organizations changed their name to include nano (e.g., InfectTech Inc. changed their name to NanoLogix Inc. given the great worldwide interest in this area), and there are many organizations that have nano in their name but do not really engage in nano activities. There is also a robust merger and acquisition activity (e.g., the merger of Carbon Nanotechnologies Inc., or CNI, and Unidym in 2007).

Universities and government-sponsored laboratories

Worldwide there are more than 185 academic institutions with major efforts in nanotechnology. Turning academic research into products, however, remains a challenge because while universities produce extraordinary technologies they still need to be developed to fit the market or solve a compelling problem (Thayer, 2008). Creating and maintaining research facilities and instrumentation is one of eight areas of investment highlighted in the NNI (2009) strategic plan (see previous section). Since 2005 this component has accounted for about 11% of annual spending or about $161.3 million in the proposed 2009 NNI budget. Combined federal, state, and local investment in facilities approaches several hundred million dollars per year. More than 60 centers, networks, and facilities funded by at least 7 agencies reporting under the National Nanotechnology Initiative are in operation or soon to open. Some estimates place the total number of center initiatives including those under the NNI at more than 120 (Thayer, 2007). The Cornell Nanobiotechnology Center, the NSF Science & Technology Center, and the Rensselaer Polytechnic Institute (RPI) Center for Directed Assembly of Nanostructures are examples of NSF-funded centers. Northwestern University has one of eight National Cancer Institute Centers for Cancer Nanotechnology Excellence and UC Berkeley's Molecular Foundry at Lawrence Berkeley National Laboratory is one of five national lab–based user facilities. The Nanotechnology University at Albany is an example of

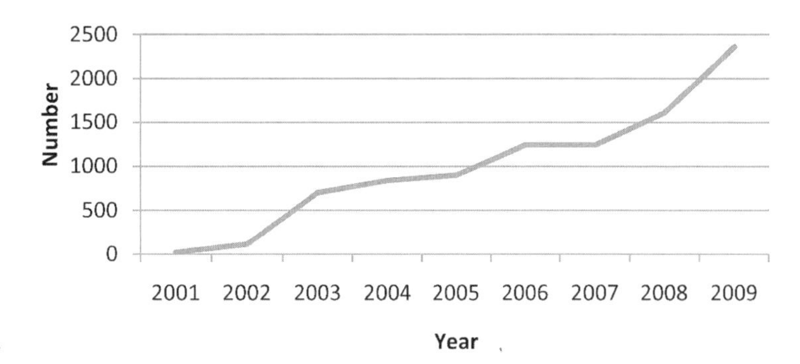

Figure 57.8. Number of nanotechnology-related organizations worldwide by year. *Source*: Peters (2009) and nano institution databases.

a large state-funded initiative. The Center for Functional Nanomaterials at Brookhaven National Laboratory is an example of a new Department of Energy center The Cornell NanoScale Science & Technology Facility is an example of a retrofit of an old center founded in 1977 that originally focused on sub-micrometer work. Today it is one of 13 NSF nanofabrication user facilities in the National Nanotechnology Infrastructure Network (NNIN).

Multinational corporations and high-tech startups

A total of 1,186 firms identified themselves as belonging to the nanotechnology industry as of March 2008. Some of these companies are multinational companies, such as IBM, DuPont, PPG, Kodak, who now have big investments in nanotechnology platforms. Others are companies who have renamed themselves or redirected themselves to take advantage of nanotechnology's commercial potential. Many multinational companies are active in the nanotechnology domain mainly to enhance their existing businesses. An example is DuPont's Light Stabilizer, which provides sun protection for plastics. Others are called "pure play" or "nanotechnology specialist" companies because they focus only on nanoproducts. These are the companies that produce single and multi-walled nanotubes, fullerenes, quantum dots, and other new materials. Examples of such companies are provided in Table 57.3. These are the types of companies that are more likely to be seeking to capture value from the more revolutionary aspects of nanotechnology. However, it should be noted from Table 57.3 that large multinational companies are also key players in each of the identified nanotechnology platforms. By March 2009 there were 114 nanotechnology stock companies. Of these 14% were multinational corporations and 66% were small high-tech startups. This is consistent with other estimates of the percentage small companies represent of total organizations listed in the nanotechnology sector.

The Project on Emerging Nanotechnologies (PEN) has mapped 800 U.S. organizations which include 637 companies, 138 university and government laboratories, and 45 other types of organizations. The resulting interactive map can be accessed at *http://www.nanotechproject.org/inventories/map/* (last accessed March 21, 2009). Their analysis concludes the following:

- The top-four nanotechnology states are California, Massachusetts, New York, and Texas (each with over 50 entries).
- The top-five "nano metro" areas are San Jose, CA; Boston, MA; San Francisco, CA; Oakland, CA; and Middlesex-Essex, MA (each with over 20 entries).
- Nanotechnology companies are working in three main sectors—materials; medicine and health; and tools and instruments—each with over 100 entries.
- The number of U.S. universities and government laboratories working on some aspect of nanotechnology is significant, with 138 identified.
- In all, 47 of 50 states and the District of Columbia contain at least one company, university, government laboratory, or organization working in nanotechnology, showing that nanotechnology activity is occurring throughout the U.S.

Economic foundations of nanotechnology

Differing perspectives

There is a continuing debate about whether nanotechnology is an industry or technology platform. This goes back to 2001 when the CEO of Científica, Tim Harper, predicted there would not be a nanotechnology industry. According to Científica's (2008) *Nanotechnology Opportunity Report* there is evidence to support this view because of the many failed nano initial public offerings (IPOs) and the collapse of pure play nanotechnology companies. In alignment with this perspective, some suggest that it is inappropriate to think of nanotechnology as an industry because it is really a collection of tools and techniques that enables technology and commercial development across many industries. Also consistent with this view is the

Table 57.3. Nanotechnology platforms.

Nanotechnology platform	*Definition*	*Number of patents using terms and nanotechnology classification 977 in the USPTO database*	*Examples of companies with expertise in the identified platform domain*
Carbon nanotubes	Chickenwire-like tubes of carbon that can act either as semiconductors or as metals. They can be single or multi-walled. They exhibit extraordinary strength and unique *electrical* properties, and are efficient *conductors*. They are members of the fullerene structural family	1,284	IBM, Carbon Nanotechnologies Inc. (CNI) (now Unidym after merger with that company), Intel, NEC
Quantum dots	Nanometer-scale crystals of semiconductors, such as cadmium selenide	457	TI, Quantum Dot Corporation, Evident Technologies
Fullerenes	Spherical cages of carbon such as C_{60}, also called buckyballs	400	
Nanowires	Wires and rods with the diameter of the order of 1 nanometer. They can be made of one or more semiconductors, such as gallium arsenide or indium phosphate or metals, including metallic (e.g., Ni, Pt, Au), insulator (e.g., SiO_2,TiO_2), or repeating molecular units either organic (e.g., DNA) or inorganic (e.g., $Mo_6S_{9-x}I_x$)	304	Nanosys, Nanolane Agere Systems (merged in 2007 with LSI Corporation), Hewlett Packard, Samsung Electronics, Sharp, Intel
Dendrimers	Repeatedly branched *molecules* or spherical polymers used experimentally for drug delivery and as light emitters	115	Dow, Exxon Mobile, 3M, GE, Dendritic Nanotechnologies, Nano Cure Corporation, Starpharma
Nanocomposites	Multiphase solid materials where one of the phases has a dimension of less than 100 *nanometers* (nm)	76	Eastman Kodak, GM, 3M, GE, ExxonMobil, Lucent, Goodyear
Nanoparticles	Particles sized between 1 and 100 nanometers which may or may not have special properties (includes metal and ceramic nanoparticles)	1,083	3M, Nanosys, Eastman Kodak, Micron Technology, GM, Xerox, Entegris Corning, Sirtris Pharmaceutical (acquired by Glaxo Smith Kline in 2008)
Aerogels	Low-density *solid-state* material derived from *gel* in which the liquid component of the gel has been replaced with gas. The result is an extremely low density solid with several remarkable properties, most notably its effectiveness as a thermal insulator	69	Nanosys, Novellus Systems Inc., NanoProducts Corporation, ExxonMobil, Nicron Technology, Cabot

Sources: Foley & Lardner LLP (2007), Moore (2005), Peters (2009), USPTO.

claim that nanotechnology is potentially a general purpose technology (Palmberg, 2008; Pandze and Holt, 2007), meaning that it is an innovation leap that will redefine society and therefore be important for all industries. Other examples of general purpose technologies are the steam engine, railroad, electricity, electronics, the automobile, the computer, and the Internet.

Nanotechnology platform definitions and patent numbers are identified in Table 57.3 and indicate that both small companies and large established companies are key players.

Nevertheless, there are ample predictions of the 20-year growth of nanotechnology as an industry. A justification might be that carbon nanotubes and other new materials, such as quantum dots, nanowires, etc., are the foundations of an industry sub-sector of new materials. Also if you consider an industry in terms of a value system of suppliers, buyers, rivals, etc., there is justification for a nanotech industry. Certainly a case can be made for an industry ecosystem of enterprises seeking to create value for society through nanotechnology.

Part of this debate relates to the varying definitions of nanotechnology (e.g., something that just happens to have dimensions of a few nanometers). Under such a definition the revolutionary aspect of nanotechnology—which in fact could create a "new industry sector" whether through new materials or molecular manufacturing—is dwarfed.

Impact on existing industries

The main industry segments and sub-sectors impacted by nanotechnology are as follows (Glapa, 2002; Peters and Sundarajaran, 2008):

- Tools and devices: these include imaging and fabrication devices, molecular switches, nanodevices and systems. An example is the atomic force microscope (AFM),
- Materials: this includes new materials such as nanotubes, fullerenes, nano-powders, ceramics, and new chemical manufacturing know-how that allows production of thin film, coatings, nanocomposites, etc.
- Biotechnology: which includes nanobio a connection between molecular biology and nanotechnology that involves nano approaches to drug delivery, diagnostics, molecular biology, bionanodevices and systems, etc.
- Other areas: such as modeling and software that includes nano-simulation (Nano Cad), virtual reality (CAVE), etc.

Market estimates of the impact for different industry segments are given in Table 57.4. Lately much has been made about nanotechnology activity in the energy and environment sector but it barely makes a dent in total emerging nanotech revenue, amounting to just $876 million or 0.6% of total 2007 nanotechnology revenues.

The breadth of the industries that will be affected by nanotechnology is very large, and it is expected that some platform nanotechnologies will serve more industries in improved manufacturing and product performance enhancement than others. The problem that this complexity imposes for market forecasters is multifold. Market forecast focused on a certain industry will not reveal the entire demand picture for a product. It seems that it would be appropriate to integrate cross-industry factors in order to find the aggregate demand for the different nanotechnology markets by building value chain verticals in each market. The difficulty, however, is that products in certain markets have different values associated with nanotechnology, and different industries are expected to have different values for the same technology.

Table 57.4. Industry impact (millions of dollars).

Sub-sectors	*Rank market size*	*Market size 2004*	*Rank growth*	*Growth (%) 2015*
Materials	1	440	2	23.8
Tools and devices	2	370	1	47
Nanobio	3	210	3	20
Others	4	65	NA	NA

Sources: Murdock (2002).

The nanotechnology value chain

A value chain based on the current state of nanotechnology identifies three key segments that are underpinned by a tools and instruments sector (Figure 57.9). It is nanotechnology intermediates that are projected to have the most value. In 2007 nano-intermediates garnered an aggregate net profit margin of 9%, almost twice that of nanomaterials and nano-enabled products, a figure that will expand to 15% in 2015 according to Lux Research (2009a).

As mentioned earlier there are large discrepancies over the definition of nanotechnology and subsequently its markets (i.e., whether it is nano-sized particles or any product influenced by nanotechnology). Based on the latter very broad definition and a long-range view of nanotechnology we see (Figure 57.10) a potential value chain of the future nanotechnology industry.

Nanotechnology products

Current status

According to Lux Research, nano-enabled products in 2006 resulted in $50 billion of product sales and $147 billion in 2007 (Lux Research, 2009a), while Científica place the value of nanotechnology products in 2008 at $166 billion. Products that currently incorporate nano-

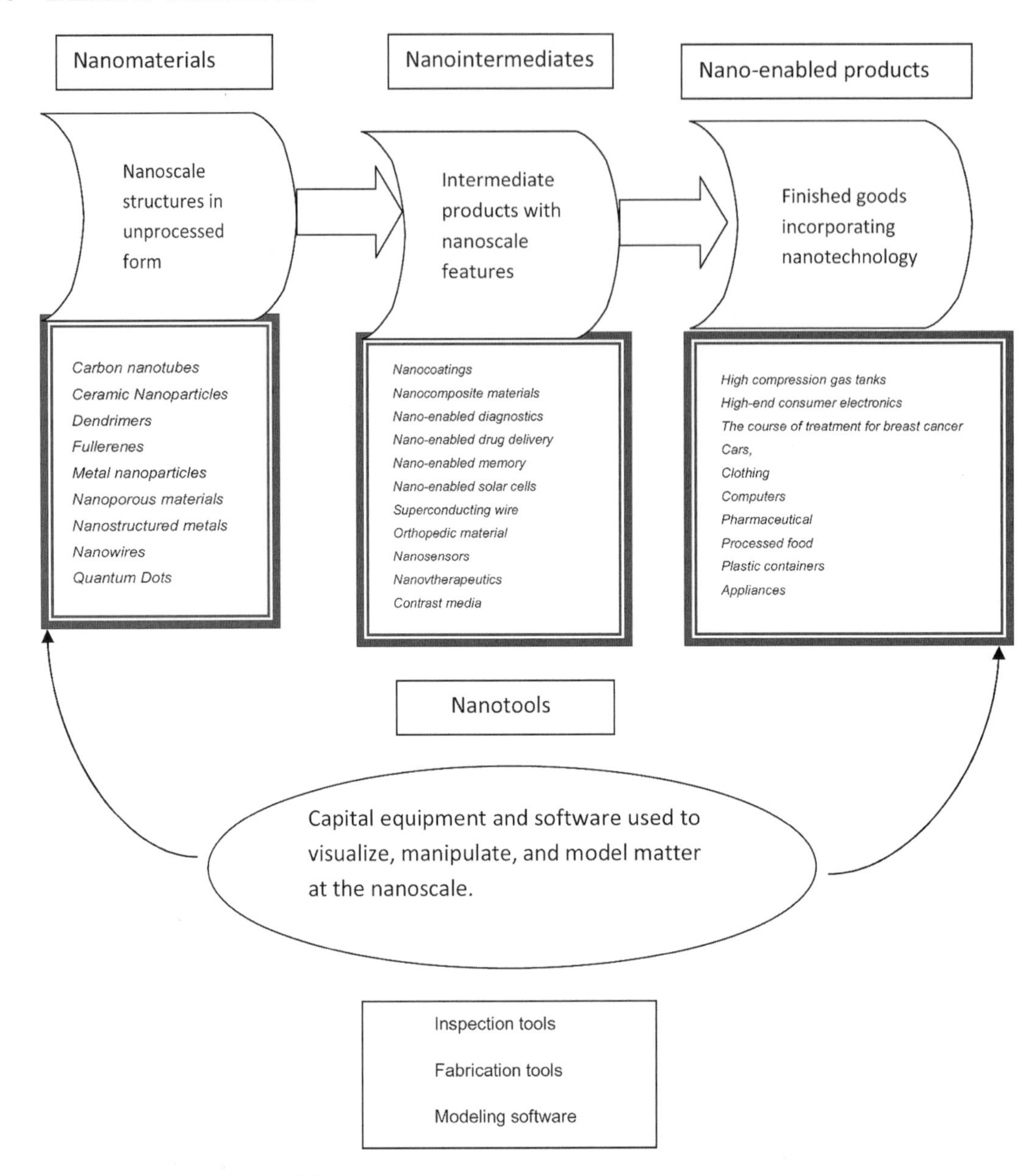

Figure 57.9. Nanotechnology value chain.
Source: adapted from Lux Research.

technology are numerous, and include stain-repellent and wrinkle-resistant threads, high-performance ski wax, deep-penetrating skin care, OLED digital cameras, high-performing sun-glasses, nano-socks, and high-tech tennis rackets and balls. A public inventory of nanotechnology consumer products (PEN, 2009) reveals that as of August 21, 2008 the majority of 903 products fall in the category of health and fitness and that the U.S. is the region of origin of the greatest number of products, 426, followed by East Asia with 227.

Future projections of nanotechnology products

The market growth for all nanotech products is estimated to be 44% over the next 12–15 years (Figure 57.11). Nanotechnology systems and products with high-tech complexity and regulatory challenges are further along in the time horizon.

Market growth

In 2000 the NSF predicted that by 2015 the nanotech market worth would be $1 trillion. Since that time this

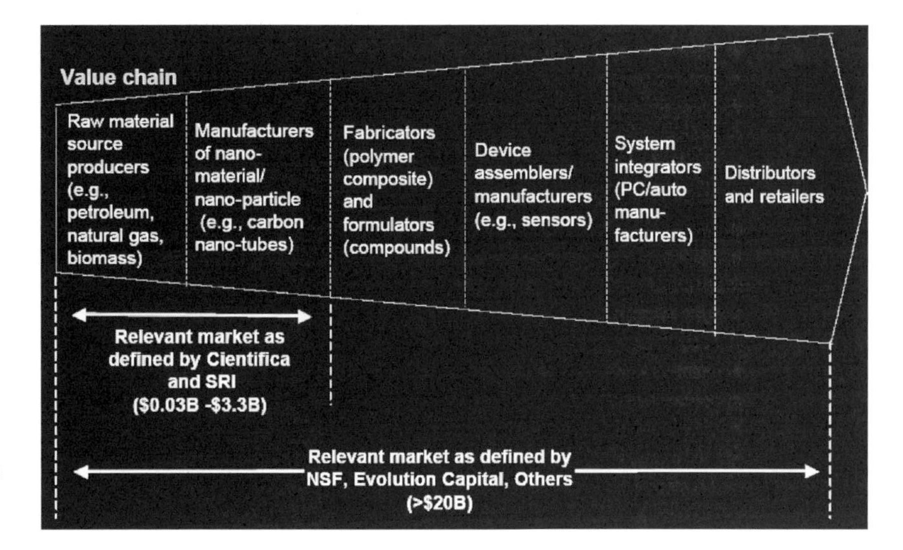

Figure 57.10. Based on the nanotech market definition eect on market forecasting.
Source: Sanghvi (2003).

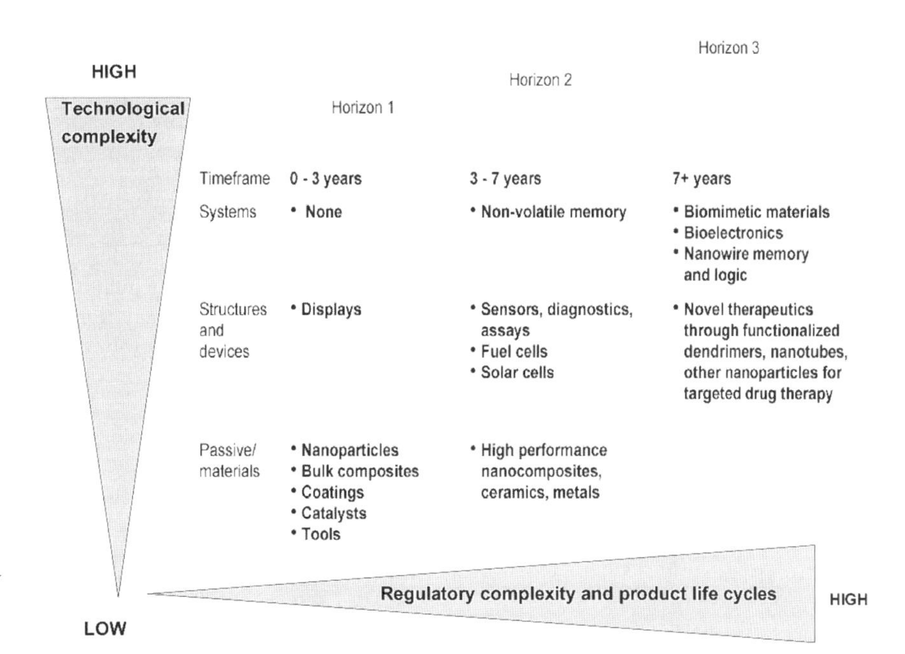

Figure 57.11. Nanotechnology market horizons.
Source: Murdock (2002).

figure has been much repeated and criticized. Forecasts for the nanotech market in 2015 range from $750 billion (WGR, 2004) to $3.1 trillion (Lux Research, 2009b). Científica (2008) states that the market will be $1.5 trillion in 2015 if you exclude semiconductors and $2.95 trillion if you include them.

Finding the right growth rate is not trivial, since the numbers vary widely from more optimistic growth rates of 100% in the chemical-manufacturing industry to the more pessimistic growth rates of 20% to 30% in other industries. It is a lot easier to forecast anything that is evolutionary since evolutionary nanotechnology is much closer to commercialization because the only barriers to adoption are materials performance, scalability, and price. However, as seen earlier, important breakthroughs in molecular/nano manufacturing are imperative in order to bring about any form of significant revolutionary changes in the industry. A collaborative effort between Battelle and the Foresight Institute resulted in a *Technology Roadmap for Productive Nanosystems* released in January 2007. It provides a foundation for moving towards atomically precise manufacturing and has contributed to increased recognition that engineering of functional systems at the molecular scale will be a reality in the foreseeable future.

Issues and challenges

There are five major challenges to nanotechnology development: technical; environment, health and safety; legal; education and training; and addressing societal fears.

Technical

With regard to such nanomaterials as carbon nanotubes the technical challenges include cost reduction, reliable yields, ability to control desired sub-assemblies, outcomes, and scalability. For example, carbon nanotubes tend to clump, whereas uniform dispersion is often required. The technical challenges to molecular manufacturing are still considerable and the few companies (e.g., Zyvex) involved are merely laying a foundation for its future.

Environment, health, and safety (EHS)

Recent research on the environmental and health impact of nanotechnology has tempered the excitement of some firms about the economic potential of nanotechnology and raised a call for a greater number and more rigorous studies of the environmental and health impact of nanotechnology (ICON, 2009).

In 2007 both the EU and the U.S. produced strategic nanotechnology plans highlighting the importance of increased research spending on health, safety, and environmental nanotechnology impacts. Understanding and managing the risks of nanotechnology accounted for $13 million in 2006 in the U.S. while European nations invested nearly $24 million (congressional testimony of Dr. André Naynard, April 16, 2008, in testimony to the House Science & Technology Committee hearing on the National Nanotechnology Initiative Amendment Act 2008.PEN). Analysis also highlights a substantial overinflation of the Federal Government's nanotechnology risk research investment figures for the U.S. A report by the National Research Council (NRC, 2008) was critical of U.S. investment in EHS and risk research (NRC, 2008). Legislation proposed by U.S. House of Representatives Science Committee Chair Bart Gordon (D-TN) amended the NNI act to include a minimum 10% mandate for the nanotechnology federal R&D budget devoted to EHS research in the future, amounting to approximately $150 million annually.

Legal

Within the legal domain there are significant challenges to intellectual property rights and developing useful regulations and standards. In several places we have called attention to the lack of consensus in what is covered by nanotechnology and its definition. This can be a problem for determining patent claims. It is generally accepted that the properties of matter and other fundamental scientific discoveries are not patentable. Hence a challenge is how to obtain a patent based on the discovery of the inherent properties of materials. Showing additional novelty or utility is key and this can relate to how nanotechnology is conceptualized. The cross-disciplinary nature of nanotechnology and its complexity requires scientific collaboration, and therefore patent filing will often involve teams of scientists both domestic and global, which can complicate patent ownership. Nanotechnology applications do not conform to existing classifications of intellectual property and therefore getting a patent to the right examiner can be a challenge. Furthermore, the USPTO is having trouble keeping pace with the nano-patent boom giving rise to uncertainty about intellectual property rights. Crowded and vulnerable patents in areas such as carbon nanotubes, quantum dots, dendrimers, etc. indicate there will be a lot of legal wrangling in the future. The fast-paced tempo of nanotechnology development and uncertainty about environmental health and safety realities makes it difficult to write regulatory laws and develop appropriate guidelines ensuring minimal threats to society.

Education and training

There are three important issues within this domain, one is attracting students to the fields that underpin nanotechnology, a second is building programs that can effectively deliver the broad cross-disciplinary education required to be effective in the nanotechnology domain. A third is eeducating the public. Research indicates that educating the public about nanotechnology causes people to become more worried and cautious about the technology (*http://www.azonano.com/news.asp?newsID =68520*, last accessed March 22, 2009).

Societal fears

Eric Drexler, author of *Engines of Creation* (1986), an early book about molecular manufacturing, coined the term "grey goo" in describing the potential danger of molecular manufacturing. *Engines of Creation* raised fears about nanotechnology research in general and set the stage for misunderstandings about nanotechnology which continue today. The "grey goo" concept was further sensationalized in the media (e.g., as body-altering "nanoprobes" used by *Star Trek*'s Borg or the malevolent self-replicating "nanobots" in Michael Crichton's 2002 novel. Drexler has since distanced himself from his initial conceptualization of nanobots overtaking the earth, stating that self-replication is not essential to molecular manufacturing and pointing out that focusing on "grey goo" draws attention away from much more pressing issues such as civic engagement, weapons competition, and arms control (Phoenix and Drexler, 2004). But the earlier Drexler version of advanced nanotechnology continues to spur public fears about nanotechnology centered on the idea that the technology could spiral out of control and take over the earth. Indeed, there may be potentially serious consequences from the development of nanotechnology including the risk of severe economic disruption, the possibly dehumanizing effect of using nanotechnology on ourselves, and potential criminal, military, or terrorist use of advanced nanotechnology. Many organizations are paying attention to this alongside investment in technical R&D.

Conclusion

While nanotech, on the one hand, is a very new area, claims have been made that it has already impacted markets (e.g., products like stain repellents and high-tech tennis rackets) and even that it has been around for 30 or more years (e.g., the first molecular electronic device was invented 30 years ago). More importantly, new discoveries such as carbon nanotubes and quantum dots are fueling excitement about future nanotech markets. A review of the growth of patents, company activities, interviews with company representatives, and discussions with nanotechnology scientists and engineers confirm that this excitement is grounded in reality. The reviews and discussions further suggest that nanotech's first impact will be in the areas of coatings, films, and sensors and thus rightfully ranking the materials sub-sector as having the largest market followed by the devices and then the nanobio sectors. Most agree that the impact will be significant and could be revolutionizing. The economic risk and potential of nanotechnology remains high. High returns will likely come from investment in new materials. Companies seeking to gain extraordinary value from nanotechnology need to develop capabilities for dealing with the uncertainties surrounding the generation of resources as opposed to focusing on the transactional capabilities associated with picking resources.

References

Battelle/Foresight Institute. (2007). Foresight Nanotech Institute and Battelle Memorial Institute. *Technology Roadmap for Productive Nanosystems*; *http://www.foresight.org/roadmaps/index.html* (last accessed December 14, 2009).

Chen, Hsinchun; Roco, M. C.; Lin, Xin; and Lin, Yiling. (2008). "Trends in nanotechnology." *Nature Nanotechnology*, **3**, 123–125.

BIAC. (2009). *Responsible Development of Nanotechnology: Turning Vision into Reality*. Paris: Business and Industry Advisory Committee, Organization for Economic Cooperation & Development; *http://www.biac.org/statements/nanotech/FIN09-01_Nanotechnology_Vision_Paper.pdf* (last accessed March 23, 2009).

Científica. (2008). *The Nanotechnology Opportunity Report (NOR)* (Third Edition). London: Científica Ltd.

Crichton, M. (2002). *Prey*. New York: Harper Collins.

Drexler, E. (1986). *Engines of Creation*. New York: Anchor Books.

Feynman, R. (1959). "There is plenty of room at the bottom." Lecture given by Richard Feynman to the American Physical Society at the California Institute of Technology December 29, 1959. Published in *Engineering Science*, February 1960, by the California Institute of Technology.

Foley & Lardner LLP. (2007). *Top Nanotechnology Patents: A Baker's Dozen*. Washington, D.C.: Foley & Lardner LLP; *http://www.foley.com/files/tbl_s31Publications/FileUpload137/3869/Top%20Nano%20Patents.pdf* (last accessed December 14, 2009).

Frost & Sullivan. (2005). *http://www.frost.com/prod/servlet/report-brochure.pag?id=F175-01-00-00-00* (last accessed March 23, 2009).

Glapa, S. (2002). "A critical investor's guide to nanotechnology." In: L. S. Peters and M. Sundarajaran (Eds.), *CEG Nanotechnology Technology Road Map*; *http://ceg.org/assets/files/PDFs/Nanotechnology%20executive%20summary.pdf* (last accessed March 23, 2009).

Huang, Z., Chen, H., Yip, A., Ng, G., Guo, F., Chen, Z.-K., and Roco, M. C. (2003). "Longitudinal patent analysis for nanoscale science and engineering: Country, institution and technology field." *Journal of Nanoparticle Research*, **5**(3/4).

ICON. (2008). *Towards Predicting Nano-biointeractions: An International Assessment of Nanotechnology Environment, Health and Safety Research Needs*. Texas: International Council of Nanotechnology; *http://icon.rice.edu/* (last accessed December 14, 2009).

Jayarajah, A. (2008). "Full steam ahead." *ICIS Chemical Business*, January 14, **273**(2), 28–29.

Jones, R. (2004). *Soft Machines: Nanotechnology and Life*. New York: Oxford University Press

Lux Capital. (2003). *The Nanotech Report* (Fifth Edition). New York: Lux Capital.

Lux Research. (2009a). *Press Releases* (1/22/09 and 7/22/08); *http://www.luxresearchinc.com/2003-2009%20last%20accessed%20March%2023* (last accessed March 23, 2009).

Lux Research. (2009b). *Lux Nanotech Index*. New York: Lux Research; *http://www.lux researchinc.com/pxn.php%20last%20accessed%20March%2023* (last accessed March 23, 2009).

Moore, S. K. (2005). "Nanotech patent trap.". *IEEE Spectrum*, July.

Murdock, S. (2002). "A disruptive technology with the potential to create new winners and redefine industry boundaries." Presentation about the nanotechnology business roadmap for the industry given at *AtomWorks*.

NNI. (2009). National Nanotechnology Initiative; *http://www.nano.gov/* (last accessed December 14, 2009).

NRC. (2008). *Review of Federal Strategy for Nanotechnology-related Environmental, Health, and Safety Research* (97 pp.). Washington, D.C.: National Research Council, National Academy of Sciences.

NSB. (2008). *NSF S&E Indicators*. Washington, D.C.: National Science Board/Government Printing Office; *http://www.nsf.gov/statistics/seind08/*

NSF. (2001). *Societal Implications of Nanoscience and Nanotechnology* (report). Washington, D.C.: National Science Foundation.

Palmberg, C. (2008). "The transfer and commercialization of nanotechnology: A comparative analysis of university and company research." *Journal of Technology Transfer*, **33**, 631–652.

Pandze, K., and Holt, R. (2007). "Absorptive and transformative capacities in nanotechnology innovation systems." *Journal of Engineering Technology Management*, **24**(4), 347–365.

Phoenix, C., and Drexler, E. (2004). "Safe exponential manufacturing." *Journal of Nanotechnology*, **15**, 869–872.

Peters, L. (2009). *Nanotechnology Databases.* Troy, NY: RPI.

Peters, L., and Sundarajaran. (2008). *Drivers of Growth for Firms in Different Sub-sectors of the Nanotechnology Industry: Implications for R&D Management* (Working Paper). Troy, NY: RPI.

PEN. (2009). "Project on emerging nanotechnologies"; *http://www.nanotechproject.org/consumerproducts%20last%20accessed%20March%2023* (last accessed March 23, 2009).

Roco, W., and Alivisatos. (Eds.). (1999). *Nanotechnology Research Directions* (316 pp.). Washington, D.C.: U.S. National Science & Technology Council; *http://nano.gov* [also published by Kluwer Academic, Boston, MA in 2000].

Russell, R. M. (2007). *Speech Delivered at the University of Missouri in September 2007.* Missouri: Technology Office of Science &Technology Policy

Sanghvi, S. (2003). "Nanotech: Positioning today for long term value creation." *Nanocommerce*. Chicago, IL: University of Chicago Press.

Siegel, H., and Roco, W. (1999). *International Council of Nanotechnology, 2008.* Washington, D.C.: Woodrow Wilson Center for International Scholars.

Stark, D. (2007). *Nanotechnology in Europe: Ensuring the EU Competes Effectively on the World Stage* (Nanoforum Dusseldorf Report). Dusseldorf, Germany: European Nanotechnology Trade Alliance.

Thayer, A. M. (2008). "Building businesses: Turning university research into products takes time, money, and initiative as four nanotechnology companies' experiences show." *Chemical and Engineering News*, **86**(13), 10–15.

Thayer, A. M. (2007). "Building up nanotech research: Investments in centers and institutes underpin interdisciplinary efforts." *Chemical & Engineering News*, April 9, **85**(15), 15–21.

Trader, M. (2007). *Russian and Nanotechnology Institute for Ethics and Emerging Technologies*; *http://ieet.org/index.php/IEET/more/treder20070505/Mayevels* (last accessed March 2009).

Tullis, T. K. H. (2005). Comment on "Application of the government license defense to federally funded nanotechnology research: The case for a limited patent compulsory licensing regime." *52 UCLA Law Review*, **11**, 279–280.

WGR. (2009). *2004 Report on Nanotechnology*. Winter Green Research; *http://www.wintergreenresearch.com/nanotech_reports.htm%20accessed%20March%2021* (last accessed March 21, 2009).

WWCIS. (2009). *Project on Emerging Technologies.* Washington, D.C.: Woodrow Wilson Center for International Scholars; *http://www.wilson center.org/* (last accessed December 14, 2009).

58

Emerging Web Technologies

Murugan Anandarajan, Bay Arinze,* Chittibabu Govindarajulu,[†] and Maliha Zaman**

*Drexel University and [†]Delaware State University

Introduction

The Internet has radically transformed the way millions of people work and play, as well as all types of industries across the board. The World Wide Web (or simply the Web), is a graphical HTML-based user interface that over a billion users worldwide now use and provides the Internet user access to a multitude of applications and services.

Evolving from a rudimentary network with military origins and used by a small number of scientists to transfer files between different destinations, most businesses now depend critically on the Internet and the Web to execute transactions, perform general business processes, exchange information, and connect with customers and other business partners.

Evolution of Web technologies

The Web is a network of interlinked nodes (HTML documents linked by hypertext) and is itself built upon the technologies and protocols of the Internet (routers, DNS, TCP/IP, etc.) which form a telecommunications network. There are over a billion people online and as these technologies mature and grow in size and scale, the implications of working with these kinds of networks are beginning to be explored in detail. Understanding the topology of the Web and the Internet, its shape, and interconnectedness is therefore increasingly important.

Over the last two decades, the Web has evolved significantly, its stages being referred to generally as Web 1.0, Web 2.0, and in the immediate future, Web 3.0. Web 1.0 was about client/server-based ecommerce, Web 2.0 is about peer/peer-based and user-generated content. Sites such as Amazon, eBay, and thousands of others flourished under Web 1.0, whereas YouTube, Flickr, and Wikipedia are prime examples of Web 2.0. This evolution has resulted from advances in several areas, namely new development tools and user interfaces, new computing paradigms, and improved bandwidth and processing power. Web 3.0, a phrase coined by John Markoff of the *New York Times* in 2006, refers to an evolution of Web 2.0 services that collectively comprise what might be called "the intelligent Web" such as those using natural language search and artificial intelligence technologies that emphasize machine-facilitated understanding of information in order to provide a more productive user experience.

While the origins of Internet (ARPANET) can be traced to the need to share information among scientists dispersed geographically, those early command line–based rudimentary technologies were difficult to use and were embraced only by technically savvy researchers. With the advent of the WWW, commercialization has gained prominence in the past two decades.

Web 1.0 technologies

Web 1.0 was the earliest form of the Web, led mainly by companies seeking a web presence. Content would be posted mostly by companies (by some individuals via personal web pages) and this content was typically accessed in read-only mode. These websites typically featured static content or database content from catalogs that were presented to the user, but with little user-provided content possible. The emphasis of Web 1.0 was "publishing" content onto the Web for users, not user-provided content. Other aspects of Web 1.0 include less sophisticated web browsers and slower Internet access, often through analog speeds of 56K or less. Figure 58.1 provides a timeline of web technologies.

In the early days of the Web, corporations used the Internet platform purely for commerce. Typically, customers would use the Web to search for products and/or buy them online. While the Internet was growing in popularity, telecommunications technology was advancing in leaps and bounds giving way to several new technologies

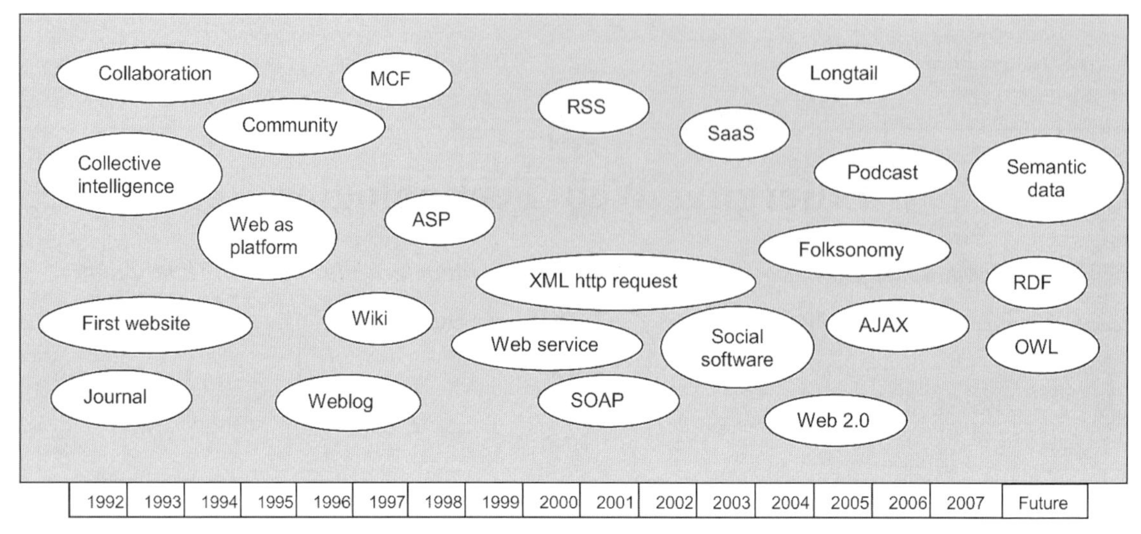

Figure 58.1. Evolution of web technologies.

such as XML web services, AJAX, and podcasting. XML web services provide a way for companies that specialize in a specific set of services to offer those services to other companies through a standard set of web protocols such as SOAP (Simple Object Access Protocol). AJAX (Asynchronous Javascript and XML), a more recent technology, helps in providing updating only parts of a web page instead of reloading the whole web page. It is important to recognize three elements in the growth of the Internet: the concepts (Web 1.0, 2.0, and 3.0), the technology (HTML, Java, AJAX, XML web services, etc.), and telecommunications. In the beginning stages of the Web, telecommunications and web technology was at its infancy and hence interactions were limited to searching and buying online using low-speed telecom equipment such as the now-outdated 56K modems and hubs. As technologies grew, primarily in the telecom industry through cable modems and fiber-optic technology, Web 2.0 gained prominence by way of user-generated content (Flickr, MySpace, Facebook, YouTube, blogs, etc.). Even companies such as Amazon and eBay that have origins in the Web 1.0 era have facilitated user-generated content through the use of feedback by customers on products and sellers. As technologies continue to evolve, Web 3.0 will provide rich user interfaces, natural language-based queries, intelligence searching, etc in all devices making the Web truly ubiquitous.

Web 2.0 technologies

While Web 1.0 revolved mostly around ecommerce, Web 2.0 architectures focus on encouraging user participation. The key aspects of Web 2.0 are the interconnectivity and interactivity of web-delivered content. As shown in

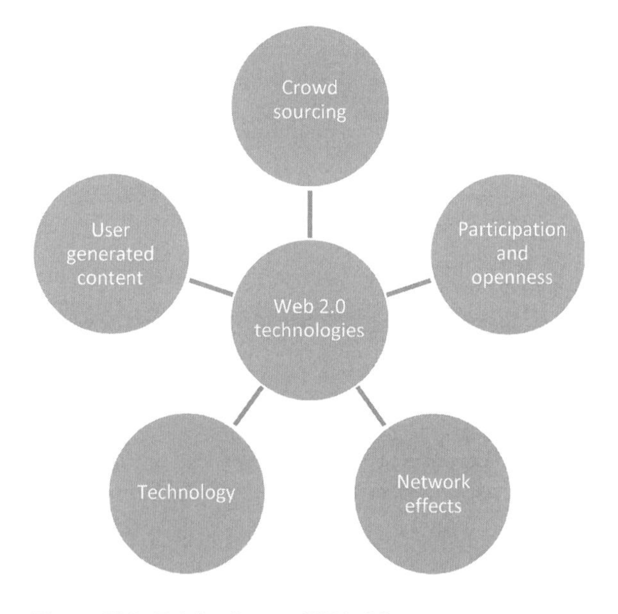

Figure 58.2. Driving forces of Web 2.0.

Figure 58.2, there are six forces which drive Web 2.0 technologies and these are discussed below.

(a) User-generated content

User-generated content refers to any material created and uploaded to the Internet by non-media professionals. This includes user comments to eBay or a video uploaded to YouTube, or a professor's profile on MyNetResearch. While user-generated content has been around since the early days of the Internet (e.g., Usenet in the 1980s), it was the advent of broadband Internet access and search technologies that made it a dominant form of global media. User-generated content (UGC) is now one of the fastest growing forms of content on the Internet. It fundamentally changes how individuals interact with the Internet, and how advertisers reach the market. In 2006, UGC sites attracted 69 million users in the U.S. alone, and in 2007 generated $1 billion in advertising revenue. By 2011, UGC sites are projected to attract 101 million users in the U.S. and earn $4.3 billion in ad revenue.

(b) Crowdsourcing

Crowdsourcing occurs when a company or institution takes a function once performed by employees and outsources it to an undefined and generally large network of people in the form of an open call for proposals.

In a crowdsourcing, the crowd is the collective of users who participate in the problem-solving process. Since it takes place through the Web, the crowd comprises Web users (i.e., individuals or organizations that posit solutions to the problem). It is in this composite or aggregate of ideas, rather than in the collaboration of individuals, where the strength of this approach lies. According to Surowiecki (2004), under the right circumstances, groups are remarkably intelligent and are often smarter than the smartest people in them. Furthermore, Surowiecki argues that in order to provide the best mix of solutions to a problem, the crowd must be diverse in opinion. The crucial prerequisite is the use of the open-call format and the large network of diverse potential web users.

While crowdsourcing has proven its worth in for–profit contexts, many have hopes for crowdsourcing as a far-reaching problem-solving model that can harness the collective intelligence of researchers across the world to benefit government and non-profit projects. No matter the purpose—for business, research, or the social good—the use of crowdsourcing appears set to grow dramatically.

(c) Participation and openness

The concept of participation and openness has its foundations in the open-source software development communities. According to O'Reilly (2003), these communities organize themselves so that there are lowered barriers to participation and a real market for new ideas and suggestions that are adopted by popular acclamation.

The most successful web-based services are those that encourage mass participation and provide an architecture featuring ease of use, toolboxes, etc. that lowers the barriers to participation. As a Web 2.0 concept, the idea of opening up collaborations and encouraging participation goes beyond the open-source software idea of opening up code to developers, to opening up content production to all users and exposing data for re-use and combination in mashups. A mashup is the result of combining data from different sources. It is comprised of three parts: (a) the content provider—who has the data and makes it available through web services, (b) the mashup site that uses data from different content providers to create a new service (the mashup), and (c) a client web browser that helps to mash up content using client-side scripts. It should be noted that embedding a YouTube video in a site is not a mashup. A true mashup site will access third-party data itself using an API or web service and process the data in a way that enhances the data's value to visitors of the site.

(d) Network effects

The network effect is a general economic term used to describe the increase in value to existing users of a service in which there is some form of interaction with others, as more and more people start to use it (Klemperer, 2006). The concept is commonly used when describing the extent of the increase in usefulness of a social system as more and more users join it. For example, as a new person joins a social networking site, other users of the site also benefit. Once the network effect begins to build and people become aware of the increase in a service's popularity, a product often takes off very rapidly in a marketplace. Given the social nature of Web 2.0 technologies, they rely heavily on the network effect for their adoption.

(e) Technology

Many technologies contribute to Web 2.0. Hardware-related developments include pervasive broadband, especially within developed countries where the majority of users now possess broadband connections. These enable the use of graphics-rich user interfaces that characterize Web 2.0 websites. Others include weblogs (or blogs), wikis, podcasts, RSS feeds (and other forms of many-to-many publishing), social software, and web application programming interfaces (APIs). These open APIs—with "open" signifying their availability to anyone—allow other developers and even users to develop their own applications or "applets" that can work in concert with Web 2.0 websites.

Web-based services resulting from Web 2.0

There are a number of web-based services and applications that demonstrate the foundations of the Web 2.0 concept. These are not really technologies as such, but services (or user processes) built using the building blocks of the technologies and open standards that underpin the Internet and the Web. These include blogs, wikis, multimedia-sharing services, content syndication, podcasting, and content-tagging services. Many of these applications of web technology are relatively mature, having been in use for a number of years, although new features and capabilities are being added on a regular basis. It is worth noting that many of these newer technologies are concatenations (i.e., they make use of existing services).

Table 58.1. Examples of popular blogs.

Examples of blog sites	*Functionality*
http://radar.oreilly.com/	Educational blog—innovative technology
http://www.techcrunch.com/	Educational blog—reviewing Internet products
http://www.instapundit.com/	Educational blog—law-related
http://wordpress.org/	Blog software—provider of WordPress
http://www.sixapart.com/typepad/	Blog software—provider of TypePad
http://www.blogger.com/start	Blog site—Blogger
http://radio.userland.com/	Blog software—provider of Radio Userland
http://www.bblog.com/	PHP blog software—provider of bBlog
http://technorati.com/	Blog service—miscellaneous current happenings
http://www.gnosh.org/	Blog service—meta search engine—Gnosh
http://blogsearch.google.com/	Blog service—Google's blog search engine
http://www.weblogs.com/about.html	Blog service—Ping service to notify users of new blogs

These applications and their collaborative capabilities are discussed next.

Blogs

Blog (short form of weblog) is a website maintained by an individual to express his/her opinion on a specific matter that could be political, social, religious, or scientific in nature. Most blogs allow visitors to post comments to encourage user participation. While not all sites are popular, blogging has encouraged professionals and nonprofessionals alike to express their opinions freely using the Internet. Table 58.1 provides a list of popular blogs.

Wikis

A wiki is a collection of web pages created using the collective knowledge of visitors. Wiki users contribute and edit the content of these web pages. In the words of Ward Cunningham, creator of Wiki software, it is "the simplest online database that could possibly work." A wiki enables documents to be written collaboratively, in a simple markup language using a web browser. A single page in a wiki website is known as a "wiki page", while the entire collection of pages, which are usually well interconnected by hyperlinks, is "the wiki".

A wiki is essentially a database for creating, browsing, and searching through information. Wikipedia.org, the online user-created encyclopedia, is an excellent example of the power of wikis to enable collaboration. Most wikis, except private corporate ones, are open to the public and anyone can edit their contents. However, because of this open nature of wikis and the lack of review before modifications are accepted, they are subject to easy tampering by vandals. The degree of tampering depends on the degree of openness of the wiki. Secure wikis that require user authentication and moderation grow at a slower pace than open Wikis. Wikis can be specialized (e.g., for species: *http://species.wikimedia.org/wiki/Main_Page*) to address certain areas, and the best way to find them is to follow wiki nodes or bus tours (*http://london.openguides.org/wiki/?TourBusStop*).

Multimedia-sharing services

The explosive growth in affordable broadband has led to rich bandwidth-intensive Web 2.0 multimedia-sharing services. Users now share photos with friends and family or even publish in the public domain using popular services such as flickr.com. They can upload home and other videos to YouTube or publish audio content on various podcasting sites. To add value, several of these sites also allow the multimedia content to be tagged to make searching easier.

Content syndication

The Internet is rich in content that is constantly updated. It is difficult to keep abreast of new information from various sites on a regular basis. This is where content syndication is really helpful. Created by Netscape in 1999 to get into portal business, RSS (Really Simple Syndication) helps users stay informed by aggregating information from several sites. RSS is based on XML and helps both the sender and receiver of information. For the sender, it helps to display content on many websites, thus increasing exposure; and for the receivers, it helps them to stay current with latest news and information. RSS feeds can be read using either a standalone reader or embedded readers in supported browsers or email clients. Content syndication can be a very useful tool to researchers to keep up to date in their respective fields.

Table 58.2. Example of podcasts.

	Podcast sites	*Podcast focus*
Education Podcast Network	*http://epnweb.org/index.php*	Educational—school level
Bowdoin College	*http://www.bowdoin.edu/podcasts/*	Educational—university
Time magazine	*http://www.time.com/time/podcast/*	News magazine
WeatherBug	*http://weather.weatherbug.com/labs/weather-podcasts.html*	Weather
USA.gov	*http://www.usa.gov/Topics/Reference_Shelf/Libraries/Podcasts.shtml*	Government

Table 58.3. Example of content-tagging sites.

Example of tagging services	*Functionality*
http://www.connotea.org/	Reference: management software
http://www.citeulike.org/	Net-based citation software
http://www.librarything.com/	Book management software and networking
http://del.icio.us/	Book management software and networking
http://www.sitebar.org	Net-based bookmarker software
http://www.furl.net/index.jsp	Social bookmarking
http://www.stumbleupon.com/	Website discovery service
http://www.blinklist.com/	Website discovery service
http://www.digg.com/	Share content over Web
http://www.rawsugar.com	Advanced Web content search

Podcasting

Podcasting is the distribution of digital media (mostly audio) files using syndication technologies for playback on computers and handheld devices such as mp3 players and iPods. It is increasingly becoming popular in education to distribute lectures. The author of a podcast is a podcaster. Podcasting is also popular in providing audio tours of museums, conference meeting alerts, and in providing public alert messages by police departments. Given the ubiquity of broadband, such technologies are gaining widespread acceptance in society. It is a Web 2.0 technology that focuses on audio distribution while sites such as YouTube provide platforms for user-generated video content. Table 58.2 provides a list of various types of podcasts

Content tagging

A tag is a relevant keyword or term associated with or assigned to a piece of information (a picture, a geographic map, a blog entry, a video clip, etc.), thus describing the item and enabling keyword-based classification and search of information (Wikipedia). In 2005, Technorati launched a keyword classification scheme for web publishers to provide reference terms for any digital content they published online. Similar to the metatags of Web 1.0 that helped to index web pages, tagging content is a popular Web 2.0 concept to help categorize user-generated content. While the metatags of Web 1.0 were invisible to users, tags for user-generated content are supplied by the authors and users of the content. This practice is also known as collaborative tagging or social tagging. Typically, freely chosen words are used instead of words from a predefined vocabulary. Table 58.3 provides a list of tagging services.

Beyond Web 2.0 technologies and the future

So what comes after Web 2.0? For Berners-Lee, Hendler, and Lassila (2001), it is the shift from documents to data—the transformation of a space consisting largely of human-readable, text-oriented documents to an information space in which machine-readable data, imbued with some sense of "meaning", is being exchanged and acted upon. This is commonly referred to as the Semantic Web. The Semantic Web is the symbiosis of web technologies and knowledge representation concerned with constructing and maintaining (potentially complex) models of the world that enable reasoning about themselves and their associated information.

In the last few years, the Semantic Web has begun to emerge into a mature set of standards that support "open" data and a view of information processing that emphasizes information rather than processing. This has been largely attributed to the maturing of RDF (Resource Description Framework) languages and the technologies that support them. According to the Microsoft (2006) Connected Services Framework, there are two main benefits offered by a profile store that has been created by using RDF. The first is that RDF enables you to store data in a flexible schema so you can store additional types of information that you might have been unaware of when you originally designed the schema. The second is that it helps you to create web-like relationships between data, which is not easily done in a typical relational database.

Various organizations are now focusing in different areas of the Semantic Web space. Garlik (*www.garlik.com*), for example, uses Semantic Web technologies for the "control of personal data in the digital world." Specifically, the company is working to let users discover what's known about them on the Web to see what the aggregation of this information (exposed via an RDF store) reveals. The Yahoo food site (http://food.yahoo.com) is being powered by OWL (Web Ontology Language) and RDF, as well as several other technologies. Teranode (www.teranode.com), among others, is exploring technologies for scientific data integration, particularly in the biology sector. Joost (www.joost.com), the new Internet TV platform that made big news in February in announcing a partnership with Viacom, uses RDF extensively.

An issue that has dogged the development of the Semantic Web is the need to develop ontologies for a multitude of domains, which could have considerable resource costs. As part of this process there are several areas where developments in the Semantic Web and those within social software are beginning to be explored.

Semantic wikis

This is a developing research area, but in essence researchers are looking at ways to annotate wiki content with semantic information. A Semantic wiki allows users to make formal descriptions of things in a manner similar to Wikipedia, and also annotate these pages with semantic information using formal languages such as RDF and OWL.

Semantic blogging

Blogs can be more than an easy-to-use publishing tool. Their ability to also generate machine-readable RSS and Atom feeds means that they can also be used to distribute machine-readable summaries of their content and thus facilitate the aggregation of similar information from a number of sources (Cayzer, 2004). Traditionally, these-feeds are used for headlines from blog postings, but by combining the ideas behind the Semantic Web with blogging software—semantic blogging—it may be possible to develop new information management systems. For example, RDF semantic data can be used to represent and export blog metadata, which can then be processed by another machine. In the long run the inclusion of this semantic information, by instilling some level of meaning, will allow queries such as "Who in the blogosphere agrees/disagrees with me on this point?"

Semantic desktop

It is envisaged that combining the ideas of Semantic Web and Web 2.0 services with traditional desktop applications and the data they hold (such as wordprocessor files, emails, and photos) on your local computing device will facilitate a more personalized way of working. In theory, this should create a more focused information and knowledge management environment, helping to find a way through personal "data swamps". Research work is at an early stage, but IBM is working on QED Wiki, a wiki-based application framework for collaboration working which enables the creation of enterprise mashups.

Collaboration has remained virtually identical in its conduct and organization over the last few decades. Disparate groups of individuals have worked on their ideas, projects, and inventions in isolated clusters, with little sharing of information and synergies from collaboration. The advent of social networks and Web 2.0 technologies have led to the creation of new networks that dramatically reduce the barriers and obstacles to collaboration by individuals who are geographically and organizationally distant.

Web 2.0 technologies use broadband connections, improved browsers, "rich" multimedia in concert with a new generation of websites that encourage user contributions to content. Blogs, forums, wikis, and other forms of user-generated content are in many cases the major source of content for these websites.

The current tens of thousands and eventually millions of individuals who will use networks are ushering in a new paradigm for collaboration. In this paradigm, collaboration is made much easier, sharing of knowledge is instant, and synergies from routine collaboration will yield huge advances in productivity and innovation. The challenge for administrators in both industry and academia is to understand how Web 2.0 technologies are changing the practice of collaboration and decide how best to embrace such technologies and use them to their best advantage.

Appendix 58.A. Summary of web technology.

Categorization	*Explanation*	*Examples of service*
Social networking	Professional and social networking sites that facilitate meeting people, finding like minds, sharing content—uses ideas from harnessing the power of the crowd, network effect and individual production/user generated content	Professional networking: *http://www.siphs.com/aboutus.jsp* *https://www.linkedin.com/* *http://www.zoominfo.com/* Social networking: *www.myspace.com* *www.facebook.com* *http://fo.rtuito.us/* *http://www.spock.com/* *http://www.flock.com/* *http://www.bebo.com/*
Aggregation services	Gather information from diverse sources across the Web and publish in one place. Includes news and RSS feed aggregators and tools that create a single webpage with all your feeds and email in one place—uses ideas from individual production/user generated content Collect and aggregate user data, user and intentions—uses ideas from the architecture of participation, data on epic scale and power of the crowd	*http://www.techmeme.com/* *http://www.google.co.uk/nwshp?hl=en* *http://www.blogbridge.com/* *http://www.suprglu.com/* *http://www.netvibes.com/* *http://www.attentiontrust.org/* *http://www.digg.com/*
Data "mash-ups"	Web services that pull together data from different sources to create a new service (i.e., aggregation and recombination). Uses, for example, ideas from data on epic scale and openness of data	*http://www.housingmaps.com/* *http://darwin.zoology.gla.ac.uk/~rpage/ispecies/* *http://www.rrove.com/set/item/59/top-11-us-universities*
Tracking and filtering content	Services that keep track of, filter, analyse and allow search of the growing amounts of Web 2.0 content from blogs, multimedia sharing services, etc. Uses ideas from e.g. data on epic scale	*http://technorati.com/about/* *http://www.digg.com/* *http://www.blogpulse.com* *http://cloudalicio.us/about/*
Collaborating	Collaborative reference works (like Wikipedia) that are built using wiki-like software tools. Uses ideas from harnessing the power of the crowd Collaborative, Web-based project and work group productivity tools. Uses architecture of participation	*http://www.squidoo.com/* *http://wikia.com/wiki/Wikia* *http://www.MyNetResearch.com* *http://vyew.com/always-on/collaboration/* *http://www.37signals.com/*
Replicate office-style software in the browser	Web-based desktop application/document tools. Replicate desktop applications. Based on technological developments	*http://www.google.com/google-d-s/tour1.html* *http://www.stikkit.com/* *http://www.backpackit.com/tour*
Source ideas or work from the crowd	Seek ideas, solutions to problems or get tasks completed by outsourcing to users of the Web. Uses the idea of power of the crowd	*http://www.mturk.com/mturk/welcome* *http://www.innocentive.com/*

Source: adapted from JISC Technology and Standards Watch, Feb. 2007 Web 2.0.

References

Berners-Lee, T., Hendler, J., and Lassila, O. (2001). "The Semantic Web: A new form of Web content that is meaningful to computers will unleash a revolution of new possibilities." *Scientific American.*

Cayzer, S. (2004). "Semantic blogging and decentralized knowledge management." *Communications of the ACM*, **47**(12), 47–52.

JISC. (2007). *Technology and Standards Watch: Web 2.0*; *http://www.jisc.ac.uk/whatwedo/topics/web2.aspx*

Klemperer, P. (2006). *Network Effects and Switching Costs: Two Short Essays for the New Palgrave* (Economics Papers 2006-W06). Oxford, U.K.: Economics Group, Nuffield College, University of Oxford.

Markoff, J. (2006). "Entrepreneurs see a Web guided by common sense." *New York Times*, November.

O'Reilly; *http://www.oreillynet.com/pub/wlg/3017*

Surowiecki, J. (2004). *The Wisdom of Crowds*. New York: Random House Large Print.

Glossary

Absorptive capacity: the ability of a firm to recognize, assimilate, value, or make productive use of new information such as research discoveries or new manufacturing methods. Absorptive capacity is seen as a function of internal expertise; where firms' internal capabilities obsolesce, they may be unable to make use of others' discoveries, or even recognize their importance.

Action agendas: programs to develop joint industry–government initiatives in specific sectors.

Adopter categories: market segments defined in terms of how quickly end-users purchase the new product.

America COMPETES Act: U.S. enabling legislation of 2007 that enacted but did not fund a range of policies to support U.S. R&D and innovation, including doubling over 10 years funding for innovation-enabling research at federal agencies, modernizing research and experimentation tax credits, strengthening K-12 math and science education, reforming the workforce training system, and reforming immigration in order to retain highly skilled foreign workers in the U.S.; the federal budget of December 2007 funded only a minor fraction of the initiative.

American Competitiveness Initiative—*see* **America COMPETES Act**

Angels: typically private, wealthy investors seeking to invest personal funds to advance innovation and development in areas outside their own business framework.

Antibody: protein produced by humans and higher animals in response to the presence of a specific antigen.

Appropriability regime: this is a term used in economics that describes the degree to which a firm is able to secure the profits from its innovation without the innovation being imitated by others. Patents, copyrights, and trademarks are some mechanisms that assist a firm in creating a "tight" appropriability regime. A weak appropriability regime is characterized by relatively quick competitor imitation.

ARC (Australian Research Council): a competitive funding scheme for universities.

Architectural innovation: reconfigures the *linkages* between components of established products in new ways while leaving the core design elements untouched.

Asset-based lending: asset-based loans base the terms of the contract on the assets that can back the loan rather than financial statement fundamentals.

Barriers to diffusion: product characteristics that will influence and possibly slow down its rate of diffusion. These include relative advantage, compatibility, complexity, divisibility, and observability of results.

Bass model: a diffusion model for durable goods that estimates peak sales and length of time required to reach peak sales.

Bayh–Dole Act: a 1980 U.S. federal law that allowed universities to retain patent rights from federally funded R&D.

Bell Laboratories: the research organization of AT&T (American Telephone & Telegraph). Formed in 1925, its principal focus was on design and support of equipment supplied to the Bell System operating companies. Bell Laboratories emerged as one of the nation's leading R&D labs during the middle decades of the 20th century.

Bioprocess: a process in which living cells, or components thereof, are used to produce a desired product.

Bioreactor: a vessel used for bioprocessing.

Bioremediation: the use of microorganisms to remedy environmental problems, rendering hazardous wastes non-hazardous.

BITS (Building Information Technology Strengths): an incubator support program.

Blog: a website containing opinion, events, grips, fun facts reflecting the view of the individual who is in charge of the blog. The website provides opportunities for visitors to comment on issues raised on the site.

Bootstrapping: the technique of obtaining funds for innovation or advancement without resorting to external sources, such as credit cards, second mortgages, personal savings, personal loans, or resources from friends and family.

Breakthrough innovation (also called **radical innovation**): innovations with features offering dramatic improvements in performance or cost, which result in transformation of existing markets or creation of new ones. They involve fundamental technological discoveries for the firm, and thus are new to the firm and/or industry, and offer substantially new benefits and higher performance to customers.

Business model innovation: design of a fundamentally different business model in an existing mature industry.

Career path: a planned progression of jobs within an organization or in a professional field.

Catalyst: an agent (such as an enzyme or a metallic complex) that facilitates a reaction but is not itself changed during the reaction.

Cell culture: growth of cells under laboratory conditions.

Chasm: in Moore's diffusion model, the difference in expectations between the visionaries and pragmatists, which accounts for the fact that pragmatists might not ever adopt the product even if the visionaries do.

Chromosomes: threadlike components in the cell that contain DNA and proteins. Genes are carried on the chromosomes.

Civilian Industrial Technology Program: proposed but defeated U.S. legislation of 1964, that would have supported industrial research at universities and provided incentives for industry to take on costly and risky R&D initiatives.

Clone: a term that is applied to genes, cells, or entire organisms that are derived from—and are genetically identical to—a single common ancestor gene, cell, or organism, respectively. Cloning of genes and cells to create many copies in the laboratory is a common procedure essential for biomedical research. Note that several processes commonly described as cell "cloning" give rise to cells that are almost but not completely genetically identical to the ancestor cell. Cloning of organisms from embryonic cells occurs naturally in nature (e.g., identical twins). Researchers have achieved laboratory cloning using genetic material from adult animals of several species, including mice, pigs, and sheep.

Co-evolution: when two technologies evolve together in such a way that developments in one influence the course of development of the other and *vice versa*.

COMET (COMmercializing Emerging Technologies): an early-stage venture support program.

Commission on Industrial Competitiveness: a U.S. commission, under President Ronald Reagan, charged with reviewing federal R&D and innovation policies.

Compatibility: how well the new product fits with the end-user's experiences and values. Low compatibility can become a barrier to diffusion.

Competence (and competencies): a set of organizational capabilities (such as managerial processes or shared collective technical/technological know-how) to achieve something.

Competence gap: the gap between a firm's portfolio of technology and the set of competences required by a new technology.

Competitive intelligence: the activity of information gathering and scanning of the environment in search of inputs for the activities of the firm. Technology intelligence is the scientific/technical/technological part of CI.

Complementary assets: the other organizational and functional elements that a firm needs to commercialize its innovation. For example, a company might have a product design but it will also require appropriate manufacturing assets, marketing assets, and distribution assets to actually bring that design to the marketplace.

Complexity: how difficult the new product will be perceived by the end-user. High complexity can become a barrier to diffusion.

Component innovation—*see* **Modular innovation**

Connecticut Innovations—*see* **Connecticut Product Development Corporation**

Connecticut Product Development Corporation: one of many product development corporations established by U.S. states to provide grants to companies to aid development of new products, established in 1973, became Connecticut Innovations *circa* 1989.

Content tagging: on the Internet, a tag is a relevant keyword or term associated with or assigned to a piece of information (a picture, a geographic map, a blog entry, a video clip, etc.).

Continuous improvement: a key process in total

quality management, where specific routines are embedded in organizations to generate incremental innovations regularly.

Copyright: a statutory right that gives the author of an artistic work, for a limited period, the exclusive privilege to make copies of the work and publish and sell the copies.

Copyright Term Extension Act: U.S. legislation of 1998 that extended new and existing copyrights by 20 years, effectively taking the right to use materials away from the general public and, for new copyrights, expanding the incentives to produce copyrighted materials by a tiny amount.

Court of Appeals for the Federal Circuit: a U.S. federal appeals court created in 1982 to handle, in part, patent-related cases; the court made patents easier to obtain and defend until a partial reversal by the Supreme Court in 2006 and 2007.

CRADAs (Cooperative R&D Agreements): U.S. collaborative agreements between a federal government laboratory and a non-federal partner for the purpose of conducting joint R&D.

Creative destruction: the process of industrial transformation by which successive waves of innovation create value while undermining the basis of established companies.

CSIRO (Commonwealth Scientific & Industrial Research Organization): a major multisectoral public-sector research organization.

CTO (Chief Technology Officer): the executive function in charge of covering technology strategy and thus TI. By extension it may also cover R&D management, the promotion of innovation, and technical support and capitalization. CTO may be seen as a modern name for the R&D VP.

Defense Advanced Research Projects Agency: an agency of the U.S. Department of Defense that develops new technology for use by the U.S. military, controlling 2.3% (in 2007) of the nation's federal R&D budget.

Demand-side sources of innovation: innovations that are developed in response to expressed or latent consumer needs.

Department of Agriculture: a U.S. government department responsible for agriculture, controlling 1.6% (in 2007) of the nation's federal R&D budget (half as much as the U.S. National Science Foundation).

Department of Defense: the U.S. government department responsible for national defense, controlling 55% (in 2007) of the nation's federal R&D budget.

Department of Energy: the U.S. government department dealing with energy policy and issues, controlling 6.3% (in 2007) of the nation's federal R&D budget.

Department of Health & Human Services: the U.S. government department responsible for health and human welfare, controlling 21% (in 2007) of the nation's federal R&D budget.

Derivative innovation—*see* **Incremental innovation**

Diamond v. Chakrabarty: a U.S. Supreme Court case that made possible patenting of engineered organisms.

Diffusion curve: a graph showing the percentage of a population of potential adopters who have selected a given technology as a function of time since the introduction of the technology.

Diffusion of innovation: the process by which an innovative product category spreads through the market, over several categories of adopters.

Directed placements: target specific institutional investors offering shares at a discount prior to being publicly traded.

Disruptive technology: a specific type of technological innovation that initially appeals only to a marginal market segment, but because of its improvement over time it can eventually satisy the mainstream market.

Divisibility: how easily the new product can be purchased and/or used in smaller quantities. Low divisibility can become a barrier to diffusion.

DNA (DeoxyriboNucleic Acid): the molecule that carries the genetic information for most living systems. The DNA molecule consists of four bases (adenine, cytosine, guanine, and thymine) and a sugar–phosphate backbone, arranged in two connected strands to form a double helix.

Domestic Policy Review: a roadmap established by President Jimmy Carter in 1979 to optimize U.S. innovation and competitive advantage, and to respond to Japan's industrial policy. It resulted in the Industrial Innovation Initiatives to promote cooperative research between government, industry, and universities by: enhancing technology transfer, increasing technical knowledge, improving the patent system, clarifying antitrust policy, fostering development of smaller innovative firms, opening federal procurement to innovations, improving the regulatory system, facilitating labor–management adjustments, and maintaining a supportive innovation climate by removing barriers to innovation.

Dominant design: a common form of a product upon which all companies will converge in a particular industry. For example, the personal computer (PC) is a dominant design in the computer industry. All makers produce their computers according to a dominant architecture.

Downsizing: reduction of staff, elimination of units or functions in a business, often by substituting new technology or external sourcing.

Dual-ladder career development systems: career development alternatives for technical employees; typically provides a choice of a managerial track or a technical track.

Dual-technology tree: a mapping technique to visualize competing technologies, the technological trajectories, and the competence gap. The process/product duality appears visually on the map. A technology is a branch of the tree.

E-collaboration: collaboration among people or organizations made possible by means of electronic technologies such as the Internet, video conferencing, and wireless devices.

Early majority: one of the adopter categories—the first half of the majority of end-users to adopt the product.

Early-adopters: after the innovators, the second segment of end-users to adopt the product.

eBay v. MercExchange: a U.S. Supreme Court ruling of May 15, 2006 that effectively made it more difficult for a patent holder to block the activities of another firm by alleging in court that the patent was infringed.

Economic Tax Recovery Act: U.S. legislation introduced by President Reagan and approved by the U.S. Congress in 1981 which reduced business taxes and lowered the maximum marginal tax rate to 50% and capital gains tax to 20%.

Economies of agglomeration: cost savings that arise when an economic activity congregates in or close to a single location, rather than being spread out uniformly over space.

Ecosystem: the set of external partners, suppliers, clients, consultants, engineers, etc. that relate to an organization thus feeding in their vision, knowledge, technologies, ideas, etc., while being nurtured in return by the organization.

Entrepreneurial capitalism: capitalism in which entrepreneurs play a predominant role as a fraction of overall economic activity.

Enzyme: a protein catalyst that facilitates specific chemical or metabolic reactions necessary for cell growth and reproduction.

Evolutionary—*see both* **Evolutionary market** and **Evolutionary technical**

Federal laboratory: a research laboratory funded and sometimes managed by a federal-level government agency. Examples include the Department of Energy's Los Alamos National Laboratory and National Renewable Energy Laboratory.

Federal Technology Transfer Act of 1986: passed by the U.S. Congress to facilitate the commercialization of inventions developed in government-owned, government-operated federal laboratories (GOGOs). This act permitted the formation of CRADAs.

Fermentation: the process of growing microorganisms for the production of various chemical or pharmaceutical compounds. Microbes are normally incubated under specific conditions in the presence of nutrients in large tanks called fermentors.

First to discover: a system of awarding priority in patent or copyright claims to the original discoverer, in contrast to first to file. Priority of discovery can be documented by dated notes, records of experiments, and the like.

First to file: a system of awarding priority in patent or copyright claims to the first party to file for ownership recognition, in contrast to first to discover. First to file is a much easier system to manage, as the filing date is typically indisputable.

Foresight: the activity of describing what the variety of potential futures of a system or context could look like. Foresight exercises lead to building scenarios to capture the variety of potential outcomes as a way of helping decision-makers cope with uncertainty and strategize.

Functional need: the specifications of the needs of those being served by the technologies (i.e., the users, clients in the markets, or internal users in the manufacturing or distribution processes).

Gene: a segment of chromosome. Some genes direct the syntheses of proteins, while others have regulatory functions.

Gene sequencing: determination of the sequence of nucleotide bases in a strand of DNA. *See also* **Sequencing.**

Gene therapy: the replacement of a defective gene in an organism suffering from a genetic disease. Recombinant DNA techniques are used to isolate the functioning gene and insert it into cells. More than 300 single-gene genetic disorders have been identified in humans. A significant percentage of these may be amenable to gene therapy.

Genetic code: the code by which genetic information in DNA is translated into biological function. A set of three nucleotides (codons), the building blocks of DNA, signify one amino acid, the building blocks of proteins.

Genetic modification: a number of techniques, such as selective breeding, mutagenesis, transposon insertions, and recombinant DNA technology, that are used to alter the genetic material of cells in order to make them capable of producing new substances, performing new functions or blocking the production of substances.

Genetic testing: analysis of an individual's genetic material. Genetic testing can be used to gather information on an individual's genetic predisposition to a particular health condition, or to confirm a diagnosis of genetic disease.

Genome: the total hereditary material of a cell, comprising the entire chromosomal set found in each nucleus of a given species.

Genomics: the study of genes and their function. Recent advances in genomics are bringing about a revolution in our understanding of the molecular mechanisms of disease, including the complex interplay of genetic and environmental factors. Genomics is also stimulating the discovery of breakthrough health care products by revealing thousands of new biological targets for the development of drugs and by giving scientists innovative ways to design new drugs, vaccines, and DNA diagnostics. Genomic-based therapeutics may include "traditional" small chemical drugs, as well as protein drugs and gene therapy.

Gross Domestic Product (GDP): the monetary value of all final goods and services produced within the boundaries of a country during a set period of time such as a calendar quarter or year.

H-1B visa: temporary work visas allowing non-U.S. citizens to enter the U.S. and work in sponsoring U.S. businesses in specialized skilled occupations.

Hold-up (by patents): the use of patents to threaten prevention of an allegedly infringing company's ability to produce a product; firms are claimed in many cases to have used such tactics even with frivolous patents to extract monetary gains from large companies.

IIF (nnovation Investment Funds): subsidy incentives for venture capital investment.

Imitation rate: one of the estimates used in the Bass model; refers to how quickly the rest of the market follows the innovators. *See* **Innovation rate**.

Incremental innovation: continuous innovation where the competence gap is minimum.

Infratechnology: technologies that facilitate technological infrastructure for R&D, production, and marketing in industries.

Initial Public Offering (IPO): the first sale of common stock or shares in a company, when it becomes listed on a public exchange.

Innovation: application of the existing stock of knowledge in new combinations and new ways to meet some human need.

Innovation rate: one of the estimates used in the Bass model; refers to the proportion of innovators (adopters in the first time period) in the population.

Innovator traits: identifiable characteristics of opinion leaders; one list includes venturesomeness, social integration, cosmopolitanism, social mobility, and priviligedness.

Innovators: the first end-users to adopt the new product, comprising about 5% of the market.

Intellectual capital: intangible assets of value in commercial activity, as the base to create intellectual property.

Intellectual property: formally recognized ownership rights in knowledge, inventions, trademarks, logos, and other knowledge. Contemporary usage expands consideration to include aspects of knowledge that may not be formally be acknowledged by government as intellectual capital.

Invention: the discovery of new knowledge about natural phenomena.

Know-how: practical expertise acquired from study, training and experience.

Knowledge economy: commercial activity resting on knowledge, insight, judgment, discovery, and innovation. The contemporary global economy is increasingly characterized as knowledge-based, according increasing importance to intellectual capital of a wide array of forms.

Kondratiev waves: recurrent cycles or wave-like fluctuations in economic aggregates with a period of approximately 50 to 60 years.

KSR International v. Teleflex: a U.S. Supreme Court ruling of April 30, 2007 that made U.S. patents more difficult to obtain by returning to more traditional views (as opposed to practices introduced by the Court of Appeals for the Federal Circuit) that patented technologies must be non-obvious.

Labor productivity: the quantity of output that is produced per unit of labor input.

Laggards: one of the adopter categories, the last segment of end-users to adopt the product.

Large-firm capitalism: capitalism in which large corporations play a predominant role as a fraction of overall economic activity.

Late majority: one of the adopter categories: the second half of the majority of end-users to adopt the product.

Learning-by-doing: improvements in productivity resulting from experience gained through producing a given product.

Learning-by-using: improvements in productivity resulting from experience gained using a technologically complex product.

Living standards: the level of well-being that can be achieved by an individual or household per unit of labor effort.

Local/State/Federal programs: provide financial support to innovators in the form of grants or loans which allow for the development of technology. The purpose of such agencies is to promote economic development and competitive positioning.

Lock-in: a situation in which a particular technology becomes dominant despite other potentially more economically efficient alternatives. Lock-in may occur either because of sunk costs or because of external economies and a lack of coordination mechanisms that prevent individuals from switching to the superior technology.

Managerial career ladder: dual-ladder career path alternative that focuses on organizational advancement through increased managerial responsibilities involving planning, organizing, leading, and controlling.

Manhattan Project: the U.S. government's World War II project that developed nuclear weapons.

Market breakthrough innovation—*see* **Breakthrough innovation**

Mergers and acquisitions: synergies created for research and development by merging or acquiring firms, such as the proliferation of new ideas and access to finance and other resources that foster the development process.

Microsoft v. AT&T: a U.S. Supreme Court ruling of April 30, 2007 that held that, for software developed in the U.S. and exported on a disk used outside the U.S. to make copies onto devices sold outside the U.S., the software copied onto those devices does not infringe U.S. patents with technologies embodied in the software.

Modular innovation: the introduction of new technology to specific modules of a product that displaces *core design concepts* while leaving the established linkages between components relatively untouched.

Monoclonal antibody (MAb): highly specific, purified antibody that is derived from only one clone of cells and recognizes only one antigen.

Moore diffusion model: a model of diffusion of discontinuous products which views end-users in two segments (visionaries and pragmatists), and suggests that the two segments have different expectations about the new product.

Moral rights: the right of an author to prohibit others from tampering with a copyrighted work.

Mutation: a change in the genetic material of a cell.

National Advisory Committee on Aeronautics: a U.S. federal agency, from 1915 to 1958, that carried out aeronautical research; forerunner to the National Aeronautics and Space Administration.

National Aeronautics and Space Administration: a U.S. federal agency that carries out the U.S.'s publicly-funded space program, controlling 8% (in 2007) of the nation's federal R&D budget.

National Cancer Institute: part of the U.S. National Institutes of Health, established in 1937 and charged with cancer-related research.

National Center for Manufacturing Sciences: a U.S. federally-funded program initiated by President Ronald Reagan to enhance civilian technological strength to support U.S. defense and related high technology that relied on commercial technological capabilities; in 2008 it remains the largest not-for-profit, cross-industry collaborative research consortium of North American corporations, with an annual R&D portfolio of over $80 million.

National Competitiveness Act of 1989: passed by the U.S. Congress, this act extended the provisions of the Federal Technology Transfer Act of 1986 to government-owned, contractor-operated federal laboratories (GOCOs). This act permitted the formation of CRADAs.

National Cooperative Research Act: U.S. legislation of 1984 that relaxed antitrust regulations to more readily allow firms to carry out joint R&D activities.

National Innovation Act: proposed but not enacted U.S. legislation of 2005 that would have enacted a range of policies to support U.S. R&D and innovation, including a President's Council on Innovation, an Innovation Acceleration Grants Program, a near doubling of U.S. National Science Foundation research funding by 2011, and making permanent the U.S.'s sporadic research and experimentation tax credit.

National Institutes of Health: a U.S. federal agency that conducts and supports medical research.

National Science Foundation: a U.S. federal agency

that funds about 20% (*circa* 2007) of federally supported basic research in U.S. universities and controls 3.1% (in 2007) of the nation's federal R&D budget.

National Science Foundation Act: U.S. legislation that in 1950 established the U.S. National Science Foundation.

National treatment: once goods are legally imported, they must be treated no differently than domestic goods, with no additional requirements imposed based on foreign source.

Network effects or externalities: the mechanism through which the value of a service becomes more valuable to each user as additional users are added.

New product: general terminology for tangible innovations.

New Technologies Opportunity Program: a U.S. effort launched by President Richard Nixon in 1970–1971 to identify new technological opportunities under government auspices that could be transferred to the private sector. The program received much opposition and criticism and eventually stalled due to fiscal and ideological constraints—that subsidizing research and development might reduce incentives for privately funded R&D.

NICTA (National ICT Australia): a largely government-funded research center.

Non-rival good: a product whose use by one individual does not affect its value or utility to other consumers. For example, each listener to a radio broadcast receives the same benefit regardless of the number of other listeners.

Nucleic acids: large molecules, generally found in the cell's nucleus and/or cytoplasm, that are made up of nucleotides. The two most common nucleic acids are DNA and RNA.

Nucleotides: the building blocks of nucleic acids. Each nucleotide is composed of sugar, phosphate, and one of four nitrogen bases. The sugar in DNA is deoxyribose and RNA's sugar is ribose. The sequence of the bases within the nucleic acid determines the sequence of amino acids in a protein.

Observability of results: the extent to which the end-user can see the benefits of the new product. Low observability can become a barrier to diffusion.

Office of Scientific Research & Development: a U.S. federal agency during World War II that coordinated scientific research to support the military during the war.

Oligonucleotide: a polymer consisting of a small number (about 2 to 10) of nucleotides.

Omnibus Trade & Competitiveness Act: U.S. legislation of 1988, sporadically renewed and then expired, that reacted to U.S. trade deficits by requiring investigation of certain U.S. trade deficits and in some cases enactment of trade barriers.

Opinion leaders: end-users who have a strong influence on whether the majority of the market ever adopts the product. These are equivalent to the first two adopter categories (innovators and early-adopters).

Outsourcing: external purchase of (formerly internally performed) business functions. Where external providers offer lower cost or better quality, particularly in specialty items (e.g., tax services, components), or where external sourcing permits the firm to focus attention on its core competence, outsourcing is seen as desirable because it is thought to enhance economic efficiency. Outsourcing may, however, produce undesirable declines in customer experience or market knowledge—as, for example, when customer service call centers are remotely located in low-cost countries.

Patent: a statutory right that gives the inventor, for a limited period, the exclusive right to use or sell a patented product or to use a patented method or process.

Patent strategies: concerted and integrated approaches to patenting, patent enforcement, negotiation, and other uses of IP and intellectual capital to generate competitive advantage for the firm.

Patenting around: patenting a modified technology that achieves the same purpose as an existing patent, but that takes a different enough approach as to circumvent the existing patent's intellectual property claims.

Path dependence: the idea that decisions taken in the past limit the scope of contemporary choices.

Pecuniary rights: the right of an author to exploit a copyrighted work for economic gain.

Penicillin: an antibiotic discovered by Alexander Fleming and developed to production by U.S. government and U.S. firms in a major project during World War II.

Performance guarantees: a specified rate of return on an investment over some period of time (usually 1 year).

Pharmacogenetics: the study of inherited differences (variation) in drug metabolism and response. *See also* **Pharmacogenomics**.

Pharmacogenomics: the science that examines the inherited variations in genes that dictate drug response and explores the ways these variations can be used to predict whether a patient will have a good response to a drug, a bad response to a drug, or no response at all. *See also* **Pharmacogenetics**.

PIPE (Private Investment in Public Entities): PIPE transactions tender equity or equity-linked investments to a select group of investors, while removing the burdens of regulation requirements on typical equity offerings.

Plant Patent Act: U.S. legislation of 1930 that allowed patenting of new varieties of plants, excepting sexual and tuber-propagated plants.

Podcast: an audio or video file distributed over the Internet on to portable media devices and personal computers.

Polymerase chain reaction (PCR): a technique to amplify a target DNA sequence of nucleotides by several hundred thousandfold.

Pragmatists: one category of end-user in Moore's model, roughly equivalent to the early and late majority and the laggards.

Private placement: the obtaining of funds through the sale of equity shares without the regulatory burden of the public market.

Product life cycle: a model of the sales and profits of a product category from its introduction until its decline and disappearance from the market; focuses on the appropriate strategies at each stage.

Program for International Student Assessment: international assessments of 15-year-olds' literacy in reading, mathematics, science, and other skills.

Proteomics: each cell produces thousands of proteins, each with a specific function. This collection of proteins in a cell is known as the proteome, and, unlike the genome, which is constant irrespective of cell type, the proteome varies from one cell type to the next. The science of proteomics attempts to identify the protein profile of each cell type, assess protein differences between healthy and diseased cells, and uncover not only each protein's specific function but also how it interacts with other proteins.

Purchasing power: the quantity of goods and services that can be purchased with a given quantity of money.

R&D start: a program to provide direct grants to firms for innovation projects.

R&D tax credit: a national or regional corporate tax benefit that encourages business to carry out R&D.

Radical innovation: a change where the competence gap is major. It is a paradigmatic shift, a change in technological trajectory. It happens in case of a breakthrough. *See also* **Breakthrough innovation**.

Radically new innovation—*see* **Radical innovation** and **Breakthrough innovation**

RDCs (Rural Research & Development Corporations): funded by grower levies and government support.

Real options: these stem from decision theory, keeping alternate options open and continuously assessing their value. This is applied here to technological options visualized as branches on the dual-technology tree.

Recombinant DNA (rDNA): the DNA formed by combining segments of DNA from two different sources.

Relative advantage: a barrier to diffusion in which the new product offers nothing better than other products with which it is competing.

Research parks: an option for financing technology especially to early-phase research and development projects or startup technology companies. Research parks are typically aligned with governments or universities for the purpose of transferring intellectual property that arises from technological innovation. While research parks do not necessarily provide funding in the traditional sense, they offer technology starts facilities and services to foster the development process.

Research Triangle Park: a large U.S. commercial research park, located in North Carolina.

Revolutionary innovation—*see* **Breakthrough innovation**

RiboNucleic Acid (RNA): a molecule similar to DNA that delivers DNA's genetic message to the cytoplasm of a cell where proteins are made.

Roadmaps: techniques that capture market trends, product launches, technology development, and competence building over time in a multilayer, consistent framework.

Rogers' diffusion model: an important early model that characterizes the diffusion of a product through several adopter categories (innovators, early-adopters, etc.).

Route 128: a highway around Boston, Massachusetts with a high concentration of computer-related businesses.

Royalty financing: an alternative to regular debt-and-equity financing whereby funds contributed by outside investors are repaid out of future earnings as determined by a pre-set formula.

Sarbanes–Oxley Act: U.S. legislation of 2002 that mandates strict accounting standards, and may have decreased formation of public (shareholder-held) firms because of its substantial compliance costs.

Scenario: a description of any typical potential future (not only the most probable).

Sematech: a non-profit research consortium for basic research in semiconductor manufacturing, formed in 1987 as an exclusive program of U.S. Government and semiconductor manufacturers, but changed after 1996 to a non-governmental international program.

Sequencing: decoding a strand of DNA or gene into the specific order of its nucleotides: adenine, cytosine, guanine, and thymine. This analysis can be done manually or with automated equipment. Sequencing a gene requires analyzing an average of 40,000 nucleotides.

Service mark: a mark or symbol used to identify a person or entity which provides services.

Silicon Valley: a region south of San Francisco, California with a high concentration of computer-related and high-tech R&D and manufacture.

siRNA (silencing or short interfering RNA: mechanism for controlling production of specific proteins by interfering with gene transcription.

Small Business Act: U.S. legislation of 1953 that established the U.S. Small Business Administration.

Small Business Administration: a U.S. Government agency that supports small businesses.

Small Business Innovation Development Act: U.S. legislation that established the Small Business Innovation Research program

Small Business Innovation Research program: a program in which U.S. federal agencies with extramural R&D budgets over $100 million must set aside 2.5% of their budget for innovative research conducted at small companies.

Small Business Technology Transfer program: a U.S. federal program supporting joint R&D of small businesses with non-profit research institutions.

Spatial clustering: the concentration of technological innovations or other economic activities in certain geographically confined locations in greater numbers than would be expected based on the distribution of population or economic activity alone.

Spinouts: separate entities created exclusively for the research and development portion of a business.

Splicing: the removal of introns and joining of exons to form a continuous coding sequence in RNA.

Standards: explicit or implicit specifications or rules regarding important aspects of a technology which facilitate interoperability, safety, reliability, or other characteristics of a technology.

Strategic alliance: collaboration among partners seeking to achieve critical longer term or unique advantages by cooperation, as, for example, proprietary products or capabilities arising from their partnership while nevertheless remaining independent firms.

Strategic alliances/partnerships: partnering or allying with other institutions in order to gain access to capital, markets, and other technologies provides a highly effective capital-raising strategy. Innovators are able to spread the risk of the technology development. Collaboration is the key element to these relationships and the affiliation offers flexibility in terms of design and extent of the association

Superconducting Super Collider: a cancelled U.S. particle accelerator for basic physics research, with major construction begun in 1991 but cancelled in 1993 given multibillion dollar–projected cost overruns.

Supply-side sources of innovation: innovations that originate by first creating a new product and then generating consumer demand.

Sustaining innovations: these improve the performance levels of established products (such as video-capable cellphones) and provide incumbent firms an opportunity to reinfore their core competencies.

SWORD (Stock Warrant Off-balance-sheet Research and Development): generally defined as a set of contracts that define the relationship among parties and the rights accruing to those parties. In a SWORD agreement a separate entity is set up to house a research and development effort.

Synchronous innovation: the simultaneous adoption of process innovation technology and organizational innovations designed specifically to successfully cature the benefits of this new technology, which is typically available for competitors and precipitates weak appropriation conditions.

Systems biology: a hypothesis-driven field of research that creates predictive mathematical models of complex biological processes or organ systems.

Technical career ladder: a dual-ladder career path alternative that focuses on organizational advancement through increasingly important and difficult technical tasks.

Technical empirical know-how: a technique that works in a specific context although its fundamental underpinnings may not be well understood.

Technical standards: these are specifications to insure that different components of the same technological system are compatible. Technical standards allow for multiple firms to compete in various segments of an industry value chain. For example, USB slots on computers are standard so peripherals can be manufactured to fit.

Technological dynamism: this refers to how quickly technological advances move in a particular industry. The pace of innovation will vary among industries, with some industries having much longer technology life cycles than others. Typically, the underlying scientific and technological advancements in an industry will determine how technologically dynamic it is.

Technological frontier: the most advanced instances of a particular technology.

Technological paradigm: a state of scientific knowledge and technical know-how (i.e., the state of the art at a moment in time) to fulfill some generic market need.

Technological trajectory: the path followed over time to explore the technological paradigm.

Technology: a technique that is, at least in part, understood via scientific knowledge.

Technology Transfer Act: a U.S. legislation of 1986 that encouraged transfer of technologies from U.S. federal laboratories to businesses, where possible, for manufacture in the U.S.

Temporal clustering: the bunching of innovation at certain points in time and not at others.

Thickets: large numbers of patents in a technological area, with protection to patent-holders provided by the large number of patents held not by any one patent-holder.

Trade name: a mark or symbol used to identify a manufacturer or merchant.

Trade secret: specific information or know-how that is kept secret, protected in some countries by trade secrecy laws.

Trademark: a mark or symbol used to identify goods of a particular manufacturer or merchant.

Transcription: synthesis of messenger (or any other) RNA on a DNA template.

Transilience: the competencies that will survive a technological change. It results from the contraction of two words, transition and resilience.

Trends in International Mathematics and Science Study: a periodic international assessment of youth skills in mathematics and science.

TRIPS: a multilateral and comprehensive set of rights and obligations governing international trade in intellectual property, set forth in an annex to the agreement establishing the WTO.

U.S. Patent and Trademark Office (USPTO): a U.S. agency that issues U.S. patents, trademarks, and copyrights.

U.S. patent types
Business process patents: patents on methods for doing business (e.g., Amazon's "one-click" ordering system) that prevent others from utilizing the method without license from the patent-holder. Business process patents are especially important in ecommerce, informated business operations, and information-laden businesses like insurance, banking, and finance. Business methods (as discrete from methods linked to a device or technology) cannot be patented in Canada or Europe.
Design patents: these "may be granted to anyone who invents a new, original, and ornamental design for an article of manufacture."
Plant patents: these "may be granted to anyone who invents or discovers and asexually reproduces any distinct and new variety of plant."
Utility patents: these cover "any new and useful process, machine, article of manufacture, or compositions of matters, or any new useful improvement thereof."

Utility patent: in the U.S., the main type of patent, issued for the "invention of a new and useful process, machine, manufacture, or composition of matter, or a new and useful improvement thereof"; not to be confused with a "utility model" which in some nations refers to a short-duration and more informally granted equivalent of a patent.

Utterback–Abernathy life cycle model: this model identifies three phases in an innovation's life cycle—fluid, transitional, and specific—and suggests that industries evolve predictably from one phase to the other.

Venture capital (VC): funds availed to innovators by organizations established typically as a general partnership with committed funds partners such as institutional investors, corporations, and wealthy individuals. The contributed funds ultimately convert to an ownership position.

Venture philanthropy: also called social venture philanthropy, this refers to non-profit firms willing to contribute capital to a technological effort which coincides with the goals and mission of the philanthropic organization.

Virtual teams: teams composed of members who collaborate electronically in accomplishment of some common task, often via the Internet. Some virtual teams meet face to face from time to time, or initially; others are never co-located.

Visionaries: one category of end-user in Moore's model, roughly equivalent to the innovators and early-adopters.

Web 1.0: a system of Internet servers that support specially formatted documents. The documents are formatted in a markup language called HTML (HyperText Markup Language) that supports links to other documents, as well as graphics, audio, and video files.

Web 2.0: describes a second generation of the World Wide Web that is focused on the ability for people to collaborate and share information online. Web 2.0 basically refers to the transition from static HTML web pages to a more dynamic Web that is more organized and is based on serving web applications to users.

Web 3.0: an extension of the current Web that provides an easier way to find, share, reuse, and combine information more easily. It also provides a common language for recording how the data relate to real-world objects, allowing a person, or a machine, to start off in one database and then move through an unending set of databases which are connected not by wires but by being about the same thing.

Wiki: a wiki is a piece of server software that allows users to freely create and edit web page content using any web browser.

WTO (World Trade Organization): the intergovernmental organization responsible for implementing and administering the WTO Agreement and its annexes and serving as a tribunal for resolving disputes.

Yeast: a general term for single-celled fungi that reproduce by budding. Some yeasts can ferment carbohydrates (starches and sugars) and thus are important in brewing and baking.

Index